Lecture Notes in Artificial Intelligence 3056

Edited by J. G. Carbonell and J. Siekmann

Subseries of Lecture Notes in Computer Science

Springer
Berlin
Heidelberg
New York
Hong Kong
London
Milan
Paris
Tokyo

Honghua Dai Ramakrishnan Srikant
Chengqi Zhang (Eds.)

Advances in Knowledge Discovery and Data Mining

8th Pacific-Asia Conference, PAKDD 2004
Sydney, Australia, May 26-28, 2004
Proceedings

Springer

Series Editors

Jaime G. Carbonell, Carnegie Mellon University, Pittsburgh, PA, USA
Jörg Siekmann, University of Saarland, Saarbrücken, Germany

Volume Editors

Honghua Dai
Deakin University
School of Information Technology
221 Burwood Highway, Burwood Vic 3125, Australia
E-mail: hdai@deakin.edu.au

Ramakrishnan Srikant
IBM Almaden Research Center
K55,B1, 650 Harry Road, San Jose, CA 95120, USA
E-mail: srikant@almaden@ibm.com

Chengqi Zhang
University of Technology, Sydney
Faculty of Information Technology
04.554, Building 10, City Campus, Broadway, NSW 2007, Australia
E-mail: chengqi@it.uts.edu.au

Library of Congress Control Number: 2004105537

CR Subject Classification (1998): I.2, H.2.8, H.3, H.5.1, G.3, J.1, K.4

ISSN 0302-9743
ISBN 3-540-22064-X Springer-Verlag Berlin Heidelberg New York

Springer-Verlag is a part of Springer Science+Business Media

springeronline.com

Printed in Germany

Typesetting: Camera-ready by author, data conversion by PTP-Berlin, Protago-TeX-Production GmbH
Printed on acid-free paper SPIN: 11007623 06/3142 5 4 3 2 1 0

Preface

The Pacific-Asia Conference on Knowledge Discovery and Data Mining (PAKDD) has been held every year since 1997. This year, the eighth in the series (PAKDD 2004) was held at Carlton Crest Hotel, Sydney, Australia, 26–28 May 2004. PAKDD is a leading international conference in the area of data mining. It provides an international forum for researchers and industry practitioners to share their new ideas, original research results and practical development experiences from all KDD-related areas including data mining, data warehousing, machine learning, databases, statistics, knowledge acquisition and automatic scientific discovery, data visualization, causal induction, and knowledge-based systems.

The selection process this year was extremely competitive. We received 238 research papers from 23 countries, which is the highest in the history of PAKDD, and reflects the recognition of and interest in this conference. Each submitted research paper was reviewed by three members of the program committee. Following this independent review, there were discussions among the reviewers, and when necessary, additional reviews from other experts were requested. A total of 50 papers were selected as full papers (21%), and another 31 were selected as short papers (13%), yielding a combined acceptance rate of approximately 34%.

The conference accommodated both research papers presenting original investigation results and industrial papers reporting real data mining applications and system development experience. The conference also included three tutorials on key technologies of knowledge discovery and data mining, and one workshop focusing on specific new challenges and emerging issues of knowledge discovery and data mining. The PAKDD 2004 program was further enhanced with keynote speeches by two outstanding researchers in the area of knowledge discovery and data mining: Philip Yu, Manager of Software Tools and Techniques, IBM T.J. Watson Research Center, USA and Usama Fayyad, President of DMX Group, LLC, USA.

We would like to thank everyone who participated in the development of the PAKDD 2004 program. In particular, we would give special thanks to the Program Committee. We asked a lot from them and were repeatedly impressed with their diligence and deep concern for the quality of the program, and also with their detailed feedback to the authors. We also thank Graham Williams for chairing the Industry Portal, Chinya V. Ravishankar and Bing Liu for chairing the workshop program, and Kai Ming Ting and Sunita Sarawagi for chairing the tutorials program. We are very grateful to Gang Li who put a lot of effort into the PAKDD 2004 Web site. We also thank Hongjun Lu, Hiroshi Motoda, and the other members of the PAKDD Steering Committee for providing valuable support and advice on many issues. The task of selecting papers was facilitated by Microsoft's Conference Management Tool, and by the tireless efforts of Mark Lau. The general organization of the conference also relied on the efforts of many. Simeon J. Simoff did an impressive job of ensuring that the logistics for the conference site were executed flawlessly. Li Liu and Guangquan Zhang made sure that the registration process flowed smoothly. Sanjay Chawla successfully publicized the conference and Peter O'Hanlon obtained sponsorships for the conference. We would also like to thank our financial sponsors SAS, UTS, Deakin University, and NICTA.

Finally and most importantly, we thank all the authors, who are the primary reason why the PAKDD conference continues to be so exciting, and to be the foremost place to learn about advances in both theoretical and applied research on KDD. Because of your work, PAKDD 2004 was a great success.

May 2004

Honghua Dai, Ramakrishnan Srikant
Chengqi Zhang, Nick Cercone

PAKDD 2004 Conference Committee

Honorary Chair
Ramamohanarao Kotagiri — University of Melbourne, Australia

Conference Co-chairs
Chengqi Zhang — University of Technology, Sydney, Australia
Nick Cercone — Dalhousie University, Canada

Program Co-chairs
Honghua Dai — Deakin University, Australia
Ramakrishnan Srikant — IBM Almaden Research Center, USA

Organizing Chair
Simeon J. Simoff — University of Technology, Sydney, Australia

Workshop Co-chairs
Chinya V. Ravishankar — University of California, USA
Bing Liu — University of Illinois, USA

Tutorial Co-chairs
Kai Ming Ting — Monash University, Australia
Sunita Sarawagi — IIT Bombay, India

Publicity Chair
Sanjay Chawla — The University of Sydney, Australia

Industrial Chair
Graham Williams — CSIRO, Australian National University, Australia

Sponsorship Chair
Peter O'Hanlon — SAS Australia & New Zealand

Treasurer
Guangquan Zhang — University of Technology, Sydney, Australia

Web Master
Gang Li — Deakin University, Australia

Conference Secretary
Li Liu — University of Technology, Sydney, Australia

PAKDD Steering Committee

Hiroshi Motoda (Chair)	Osaka University, Japan
David W. Cheung (Co-chair)	The University of Hong Kong, Hong Kong
Hongjun Lu	Hong Kong University of Science & Technology, Hong Kong
Arbee L. P. Chen	National Tsinghua University, Taiwan
Ming-Syan Chen	National Taiwan University, Taiwan
Jongwoo Jeon	Seoul National Unversity, Korea
Masaru Kitsuregawa	The University of Tokyo, Japan
Rao Kotagiri	University of Melbourne, Australia
Huan Liu	Arizona State University, USA
Takao Terano	University of Tsukuba, Japan
Kyu-Young Whang	Korea Adv. Inst. of Sci. & Tech., Korea
Graham Williams	CSIRO, Australia
Ning Zhong	Maebashi Institute of Technology, Japan
Lizhu Zhou	Tsinghua University, China

PAKDD 2004 Program Committee

Raj Acharya	The Pennsylvania State University, USA
Bharat Bhasker	Indian Institute of Management, India
Paul Bradley	Apollo Data Technologies, USA
Arbee L.P. Chen	National Tsing Hua University, Taiwan
Ming-Syan Chen	National Tsing Hua University, Taiwan
Yi-Ping Phoebe Chen	Deakin University, Australia
David Cheung	The University of Hong Kong, China
Vic Ciesielski	RMIT University, Australia
Jirapun Daengdej	Assumption University, Thailand
Guozhu Dong	Wright State University, USA
Usama Fayyad	DMX Group, USA
Eibe Frank	University of Waikato, New Zealand
Joydeep Ghosh	University of Texas, USA
Sudipto Guha	University of Pennsylvania, USA
Dimitrios Gunopulos	University of California, USA
Jingyu Hou	Deakin University, Australia
San-Yih Hwang	University of Minnesota, USA
Edward Ip	Wake Forest University, USA
Myoung Ho Kim	KAIST, Korea
Minos Garofalakis	Bell labs, USA
Inna Kulyshkina	Pricewaterhouse Coopers, Australia
Shyam Kumar Gupta	IIT Delhi, India

Howard J. Hamilton	University of Regina, Canada
Xiaoshu Hang	Deakin University, Australia
Joshua Zhexue Huang	University of Hong Kong, China
Soon J. Hyun	ICU, Korea
Jongwoo Jeon	SNU, Korea
Myoung Ho Kim	KAIST, Korea
Nick Koudas	AT&T, USA
Vipin Kumar	University of Minnesota, USA
Sang Ho Lee	SSU, Korea
Aleksandar Lazarevic	University of Minnesota, USA
Gang Li	Deakin University, Australia
Xue Li	University of Queensland, Australia
Xuemin Lin	University of New South Wales, Australia
Yuefeng Li	Queensland University of Technology, Australia
Bing Liu	UIC, USA
Huan Liu	Arizona State University, USA
James Liu	HKPU, China
Hong-Jun Lu	HKUST, China
Xiaofeng Meng	Renmin University of China, China
Hiroshi Motoda	Osaka University, Japan
Rajeev Motwani	Stanford, USA
Douglas Newlands	Deakin University, Australia
Raymond Ng	University of British Columbia, USA
Tom Osborn	Nuix Pty Ltd., Australia
Rajeev Rastogi	Bell Laboratories, USA
John Roddick	Flinders University, Australia
Kyuseok Shim	SNU, Korea
Simeon J. Simoff	University of Technology, Sydney, Australia
Andrew Skabar	Deakin University, Australia
Kate A. Smith	Monash University, Australia
Takao Terano	Tsukuba University, Japan
Hannu Toivonen	University of Helsinki, Finland
Dianhui Wang	La Trobe University, Australia
Lipo WANG	Nanyang Technological University, Singapore
Zhihai Wang	Monash University, Australia
Kyu Young Whang	KAIST, Korea
Graham Williams	CSIRO, Australia
Xindong Wu	University of Vermont, USA
Qiang Yang	HKUST, China
Min Yao	Zhejiang University, China
Yangdong Ye	Zhengzhou University, China
Philip S. Yu	IBM T.J. Watson Research Center, USA

Mohammed J. Zaki	Rensselaer Polytechnic Institute, USA
Shichao Zhang	University of Technology, Sydney, Australia
Aoying Zhou	Fudan University, China
Lizhu Zhou	Tsinghua University, China
Ning Zhong	Maebashi Institute of Technology, Japan
Zhi-Hua Zhou	Nanjing University, China

PAKDD 2004 External Reviewers

Aaron Ceglar
Aleksandar Lazarevic
Alex Liu
Alin Dobra
Amit Mandvikar
Ashwin Shriram
Aysel Ozgur
Bi-Ru Dai
Carl Mooney
Chase Krumpelman
Chulyun Kim
Chunyu Jiang
Daling Wang
Damien McAullay
Daofeng Luo
Ding-Yi Chen
Dongdong Hu
Dragoljub Pokrajac
Ehtesham Haque
Eric Eilertso
Gabriel P.c. Fung
Ganesh Ramesh
Guichong Li
Gunjan Gupta
Gyoergy J. Simon
Gyorgy Simon
Haixun Wang
Hongxing He
Hui Xiong
Hung-Chen Chen
Hyoungmin Park
Injae Sung
Jaana Heino
Jaehyok Chong
Jeff Riley
Jie Chen
Jigar Mody
Jinseog Kim
Jiyuan An
Juggapong Natwichai
Jun Hu
K. Srikumar
Kamran Karimi
Kari Laasonen
Ken-Hao Liu
Kouzou Ohara
Lance Parsons
Levent Ertoz
Lijun Chen
Liqiang Geng
Madhukar Suryaprakash
Mika Raento
Min-Jae Lee
Mintz Hsieh
Narasimha D. Kolippakkam
Neoklis Polyzotis
Ning-Han Liu
Pavanreddy Sannapareddy
Petteri Hintsanen
Phu Chien Nguyen
Pusheng Zhang
Qing Liu
Qing Zhang
Ralf Rantzau
Ran Wolff
Ravi Janga
S. Merugu
Sheng-Tang Wu
Shihong Mao
Shu-Chuan Chu
Soohwan Chung
Srihari Venkatesan
Suju Rajan
Supawan Prompramote
Taneli Mielikäinen
Tetsuya Yoshida
Varun Chandola
Wan-Shiou Yang
Warren Jin
Wei-Guang Teng
Weining Qian
Xin Wang
Xin Yan
Xingquan Zhu
Xu Jian
Yan Chen
Yang-Sae Moon
Yidong Yuan
Yi-Hung Wu
Ying Sun
Ying Yang
Yiqing Tu
Yongdai Kim
Yue Xu
Yuelong Jiang
Yuwon Kim
Zhiheng Li

Sponsors

Table of Contents

Session 2A: Novel Algorithms (I)

Session 2B: Association (II)

Session 2C: Classification (II)

Session 3A: Event Mining, Anomaly Detection, and Intrusion Detection

Session 3B: Ensemble Learning

Session 3C: Bayesian Network and Graph Mining

Session 3D: Text Mining (I)

Session 4A: Clustering (II)

Session 4B: Association (III)

Session 4C: Novel Algorithms (II)

Session 4D: Multimedia Mining

Session 5A: Text Mining and Web Mining (II)

Session 5B: Statistical Methods, Sequential Data Mining, and Time Series Mining

Session 5C: Novel Algorithms (III)

Session 5D: Biomedical Mining

Mining of Evolving Data Streams with Privacy Preservation

Philip S. Yu

Manager, Software Tools and Techniques
IBM T.J. Watson Research Center, USA
psyu@us.ibm.com

Abstract. The data stream domain has become increasingly important in recent years because of its applicability to a wide variety of applications. Problems such as data mining and privacy preservation which have been studied for traditional data sets cannot be easily solved for the data stream domain. This is because the large volume of data arriving in a stream renders most algorithms to inefficient as most mining and privacy preservation algorithms require multiple scans of data which is unrealistic for stream data. More importantly, the characteristics of the data stream can change over time and the evolving pattern needs to be captured. In this talk, I'll discuss the issues and focus on how to mine evolving data streams and preserve privacy.

H. Dai, R. Srikant, and C. Zhang (Eds.): PAKDD 2004, LNAI 3056, p. 1, 2004.

Data Mining Grand Challenges

Usama Fayyad

President, DMX Group, LLC, USA
fayyad@aig.jpl.nasa.gov

Abstract. The past two decades has seen a huge wave of computational systems for the "digitization" of business operations from ERP, to manufacturing, to systems for customer interactions. These systems increased the throughput and efficiency of conducting "transactions" and resulted in an unprecedented build-up of data captured from these systems. The paradoxical reality that most organizations face today is that they have more data about every aspect of their operations and customers, yet they find themselves with an ever diminishing understanding of either. Data Mining has received much attention as a technology that can possibly bridge the gap between data and knowledge.

While some interesting progress has been achieved over the past few years, especially when it comes to techniques and scalable algorithms, very few organizations have managed to benefit from the technology. Despite the recent advances, some major hurdles exist on the road to the needed evolution. Furthermore, most technical research work does not appear to be directed at these challenges, nor does it appear to be aware of their nature. This talk will cover these challenges and present them in both the technical and the business context. The exposition will cover deep technical research questions, practical application considerations, and social/economic considerations. The talk will draw on illustrative examples from scientific data analysis, commercial applications of data mining in understanding customer interaction data, and considerations of coupling data mining technology within database management of systems. Of particular interest is the business challenge of how to make the technology really work in practice. There are many unsolved deep technical research problems in this field and we conclude by covering a sampling of these.

H. Dai, R. Srikant, and C. Zhang (Eds.): PAKDD 2004, LNAI 3056, p. 2, 2004.

Evaluating the Replicability of Significance Tests for Comparing Learning Algorithms

Remco R. Bouckaert[1,2] and Eibe Frank[2]

[1] Xtal Mountain Information Technology
215 Three Oaks Drive, Dairy Flat, Auckland, New Zealand
rrb@xm.co.nz

[2] Computer Science Department, University of Waikato
Private Bag 3105, Hamilton, New Zealand
{remco,eibe}@cs.waikato.ac.nz

Abstract. Empirical research in learning algorithms for classification tasks generally requires the use of significance tests. The quality of a test is typically judged on Type I error (how often the test indicates a difference when it should not) and Type II error (how often it indicates no difference when it should). In this paper we argue that the replicability of a test is also of importance. We say that a test has low replicability if its outcome strongly depends on the particular random partitioning of the data that is used to perform it. We present empirical measures of replicability and use them to compare the performance of several popular tests in a realistic setting involving standard learning algorithms and benchmark datasets. Based on our results we give recommendations on which test to use.

1 Introduction

Significance tests are often applied to compare performance estimates obtained by resampling methods—such as cross-validation [1]—that randomly partition data. In this paper we consider the problem that a test may be very sensitive to the particular random partitioning used in this process. If this is the case, it is possible that, using the same data, the same learning algorithms A and B, and the same significance test, one researcher finds that method A is preferable, while another finds that there is not enough evidence for this. Lack of replicability can also cause problems when "tuning" an algorithm: a test may judge favorably on the latest modification purely due to its sensitivity to the particular random number seed used to partition the data. In this paper we extend previous work on replicability [2,3] by studying the replicability of some popular tests in a more realistic setting based on standard benchmark datasets taken from the UCI repository of machine learning problems [4].

The structure of the paper is as follows. In Section 2 we review how significance tests are used for comparing learning algorithms and introduce the notion of replicability. Section 3 discusses some popular tests in detail. Section 4 contains empirical results for these tests and highlights the lack of replicability of some of them. Section 5 summarizes the results and makes some recommendations based on our empirical findings.

H. Dai, R. Srikant, and C. Zhang (Eds.): PAKDD 2004, LNAI 3056, pp. 3–12, 2004.

2 Evaluating Significance Tests

We consider a scenario where we have a certain application domain and we are interested in the mean difference in accuracy between two classification algorithms in this domain, given that the two algorithms are trained on a dataset with N instances. We do not know the joint distribution underlying the domain and consequently cannot compute the difference exactly. Hence we need to estimate it, and, to check whether the estimated difference is likely to be a "true" difference, perform a significance test. To this end we also need to estimate the variance of the differences across different training sets.

Obtaining an unbiased estimate of the mean and variance of the difference is easy if there is a sufficient supply of data. In that case we can sample a number of training sets of size N, run the two learning algorithms on each of them, and estimate the difference in accuracy for each pair of classifiers on a large test set. The average of these differences is an estimate of the expected difference in generalization error across all possible training sets of size N, and their variance is an estimate of the variance. Then we can perform a paired t-test to check the null hypothesis that the mean difference is zero. The Type I error of a test is the probability that it rejects the null hypothesis incorrectly (i.e. it finds a "significant" difference although there is none). Type II error is the probability that the null hypothesis is not rejected when there actually is a difference. The test's Type I error will be close to the chosen significance level α.

In practice we often only have one dataset of size N and all estimates must be obtained from this one dataset. Different training sets are obtained by subsampling, and the instances not sampled for training are used for testing. For each training set S_i, $1 \leq i \leq k$, we get a matching pair of accuracy estimates and the difference x_i. The mean and variance of the differences x_i is used to estimate the mean and variance of the difference in generalization error across different training sets. Unfortunately this violates the independence assumption necessary for proper significance testing because we re-use the data to obtain the different x_i's. The consequence of this is that the Type I error exceeds the significance level. This is problematic because it is important for the researcher to be able to control the Type I error and know the probability of incorrectly rejecting the null hypothesis. Several heuristic versions of the t-test have been developed to alleviate this problem [5,6].

In this paper we study the *replicability* of significance tests. Consider a test based on the accuracy estimates generated by cross-validation. Before the cross-validation is performed, the data is randomized so that each of the resulting training and test sets exhibits the same distribution. Ideally, we would like the test's outcome to be independent of the particular partitioning resulting from the randomization process because this would make it much easier to replicate experimental results published in the literature. However, in practice there is always a certain sensitivity to the partitioning used. To measure replicability we need to repeat the same test several times on the same data with different random partitionings—in this paper we use ten repetitions—and count how often the outcome is the same. Note that a test will have greater replicability than another test with the same Type I and Type II error if it is more consistent in its outcomes for each individual dataset.

We use two measures of replicability. The first measure, which we call *consistency*, is based on the raw counts. If the outcome is the same for every repetition of a test on the same data, we call the test *consistent*, and if there is a difference at most once, we call it *almost consistent*. This procedure is repeated with multiple datasets, and the fraction of outcomes for which a test is consistent or almost consistent is an indication of how replicable the test is. The second measure, which we call *replicability*, is based on the probability that two runs of the test on the same data set will produce the same outcome. This probability is never worse than 0.5. To estimate it we need to consider pairs of randomizations. If we have performed the test based on n different randomizations for a particular dataset then there are $\binom{n}{2}$ such pairs. Assume the tests rejects the null hypothesis for k $(0 \leq k \leq n)$ of the randomizations. Then there are $\binom{k}{2}$ rejecting pairs and $\binom{n-k}{2}$ accepting ones. Based on this the above probability can be estimated as $R(k,n) = (\binom{k}{2} + \binom{n-k}{2})/\binom{n}{2} = \frac{k(k-1)+(n-k)(n-k-1)}{n(n-1)}$. We use this probability to form a measure of replicability across different datasets. Assume there are m datasets and let i_k $(0 \leq k \leq n)$ be the number of datasets for which the test agrees k times (i.e. $\sum_{k=0}^{n} i_k = m$). Then we define replicability as $R = \sum_{k=0}^{n} \frac{i_k}{m} R(k,n)$. The larger the value of this measure, the more likely the test is to produce the same outcome for two different randomizations of a dataset.

3 Significance Tests

In this section we review some tests for comparing learning algorithms. Although testing is essential for empirical research, surprisingly little has been written on this topic.

3.1 The 5x2cv Paired t-Test

Dietterich [5] evaluates several significance tests by measuring their Type I and Type II error on artificial and real-world data. He finds that the paired t-test applied to random subsampling has an exceedingly large Type I error. In random subsampling a training set is drawn at random without replacement and the remainder of the data is used for testing. This is repeated a given number of times. In contrast to cross-validation, random subsampling does not ensure that the test sets do not overlap. Ten-fold cross-validation can be viewed as a special case of random subsampling repeated ten times, where 90% of the data is used for training, and it is guaranteed that the ten test sets do not overlap. The paired t-test based on ten-fold cross-validation fares better in the experiments in [5] but also exhibits an inflated Type I error. On one of the real-world datasets its Type I error is approximately twice the significance level.

As an alternative [5] proposes a heuristic test based on five runs of two-fold cross-validation, called "5x2cv paired t-test". In an r-times k-fold cross-validations there are r, $r > 1$, runs and k, $k > 1$, folds. For each run j, $1 \leq j \leq r$, the data is randomly permutated and split into k subsets of equal size.[1] We call these i, $1 \leq i \leq k$, subsets the k folds of run j. We consider two learning schemes A and B and measure their respective accuracies a_{ij} and b_{ij} for fold i and run j. To obtain a_{ij} and b_{ij} the corresponding learning scheme is trained on all the data excluding that in fold i of run j and tested on

[1] Of course, in some cases it may not be possible to split the data into subsets that have exactly the same size.

the remainder. Note that exactly the same pair of training and test sets is used to obtain both a_{ij} and b_{ij}. That means a paired significance test is appropriate and we can consider the individual differences in accuracy $x_{ij} = a_{ij} - b_{ij}$ as the input for the test.

Let $x_{.j}$ denote the mean difference for a single run of 2-fold cross-validation, $x_{.j} = (x_{1j} + x_{2j})/2$. The variance is $\hat{\sigma}_j^2 = (x_{1j} - x_{.j})^2 + (x_{2j} - x_{.j})^2$. The 5x2cv paired t-test uses the following test statistic:

$$t = \frac{x_{11}}{\sqrt{\frac{1}{5}\sum_{j=1}^{5}\hat{\sigma}_j^2}}$$

This statistic is plugged into the Student-t distribution with five degrees of freedom. Note that the numerator only uses the term x_{11} and not the other differences x_{ij}. Consequently the outcome of the test is strongly dependent on the particular partitioning of the data used when the test is performed. Therefore it can be expected that the replicability of this test is not high. Our empirical evaluation demonstrates that this is indeed the case.

The empirical results in [5] show that the 5×2cv paired t-test has a Type I error at or below the significance level. However, they also show that it has a much higher Type II error than the standard t-test applied to ten-fold cross-validation. Consequently the former test is recommended in [5] when a low Type I error is essential, and the latter test otherwise.

The other two tests evaluated in [5] are McNemar's test and the test for the difference of two proportions. Both of these tests are based on a single train/test split and consequently cannot take variance due to the choice of training and test set into account. Of these two tests, McNemar's test performs better overall: it has an acceptable Type I error and the Type II error is only slightly lower than that of the 5×2cv paired t-test. However, because these two tests are inferior to the 5×2cv test, we will not consider them in our experiments.

3.2 Tests Based on Random Subsampling

As mentioned above, Dietterich [5] found that the standard t-test has a high Type I error when used in conjunction with random subsampling. Nadeau and Bengio [6] observe that this is due to an underestimation of the variance because the samples are not independent (i.e. the different training and test sets overlap). Consequently they propose to correct the variance estimate by taking this dependency into account.

Let a_j and b_j be the accuracy of algorithms A and B respectively, measured on run j ($1 \leq j \leq n$). Assume that in each run n_1 instances are used for training, and the remaining n_2 instances for testing. Let x_j be the difference $x_j = a_j - b_j$, and $\hat{\mu}$ and $\hat{\sigma}^2$ the estimates of the mean and variance of the n differences. The statistic of the "corrected resampled t-test" is:

$$t = \frac{\frac{1}{n}\sum_{j=1}^{n} x_j}{\sqrt{(\frac{1}{n} + \frac{n_2}{n_1})\hat{\sigma}^2}}$$

This statistic is used in conjunction with the Student-t distribution and $n - 1$ degrees of freedom. The only difference to the standard t-test is that the factor $\frac{1}{n}$ in the denominator has been replaced by the factor $\frac{1}{n} + \frac{n_2}{n_1}$. Nadeau and Bengio [6] suggest that "... normal usage would call for n_1 to be 5 or 10 times larger than n_2, ...".

Empirical results show that this test dramatically improves on the standard resampled *t*-test: the Type I error is close to the significance level, and, unlike McNemar's test and the 5×2cv test, it does not suffer from high Type II error [6].

3.3 Tests Based on Repeated k-Fold Cross Validation

Here we consider tests based on r-times k-fold cross-validation where r and k can have any value. As in Section 3.1, we observe differences $x_{ij} = a_{ij} - b_{ij}$ for fold i and run j. One could simply use $m = \frac{1}{k.r}\sum_{i=1}^{k}\sum_{j=1}^{r} x_{ij}$ as an estimate for the mean and $\hat{\sigma}^2 = \frac{1}{k.r-1}\sum_{i=1}^{k}\sum_{j=1}^{r}(x_{ij} - m)^2$ as an estimate for the variance. Then, assuming the various values of x_{ij} are independent, the test statistic $t = m/\sqrt{(1/k.r)\hat{\sigma}^2}$ is distributed according to a *t*-distribution with $df = k.r - 1$ degrees of freedom. Unfortunately, the independence assumption is highly flawed, and tests based on this assumption show very high Type I error, similar to plain subsampling.

However, the same variance correction as in the previous subsection can be performed here because cross-validation is a special case of random subsampling where we ensure that the test sets in one run do not overlap. (Of course, test sets from different runs will overlap.) This results in the following statistic:

$$t = \frac{\frac{1}{k.r}\sum_{i=1}^{k}\sum_{j=1}^{r} x_{ij}}{\sqrt{(\frac{1}{k.r} + \frac{n_2}{n_1})\hat{\sigma}^2}}$$

where n_1 is the number of instances used for training, and n_2 the number of instances used for testing. We call this test the "corrected repeated k-fold cv test".

4 Empirical Evaluation

To evaluate how replicability affects the various tests, we performed experiments on a selection of datasets from the UCI repository [4]. We used naive Bayes, C4.5 [7], and the nearest neighbor classifier, with default settings as implemented in Weka[2] version 3.3 [1]. For tests that involve multiple folds, the folds were chosen using stratification, which ensures that the class distribution in the whole dataset is reflected in each of the folds. Each of the tests was run ten times for each pair of learning schemes and a 5% significance level was used in all tests unless stated otherwise.

4.1 Results for the 5x2cv Paired *t*-Test

Table 1 shows the datasets and their properties, and the results for the 5x2 cross validation test. The three right-most columns show the number of times the test does not reject the null hypothesis, i.e, the number of times the 5x2 cross validation test indicates that there

[2] Weka is freely available with source from http://www.cs.waikato.ac.nz/ml.

Table 1. The number of cases (#inst.), attributes (#atts.), and classes (#cl.) for each dataset; and the number of draws for each pair of classifiers based on the 5x2 cross validation test (NB = naive Bayes, NN = nearest neighbor).

dataset	#inst.	#atts.	#cl.	NB vs C4.5	NB vs NN	C4.5 vs NN
anneal	898	38	5	4	4	10
arrhythmia	452	280	13	9	9	2
audiology	226	69	24	5	10	8
autos	205	25	6	10	7	10
balance-scale	625	4	3	1	4	7
breast-cancer	286	9	2	10	9	8
credit-rating	690	16	2	6	8	10
ecoli	336	8	8	7	10	10
German credit	1000	20	2	9	6	10
glass	214	9	6	6	6	9
heart-statlog	270	13	2	4	5	9
hepatitis	155	19	2	9	10	10
horse-colic	368	22	2	8	10	7
Hungarian heart disease	294	13	2	10	10	10
ionosphere	351	34	2	10	10	8
iris	150	4	3	10	10	10
labor	57	16	2	8	10	10
lymphography	148	18	4	9	10	10
pima-diabetes	768	8	2	10	6	7
primary-tumor	339	17	21	7	3	10
sonar	208	60	2	10	9	6
soybean	683	35	19	8	8	9
vehicle	846	18	4	0	0	9
vote	435	16	2	4	9	7
vowel	990	13	11	4	0	0
Wisconsin breast cancer	699	9	2	8	9	10
zoo	101	16	7	10	10	8
			Consistent:	9	12	13
			Almost consistent:	14	17	17
			Replicability (R):	0.737	0.783	0.816

is no difference between the corresponding pair of classifiers. For example, for the anneal dataset, the test indicates no difference between naive Bayes and C4.5 four times, so six times it does indicate a difference. Note that the same dataset, the same algorithm, the same settings, and the same significance test were used in each of the ten experiments. The only difference was in the way the dataset was split into the 2 folds in each of the 5 runs. Clearly, the test is very sensitive to the particular partitioning of the anneal data.

Looking at the column for naive Bayes vs. C4.5, this test could be used to justify the claim that the two perform the same for all datasets except the vehicle dataset just by choosing appropriate random number seeds. However, it could just as well be used to support the claim that the two algorithms perform differently in 19 out of 27 cases.

For some rows, the test consistently indicates no difference between any two of the three schemes, in particular for the iris and Hungarian heart disease datasets. However, most rows contain at least one cell where the outcomes of the test are not consistent.

The row labeled "consistent" at the bottom of the table lists the number of datasets for which all outcomes of the test are the same. These are calculated as the number of 0's and 10's in the column. For any of the compared schemes, less than 50% of the results turn out to be consistent.

Note that, it is possible that, when comparing algorithms A and B, sometimes A is preferred and sometimes B if the null hypothesis is rejected. However, closer inspection of the data reveals that this only happens when the null hypothesis is accepted most of the time, except for 2 or 3 runs. Consequently these cases do not contribute to the value of the consistency measure.

If we could accept that one outcome of the ten runs does not agree with the rest, we get the number labeled "almost consistent" in Table 1 (i.e. the number of 0's, 1's, 9's and 10's in a column). The 5x2 cross validation test is almost consistent in fewer than 66% of the cases, which is still a very low rate.

The last row shows the value of the replicability measure R for the three pairs of learning schemes considered. These results reflect the same behaviour as the consistency measures. The replicability values are pretty low considering that R cannot be smaller than 0.5.

4.2 Results for the Corrected Resampled t-Test

In the resampling experiments, the data was randomized, 90% of it used for training, and the remaining 10% used to measure accuracy. This was repeated with a different random number seed for each run. Table 2 shows the results for the corrected resampled t-test. The number of runs used in resampling was varied from 10 to 100 to see the effect on the replicability.

The replicability increases with the number of runs almost everywhere. The only exception is in the last row, where the "almost consistent" value decreases by one when increasing the runs from 10 to 20. This can be explained by random fluctuations due to the random partitioning of the datasets. Overall, the replicability becomes reasonably acceptable when the number of runs is 100. In this case 80% of the results are "almost consistent", and the value of the replicability measure R is approximately 0.9 or above.

4.3 Results for Tests Based on (Repeated) Cross Validation

For the standard t-test based on a single run of 10-fold cross validation we observed consistent results for 15, 16, and 14 datasets, comparing NB with C4.5, NB with NN, and C4.5 with NN respectively. Contrasting this with corrected resampling with 10 runs, which takes the same computational effort, we see that 10-fold cross validation is at least as consistent. However, it is substantially less consistent than (corrected) resampling at 100 runs. Note also that this test has an inflated Type I error [5].

Table 2. Replicability for corrected resampled t-test.

	#Runs			
	10	20	50	100
NB vs C4.5				
consistent	15	14	19	21
almost consistent	16	19	22	23
replicability (R)	0.801	0.843	0.892	0.922
NB vs NN				
consistent	12	15	18	20
almost consistent	20	21	22	23
replicability (R)	0.835	0.865	0.882	0.899
C4.5 vs NN				
consistent	12	14	18	23
almost consistent	18	17	22	24
replicability (R)	0.819	0.825	0.878	0.935

Performing the same experiment in conjunction with the standard t-test based on the 100 differences obtained by 10-times 10-fold cross validation, produced consistent results for 25, 24, and 18 datasets, based on NB with C4.5, NB with NN, and C4.5 with NN respectively. This looks impressive compared to any of the tests we have evaluated so far. However, the Type I error of this test is very high (because of the overlapping training and test sets) and therefore it should not be used in practice.

To reduce Type I error it is necessary to correct the variance. Table 3 shows the same results for the corrected paired t-test based on the paired outcomes of r-times 10-fold cross validation. Comparing this to Table 2 (for corrected resampling) the consistency is almost everywhere as good and often better (assuming the same computational effort in both cases): the column with 1 run in Table 3 should be compared with the 10 runs column in Table 2, the column with 2 runs in Table 3 with the column with 20 runs in Table 2, etc. The same can be said about the replicability measure R. This indicates that repeated cross validation helps to improve replicability (compared to just performing random subsampling).

To ensure that the improved replicability of cross-validation is not due to stratification (which is not performed in the case of random subsampling), we performed an experiment where resampling was done with stratification. The replicability scores differed only very slightly from the ones shown in Table 2, suggesting the improved replicability is not due to stratification.

Because the corrected paired t-test based on 10-times 10-fold cross validation exhibits the best replicability scores, we performed an experiment to see how sensitive its replicability is to the significance level. The results, shown in Table 4, demonstrate that the significance level does not have a major impact on consistency or the replicability measure R. Note that the latter is greater than 0.9 in every single case, indicating very good replicability for this test.

Table 3. Replicability for corrected rx10 fold cross-validation test.

	#Runs			
	1	2	5	10
NB vs C4.5				
consistent	16	20	21	24
almost consistent	18	21	23	25
replicability (R)	0.821	0.889	0.928	0.962
NB vs NN				
consistent	18	20	23	23
almost consistent	19	21	23	24
replicability (R)	0.858	0.890	0.939	0.942
C4.5 vs NN				
consistent	13	19	22	22
almost consistent	18	23	24	24
replicability (R)	0.814	0.904	0.928	0.928

Table 4. Replicability of corrected 10x10 fold cross-validation test for various significance levels.

	Significance level			
	1%	2.5%	5%	10%
NB vs C4.5				
consistent	22	23	24	21
almost consistent	23	24	25	22
replicability (R)	0.927	0.936	0.962	0.915
NB vs NN				
consistent	23	24	23	23
almost consistent	24	27	24	23
replicability (R)	0.939	0.978	0.942	0.939
C4.5 vs NN				
consistent	23	24	22	20
almost consistent	23	24	24	24
replicability (R)	0.943	0.953	0.928	0.919

Table 5. Results for data sources 1 to 4: the difference in accuracy between naive Bayes and C4.5 (in percent) and the consistency of the tests (in percent).

Source	1	2	3	4	
Δ accuracy	0.0	2.77	5.83	11.27	min.
5x2 cv	72.3	71.2	63.5	16.9	16.9
10 x resampling	65.5	44.0	26.0	48.8	26.0
100 x resampling	90.9	73.2	66.8	97.2	66.8
10-fold cv	49.7	47.6	33.2	90.8	33.2
corrected 10x10 fold cv	91.9	80.3	76.7	98.9	76.7

4.4 Simulation Experiment

To study the effect of the observed difference in accuracy on replicability, we performed a simulation study. Four data sources were selected by randomly generating Bayesian networks over 10 binary variables where the class variable had 0.5 probability of being zero or one. A 0.5 probability of the class variable is known to cause the largest variability due to selection of the training data [5]. The first network had no arrows and all variables except the class variables were independently selected with various different probabilities. This guarantees that any learning scheme will have 50% expected accuracy on the test data. The other three data sources had a BAN structure [8], generated by starting with a naive Bayes model and adding arrows while guaranteeing acyclicity.

Using stochastic simulation [9], a collection of 1000 training sets with 300 instances each was created. Naive Bayes and C4.5 were trained on each of them and their accuracy measured on a test set of 20,000 cases, generated from each of the data sources. The average difference in accuracy is shown in Table 5 in the row marked Δ accuracy, and it ranges from 0% to 11.27%.

Each of the tests was run 10 times on each of the 4×1000 training sets. Table 5 shows, for each of the tests and each data source, the percentage of training sets for which the test is consistent (i.e., indicates the same outcome 10 times). The last column shows the minimum of the consistency over the four data sources.

Again, 5×2 cross validation, 10 times resampling, and 10 fold cross validation show rather low consistency. Replicability increases dramatically with 100 times resampling, and increases even further when performing 10 times repeated 10 fold cross validation. This is consistent with the results observed on the UCI datasets.

Table 5 shows that the tests have fewer problems with data sources 1 and 4 (apart from the 5×2 cv test), where it is easy to decide whether the two schemes differ. The 5×2 test has problems with data source 4 because it is a rather conservative test (low Type I error, high Type II error) and tends to err on the side of being too cautious when deciding whether two schemes differ.

5 Conclusions

We considered tests for choosing between two learning algorithms for classification tasks. We argued that such a test should not only have an appropriate Type I error and low Type II error, but also high replicability. High replicability facilitates reproducing published results and reduces the likelihood of oversearching. In our experiments, good replicability was obtained using 100 runs of random subsampling in conjunction with Nadeau and Bengio's corrected resampled t-test, and replicability improved even further by using 10-times 10-fold cross-validation instead of random subsampling. Both methods are acceptable but for best replicability we recommend the latter one.

Acknowledgment. Eibe Frank was supported by Marsden Grant 01-UOW-019.

References

1. I.H. Witten and E. Frank. Data Mining: Practical Machine Learning Tools and Techniques with Java Implementations. Morgan Kaufmann, 2000.
2. R.R. Bouckaert. Choosing between two learning algorithms based on calibrated tests. Proc 20th Int Conf on Machine Learning. Morgan Kaufmann, 2003.
3. R.R. Bouckaert. Choosing learning algorithms using sign tests with high replicability. Proc 16th Australian Joint Conference on Artificial Intelligence. Springer-Verlag, 2003.
4. C.L. Blake and C.J. Merz. UCI Repository of machine learning databases. Irvine, CA, 1998. http://www.ics.uci.edu/~mlearn/MLRepository.html
5. T. G. Dietterich. Approximate Statistical Tests for Comparing Supervised Classification Learning Algorithms. Neural Computation, 10(7) 1895–1924, 1998
6. C. Nadeau and Y. Bengio. Inference for the generalization error. In Machine Learning 52:239–281, 2003
7. R. Quinlan. C4.5: Programs for Machine Learning, Morgan Kaufmann, 1993.
8. N. Friedman, D. Geiger, and M. Goldszmidt. Bayesian Network Classifiers. In Machine Learning 29:131–163, 1997
9. J. Pearl: Probabilistic Reasoning in Intelligent Systems. Morgan Kaufmann, 1988.

Spectral Energy Minimization for Semi-supervised Learning

Chun-hung Li* and Zhi-li Wu

Department of Computer Science, Hong Kong Baptist University
{chli,vincent}@comp.hkbu.edu.hk

Abstract. Data mining problems often involve a large amount of unlabeled data and there is often very limited known information on the dataset. In such scenario, semi-supervised learning can often improve classification performance by utilizing unlabeled data for learning. In this paper, we proposed a novel approach to semi-supervised learning as as an optimization of both the classification energy and cluster compactness energy in the unlabeled dataset. The resulting integer programming problem is relaxed by a semi-definite relaxation where efficient solution can be obtained. Furthermore, the spectral graph methods provide improved energy minimization via the incorporation of additional criteria. Results on UCI datasets show promising results.

1 Introduction

In this paper, we adopt an energy minimization approach to semi-supervised learning where the energy function is composed of two distinct parts: the supervised energy function and the unsupervised energy function. The unsupervised energy incorporates measures on cluster compactness. The supervised energy is specified as function of classification probability of any chosen classifier.

The modeling of the semi-supervised learning process as two separate energy functions allows considerable flexibility in incorporating existing clustering model and selection of suitable classifier. While the estimation of unknown labels via minimization of energy function is an integer programming problem which often leads to solutions of combinatorial nature. We propose to adopt a semi-definite relaxation of the integer programming to obtain solution in polynomial time. Using results from spectral graph theory, minimum energy solutions can be attained that satisfiy different desired constraint.

This paper is organized as follows. In section 2, we discuss the energy minimization approach to semi-supervised learning. In section 3, we describe the relationships with graph. In section 4, we discuss the spectral methods for energy minimization. Section 5 gives experimental results on datasets from the UCI repostory.

* The funding from HKBU FRG/01-02/II-67 is acknowledged

H. Dai, R. Srikant, and C. Zhang (Eds.): PAKDD 2004, LNAI 3056, pp. 13–21, 2004.

2 Energy Function for Semi-supervised Learning

2.1 Energy Minimization

Given dataset $D = \{(\mathbf{x}_1, y_1), \ldots, (\mathbf{x}_n, y_n)\} \in \mathbf{R}^N \times \{\pm 1\}$ and n is the total number of data, D can also be partitioned into two sets $D = \{L \cup U\}$ where L is the set of labeled data with known classification labels and U is the set of unlabeled data with unknown classification labels. Classification algorithms often relies primarily on the labeled dataset L for learning rule or mapping to classify any unlabeled dataset. On the other hand, clustering algorithms often relies on compactness measures or probabilistic assumptions on the unlabeled dataset U to infer the possible groupings and hence the values of the labels y_m. In general, learning from labeled dataset L and unlabeled dataset U can be a simultaneous optimization of these two criteria

$$E(D) = E_{sup}(D) + \lambda E_{unsup}(D) \tag{1}$$

where E_{sup} measures the agreement between the entire dataset with the training dataset, E_{unsup} measures the compactness of the entire dataset and λ is a real positive constant that balances the importance of the two criteria. As the simultaneous optimization of the above criteria often leads to combinatorial problems which cannot be solved in polynomial time, the development of the energy function is emphasized on arriving at a continuous variations where solutions can be obtained in polynomial time.

2.2 Unsupervised Learning

There are various forms of energy functions for unsupervised learning. The primary objective is to find a partitioning of the dataset such that the labels are assigned to minimize the within-class distance or to minimize the between-class similarity. A simple quadratic energy for minimizing the within-class distance is $\sum_{i,j:y_i=y_j} ||\mathbf{x}_i - \mathbf{x}_j||^2$ where the Euclidean distance between $\mathbf{x}_i$ and $\mathbf{x}_j$ is calculated only if both have the same labels. However, recent advances in clustering and kernel based learning motivates the use of similarity measures in the form of a dot-product. Let $K_{i,j}$ represents the similarity between $\mathbf{x}_i$ and $\mathbf{x}_j$, the between-class similarity can be specified as

$$E_{unsup}(D) = \sum_{i,j:y_i \neq y_j} K_{i,j} \tag{2}$$

which is equivalent to the following form where the discrete variables $y \in +1, -1$ are explicitly specified in a product form

$$E_{unsup}(D) = \sum_{i,j} K_{i,j}(y_i - y_j)^2 \tag{3}$$

While the above equations are equivalent when the variable y takes only the value {+1,-1}, Eqn 3 caters for the case where y is allowed to take a continuous value.

2.3 Supervised Learning

We consider a generic model of classifier using the class output probability model of a classifier in this paper. Although only a small number of classifiers have class probability model, classification scores can often be given to classifiers that reflects some measures of confidence on the classification. Assuming a generic classifier C which output conditional probability $P(y_i|\mathbf{x}_i)$, the supervised energy E_{sup} can be specified as

$$E_{sup}(D) = \sum_i P(+1|\mathbf{x}_i)(y_i - (+1))^2 + \sum_i P(-1|\mathbf{x}_i)(y_i - (-1))^2 \tag{4}$$

Without additional constraint, this criteria will give labels in accordance with the conditional probability. The second step is to further relax on the fixed labels $\{+1, -1\}$ inside the quadratic term to allow it to take two continuous values to be denoted by $\{y_{Z+}, y_{Z-}\}$ where

$$E_{sup}(D) = \sum_i P(+1|\mathbf{x}_i)(y_i - y_{Z+})^2 + \sum_i P(-1|\mathbf{x}_i)(y_i - y_{Z-})^2 \tag{5}$$

As y_i is allowed to take continuous value, we need to subject the optimization to feasible solution space. Suitable constraint includes the symmetric constraint where $\sum_i y_i = 0$ and the normalization constraint $\sum_i y_i^2 = k$ which set a scale on the solution. In particular, the normalization constraint of setting $k = 1$ is usually taken.

Construct a symmetric matrix H with positive entries by

$$H_{i,j} = \begin{cases} K_{i,j} & \text{if } 1 \leq i,j \leq n \\ \frac{\lambda}{2} P(+1|\mathbf{x}_j) & \text{if } i = n+1 \text{ and } 1 \leq j \leq n \\ \frac{\lambda}{2} P(+1|\mathbf{x}_i) & \text{if } j = n+1 \text{ and } 1 \leq i \leq n \\ \frac{\lambda}{2} P(-1|\mathbf{x}_j) & \text{if } i = n+2 \text{ and } 1 \leq j \leq n \\ \frac{\lambda}{2} P(-1|\mathbf{x}_i) & \text{if } j = n+2 \text{ and } 1 \leq i \leq n \\ 0 & \text{if } n+1 \leq i,j \leq n+2 \end{cases} \tag{6}$$

and letting y_{n+1} denotes the value y_{Z+} and y_{n+2} denotes the value y_{Z-}, enables the total energy function to be represented in the following form

$$E(D) = \sum_{i,j} H_{i,j}(y_i - y_j)^2 \tag{7}$$

subject to the constraint where $\sum_i y_i^2 = 1$ and $\sum_i y_i = 0$. This optimization problem can then be solved via standard spectral methods. Before going into details of using different spectral methods for solving the above optimization problems, the relationship with graph-based method is first discussed.

3 Graph-Based Modeling

Constrained minimum cuts incorporating labeled data into similarity graph has been proposed for semi-supervised learning (Li, 2001). The minimization of the constrained cut is approximated via a variation of the randomized contraction algorithm of (Karger & Stein, 1996) and the class labels are estimated via computing the class probability via averaging over an ensemble of approximate minimum cuts partitions. Another type of similarity graph has been proposed by (Blum & Chawla, 2001) which provide several theoretical justification in terms of LOOCV errors of nearest neighbor classifier and a generative model from an underlying distribution.

We model the data on an undirected weighted graph $G = (V, E)$ where V is the set of vertices and E is the set of edges. The weight on edge represents the similarity between the data. The similarity matrix K_{ij} represents the similarity between the vector $\mathbf{x}_i$ and $\mathbf{x}_j$. The exponential similarity

$$k(\mathbf{x}_i, \mathbf{x}_j) = \exp\,(-\frac{||\mathbf{x}_i - \mathbf{x}_j||^2}{\sigma^2}) \tag{8}$$

is adopted in this work.

3.1 Constrained Minimum Cut for Semi-supervised Learning

Given labeled data of two classes stored in the sets a,b where $i \in a$ if $y_i = +1$ and $i \in b$ if $y_i = -1$, and $a \cap b = \phi$. The constrained minimum cut is a constrained combinatorial minimization for finding a disjoint partition of V into A and B such that,

$$A, B = \arg\min cut(A, B) \quad \text{where} \quad a \subset A \quad \text{and} \quad b \subset B. \tag{9}$$

This constrained minimum cut partitions the unlabeled data into sets with smallest similarity while each of the partitions contains only vertices having identical training labels.

The above optimization involves hard constraint where the labeled data act as constraint for the minimum cut optimization. If the hard constraint is replaced by fixed cost function, the criterion function for semi-supervised learning is

$$A, B = \arg\min cut(A, B) + \rho g(A, B, a, b) \tag{10}$$

where ρ is a positive integer balancing the compactness of the segmented graph and the agreement with the labeled samples and $g(a, b)$ is a function measuring the disagreement of the labeled samples with the estimated classification. Setting the balancing constant ρ to approach infinity would be equivalent to the constrained minimum cut formulation in Eqn. 9. The number of labeled samples that is assigned to the wrong class can be used to construct g,

$$g(A, B, a, b) = N(a \cap B) + N(b \cap A) \tag{11}$$

where $N(.)$ is the number of elements in the set. Thus the function g measures the number of labeled data that is assigned to other class. The optimization problem in Eqn.10 can be simplified by integrating the cost function g into the original graph G. The augmented graph G' can be constructed by adding new edges and new nodes to the original graph G. We can construct two extra nodes l_{-1}, l_{+1}, each of which represents the two class labels. The cost function of labeled samples can then be embedded in the augmented graph by adding edges connecting the above nodes with the labeled samples,

$$\begin{aligned} K'_{il_{+1}} &= \rho \quad \text{if } i \in a \\ K'_{il_{-1}} &= \rho \quad \text{if } i \in b \end{aligned} \tag{12}$$

It is straightforward to observe that the minimum cut on the augmented graph is equivalent to minimizing the criterion function in Eqn. 10 under the assumption that the label nodes l_{-1}, l_{+1} are not merged into a single class.

4 Spectral Energy Minimization

Graph spectral partitioning is an effective approximation for graph partitioning (Kannan et al., 2000), (Drineas et al., 1999), (Ding et al., 2001), (Cristianini et al., 2002). Furthermore, studies in using spectral methods for optimal dimension embedding provides additional motivation for its natural capabilities of representing high dimension data in low dimension (Belkin & Niyogi, 2002).

In this section, we discuss the spectral methods for minimizing the energy function in Eqn.7. First we notice that the energy minimization has a trial solution at y with values all $1s$ or $-1s$. We set $y^T e = 0$ to restrict the minimization to a balanced partitioning

$$E = \frac{1}{2}(\sum_{i,j} H_{ij} - \mathbf{y}'\mathbf{H}\mathbf{y}) = \frac{1}{2}\mathbf{y}'\mathbf{L}\mathbf{y}, \tag{13}$$

where $\mathbf{L} = \mathbf{D} - \mathbf{H}$ and $\mathbf{D} = \text{diag}(d_1, \dots, d_n)$ with $d_i = \sum_{j=1}^{n} H_{ij}$. Let the eigenvector of L associated with the second smallest eigenvalue be denoted as $\mathbf{y}_2$, which is orthogonal to the eigenvector, e, corresponding to the smallest eigenvalue, 0.

The normalized cut alternatively minimizes the following objective cut under the constraint $yDe = 0$.

$$\begin{aligned} & E \\ = & \textstyle\sum_{i\in A, j\in B} H_{ij}\left(\frac{1}{\sum_{i\in A, j\in A\cup B} H_{ij}} + \frac{1}{\sum_{i\in B, j\in A\cup B} H_{ij}}\right) \\ \propto & \frac{\mathbf{y}'\mathbf{L}\mathbf{y}}{\mathbf{y}'\mathbf{D}\mathbf{y}}. \end{aligned}$$

And by solving the generalized eigen-system

$$(\mathbf{D} - \mathbf{H})V = \lambda \mathbf{D} V, \tag{14}$$

the eigenvector corresponding to the second smallest eigenvalue y_2 is used as the average cut case.

The algorithm for the spectral energy minimization is as follows:

1. Construct similarity matrix K with all samples
2. Construct probability+similarity matrix H by adding probability of classification
3. Compute the k-th smallest eigenvalues and the associated eigenvectors by solving the suitable eigenvalue problem
4. Perform classification on the k-th dimension subspace with the k-th dimension eigenvector using labeled data

In previous work in clustering, the choice of the number of eigenvectors is often limited to observations and intuitions. There is considerable debate on whether the second eigenvector is sufficient and whether the more eigenvectors the better is the clustering. However, in the case of semi-supervised learning, the errors on the training samples can be used to select the number of dimensions to be employed.

A simple classifier based on weighted distance constructed by weighting all training samples with exponential distances has been adopted. Assuming the similarity is constructed as in Eqn.8, the classifier's output can be represented via the following

$$H_{i,n+1} = \frac{\sum_t K_{i,t} y_t}{\sum_t K_{i,t}} \tag{15}$$

where t indices only the labeled data and $H_{i,n+2} = 1 - H_{i,n+1}$. Similarly, the entries on the transpose is constructed. Although this classifier has relatively poor performance on its own, the total energy minimization does achieves very good performance.

5 Experiments

The spectral method is tested on the following datasets: crab, Wisconsin breast cancer, pima indian diabetes, vote, and ionosphere. For the spectral method, the scale constant σ in the radial basis function is set as the average value of eight nearest-neighbour distance. The balance constant λ controlling the fidelity to training labels and cluster compactness is set as a fixed value of 2π for all datasets. The choice of the value of 2π is only arbitrary as the algorithm is not sensitive to the selection of scale constant. All experiments are repeated ten times and the average classification error on the unlabeled dataset are reported. The crab data consists of two species, each species is further broken down into two sexes (Ripley, 1996). The classification of the species is relatively easy and the distinction of the two sexes is more challenging. In this experiment, the task is to classify the sex of the crab given a limited number of training samples.

5.1 Comparison with the Semi-supervised kmeans Algorithm

Figure 1(a) shows the classification error of the spectral method compared with the semi-supervised k-means for the crab dataset (Basu et al., 2002). It is clear from the figure that the proposed spectral method is capable of learning from both labeled and unlabeled data and perform better than both the constrained k-mean and the seeded k-mean algorithm.

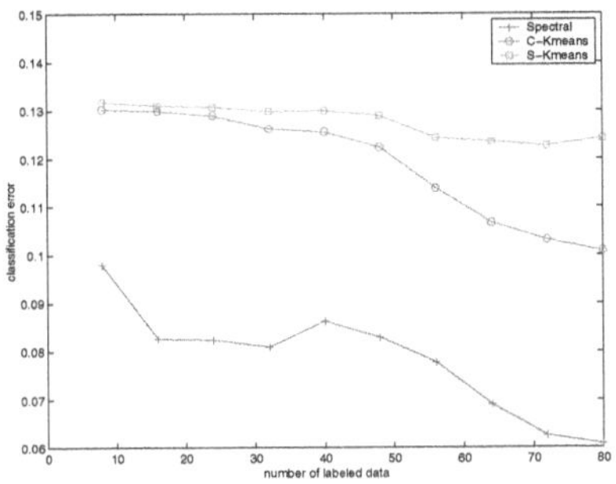

(a) Crab

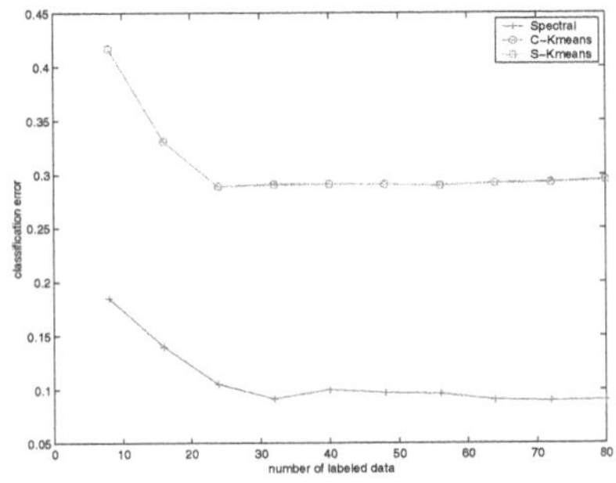

(b) Inonsphere

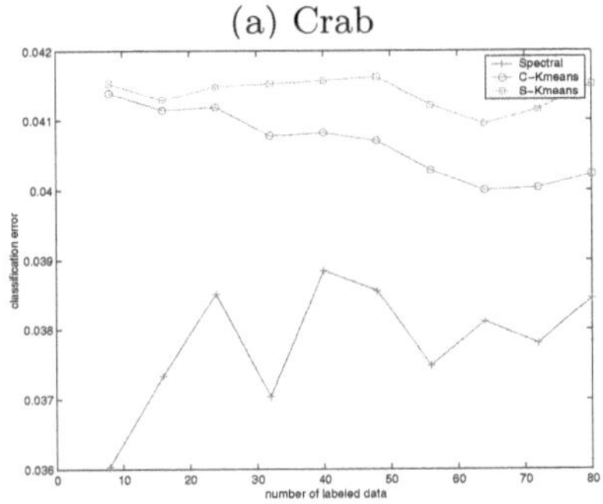

(c) Breast Cancer

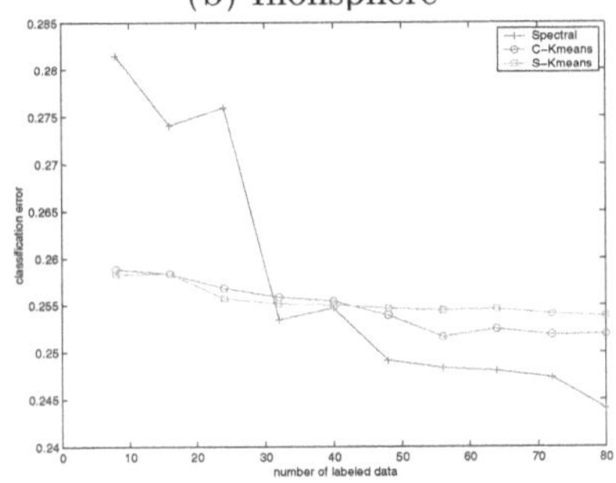

(d) Diabetes

Fig. 1. Average Classification Error

Figure 1(b) shows the classification error of the spectral method compared with the semi-supervised k-means for the ionosphere dataset. The spectral method performs also very good on this dataset compared with the semi-supervised k-means algorithm. Furthermore, comparing the semi-supervised graph mincut's accuracy 81.6% using 50 training samples, the spectral algorithm achieve an accuracy of 88.3% at 48 training samples (Blum & Chawla, 2001).

Figure 1(c) shows the classification error of the spectral method compared with the semi-supervised k-means for the breast cancer dataset. The spectral method performs also very good on this dataset compared with the semi-supervised k-means algorithm. Although there are fluctuations in the classification errors of the spectral algorithm, the range of the fluctuations are within 0.5% which is reasonable given the randomly drawn training samples. Figure 1(d) shows the classification error of the spectral method compared with the semi-supervised k-means for diabetes dataset. All data with missing values are removed from the dataset. The spectral method has larger errors when there is a smaller number of training data while the semi-supervised k-means does not gain a lot of performance with more training data.

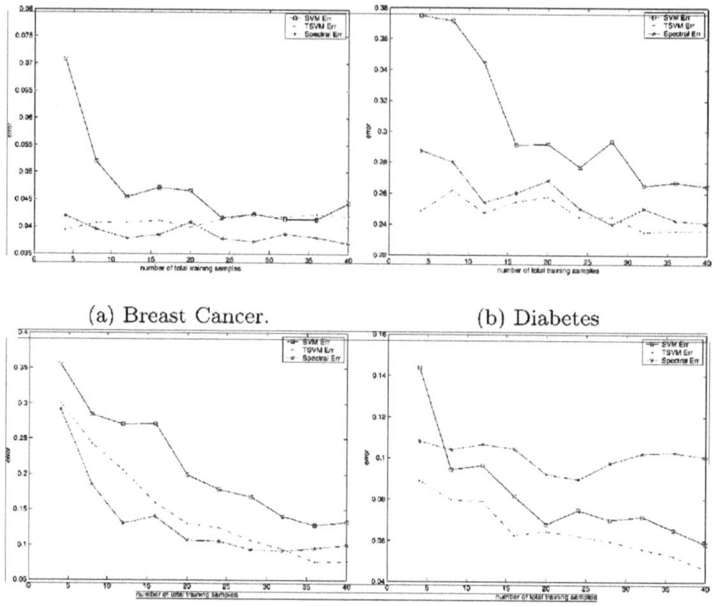

(a) Breast Cancer. (b) Diabetes

(c) Ionosphere (d) Votes

Fig. 2. Average Classification Error c.f. TSVM

5.2 Comparison with the Transductive Support Vector Machine

In this section, comparisons are performed with the transductive support vector machine (TSVM) and support vector machine (Joachims, 1999). The tests with the TSVM are difficult to be performed as there are different parameters to be estimated and the algorithm does not always converge. Effort has been spent on obtaining the best possible results using leave-one-out cross validation (LOOCV) for selecting the parameters of the TSVM. For each set of training data, a pair of σ and C that leads to the best LOOCV accuracy is obtained through grid search on the region spanned by $\sigma \in \{2^i | i = -5, \ldots, 5\}$ and $C \in \{2^i | i = -5, \ldots, 10\}$. However, the computational requirement of such a process is order of magnitude larger than the spectral approach and a lot of manual effort has been spend in setting the parameters in optimization.

Figure 5.1(a) to Figure 5.1(d) shows the classification error of the spectral method compared with SVM and TSVM. The spectral method with fixed parameter and simple classifier probability estimation is able to achieve better performance than SVM and comparable performance to TSVM in some datasets. It is not surprising the TSVM has in general superior performance especially when there are more training data.

6 Conclusion

Posing the semi-supervised learning as the optimization of both the classification energy and the cluster compactness energy is a flexible representations where different classifiers and different similarity criteria can be adopted. The resulting integer programming can be relaxed by a semi-definite relaxation which allows efficient solutions to be obtained via spectral graph methods. The spectral energy minimization method perform very well when compared with constrained k-means and SVM. Although results from spectral energy minimization do not exceed the performance of the best result achieved by the transductive SVM, it provides a fast solution without difficult parameters estimations and convergence problems.

References

Basu, S., Banerjee, A., & Mooney, R. J. (2002). Semi-supervised clustering by seeding. *Proceedings of the Nineteenth International Conference on Machine Learning (ICML-2002)* (pp. 19-26).

Belkin, M., & Niyogi, P. (2002). Laplacian eigenmaps for dimensionality reduction and data representation. *Technical Report, TR-2002-1.*

Blum, A., & Chawla, S. (2001). Learning from labeled and unlabeled data using graph mincuts. *The Eighteenth International Conference on Machine Learning.*

Cristianini, N., Shawe-Taylor, J., & Kandola, J. (2002). Spectral kernel methods for clustering. *Neural Information Processing Systems.*

Ding, C.H.Q., He, X., Zha, H., Gu, M., & Simon, H. D. (2001). A min-max cut algorithm for graph partitioning and data clustering. *Proceedings of ICDM* (pp. 107–14)

Drineas, Frieze, Kannan, Vempala, & Vinay (1999). Clustering in large graphs and matrices. *ACM-SIAM Symposium on Discrete Algorithms (A Conference on Theoretical und Experimental Analysis of Discrete Algorithms).*

Joachims, T. (1999). Transductive inference for text classification using support vector machines. *International Conference on Machine Learning (ICML)* (pp. 200-209). Morgan-Kaufman.

Kannan, R., Vempala, S., & A., V. (2000). On clusterings-good, bad and spectral. *Proceedings of the 41st Annual Symposium on Foundations of Computer Science.*

Karger, D. R., & Stein, C. (1996). A new approach to the minimum cut problem. *Journal of the ACM, 43,* 601-640.

Li, C. H. (2001). Constrained minimum cut for classification using labeled and unlabeled data. *Proceedings of the IEEE Conference on Computer Vision and Pattern Recognition.* IEEE Computer Society Press.

Ripley, B. (1996). *Pattern recognition and neural networks.* Cambridge, UK: Cambridge University Press.

Discriminative Methods for Multi-labeled Classification

Shantanu Godbole and Sunita Sarawagi

KReSIT, IIT Bombay Powai, Mumbai, 400076, India
{shantanu,sunita}@it.iitb.ac.in

Abstract. In this paper we present methods of enhancing existing discriminative classifiers for multi-labeled predictions. Discriminative methods like support vector machines perform very well for uni-labeled text classification tasks. Multi-labeled classification is a harder task subject to relatively less attention. In the multi-labeled setting, classes are often related to each other or part of a is-a hierarchy. We present a new technique for combining text features and features indicating relationships between classes, which can be used with any discriminative algorithm. We also present two enhancements to the margin of SVMs for building better models in the presence of overlapping classes. We present results of experiments on real world text benchmark datasets. Our new methods beat accuracy of existing methods with statistically significant improvements.

1 Introduction

Text classification is the task of assigning documents to a pre-specified set of classes. Real world applications including spam filtering, e-mail routing, organizing web content into topical hierarchies, and news filtering rely on automatic means of classification. Text classification can be broadly categorized into discriminative techniques, typified by support vector machines [1] (SVMs), decision trees [2] and neural networks; and generative techniques, like Naïve Bayes (NB) and Expectation Maximization (EM) based methods. From a performance point of view, NB classifiers are known to be the fastest, learning a probabilistic generative model in just one pass of the training data. Their accuracy is however relatively modest. At the other end of the spectrum lie SVMs based on elegant foundations of statistical learning theory [3]. Their training time is quadratic to the number of training examples, but they are known to be the most accurate.

The simplest task in text classification is to determine whether a document belongs to a class of interest or not. Most applications require the ability to classify documents into one out of many (> 2) classes. Often it is not sufficient to talk about a document belonging to a single class. Based on the granularity and coverage of the set of classes, a document is often about more than one topic. A document describing the politics involved in the sport of cricket, could be classified as `Sports/Cricket`, as well as `Society/Politics`. When a document can belong to more than one class, it is called multi-labeled. Multi-labeled classification is a harder problem than just choosing one out of many classes.

H. Dai, R. Srikant, and C. Zhang (Eds.): PAKDD 2004, LNAI 3056, pp. 22–30, 2004.

In this paper, we present algorithms which use existing discriminative classification techniques as building blocks to perform better multilabeled classification. We propose two enhancements to existing discriminative methods. First, we present a new algorithm which exploits correlation between related classes in the label-sets of documents, by combining text features and information about relationships between classes by constructing a new kernel for SVMs with heterogeneous features. Next, we present two methods of improving the margin of SVMs for better multilabeled classification. We present experiments comparing various multilabeled classification methods. Following this, we review related work and conclude with future research directions.

2 Multi-labeled Classification Using Discriminative Classifiers

Suppose we are given a vector space representation of n documents. In the bag-of-words model, each document vector d_i has a component for each term feature which is proportional to it's importance (term frequency or TFIDF are commonly used). Each document vector is normalized to unit L_2 norm and is associated with one of two labels, +1 or −1. The training data is thus $\{(d_j, c_i), j = 1, \ldots, n\}$, $c_i \in \{-1, +1\}$.

A linear SVM finds a vector $\mathbf{w}$ and a scalar constant b, such that for all i, c_i $(\mathbf{w}_{c_i} \cdot d_j + b) \geq 1$, and $||\mathbf{w}||$ is minimized. This optimization corresponds to fitting the thickest possible slab between the positive ($c = +1$) and negative ($c = -1$) documents.

Most discriminative classifiers, including SVMs, are essentially twoclass classifiers. A standard methods of dealing with multi-class problems is to create an *ensemble* of yes/no binary classifiers, one for each label. This method is called *one-vs-others*. For each label l_i, the positive class includes all documents which have l_i as one of their labels and the negative side includes all other documents. During application, the set of labels associated with a document d_j is $\{k\}$, such that $\mathbf{w}_k \cdot d_j + b_k > 0$. This is the basic SVM method (denoted SVM) that serves as a baseline against which we compare other methods.

2.1 Limitations of the Basic SVM Method

Text classification with SVMs is faced with one issue; that of all classifiers in an ensemble rejecting instances. In one-vs-others, all constituents of the ensemble emit a $(\mathbf{w}_c \cdot d + b_c)$ score; for multi-labeled classification we admit all classes in the predicted set, whose score $\mathbf{w}_c \cdot d + b_c > 0$. However, in practice, we find that a significant fraction of documents get negative scores by all the classifiers in the ensemble.

Discriminative multi-class classification techniques, including SVMs, have historically been developed to assign an instance to exactly one of a set of classes that are assumed to be disjoint. In contrast, multi-labeled data, by its very nature, consists of highly correlated and overlapping classes. For instance, in the

Reuters-21578 dataset, there are classes like *wheat–grain*, *crude–fuel*, where one class is almost a parent of the other class although this knowledge is not explicitly available to the classifier. Such overlap among classes hurts the ability of discriminative methods to identify good boundaries for a class. We devise two techniques to handle this problem in Section 4. Correlation between classes can be a boon as well. We can exploit strong mutual information among subsets of classes to "pull up" some classes when the term information is insufficient. In the next section, we present a new method to directly exploit such correlation among classes to improve multi-label prediction.

3 Combining Text and Class Membership Features

The first opportunity for improving multi-labeled classification is provided by the co-occurrence relationships of classes in label sets of documents. We propose a new method for exploiting these relationships.

If classification as class C_i is a good indicator of classification as class C_j , one way to enhance a purely text-based SVM learner is to augment the feature set with $|C|$ extra features, one for each label in the dataset. The cyclic dependency between features and labels is resolved iteratively.

Training: We first train a normal text-based SVM ensemble $S(0)$. Next, we use $S(0)$ to augment each document $d \in D$ with a set of $|C|$ new columns corresponding to scores $\mathbf{w}_{c_i} \cdot d + b_{c_i}$ for each class $c_i \in C$. All positive scores are transformed to $+1$ and all negative scores are transformed to -1. In case all scores output by $S(0)$ are negative, the least negative score is transformed to $+1$. The text features in the original document vector are scaled to $f(0 \leq f \leq 1)$, and the new "label dimensions" are scaled to $(1-f)$. Documents in D thus get a new vector representation with $|T|+|C|$ columns where $|T|$ is the number of term features. They also have a supervised set of labels. These are now used to train a new SVM ensemble $S(1)$. We call this method *SVMs with heterogeneous feature kernels* (denoted SVM-HF). The complete pseudo-code is shown in Figure 1. This approach is directly related to our previous work on Cross- Training [4] where label mappings between two different taxonomies help in building better classification models for each of the taxonomies.

Testing: During application, all test documents are classified using $S(0)$. For each document, the transformed scores are appended in the $|C|$ new columns with appropriate scaling. These document are then submitted to $S(1)$ to obtain the final predicted set of labels.

The scaling factor: The differential scaling of term and feature dimensions has special reasons. This applies a special kernel function to documents during training $S(1)$. The kernel function in linear SVMs gives the similarity between two document vectors, $K_T(d_i, d_j) = \frac{\langle d_i \cdot d_j \rangle}{|d_i||d_j|}$. When document vectors are scaled to unit L_2 norm, this becomes simply the $\cos\theta$ of the angle between the two document vectors, a standard IR similarity measure. Scaling the term and label dimensions sets up a new kernel function given by $K(d_i, d_j) = f \cdot K_T(d_i, d_j) + (1-f) \cdot K_L(d_i, d_j)$, where K_T is the usual dot product kernel between terms

```
Represent each document as a vector d in term space and ||d|| = 1
Build one-vs-rest SVM classifier S(0) using text tokens only
for each document d ∈ D do
    Apply S(0) to d, getting a vector γ_C(d) of |C| scores (see text)
    Concatenate vectors d and γ_C(d) into a single training vector with label
    carried over from S(0), with relative term-label weight determined by f,
    maintaining ||d|| = 1
    Add this vector into the training set of S(1)
end for
Induce a new one-vs-rest SVM classifier S(1) for all d ∈ D
```

Fig. 1. SVMs with heterogeneous feature kernels

and K_L is the kernel between the label dimensions. The tunable parameter f is chosen through cross-validation on a held out validation set. The label dimensions interact with each other independent of the text dimensions in the way we set up the modified kernel. Just scaling the document vector suitably is sufficient to use this kernel and no change in code is needed.

4 Improving the Margin of SVMs

In multi-labeled classification tasks, the second opportunity for improvement is provided by tuning the margins of SVMs to account for overlapping classes. It is also likely that the label set attached with individual in- stances is incomplete. Discriminative methods work best when classes are disjoint. In our experience with the Reuters-21578 dataset, multi-labeled instances often seem to have incomplete label sets. Thus multi-labeled data are best treated as 'partially labeled'. Therefore, it is likely that the 'others' set includes instances that truly belong to the positive class also. We propose two mechanisms of removing examples from the large negative set which are very similar to the positive set. The first method does this at the document level, the second at the class level.

4.1 Removing a Band of Points around the Hyperplane

The presence of very similar negative training instances on the others side for each classifier in an SVM ensemble hampers the margin, and re-orients the separating hyperplanes slightly differently than if these points were absent. If we remove these points which are very close to the resultant hyperplane, we can train a better hyperplane with a wider margin. The algorithm to do this consists of two iterations:

1. In the first iteration, train the basic SVM ensemble.
2. For each SVM trained, remove those negative training instances which are within a threshold distance (band) from the learnt hyperplane. Re-train the ensemble.

We call this method the *band-removal* method (denoted BandSVM). When selecting this band, we have to be careful not to remove instances that are crucial in defining the boundary of the others class.We use a heldout validation dataset to choose the band size. An appropriate band-size tries to achieve the fine balance between large-margin separation, achieved by removing highly related points, and over-generalization, achieved by removing points truly belonging to the negative class.

4.2 Confusion Matrix Based "Others" Pruning

Another way of countering very similar positive and negative instances, is to completely remove all training instances of 'confusing' classes. Confusing classes are detected using a confusion matrix quickly learnt over held out validation data using any moderately accurate yet fast classifier like naïve Bayes [5]. The confusion matrix for a n-class problem is nXn matrix M, where the ij^{th} entry, M_{ij}, gives the percentage of documents of class i which were misclassified as class j. If M_{ij} is above a threshold β, we prune away all confusing classes (like j) from the 'others' side of i when constructing a i-vs-others classifier. This method is called the *confusionmatrix based pruning method* (denoted ConfMat). This two-step method is specified as:

1. Obtain a confusion matrix M over the original learning problem using any fast, moderately accurate classifier. Select a threshold β.
2. Construct a one-vs-others SVM ensemble. For each class i, leave out the entire class j from the 'others' set if $M_{ij} > \beta$.

If the parameter β is very small a lot of classes will be excluded from the others set. If it is too small, none of the classes may be excluded resulting in the original ensemble. β is chosen by cross-validation.

ConfMat is faster to train than BandSVM, relying on a confusion matrix given by a fast NB classifier, and requires only one SVM ensemble to be trained. The user's domain knowledge about relationships between classes (e.g. hierarchies of classes) can be easily incorporated in ConfMat.

5 Experiments

We describe experiments with text classification benchmark datasets and report the results of a comparison between the various multi-labeled classification methods.We compare the baseline SVM method with ConfMat, BandSVM, and SVM-HF.

All experiments were performed on a 2-processor 1.3GHz P3 machine with 2GB RAM, running Debian Linux. *Rainbow*[1] was used for feature and text processing and SVMLight[2] was used for all SVM experiments.

[1] http://www.cs.cmu.edu/ mccallum/bow/

[2] http://svmlight.joachims.org

5.1 Datasets

Reuters-21578: The Reuters-21578 Text Categorization Test Collection is a standard text categorization benchmark. We use the Mod-Apte split and evaluate all methods on the given train/test split with 135 classes. We also separately use random 70–30 train/test splits (averaged over 10 random splits), to test statistical significance, for a subset of 30 classes. We did feature selection by using stemming, stopword removal and only considered tokens which occurred in more than one document at least once, and selected the top 1000 features by mutual information.

Patents: The Patents dataset is another text classification benchmark. We used the wipo–alpha collection which is an English language collection of patent applications classified into a hierarchy of classes with subclasses and groups. We take all 114 sub-classes of the top level (A to H) using the given train/test split. We also report average over 10 random 70–30 train/test splits for the F sub-hierarchy. We consider only the text in the abstract of the patent for classification and feature selection is the same as that for the Reuters dataset.

5.2 Evaluation Measures

All evaluation measures discussed are on a per instance basis and the aggregate value is an average over all instances. For each document d_j, let T be the true set of labels, S be the predicted set of labels. Accuracy is measured by the Hamming score which symmetrically measures how close T is to S. Thus, Accuracy $(d_j) = |T \cap S|/|T \cup S|$. The standard IR measures of Precision (P), Recall (R) and F_1 are defined in the multilabeled classification setting as $P(d_j) = |T \cap S|/|S|$, $R(d_j) = |T \cap S|/|T|$, and $F_1(d_j) = 2P(d_j)R(d_j)/(P(d_j) + R(d_j))$.

5.3 Overall Comparison

Figures 2 and 3 shows the overall comparison of the various methods on the Reuters and Patents datasets. Figure 2 shows comparison on all 135 classes of Reuters as well as results of averaging over 10 random train/test splits on a subset of 30 classes. Figure 3 shows the comparison for all 114 subclasses of Patents and average over 10 random train/test splits on the F class sub-hierarchy. For both datasets we see that SVM-HF has the best overall accuracy. SVM has the best precision and ConfMat has the best recall. We also observe that BandSVM and SVM-HF are very comparable for all measures.

We did a directional *t–test* of statistical significance between the SVM and SVM-HF methods for the 30 class subset and the F sub-hierarchy. The accuracy and F1 scores of SVM-HF were 2% better than SVM, being a small but significant difference at 95% level of significance. The t values were 2.07 and 2.02 respectively over the minimum required value of 1.73 for $d_f = 18$.

Method	30 class subset				All 135 classes			
	Accuracy	Precision	Recall	F1	Accuracy	Precision	Recall	F1
SVM	82.02	**92.65**	82.47	87.26	81.26	**92.41**	82.45	87.15
ConfMat	76.16	81.64	**88.00**	84.7	80.92	87	**88.37**	87.68
BandSVM	83.18	89.87	87.41	88.63	81.73	88.44	87.54	**87.99**
SVM-HF	**84.25**	91.56	86.94	**89.19**	**82**	88.66	87.27	**87.96**

Fig. 2. The Reuters-21578 dataset

Method	F class sub-hierarchy				All 114 subclasses			
	Accuracy	Precision	Recall	F1	Accuracy	Precision	Recall	F1
SVM	66.65	**73.65**	67.57	70.48	42.47	**56.76**	43.37	49.16
ConfMat	66.62	69.70	**70.63**	70.16	41.67	53.40	**51.65**	**52.51**
BandSVM	67.30	72.09	68.90	70.45	43.30	55.24	48.61	51.70
SVM-HF	**68.86**	72.06	69.78	**70.90**	**44.41**	55.35	49.84	**52.45**

Fig. 3. The Patents dataset

5.4 Interpreting Co-efficients

With all documents scaled to unit L_2 norm, inspecting the components of w along the label dimensions derived by SVM-HF gives us some interesting insights into various kinds of mappings between the labels. The signed components of w along the label dimensions represent the amount of positive or negative influence the dimension has in classifying documents. As an example for the Reuters dataset, the label dimension for *grain* (+8.13) is highly indicative of the class *grain*. *Wheat* (+1.08) also has a high positive component for *grain*, while *money-fx* (−0.98) and *sugar* (−1.51) have relatively high negative components. This indicates that a document getting classified as *wheat* is a positive indicator of the class *grain*; and a document classified as *sugar* or *money-fx* is a negative indicator of the class *grain*.

5.5 Comparing Number of Labels

Figure 4 shows the size of the true set of labels T, and the predicted set S. We fix $|S|$ to be 1, 2, 3 for each $|T| = 1, 2, 3$. For instance, for $|T| = 1$, $|S| = 1$ for 99% of the instances for the SVM method, and only 1% of the instances are assigned $|S| = 2$. For singleton labels, SVM is precise and admits only one label whereas other methods admit a few extra labels.

When $|T| = 2, 3$, we see that SVM still tends to give lesser number of predictions, often just one, compared to the other methods which have a high percentage of instances in the $|T| = |S|$ column. One reason for this is the way one-vs-others is resolved. All negative scores in one-vs-others are resolved by choosing the least negative score and treating this as positive. This forces the prediction set size to be 1 and the semantics of least negative is unclear. The percentages of documents assigned all negative scores by SVM is 18% for 30

	T=1			T=2			T=3		
Corresponding S=	1	2	3	1	2	3	1	2	3
SVM	0.99	0.01	0	0.5	0.5	0	0.52	0.35	0.13
ConfMat	0.83	0.14	0.03	0.27	0.63	0.1	0.17	0.3	0.48
BandSVM	0.89	0.09	0.01	0.32	0.64	0.03	0.22	0.3	0.43
SVM-HF	0.94	0.06	0.01	0.34	0.63	0.02	0.3	0.26	0.39

Fig. 4. Percentage of instances with various sizes of S for T=1,2,3 with 30 classes of Reuters. Here, 68% of all test instances in the dataset had T=1; 22% had T=2; 8% had T=3; others had T greater than 3.

classes of Reuters, while ConfMat, BandSVM, and SVM-HF assign all negative scores to only 4.94%, 6.24%, and 10% of documents respectively.

6 Related Work

Limited work has been done in the area of multi-labeled classification. Crammer *et al.* [6] propose a one-vs-others like family on online topic ranking algorithms. Ranking is given by $\mathbf{w}_{c_i} \cdot x$ where the model for each class $\mathbf{w}_{c_i}$ is learnt similar to perceptrons, with an update of $\mathbf{w}_{c_i}$ in each iteration, depending on how imperfect ranking is compared to the true set of labels. Another kernel method for multi-labeled classification tested on a gene dataset is given by Elisseeff *et al.* [7]. They propose a SVM like formulation giving a ranking function along with a set size predictor. Both these methods are topic ranking methods, trying to improve the ranking of all topics. We ignore ranking of irrelevant labels and try to improve the quality of SVM models for automatically predicting labels. The ideas of exploiting correlation between related classes and improving the margin for multi-label classification are unique to our paper.

Positive Example Based Learning–PEBL [8] is a semi-supervise learning method similar to BandSVM. It also uses the idea of removing selected negative instances. A disjunctive rule is learned on features of strongly positive instances. SVMs are iteratively trained to refine the positive class by selectively removing negative instances. The goal in PEBL is to learn from a small positive and a large unlabeled pool of examples which is different from multi-labeled classification.

Multi-labeled classification has also been attempted using generative models, although discriminative methods are known to be more accurate. McCallum [9] gives a generative model where each document is probabilistically generated by all topics represented as a mixture model trained using EM. The class sets which can generate each document are exponential in number and a few heuristics are required to efficiently search only a subset of the class space. The Aspect model [10] is another generative model which can be naturally employed for multi-labeled classification, though no current work exists. Documents are probabilistically generated by a set of topics and words in each document are generated by members of this topic set. This model is however used for unsupervised clustering and not for supervised classification.

7 Conclusions

We have presented methods for discriminative multi-labeled classification. We have presented a new method (SVM-HF) for exploiting co-occurrence of classes in label sets of documents using iterative SVMs and a general kernel function for heterogeneous features. We have also presented methods for improving the margin quality of SVMs (BandSVM and ConfMat). We see that SVM-HF performs 2% better in terms of accuracy and F_1 than the basic SVM method; a small but statistically significant difference. We also note that SVM-HF and BandSVM are very comparable in their results, being better than ConfMat and SVM. ConfMat has the best recall, giving the largest size of the predicted set; this could help a human labeler in the data creation process by suggesting a set of closely related labels.

In future work, we would like to explore using SVMs with the positive set containing more than one class. The composition of this positive set of related candidate classes is as yet unexplored. Secondly, we would like to theoretically understand the reasons for accuracy improvement in SVM-HF given that there is no extra information beyond terms and linear combinations of terms. Why should the learner pay attention to these features if all the information is already present in the pure text features? We would also like to explore using these methods in other application domains.

Acknowledgments. The first author is supported by the Infosys Fellowship Award from Infosys Technologies Limited, Bangalore, India. We are grateful to Soumen Chakrabarti for many helpful discussions, insights and comments.

References

1. T. Joachims. Text categorization with support vector machines: learning with many relevant features. In *Proceedings of ECML*-1998.
2. R. E. Schapire and Y. Singer. Boostexter: A boosting-based system for text categorization. *Machine Learning*, 39(2/3):135–168, 2000.
3. V. Vapnik. *Statistical Learning Theory*, John Wiley, 1998.
4. S. Sarawagi, S. Chakrabarti, and S. Godbole. Cross training: learning probabilistic mappings between topics. In *Proceedings of the ACM SIGKDD*-2003.
5. S. Godbole, S. Sarawagi, and S. Chakrabarti. Scaling multi-class support vector machines using inter-class confusion. In *Proceedings of ACM SIGKDD*-2002.
6. K. Crammer and Y. Singer. A family of additive online algorithms for category ranking. *Journal of Machine Learning Research*, 1025–1058, 2003.
7. A. Elisseeff and J. Weston. Kernel methods for multi-labelled classification and categorical regression problems. Technical Report, BioWulf Technologies, 2001.
8. H. Yu, J. Han, and K. C-C. Pebl: Positive example-based learning for web page classification using SVM. In *Proceedings of ACM SIGKDD*-2002.
9. A. McCallum. Multi-label text classification with a mixture model trained by EM. *AAAI Workshop on Text Learning*-1999.
10. T. Hofmann and J. Puzicha. Unsupervised learning from dyadic data. *Technical Report TR-98-042*, Berkeley, 1998.

Subspace Clustering of High Dimensional Spatial Data with Noises

Chih-Ming Hsu and Ming-Syan Chen

Electrical Engineering Department
Naional Taiwan University
Taipei, Taiwan, ROC
{ming, mschen}@arbor.ee.ntu.edu.tw

Abstract. Clustering a large amount of high dimensional spatial data sets with noises is a difficult challenge in data mining. In this paper, we present a new subspace clustering method, called SCI (Subspace Clustering based on Information), to solve this problem. The SCI combines Shannon information with grid-based and density-based clustering techniques. The design of clustering algorithms is equivalent to construct an equivalent relationship among data points. Therefore, we propose an equivalent relationship, named density-connected, to identify the main bodies of clusters. For the purpose of noise detection and cluster boundary discovery, we also use the grid approach to devise a new cohesive mechanism to merge data points of borders into clusters and to filter out the noises. However, the curse of dimensionality is a well-known serious problem of using grid approach on high dimensional data sets because the number of the grid cells grows exponentially in dimensions. To strike a compromise between the randomness and the structure, we propose an automatic method for attribute selection based on the Shannon information. With the merit of only requiring one data scan, algorithm SCI is very efficient with its run time being linear to the size of the input data set. As shown by our experimental results, SCI is very powerful to discover arbitrary shapes of clusters.

Index Terms: Data clustering, subspace clustering, Shannon information

1 Introduction

Data mining has attracted a significant amount of research attention to discover useful knowledge and information. Data clustering is an important technique on data mining with the aim of grouping data points into meaningful subclasses. It has been widely used in many applications such as data analysis, pattern recognition, and image processing.

Spatial data possesses much information related to the space. Increasingly large amounts of data are obtained from satellite images, medical equipments, multimedia databases, etc. Therefore, automated knowledge discovery in spatial data sets becomes very important. It is known that most existing clustering algorithms, such as k-means, k-medoids single-link and complete-link, have several drawbacks when being applied to high dimensional large spatial data sets [10].

H. Dai, R. Srikant, and C. Zhang (Eds.): PAKDD 2004, LNAI 3056, pp. 31–40, 2004.

The grid approach partitions the data space into disjoint grid cells. There is high probability that all data points belong to the same cluster within the same grid cell. Therefore, all data points that fall into the same cell can be aggregated and treated as one object. All of the clustering operations, which are performed on the grid structure, can be improved in terms of the processing time [1] [9]. Note that the total number of the grid cells is independent of the number of data points. On the other hand, each cluster has a typical density of data points that is higher than that of those points outside of the cluster [6]. In addition, there are some space relationships among the data points of spatial data sets. The data points inside a neighborhood for a given radius of a certain point shall be more similar to each other than those outside the neighborhood. Hence, the density-based approach can capture the main bodies of clusters [9] [14]. Another benefit of those approaches is the independence of the input order of the data points, because the merging of cells into cluster depends only on their space relationships and the densities. Some clustering algorithms, such as Grid clustering[13], STING[15], CLIQUE[1], and WaveCluster[14], are proposed to integrate the density-based and grid-based clustering techniques. However, there still exist some problems to solve, such as overfitting phenomenon, the confusion between noises (or outliers) and border points of clusters, and curse of dimensionality, to name a few.

In most related grid-based and density-based clustering algorithms, such as CLIQUE and STING, the two grid cells u_1, u_2 are connected if they have a common face or there exists another grid cell u_3 such that u_1 is connected to u_3 and u_2 is connected to u_3 [1]. A cluster is defined as a maximal set of connected dense grid cells. However, those methods suffer from some problems. The first is the overfitting problem. These methods are likely to divide some meaningful clusters into many subclusters, especially while the data points spread along the direction are not parallel to the axes. For example, consider the scenario of a 2-dimensional data set with one cluster, as shown in Fig. 1(a). The output of the number of clusters on this case is 3 by using CLIQUE or STING. The result is deemed unsuccessful since the phenomenon of overfitting occurs. To remedy this, we shall redefine the meaning of the connection between two grid cells in our work, as will be described in detail later.

The second problem is the detection of the boundary of clusters and noises. The prior grid-based and density-based clustering algorithms are designed in particular to deal with dense grid cells. Sometimes the points of cluster's boundary may not fall into any dense cells, as in the example shown in Fig. 1(b). Hence, we cannot separate the cluster boundaries from noises. Then, most of the cluster boundaries are parallel or vertical to the axis, and some points of borders may be lost. As a result, we design a new cohesive mechanism, based on the grid structure, to detect the boundaries of clusters and to filter out the outliers. In addition, the grid approach is vulnerable to high dimensional data sets, because the number of grid cells is of the exponential order of the dimensions. We offer a new approach to solve this problem by using Shannon information. The Shannon information, also referred to as entropy, is a measure of the uncertainty [2]. Note that structure diminishes uncertainty and randomness confuses the borders. Clustering algorithms group data points into clusters usually in such a way that the structural relationship between those assigned to the same cluster tends to be stronger than those in different clusters. The data points should be of less randomness within the same cluster, and of less structural relationship to those data points in other clusters. Therefore, the attribute is considered redundant if its entropy is too large or too small.

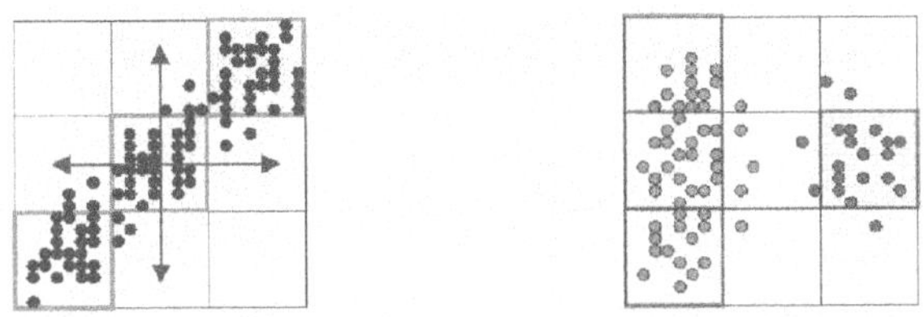

(a) (b)

Fig. 1. (a) The overfitting phenomenon by CLIQUE or STING, (b) the boundary is difficult to detect.

To strike a compromise between the randomness and the structure, we propose an automatic method for attribute selection based on the Shannon information. The proposed subspace clustering method is referred to as algorithm SCI, standing for Subapace Clustering based on Information. With the merit of only requiring one data scan, algorithm SCI is very efficient with its run time being linear to the size of the input data set. In addition, SCI has several advantages, including being scalable to the number of attributes, robust to noises, and independent of the order of input. As shown by our experimental results, SCI is very powerful to discover arbitrary shapes of clusters. In a specific comparison with algorithms CLIQUE, CURE, and k-means, algorithm SCI is able to achieve the best clustering quality while incurring even shorter execution time.

The rest of the paper is organized as follows. Algorithm SCI is developed in Section 2. In Section 3, we present some experimental results. The conclusion is given in Section 4.

2 Subspace Clustering Based on Information

As mentioned before, some conventional grid-based and density-based clustering algorithms are likely to have the following disadvantages: (1) overfitting phenomenon, (2) the confusion between noises (or outliers) and border points of clusters, and (3) the curse of dimensionality. To solve those problems, we propose a new efficient and effective subspace clustering method on high dimensional large data sets with noises, called SCI (Subspace Clustering based on Information). In essence, algorithm SCI is a hierarchical and single scan subspace clustering algorithm. A large dataset is compressed into a highly condensed table. Each cell of this table just stores the count of data points in the corresponding grid cell of data space. The main memory only stores this data table (called summary table) and the data points of spare cells. Such implementation enables this clustering algorithm to be effective and scalable.

2.1 Definitions

Given a set of d-dimensional data set D, our main task is to partition this set into meaningful subclasses. Assume that the spatial data set D has d attributes $(X_1, X_2, \cdots, X_d)$, and each attribute domain is bounded and totally ordered. Without loss of generality, we assume that the domain of attribute is a half open interval $[L_j, U_j)$ for some numbers $L_j, U_j, j = 1, 2, \cdots, d$. Therefore, the d-dimensional numerical space $\mathcal{X} = [L_1, U_1) \times [L_2, U_2) \times \cdots \times [L_d, U_d)$ denotes a data space.

Definition 1. *The k-regular partition is a partition, which divides the data space $\mathcal{X}$ into disjoint rectangular cells. The cells are obtained by partitioning every attribute domain into k half open intervals of equal length. For simplify, we let such intervals be right-open.*

The k-regular partition divides the d dimensional spatial data set D into k^d cells with the same volume. The domain of each attribute X_i is divided into bins of length $\delta_i = (U_i - L_i)/k$. The attribute X_i is the union of the k disjoint right-open intervals $I_{i,j} = [L_i + (j-1)\delta_i, L_i + j\delta_i)$, $j = 1, \cdots, k$. Each grid cell has a unique index. The $(i_1, i_2, \cdots, i_d)$th cell, $B(i_1, i_2, \cdots, i_d)$, is $I_{1,i_1} \times I_{2,i_2} \times \cdots \times I_{d,i_d}$. Since each cell has the same volume, the number of data points inside it can be used to approximate the density of the cell. For ease of exposition, we henceforth assume that the data space of the data set D has been partitioned into disjoint equal volume grid cells by using k-regular partition.

Definition 2. *The neighboring cells of the cell B are the cells that are adjacent to B. The cell B_i neighbors with B_j if B_j is one of the neighbor cells of B_i.*

The set of neighbor cells of B contains two classes: (1) those have a common boundary side with B and (2) those are adjacent with B on single point. Each grid cell has 2×2^d neighbor cells except for the cells of boundary of the data space.

Definition 3. *Let m_sup be the predetermined density threshold. A grid cell is called (k, m_sup)-dense cell if it contains at least m_sup data points. Otherwise, we call that cell (k, m_sup)-sparse cell.*

Definition 4. *Each (k, m_sup)-dense cell is density-connected with itself. Two (k, m_sup)-dense cells B_i and B_j are density-connected if they are neighbors of each other or if there exists another (k, m_sup)-dense cell B_k such that B_i is density-connected to B_k and B_k is density-connected to B_j.*

The equivalent relationship in [7] and the density-connected relationship lead to the following theorem.

Theorem 1. *The density-connected relationship is an equivalent relationship on the set of dense cells.*

An equivalence relation on a set induces a partition on it, and also any partition induces an equivalence relation [7]. The main work of the clustering algorithm is to assign each data point to exactly one of the cluster. Hence, to implement a clustering algorithm is equivalent to construct a partition function. Density-connected relationship is an equivalent relationship. We can use this equivalent relationship to partition the set of dense cells into some disjoint subclasses. In addition, the dense cells will cover most points of the data sets, and the use of density-connected relationship is helpful to classify and to identify the main body of each cluster. The remaining work is hence to discover the boundaries of clusters.

Definition 5. *A (k, m_sup)-sparse cell is called isolated if its neighboring cells are all (k, m_sup)-sparse cells. The data points, which are contained in some isolated sparse cells, are defined as noises.*

A proper density-connected equivalent subclass usually contains most data points of each cluster. The data points of sparse cell may be either the boundary points of some cluster or outliers. The property for each sparse cell to be isolated or not is an important key to recognize border points and outliers. Outliers are essentially those points of isolated sparse cell. The data points in non-isolated sparse cells are border points of clusters. They might belong to distinct clusters but lie inside the same sparse cell. We link those data points of non-isolated sparse cell to their nearest dense neighbor cell, point by point. We shall assign each data point of non-isolated sparse cells into the cluster which contains its nearest dense neighborhood cell. To this end, we have to define the nearest neighborhood cell of data point.

If v_j $j = 1, 2, \cdots, 2^d$ are the vertices of grid cell B, then $c_B^* = \sum_{j=1}^{2^d} v_j / 2^d$ is its geometrical center point. Note that the center point of some grid cell may not be the distribution center (mean) of data points on this cell.

Definition 6. *Given a distance metric d of data points, the distance function between data point x and the cell B is defined as the distance between x and the geometrical center point of B, i.e. $d_B(x, B) \equiv d(x, c_B^*)$ where c_B^* is the geometrical center point of B. The nearest neighbor cell of a data point x is the cell whose distance to x is the minimum among all neighbor cells of x.*

2.2 Algorithm SCI

Formally, algorithm SCI consists of the following four procedures:
1. Partition: Create the regular grid structure.
2. Counting: Count the number of data points in each grid cell.
3. Purify: remove the redundant attributes.
4. GD clustering: identify the clusters.

Partition Stage: The partition process partitions the data space into cells using k-regular partition. For each data point, by utilizing the attribute domain and the value of k, we can obtain the index of the corresponding grid cell.

Table 1. Algorithm SCI

Algorithm: SCI
Inputs: m_sup, k, I^*

1.Partition: We partition the data space, using k-regular partition, into disjoint rectangular cells.
2.Counting: for each $i_j = 1, \cdots, k$; $j = 1, \cdots, d$
  a.Count the number of data points that are contained in the cell $B(i_1, i_2, \cdots, i_d)$.
  b.Assume that $B(i_1, i_2, \cdots, i_d)$ has $C_{i_1 i_2 \cdots i_d}$ data points.
  We mark $B(i_1, i_2, \cdots, i_d)$ as (k, m_sup)-dense cell if $C_{i_1 i_2 \cdots i_d} \geq m\_sup$, otherwise mark it as (k, m_sup)-sparse cell.

3.Purify:
  a.Compute the information of each attribute.
  b.Prune the attribute X_j if $Info(X_j) \geq log_2 k - I^*$ or $Info(X_j) \leq I^*$.

4.GD clustering:
  4.1 Noises and boundary detection:
  a.Link the data points of non-isolated (k, m_sup)-sparse cells to their nearest dense neighbor.
  b.Mark the data points of isolated (k, m_sup)-sparse cells as outlier.
  4.2 Clustering: Find clusters.

Counting Stage: On the counting process, we scan the dataset once and count the number of data points of each cell. A large dataset is compressed into a highly condensed summary table. Each cell of this table just stores the count of data points in the corresponding grid cell of data space. We classify these grid cells of data space as either dense or sparse, and further classify sparse as either isolated or non-isolated.
Purify Stage: Note that we are interested in automatically identifying subspace of a high dimensional data space that allows better clustering of the data points than the original space. By only using the summary table of data sets, we can avoid multiple database scans. Let N be the number of input data points. After the k-regular partition process, the data space will be partitioned into k^d grid cells $\{B(i_1, i_2, \cdots, i_d) \mid i_j = 1, \cdots, k \; j = 1, \cdots, d\}$. For each j, the domain of attribute X_j is divided into k disjoint equal width intervals $\{I_{ji} \mid i = 1, \cdots, k\}$. If the grid cell $B(i_1, i_2, \cdots, i_d)$ has $C_{i_1 i_2 \cdots i_d}$ data points, then $C_{jl} = \sum_{\substack{(i_1, \cdots, i_d) \\ i_j = l}} C_{i_1 \cdots i_d}$ data points are projected on the interval I_{jl} under X_j dimension. (Note that the value of C_{jl} can also be counted at the scanning time.)

The entropy relationship [2] leads to the following useful lemma for our work.

Lemma 1. *a. The X_j dimension has Shannon information* $Info(X_j) = -\sum_{l=1}^{k} \frac{C_{jl}}{N} log_2(\frac{C_{jl}}{N})$ *bits.*

b. $0 \leq Info(X_j) \leq log_2(k)$.
c. $Info(X_j) = 0 \Leftrightarrow \exists l$ *such that* $\frac{C_{jl}}{N} = 1$ *and* $\frac{C_{jr}}{N} = 0 \; \forall r \neq l$.
d. $Info(X_j) = log_2(k) \;\; \Leftrightarrow \;\; \forall l \; \frac{C_{jl}}{N} = \frac{1}{k}$.

If the data points projected on some attribute dimension are spatially close to one another, then we cannot significantly discern among them statistically. This attribute holds less information. If we prune this attribute and cluster again under the lower dimension, the clustering result would not be much changed. On the other hand, if the distribution of the data points projected on some attribute dimension is uniform, then this attribute has more information but less group structure. Therefore, such an attribute is also useless for clustering and is subject to be pruned. Explicitly, recall that structure diminishes uncertainty and randomness confuses the borders of clusters. The Shannon information is a commonly used measure of uncertainly. Attributes of less Shannon information carry less boundary information among groups, but more group structure information. In contrast, attributes of more Shannon information carry more boundary information, but possess less group structure information. Then, we can prune those attributes whose Shannon information is either too large or too small.

Example 1. In Fig. 2(a), the distribution of data points projected on dimension X is compact, and dimension X hence has less Shannon information. In Fig. 2(b), the projection of data points on dimension X is uniform, and then dimension X has more Shannon information, which, however implies less group structure information. In both scenarios, the subspace with dimension X pruned allows better clustering of the data points than the original space.

To decide which subspace is interesting, we apply Lemma 1 to our work. Assume that I^* is a predetermined small positive threshold (called significant level), then we prune the attribute X_j if $Info(X_j) \geq log_2 k - I^*$ or $Info(X_j) \leq I^*$.

GD Clustering Stage: The GD clustering satge consists of the following two sub-processes:

1. Noises and boundary detection: Classify the clustering boundary points and noises.
2. Clustering: Find clusters.

We categorize the data points of a cluster into two kinds, points inside some dense cells and points inside some sparse cells. The core points in a cluster are a set of points inside some density-connected cells. Some border points may be falling in some spare cell, and near the core points of cluster. A cluster is a maximal of density-connected cells with some linked points of their spare neighbors. The main task of outlier detection process is to classify the data points of sparse cells into border points or outliers. The final process is to determine the maximal density-connected equivalent classes by merging the dense neighbors. Each equivalent class integrating with the linking border points is a cluster.

Note that the centroid of cluster may be out of cluster. An illustrative example is shown in Fig. 2(c). Similarly, the centroid of the region of density-

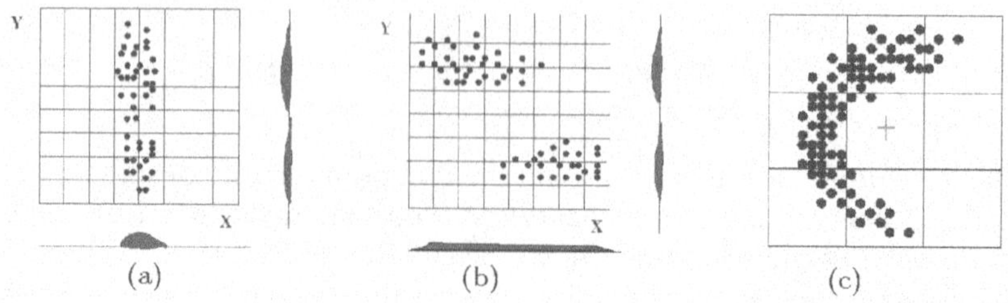

Fig. 2. (a) Dimension X has little Shannon information, (b) dimension X has much Shannon information but less group structure information, (c) the centroid of cluster is out of cluster.

connected equivalent class may be out of region. If we merge border points to the density-connected equivalent class whose centroid is nearest, we are likely to have incorrect merges sometimes. To solve this problem, we shall merge border points to their nearest neighbor cell. Moreover, to avoid the computations of cell center points, we can merge border points to the cell with nearest boundary distance among neighbors without compromising the quality of results.

3 Experimental Evaluation

To assess the performance of algorithm SCI, we have conducted a series of experiments. We compare the cluster quality and the execution time of SCI with several well-known clustering methods, including CLIQUE, CURE, and k-means algorithms. The synthetic sample data sets of our experiments are shown in Fig. 3. We added some random noises for evaluating the noises detection mechanism. For algorithms CLIQUE and SCI, the granularity of partition k is set to 25 and the density threshold m_sup is 1.1 times the expectation of data points of each grid cell under uniform distribution, i.e. $1.1 \times (N/25^2)$. The clustering results of those algorithms are shown in http://arbor.ntu.edu.tw/~ming/sci/ where algorithm SCI is able to successfully group these data sets. Further, the noises detection mechanism has been automatically triggered under algorithm SCI. From these results, the CURE and k-means algorithm are not able to capture the spatial shapes of clusters successfully. In fact, only algorithm SCI is able to detect noises successfully. Algorithm CLIQUE tends to remove too many noises. Because algorithm CLIQUE is not equipped with any detection mechanism to depart noises from the sparse cells, some border points are classified as noises. Also, parts of the cluster boundaries of CLIQUE are parallel or vertical to the axis. As shown in Fig. 4(b) and Fig. 4(d), there are overfitting phenomena on the resulting clusters of CLIQUE. From these results, it is shown that SCI is very powerful in handing any sophisticated spatial data sets.

The average execution times for those data sets are shown in Table 2 where it can be seen that grid-based and density-based approaches, i.e. CLIQUE and SCI, are more efficient than others. With the refinement on the definition of connected neighbor for grid cells, SCI is able to attain better clustering results than CLIQUE, as well as to improve the search of connected components. From these experiments, it is noted that most execution time of SCI is on the noises and boundary detection, which indeed leads to better results.

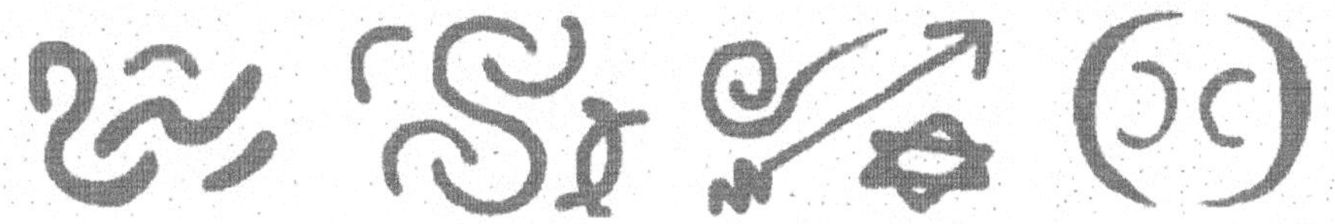

(a) Data Set 1 (b) Data Set 2 (c) Data Set 3 (d) Data Set 4

Fig. 3. Data sets used in experiments.

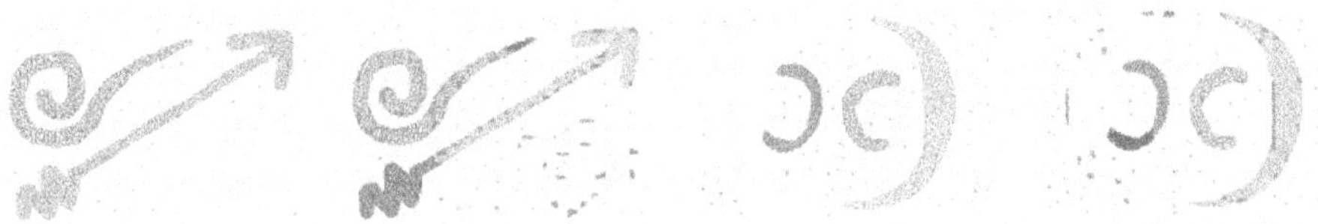

(a) SCI (b) CLIQUE (c) SCI (d) CLIQUE

Fig. 4. Some clustering results of SCI and CLIQUE.

Table 2. Execution time (in seconds) for the clustering algorithms.

(Note that CLIQUE does not have outliers detection.)

	Data Set 1	Data Set 2	Data Set 3	Data Set 4
Number of data	12,607	14,348	16,220	11,518
SCI	0.87	1.04	0.92	0.83
CLIQUE	0.33	0.28	0.38	0.33
CURE	487.08	624.29	797.35	406.05
k-means	1.15	1.93	0.94	2.75

4 Conclusions

In this paper, we developed algorithm SCI which integrates Shannon information and grid-based and density-based clustering to form a good solution for subspace clustering of high dimensional spatial data with noises. With the merit of only requiring one data scan, algorithm SCI is very efficient with its run time being linear to the size of the input data set. In addition, SCI was shown to have several advantages, including being scalable to the number of attributes, robust to noises, and independent of the order of input. As shown by our experimental results, SCI is very powerful to discover arbitrary shapes of clusters and able to solve the overfitting problem and avoid the confusion between outliers (noises) and cluster boundary points.

References

1. R. Agrawal, J. Gehrke, D. Gunopulos, and P. Raghavan. Automatic subspace clustering of high dimensional data for data mining applications. SIGMOD'98. 1998.
2. T. M. Cover and J. A. Thomas. Elements of Information Theory. Wiley. 1991.
3. B.-R. Dai, C.-R. Lin, and M.-S. Chen. On the techniques for data clustering with numerical constraints. In Proc. Of the 3rd SIAM Intern'l Conference on Data Mining. 2003.
4. A. K. Jain, R C. Dubes. Algorithms for Clustering Data. Prentice Hall. 1988.
5. A. K. Jain, M. N. Murty and P. J. Flynn. Data Clustering: A Review. ACM Computing Surveys. 1999.
6. M. Ester, H. P. Kriegel, J. Sander and X. Xu. A density-based algorithm for discovery in large spatial databases with noise. Proc. 2nd Int. Conf. Knowledge Discovery and Data Mining (KDD-96), pages 67-82, Poland, Maine, August 1996.
7. J. B. Fraleigh. A First Course in Abstract Algebra, 6th edition. Addison Wesley. 1999.
8. A. Hinneburg and D. A. Keim. An efficient approach to clustering in large multimedia databases with noise. Proc. 1998 Int. Conf. Knowledge Discovery and Data Mining (KDD-98), pages 58-65, New York, August 1998.
9. A. Hinneburg and D. A. Keim. Optimal Grid-Clustering: Towards Breaking the curse of Dimensionality in High-Dimensional Clustering. VLDB, 1999.
10. J. Han and M. Kamber. Data Mining: Concepts and Techniques. Morgan Kaufmann. 2000.
11. J. Han, M. Kamber, and A. Tung. Spatial Clustering Methods in Data Mining: A Survey. In Miller, H., and Han, J., eds., Geographic Data Mining and Knowledge Discovery. Taylor and Francis. 21, 2001.
12. C.-R. Lin and M.-S. Chen. A robust and efficient clustering algorithm based on cohesion self-merging. In Proc. Of the 8th ACM SIGKDD Internal Conf. On Knowledge Discovery and Data Mining, August 2002.
13. E. Schikuta. Grid-Clustering: A Fast Hierarchical Clustering Method for Very Large Data Sets. In Proc. 13th Int. Conf. On Pattern Recognition, Volume 2 , pages 101-105, IEEE Computer Society, 1996.
14. G. Sheikholeslami, S. Chatterjee, and A. Zhang. WaveCluster: A multi-resolution clustering approach for very large spatial databases. VLDB, 1998.
15. W. Wang, J. Yang, and R. Muntz. STING: A Statistical Information grid Approach to Spatial Data Mining. VLDB, 1997.
16. O. R. Zaiane, A. Foss, C.-H. Lee, and W. Wang. On data clustering analysis: Scalability, constraints, and validation. In PAKDD 2002, pages 28-39, 2002.

Constraint-Based Graph Clustering through Node Sequencing and Partitioning

Yu Qian[1], Kang Zhang[1], and Wei Lai[2]

[1]Department of Computer Science
The University of Texas at Dallas, Richardson, TX 75083-0688, USA
{yxq012100, kzhang}@utdallas.edu
[2]School of Information Technology
Swinburne University of Technology, Hawthorn, Victoria, 3122, Australia
wlai@it.swin.edu.au

Abstract. This paper proposes a two-step graph partitioning method to discover constrained clusters with an objective function that follows the well-known min-max clustering principle. Compared with traditional approaches, the proposed method has several advantages. Firstly, the objective function not only follows the theoretical min-max principle but also reflects certain practical requirements. Secondly, a new constraint is introduced and solved to suit more application needs while unconstrained methods can only control the number of produced clusters. Thirdly, the proposed method is general and can be used to solve other practical constraints. The experimental studies on word grouping and result visualization show very encouraging results.

1 Introduction

As a widely recognized technique for data analysis, clustering aims at gathering closely related entities together in order to identify coherent groups, i.e., clusters. Clustering methods have proven to be very useful in many application areas including data mining, image processing, graph drawing, and distributed computing. This paper presents a novel graph theoretic partitioning approach to constrained clustering, analyzes and demonstrates the advantages of such an approach.

For constrained clustering, grouping similar units into clusters has to satisfy some additional conditions. Such additional conditions come from two kinds of knowledge: background knowledge and user requirements. While there have been some works investigating the use of background knowledge in clustering process, little research, however, performs in-depth analysis on the role of user-inputs in the process of clustering. According to the role of user-input in the clustering process, clustering criteria can be classified into two categories: user-centric and data-centric. The former involves and solves practical constraints in clustering process while the latter meets only the application-independent requirements such as high cohesiveness, low coupling, less noise, and etc. A practical clustering method should be both user-centric and data-centric, e.g., in most clustering algorithms the number of the clusters to be discovered is a necessary user-input while the application-independent min-max clustering principle must hold: the similarity between two clusters is significantly less than the similarity within each cluster [3].

H. Dai, R. Srikant, and C. Zhang (Eds.): PAKDD 2004, LNAI 3056, pp. 41–51, 2004.

The number of user-input constraints varies for different applications. In spatial data clustering, user-input can be minimized because there are naturally-defined potential clusters contained in the given data and finding these clusters is exactly the final purpose. In many other cases, however, finding natural clusters is far from the destination, which makes it not enough to meet only one constraint on the number of the clusters. Let us use a simple example in Fig. 1 to demonstrate the necessity of involving more constraints.

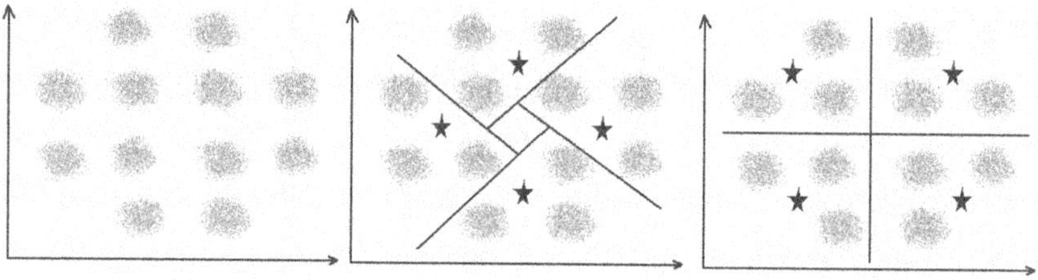

Fig. 1. (a) A population-density map with 12 communities. (b) A 4-clustering of the 12 communities. (c) A 4-clustering with a constraint on the distance between cluster centers.

Fig. 1 (a) is a map that describes the population distribution in a city. The denser the color is the more crowded the people are. A builder is planning to build some supermarkets in this city. He wants to choose four profitable locations of the supermarkets according to this map. So he uses some clustering algorithms to discover the clusters of people in order to put his supermarkets at the centers of clusters. In Fig. 1 (a), we can see that a correct and "good" data-centric clustering algorithm without any user-input would produce the twelve communities in this city as twelve clusters. Such a clustering result, unfortunately, is of little use because the builder can afford only four supermarkets. Fig. 1 (b) illustrates a feasible result after accepting the constraint on the number of clusters: the four stars indicate the best locations of the four supermarkets. The result shown in Fig. 1 (b), however, cannot satisfy the builder either. The builder needs a bigger distance between two supermarkets so that they can cover more area. After accepting one more constraint about the minimal allowable distance between two supermarkets, Fig. 1 (c) shows a desirable result. Fig. 1 has revealed an important phenomenon in many clustering applications: the clustering results that satisfy only the constraint on the number of clusters may not meet the practical requirements. It is necessary to involve more constraints into clustering process.

This paper introduces a new constraint into the generic clustering problem: the upper bound of the quantified similarity between two clusters. The similarity may have different quantified values in different application domains. In the above supermarket example, the similarity between two clusters is represented by the distance between two clusters: the nearer, the more similar. The upper bound of the similarity between two clusters is the minimum allowable distance between their centers. We believe that the upper bound of similarity is a general constraint which is required by many clustering applications. To support the argument, we further provide an example on parallel task partitioning: as we know, computers in a distributed environment lack global addressing space, communication has to be inserted whenever a processor needs to access non-local data. For example, on the Intel Paragon the processor cycle time is 20 nanoseconds whereas the remote memory access time is between 10000 and 30000 nanoseconds [8], depending on the distance between communicating processors. Therefore, it is imperative that the frequency and volume of non-local accesses are reduced as much as possible. Suppose that the whole computing task is composed of n smaller tasks. Given k machines/processors, the attempt to assign the n small tasks to the k machines/processors is a k-clustering problem, which has several concerns: firstly, there is a communication bottleneck between the machines/processors. Secondly, it may not be worth to parallelize the computing task when the ratio of the communication cost to the total cost exceeds a preset threshold. Thirdly, in most cases, the number of available machines/processors is not always fixed but flexible. The three concerns can be exactly mapped into three corresponding constraints in clustering problem: the upper bound constraint, the min-max principle, and the number of clusters.

Motivated by the above examples, this paper proposes a graph theoretic model that can represent the above three requirements (two user-input constraints and one application-independent min-max principle): 1) the desired number of clusters; 2) the objective function of clustering that reflects the min-max principle, and 3) the upper bound of the similarity between two clusters. In particular, the desired number of clusters in the proposed model is represented with a range (K_{min}, K_{max}), minimum and maximum allowable number of clusters, respectively. The objective function is defined as the ratio of the similarity within the cluster to the similarity between clusters. Maximizing the proposed objective function not only follows the min-max principle but also meets certain practical requirements, e.g., in parallel task partitioning, the ratio decides if it is worth to parallelize the computation task; in spatial data clustering, the objective function is the ratio of the density inside the cluster to the density outside the cluster when representing the spatial data sets as sparse graphs. Based on the three requirements, this paper proposes a two-step graph partitioning methodology to discover constrained clusters. The basic idea involves node sequencing and then partitioning the node sequence according to the constraints. In the sequencing process, the graph nodes are sorted using existing algorithms. In the partitioning process we find an appropriate cut point according to the objective function along the ordered node sequence so that all points on one side will be output as a cluster while all points on the other side will remain for further partitioning until all constraints are satisfied or the number of produced clusters exceeds K_{max}.

The rest of this paper is organized as follows. Related work on graph partitioning and constrained clustering is introduced in Section 2. Section 3 proposes our two-step methodology. The experimental studies on word grouping and result visualization are presented in Section 4. Section 5 concludes the paper.

2 Related Work

Since this paper focuses on a constrained graph partitioning for data clustering, the related work can be categorized into two parts: graph partitioning and constraint-based data clustering.

2.1 Graph Partitioning

Numerous graph partitioning criteria and methods have been reported in the literature. We consider matrix transformation and ordering since our method proposes similar techniques. The optimal solution to the graph partitioning problem is NP-complete due to the combinatoric nature of the problem [3, 4]. The objective of graph partitioning is to minimize the cut size [3], i.e., the similarity between the two subgraphs with the requirement that the two subgraphs have the same number of nodes. It is important to note that minimizing the cut size for each partitioning step cannot guarantee a minimal upper bound of the cut size for the whole clustering process.

Hagen and Kahng [6] remove the requirement on the sizes of the subgraphs and show that the Fiedler vector provides a good linear search order to the ratio cut (*Rcut*) partitioning criteria, which is proposed by Cheng and Wei [2]. The definition of *Rcut* is: $Rcut=cut(A,B)/|A|+cut(A,B)/|B|$, where $G=(V,E)$ is a weighted graph with node set V and edge set E, $cut(A,B)$ is defined as the similarity between the two subgraphs A and B and $|A|$, $|B|$ denote the size of A, B, respectively.

Shi and Malik [10] propose the normalized cut by utilizing the advantages of normalized Laplacian matrix: $Ncut=cut(A,B)/deg(A)+cut(A,B)/deg(B)$, where $deg(A)$ is the sum of node degrees, which is also called the volume of subgraph A, in contrast to the size of A. Ding *et al.* [3] propose a min-max cut algorithm for graph partitioning and data clustering with a new objective function called $Mcut=cut(A,B)/W(A)+cut(A,B)/W(B)$, where $W(A)$ is defined as the sum of the weights of the edges belong to subgraph A.

All objective functions above are designed for their algorithms to find an appropriate partitioning. They are algorithm-oriented and cannot reflect the practical requirements or physical meanings, which make them infeasible to serve a constrained clustering.

2.2 Constraint-Based Data Clustering

According to Tung *et al.*, constraint-based clustering [11] is defined as follows. Given a data set D with n objects, a distance function $df: D \times D \rightarrow R$, a positive integer k, and a set of constraints C, find a k-clustering $(Cl_1, Cl_2, \ldots Cl_k)$ such that

$$DISP=\sum_{i=1}^{k} disp(Cl_i, rep_i)$$

is minimized, and each cluster Cl_i satisfies the constraints C, denoted as $Cl_i|=C$, where $disp(Cl_i, rep_i)$ measures the total distance between each object in Cl_i and the representative point rep_i of Cl_i. The representative of a cluster Cl_i is chosen such that $disp(Cl_i, rep_i)$ is minimized. There are two kinds of constraint-based clustering methods, aiming at different goals: one category aims at increasing the efficiency of the clustering algorithm while the other attempts to incorporate domain knowledge using constraints. Two instance-level constraints: must-link and cannot-link constraints have been introduced by Wagstaff and Cardie [12], who have shown that the two constraints can be incorporated into COBWEB [5] to increase the clustering accuracy while decreasing runtime. Bradley *et al.* propose a constraint-based k-means algorithm [1] to avoid local solutions with empty clusters or clusters with very few data points that can often be seen when the value of k is bigger than 20 and the number of dimensions is bigger than 10.

The proposed method is different from the aforementioned approaches: firstly, we combine graph partitioning and constrained-based clustering together. Secondly, the proposed partitioning process is different from the traditional partitioning in that for each partitioning step we produce only one cluster instead of two subgraphs. The remaining part will be further partitioned until all constraints are satisfied or the number of produced clusters exceeds K_{max}. Thirdly, the upper bound constraint we introduce is new and its involvement does not aim at improving the efficiency of the clustering process but aim at encoding the user's requirements. Finally, we accept both unweighted and weighted graphs. Section 3 will introduce the proposed two-step methodology.

3 A Two-Step Methodology

This section first describes the process of node sequencing, and then introduces the node partitioning method.

3.1 Node Sequencing Method (NSM)

The process of node sequencing transforms a two-dimensional graph into a one-dimensional node sequence. The sequencing method used here is proposed in our previous paper [9]. Due to the space limitation, we only provide a brief introduction. The whole sequencing includes two steps: coarse-grained sequencing, which partitions the given graph into several parts and sorts these parts, and fine-grained sequencing, which sorts the nodes inside each part produced in the coarse-grained step. As a result, we obtain a sequence of nodes in which the nodes belonging to the same cluster will be put together. Then we can apply the node partitioning method to the node sequence to find the boundary points of clusters with constraints.

3.2 Node Partitioning Method (NPM)

This section proposes a novel node partitioning method. We first introduce the algorithm parameters used in the algorithm.

3.2.1 Algorithm Parameters

The algorithm parameters used in NPM include *alpha1*, *alpha2*, *beta*, and E_{inter}, which are defined as follows. Given a node sequence S of n nodes, and a cut at ith node separates S into two sub-sequences, say S_1, and S_2 where S_1 contains the nodes from 1 to i, and S_2 contains the nodes from $i+1$ to n. Let E_1 denote the number of edges inside S_1 and E_2 the number of edges inside S_2, and E_{inter} the number of edges between S_1 and S_2, we have the following algorithm parameters, as intuitively shown in Fig. 2.

$$alpha1(i) = E_1 / (i(i-1)/2) \quad (1)$$

$$alpha2(i) = E_2/((n-i)(n-i-1)/2) \quad (2)$$

$$beta(i) = E_{inter} / (((n-i) \cdot i)/2) \quad (3)$$

As Fig. 2 shows, the big square represents an adjacency matrix of the given graph, and the cut at i separates the node sequence $1...n$ into $1...i$, and $i+1...n$. *alpha1* represents the density of the upper left square, *alpha2* represents the density of the lower right square, and *beta* represents the density of the upper right or lower left square.

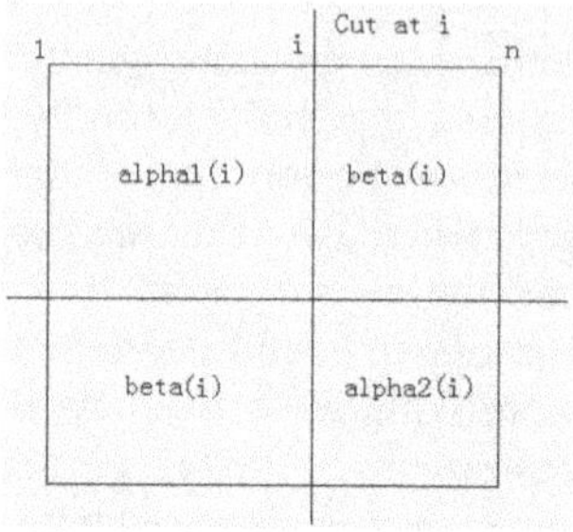

Fig. 2. The physical meanings of the algorithm parameters

Given a node sequence S of n nodes, for every node i, $1<i<n$, "*cut at i*" means partitioning the sequence at the ith node. A cut at i is *acceptable* only if its corresponding objective function *cutvalue(i)* is a peak value. A *cutvalue(i)* is a peak value means that *cutvalue(i)* is bigger than any other *cutvalue* bwteen i-ε and $i+\varepsilon$ where ε is a threshold defined as $n/2k$ and k is the number of desired clusters. *Cutvalue* measures the ratio of the similarity within the cluster and the similarity between clusters and helps discovering where we should separate the node sequence. Its definition is:

$$cutvalue(i) = \begin{cases} alpha\ 1(i)\ /\ beta\ (i) & beta\ (i) <> 0 \\ MAX\ +\ alpha\ 1(i) & beta\ (i) = 0 \end{cases} \tag{4}$$

The *MAX* in formula (4) is a very big constant used to distinguish the nodes when *beta(i)* is zero. According to the definitions, *beta(i)* represents the density of inter-cluster area while *alpha1(i)* is the density of intra-cluster area. The physical meaning of using the ratio of *alpha1(i)* to *beta(i)* is to effectively reflect the relative density. If *cutvalue* increases significantly, the corresponding cut point is more possibly located at the boundary of two clusters.

Now let us define the upper bound constraint for the similarity between two clusters. For clusters C_i and C_j, and nodes $u \in C_i$, $v \in C_j$, if $(u,v) \in E$, we say (u,v) is an edge between clusters C_i and C_j. Let *inter(i,j)* denote the number of edges between clusters C_i and C_j and *sum_inter(i,j)* the sum of the weights of the edges between clusters C_i and C_j. We have the following definitions:

Definition 3.1 (Coupling Bound, Bound Constraint)
Coupling Bound is the biggest number of inter-cluster edges for a clustering result, represented by an integer U_{inter}: $U_{inter} = Max(inter(i,j)), \forall\ i,j \in \{1,2,....,k\},\ i \neq j$. For weighted graphs, *coupling bound* is the maximal sum of the weights of inter-cluster edges, denoted by an integer $U_{inter\text{-}w}$: $U_{inter\text{-}w} = Max\ (sum\_inter(i,j)),\ \forall i,j \in \{1,2,...,k\}$, $i \neq j$. *Bound Constraint*, denoted by U, is a user-input constraint on the maximum allowable coupling bound. For a satisfactory clustering result, the formula $Uinter < U$ must hold (for weighted graph, the formula is $U_{inter\text{-}w} < U$).
Definition 3.2 (Granularity, G-constraint)
Granularity is defined as the number clusters for a clustering result, denoted by an integer k. G-constraint is a pair of integers (K_{min}, K_{max}) input by the user. For a satisfactory clustering result, the formula K_{max} k K_{min} must hold.

```
Algorithm ComputeCut(Graph g, Integer n)
begin
  for each i from 2 to n-2
     compute alpha1(i), beta(i), and
     cutvalue according to (4)
  for each node i from 1 to n do
     cut ←getFirstPeak(cutvalue[i]);
  return cut;
end

Algorithm Npm (NodeSequence seq) {seq is
the result after node sequencing}
begin
  t←0, i←0;
  while (i<Kmin) do
   begin {the partitioning procedure}
     Remove the nodes before Node[t] from
     seq; n ←n-t;
     Create the residual graph;
     t←ComputeCut(g,n);  i←i+1;
   end
  while (Uinter>U) && i<Kmax) do
     repeat the partitioning procedure;
  if (Uinter>U) return clustering result;
  else return ("No such kind of clustering");
end
```

Fig. 3. The *ComputeCut* Algorithm and the whole *NPM* algorithm

Fig. 3 describes the algorithms that computes the *cutvalue* for each cut at *i* and returns the cut with the first peak value and the whole NPM algorithm.

4 Experimental Studies

Our experimental studies consist of three parts: the first part evaluates min-max principle and demonstrates that the parameter values represent the data distribution precisely. The second part of our experiments evaluates the ability of our approach on constraint satisfaction. Since there are two kinds of constraints involved in our algorithm: the upper bound constraint and the range of the desired number of clusters, the second part uses a set of synthetic constraints to evaluate if our algorithm can find the constrained clustering results for the given data set. The third part of our experiments visualizes the clustering results intuitively and compares our results with the one produced by Kamada and Kawai's method, a well-known force-directed (spring) graph clustering algorithm.

The clustering task in our experiments is word grouping, a basic technique of text mining and document clustering. The testing data are generated from the subsets of an English dictionary. In the first part of our experiments, two data sets: DS1 and DS2 are used. In the second part of our experiments, three data sets: DS3, DS4, and DS5 are used. The properties of their corresponding graphs are shown in Table 1. The similarity between two words is defined on their edit distance except for DS1. Each word is regarded as a graph node and if the edit distance between two words is less than a fixed threshold, there is an edge between the two corresponding nodes. For DS1, there is an edge between the two nodes if and only if the two corresponding words have the same length.

4.1 Parameter Effectiveness

Fig. 4 (a) shows the computed values of the two algorithm parameters: *alpha1* and *alpha2* for DS1. According to the similarity definition of DS1, the words with the

Table 1. The properties of the five testing data sets

Testing Data	Number of Nodes	Number of Edges	Threshold of Edit Distance	Result shown in
DS1	3419	146284	N/A	Fig. 4
DS2	472	428	1	Fig. 5
DS3	300	953	2	Fig. 6 (a)
DS4	1000	4908	2	Fig. 6 (b)
DS5	5000	312834	2	Fig. 6 (c)

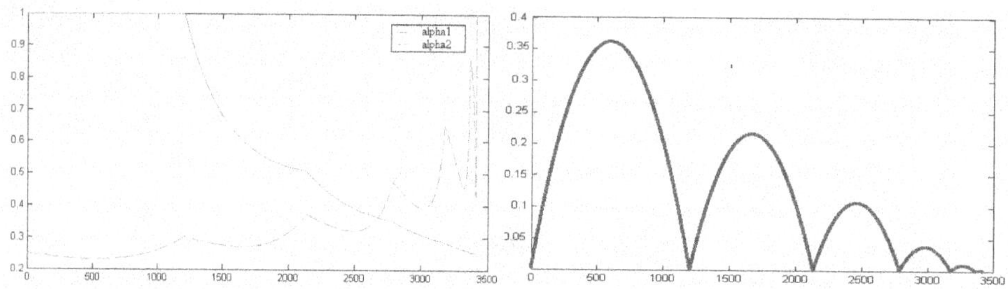

(a) (b)

Fig. 4. (a) Values of alpha1 and alpha2 for DS1 in the first partitioning step (b) Values of beta for DS1 in the first partitioning step

same length form a complete graph; the whole graph contains 6 complete subgraphs that are isolated from each other. The boundaries of the 6 subgraphs/clusters are clearly shown where the values of alpha1 drops dramatically.

Fig. 4 (b) shows the computed values of beta. We can see that the values of beta reach minimum at the boundaries of clusters while the values of *alpha1* reach maximum at the boundaries. According to the definition of *cutvalue*, the value of *alpha1(i)/beta(i)* would be significantly bigger when *i* is the index of a boundary point, i.e., the cutting point can be correctly found.

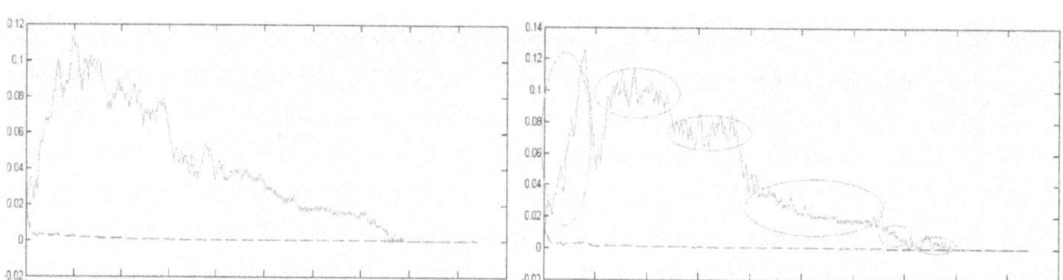

Fig. 5. The values of beta (a) before applying BEA and (b) after applying BEA for DS2.

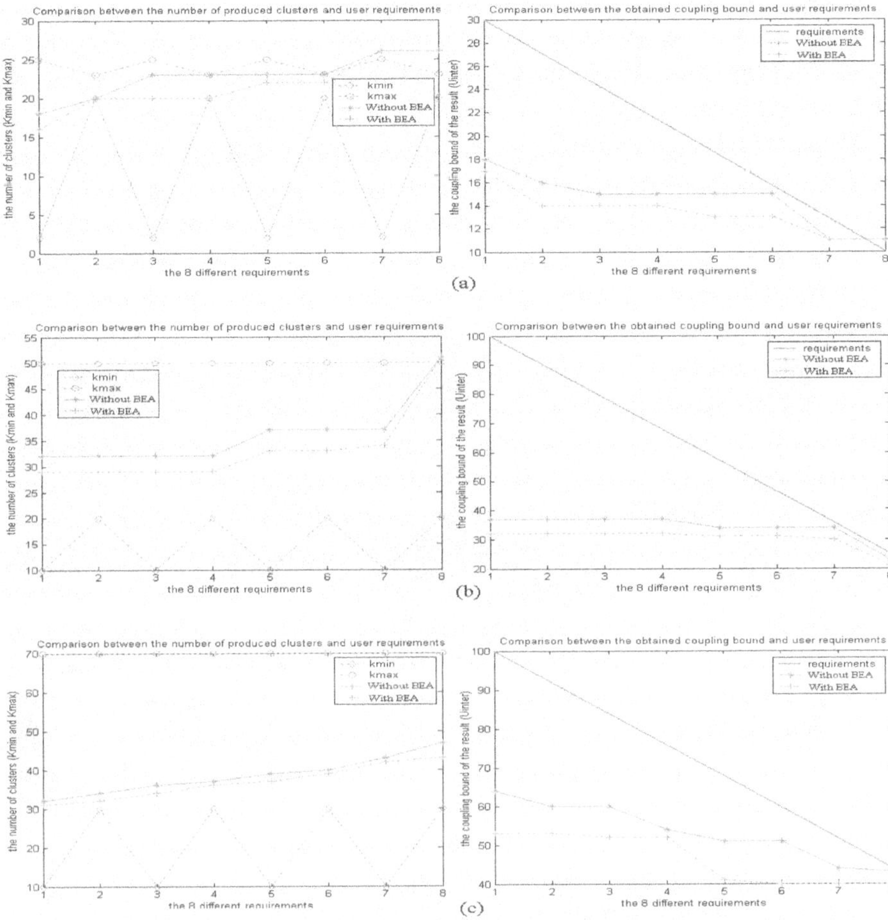

Fig. 6. The constraint satisfaction process for (a) DS3 (b) DS4 and (c) DS5

Fig. 5 shows the difference on the values of beta before and after applying the BEA. Fig. 5(a) shows that before applying the BEA the couplings between the data points are not very different and we cannot find a cut point to partition the node sequence. After applying BEA to the same graph, we find that the difference of couplings of different clusters is sharpened while the distribution of the coupling values inside the cluster becomes smoother. As clearly shown in Fig 5 (b), the graph contains six clusters.

4.2 Constraint Satisfaction

This part of experiments evaluates whether the proposed partitioning algorithm can produce satisfactory results according to different user-input constraints. Each testing graph is evaluated against 8 different clustering requirements and each requirement contains two kinds of constraints, as defined in Section 3. Each of Fig. 6 (a), Fig. 6 (b), and Fig. 6 (c), contains two sub-figures, corresponding to the two kinds of constraints respectively. The 8 requirements are designed from loose to strict, i.e., the first requirement is the easiest to be satisfied and the last is the hardest. If the produced granularity is between the minimum and maximum requirement, the corresponding clustering result is satisfactory on granularity; if the coupling bound of the produced result is below the upper bound constraint, the corresponding clustering result is satisfactory on coupling. If both kinds of constraints are satisfied, the clustering result is satisfactory.

4.3 Result Visualization

Another way to evaluate our approach is to visualize the results. Fig. 7(a) shows the original graph with 320 nodes. The graph nodes belong to the same cluster are in the same gray level. We apply a popular force-directed graph clustering algorithm: Kamada and Kawai's method [7] to the graph and its result is shown in Fig. 7 (b) while our results are shown in Fig. 7 (c) and (d). Although our approach does not compete with Kamada and Kawai's method on the quality of graph layout, it separates clusters clearly. In Fig. 7(b) many graph nodes belong to different clusters are mixed up while our method can discover all clusters correctly. Fig. 7 (c) and (d) show that our method can produce different numbers of clusters according to the user input. Apart from the advantage on effectiveness, our method is faster than Kamada and Kawai's method. It takes only several minutes for our program to generate the results in Fig. 7 (c) and (d) while Kamada and Kawai's method needs more than 1 hour to reach a stable layout.

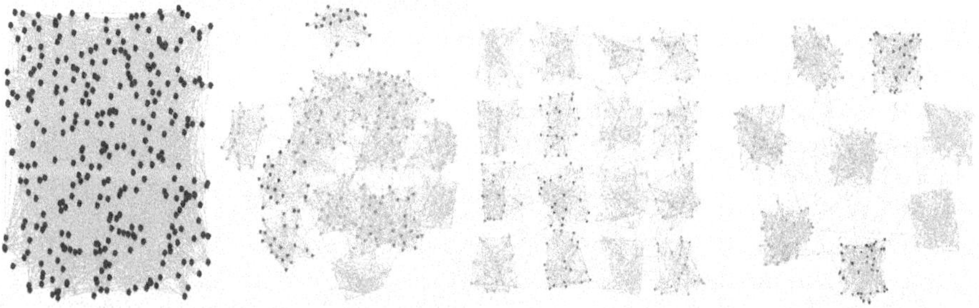

(a) (b) (c) (d)

Fig. 7. (a) The original graph before clustering (b) the same graph after applying Kamada and Kawai's method. The same graph after applying our algorithm with (c) 16 clusters (d) 8 clusters

5 Conclusions

This paper has presented a novel graph partitioning method for constrained data clustering. A new constraint: upper bound of the similarity between two clusters is introduced and solved with the proposed graph partitioning method. The method consists of two steps: sequencing the given set of graph nodes, and then partition the node sequence into final clusters. This method has at least two advantages: first, the objective function not only follows the theoretical min-max principle but also reflects certain practical requirements. Second, new constraints from practical clustering problems are introduced and solved so that the clustering results can be tailored to more application needs while unconstrained methods can only control the number of produced clusters. Our experimental studies have visualized the clustering results intuitively and demonstrated that the combination of graph partitioning and constrained data clustering is successful. Future work will explore whether it is possible to locate the feasible range of the constraints for a given clustering task so that the user can be guided on constraint input.

References

1. Bradley, P. S., Bennett, K. P., and Demiriz, A. Constrained *K*-Means Clustering, In MSR-TR-2000-65, Microsoft Research. (2000)
2. Cheng, C-K., and Wei, Y. A. An improved two-way partitioning algorithm with stable performance. IEEE. Trans. on Computed Aided Design. 10 (1991), pp. 1502-1511.
3. Ding, H. Q. C., He, X., Zha, H., Gu, M., and Simon, H. A Min-Max Cut Algorithm for Graph Partitioning and Data Clustering. Proc. of International Conf on Data Mining, (2001), pp. 107-114.
4. Donath, W. E. and Hoffman, A. J. Lower bounds for partitioning of graphs. IBM J. Res. Develop., 17 (1973), pp. 420-425.
5. Fisher, D. Knowledge acquisition via incremental conceptual clustering, Machine Learning, 2, (1987), pp. 139-172.
6. Hagen, L. and Kahng, A. B. New spectral methods for ratio cut partitioning and clustering. IEEE Trans. on Computed Aided Design, 11(1992), pp. 1074-1085.
7. Kamada, T. and Kawai, S. An algorithm for drawing general undirected graphs. Information Processing Letters, 31(1989), pp. 7-15.
8. Kandemir, M., Banerjee, P., Ramanujam, J., and Shenoy, N. A global communication optimization technique based on data-flow analysis and linear algebra. ACM Transactions on Programming Languages and Systems, Vol. 21, No. 6, (2000) pp.1251-1297.
9. Qian, Y. and Zhang, K.: A Customizable Hybrid Approach to Data Clustering. Proc. of the 2003 ACM Symposium on Applied Computing, (2003) 485-489.
10. Shi, J. and Malik, J.: Normalized cuts and image segmentation. IEEE Trans. on Pattern Analysis and Machine Intelligence. Vol. 22, No. 8, (2000), pp. 888-905.
11. Tung, A. K. H., Han, J., Lakshmanan, L. V. S., and Ng, R. T. Constrained-based clustering in large databases, Proc. 8th Intl. Conf. on Database Theory (ICDT'01), London, UK, (2001), pp. 405-419.
12. Wagstaff, K. and Cardie, C. Clustering with instance-level constraints, Proc. of the 17th Intl. Conf. on Machine Learning, (2000), pp. 1103-1110.

Mining Expressive Process Models by Clustering Workflow Traces

Gianluigi Greco[1], Antonella Guzzo[1], Luigi Pontieri[2], and Domenico Saccà[1,2]

[1] DEIS, University of Calabria, Via Pietro Bucci 41C, 87036 Rende, Italy
[2] ICAR, CNR, Via Pietro Bucci 41C, 87036 Rende, Italy
{ggreco,guzzo}@si.deis.unical.it, {pontieri,sacca}@icar.cnr.it

Abstract. We propose a general framework for the process mining problem which encompasses the assumption of workflow schema with local constraints only, for it being applicable to more expressive specification languages, independently of the particular syntax adopted. In fact, we provide an effective technique for process mining based on the rather unexplored concept of clustering workflow executions, in which clusters of executions sharing the same structure and the same unexpected behavior (w.r.t. the local properties) are seen as a witness of the existence of global constraints.
An interesting framework for assessing the similarity between the original model and the discovered one is proposed, as well as some experimental results evidencing the validity of our approach.

1 Introduction

Even though workflow management systems (WfMS) are more and more utilized in enterprises, their actual impact in automatizing complex process is still limited by the difficulties encountered in the designing phase. In fact, processes have complex and often unexpected dynamics, whose modelling requires expensive and long analysis which may eventually result unviable under an economic viewpoint.

Recent research faced this problem, by exploiting some strategies, called *process mining* techniques, for using the information collected during the enactment of a process not yet supported by a WfMS, such as the transaction logs of ERP systems like SAP, in order to derive a model explaining the events recorded. Then, the output of these techniques, i.e., the "mined" synthetic model, can be profitably used to (re)design a detailed workflow schema, capable of supporting automatic enactments of the process.

Several approaches for process mining have been proposed in the literature (see, e.g., [1,16,4,12]), that aim at reconstructing the structure of the process, by exploiting graphical models based on the notion of *control flow graph*. This is an intuitive way of specifying a process through a directed graph, where nodes correspond to the activities in the process and edges represent the potential flow of work, i.e., the relationships of precedence among the activities.

However, despite its intuitiveness, the control flow completely lacks in the ability of formalizing complex *global constraints* on the executions, which often occurs while

H. Dai, R. Srikant, and C. Zhang (Eds.): PAKDD 2004, LNAI 3056, pp. 52–62, 2004.

modelling real scenarios, for it being able to prescribe only *local constraints* in terms of relationships of precedence.

In this paper, we extend previous approaches to process mining, by proposing an algorithm which is able to discover not only the control flow of a given process, but also some interesting global constraints, in order provide the designer with a refined view of the process. The main contribution are as follows.

In Section 2, we formalize the process model discovery problem, in a context in which the target workflow schema may be enriched with some global constraints, denoted by $\mathcal{C}_G$. In order to decouple the approach from the particular syntax adopted for expressing $\mathcal{C}_G$, we exploit the observation that each global constraint leads to instances with a specific structure (short. pattern); then, a workflow schema $\mathcal{WS}^{\vee}$, accounting for global constraints, is the union of several schemas $\mathcal{WS}^1, ..., \mathcal{WS}^k$ (without global constraints), each one supporting the execution of one pattern, only.

Different patterns of executions (and, hence, $\mathcal{WS}^{\vee}$) are identified by means of an algorithm for clustering workflow traces, presented in Section 3, which is based on the projection of the traces on a suitable set of properly defined *features*. The approach is similar in the spirit to the proposals of clustering sequences using frequent itemsets, but technically more complex, for it deriving a hierarchical clustering. The theoretical properties of the algorithm are investigated as well.

In Section 3.1, we propose a level-wise algorithm for the identification of the set of features $\mathcal{F}$ for the clustering, and we study the problem of selecting the most 'representative' subset of $\mathcal{F}$, by showing its intrinsic difficulty. Therefore, we propose a greedy heuristic for quickly computing a set of features approximating the optimal solution.

Finally, we experiment an implementation of the proposed technique, by showing its scalability. An interesting framework for assessing the similarity between the original model and the discovered one is proposed in Section 4, thus, providing a quantitative way for testing the validity of the approach.

2 Formal Framework

In this section we formalize the mining problem addressed in the paper, which can be roughly described as the problem of (re)constructing a workflow model of an unknown process P, on the basis of log data related to some executions of the process.

The *control flow graph* of a process P is a tuple $\mathcal{CF}(P) = \langle A, E, a_0, F \rangle$, where A is a finite set of *activities*, $E \subseteq (A - F) \times (A - \{a_0\})$ is a relation of precedences among activities, $a_0 \in A$ is the starting activity, $F \subseteq A$ is the set of final activities.

Any connected subgraph $I = \langle A_I, E_I \rangle$ of the control flow graph, such that $a_0 \in A_I$ and $A_I \cap F \neq \emptyset$ is a *potential instance* of P. In order to model restrictions on the possible instances, the description of the process is often enriched with some additional *local* or *global* constraints, requiring, e.g., that an activity must (or may not) directly (or indirectly) follow the execution of a number of other activities.

For instance, local constraints are that an *and-join* activity can be executed only after all its predecessors are completed, and that an *or-join* activity can be executed as soon as one of its predecessors is completed. Other examples are than an *and-split* activity activates all of its successor activities, while a *xor-split* activates exactly one of its outgoing arcs.

Global constraints are, instead, richer in nature and their representation strongly depends on the particular application domain of the modelled process. Thus, they are often expressed using other complex formalisms, mainly based on a suitable logic with an associated clear semantics.

Let P be a process. A *workflow schema* for P, denoted by $\mathcal{WS}(P)$, is a tuple $\langle \mathcal{CF}(P), \mathcal{C}_L(P), \mathcal{C}_G(P)\rangle$, where $\mathcal{CF}(P)$ is the control flow graph of P, and $\mathcal{C}_L(P)$ and $\mathcal{C}_G(P)$ are sets of local and global constraints, respectively. Given a subgraph I of $\mathcal{CF}(P)$ and a constraint c in $\mathcal{C}_L(P) \cup \mathcal{C}_G(P)$, we write $I \models c$ whenever I satisfies c in the associated semantics. Moreover, if $I \models c$ for all c in $\mathcal{C}_L(P) \cup \mathcal{C}_G(P)$, I is called an *instance* of $\mathcal{WS}(P)$, denoted by $I \models \mathcal{WS}(P)$. When the process P is clear from the context, a workflow schema will be simply denoted by $\mathcal{WS} = \langle \mathcal{CF}, \mathcal{C}_L, \mathcal{C}_G\rangle$.

2.1 The Process Model Discovery Problem

Let A_P be the set of task identifiers for the process P. We assume the actual workflow schema $\mathcal{WS}(P)$ for P to be unknown, and we consider the problem of properly identifying it, in the set of all the possible workflow schemas having A_P as set of nodes. In order to formalize this problem we need some preliminarily definitions and notations.

A *workflow trace s over A_P* is a string in A_P^*, representing a task sequence. Given a trace s, we denote by $s[i]$ the i-th task in the corresponding sequence, and by $lenght(s)$ the length of s. The set of all the tasks in s is denoted by $tasks(s) = \bigcup_{1 \leq i \leq lenght(s)} s[i]$. Finally, a *workflow log for P*, denoted by $\mathcal{L}_P$, is a bag of traces over Σ_P: $\mathcal{L}_P = [\, s \mid s \in A_P^* \,]$ and is the only input from which inferring the schema $\mathcal{WS}(P)$.

In order to substantiate the problem of mining $\mathcal{WS}(P)$, one must specify which language is to be adopted for expressing the global constraints in $\mathcal{C}_G$. In order to devise a general approach, it is convenient to find an alternative (syntax-independent) way for evidencing global constraints. The solution adopted in this paper is to replace a unique target schema $\mathcal{WS}(P)$ with a variety of alternative schemata having no global constraints but directly modelling the various execution patterns prescribed by global constraints. The basic idea is to first derive from the trace logs an initial workflow schema whose global constraints are left unexpressed and, then, to stepwise refine it into a number of specific schemas, each one modelling a class of traces having the same characteristics w.r.t. global constraints.

Definition 1. Let P be a process. A *disjunctive workflow schema* for P, denoted by $\mathcal{WS}^\vee(P)$, is a a set $\{\mathcal{WS}^1, ..., \mathcal{WS}^m\}$ of workflow schemata for P, with $\mathcal{WS}^j = \langle \mathcal{CF}^j, \mathcal{C}_L^j, \emptyset\rangle$, for $1 \leq j \leq m$. The *size* of $\mathcal{WS}^\vee(P)$, denoted by $|\mathcal{WS}^\vee(P)|$, is the number of workflow schemata it contains. An instance of any $\mathcal{WS}^j$ is also an instance of $\mathcal{WS}^\vee$, denoted by $I \models \mathcal{WS}^\vee$. □

Given $\mathcal{L}_P$, we aim at discovering a disjunctive schema $\mathcal{WS}^\vee$ as "close" as possible to the actual unknown schema $\mathcal{WS}(P)$ that had generated the logs. This intuition can be formalized by accounting for two criteria, namely *completeness* and *soundness*, constraining the discovered workflow to admit exactly the traces of the log. Obviously, we preliminary need some mechanisms for deciding whether a given trace in $\mathcal{L}_P$ can be actually derived from a real instantiation of a workflow $\mathcal{WS}^\vee$. Ideally, we might exploit the following definition.

Definition 2. Let s be a trace in $\mathcal{L}_P$, $\mathcal{WS}^\vee$ be a disjunctive workflow schema, and $I = \langle A_I, E_I \rangle$ be an instance of it. Then, s is *compliant with* $\mathcal{WS}^\vee$ *through* I, denoted by $s \models^I \mathcal{WS}^\vee$, if s is a topological sort of I, i.e., s is an ordering of the activities in A_I s.t. for each $(a, b) \in E_I$, $i < j$ where $s[i] = a$ and $s[j] = b$. Moreover, s is simply said to be compliant with $\mathcal{WS}^\vee$, denoted by $s \models \mathcal{WS}^\vee$, if there exists I with $s \models^I \mathcal{WS}^\vee$. □

We are now ready to introduce, for a disjunctive workflow schema and for a trace log, the notions of soundness (i.e., every instance must be witnessed by some trace in the log) and of completeness (all traces are compliant with some instance). As the schema is not given but discovered from the analysis of the trace log, the two notions are given with a certain amount of uncertainty.

Definition 3. Let $\mathcal{WS}^\vee$ be a disjunctive workflow model, and $\mathcal{L}_P$ be a log for process P. We define:

- $soundness(\mathcal{WS}^\vee, \mathcal{L}_P) = \frac{|\{I | I \models \mathcal{WS}^\vee \wedge \nexists s \in \mathcal{L}_P \text{ s.t. } s \models^I \mathcal{WS}^\vee\}|}{|\{I | I \models \mathcal{WS}^\vee\}|}$, i.e., the percentage of instances having no corresponding traces in the log;
- $completeness(\mathcal{WS}^\vee, \mathcal{L}_P) = \frac{|\{s | s \in \mathcal{L}_P \wedge s \models \mathcal{WS}^\vee\}|}{|\{s | s \in \mathcal{L}_P\}|}$, i.e., the percentage of traces that are compliant with some trace in the log.

Given two real numbers α and σ between 0 and 1 (typically α is small whereas σ is close to 1) we say that $\mathcal{WS}^\vee$ is

- *α-sound* w.r.t. $\mathcal{L}_P$, if $soundness(\mathcal{WS}^\vee, \mathcal{L}_P) \leq \alpha$, i.e. the smaller the sounder;
- *σ-complete* w.r.t. $\mathcal{L}_P$, if $completeness(\mathcal{WS}^\vee, \mathcal{L}_P) \geq \sigma$, i.e., the larger the more complete. □

We want to discover a disjunctive schema $\mathcal{WS}^\vee$ for a given process P which is *α-sound* and *σ-complete*, for some given α and σ. However, it is easy to see that a trivial schema satisfying the above conditions always exists, consisting in the union of exactly one workflow (without global constraints) modelling each of the instances in $\mathcal{L}_P$. However, such model would be not a syntectic view of the process P, for its size being $|\mathcal{WS}^\vee| = |\mathcal{L}_P|$, where $|\mathcal{L}_P| = |\{s \mid s \in \mathcal{L}\}|$. We therefore introduce a bound on the size of $\mathcal{WS}^\vee$.

Definition 4. (Minimal Process Discovery) Let $\mathcal{L}_P$ be a workflow log for the process P. Given a real number σ and a natural number m, the *Minimal Process Discovery problem*, denoted by MPD(P,σ, m), consists in finding a σ-complete disjunctive workflow schema $\mathcal{WS}^\vee$, such that $|\mathcal{WS}^\vee| \leq m$ and $soundness(\mathcal{WS}^\vee, \mathcal{L}_P)$ is minimal. □

The problem is obviously solvable as one may sacrifice enough portions of soundness to get a result. But, as it is shown next, the problem is untractable. W.l.o.g., let us assume that the values representing soundness are suitably discretized as positive integers so that we can represent MPD as an NP optimization problem.

Theorem 1. MPD(P,σ,m) *is an* NP*-complete optimization problem whose set of feasible solution is not empty.*

Armed with the above result, we turn to the problem PD(P,σ,m) of greedily finding a suitable approximation, that is a σ-complete workflow schema $\mathcal{WS}^{\vee}$, with $|\mathcal{WS}^{\vee}| \leq m$, which is as sound as possible. In the rest, we shall propose an efficient technique for solving this problem.

3 Clustering Workflow Traces

In order to mine the underlying workflow schema of the process P (problem PD(P,σ,m)) we exploit the idea of iteratively and incrementally refining a schema, by mining some *global constraints* which are then used for discriminating the possible executions, starting with a preliminary disjunctive model $\mathcal{WS}^{\vee}$, which only accounts for the dependencies among the activities in P.

The algorithm `ProcessDiscover`, shown in Figure 1, which computes $\mathcal{WS}^{\vee}$ through a hierarchical clustering, first mines a control flow $\mathcal{CF}_{\sigma}$, according to the threshold σ[1], through the procedure *minePrecedences*, which mainly exploits techniques already presented in the literature (see, e.g., [1,18], and, therefore, it is not illustrated in more details. Each workflow schema $\mathcal{WS}_i^j$, eventually inserted in $\mathcal{WS}^{\vee}$, is identified by the number i of refinements needed, and an index j for distinguishing the schemas at the same refinement level. Moreover, we denote by $\mathcal{L}(\mathcal{WS}_i^j)$ the set of traces in the cluster defined by $\mathcal{WS}_i^j$. Notice that preliminarily $\mathcal{WS}_0^1$, containing all the logs in $\mathcal{L}_P$, is inserted in $\mathcal{WS}^{\vee}$, and in Step 3 we refine the model by mining some local constraints, too.

The algorithm is also guided by a greedy heuristic that at each step selects a schema $\mathcal{WS}_i^j \in \mathcal{WS}^{\vee}$, for being refined with the function *refineWorkflow*, by preferring the schema which can be most profitably refined. In practice, we refine the the least sound schema among the ones already discovered; however, some experiments have been also conduced refining the schema $\mathcal{WS}_i^j$ with the maximum value of $|\mathcal{L}(\mathcal{WS}_i^j)|$.

In order to reuse well know clustering methods, and specifically in our implementation the k-*means* algorithm, the procedure *refineWorkflow* translates the logs $\mathcal{L}(\mathcal{WS}_i^j)$ to relational data with the procedures ***identifyRelevantFeatures*** and ***project***, which will be discussed in the next section. Then, if more than one feature is identified, it computes the clusters $\mathcal{WS}_{i+1}^{j+1}, ..., \mathcal{WS}_{i+1}^{j+k}$, where j is the maximum index of the schemas already inserted in $\mathcal{WS}^{\vee}$ at the level $i+1$, by applying the k-*means* algorithm on the traces in $\mathcal{L}(\mathcal{WS}_i^j)$, and put inserts them into the disjunctive schema $\mathcal{WS}^{\vee}$. Finally, for each schema inserted $\mathcal{WS}^{\vee}$ the procedure *mineLocalConstraint* is applied, in order to identify local constraints as well.

The algorithm `ProcessDiscover` converges in at most m steps (see Step 4), and exploits the following interesting property of the procedure *refineWorkflow*. We observe that at each step of workflow refinement the value of soundness decreases, thus the algorithm gets closer to the optimal solution.

[1] Roughly, the edges in $\mathcal{CF}_{\sigma}$ represent a minimal set of precedences with at least a given support

Input: Problem PD(P,σ,m), natural number $maxFeatures$.
Output: A process model.
Method: Perform the following steps:

$\mathcal{CF}_\sigma(\mathcal{WS}_0^1)$:=*minePrecedences*($\mathcal{L}_\mathcal{P}$); *//See Section 3.1*
let $\mathcal{WS}_0^1$ be a schema, with $\mathcal{L}(\mathcal{WS}_0^1) = \mathcal{L}_P$;
mineLocalConstraints($\mathcal{WS}_0^1$); *//See Section 3.1*
$\mathcal{WS}^\vee := \mathcal{WS}_0^1$; *//Start clustering with the dependency graph only*
while $|\mathcal{WS}^\vee| < m$ **do**
  $\mathcal{WS}_i^j$:=*leastSound*($\mathcal{WS}^\vee$);
  $\mathcal{WS}^\vee := \mathcal{WS}^\vee - \{\mathcal{WS}_i^j\}$;
  refineWorkflow(i,j);
end while
return $\mathcal{WS}^\vee$;

Procedure *refineWorkflow*(i: step, j: schema);
$\mathcal{F}$:=***identifyRelevantFeatures***$(\mathcal{L}(\mathcal{WS}_i^j), \sigma, maxFeatures, \mathcal{CF}_\sigma)$; *//See Section 4.1*
$\mathcal{R}(\mathcal{WS}_i^j)$:=***project***$(\mathcal{L}(\mathcal{WS}_i^j), \mathcal{F})$; *//See Section 4.2*
$k := |\mathcal{F}|$;
if $k > 1$ **then**
  $j := \max\{j \mid \mathcal{WS}_{i+1}^j \in \mathcal{WS}^\vee\}$;
  $\langle \mathcal{WS}_{i+1}^{j+1}, ..., \mathcal{WS}_{i+1}^{j+k} \rangle := k\text{-}means(\mathcal{R}(\mathcal{WS}_i^j))$;
  for each $\mathcal{WS}_{i+1}^h$ **do**
    $\mathcal{WS}^\vee = \mathcal{WS}^\vee \cup \{\mathcal{WS}_{i+1}^h\}$;
    $\mathcal{CF}_\sigma(\mathcal{WS}_{i+1}^h)$:=*minePrecedences*($\mathcal{L}(\mathcal{WS}_{i+1}^h)$);
    mineLocalConstraints($\mathcal{WS}_{i+1}^h$);
  end for
else *//Leave of the tree*
  $\mathcal{WS}^\vee = \mathcal{WS}^\vee \cup \{\mathcal{WS}_i^j\}$; *//See Theorem 2.2*
end if;

Fig. 1. Algorithm `ProcessDiscover`

Theorem 2. *Given a disjunctive schema $\mathcal{WS}^\vee$, with $\mathcal{WS}_i^j \in \mathcal{WS}^\vee$, the disjunctive workflow schema $\mathcal{WS}_+^\vee$, obtained by refining $\mathcal{WS}^\vee - \{\mathcal{WS}_i^j\}$ with the procedure refineWorkflow(i,j), is such that $soundness(\mathcal{WS}_+^\vee) \leq soundness(\mathcal{WS}^\vee)$.*

A main point of the algorithm is fixing the number k of new schemata to be added at each refinement step. The range of k goes from a minimum of 2, which will require several steps for the computation, to an unbounded value, which will return the result in only one step. One could then expect that the latter case is most efficient. This is not necessarily true: the clustering algorithm could run slower with a larger number of classes thus loosing the advantage of a smaller number of iterations. In contrast, there is an important point in favor of a small value for k: the representation of the various schemata can be optimized by preserving the tree structure and storing for each node only the differences w.r.t. the schema of the father node. The tree representation is relevant not only because of the space reduction but also because it give more insights on the properties of the modelled workflow instances and provides an intuitive and expressive description of global constraints.

3.1 Dealing with Relevant Features

The crucial point of the algorithm for clustering workflow traces lies in the formalization of the procedures ***identifyRelevantFeatures*** and ***project***. Roughly, the former identifies a set $\mathcal{F}$ of relevant features [10,11,14], whereas the latter projects the traces into a vectorial space whose components are, in fact, these features.

Some works addressing the problem of clustering complex data considered the most frequent common structures (see e.g. [2,3,7]), also called frequent patterns, to be the relevant features for the clustering. Since we are interested in features that witness some kind of global constraints, we instead exploit the more involved notion of unexpected (w.r.t. the local properties) frequent rules.

Let $\mathcal{L}$ be a set of traces, $\mathcal{CF}_\sigma$ be a mined control flow, for threshold σ, and E_σ be the edge set of $\mathcal{CF}_\sigma$. Then a sequence $[a_1...a_h]$ of tasks is σ-frequent in $\mathcal{L}$ if $|\{s \in \mathcal{L} \mid a_1 = s[i_1], ..., a_h = s[i_h] \wedge i_1 < ... < i_h\}|/|\mathcal{L}| \geq \sigma$. We say that $[a_1...a_h]$ *σ-precedes* a in $\mathcal{L}$, denoted by $[a_1...a_h] \rightarrow_\sigma a$, if both $[a_1...a_h]$ and $[a_1...a_h a]$ are σ-frequent in $\mathcal{L}$.

Definition 5 (Discriminant Rules). A *discriminant rule (feature)* ϕ is an expression of the form $[a_1...a_h] \not\rightarrow_\sigma a$, s.t. (i) $[a_1...a_h]$ is σ-frequent in $\mathcal{L}$, (ii) $(a_h, a) \in E_\sigma$, and (iii) $[a_1...a_h] \rightarrow_\sigma a$ does not hold. Moreover, ϕ is *minimal* if (iv) there is no b, s.t. $[a_1...a_h] \not\rightarrow_\sigma b$ and $[b] \rightarrow_\sigma a$, and (v) there is no j, s.t. $j > 1$ and $[a_j...a_h] \not\rightarrow_\sigma a$. □

The identification of discriminant rules can be carried out by means of the level-wise algorithm shown in Figure 2. At each step k of the computation, we store in L_k all the σ-frequent sequences whose size is k. Specifically, in the Steps 5–9, the set of potential sequences M to be included in L_{k+1} are obtained by combining those in L_k with the relationships of precedences in L_2 — notice that Step 7 prevents the computation of not minimal unexpected rules. Then, only σ-frequent pattern in M are included in L_{k+1} (Step 11), while all the others will determine unexpected rules (Step 12). The process is repeated until no other frequent traces are found. The correctness of the algorithm can be easily proven.

Theorem 3. *In the algorithm of Figure 2, before its termination (Step 16):*

1. *the set R contains exactly all the σ-frequent sequences of tasks, and*
2. *the set $\mathcal{F}$ contains exactly all the minimal discriminant rules.*

Notice that the algorithm *IdentifyRelevantFeatures* does not directly output $\mathcal{F}$, but call the procedure *mostDiscriminantFeatures*, whose aim is to find a proper subset of $\mathcal{F}$ which better discriminates the traces in the log.

This intuition can be formalized as follows. Let ϕ be a discriminant rule of the form $[a_i, ..., a_j] \not\rightarrow_\sigma b$, then *the witness of ϕ in $\mathcal{L}$*, denoted by $w(\phi, \mathcal{L})$, is the set of logs in which the pattern $[a_i, ..., a_j]$ occurs.

Moreover, given a set of rules R, then the witness of R in $\mathcal{L}$ is $\bigcup_{\phi \in R} w(\phi, \mathcal{L})$. For a fixed k, R is the most discriminant k-set of features if $|R| = k$ and there exists no R' with $|w(R', \mathcal{L})| > |w(R, \mathcal{L})|$, and $|R'| = k$. Notice that the most discriminant k-set of features can be computed in polynomial time by considering all the possible combinations of features of R, with k element.

The minimum k, for which the most discriminant k-set of features, say S, covers all the logs, i.e., $w(S, \mathcal{L}) = \mathcal{L}$, is called *dimension* of $\mathcal{L}$, whereas S is the *most discriminant set of features*.

Input: A log $\mathcal{L}$, a threshold σ, the max nr. of features $maxFeatures$, the control flow graph $\mathcal{CF}_\sigma$, with edge set E_σ.
Output: A set of minimal discriminant rules.
Method: Perform the following steps:

$L_2 := \{[ab] \mid (a,b) \in E_\sigma\}$;
$k := 1, R := L_2, \mathcal{F} := \emptyset$;
repeat
  $M := \emptyset$; $k := k+1$;
  forall $[a_i...a_j] \in L_k$ **do**
    forall $[a_j b] \in L_2$ **do**
      if $[a_{i+1}...a_j] \not\rightarrow_\sigma b$ is not in $\mathcal{F}$ **then**
        $M := M \cup [a_i...a_j b]$;
  end for
  forall $p \in M$ of the form $[a_i...a_j b]$ **do**
    if p is σ-frequent in $\mathcal{L}$ **then** $L_{k+1} := \{p\}$;
    else $\mathcal{F} := \mathcal{F} \cup \{[a_i...a_j] \not\rightarrow_\sigma b\}$;   *//See Theorem 3.2*
  end for
  $R := R \cup L_{k+1}$;   *//See Theorem 3.1*
until $L_{k+1} = \emptyset$;
return *mostDiscriminant*($\mathcal{F}$);

Procedure *mostDiscriminantFeatures*($\mathcal{F}$: set of unexpected rules): set of unexpected rules;

1 $S' := \mathcal{L}$; $\mathcal{F}' := \emptyset$;
2 **do**
3   **let** $\phi = \texttt{argmax}_{\phi' \in \mathcal{F}} |w(\phi', S')|$;
4   $\mathcal{F}' := \mathcal{F}' \cup \{\phi\}$;
5   $S' := S' - w(\phi, S')$;
6 **while** $(|S'|/|\mathcal{L}_P| > \sigma)$ and $(\mathcal{F}' < maxFeatures)$;
7 **return** $\mathcal{F}'$;

Fig. 2. **Algorithm** *IdentifyRelevantFeatures*

Theorem 4. *Let $\mathcal{L}$ be a set of traces, n be the size of $\mathcal{L}$ (i.e., the sum of the lengths of all the traces in $\mathcal{L}$), and $\mathcal{F}$ be a set of features. Then, computing any most discriminant set of features is* NP *hard.*

Due to the intrinsic difficulty of the problem, we turn to the computation of a suitable approximation. In fact, the procedure *mostDiscriminantFeatures*, actually implemented in the algorithm for identifying relevant features, computes a set $\mathcal{F}'$ of discriminant rules, guided by the heuristics of greedily selecting a feature ϕ covering the maximum number of traces, among the ones (S') not covered by previous selections.

Finally, the set of relevant features $\mathcal{F}$, can be used for representing each trace s as a point in the vectorial space $\mathbb{R}^{|\mathcal{F}|}$, denoted by $\overrightarrow{s}$. Then, the procedure ***project*** maps traces in $\mathbb{R}^{|\mathcal{F}|}$, where *k-means* algorithm can operate. Due to its simplicity we do not report the code here.

4 Experiments

In this section we study the behavior of the `ProcessDiscover` algorithm for evaluating both its effectiveness and its scalability, with the help of a number of tests performed on synthetic data. The generation of such data can be tuned according to: (i) the size of $\mathcal{WS}$, (ii) the size of $\mathcal{L}_P$, (iii) the number of global constraints in $\mathcal{C}_G$, and (iv) the probability p of choosing any successor edge, in the case of *nondeterministic fork* activities. The ideas adopted in generating synthetic data are essentially inspired by [3], and the generator we exploited is an extension of the one described in [6].

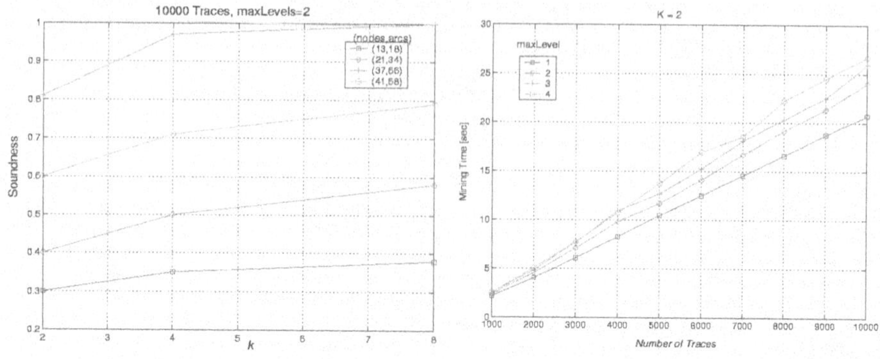

Fig. 3. Fixed Schema. Left: Soundness w.r.t. levels. Right: Scaling w.r.t. number of traces.

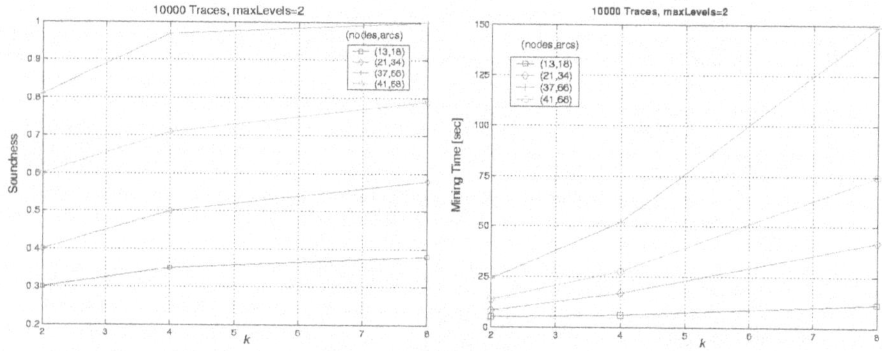

Fig. 4. Variable Workflow Schema. Left: Soundness w.r.t. k. Right: Scalability w.r.t. k.

Test Procedure. In order to asses the effectiveness of the technique, we adopted the following test procedure. Let $\mathcal{WS}(I)$ be a workflow schema for the input process I, and $\mathcal{L}_I$ a log produced with the generator. The quality of any workflow $\mathcal{WS}^{\vee}(O)$, extracted by providing the mining algorithm with $\mathcal{L}_I$, is evaluated, w.r.t. the original one $\mathcal{WS}(I)$, essentially by comparing two random samples of the traces they respectively admit. This allow us to compute an estimate of the actual soundness and completeness. Moreover, in order to avoid statistical fluctuations in our results, we generate a number of different training logs, and hence, whenever relevant, we report for each measure its mean value together with the associated standard deviation. In the test described here, we focus on the influence of two major parameters of the method: (i) the branching factor k and (ii) the maximum number ($maxLevels$) of levels in the resulting disjunctive scheme. Notice that the case $k = 1$ coincides with traditional algorithms which do not account for global constraints. All the tests have been conduced on a 1600MHz/256MB Pentium IV machine running *Windows XP Professional*.

Results. In a first set of experiments we considered a fixed workflow schema and some randomly generated instances. Figure 3 (on the left) reports the mean value and the standard deviation of the soundness of the mined model, for increasing values of $|\mathcal{L}_I|$

by varying the factor k. Notice that for $k = 1$, the algorithm degenerates in computing a unique schema, and in fact, the soundness is not affected by the parameter $maxLevel$ — this is the case of any algorithm accounting of local constraints only. Instead, for $k > 1$, we can even rediscover exactly the underlying schema, after a number of iterations. These experiments have been conduced on an input log of 1000 instances. Then, on the right, we report the scaling of the approach at the varying of the number of logs in $\mathcal{L}_I$.

In a second set of experiments we also consider variable schemas. In Figure 4 we report the results for four different workflow schemas. Observe (on the left) that for a fixed value of k, the soundness of the mined schema tends to be low at the increasing of the complexity of the schemas, consisting of many nodes and possibly many constraints. This witness the fact that on real processes, traditional approaches (with $k = 1$) performs poorly, and that for having an effective reconstruction of the process it is necessary not only to fix $k > 1$, but also to deal with several levels of refinements. Obviously, for complex schemas, the algorithm takes more time, as shown in the same figure on the right.

5 Conclusions

In this paper, we have continued on the way of the investigation of data mining techniques for process mining, by providing a method for discovering global constraints, in terms of the patterns of executions they impose. This is achieved through a hierarchical clustering of the logs, in which each trace is seen as a point of a properly identified space of features. The precise complexity of the task of constructing this space is provided, as well as a practical efficient algorithm for its solution.

References

1. R. Agrawal, D. Gunopulos, and F. Leymann. Mining process models from workflow logs. In *Proc. 6th Int. Conf. on Extending Database Technology (EDBT'98)*, pages 469–483, 1998.
2. R. Agrawal, T. Imielinski, and A. Swami. Mining Association Rules between Sets of Items in Large Databases. In *Proc. of SIGMOD'93*, pages 207–216, 1993.
3. R. Agrawal and R. Srikant. Fast Algorithms for Mining Association Rules. In *Proc. of the 20th Int'l Conference on Very Large Databases*, pages 487–499, 1994.
4. J.E.Cook and A.L. Wolf. Automating process discovery through event-data analysis. In *Proc. 17th Int. Conf. on Software Engineering (ICSE'95)*, pages 73–82, 1995.
5. H. Davulcu, M. Kifer, C.R. Ramakrishnan, and I.V. Ramakrishnan. Logic based modeling and analysis of workflows. In *Proc. of PODS'98*, pages 25–33, 1998.
6. G.Greco, A.Guzzo, G.Manco, and D. Saccà. Mining Frequent Instances on Workflows. In *Proc. of PAKDD'03*, pages 209–221, 2003.
7. J. Han, J. Pei, and Y. Yi. Mining frequent patterns without candidate generation. In *Proc. Int. ACM Conf. on Management of Data (SIGMOD'00)*, pages 1–12, 2000.
8. Y. Kim, W. Nick Street, and F. Menczer. Feature selection in unsupervised learning via evolutionary search. In *Proceedings of the KDD'00*, pages 365–369, 2000.
9. V. I. Levenshtein. Binary codes capable of correcting deletions, insertions and reversals. *Doklady Akademii Nauk SSSR 163(4)* page 845-848, 1965.

10. N.Lesh, M.J. Zaki, and M.Ogihara. Mining features for sequence classification. In *Proc. of KDD'99*, pages 342–346, 1999.
11. H.Motoda and H.Liu. Data reduction: feature selection. pages 208–213, 2002.
12. P. Muth, J. Weifenfels, M.Gillmann, and G. Weikum. Integrating light-weight workflow management systems within existing business environments. In *Proc. 15th IEEE Int. Conf. on Data Engineering (ICDE'99)*, pages 286–293, 1999.
13. J. Pei, J. Han, B. Mortazavi-Asl, H. Pinto, Q. Chen, U. Dayal, and M. Hsu. Prefixspan: Mining sequential patterns by prefix-projected growth. In *Proc. of ICDE'2001*, pages 215–224, 2001.
14. B.Padmanabhan and A.Tuzhilin. Small is beautiful: discovering the minimal set of unexpected patterns. In *Proc. of the 6th ACM SIGKDD*, pages 54–63, 2000.
15. W.M.P. van der Aalst. The application of petri nets to worflow management. *Journal of Circuits, Systems, and Computers*, 8(1):21–66, 1998.
16. W.M.P. van der Aalst, A. Hirnschall, and H.M.W. Verbeek. An alternative way to analyze workflow graphs. In *Proc. 14th Int. Conf. on Advanced Information Systems Engineering*, pages 534–552, 2002.
17. W.M.P. van der Aalst and K.M. van Hee. *Workflow Management: Models, Methods, and Systems*. MIT Press, 2002.
18. W.M.P. van der Aalst, B.F. van Dongen, J. Herbst, L. Maruster, G.Schimm, and A.J.M.M. Weijters. Workflow mining: A survey of issues and approaches. *Data and Knowledge Engineering*, 47(2):237–267, 2003.

CMTreeMiner: Mining Both Closed and Maximal Frequent Subtrees*

Yun Chi, Yirong Yang, Yi Xia, and Richard R. Muntz

University of California, Los Angeles, CA 90095, USA
{ychi,yyr,xiayi,muntz}@cs.ucla.edu

Abstract. Tree structures are used extensively in domains such as computational biology, pattern recognition, XML databases, computer networks, and so on. One important problem in mining databases of trees is to find frequently occurring subtrees. However, because of the combinatorial explosion, the number of frequent subtrees usually grows exponentially with the size of the subtrees. In this paper, we present *CMTreeMiner*, a computationally efficient algorithm that discovers all closed and maximal frequent subtrees in a database of rooted unordered trees. The algorithm mines both closed and maximal frequent subtrees by traversing an enumeration tree that systematically enumerates all subtrees, while using an enumeration DAG to prune the branches of the enumeration tree that do not correspond to closed or maximal frequent subtrees. The enumeration tree and the enumeration DAG are defined based on a canonical form for rooted unordered trees–the depth-first canonical form (DFCF). We compare the performance of our algorithm with that of *PathJoin*, a recently published algorithm that mines maximal frequent subtrees.

Keywords: Frequent subtree, closed subtree, maximal subtree, enumeration tree, rooted unordered tree.

1 Introduction

Tree structures are used extensively in domains such as computational biology, pattern recognition, XML databases, computer networks, and so on. Trees in real applications are often *labeled*, with labels attached to vertices and edges where these labels are not necessarily unique. In this paper, we study one important issue in mining databases of labeled rooted unordered trees–finding frequently occurring subtrees, which has much practical importance [5]. However, as we have discovered in our previous study [5], because of the combinatorial explosion, the number of frequent subtrees usually grows exponentially with the tree size. This is the case especially when the transactions in the database are strongly correlated. This phenomenon has two effects: first, there are too many frequent subtrees for users to manage and use, and second, an algorithm that discovers all frequent subtrees is not able to handle frequent subtrees with large size. To solve

* This work was supported by NSF under Grant Nos. 0086116, 0085773, and 9817773.

H. Dai, R. Srikant, and C. Zhang (Eds.): PAKDD 2004, LNAI 3056, pp. 63–73, 2004.

this problem, in this paper, we propose *CMTreeMiner*, an efficient algorithm that, instead of looking for all frequent subtrees, only discovers both closed and maximal frequent subtrees in a database of labeled rooted unordered trees.

Related Work Recently, there has been growing interest in mining databases of labeled trees, partly due to the increasing popularity of XML in databases. In [9], Zaki presented an algorithm, TREEMINER, to discover all frequent embedded subtrees, i.e., those subtrees that preserve ancestor-descendant relationships, in a forest or a database of rooted ordered trees. The algorithm was extended further in [10] to build a structural classifier for XML data. In [2] Asai *et al.* presented an algorithm, FREQT, to discover frequent rooted ordered subtrees. For mining rooted unordered subtrees, Asai *et al.* in [3] and we in [5] both proposed algorithms based on enumeration tree growing. Because there could be multiple ordered trees corresponding to the same unordered tree, similar canonical forms for rooted unordered trees are defined in both studies. In [4] we have studied the problem of indexing and mining free trees and developed an Apriori-like algorithm, *FreeTreeMiner*, to mine all frequent free subtrees. In [8], Xiao *et al.* presented an algorithm called *PathJoin* that rewrites a database into a compact in-memory data structure–*FST-Forest*, for the mining purpose. In addition, to the best of our knowledge, *PathJoin* is the only algorithm that mines *maximal* frequent subtrees.

Our Contributions The main contributions of this paper are: (1) We introduce the concept of *closed frequent subtrees* and study its properties and its relationship with *maximal frequent subtrees.* (2) In order to mine both closed and maximal frequent rooted unordered subtrees, we present an algorithm–*CMTreeMiner*, which is based on the canonical form and the enumeration tree that we have introduced in [5]. We develop new pruning techniques based on an enumeration DAG. (3) Finally, we have implemented our algorithm and have carried out experimental study to compare the performance of our algorithm with that of *PathJoin* [8].

The rest of the paper is organized as follows. In section 2, we give the background concepts. In section 3, we present our *CMTreeMiner* algorithm. In section 4, we show experiment results. Finally, in Section 5, we give the conclusion.

2 Background

2.1 Basic Concepts

In this section, we provide the definitions of the concepts that will be used in the remainder of the paper. We assumed that the readers are familiar with the notions such as *rooted unordered tree*, *ancestor/descendant*, *parent/child*, *leaf*, etc. In addition, a rooted tree t is a (proper) *subtree* of another rooted tree s if the vertices and edges of t are (proper) subsets of those of s. If t is a (proper) subtree of s, we say s is a (proper) *supertree* of t. Two labeled rooted unordered trees t and s are *isomorphic* to each other if there is a one-to-one mapping from the vertices of t to the vertices of s that preserves vertex labels, edge labels, adjacency, and the root. A *subtree isomorphism* from t to s is an isomorphism from t to some subtree of s. For convenience, in this paper we call a rooted tree with k vertices a k-tree.

Let D denote a database where each transaction $s \in D$ is a labeled rooted unordered tree. For a given pattern t, which is a rooted unordered tree, we say t *occurs* in a transaction s if there exists at least one subtree of s that is isomorphic to t. The *occurrence* $\delta_t(s)$ of t in s is the number of distinct subtrees of s that are isomorphic to t. Let $\sigma_t(s) = 1$ if $\delta_t(s) > 0$, and 0 otherwise. We say s *supports* pattern t if $\sigma_t(s)$ is 1 and we define the *support* of a pattern t as $supp(t) = \sum_{s \in D} \sigma_t(s)$. A pattern t is called *frequent* if its support is greater than or equal to a *minimum support* (*minsup*) specified by a user. The frequent subtree mining problem is to find all frequent subtrees in a given database.

One nice property of frequent trees is the *a priori* property, as given below:

Property 1. Any subtree of a frequent tree is also frequent and any supertree of an infrequent tree is also infrequent.

We define a frequent tree t to be *maximal* if none of t's proper supertrees is frequent, and *closed* if none of t's proper supertrees has the same support that t has. For a subtree t, we define the *blanket* of t as the set of subtrees $B_t = \{t' |$removing a leaf or the root from t' can result in $t\}$. In other words, the blanket B_t of t is the set of all supertrees of t that have one more vertex than t. With the definition of blanket, we can define maximal and closed frequent subtrees in another equivalent way:

Property 2. A frequent subtree t is maximal iff for every $t' \in B_t$, $supp(t') < minsup$; a frequent subtree t is closed iff for every $t' \in B_t$, $supp(t') < supp(t)$.

For a subtree t and one of its supertrees $t' \in B_t$, we define the *difference* between t' and t ($t' \backslash t$ in short) as the additional vertex of t' that is not in t. We say $t' \in B_t$ and t are *occurrence-matched* if for each occurrence of t in (a transaction of) the database, there is at least one corresponding occurrence of t'; we say $t' \in B_t$ and t are *support-matched* if for each transaction $s \in D$ such that $\sigma_t(s) = 1$, we have $\sigma_{t'}(s) = 1$. It is obvious that if t' and t are occurrence-matched, it implies that they are support-matched.

2.2 Properties of Closed and Maximal Frequent Subtrees

The set of all frequent subtrees, the set of closed frequent subtrees and the set of maximal frequent subtrees have the following relationship.

Property 3. For a database D and a given *minsup*, let $\mathcal{F}$ be the set of all frequent subtrees, $\mathcal{C}$ be the set of closed frequent subtrees, and $\mathcal{M}$ be the set of maximal frequent subtrees, then $\mathcal{M} \subseteq \mathcal{C} \subseteq \mathcal{F}$.

The reason why we want to mine closed and maximal frequent subtrees instead of all frequent subtrees is that usually, there are much fewer closed or maximal frequent subtrees compared to the total number of frequent subtrees

[7]. In addition, by mining only closed and maximal frequent subtrees, we do not lose much information because the set of closed frequent subtrees maintains the same information (including support) as the set of all frequent subtrees and the set of maximal frequent subtrees subsumes all frequent subtrees:

Property 4. We can obtain all frequent subtrees from the set of maximal frequent subtrees because any frequent subtree is a subtree of one (or more) maximal frequent subtree(s); similarly, we can obtain all frequent subtrees with their supports from the set of closed frequent subtrees with their supports, because for a frequent subtree t that is not closed, $supp(t) = \max_{t'}\{supp(t')\}$ where t' is a supertree of t that is closed.

2.3 The Canonical Form for Rooted Labeled Unordered Trees

From a rooted unordered tree we can derive many rooted ordered trees, as shown in Figure 1. From these rooted ordered trees we want to uniquely select one as the canonical form to represent the corresponding rooted unordered tree. Notice that if a labeled tree is rooted, then without loss of generality we can assume that all edge labels are identical: because each edge connects a vertex with its parent, so we can consider an edge, together with its label, as a part of the child vertex. So for all running examples in the following discussion, we assume that all edges in all trees have the same label or equivalently, are unlabeled, and we therefore ignore all edge labels.

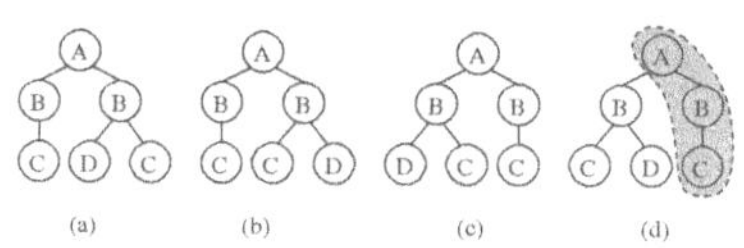

Fig. 1. Four Rooted Ordered Trees Obtained from the Same Rooted Unordered Tree

Without loss of generality, we assume that there are two special symbols, "\$" and "#", which are not in the alphabet of edge labels and vertex labels. In addition, we assume that (1) there exists a total ordering among edge and vertex labels, and (2) "#" sorts greater than "\$" and both sort greater than any other symbol in the alphabet of vertex and edge labels. We first define the *depth-first string encoding* for a rooted ordered tree through a *depth-first* traversal and use "\$" to represent a backtrack and "#" to represent the end of the string encoding. The depth-first string encodings for each of the four trees in Figure 1 are for (a) $ABC\$\$BD\$C\#$, for (b) $ABC\$\$BC\$D\#$, for (c) $ABD\$C\$\$BC\#$, and for (d) $ABC\$D\$\$BC\#$. With the string encoding, we define the *depth-first canonical string (DFCS)* of the rooted unordered tree as the minimal one among all possible depth-first string encodings, and we define the *depth-first canonical form (DFCF)* of a rooted unordered tree as the corresponding rooted ordered tree that gives the minimal DFCS. In Figure 1, the depth-first string encoding for *tree (d)* is the DFCS, and *tree (d)* is the DFCF for the corresponding labeled rooted unordered tree. Using a tree isomorphism algorithm given by Aho *et al.* [1,4,5], we can construct the DFCF for a rooted unordered tree in $O(ck \log k)$ time, where k is the number of vertices the tree has and c is the maximal degree of the vertices in the tree.

For a rooted unordered tree in its DFCF, we define the *rightmost leaf* as the last vertex according to the depth-first traversal order, and *rightmost path* as the path from the root to the rightmost leaf. The rightmost path for the DFCF of the above example (Figure 1(d)) is the path in the shaded area and the rightmost leaf is the vertex with label C in the shaded area.

3 Mining Closed and Maximal Frequent Subtrees

Now, we describe our *CMTreeMiner* algorithm that mines both closed and maximal frequent subtrees from a database of labeled rooted unordered trees.

3.1 The Enumeration DAG and the Enumeration Tree

We first define an enumeration DAG that enumerates all rooted unordered trees in their DFCFs. The nodes of the enumeration DAG consist of all rooted unordered trees in their DFCFs and the edges consist of all ordered pairs (t, t') of rooted unordered trees such that $t' \in B_t$. Figure 2 shows a fraction of the enumeration DAG. (For simplicity, we have only shown those trees with A as the root.)

Next, we define a unique enumeration tree based on the enumeration DAG. The enumeration tree is a spanning tree of the enumeration DAG so the two have the same set of the nodes. The following theorem is key to the definition of the enumeration tree.

Theorem 1. *Removing the rightmost leaf from a rooted unordered (k+1)-tree in its DFCF will result in the DFCF for another rooted unordered k-tree.*[1]

Based on the above theorem we can build an enumeration tree such that the parent for each rooted unordered tree is determined uniquely by removing the rightmost leaf from its DFCF. Figure 3 shows a fraction of the enumeration tree for the enumeration DAG in Figure 2. In order to grow the enumeration tree, starting from a node v of the enumeration tree, we need to find all valid children of v. Each child of v is obtained by adding a new vertex to v so that the new vertex becomes the new rightmost leaf of the new DFCF. Therefore, the possible positions for adding the new rightmost leaf to a DFCF are the vertices on the rightmost path of the DFCF.

[1] Proofs for all the theorems in this paper are available in [6].

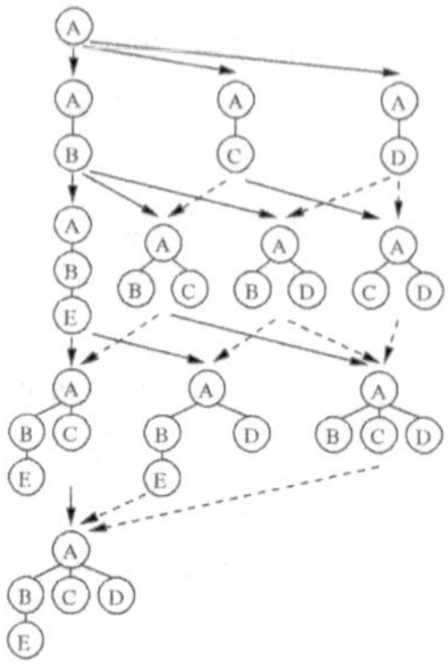

Fig. 2. The Enumeration DAG for Rooted Unordered Trees in DFCFs

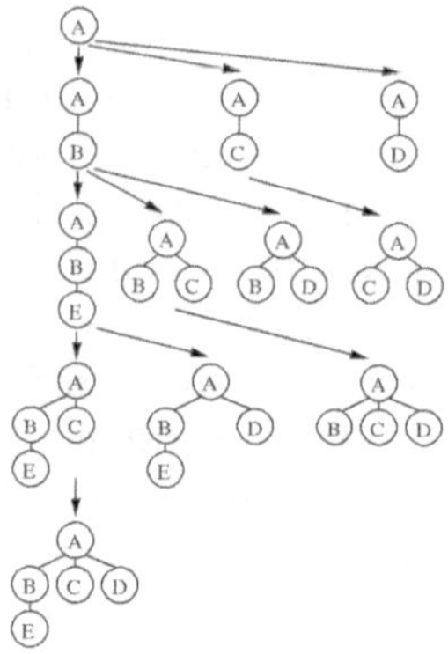

Fig. 3. The Enumeration Tree for Rooted Unordered Trees in DFCFs

3.2 The CMTreeMiner Algorithm

In the previous section, we have used the enumeration tree to enumerate all (frequent) subtrees in their DFCFs. However, the final goal of our algorithm is to find all closed and maximal frequent subtrees. As a result, it is not necessary to grow the complete enumeration tree, because under certain conditions, some branches of the enumeration tree are guaranteed to produce no closed or maximal frequent subtrees and therefore can be pruned. In this section, we introduce techniques that prune the unwanted branches with the help of the enumeration DAG (more specifically, the blankets).

Let us look at a node v_t in the enumeration tree. We assume that v_t corresponds to a frequent k-subtree t and denote the blanket of t as B_t. In addition, we define three subsets of B_t:

$$
\begin{aligned}
B_t^F &= \{t' \in B_t | t' \text{ is frequent}\} \\
B_t^{SM} &= \{t' \in B_t | t' \text{ and } t \text{ are support-matched}\} \\
B_t^{OM} &= \{t' \in B_t | t' \text{ and } t \text{ are occurrence-matched}\}
\end{aligned}
$$

From Property 2 we know that t is closed iff $B_t^{SM} = \emptyset$, that t is maximal iff $B_t^F = \emptyset$, and that $B_t^{OM} \subseteq B_t^{SM}$. Therefore by constructing B_t^{SM} and B_t^F for t, we can know if t is closed and if t is maximal. However, there are two problems. First, knowing that t is not closed does not automatically allow us to prune v_t from the enumeration tree, because some descendants of v_t in the enumeration tree might be closed. Second, computing B_t^F is time and space consuming, because we have to record all members of B_t and their support. So we want to avoid computing B_t^F whenever we can. In contrast, computing B_t^{SM} and B_t^{OM} is not that difficult, because we only need to record the intersections of all occurrences.

To solve the first problem mentioned above, we use B_t^{OM}, instead of B_t^{SM}, to check if v_t can be pruned from the enumeration tree. For a $t^o \in B_t^{OM}$ (i.e.,

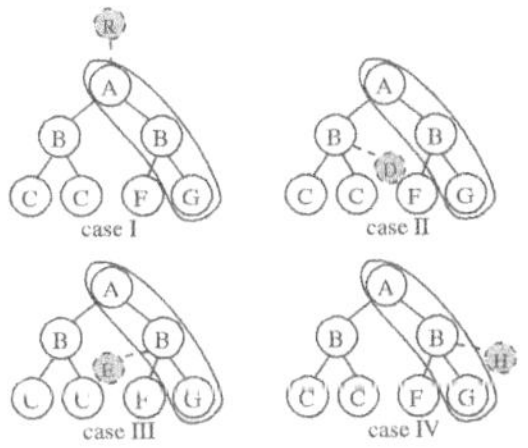

Fig. 4. Locations for an Additional Vertex to Be Added To a Subtree

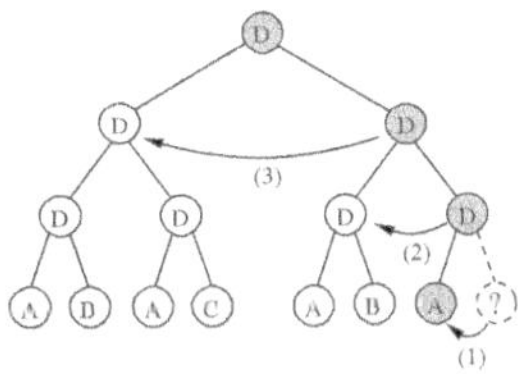

Fig. 5. Computing the Range of a New Vertex for Extending a Subtree

t^o and t are occurrence-matched), the new vertex $t^o \backslash t$ can occur at different locations, as shown in Figure 4. In Case I of Figure 4, $t^o \backslash t$ is the root of t^o; in Case II $t^o \backslash t$ is attached to a vertex of t that is not on the rightmost path; in Case III and case IV, $t^o \backslash t$ is attached to a vertex on the rightmost path. The difference between Case III and Case IV is whether or not $t^o \backslash t$ can be the new rightmost vertex of t^o.

To distinguish Case III and Case IV in Figure 4, we compute the range of vertex labels that could possibly be the new rightmost vertex of a supertree in B_t. Notice that this information is also important when we extend v_t in the enumeration tree–we have to know what are the valid children of v_t. Figure 5 gives an example for computing the range of valid vertex labels at a given position on the rightmost path. In the figure, if we add a new vertex at the given position, we may violate the DFCF by changing the order between some ancestor of the new vertex (including the vertex itself) and its immediate left sibling. So in order to determine the range of allowable vertex labels for the new vertex (so that adding the new vertex will guarantee to result in a new DFCF), we can check each vertex along the path from the new vertex to the root. In Figure 5, the result of comparison (1) is that the new vertex should have label greater than or equal to A, comparison (2) increases the label range to be greater than or equal to B, and comparison (3) increases the label range to be greater than or equal to C. As a result, before start adding new vertices, we know that adding any vertex with label greater than or equal to C at that specific position will surely result in a DFCF. Therefore, at this given location, adding a new vertex with label greater than or equal to C will result in case IV (and therefore the new vertex becomes the new rightmost vertex), and adding a new vertex with label less than C will result in case III in Figure 4.

Now we propose a pruning technique based on B_t^{OM}, as given in the following theorem.

Theorem 2. *For a node v_t in the enumeration tree and the corresponding subtree t, assume that t is frequent and $B_t^{OM} \neq \emptyset$. If there exists a $t^o \in B_t^{OM}$ such that $t^o \backslash t$ is at location of Case I, II, or III in Figure 4, then neither v_t nor any descendant of v_t in the enumeration tree correspond to closed or maximal frequent subtrees, therefore v_t (together with all of its descendants) can be pruned from the enumeration tree.*

For the second problem mentioned above, in order to avoid computing B_t^F as much as possible, we compute B_t^{OM} and B_t^{SM} first. If some $t^o \in B_t^{OM}$ is of Case I, II, or III in Figure 4, we are lucky because v_t (and all its descendants) can be pruned completely from the enumeration tree. Even if this is not the case, as long as $B_t^{SM} \neq \emptyset$, we only have to do the regular extension to the enumeration tree, with the knowledge that t cannot be maximal. To extend v_t, we find all potential children of v_t by checking the potential new rightmost leaves within the range that we have computed as described above. Only when $B_t^{SM} = \emptyset$, before doing the regular extension to the enumeration tree, do we have to compute B_t^F to check if t is maximal. Putting all the above discussion together, Figure 6 gives our *CMTreeMiner* algorithm.

```
Algorithm CMTreeMiner(D,minsup)
CL ← ∅, MX ← ∅;
C ← frequent 1-trees;
CM-Grow(C,CL, MX, minsup);
return CL, MX;
```

```
Algorithm CM-Grow(C,CL, MX, minsup)
for i ← 1,...,|C| do
    E ← ∅;
    compute B_{c_i}^{OM}, B_{c_i}^{SM};
    if ∃c^o ∈ B_{c_i}^{OM} that is of case I,II, or III then continue;
    if B_{c_i}^{SM} = ∅ then
        CL ← CL ∪ c_i;
        compute B_{c_i}^F;
        if B_{c_i}^F = ∅ then MX ← MX ∪ c_i;
    for each vertex v_n on the rightmost path of c_i do
        for each valid new rightmost vertex v_m of c_i do
            e ← c_i plus vertex v_m, with v_n as v_m's parent;
            if supp(e) ≥ minsup then E ← E ∪ e;
    if E ≠ ∅ then CM-Grow(E,CL,MX,minsup);
return;
```

Fig. 6. The CMTreeMiner Algorithm

We want to point out two possible variations to the *CMTreeMiner* algorithm. First, the algorithm mines both closed frequent subtrees and maximal frequent subtrees at the same time. However, the algorithm can be easily changed to mine only closed frequent subtrees or only maximal frequent subtrees. For mining only closed frequent subtrees, we just skip the step of computing B_t^F. For mining only maximal frequent subtrees, we just skip computing B_t^{SM} and use B_t^{OM} to prune the subtrees that are not maximal. Notice that this pruning is indirect: B_t^{OM} only prunes the subtrees that are not closed, but if a subtree is not closed then it cannot be maximal. If $B_t^{OM} = \emptyset$, for better pruning effects, we can still compute B_t^{SM} to determine if we want to compute B_t^F. If this is the case, although we only want the maximal frequent subtrees, the closed frequent subtrees are the byproducts of the algorithm. For the second variation, although our enumeration tree is built for enumerating all rooted *unordered* subtrees, it can be changed easily to enumerate all rooted *ordered* subtrees–the rightmost expansion is still valid for rooted ordered subtrees and we only have to remove the canonical form restriction. Therefore, our *CMTreeMiner* algorithm can handle databases of rooted ordered trees as well.

4 Experiments

We performed extensive experiments to evaluate the performance of the *CMTreeMiner* algorithm using both synthetic datasets and datasets from real applications. Due to the space limitation, here we only report the results for a synthetic dataset. We refer interested readers to [6] for other results. All experiments were done on a 2GHz Intel Pentium IV PC with 1GB main memory, running RedHat Linux 7.3 operating system. All algorithms were implemented in C++ and compiled using the g++ 2.96 compiler.

As far as we know, *PathJoin* [8] is the only algorithm for mining maximal frequent subtrees. *PathJoin* uses a subsequent pruning that, after obtaining all frequent subtrees, prunes those frequent subtrees that are not maximal. Because *PathJoin* uses the paths from roots to leaves to help subtree mining, it does not allow any siblings in a tree to have the same labels. Therefore, we have generated a dataset that meets this special requirement. We have used the data generator given in [5] to generate synthetic data. The detailed procedure for generating the dataset is described in [5] and here we give a very brief description. A set of $|N|$ subtrees are sampled from a large base (labeled) graph. We call this set of $|N|$ subtrees the *seed trees*. Each seed tree is the starting point for $|D| \cdot |S|$ transactions where $|D|$ is the number of transactions in the database and $|S|$ is the minimum support. Each of these $|D| \cdot |S|$ transactions is obtained by first randomly permuting the seed tree then adding more random vertices to increase the size of the transaction to $|T|$. After this step, more random transactions with size $|T|$ are added to the database to increase the cardinality of the database to $|D|$. The number of distinct vertex labels is controlled by the parameter $|L|$. The parameters for the dataset used in this experiment are: $|D|$=100000, $|N|$=90, $|L|$=1000, $|S|$=1%, $|T|$=$|I|$, and $|I|$ varies from 5 to 50. (For $|I| > 25$, *PathJoin* exhausts all available memory.)

Figure 7 compares the performance of *PathJoin* with that of *CMTreeMiner*. Figure 7(a) gives the number of all frequent subtrees obtained by *PathJoin*, the number of subtrees checked by *CMTreeMiner*, the number of closed frequent subtrees, and the number of maximal frequent subtrees. As the figure shows, the number of subtrees checked by *CMTreeMiner* and the number of closed subtrees grow in polynomial fashion. In contrast, the total number of all frequent subtrees (which is a lower bound of the number of subtrees checked by *PathJoin*) grows exponentially. As a result, as demonstrated in Figure 7(b), although *PathJoin* is very efficient for datasets with small tree sizes, as tree sizes increases beyond some

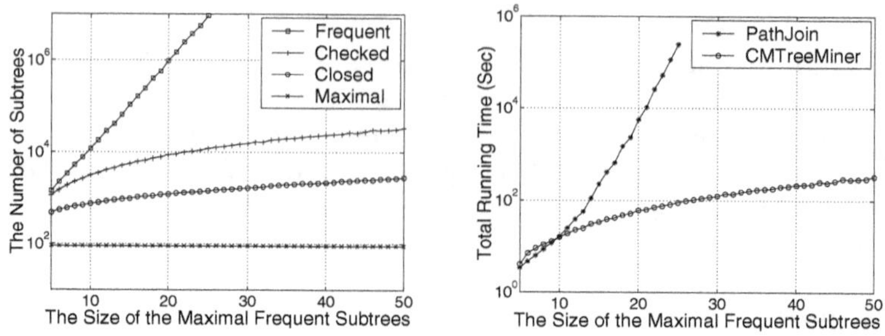

(a) (b)

Fig. 7. CMTreeMiner vs. PathJoin

reasonably large value (say 10), it becomes obvious that *PathJoin* suffers from exponential explosion while *CMTreeMiner* does not. (Notice the logarithmic scale of the figure.) For example, with the size of the maximal frequent subtrees to be 25 in the dataset, it took *PathJoin* around 3 days to find all maximal frequent subtrees while it took *CMTreeMiner* only 90 seconds!

5 Conclusion

In this paper, we have studied the issue of mining frequent subtrees from databases of labeled rooted unordered trees. We have presented a new efficient algorithm that mines both closed and maximal frequent subtrees. The algorithm is built based on a canonical form that we have defined in our previous work. Based on the canonical form, an enumeration tree is defined to enumerates all subtrees and an enumeration DAG is used for pruning branches of the enumeration tree that will not result in closed or maximal frequent subtrees. The experiments showed that our new algorithm performs in polynomial fashion instead of the exponential growth shown by other algorithms.

Acknowledgements. Thanks to Professor Y. Xiao at the Georgia College and State University for providing the *PathJoin* source codes and offering a lot of help.

References

1. A. V. Aho, J. E. Hopcroft, and J. E. Ullman. *The Design and Analysis of Computer Algorithms.* Addison-Wesley, 1974.
2. T. Asai, K. Abe, S. Kawasoe, H. Arimura, H. Satamoto, and S. Arikawa. Efficient substructure discovery from large semi-structured data. In *Proc. of the 2nd SIAM Int. Conf. on Data Mining*, 2002.

3. T. Asai, H. Arimura, T. Uno, and S. Nakano. Discovering frequent substructures in large unordered trees. In *Proc. of the 6th Intl. Conf. on Discovery Science*, 2003.
4. Y. Chi, Y. Yang, and R. R. Muntz. Indexing and mining free trees. In *Proc. of the 2003 IEEE Int. Conf. on Data Mining*, 2003.
5. Y. Chi, Y. Yang, and R. R. Muntz. Mining frequent rooted trees and free trees using canonical forms. Technical Report CSD-TR No. 030043, ftp://ftp.cs.ucla.edu/tech-report/2003-reports/030043.pdf, UCLA, 2003.
6. Y. Chi, Y. Yang, Y. Xia, and R. R. Muntz. CMTreeMiner: Mining both closed and maximal frequent subtrees. Technical Report CSD-TR No. 030053, ftp://ftp.cs.ucla.edu/tech-report/2003-reports/030053.pdf, UCLA, 2003.
7. N. Pasquier, Y. Bastide, R. Taouil, and L. Lakhal. Discovering frequent closed itemsets for association rules. *Lecture Notes in Computer Science*, 1540:398–416, 1999.
8. Y. Xiao, J-F Yao, Z. Li, and M. Dunham. Efficient data mining for maximal frequent subtrees. In *Proc. of the 2003 IEEE Int. Conf. on Data Mining*, 2003.
9. M. J. Zaki. Efficiently mining frequent trees in a forest. In *8th ACM SIGKDD International Conference on Knowledge Discovery and Data Mining*, 2002.
10. M. J. Zaki and C. C. Aggarwal. XRules: An effective structural classifier for XML data. In *Proc. of the 2003 Int. Conf. Knowledge Discovery and Data Mining*, 2003.

Secure Association Rule Sharing

Stanley R.M. Oliveira[1], Osmar R. Zaïane[1], and Yücel Saygin[2]

[1] Department of Computing Science
University of Alberta, Edmonton, Canada, T6G 2E8
{oliveira, zaiane}@cs.ualberta.ca
[2] Faculty of Engineering and Natural Sciences
Sabanci University, Orhanli, 34956, Istanbul, Turkey
ysaygin@sabanciuniv.edu

Abstract. The sharing of association rules is often beneficial in industry, but requires privacy safeguards. One may decide to disclose only part of the knowledge and conceal strategic patterns which we call restrictive rules. These restrictive rules must be protected before sharing since they are paramount for strategic decisions and need to remain private. To address this challenging problem, we propose a unified framework for protecting sensitive knowledge before sharing. This framework encompasses: (a) an algorithm that sanitizes restrictive rules, while blocking some inference channels. We validate our algorithm against real and synthetic datasets; (b) a set of metrics to evaluate attacks against sensitive knowledge and the impact of the sanitization. We also introduce a taxonomy of sanitizing algorithms and a taxonomy of attacks against sensitive knowledge.

Keywords: Privacy preserving data mining, Protecting sensitive knowledge, Sharing association rules, Data sanitization, Sanitizing algorithms.

1 Introduction

Protecting against inference learning in data mining sense has begun to receive attention. In particular, the problem of privacy preservation, when sharing data while wanting to conceal some restrictive associations has been addressed in the literature [1,2,7,4,5]. The proposed solutions consist in transforming a transactional database to be shared in such a way that the restrictive rules cannot be discovered. This process is called data sanitization [1]. The effectiveness of the data sanitization is measured by the proportion of restrictive rules effectively hidden (hiding failure), the proportion of rules accidentally hidden (misses cost) and the amount of artifactual rules created by the process [4]. The problem we address here is different and more practical. It is the problem of *rule sanitization.* Rather than sharing the data, collaborators prefer to mine their own data and share the discovered patterns.

Let us consider a motivating example based on a case discussed in [3]. Suppose we have a server and many clients in which each client has a set of sold items

H. Dai, R. Srikant, and C. Zhang (Eds.): PAKDD 2004, LNAI 3056, pp. 74–85, 2004.

(e.g. books, movies, etc). The clients want the server to gather statistical information about associations among items in order to provide recommendations to the clients. However, the clients do not want the server to know some restrictive association rules. In this context, the clients represent companies and the server is a recommendation system for an e-commerce application, for example, fruit of the clients collaboration. In the absence of rating, which is used in collaborative filtering for automatic recommendation building, association rules can be effectively used to build models for on-line recommendation. When a client sends its frequent itemsets or association rules to the server, it sanitizes some restrictive itemsets according to some specific policies. The server then gathers statistical information from the sanitized itemsets and recovers from them the actual associations.

The simplistic solution to address the motivating example is to implement a filter after the mining phase to weed out/hide the restricted discovered rules. However, we claim that trimming some rules out does not ensure full protection. The sanitization applied to the set of rules must not leave a trace that could be exploited by an adversary. We must guarantee that some inference channels have been blocked as well.

This paper introduces the notion of rule sanitization. The main contribution of this paper is a novel framework for protecting sensitive knowledge before sharing association rules. This framework encompasses: (a) a sanitizing algorithm called Downright Sanitizing Algorithm (DSA). This algorithm sanitizes a set of restrictive rules while blocking some inference channels; (b) a set of metrics to evaluate attacks against sensitive knowledge and the impact of the sanitization. Another contribution is a taxonomy of existing sanitizing algorithms. Finally, we present a taxonomy of attacks against sensitive knowledge. To our best knowledge, the investigation of attacks against sensitive knowledge, notably in the context of data or rule sanitization, has not been explored in any detail.

This paper is organized as follows. Related work is reviewed in Section 2. The problem definition is stated in Section 3. In Section 4, we present our framework for protecting sensitive knowledge. In Section 5, we introduce our Downright Sanitizing Algorithm (DSA). The experimental results and discussion are presented in Section 6. Finally, Section 7 presents our conclusions and a discussion of future work.

2 Related Work

Some effort has been made to address the problem of protecting sensitive knowledge in association rule mining by data sanitization. The existing sanitizing algorithms can be classified into two major classes: *Data-Sharing approach* and *Pattern-Sharing approach*, as can be seen in Figure 1A. In the former, the sanitization process acts on the data to remove or hide the group of restrictive association rules that contain sensitive knowledge. To do so, a small number of transactions that contain the restrictive rules have to be modified by deleting one or more items from them or even adding some noise, i.e., new items not originally present in such transactions. In the latter, the sanitizing algorithm acts on the rules mined from a database, instead of the data itself. The only known algorithm in this category is our DSA algorithm herein presented. The algorithm removes all restrictive rules before the sharing process.

Among the algorithms of the Data-Sharing approach, we classify the following categories: *Item Restriction-Based*, *Item Addition-Based*, and *Item Obfuscation-Based*.

Item Restriction-Based: These algorithms [2] remove one or more items from a group of transactions containing restrictive rules. In doing so, the algorithms hide restrictive rules by reducing either their support or confidence below a privacy threshold. Other algorithms [4,5,6], that lie in this category, hide rules by satisfying a disclosure threshold ψ controlled by the database owner. This threshold basically expresses how relaxed the privacy preserving mechanisms should be. When $\psi = 0\%$, no restrictive association rules are allowed to be discovered. When $\psi = 100\%$, there are no restrictions on the restrictive association rules.

Item Addition-Based: Unlike the previous algorithms, item addition-based algorithms modify existing information in transactional databases by adding some items not originally present in some transactions. The items are added to the antecedent part X of a rule $X \rightarrow Y$ in transactions that partially support it. In doing so, the confidence of such a rule is decreased. This approach [2] may generate artifacts such as artificial association rules that would not exist in the original database.

Item Obfuscation-Based: These algorithms [7] hide rules by placing a mark "?" (unknowns) in items of some transactions containing restrictive rules, instead of deleting such items. In doing so, these algorithms obscure a given set of restrictive rules by replacing known values with unknowns. Like the item reduction-based algorithms, these algorithms reduce the impact in the sanitized databases protecting miners from learning "false" rules.

The work presented here differs from the related work in some aspects, as follows: first, our algorithm addresses the issue of pattern sharing and sanitizes rules, not transactions. Second, we study attacks against sensitive knowledge in the context of rule sanitization. This line of work has not been considered so far. Most importantly, our contribution in rule sanitization and the existing solutions in data sanitization are complementary.

3 Problem Definition

The specific problem addressed in this paper can be stated as follows: Let D be a database, R be the set of rules mined from D based on a minimum support threshold σ, and R_R be a set of restrictive rules that must be protected according to some security/privacy policies. The goal is to transform R into R', where R' represents the set of non-restrictive rules. In this case, R' becomes the released set of rules that is made available for sharing.

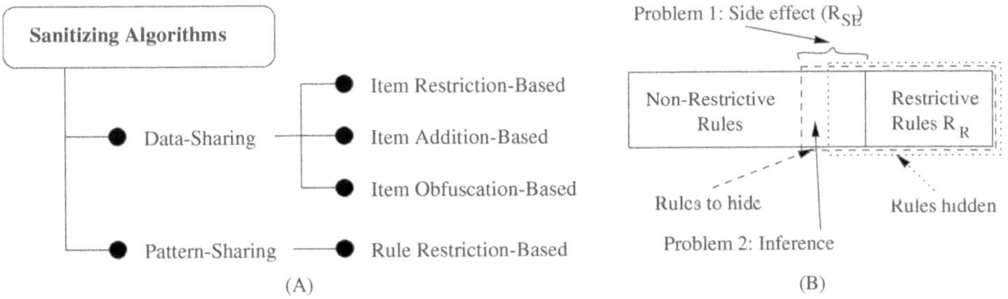

Fig. 1. (A): A Taxonomy of Sanitizing Algorithms. (B): Rule Sanitization problems.

Ideally, $R' = R - R_R$. However, there could be a set of rules r in R' from which one could derive or infer a restrictive rule in R_R. So in reality, $R' = R - (R_R + R_{SE})$, where R_{SE} is the set of non-restrictive rules that are removed as side effect of the sanitization process to avoid recovery of R_R.

Figure 1B illustrates the problems that occur during the rule sanitization process. Problem 1 conveys the non-restrictive rules that are removed as a side effect of the process (R_{SE}). We refer to this problem as *side effect.* It is related to the misses cost problem in data sanitization [4]. Problem 2 occurs when using some non-restrictive rules, an adversary may recover some restrictive ones by inference channels. We refer to such a problem as *recovery factor.*

4 Framework for Protecting Sensitive Knowledge

Before introducing our framework for protecting sensitive knowledge, we briefly review some terminology from graph theory. We present our new sanitizing algorithm in Section 5.2.

4.1 Basic Definitions

The itemsets in a database can be represented in terms of a directed graph. We refer to such a graph as *frequent itemset graph* and define it as follows:

Definition 1 (Frequent Itemset Graph). *A frequent itemset graph, denoted by $G = (C, E)$, is a directed graph which consists of a nonempty set of frequent itemsets C, a set of edges E that are ordered pairings of the elements of C, such that $\forall u, v \in C$ there is an edge from u to v if $u \cap v = u$ and if $|v| - |u| = 1$ where $|x|$ is the size of itemset x.*

Figure 2b shows a frequent itemset graph for the sample transactional database depicted in Figure 2a. In this example, the minimum support threshold σ is set to 2. As can be seen in Figure 2b, in a frequent itemset graph G, there is an ordering for each itemset. We refer to such an ordering as *itemset level* and define it as follows:

TID	List of Items
T1	A B C D
T2	A B C
T3	A C D
T4	A B C
T5	A B

(a)

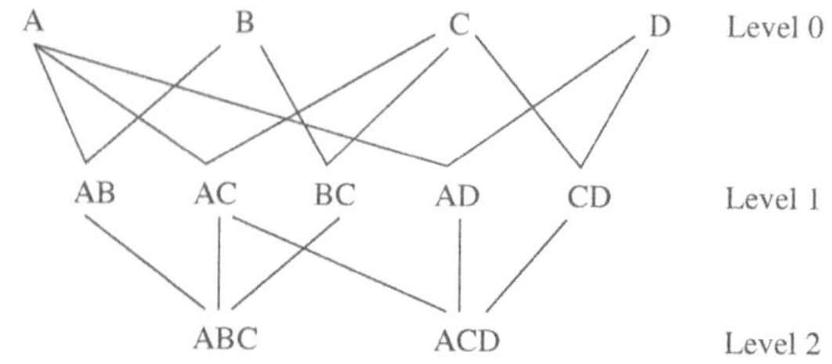

(b)

Fig. 2. (a) A transactional database. (b) The corresponding frequent itemset graph.

Definition 2 (Itemset Level). *Let $G = (C, E)$ be a frequent itemset graph. The level of an itemset u, such that $u \subset C$, is the length of the path connecting an 1-itemset to u.*

Based on Definition 2, we define the level of a frequent itemset graph G as follows:

Definition 3 (Frequent Itemset Graph Level). *Let $G = (C, E)$ be a frequent itemset graph. The level of G is the length of the maximum path connecting an 1-itemset u to any other itemset v, such that $u, v \in C$, and $u \subset v$.*

In general, the discovery of itemsets in G is the result of top-down traversal of G constrained by a minimum support threshold σ. The discovery process employs an iterative approach in which k-itemsets are used to explore $(k + 1)$-itemsets.

4.2 Taxonomy of Attacks

An attack occurs when someone mines a sanitized set of rules and, based on non-restrictive rules, deduce one or more restrictive rules that are not supposed to be discovered. We have identified some attacks against sanitized rules, as follows:

Forward-Inference Attack: Let us consider the frequent itemset graph in Figure 3A. Suppose we want to sanitize the restrictive rules derived from the itemset ABC. The naïve approach is simply to remove the itemset ABC. However, if AB, AC, and BC are frequent, a miner could deduce that ABC is frequent. A database owner must assume that an adversary can use any inference channel to learn something more than just the permitted association rules. We refer to this attack as forward-inference attack. To handle this attack, we must also remove at least one subset of ABC in the level 1 of the frequent itemset graph. This complementary sanitization is necessary. In the case of a deeper graph, the removal is done recursively up to level 1. We start removing from level 1 because we assume that the association rules recovered from the itemsets have at least 2 items. Thus, the items in level 0 of the frequent itemset graph are not shared with a second party. In doing so, we reduce the inference channels and minimize the side effect.

Backward-Inference Attack: Another type of attack occurs when we sanitize a non-terminal itemset. Based on Figure 3B, suppose we want to sanitize any rule derived from the itemset AC. If we simply remove AC, it is straightforward to infer the rules mined from AC since either ABC or ACD is frequent. We refer to this attack as backward-inference attack. To block this attack, we must remove any superset that contains AC. In this particular case, ABC and ACD must be removed as well.

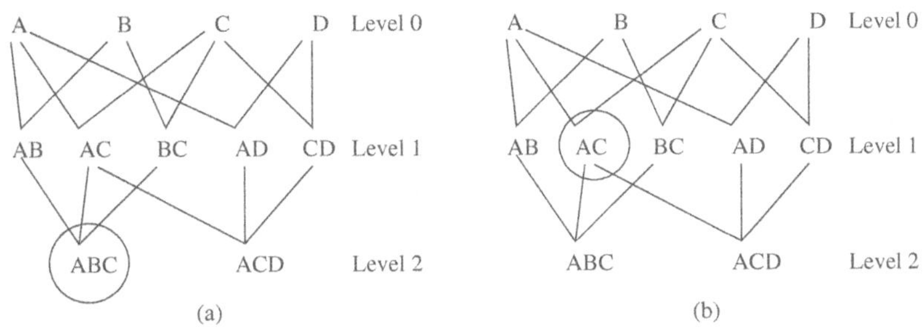

Fig. 3. (a) An example of forward-inference. (b) An example of backward-inference.

4.3 Metrics

In this section, we introduce two metrics related to the problems illustrated in Figure 1B: The *side effect* and the *recovery*.

Side Effect Factor (SEF): Measures the amount of non-restrictive association rules that are removed as side effect of the sanitization process. The side effect factor is calculated as follows: $SEF = \frac{(|R|-(|R'|+|R_R|))}{(|R|-|R_R|)}$ where R, R', and R_R represent the set of rules mined from a database, the set of sanitized rules, and the set of restrictive rules, respectively, and $|S|$ is the size of the set S.

Recovery Factor (RF): This measure expresses the possibility of an adversary recovering a restrictive rule based on non-restrictive ones. The recovery factor of one pattern takes into account the existence of its subsets. The rationale behind the idea is that all nonempty subsets of a frequent itemset must be frequent. Thus, if we recover all subsets of a restrictive itemset (rule), we say that the recovery factor for such an itemset is possible, thus we assign it the value 1. However, the recovery factor is never certain, i.e., an adversary may not learn an itemset even with its subsets. On the other hand, when not all subsets of an itemset are present, the recovery of the itemset is improbable, thus we assign value 0 to the recovery factor.

5 Rule Sanitization: The DSA Algorithm

5.1 The General Approach

Our approach has essentially three steps as follows. These steps are applied after the mining phase, i.e., we assume that the frequent itemset graph G is built. The set of all itemsets that can be mined from G, based on a minimum support threshold σ, is denoted by C.

Step1: Identifying the restrictive itemsets. For each restrictive rule in R_R, convert it to an itemset $c_i \in C$ and mark it to be sanitized.
Step2: Selecting subsets to sanitize. In this step, for each itemset c_i to be sanitized, we compute its item pairs from level 1 in G, subsets of c_i. If none of them is marked, we randomly select one of them and mark it to be removed.
Step3: Sanitizing the set of supersets of marked pairs in level 1. The sanitization of restrictive itemsets is simply the removal of the set of supersets of all itemsets in level 1 of G that are marked for removal. This process blocks inference channels.

5.2 The Downright Sanitizing Algorithm

In this section, we introduce the Downright Sanitizing Algorithm, denoted by DSA. The inputs for DSA are the frequent itemset graph G, the set of all association rules R mined from a database G, and the set of restrictive rules R_R to be sanitized. The output is the set of sanitized association rules R'.

Downright_Sanitizing_Algorithm
Input: G, R, R_R
Output: R'
Step 1. **For each** association rule $rr_i \in R_R$ **do**
  1.1. $pattern_i \leftarrow rr_i$ //Convert each rr_i into a frequent itemset $pattern_i$
Step 2. **For each** $pattern_i$ in the level k of G, where $k > 1$ **do** {
  2.1. Pairs($pattern_i$) //Compute all the item pairs of $pattern_i$
  2.2. **If** (Pairs($pattern_i$) $\cap$ $MarkedPair = \emptyset$) **then** {
    2.2.1 $p_i \leftarrow$ random(Pairs($pattern_i$)) //Select randomly a pair $p_i \in pattern_i$
    2.2.2 $MarkedPair \leftarrow MarkedPair \cup p_i$ //Update the list $MarkedPair$
  }
}
Step 3. $R' \leftarrow R$
  3.1. **For** each itemset $c_i \in G$ **do** {
    3.1.1. **If** $\exists$ a marked pair p, such that $p \in MarkedPair$ and $p \subset c_i$ **then**
      3.1.1.1. Remove(c_i) from R' //c_i belongs to the set of supersets of p
  }
End Algorithm

6 Experimental Results

In this section, we study the efficiency and scalability of DSA and the similar counterparts in the literature. There are no known algorithms for rule sanitization. However, transaction sanitization algorithms can be used for this purpose. Indeed, in order to sanitize a set of rules R to hide R_R, one can use data sanitization to transform the database D into D' to hide R_R and then mine D' to get the rules to share. We used this idea to compare our algorithm to existing data sanitization approaches.

6.1 Datasets

We validated DSA against real and synthetic datasets. The real dataset, BMS-Web-View-2 [8], placed in the public domain by Blue Martini Software. The dataset contains 22,112 transactions with 2717 distinct items, and each customer purchasing has four items on average. The synthetic dataset was generated with the IBM synthetic data generator. This dataset contains 100,000 transactions with 500 different items, and each customer purchasing has ten items on average.

6.2 Sanitizing Algorithms

For our comparison study, we selected the best sanitizing algorithms in the literature: (1) Algo2a hides restrictive rules by reducing support [2]. (2) Item Grouping Algorithm (IGA) [4] which groups restricted association rules in clusters of rules sharing the same itemsets. The shared items are removed to reduce the impact on the sanitized dataset; (3) Sliding Window Algorithm (SWA) [6] scans a group of K transactions, at a time and sanitizes the restrictive rules present in such transactions based on a disclosure threshold ψ defined by a database owner. We set the window size of SWA to 20000 transactions in both datasets; (4) An algorithm similar to DSA, called Naïve, which sanitizes restrictive itemsets and their supersets, i.e., Naïve blocks the forward-inference attack without considering blocking the backward-inference attack.

6.3 Methodology

We performed two series of experiments: the first to evaluate the effectiveness of DSA, and the second to measure its efficiency and scalability.

Our comparison study was carried out through the following steps: (1) We used the algorithms IGA, SWA, and Algo2a to sanitize both initial databases; (2) We applied the Apriori algorithm on the sanitized data to extract the rules to share. For DSA, also two steps were necessary: (1) Apply Apriori algorithm to extract rules from the two initial datasets; (2) Use DSA to sanitize these rules. The effectiveness is measured in terms of restrictive associations that can be recovered by an adversary, as well as the proportion of non-restrictive rules hidden due to the sanitization.

All the experiments were conducted on a PC, AMD Athlon 1900/1600 (SPEC CFP2000 588), with 1.2 GB of RAM running a Linux operating system. In our experiments, we selected a set of 20 restrictive association rules for the real dataset and 30 for the synthetic dataset, with rules ranging from 2 to 5 items. The real dataset has 17,667 association rules with support $\geq 0.2\%$ and confidence $\geq 50\%$, while the synthetic dataset has 20,823 rules with support $\geq 0.1\%$ and confidence $\geq 50\%$.

6.4 Measuring Effectiveness

In this section, we measure the effectiveness of DSA, IGA, SWA, and Algo2a considering the metrics introduced in Section 4.3.

In order to compare the algorithms under the same conditions, we set the disclosure thresholds ψ of the algorithms IGA and SWA, and the privacy threshold λ of algorithm Algo2a to 0%. In this case, all restrictive rules are completely sanitized. We purposely set these thresholds to zero because DSA always sanitizes all the restrictive rules. However, the value for the side effect factor differs from one algorithm to another. For instance, Figure 4A shows the side effect factor on the synthetic dataset. The lower the result the better. For this example, 1.09% of the non-restrictive association rules in the case of Naïve, 3.58% in the case of DSA, 6.48% in the case of IGA, 6.94% in the case of SWA, and 8.12% in the case of Algo2a are removed by the sanitization process.

Similarly, Figure 4B shows the side effect of the sanitization on the real dataset. In this situation, 3.2% of the non-restrictive association rules in the case of Naïve, 4.35% in the case of DSA, 11.3% in the case of IGA, 22.1% in the case of SWA, and 27.8% in the case of Algo2a are removed.

In both cases, Naïve yielded the best results, but we still need to evaluate how efficient Naïve is to block inference channels. We do so below. DSA also yielded promising results, while the sanitization performed by Algo2a impacts the database more significantly. An important observation here is the results yielded by SWA and IGA. Both algorithms benefit from shared items in the restrictive rules during the process of sanitization. By sanitizing the shared items of these restrictive rules, one would take care of hiding such restrictive rules in one step. As a result, the impact on the non-restrictive rules is minimized. In general, the heuristic of IGA is more efficient than that one in SWA. This explains the better performance of IGA over SWA in both datasets.

After identifying the side effect factor, we evaluated the recovery factor for Naïve and DSA. This measure is not applied to IGA, SWA, and Algo2a since these algorithms rely on data sanitization instead of rule sanitization. Thus, once the data is shared for mining, there is no restriction about the rules discovered from a sanitized database.

In the case of rule sanitization, some inference channels can occur, as discussed in Section 4.2. We ran a checklist procedure to evaluate how efficient is the sanitization performed by Naïve and DSA. We check the existence of any subset of the restrictive rules removed in order to identify the recovery factor. If all subsets of a rule are found, we assume the rule could be recovered. As expected,

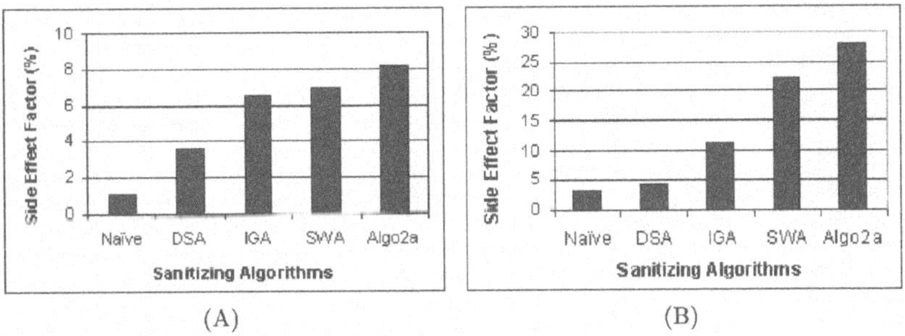

Fig. 4. (A): Side effect on the synthetic dataset. (B): Side effect on the real dataset.

Naïve blocked well the forward-inference attacks but failed to block backward-inference attacks in both datasets. On the contrary, DSA yielded the best results in all cases, i.e., DSA blocked both forward-inference and the backward-inference attacks. The results suggested that hardly can an adversary reconstruct the restrictive rules after the sanitization performed by DSA.

6.5 CPU Time for the Sanitization Process

We tested the scalability of DSA and the other algorithms vis-à-vis the number of rules to hide. We did not plot Naïve because its CPU time is very similar to that one of DSA. We varied the number of restrictive rules to hide from 20 to 100 and set the disclosure thresholds to $\psi = 0\%$. The rules were randomly selected from both datasets. Figures 5A and 5B show that the algorithms scale well with the number of rules to hide. Note that IGA, SWA, and DSA increase CPU time linearly, while the CPU time in Algo2a grows fast. This is due the fact that Algo2a requires various scans over the original database, while IGA requires two, and both DSA and SWA require only one.

Although IGA requires 2 scans, it is faster than SWA in most cases. The main reason is that SWA performs a number of operations in main memory to fully sanitize a database. IGA requires on scan to build an inverted index where the vocabulary contains the restrictive rules and the occurrences contain the transaction IDs. In the second scan, IGA sanitizes only the transactions marked in the occurrences. Another important result is that IGA and DSA yielded very similar CPU time for both datasets. In particular, IGA was better in the synthetic dataset because the transactions contain more items and IGA requires less operations in main memory.

6.6 General Discussion

Our experiments demonstrated the evidence of attacks in sanitized databases. The figures revealed that DSA is a promising solution to protect sensitive knowledge before sharing association rules, notably in the context of rule sanitization.

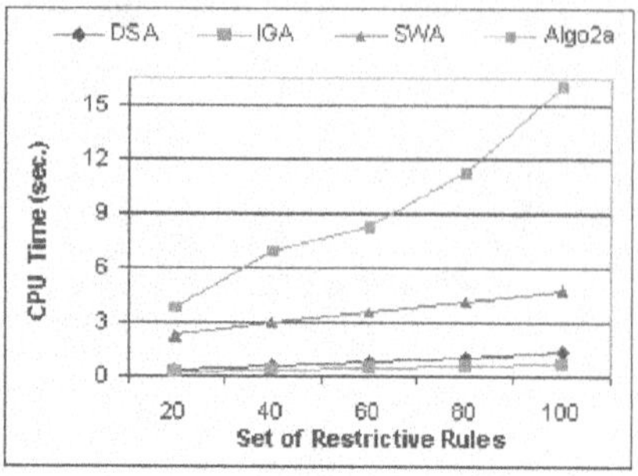

(A)

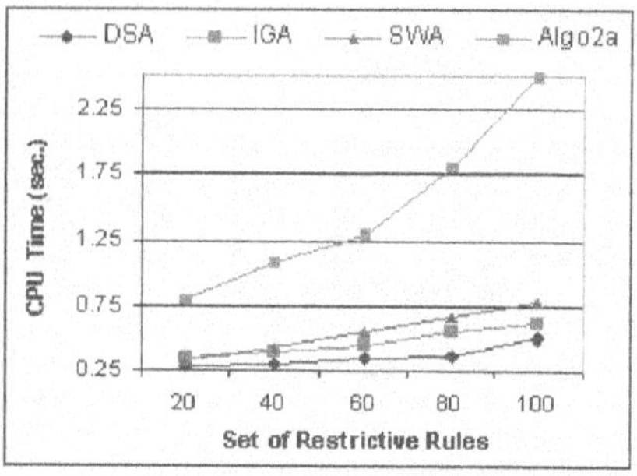

(B)

Fig. 5. (A): CPU time for the synthetic dataset. (B): CPU time for the real dataset.

DSA has a low value for side effect factor and very low recovery factor. We have identified some advantages of DSA over the previous data sanitizing algorithms in the literature as follows: (1) Using DSA, a database owner would share patterns (only rules) instead of the data itself; (2) Sanitizing rules, one reduces drastically the possibility of inference channels since the support threshold and the mining algorithm are selected previously by the database owner; and (3) Sanitizing rules instead of data results in no alteration in the support and confidence of the non-restrictive rules, i.e., the released rules have the original support and confidence. As a result, the released rules seem more interesting for practical applications. Note that the other approaches reduce support and confidence of the rules as a side effect of the sanitization process.

On the other hand, DSA reduces the flexibility of information sharing since each time a client (party) wants to try a different set of support and confidence levels, it has to request for the rules from the server.

7 Conclusions

In this paper, we have introduced a novel framework for protecting sensitive knowledge before sharing association rules.

Our contributions in this paper can be summarized as follows: First, a sanitizing algorithm called Downright Sanitizing Algorithm (DSA). DSA blocks some inference channels to ensure that an adversary cannot reconstruct restrictive rules from the non-restrictive ones. In addition, DSA reduces drastically the side effect factor during the sanitization process. Our experiments demonstrated that DSA is a promising approach for protecting sensitive knowledge before sharing association rules. Second, the framework also encompasses metrics to evaluate the effectiveness of the rule sanitization process. Another contribution is a taxonomy of existing sanitizing algorithms. We also introduced a taxonomy of attacks against sensitive knowledge.

The work presented here introduces the notion of rule sanitization, which complements the idea behind data sanitization. While data sanitization relies on protecting sensitive knowledge before sharing data, rule sanitization is concerned with the sharing of patterns. Currently, we are investigating the existence of new type of attacks against sanitized databases, and the effective response to such attacks. Another interesting issue to address is the problem of hiding rules in collective data. In our previous motivating example, if all clients share their rules in the server, but want to hide some global rules, i.e. rules that become confident with the collective support, our algorithm seems vulnerable in such context and a collaborative sanitization should be explored.

Acknowledgments. Stanley Oliveira was partially supported by CNPq, Brazil, under grant No. 200077/00-7. Osmar Zaïane was partially supported by a research grant from NSERC, Canada. Yücel Saygin was partially supported by European Commission under IST-2001-39151 CODMINE project. We would like to thank Elena Dasseni for providing us with the code of her algorithm for our comparison study.

References

1. M. Atallah, E. Bertino, A. Elmagarmid, M. Ibrahim, and V. Verykios. Disclosure Limitation of Sensitive Rules. In *Proc. of IEEE Knowledge and Data Engineering Workshop*, pages 45–52, Chicago, Illinois, November 1999.
2. E. Dasseni, V. S. Verykios, A. K. Elmagarmid, and E. Bertino. Hiding Association Rules by Using Confidence and Support. In *Proc. of the 4th Information Hiding Workshop*, pages 369–383, Pittsburg, PA, April 2001.
3. A. Evfimievski, R. Srikant, R. Agrawal, and J. Gehrke. Privacy Preserving Mining of Association Rules. In *Proc. of the 8th ACM SIGKDD Intl. Conf. on Knowlegde Discovery and Data Mining*, pages 217–228, Edmonton, AB, Canada, July 2002.
4. S. R. M. Oliveira and O. R. Zaïane. Privacy Preserving Frequent Itemset Mining. In *Proc. of the IEEE ICDM Workshop on Privacy, Security, and Data Mining*, pages 43–54, Maebashi City, Japan, December 2002.
5. S. R. M. Oliveira and O. R. Zaïane. Algorithms for Balancing Privacy and Knowledge Discovery in Association Rule Mining. In *Proc. of the 7th International Database Engineering and Applications Symposium (IDEAS'03)*, pages 54–63, Hong Kong, China, July 2003.
6. S. R. M. Oliveira and O. R. Zaïane. Protecting Sensitive Knowledge By Data Sanitization. In *Proc. of the 3rd IEEE International Conference on Data Mining (ICDM'03)*, pages 613–616, Melbourne, Florida, USA, November 2003.
7. Y. Saygin, V. S. Verykios, and C. Clifton. Using Unknowns to Prevent Discovery of Association Rules. *SIGMOD Record*, 30(4):45–54, December 2001.
8. Z. Zheng, R. Kohavi, and L. Mason. Real World Performance of Association Rules Algorithms. In *Proc. of the 7th ACM SIGKDD International Conference on Knowledge Discovery and Data Mining*, San Francisco, CA, USA, August 2001.

Self-Similar Mining of Time Association Rules

Daniel Barbará[1], Ping Chen[2], and Zohreh Nazeri[3]

[1] George Mason University, Department of Information and Software Engineering
Fairfax, VA 22303
dbarbara@gmu.edu
[2] CMS Dept. University of Houston-Downtown
1 Main St., Houston, TX 77002
chenp@uhd.edu
[3] The MITRE Corporation
1820 Dolley Madison Boulevard, McLean, VA 22102-3481
nazeri@mitre.org

Abstract. Although the task of mining association rules has received considerable attention in the literature, algorithms to find time association rules are often inadequate, by either missing rules when the time interval is arbitrarily partitioned in equal intervals or by clustering the data before the search for high-support itemsets is undertaken. We present an e.cient solution to this problem that uses the fractal dimension as an indicator of when the interval needs to be partitioned. The partitions are done with respect to every itemset in consideration, and therefore the algorithm is in a better position to find frequent itemsets that would have been missed otherwise. We present experimental evidence of the e.ciency of our algorithm both in terms of rules that would have been missed by other techniques and also in terms of its scalability with respect to the number of transactions and the number of items in the data set.

Keywords: Temporal association rules, fractal dimension, intrusion detection

1 Introduction

Association rules have received a lot of attention in the data mining community since their introduction in [1]. Association rules are rules of the form $X \rightarrow Y$ where X and Y are sets of attribute-values, with $X \bigcap Y = \emptyset$ and $||Y|| = 1$. The set X is called the antecedent of the rule while the item Y is called consequent. For example, in a market-basket data of supermarket transactions, one may find that customers who buy milk also buy honey in the same transaction, generating the rule *milk* $\rightarrow$ *honey*. There are two parameters associated with a rule: *support* and *confidence*. The rule $X \rightarrow Y$ has *support s* in the transaction set T if s% of transactions in T contain $X \cup Y$. The rule $X \rightarrow Y$ has *confidence c* if c of transactions in T that contain X also contain Y. The most difficult and dominating part of an association rules discovery algorithm is to find the

H. Dai, R. Srikant, and C. Zhang (Eds.): PAKDD 2004, LNAI 3056, pp. 86–95, 2004.

itemsets $X \bigcup Y$, that have strong support. (Once an itemset is deemed to have strong support, it is an easy task to decide which item in the itemset can be the consequent by using the confidence threshold.)

Algorithms for discovering these "classical" association rules are abundant [2, 3]. They work well for data sets drawn from a domain of values with no relative meaning (categorical values). However, when applied to interval data, they yield results that are not very intuitive. Algorithms to deal with interval data have appeared in the literature. The first algorithm ever proposed (by Srikant and Agrawal [8]), defined the notion of *quantitative association rules* as rules where the predicates may be equality predicates ($Attribute = v$) or range predicates ($v_1 \leq Attribute \leq v_2$). The problem that arises then is how to limit the number of ranges (or intervals) that must be considered: if intervals are too large, they may hide rules inside portion of the interval, and if intervals are too small they may not get enough support to be considered.

Srikant and Agrawal deal with the problem by doing an Equi-depth initial partitioning of the interval data. In other words, for a depth d, the first d values (in order) of the attribute are placed in one interval, the next d in a second interval, and so on. No other considerations, such as the density of the interval or the distance between the values that fall in the same interval are taken.

Miller and Yang in [7] point out the pitfalls of Equi-depth partitioning. In short, intervals that include close data values (e.g., time values in an interval [9:00, 9:30]) are more meaningful than those intervals involving distant values (e.g., an interval [10:00, 21:00]). Notice that both types of intervals can coexist in an Equi-depth partitioning schema, since the population of values can be more dense in the closely tight intervals than in the loosely tight intervals. As the authors in [7] point out, it is less likely that a rule involving a loosely tight interval (such as [10:00, 21:00]) will be of interest than rules that involve the closely tight intervals. Srikant and Agrawal [8] had pointed out this problem, mentioning that their Equi-depth partitioning would not work well with skewed data, since it would separate close values that exhibit the same behavior.

Miller and Yang proceed to establish a guiding principle that we include here for completeness:

Goal 1. [7] In selecting intervals or groups of data to consider, we want a measure of interval quality that re.ects the distance between data points.

To achieve this goal, Miller and Yang, de.ned a more general form of association rules with predicates expressing that an attribute (or set of attributes) ought to fall within a given subset of values. Since considering all possible subsets is intractable, they use clustering [5] to find subsets that "make sense" by containing a set of attributes that are sufficiently close. A cluster is defined in [7] as a set of values whose "diameter" (average pairwise distance between values) and whose "frequency" (number of values in the cluster) are both bounded by pre-specified thresholds. They propose an algorithm that uses BIRCH [12] to find clusters in the interval attribute(s) and use the clusters so found as "items" which with the items of categorical nature are fed into the a-priori algorithm [2] to find high-support itemsets and rules. So, an itemset found in this way could take the form ([9 : 00, 9 : 30], *bread*) meaning that people buy their bread frequently between 9:00 and 9:30 am. The interval [9 : 00, 9 : 30] would have been found as constituting a cluster by feeding all the time values in the data set to BIRCH.

We believe that Miller's technique still falls short of a desirable goal. The problem is that by clustering the interval attribute alone, without taking into consideration the other (categorical) values in the data set, one can fail to discover interesting rules. For instance, it is possible that while the itemset ([9 : 00, 9 : 30], *bread*) is a frequent one, the itemset ([9 : 00, 9 : 15], *bread, honey*) is also frequent. Using clustering in the time attribute before considering the other items in the data set may lead to decide that the interval [9 : 00, 9 : 30] is to be used as an item and therefore the second itemset above will never be found as frequent. Another problem is that invoking a clustering algorithm to cluster the interval values along the itemset that we are considering (say, for instance, *bread* and *honey*), would be prohibitively expensive.

In [11] Apriori algorithm was extended to find association rules which satisfy minimum support and confidence within a user-specified interval. In [10], the authors proposed to use a density-based clustering algorithm to find interval clusters first, then generate temporal association rules with Apriori algorithm. This method requires at least two scans of a data set, which may still be too expensive for large data sets. More seriously, a pre-clustering step assumes all association rules share the same temporal clusters, which may cause the problems similar to Miller's method as discussed above. Fortunately, we have devised a method that can do the partitioning of the interval data in an *on-line* fashion, i.e, while we are trying to compute the support of the non-interval items of the itemset (*bread* and *honey*). This method uses the notion of fractal dimension to produce a natural partitioning of the set of interval data. Although we conduct our experiments only on temporal interval data, our method can be applied to any data sets with one or multiple interval attributes.

The rest of the paper is divided as follows. Section 2 brie.y reviews the background on fractal properties of data sets. Section 3 offers a motivating example and our algorithm. In section 4 we show the experimental results of applying our technique to a real data set. Finally, Section 5 gives the conclusions and future work.

2 Our Approach

2.1 Motivation

To motivate our technique, consider the data set shown in Figure 1, where timestamped costumer transactions (baskets) are shown.

An Equi-depth algorithm such as the one presented in [8] may run the risk of breaking the obvious temporal interval [9 : 00, 9 : 30] and miss the fact that the item bread had high support for this entire period of time. Miller and Yang's algorithm will find that the attribute Time contains at least two clusters: points

Time	Items				Time	Items			
9:00	bread	honey	milk	cookies	12:45	honey	diapers		
9:01	bread	honey	milk	cookies	3:25	beer			
9:02	bread	honey	cookies	milk	14:10	poultry			
9:05	bread	honey	poultry	milk	18:00	beer	diapers		
9:07	bread	honey	cereal	cookies	18:10	beer	diapers		
9:10	bread	honey	beer	diapers	18:22	beer	diapers		
9:12	bread	honey	poultry		18:24	bread	honey	milk	cookies
9:15	bread	honey			18:25	bread	honey	milk	cookies
9:19	bread	beer	diapers		18:26	bread	honey	cookies	milk
9:20	bread	poultry			18:29	bread	honey	poultry	milk
9:22	cereal	milk			18:31	bread	honey	cereal	cookies
9:25	bread				18:34	bread	honey		
9:27	bread	honey			18:35	beer	diapers		
9:30	bread	cereal			18:36	bread	honey	poultry	
11:00	meat	beer			18:39	bread	honey		
12:00	poultry	beer			18:40	beer	diapers	honey	bread
					20:40	honey	bread		

Fig. 1. A timestamped set of baskets.

in the interval $[9:00,\ 9:30]$ and points in the interval $[18:00,\ 18:40]$. (The rest of the time points may be grouped in other clusters or considered outliers.) Using the first cluster and with support threshold equal to 0.16 ($\frac{4}{24}$), Miller and Yang's algorithm would find, among others, high support itemsets ($[9:00, 9:30]$, *bread, honey*), ($[9:00,\ 9:30]$, *milk, cookies*) and ($[18:00,\ 18:40]$, *beer, diapers, bread, honey*). The algorithm would have missed the itemset ($[9:00, 9:15]$, *bread, honey*) that has support 0.33, and the itemset $[9:00,\ 9:05]$, *milk, cookies*) with support 0.16. The reason is that this algorithm would have fed the item $[9:00,\ 9:30]$ to the a-priori algorithm, without further consideration for subintervals within it, like the $[9:00,\ 9:15]$ subinterval, which in this case is a better "fit" for the items (*bread, honey*). (Or the $[9:00,\ 9:05]$ interval for the items (*milk, cookies*).)

As pointed out before, it would be extremely expensive to invoke clustering of each itemset under consideration due to too many scans of a dataset. However, using the fractal dimension we can design a simple, incremental way of deciding when to break the interval data for any itemset under consideration. The central idea is to modify a-priori in such a way that while the algorithm is measuring the support for an itemset whose items are categorical (e.g., (*bread, honey*)), we incrementally compute the fractal dimensions of all (possibly multidimensional, since the number of interval attributes in the clusters can be more than 1) clusters that include each of the interval attributes where the itemset shows up. To illustrate, Figure 2 shows the configuration of this data set using the data of Figure 1, while considering the support of the itemset (*bread, honey*). As each of the rows in the data set of Figure 1 is considered, the rows of the cluster (so far, we only have two clusters $[9:00,\ 9:15]$ and $[18:24,\ 18:39]$ which has only one interval attribute and we are going to process point 18:40) shown in Figure

2 can be built and the fractal dimension of the cluster of rows seen so far can be computed. Figure 3 shows the two box-counting plots of the data set of Figure 2. The first plot includes all the points up to time 9:15 (the first cluster), while the second includes the first cluster plus point 18:40 currently under consideration. The first straight region shows a slope of -0.2 for a fractal dimension of 0.2. The addition of the 18:40 point decreases the fractal dimension to 0.16. Similarly we compute the fractal dimension of second cluster [18:24, 18:39], and fractal dimension of [18:24, 18:39] plus point 18:40, and the difference is smaller that 0.01. So point 18:40 is added to the second cluster and the new second cluster will be [18:24, 18:40]. When we input next transaction which have itemset (*bread, honey*) with time 20:40, which will cause big changes of fractal dimension to both existing clusters, which indicates that we should start a new cluster with the new point 20:40. A big change of fractal dimension (suppose we set the change threshold as 0.01) is a good indication for a breaking point in the interval data, and a good place to compute the support of the itemset composed by the interval considered so far ([18:24, 18:40]) and the other two items (*bread, honey*). If the support is equal or exceeds the threshold, then this itemset would be reported by the algorithm. In summary, if the minimum change of fractal dimension for all existing clusters is less than a preset threshold, we should put the new point into the cluster whose fractal dimension is changed the least, otherwise, we should start a new cluster. For the itemset (*milk, cookies*) there is a change after the point 9:07. (The itemset receives its strong support before that time.) And for the itemset *beer, diapers*, there is a clear change after the point 18:00. Indeed, after that point a new interval opens in which the itemset receives strong support. Thus, a potential high support itemset would be ([18:00, 18:40], beer, diapers). (This, in fact would reveal an interesting nugget: that this pair of items are bought frequently after work hours, for example.) Again, we could not have found these itemsets (([9:00, 9:05], *milk, cookies*), ([9:00, 9:15], *bread, honey*) and ([18:00, 18:40], *beer, diapers*)) by preclustering the time attribute.

By our method we do not require any order in interval attributes, and it can be readily applied to a data set with multiple interval attributes.

Time	9:00	9:01	9:02	9:05	9:07	9:10	9:12	9:15	18:24
	18:25	18:26	18:29	18:31	18:34	18:36	18:39	18:40	20:40

Fig. 2. The data set resulting selecting the values of the interval attribute (Time) for which the itemset (*bread, honey*) is present in the transaction.

2.2 Algorithm

The pseudo-code for the our algorithm is shown in Figure 4. Line 1 selects all the 1-itemsets, as in a-priori, but considering only the categorical (non-interval)

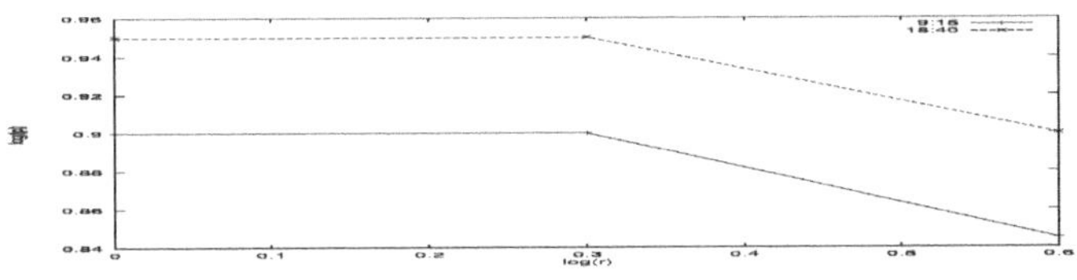

Fig. 3. Loglog plot of the data set shown in Figure 2 up to the points 9:15 and 18:40 respectively. The slope of the 9:15 curve is -0.2, indicating a fractal dimension 0.2, while the slope of the 18:40 curve is -0.16, for a fractal dimension of 0.16

items. (Every L_k will contain only categorical itemsets.) Then the algorithm iterates until no more candidates exist (Line 2), generating the next set of candidates from L_{k-1} (as in a-priori), and for every candidate I in that set initializing the multidimensional set M_I as empty and making the lower bound of the interval for I, low_I equal to $-\infty$. (This will be changed as soon as the first occurrence of I is spotted in the data set, in Lines 14-15.) For each transaction (basket) t in D, if the transaction contains I, a new row is added to the set M_I containing the interval attribute values for t in Line 13. (The notation $interval(t)$ indicates extracting the interval data from a tuple t.) Regardless of whether the item is in t or not, the overall count for the item interval $M_I.totalcount$ is increased in Line 10. This count will serve as the denominator when the support of I within the interval needs to be computed. Both, the overall count ($I.count$) and interval count ($M_I.count$) for the itemset are increased (line 12). The fractal dimension of the set M_I is then computed (Line 16), and if a significant change is found, a new interval is declared (line 17). The notation $last(M_I, interval(t))$ indicates the interval data point before the current $interval(t)$ in M_I. Line 19 tests if the itemset composed by this interval and I has enough support. Notice that both the $M_I.count$ and $M_I.totalcount$ need to be reduced by one, since the current transaction is not part of the interval being tested. If the support is greater than or equal to $minsup$, the itemset is placed in the output. (Line 19.) Line 20 re-initializes the counts. Line 22 is the classic test for the overall support in a-priori, with N being the total number of transactions in the data set.

3 Experiments

The experiments reported in this section were conducted over a SunWorkstation Ultra2 with 500 MB. of RAM, running Solaris 2.5. The experiment was done over a real data set of sniffed connections to a network, collected for intrusion detection purposes. The data is collected over MITRE's network for the purpose of intrusion detection. The attributes in the collected data are: Time (one field containing both date and time), source IP address, destination IP address, duration of the connection, type of the connection (TELNET, Login,), and name of the sensor that reported the connection. All IP addresses and sensor names in the data are changed to symbolic names such as sensor1 or 111.11.1.1 IP address.

(1)L_1 = { large 1-itemsets, from categorical values};
(2)for ($k = 2; L_{k-1} != \emptyset; k++$) do
  (3) C_k = generate(L_{k-1})
  (4) for every I in C_k
    (5) make $M_I = \emptyset$ for every I in C_k
    (6) make $low_I = -\infty$
  (7) while there are transactions in D do
    (8) consider next transaction t in D
    (9) for every candidate in I in C_k
      (10) $M_{Ic}.totalcount++$
      (11) if I is in t
        (12)make $M_I.minfdchange = +\infty$
        (13) for every interval cluster c in I in C_k
          (14) $I.count++$, $M_{Ic}.count++$
          (15) add row with $interval(t)$ to M_{Ic}
          (16) if $low_I = -\infty$
            (17) $low_I = interval(t)$
          (18) compute $F(M_{Ic}) = fd(M_{Ic})$
          (19) if the fractal dimension change is less that $M_I.minfdchange$
            (20) assign this smaller change to $M_I.minfdchange$
            (21) assign the number of the cluster with the smallest change to j
        (22) if $M_I.minfdchange$ is larger than τ
          (23) if $\frac{M_I.count - 1}{M_I.totalcount - 1} \geq minsup$
            (24) output itemset $[low_I, last(M_I, interval(t))], I$
          (25) $M_{Ij} = interval(t)$, $low_I = interval(t)$,
          $M_I.count = M_I.totalcount = 1$
  (26) end
  (27) $L_k = \{I \text{ in } C_k \mid \frac{I.count}{N} \geq minsup\}$
  (28) output items in L_k
(29) end

Fig. 4. Our algorithm.

A total of 4000 data records were used in this experiment. These records cover about 40 hours of network events (telnet, login, etc.). The Time attribute is the quantitative or interval attribute.

We tried the Equi-depth approach [8] and our algorithm on this data. For our algorithm, the parameters needed from the user are minimum support, minimum window size, and a threshold for the Fractal dimension. The parameters needed for the Equi-depth approach are minimum support, minimum confidence, and maximum support. We used the same minimum support for both algorithms. This parameter then has equal impact on both approaches and this is desirable for the comparison of the rules generated by each. For the Equi-depth approach, we used a minimum confidence of 50% and after trying different values of maximum support we picked the one that gave the best results, that is 12%. Using the maximum support in the equations suggested in Srikant and Agrawal's paper [8], the number of intervals is 8. Dividing the 24 hours by 8, our intervals are 0:00-2:59, 3:00-5:59, 6:00-8:59, and so on.

Rule Number	Rule	confidence
1.	$Time = 7/25 - [09 - 12] -> event = TELNET$	91%
2.	$Time = 7/25 - [12 - 15] -> event = TELNET$	90%
3.	$Time = 7/25 - [15 - 18] -> event = TELNET$	93%
4.	$Time = 7/26 - [09 - 12] -> event = TELNET$	90%
5.	$Time = 7/26 - [12 - 15] -> event = TELNET$	90%
6.	$dstIP = 222.22.2.2 -> srcIP = 111.11.1.1$	100%
7.	$srcIP = 111.11.1.1 -> dstIP = 222.22.2.2$	100%

Fig. 5. Results of the Equi-depth approach with minimum support = 10%, minimum confidence = 50%, and number of intervals = 8 (every 3 hours). Only first 7 rules are shown.

Rule Number	Rule(window support, overall support)
1.	event=TELNET [7/25-00:00-7/25-09:06](89%,11%),[7/25-09:06-7/25-11:13](90%,11%) [7/25-11:13-7/25-16:04(91%,22%),[7/25-16:04-7/26-00:01(93%,11%) [7/26-00:01-7/26-12:17(91%,22%),[7/26-12:17-7/26-16:22(90%,11%)
	Rule 2 through 15 are omitted due to page limit.
16.	duration=0 [7/25-00:00-7/25-09:06(19%,2%),[7/25-09:06-7/25-11:13(14%,1%) [7/25-11:13-7/25-16:04(12%,3%),[7/25-16:04-7/26-00:01(12%,1%) [7/26-00:01-7/26-12:17(13%,3%),[7/26-12:17-7/26-16:22(11%,1%)
	Rule 17 through 21 are omitted due to page limit.

Fig. 6. Results of our algorithm with minimum support = 10.00%, minimum window size = 500, FD threshold = 0.02.

For our algorithm, we tried di.erent thresholds values. The threshold specifies the amount of change that should be seen in the fractal dimension before the algorithm draws a line and starts a new window. The bigger this number is, the less number of rules are generated. The minimum window size forces the window to have at least the specified minimum number of records regardless of the change in fractal dimension. A bigger window size reduces the execution time but might miss some rules.

The results generated by each approach in our experiment are shown in Figures 5 and 8 and compared below. We show the output rules in groups so they can be matched and compared easily. For example, rules 1 through 5 in the Equidepth output are comparable to rule 1 in the Fractals output. To read the rules, Equi-depth rule 1 shows that 91% of network connections on July 25 between 9am to 12pm are TELNET connections. Fractals rule 1 shows 89% of connection from 12am (midnight) to 9:06:43 am on July 25 are TELNET connections; it also shows that this is 11% of all connections in the data set. Note that Fractal rules are all associated with the interval attribute (Time) while the Equi-depth rules show such associations only if they make the minimum support and minimum confidence. As a result, the Equi-depth rules 6 and 7 show associations between source IP= 111.11.1.1 and destination IP = 222.22.2.2 but their associations with the interval attribute, as shown by Fractals rule 2, are missed. An itemset's association with the interval attribute, Time in this case, could be an important one. For example, the fact that 26% of connections after business hours from 16:04 to midnight (as shown by third line under Fractals rule 2) have occurred between source IP address 111.11.1.1 and destination IP address 222.22.2.2, is an interesting rule which is worth further investigation (since normally one expects less number of connections after business hours).

Fractals rules 14 through 21 are associations found by the Fractals approach and not found by the Equi-depth approach. Some of these rules could be important. For example Fractals rule 16 shows connections with 0 seconds connection duration. A connection with very short duration is an important characteristic in intrusion detection. The first line under rule 16 shows 19% of the connections between midnight and 9 in the morning (on July 25) have occurred with a duration of 0. Again this is a rule that is worth further investigation. Note that this association, $[7/25 - 00:00:00 - 7/25 - 09:06:43] -> duration = 0$, has an overall support of 2% and would never make it through the filter of 10% minimum support.

4 Conclusions

We have presented here a new algorithm to efficiently find association rules for data sets that contain one dimension of interval values. Previous algorithms dealt with this problem by dividing the interval data in Equi-depth intervals, risking missing some important rules, or by clustering the interval data before using the classical a-priori approach, and using the intervals found by the clustering algorithm as items. In doing so, these intervals remain defined for the rest of the algorithm, independently on whether for certain itemsets it would have been more adequate to consider a subinterval. We have successfully implemented the code shown in Figure 4 and conducted experiments in synthetic and real data sets with it. Our results show that this technique can efficiently find frequent itemsets that were missed by previous approaches. Also, our results demonstrate that we are able to find important associations that could be missed by previous algorithms for interval data.

References

1. R. Aggrawal, T. Imielinski, and A. Swami. Mining association rules between sets in large databases. In *Proceedings of the ACM-SIGMOD International Conference on Management of Data*, Washington, DC, 1993.
2. R. Agrawal, H. Mannila, R. Srikant, H. Toivonen, and A.Inkeri. Fast Discovery of Association Rules. In *Advances in Knowledge Discovery and Data Mining,* U. Fayyad, G. Shapiro, P. Smyth, R. Uthurusamy, eds. AAAI press, 1996.
3. R. Agrawal and R. Srikant. Fast Algorithms for Mining Association Rules. In *Proceedings of the 20th Int'l Conference on Very Large Databases*, Santiago, Chile, 2000.

4. A. Belussi and C. Faloutsos. Estimating the Selectivity of Spatial Queries Using the 'Correlation' Fractal Dimension. In *Proceedings of the International Conference on Very Large Data Bases*, pp. 299–310, September 1995.
5. A. Jain, R.C. Dubes. Algorithms for Clustering Data. Prentice Hall, Englewood Cliffs, New Jersey 1988.
6. L.S. Liebovitch and T. Toth. A Fast Algorithm to Determine Fractal Dimensions by Box Counting. *Physics Letters*, 141A(8), 1989.
7. R.J. Miller and Y.Yang. Association Rules over Interval Data. In *Proceedings of the ACM-SIGMOD International Conference on Management of Data*, Phoenix, Arizona, 1997.
8. R. Srikant and R. Agrawal. Mining quantitative association rules in large relational tables. In *Proceedings of the ACM-SIGMOD International Conference on Management of Data*, Montreal, Canada, 1996.
9. C. Traina Jr., A. Traina, L. Wu and C. Faloutsos, Fast feature selection using the fractal dimension. In *Proceedings of the XV Brazilian Symposium on Databases (SBBD)*, Paraiba, Brazil, October 2000.
10. W. Wang, J. Yang, R. Muntz, TAR: Temporal Association Rules on Evolving Numerical Attributes In *Proceedings of 17th International Conference on Data Engineering April 02–06*, 2001 Heidelberg, Germany 2001.
11. X. Chen, I. Petrounias Discovering Temporal Association Rules: Algorithms, Language and System. In *Proceedings of 16th International Conference on Data Engineering San Diego*, California, February 28–March 03, 2000.
12. R. Zhang, R. Ramakrishnan and M. Livny. BIRCH: An E.cient Data Clustering Method for Very Large Databases. In *Proceedings of the ACM SIGMOD Conference on Management of Data*, Montreal, Canada, 1996. Pages 103–114.

ParaDualMiner: An Efficient Parallel Implementation of the DualMiner Algorithm

Roger M.H. Ting, James Bailey, and Kotagiri Ramamohanarao

Department of Computer Science and Software Engineering
The University of Melbourne, Australia

Abstract. Constraint based mining finds all itemsets that satisfy a set of predicates. Many constraints can be categorised as being either monotone or antimonotone. Dualminer was the first algorithm that could utilise both classes of constraint simultaneously to prune the search space. In this paper, we present two parallel versions of DualMiner. The ParaDualMiner with Simultaneous Pruning efficiently distributes the task of expensive predicate checking among processors with minimum communication overhead. The ParaDualMiner with Random Polling makes further improvements by employing a dynamic subalgebra partitioning scheme and a better communication mechanism. Our experimental results indicate that both algorithms exhibit excellent scalability.

1 Introduction

Data mining in the presence of constraints is an important problem. It can provide answers to questions such as "find all sets of grocery items that occur more than 100 times in the transaction database and the maximum price of the items in each of those sets is greater than 10 dollars". To state our problem formally, we denote an item as i. A group of items is called an itemset, denoted as S. The list of items that can exist in our database is denoted as $I = \{i_1, i_2, \ldots, i_n\}$, $S \subseteq I$. The constraints are a set of predicates $\{P_1, P_2, \ldots, P_n\}$ that have to be satisfied by an itemset S. Constraint based mining finds all sets in the powersets of I that satisfy $P_1 \wedge P_2 \wedge \ldots \wedge P_n$.

Many constraints can be categorised into being either monotone or antimonotone constraints [8,10]. There exist many algorithms that can only use one of them to prune the search space. There are also algorithms that can mine itemsets using these two constraint categories in a sequential fashion (e.g. [13]). DualMiner was the first algorithm able to interleave both classes of constraint simultaneously during mining [4]. Nevertheless, the task of finding itemsets that satisfy both classes of constraint is still time consuming. High performance computation offers a potential solution to this problem, provided that an efficient parallel version of DualMiner can be constructed.

Contributions: In this paper, we introduce two new parallel algorithms which extend the original serial DualMiner algorithm [4]. We have implemented our algorithms on a Compaq Alpha Server SC machine and they both show excellent

H. Dai, R. Srikant, and C. Zhang (Eds.): PAKDD 2004, LNAI 3056, pp. 96–105, 2004.

scalability. To the best of our knowledge, our algorithms are the first parallel algorithms that can perform constraint-based mining using both monotone and antimonotone constraints simultaneously.

2 Preliminaries

We now provide a little background on the original DualMiner algorithm [4]. Formally, given itemsets M,J and S, a constraint is *antimonotone* if

$$\forall S, J : ((J \subseteq S \subseteq M) \wedge P(S)) \Rightarrow P(J)$$

One of the most widely cited antimonotone constraints is $support(S) > c$. A constraint is *monotone* if

$$\forall S, J : ((S \subseteq J \subseteq M) \wedge Q(S)) \Rightarrow Q(J)$$

Monotone constraints are the opposite of antimonotone constraints. Therefore, a corresponding example of a monotone constraint is $support(S) < d$. A conjunction of antimonotone predicates is antimonotone and a conjunction of monotone predicates is monotone [4]. Therefore, many itemset mining problems involving multiple constraints can be reduced to looking for all itemsets that satisfy a predicate of the form $P(S) \wedge Q(S)$. Even though some constraints are not exactly monotone or antimonotone constraints, previous research indicates that they can be approximated to be either monotone or antimonotone if some assumptions are made [10]. According to previous work, the search space of all itemsets forms a *lattice*. Given a set I with n items, the number of elements in the lattice is 2^n. This is equal to the number of elements in the powerset of I, which is denoted as 2^n. By convention, the biggest itemset is at the bottom of this lattice and the smallest itemset will always be at the top. Beside that, our search space also forms a *Boolean algebra* with maximal element B and minimal element T. It has the following properties (i)$X \in \Gamma$ (ii) $B = \bigcup X$ which is the bottom element of Γ (iii) $T = \bigcap X$ which is the top element of Γ (iv) for any $A \in \Gamma$, $\overline{A} = B \backslash A$. A *subalgebra* is a collection of elements $\subseteq 2^n$ closed under $\bigcap$ and $\bigcup$. The top and bottom element of the algebra is sufficient to represent all the itemsets in between them. If the top and bottom elements satisfy both constraints, the monotone and antimonotone properties guarantee that all itemsets in between them will satisfy both constraints. A *good subalgebra* is a subalgebra which has top and bottom elements that satisfy both the antimonotone and monotone constraints.

Overview of DualMiner. DualMiner builds a dynamic binary tree when searching for all good subalgebras. A tree node represents a subalgebra, but not necessarily a good subalgebra. Each tree node τ consists of three item lists which are (i) $IN(\tau)$ representing all the items that must be in the subalgebra and the top element of current subalgebra, T, (ii) $CHILD(\tau)$ representing all the items that have not been apportioned between $IN(\tau)$ and $OUT(\tau)$ and (iii) $OUT(\tau)$

representing all the items that cannot be contained in the current subalgebra. Because our search space forms a Boolean algebra, $\overline{OUT}$ represents the bottom element of the current subalgebra, B. Note that $\overline{OUT(\tau)} = \{IN(\tau) \cup CHILD(\tau)\}$.

When DualMiner starts, it will create a root node with $IN(\alpha)$ and $OUT(\alpha)$ empty . It will start checking the top element of the current subalgebra first. If the top element does not satisfy the antimonotone constraint, every itemset below it will not satisfy the constraint too. Therefore, we can eliminate the subalgebra. If it satisfies the antimonotone constraint, DualMiner will check all the itemsets below T in the subalgebra using the antimonotone constraint. Each is of the form $IN \cup \{X\}$, where X is an item from the $CHILD$ item list. If all itemsets $IN \cup \{X\}$ satisfy the constraint, no item list will be altered. If an itemset does not satisfy the constraint, X will be put into OUT item list. This effectively eliminates the region that contains that item.

Next, the algorithm will apply the monotone constraint on B of the current subalgebra. If the maximal itemset fails, the algorithm eliminates the current subalgebra immediately. If it does not fail, DualMiner will start checking all the itemsets one level above the current bottom itemset using the monotone constraint. Each is of the form $\overline{OUT \cup \{X\}}$, where X is an item from the $CHILD$ item list. If all itemsets $\overline{OUT \cup \{X\}}$ satisfy the constraint, no item list will be altered. If an itemset does not satisfy the constraint, X will be put into the IN item list. This eliminates the region that does not contain that item.

The pruning process will continue until no pruning can be done. At the end of the pruning phase, the top itemset must satisfy the antimonotone constraint and the bottom itemset must satisfy the monotone constraint. If the top itemset also satisfies the monotone constraint and the bottom itemset also satisfies the antimonotone constraint, we have found one good subalgebra. If this is not the case, DualMiner will partition the subalgebra into two halves. This is done by firstly creating two child tree nodes and picking an item from the $CHILD$ itemset and inserting it into the IN of one child and OUT of another child. The algorithm will mark the current parent node as visited and proceed to the child nodes. The process is repeated until all nodes are exhausted.

3 Parallel DualMiner

DualMiner does not prescribe specific constraints that have to be used. The antimonotone constraint and the monotone constraint are two types of predicates over an itemset or oracle functions that return true or false. To simplify our implementation, we will use $support(S) > C$ as our antimonotone constraint and $support(S) < D$ as our monotone constraint. $C \leq D$. We represent the database with a series of bit vectors. Each bit vector represents an item in the database. The support count of an itemset can be found by performing a bitwise AND operation on the bit vector of each item in the itemset. This approach has been used by many other algorithms[5,9]. We represent the IN, $CHILD$ and OUT itemlist in each node as three bit vectors. Each item is represented as 1 bit in each of the three bit vectors. The position of the bit will indicate the item id of the item.

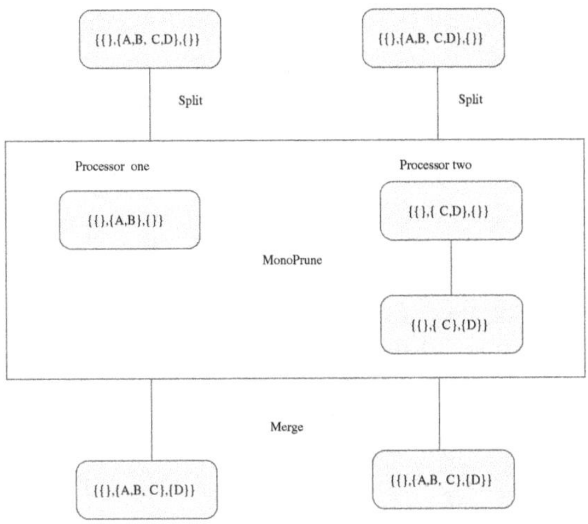

Fig. 1. ParaDualMiner with Simultaneous Pruning

ParaDualMiner with Simultaneous Pruning. In the original DualMiner paper, it was observed that the most expensive operation in any constraint based mining algorithm is the oracle function. Therefore, we can achieve great performance gain if we can distribute the call to the oracle function evenly among different processors. We notice that after DualMiner verifies that the top element of a subalgebra satisfies the antimonotone constraint, it will check whether all the elements one level below the top element satisfies the antimonotone constraint. Each of them is of the form $IN \bigcup \{X\}$, where X is any item from the $CHILD$ itemset. Since each oracle function call is independent, it is possible to partition the $CHILD$ item list and perform the oracle function call simultaneously. e.g. Given the following transaction database: Transaction 1={A,B,C,D}, Transaction 2 = {A,B,C} and suppose our constraints are $support(S) > 1$ and $support(S) < 3$, the execution of the algorithm is illustrated in figure 1. Before partitioning the $CHILD$ item list, all the processors will have the same IN,$CHILD$ and OUT item list. After the parallel algorithm distributes the antimonotone constraint checking among different processors, any itemset of the form $IN \bigcup \{X\}$ such as $\{D\}$, that does not satisfy the antimonotone constraint, will lead to an item being inserted into the OUT item list in order to prune away that part of the search space. Therefore, at the end of the simultaneous antimonotone checking, the item lists that will be altered are the $CHILD$ and OUT item list.

Before proceeding, we must merge the search space that has not been pruned away using the antimonotone constraint. It only has to perform a bitwise boolean

OR operation on the $CHILD$ and OUT item lists of all processors. This will give us the global result based on the individual processor pruning process. This simplification is the direct result of us choosing the bit vector as our item list representation. The merging operation can be done between all processors using the MPI_Allreduce function, with boolean OR as the operator.

A similar process can be applied when we are pruning the search space using the monotone constraint. The difference is the algorithm is partitioning and merging the IN and $CHILD$ item lists instead of the OUT and $CHILD$ item-lists. The partitioning operation is entirely a local operation.There is no message passing involved. Each processor will check the number of items in the $CHILD$ item list, the number of processors and its own rank to consider which part of the $CHILD$ itemlist to be processed. If it can divide the number of items evenly, each processor will perform an equal amount of oracle calls. However, this will happen only if the number of processors is a perfect multiple of the number of items in the $CHILD$ item list. If the algorithm cannot divide the $CHILD$ item list evenly, the algorithm will distribute the residual evenly to achieve optimal load balancing. Therefore, the maximum idle time for processors each time the algorithm distributes the task of oracle function call will be T_{oracle}.

ParaDualMiner with Random Polling. There are a number of parallel frequent pattern miners that use the concept of candidate partitioning and migration to distribute task among processors (e.g. [7,1]). We can see similar behaviour in DualMiner. Whenever DualMiner cannot perform any pruning on the current subalgebra using both constraints, DualMiner will split the current subalgebra into two halves by splitting the tree node. This node splitting operation is essentially a divide-and-conquer strategy. No region of the search space has been eliminated in the process. Therefore, the algorithm permits an arbitrary amount splitting of subalgebras subject to the condition that they are to be evaluated later. The number of splits permitted is equal to the number of items in the $CHILD$ item list. This intuition gives us the simplest form of a parallel subalgebra partitioning algorithm.

In the 2 processor case, the original search space is partitioned into two subalgebras. Both processors can turn off 1 bit in the $CHILD$. One puts it in the IN item list by turning on the similar bit in the IN bit vector. Another processor will put it in the OUT item list by turning on the similar bit. The two processors search two disjoint search spaces without any need for communication. After the original algebra has been partitioned, each processor will simultaneously run Dualminer locally to find itemsets that satisfy both constraints. Since our search space can be partitioned up to the number of items in the $CHILD$ itemlist, this strategy can be applied to cases with more than two processors. The number of processors that are needed must be 2^n, where n is the number of times the splitting operation has been performed. The partitioning operation is a local process. Each processor will only process one of the subalgebras according to its own rank. There is no exchange of messages.

This algorithm will only achieve perfect load balancing if the two nodes contain equal amounts of work. This is unlikely because it is unlikely that the search space of each processor is even. One of the processors may terminate earlier than the rest of processors. Without any dynamic load balancing, the processor will remain idle throughout the rest of the execution time. This leads to poor load balancing and longer execution time. To overcome this problem, we can view a node as a job parcel. Instead of letting a free processor stay idle throughout the execution time, the new algorithm can delegate one of the nodes to an idle processor to obtain better load balancing.

There are two ways to initiate task transfer between processors. They are sender-initiated and receiver-initiated methods[6]. Our study indicated that the receiver-initiated scheme outperformed the sender-initiated scheme. The reason for this is that the granularity of time when a processor is idle in the receiver-initiated scheme is large. DualMiner spends most of its time in the pruning process. Therefore, the time a processor takes before it splits a node can be very long. If a processor terminates very early at the start of its own pruning process, it has to passively wait for another processor splits a node and sends it. This greatly decreases the work done per time unit which leads to poor speedup. Instead of being passive, the idle processor should poll for a job from a busy processor. DualMiner can split a node anywhere, provided that we do the splitting and job distribution properly. e.g. Given the transaction database Transaction 1 = {A,B,C,D}, Tranaction 2= {C,D} and the constraint is $support(S) > 1$ and $support(S) < 3$,the set of itemsets that satisfies both constraints is $\{\{C\},\{D\},\{C,D\}\}$.

The original search space will firstly be split into two subalgebras as shown in figure 2. Since itemset $\{A\}$ is infrequent, processor one that processes the left node will finish earlier. Processor two that processes right node could be still within one of the pruning functions. Instead of staying idle, processor one should then poll for a job from processor two.

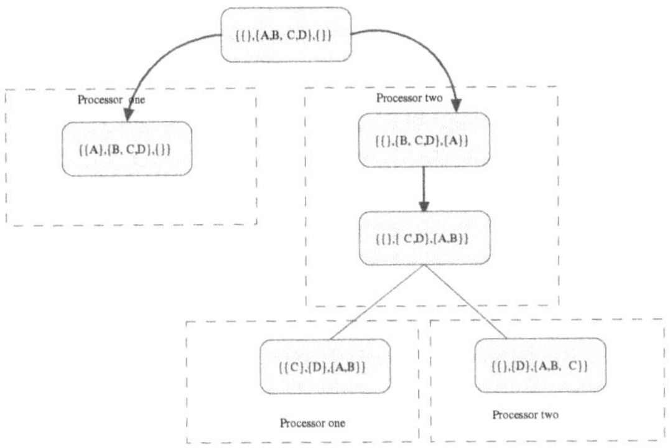

Fig. 2. Subalgebra Partitioning with Random Polling

Suppose processor two is pruning the search space using the antimonotone constraint. This implies that the top element of the current subalgebra such as $\{\}$ has already satisfied the antimonotone constraint. Otherwise, this subalgebra would have been eliminated. Therefore, while it is evaluating all the elements one level below the top element, it can check for an incoming message from processor one. Suppose it finds that processor one is free after evaluating itemset $\{B\}$, it can split the subalgebra and send it to processor one as shown in figure 2. In this case, the subalgebra is further split into two smaller subalgebras and can be distributed between these two processors. When the algorithm is pruning the search space using the antimonotone constraint, the IN itemset must have already satisfied the antimonotone constraint. Therefore, in this example, if processor two continues pruning using the antimonotone constraint, processor two should pick the right node. Likewise, if the algorithm is pruning the search space using monotone constraint, the sender processor should pick the left node. Suppose that processor two has already split a node and there is already a node or subalgebra that is yet to be processed. The algorithm should send that node to the idle processor instead of splitting the current one that it is working on. This is because the size of the subalgebra that is yet to be processed is equal to or greater than the size of the current subalgebra, if we are using a depth first or breadth first traversal strategy .

There are mainly two approaches to extend this algorithm to multiple processors. They are categorised into decentralised and centralised schemes [15]. To simplify our implementation, we have adapted the master-slave scheme. The master processor acts as an agent between all the slave processors. Each slave processor will work on the subalgebras that are assigned to it simultaneously. However, it will anticipate an incoming message from the master. Whenever a slave processor runs out of jobs locally, it will send a message to the master. The master processor will then poll for a job from a busy processor. Therefore the master has to know which processors are busy and which are idle.

For this purpose, we keep a list of processors that are busy and idle. The list can be efficiently represented as a bit vector. The bit position of the vector will then be the rank of the processor. A busy processor will be denoted as 1 in the bit vector. A free processor will be denoted as 0 in the bit vector. Whenever the master receives a message from the processor X, it will initialise bit X to zero. It will then select a processor to poll for job. There are various way to select a processor. A random selection algorithm has been found to work very well in many cases. Also, there is previous work that analyses the complexity of this kind of random algorithm [11,12,11]. Therefore, we have used this algorithm in our implementation. The master will randomly generate a number between 0 and $n-1$, where n is the number of processors. It will then send a free message to the selected processor to poll for a job. If the selected slave processor does not have any job, it will send a free message to the master processor. The master processor will then mark it as free and put it into a free CPU queue. It will continue polling until a job message is replied to it. It will then send the job to a free processor. The slave processors can detect incoming messages using the MPI_Iprobe function. The termination condition is when the number of free processors in the list is equal to the number of processors available. This implies that there is no outstanding node to be processed. The master processor will then send a termination message to all the slave processors.

4 Experiments

We implemented both serial and parallel versions of DualMiner on a 128-processor Unix Cluster.The specification of the machine is 1 Terabyte of shared file storage, 128 Alpha EV68's at 833 MHz processor, a Quadrics interconnect which has 200 Megabyte/sec bandwidth and 6 milliseconds latency and 64 Gigabytes of memory. We used databases generated from the IBM Quest Synthetic data generator. The datasets generated from it are used in various papers [4,2, 14]. The number of transactions is 10000. The dimension of the dataset is 100000, which means maximum number of distinct items in the transaction database is 100000. The length of a transaction is determined by a Poisson distribution with a parameter, average length. The average length of transactions is 10. We also scaled up the dimension of dataset by doubling the average length of transactions. This is because the computing resources demanded is significantly higher if the dataset is dense [3]. For our purpose, we define a dataset with an average transaction length of 10 to be sparse and one with an average length of 20 to be dense.

In the Random Polling version of ParaDualMiner, the master node only acts as an agent between all the slave processors. It does not run the DualMiner like the slave processors. At the start of algorithm, the original algebra is partitioned into 2^n part. This means this version of ParaDualMiner can only accept 2^n processors for the slaves and one additional processor for the master. Therefore, we studied our algorithm with 2,3,5,9,17 processors.

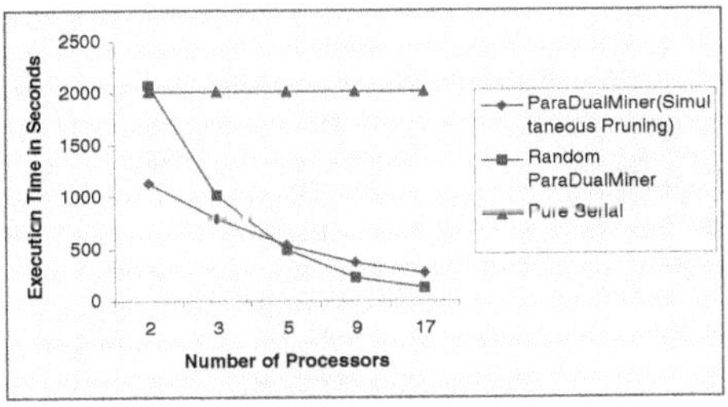

Fig. 3. Support between 0.015 and 0.03 percent and sparse dataset

Results. As shown in figure 3 and figure 4, the execution time of ParaDualMiner with Random Polling is almost identical to the serial version of DualMiner even

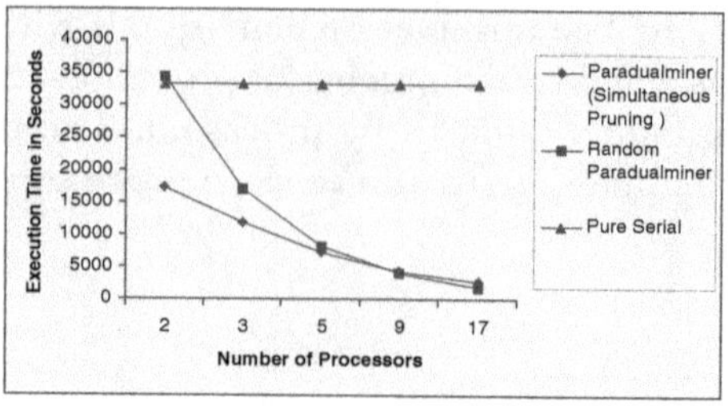

Fig. 4. Support between 0.015 and 0.03 percent and dense dataset

though it is using 2 processors. The reason is that the master in Random ParaDualMiner does not perform any task, besides acting as an agent between all the slave processors. Therefore, in the 2 processors case, there is only one processor working on the original algebra. When there are 3 processors, there will be two slave processors that work on the subalgebra distributed to them. Since the master will always poll for job from the busy processor and it is always possible to split work, the load balancing is excellent. Beside that, every communication is point to point communication and the message is relatively small. This leads to almost perfectly linear speedup after 2 processors. We also observe that there is super linear speedup in some cases. This is due to better memory usage. Whenever a processor has relatively more nodes to process, the nodes will migrate to other processors with less work load. This distributes the memory requirement among all the processors.

From figure 3 and figure 4, we can see that ParaDualMiner with Simultaneous Pruning is not as scalable as Random ParaDualMiner. The reason is that whenever there is n processors, there will be exactly n processors that will split the most computational intensive part of the algorithm which is the oracle function call. However, the algorithm will only achieve perfect load balancing if the number of items in the $CHILD$ itemlist is a perfect multiple of the number of processors. As the number of processors increases, the chances of getting a perfect multiple of the number of processors decreases. This implies the chance of having some processors stay idle for one oracle function call becomes larger. This may cause many processors to become idle too often, which impairs the parallelism that can be achieved by this algorithm. Also, the algorithm only parallelises the oracle function call. Furthermore, there is a need to have all-to-all communication to merge all the result of pruning. This is much more expensive than point-to-point communication in ParaDualMiner with Random Polling.

5 Conclusion

We have proposed two parallel algorithms for mining itemsets that must satisfy a conjunction of antimonotone and monotone constraints. There are many serial or parallel algorithms that take advantage of one of these constraints.

However, both of our parallel algorithms are the first parallel algorithms that take advantage of both constraints constraints simultaneously to perform constraint based mining. Both algorithms demonstrate excellent scalability. This is backed by our experimental result. We are currently investigating the scalability of both algorithms using hundreds of processors. Also, we are investigating how ParaDualMiner performs if we use other type of constraints. We believe that both parallel algorithms should perform well, because there is no reliance on the underlying nature of the constraints.

References

1. R. Agrawal and J. C. Shafer. Parallel mining of association rules. *IEEE Trans. On Knowledge And Data Engineering*, 8:962–969, 1996.
2. R. Agrawal and R. Srikant. Fast algorithms for mining association rules in large databases. In *Proceedings of VLDB'94*, pages 487–499, 1994.
3. R. Bayardo, R. Agrawal, and D. Gunopulos. Constraint-based rule mining in large, dense databases. In *Proceedings of ICDE'99*, pages 188–197, 1999.
4. C. Bucila, J. Gehrke, D. Kifer, and W. White. Dualminer: a dual-pruning algorithm for itemsets with constraints. In *Proceedings of ACM SIGKDD'02*, pages 42–51, 2002.
5. D. Burdick, M. Calimlim, and J. Gehrke. Mafia: A maximal frequent itemset algorithm for transactional databases. In *Proceedings of ICDE'01*, pages 443–452, 2001.
6. D. L. Eager, E. D. Lazowska, and J. Zahorjan. A comparison of receiver-initiated and sender-initiated adaptive load sharing. In *Proceedings of ACM SIGMETRICS'85*, pages 1–3, 1985.
7. E. H. Han, G. Karypis, and V. Kumar. Scalable parallel data mining for association rules. In *Proceedings of SIGMOD'97*, pages 277–288, 1997.
8. R. T. Ng, L. V. Lakshmanan, J. Han, and A. Pang. Exploratory mining and pruning optimizations of constrained association rules. In *Proceedings of SIGMOD'98*, pages 13–24, 1998.
9. S. Orlando, P. Palmerini, R. Perego, and F. Silvestri. Adaptive and resource-aware mining of frequent sets. In *Proceedings of ICDM'02*, page 338, 2002.
10. J. Pei, J. Han, and L. Lakshmanan. Mining frequent item sets with convertible constraints. In *Proceedings of ICDE'01*, pages 433–442, 2001.
11. P.Sanders. A detailed analysis of random polling dynamic load balancing. In *Proceedings of ISPAN'94*, pages 382–389, 1994.
12. P.Sanders. Asynchronous random polling dynamic load balancing. In *Proceedings of ISAAC'99,*, page 39, 1999.
13. L. D. Raedt and S. Kramer. The levelwise version space algorithm and its application to molecular fragment finding. In *Proceedings of IJCAI'01*, pages 853–8622, 2001.
14. O. Zaiane, M. El-Hajj, and P. Lu. Fast parallel association rule mining without candidacy generation. In *Proceedings of ICDM'01*, pages 665–668, 2001.
15. M. Zaki, W. Li, and S. Parthasarathy. Customized dynamic load balancing for a network of workstations. *Journal of Parallel and Distributed Computing*, 43(2):156–162, 1997.

A Novel Distributed Collaborative Filtering Algorithm and Its Implementation on P2P Overlay Network*

Peng Han, Bo Xie, Fan Yang, Jiajun Wang, and Ruimin Shen

Department of Computer Science and Engineering, Shanghai Jiao Tong University, Shanghai 200030, China
{phan,bxie,fyang,jjwang,rmshen}@sjtu.edu.cn

Abstract. Collaborative filtering (CF) has proved to be one of the most effective information filtering techniques. However, as their calculation complexity increased quickly both in time and space when the record in user database increases, traditional centralized CF algorithms has suffered from their shortage in scalability. In this paper, we first propose a novel distributed CF algorithm called PipeCF through which we can do both the user database management and prediction task in a decentralized way. We then propose two novel approaches: significance refinement (SR) and unanimous amplification (UA), to further improve the scalability and prediction accuracy of PipeCF. Finally we give the algorithm framework and system architecture of the implementation of PipeCF on Peer-to-Peer (P2P) overlay network through distributed hash table (DHT) method, which is one of the most popular and effective routing algorithm in P2P. The experimental data show that our distributed CF algorithm has much better scalability than traditional centralized ones with comparable prediction efficiency and accuracy.

1 Introduction

Collaborative filtering (CF) has proved to be one of the most effective information filtering techniques since Goldberg et al [1] published the first account of using it for information filtering. Unlike content-based filtering, the key idea of CF is that users will prefer those items that people with similar interests prefer, or even that dissimilar people don't prefer. The k-Nearest Neighbor (KNN) method is a popular realization of CF for its simplicity and reasonable performance. Up to now, many successful applications have been built on it such as GroupLens [4], Ringo [5]. However, as its computation complexity increased quickly both in time and space as the record in the database increases, KNN-based CF algorithm suffered a lot from its shortage in scalability.

One way to avoid the recommendation-time computational complexity of a KNN method is to use a model-based method that uses the users' preferences to learn a model, which is then used for predications. Breese et al utilizes clustering and Bayesian network for a model-based CF algorithm in [3]. Its results show that the clustering-based method is the more efficient but suffering from poor accuracy while

* Supported by the National Natural Science Foundation of China under Grant No. 60372078

H. Dai, R. Srikant, and C. Zhang (Eds.): PAKDD 2004, LNAI 3056, pp. 106–115, 2004 the Bayesian networks prove only practical for environments in which knowledge of user preferences changes slowly. Further more, all model-based CF algorithms still require a central database to keep all the user data which is not easy to achieve sometime not only for techniques reasons but also for privacy reasons.

An alternative way to address the computational complexity is to implement KNN algorithm in a distributed manner. As Peer-to-Peer (P2P) overlay network gains more and more popularity for its advantage in scalability, some researchers have already begun to consider it as an alternative architecture [7,8,9] of centralized CF recommender system. These methods increase the scalability of CF recommender system dramatically. However, as they used a totally different mechanism to find appropriate neighbors than KNN algorithms, their performance is hard to analyze and may be affected by many other factors such as network condition and self-organization scheme.

In this paper we solve the scalability problem of KNN-based CF algorithm by proposing a novel distributed CF algorithm called PipeCF which has the following advantage:

1. In PipeCF, both the user database management and prediction computation task can be done in a decentralized way which increases the algorithm's scalability dramatically.
2. PipeCF keeps all the other features of traditional KNN CF algorithm so that the system's performance can be analyzed both empirically and theoretically and the improvement on traditional KNN algorithm can also be applied here.
3. Two novel approaches have been proposed in PipeCF to further improve the prediction and scalability of KNN CF algorithm and reduce the calculation complexity from to where M is the user number in the database and N is the items number.
4. By designing a heuristic user database division strategy, the implementation of PipeCF on a distributed-hash-table (DHT) based P2P overlay network is quite straightforward which can obtain efficient user database management and retrieval at the same time.

The rest of this paper is organized as follows. In Section 2, several related works are presented and discussed. In Section 3, we give the architecture and key features of PipeCF. Two techniques: SR and UA are also proposed in this section. We then give the implementation of PipeCF on a DHT-based P2P overlay network in Section 4. In Section 5 the experimental results of our system are presented and analyzed. Finally we make a brief concluding remark and give the future work in Section 6.

2 Related Works

2.1 Basic KNN-Based CF Algorithm

Generally, the task of CF is to predict the votes of active users from the user database which consists of a set of votes corresponding to the vote of user *i* on item *j*. The KNN-based CF algorithm calculates this prediction as a weighted average of other users' votes on that item through the following formula:

$$P_{a,j} = \overline{v_a} + \kappa \sum_{i=1}^{n} \varpi(a,j)(v_{i,j} - \overline{v_i}) \tag{1}$$

Where $P_{a,j}$ denotes the prediction of the vote for active user a on item j and n is the number of users in user database. $\overline{v_i}$ is the mean vote for user i as:

$$\overline{v_i} = \frac{1}{|I_i|} \sum_{j \in I_i} v_{i,j} \tag{2}$$

Where I_i is the set of items on which user i has voted. The weights $\varpi(a,j)$ reflect the similarity between active user and users in the user database. κ is a normalizing factor to make the absolute values of the weights sum to unity.

2.2 P2P System and DHT Routing Algorithm

The term "Peer-to-Peer" refers to a class of systems and applications that employ distributed resources to perform a critical function in a decentralized manner. Some of the benefits of a P2P approach include: improving scalability by avoiding dependency on centralized points; eliminating the need for costly infrastructure by enabling direct communication among clients. As the main purpose of P2P systems are to share resources among a group of computers called peers in a distributed way, efficient and robust routing algorithms for locating wanted resource is critical to the performance of P2P systems. Among these algorithms, distributed hash table (DHT) algorithm is one of the most popular and effective and supported by many P2P systems such as CAN [10], Chord [11], Pastry [12], and Tapestry [13].

A DHT overlay network is composed of several DHT nodes and each node keeps a set of resources (e.g., files, rating of items). Each resource is associated with a key (produced, for instance, by hashing the file name) and each node in the system is responsible for storing a certain range of keys. Peers in the DHT overlay network locate their wanted resource by issue a lookup(*key*) request which returns the identity (e.g., the IP address) of the node that stores the resource with the certain key. The primary goals of DHT are to provide an efficient, scalable, and robust routing algorithm which aims at reducing the number of P2P hops, which are involved when we locate a certain resource, and to reduce the amount of routing state that should be preserved at each peer.

3 PipeCF: A Novel Distributed CF Algorithm

3.1 Basic PipeCF Algorithm

The first step to implement CF algorithm in a distributed way is to divide the original centralized user database into fractions which can then be stored in distributed peers. For concision, we will use the term *bucket* to denote the distributed stored fraction of user database in the following of this paper. Here, two critical problems should be considered. The first one is how to assign each bucket with a unique identifier through which they can be efficiently located. The second is which bucket should be retrieved when we need to make prediction for a particular user.

Here, we solve the first problem by proposing a division strategy which makes each bucket hold a group of users' record who has a particular <ITEM_ID, VOTE> tuple. It means that users in the same bucket at least voted one item with the same rating. This <ITEM_ID, VOTE> will then be used to a unique key as the identifier for the bucket in the network which we will describe in more detail in Section 4.

To solve the second problem, we propose a heuristic bucket choosing strategy by only retrieving those buckets whose identifiers are the same with those generated by the active user's ratings. Figure 1 gives the framework of PipeCF. Details of the function of lookup(*key*) and implemention of PipeCF on DHT-based P2P overlay network will be described in Section 4.

The bucket choosing strategy of PipeCF is based on the assumption that people with similar interests will at least rate one item with similar votes. So when making prediction, PipeCF only uses those users' records that are in the same bucket with the active user. As we can see in Figure 5 of section 5.3.1, this strategy have very high hitting ratio. Still, we can see that through this strategy we reduce about 50% calculation than traditional CF algorithm and obtain comparable prediction as shown in Figure 6 and 7 in section 5.

Algorithm: *PipeCF*
Input: rating record of the active user, target item
Output: predictive rating for target item
Method:
For Each <ITEM_ID, VOTE> tuple in the rating record of active user:
1) Generate the *key* corresponding to the <ITEM_ID, VOTE> through the hash algorithm used by DHT
2) Find the host which holds the bucket with the identifier key through the function lookup(*key*) provided by DHT
3) Copy all ratings in bucket key to current host

Use traditional KNN-based CF algorithm to calculate to predictive rating for target item.

Fig. 1. Framework of PipeCF

3.2 Some Improvement

3.2.1 Significance Refinement (SR)

In the basic PipeCF algorithm, we return all users which are in the at least one same bucket with the active user and find that the algorithm has an O(N) fetched user number where N is the total user number as Figure 7 shows. In fact, as Breese presented in [3] by the term inverse user frequency, universally liked items are not as useful as less common items in capturing similarity. So we introduce a new concept significance refinement (SR) which reduces the returned user number of the basic PipeCF algorithm by limiting the number of returned users for each bucket. We term the algorithm improved by SR as Return K which means "for every item, the PipeCF algorithm returns no more than K users for each bucket". The experimental result in Figure 7 and 8 of section 5.3.3 shows that this method reduces the returned user number dramatically and also improves the prediction accuracy.

3.2.2 Unanimous Amplification (UA)

In our experiment in KNN-based CF algorithm, we have found that some highly correlated neighbors have little items on which they vote the same rating with the active users. These neighbors frequently prove to have worse prediction accuracy than those neighbors who have same rating with active users but relatively lower correlation. So we argue that we should give special award to the users who rated some items with the same vote by amplify their weights, which we term Unanimous Amplification. We transform the estimated weights as follows:

$$w'_{a,i} = \begin{cases} w_{a,i} & N_{a,i} = 0 \\ w_{a,i} \cdot \alpha & 0 < N_{a,i} \leq \gamma \\ w_{a,i} \cdot \beta & N_{a,i} > \gamma \end{cases} \tag{3}$$

Where $N_{a,i}$ denotes the number of items which user a and user i have the same votes. A typical value for α for our experiments is 2.0, β is 4.0, and γ is 4. Experimental result in Figure 9 of section 4.3.4 shows that UA approach improves the prediction accuracy of the PipeCF algorithm.

4 Implemention of PipeCF on a DHT-Based P2P Overlay Network

4.1 System Architecture

Figure 2 gives the system architecture of our implementation of PipeCF on the DHT-based P2P overlay network. Here, we view the users' rating as resources and the system generate a unique key for each particular <ITEM_ID, VOTE> tuple through the hash algorithm, where the ITEM_ID denotes identity of the item user votes on and VOTE is the user's rating on that item. As different users may vote particular item with same rating, each key will correspond to a set of users who have the same <ITEM_ID, VOTE> tuple. As we stated in section 3, we call such set of users' record as *bucket*. As we can see in Figure 2, each peer in the distributed CF system is responsible for storing one or several buckets. Peers are connected through a DHT-based P2P overlay network. Peers can find their wanted buckets by their keys efficiently through the DHT-based routing algorithm.

As we can see from Figure 1 and Figure 2, the implementation of our PipeCF on DHT-based P2P overlay network is quite straightforward except two key pieces: how to store the buckets and fetch them back effectively in this distributed environment. We solve these problems through two fundamental DHT function: put(*key*) and lookup(*key*) which are described in Figure 3 and Figure 4 respectively.

These two functions inherit from DHT the following merits:

- *Scalability*: it must be designed to scale to several million nodes.
- *Efficiency*: similar users should be located reasonably quick and with low overhead in terms of the message traffic generated.
- *Dynamicity*: the system should be robust to frequent node arrivals and departures in order to cope with highly transient user populations' characteristic to decentralized environments.
- *Balanced load*: in keeping with the decentralized nature, the total resource load (traffic, storage, etc) should be roughly balanced across all the nodes in the system.

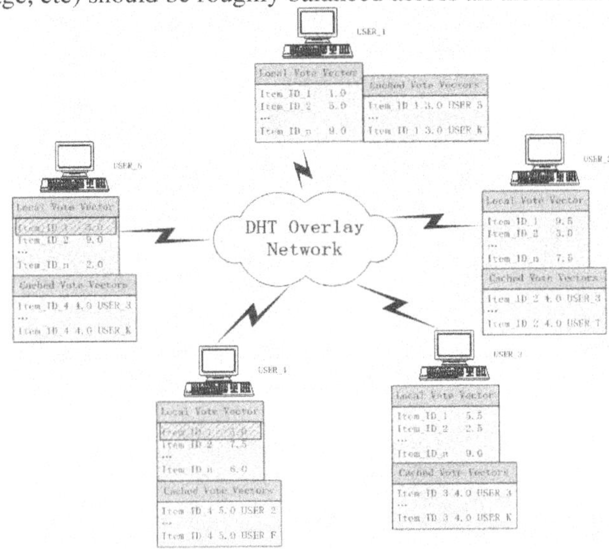

Fig. 2. System Architecture of Distributed CF Recommender System

5 Experimental Results

5.1 Data Set

We use EachMovie data set [6] to evaluate the performance of improved algorithm. The EachMovie data set is provided by the Compaq System Research Center, which ran the EachMovie recommendation service for 18 months to experiment with a collaborative filtering algorithm. The information they gathered during that period consists of 72,916 users, 1,628 movies, and 2,811,983 numeric ratings ranging from 0 to 5. To speed up our experiments, we only use a subset of the EachMovie data set.

5.2 Metrics and Methodology

The metrics for evaluating the accuracy of we used here is statistical accuracy metrics which evaluate the accuracy of a predictor by comparing predicted values with user-provided values. More specifically, we use Mean Absolute Error (MAE), a statistical accuracy metrics, to report prediction experiments for it is most commonly used and easy to understand:

$$MAE = \frac{\sum_{a \in T} | v_{a,j} - p_{a,j} |}{|T|} \tag{4}$$

Where $v_{a,j}$ is the rating given to item *j* by user *a*, is the predicted value of user a on item *j*, *T* is the test set, |*T*| is the size of the test set.

Algorithm: DHT-based CF puts a peer *P's* vote vector to DHT overlay network
Input: *P*'s vote vector
Output: NULL
Method:
For each <ITEM_ID, VOTE> in *P*'s vote vector:

1) *P* generates a unique 128-bit DHT Key K_{local} (i.e. hash the system unique username).
2) *P* hashes this <ITEM_ID, VOTE> tuple to key *K*, and routes it with *P*'s vote vector to the neighbor P_i whose local key K_{i_local} is the most similar with *K*.
3) When P_i receives the PUT message with *K*, it caches it. And if the most similar neighbor is not itself, it just routes the message to its neighbor whose local key is most similar with *K*.

Fig. 3. DHT Put(*key*) Function

Algorithm: *lookup(key)*
Input: identifier *key* of the targeted bucket
Output: targeted bucket (retrieved from other peers)
Method:

1) Routes the *key* with the targeted bucket to the neighbor P_i whose local key K_{i_local} is the most similar with *K*.
2) When P_i receives the LOOKUP message with *K*, if P_i has enough cached vote vectors with the same key *K*, it returns the vectors back to *P*, otherwise it just routes the message to its neighbor whose local key is most similar with *K*. Anyway, *P* will finally get all the records in the bucket whose identifier is *key*.

Fig. 4. DHT Lookup(*key*) Function

We select 2000 users and choose one user as active user per time and the remainder users as his candidate neighbors, because every user only make self's recommendation locally. We use the mean prediction accuracy of all the 2000 users as the system's prediction accuracy. For every user's recommendation calculation, our tests are performed using 80% of the user's ratings for training, with the remainder for testing.

5.3 Experimental Result

We design several experiments for evaluating our algorithm and analyze the effect of various factors (e.g., SR and UA, etc) by comparison. All our experiments are run on a Windows 2000 based PC with Intel Pentium 4 processor having a speed of 1.8 GHz and 512 MB of RAM.

5.3.1 The Efficiency of Neighbor Choosing

We used a data set of 2000 users and show among the users chosen by PipeCF algorithm, how many are in the top-100 users in Figure 5. We can see from the data that when the user number rises above 1000, more than 80 users who have the most similarities with the active users are chosen by PipeCF algorithm.

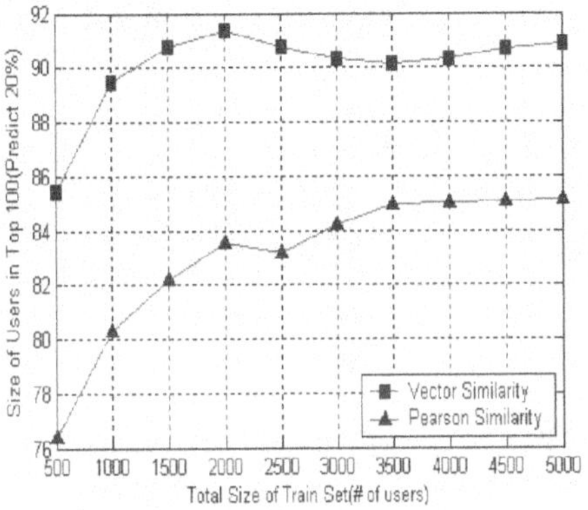

Fig. 5. How Many Users Chosen by PipeCF in Traditional CF's Top 100

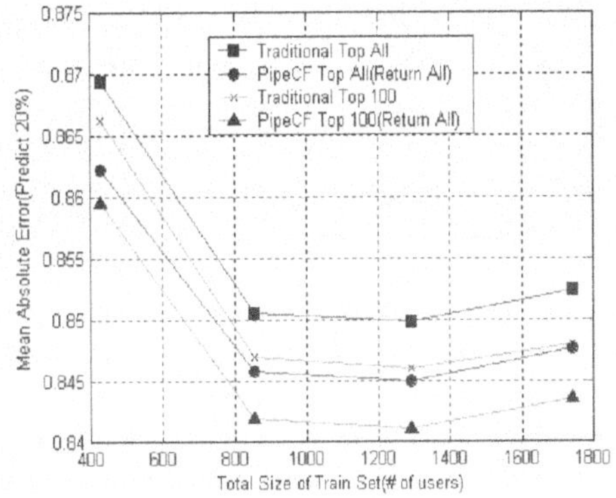

Fig. 6. PipeCF vs. Traditional CF

5.3.2 Performance Comparison

We compare the prediction accuracy of traditional CF algorithm and PipeCF algorithm while we apply both top-all and top-100 user selection on them. The results are shown as Figure 6. We can see that the DHT-based algorithm has better prediction accuracy than the traditional CF algorithm.

5.3.3 The Effect of Significance Refinement

We limit the number of returned user for each bucket by 2 and 5 and do the experiment in Section 5.3.2 again. The user for each bucket is chosen randomly. The result of the number of user chosen and the prediction accuracy is shown in Figure 7 and Figure 8 respectively. The result shows:

- "Return All" has an *O(N)* returned user number and its prediction accuracy is also not satisfying;
- "Return 2" has the least returned user number but the worst prediction accuracy;
- "Return 5" has the best prediction accuracy and the scalability is still reasonably well (the returned user number is still limited to a constant as the total user number increases).

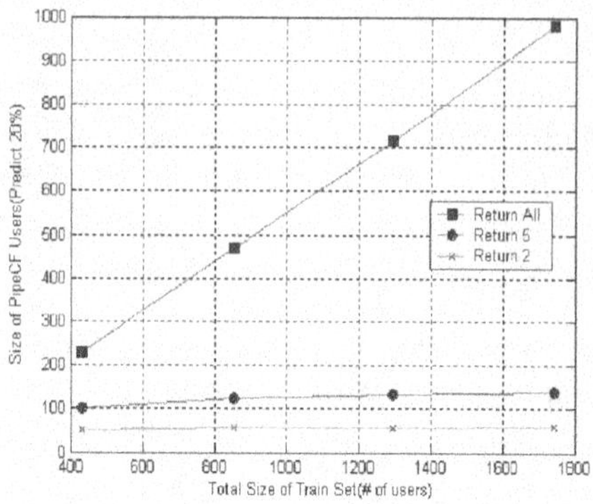

Fig. 7. The Effect on Scalability of SR on PipeCF

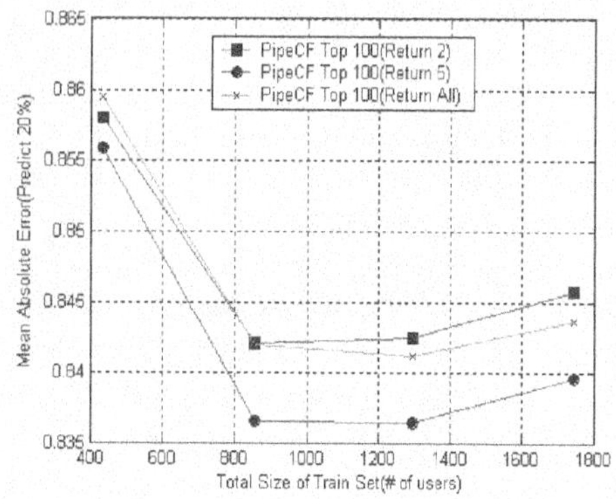

Fig. 8. The Effect on Prediction Accuracy of SR on PipeCF Algorithm

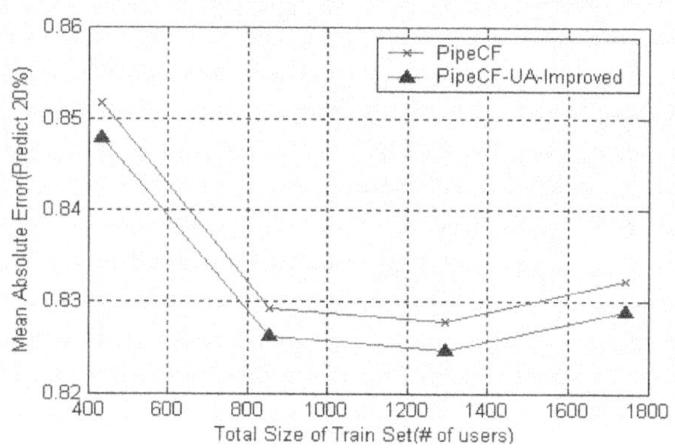

Fig. 9. The Effect on Prediction Accuracy of Unanimous Amplification

5.3.4 The Effect of Unanimous Amplification

We adjust the weights for each user by using Equation (5) while setting value for α as 2.0, β as 4.0, γ as 4 and do the experiment in Section 5.3.2 again. We use the top-100 and "Return All" selection method. The result shows that the UA approach improves the prediction accuracy of both the traditional and the PipeCF algorithm. From Figure 9 we can see that when UA approach is applied, the two kinds of algorithms have almost the same performance.

6 Conclusion

In this paper, we solve the scalability problem of KNN-based CF algorithm by proposing a novel distributed CF algorithm called PipeCF and give its implementation on a DHT-based P2P overlay network. Two novel approaches: significance refinement (SR) and unanimous amplification (UA) have been proposed to improve the performance of distributed CF algorithm. The experimental data show that our algorithm has much better scalability than traditional KNN-based CF algorithm with comparable prediction efficiency.

References

1. David Goldberg, David Nichols, Brian M. Oki, Douglas Terry.: Using collaborative filtering to weave an information tapestry, Communications of the ACM, v.35 n.12, p.61-70, Dec. 1992.
2. J. L. Herlocker, J. A. Konstan, A. Borchers, and J. Riedl.: An algorithmic framework for performing collaborative filtering. In Proceedings of the 22nd annual international ACM SIGIR conference on Research and development in information retrieval, pages 230-237, 1999.
3. Breese, J., Heckerman, D., and Kadie, C.: Empirical Analysis of Predictive Algorithms for Collaborative Filtering. Proceedings of the 14th Conference on Uncertainty in Artificial Intelligence, 1998 (43-52).
4. Paul Resnick, Neophytos Iacovou, Mitesh Suchak, Peter Bergstrom, John Riedl.: GroupLens: an open architecture for collaborative filtering of netnews, Proceedings of the 1994 ACM conference on Computer supported cooperative work, p.175-186, October 22-26, 1994, Chapel Hill, North Carolina, United States.
5. Upendra Shardanand, Pattie Maes.: Social information filtering: algorithms for automating "word of mouth", Proceedings of the SIGCHI conference on Human factors in computing systems, p.210-217, May 07-11, 1995, Denver, Colorado, United States.
6. Eachmovie collaborative filtering data set.: http://research.compaq.com/SRC/eachmovie
7. Amund Tveit.: Peer-to-peer based Recommendations for Mobile Commerce. Proceedings of the First International Mobile Commerce Workshop, ACM Press, Rome, Italy, July 2001, pp. 26-29.
8. Tomas Olsson.: "Bootstrapping and Decentralizing Recommender Systems", Licentiate Thesis 2003-006, Department of Information Technology, Uppsala University and SICS, 2003
9. J. Canny.: Collaborative filtering with privacy. In Proceedings of the IEEE Symposium on Research in Security and Privacy, pages 45--57, Oakland, CA, May 2002. IEEE Computer Society, Technical Committee on Security and Privacy, IEEE Computer Society Press.
10. S. Ratnasamy, P. Francis, M. Handley, R. Karp, and S. Shenker.: A scalable content-addressable network. In SIGCOMM, Aug. 2001
11. Stocal I et al.: Chord: A scalable peer-to-peer lookup service for Internet applications (2001). In ACM SIGCOMM, San Diego, CA, USA, 2001, pp.149-160
12. Rowstron A. Druschel P.: Pastry: Scalable, distributed object location and routing for large scale peer-to-peer systems. In IFIP/ACM Middleware, Hedelberg, Germany, 2001
13. Zhao B Y et al.: Tapestry: An infrastructure for fault-tolerant wide-area location and routing. Tech.Rep.UCB/CSB-0-114,UC Berkeley, EECS,2001

An Efficient Algorithm for Dense Regions Discovery from Large-Scale Data Streams

Andy M. Yip[1*], Edmond H. Wu[2**], Michael K. Ng[2**], and Tony F. Chan[1*]

[1] Department of Mathematics, University of California,
405 Hilgard Avenue, Los Angeles, CA 90095-1555, USA.
{mhyip,chan}@math.ucla.edu

[2] Department of Mathematics, The University of Hong Kong
Pokfulam Road, Hong Kong.
hcwu@hkusua.hku.hk, mng@maths.hku.hk

Abstract. We introduce the notion of dense region as distinct and meaningful patterns from given data. Efficient and effective algorithms for identifying such regions are presented. Next, we discuss extensions of the algorithms for handling data streams. Finally, experiments on large-scale data streams such as clickstreams are given which confirm that the usefulness of our algorithms.

1 Introduction

Besides the patterns identified by traditional data mining tasks such as clustering, classification and association rules mining, we realize that *dense regions* [1], which are two-dimensional regions defined by subsets of entities and attributes whose corresponding values are mostly constant, are another type of patterns which are of practical use and are significant. For example, dense regions may be used to evaluate discriminability of a subset of attributes, thus identifying such regions enhances the feature selection process in classification applications. Indeed, such patterns are very useful when analyzing users' behavior and improving web page or website topology design [2,3].

Our goals in this paper are to introduce the notion of dense regions and present an efficient and effective algorithm, called DRIFT (Dense Region IdentiFicaTion), for identifying such patterns. Extensions of the DRIFT algorithm for handling data streams are also discussed. Due to the lack of space, please refer to the extended version [5] for a thorough treatment on the subject.

* The research of this author is partially supported by grants from NSF under contracts DMS-9973341 and ACI-0072112, ONR under contract N00014-02-1-0015 and NIH under contract P20 MH65166.

** The research of this author is supported in part by Hong Kong Research Grants Council Grant Nos. HKU 7130/02P and HKU 7046/03P.

[1] A dense region usually refers to a subset of the space where data points are highly populated whereas dense regions discussed in this paper lie in the entity-by-attribute space. Our precise definition is given in Section 2.

H. Dai, R. Srikant, and C. Zhang (Eds.): PAKDD 2004, LNAI 3056, pp. 116–120, 2004.

To the best of the authors' knowledge, dense region patterns have not been previously studied. A similar but not identical notion is error-tolerant frequent itemset introduced by Yang et al in [4] which focuses on mining association rules.

2 Definition of Dense Regions

Let X be a given n-by-p data matrix where n is the number of entities and p is the number of attributes of each entity. Denote by $X(R,C)$ the submatrix of X defined by a subset of rows R and a subset of columns C. We also identify $X(R,C)$ by the set $D = R \times C$.

Definition 1 (Dense regions (DRs)). *A submatrix $X(R,C)$ is called a maximal dense region with respect to v, or simply a dense region with respect to v, if $X(R,C)$ is a constant matrix whose entries are v (density), and, any proper superset of $X(R,C)$ is a non-constant matrix (maximality).*

Example 1. Let X be a data matrix given by the first matrix below. The DRs of X with value 1 are given by the four matrices in the brace.

$$\begin{pmatrix} 1\,0\,0\,1 \\ 1\,1\,0\,1 \\ 1\,1\,0\,1 \\ 1\,1\,2\,0 \end{pmatrix}; \left\{ \begin{pmatrix} 1 * * * \\ 1 * * * \\ 1 * * * \\ 1 * * * \end{pmatrix}, \begin{pmatrix} 1 * * 1 \\ 1 * * 1 \\ 1 * * 1 \\ * * * * \end{pmatrix}, \begin{pmatrix} * * * * \\ 1\,1 * 1 \\ 1\,1 * 1 \\ * * * * \end{pmatrix}, \begin{pmatrix} * * * * \\ 1\,1 * * \\ 1\,1 * * \\ 1\,1 * * \end{pmatrix} \right\}.$$

Alternatively, we may denote the above DRs by $\{1,2,3,4\} \times \{1\}$, $\{1,2,3\} \times \{1,4\}$, $\{2,3\} \times \{1,2,4\}$ and $\{2,3,4\} \times \{1,2\}$ respectively.

3 The DRIFT Algorithm

The BasicDRIFT Algorithm. This starts from a given point (s,t) containing the target value v and returns two regions containing (s,t) where the first one is obtained by a vertical-first-search; the other is by a horizontal-first-search. It is proven in [5, Theorem 1] that the two returned regions are in fact DRs.

Algorithm: BasicDRIFT(X, s, t)
$R_v \leftarrow \{1 \le i \le n | X_{it} = X_{st}\}$, $C_v \leftarrow \{1 \le j \le p | X_{ij} = X_{it} \forall i \in R_v\}$
$C_h \leftarrow \{1 \le j \le p | X_{sj} = X_{st}\}$, $R_h \leftarrow \{1 \le i \le n | X_{ij} = X_{sj} \forall j \in C_h\}$
Return $\{R_v \times C_v, R_h \times C_h\}$

To determine the time complexity, we suppose the two resulting DRs have dimensions n_v-by-p_v and n_h-by-p_h respectively. The number of computations required by the algorithm is $n + n_v p + p + p_h n$. Moreover, in practice, $n_v, n_h \ll n$ and $p_v, p_h \ll p$. In this case, the complexity is of $O(n+p)$ essentially.

The ExtendedBasicDRIFT Algorithm. The BasicDRIFT algorithm is very fast but it may miss some DRs. To remedy such a deficiency, we introduce the ExtendedBasicDRIFT algorithm which performs similarly to the BasicDRIFT but after we obtain the set R_v (respectively C_h) as in the BasicDRIFT algorithm starting at (s,t), we perform a horizontal (respectively vertical) search over all possible subsets of $R_v \setminus \{s\}$ (respectively $C_h \setminus \{t\}$) and return the maximal ones.

This algorithm returns all DRs containing (s,t). However, since it requires more computations than the BasicDRIFT does, we only invoke it to find DRs that the BasicDRIFT misses. The question now becomes how to combine the two algorithms in an effective way which is the purpose of our next algorithm.

The DRIFT Algorithm. We begin by introducing a key concept, called *isolated point*, which allows us to fully utilize the fast BasicDRIFT algorithm to find as many regions as possible while minimizes the use of the more expensive ExtendedBasicDRIFT algorithm.

Definition 2 (Isolated points). *A point (i,j) in a dense region D is isolated if it is not contained in any other dense region.*

By Theorem 3 in [5], (s,t) is an isolated point iff the two DRs obtained from the BasicDRIFT are identical. Moreover, each isolated point belongs only to one DR, hence, after we record this region, the removal of such a point does not delete any legitimate DR but enhances the search for other DRs by the BasicDRIFT. After we remove all the isolated points recursively, the ExtendedBasicDRIFT is run on the reduced data matrix to find all remaining DRs.

Algorithm: DRIFT(X, v)
Repeat
    Start the BasicDRIFT at every point having value v
    Record all the regions found that are legitimate DRs
    Set the entries in X corresponding to the identified isolated points to be ∞
Until no further isolated point is found
Start the ExtendedBasicDRIFT at every point in the updated X having value v
Record all the regions found that are legitimate DRs

We remark that, a DR is "legitimate" if it is not a subset of any previously found DR. Moreover, one might want to discard DRs with small size. To do so, one may define a DR to be "illegitimate" if its size is below a user-specified threshold and thus it is not inserted into the output sequence.

4 Extending the DRIFT for Data Streams

In applications where data are generated in the form of continuous, rapid data streams, such as clickstreams, an important consideration is how to update the changing dense regions efficiently under such a stream environment. Our data model is as follows. The entity-by-attribute table has its size fixed for all time but its entries are changing. Moreover, at each instant, only one entry in the table is allowed to change.

A naive way to obtain an updated set of DRs is to simply apply the DRIFT algorithm to the most recent table (according to a prescribed time schedule), we call this approach Snapshot Update (SSU). Obviously, such a method requires massive amount of computations where, very often, many of them are redundant.

To design a better way for updating, we first observe that in data stream environments, only newly formed DRs are interesting. Thus, we only consider a change where $X(s,t)$ is changed from not-equal-to-v to v. Moreover, only the DRs containing the entry (s,t) is updated and output. Such an approach greatly reduces the update cost and makes it more possible to be point-triggered, i.e., find new DRs every time when an entry is changed.

Algorithm: PointTriggerUpdate(s,t) /* input a changed point of X */
Run BasicDRIFT(X,s,t) to obtain two (may be identical) new dense regions
If (s,t) is not an isolated point
    Run ExtendedBasicDRIFT(X,s,t) to obtain the remaining DRs
EndIf
Return All dense regions containing the changed entry (s,t)

---

5 Experimental Results

We employ the web-log data from http://espnstar.com.cn to test the performance of the DRIFT and the PointTriggerUpdate (PTU) algorithms. We use two months' web-log data to do the experiments. Table 1 lists the datasets for our experiments where ES1 and ES2 are the log datasets during December, 2002. Each data stream contains a set of continuous access sessions during some period of time. Using the cube model purposed [1], we can convert original web-log streams into access session data streams for dense regions discovery.

Table 1. Characteristics of the data streams ES1 and ES2.

Data stream	# Accesses	# Sessions	# Visitors	# Pages
ES1	78,236	5,000	120	236
ES2	7,691,105	669,110	51,158	1,609

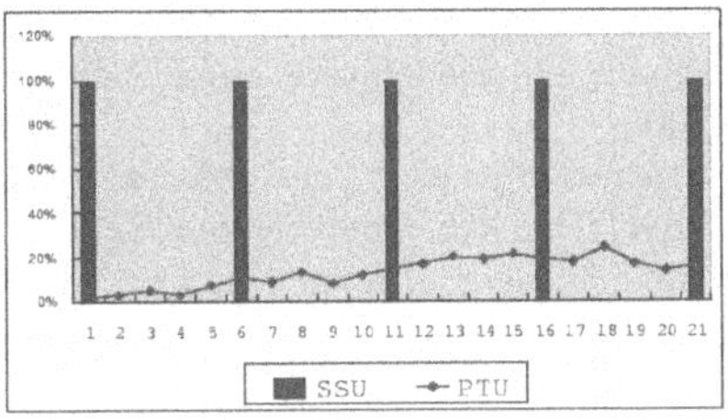

Fig. 1. Percentage of DRs updated by PTU and SSU during a 4-hour period.

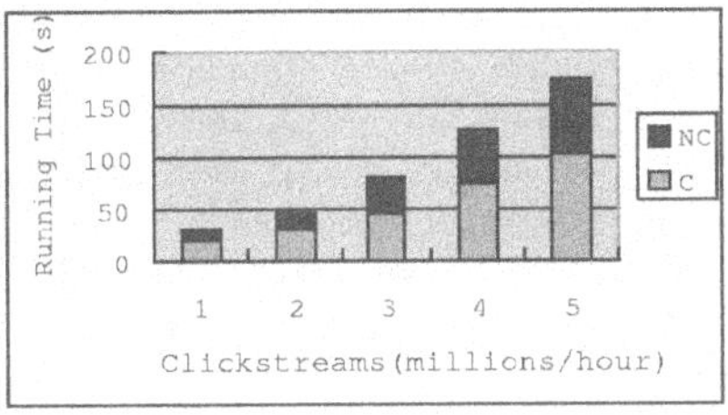

Fig. 2. Running time (in sec.) of PTU vs. stream size.

Experiment 1. This experiment is to compare the cost of the two update methods discussed in Section 4. We simulate data stream updates by using the ES1 dataset which contains continuous user accesses during a peak visiting period of four hours. SSU is employed only at the beginning of each hour and at the end of the 4th hour while PTU is performed whenever there is a new arrival tuple. What we compare is the percentage of DRs needed to be regenerated. Obviously, SSU updates all the DRs no matter whether they are changed or not. The experimental results suggeste that PTU is more adaptive than SSU to update DRs in a data stream environment, see Fig.1. On average, the update cost by PTU is just around 16% of that by SSU meaning that most of the updates by SSU are redundant and wasted. Moreover, PTU can response to the peak period (the 18th time slot in Fig.1) for updates while SSU has to wait until the 21st time slot.

Experiment 2. The next experiment is to employ the largest dataset ES2 to test the scalability of the DRIFT with PTU to update the continuous arriving clickstream data. The experimental results show that both the searching time on dense regions without isolated points ($\mathcal{C}$) and with isolated points ($\mathcal{NC}$) are acceptable, even for several millions of clickstreams per hour, see Fig. 2.

6 Conclusion

We demonstrate that dense regions are significant patterns which are useful in knowledge discovery. Efficient and effective algorithms are presented for finding and updating DRs. We refer the readers to [5] for theoretical treatments on the subject. Our experiments validate that the DRIFT algorithm is very useful in data stream applications such as online web usage mining. As future works, we would like to develop some further theoretical results characterizing dense regions from which more efficient algorithms may be derived and explore the use of dense regions in other data mining applications.

References

1. J. Huang, M. Ng, W. Ching, J. Ng, and D. Cheung, *A cube model and cluster analysis for web access sessions*, Proc. of the WEBKDD 2001 Workshop, San Francisco, USA, 2001.
2. E. H. Wu, M. K. Ng, and J. Z. Huang, *On improving website connectivity by using web-log data streams*, Proc. of the 9th Intl. Conf. on Database Systems for Advanced Applications (DASFAA 2004), Jeju, Korea, 2004.
3. E. H. Wu and M. K. Ng, *A graph-based optimization algorithm for Website topology using interesting association rules*, Proc. of PAKDD 2003, Seoul, Korea, 2003.
4. C. Yang, U. Fayyad, and P. S. Bradley, *Efficient discovery of error-tolerant frequent itemsets in high dimensions*, Proc. of ACM SIGKDD Intl. Conf. on Knowledge Discovery and Data Mining: San Francisco, California, pp. 194–203, 2001.
5. A. M. Yip, E. H. Wu, M. K. Ng, and T. F. Chan, *An efficient algorithm for dense regions discovery from large-scale data streams* (extended version), UCLA CAM Reports 03-76, Math. Dept., University of California, Los Angeles, CA, 2003.

Blind Data Linkage Using n-gram Similarity Comparisons

Tim Churches[1] and Peter Christen[2*]

[1] Centre for Epidemiology and Research, New South Wales Department of Health, Locked Mail Bag 961, North Sydney NSW 2059, Australia, tchur@doh.health.nsw.gov.au

[2] Department of Computer Science, Australian National University, Canberra ACT 0200, Australia, peter.christen@anu.edu.au

Abstract. Integrating or linking data from different sources is an increasingly important task in the preprocessing stage of many data mining projects. The aim of such linkages is to merge all records relating to the same entity, such as a patient or a customer. If no common unique entity identifiers (keys) are available in all data sources, the linkage needs to be performed using the available identifying attributes, like names and addresses. Data confidentiality often limits or even prohibits successful data linkage, as either no consent can be gained (for example in biomedical studies) or the data holders are not willing to release their data for linkage by other parties. We present methods for confidential data linkage based on hash encoding, public key encryption and n-gram similarity comparison techniques, and show how *blind data linkage* can be performed.

Keywords: Privacy preserving data mining, hash encoding, data matching, public key infrastructure, n-gram indexing.

1 Introduction and Related Work

The ability to find matches between confidential data items held in two (or more) separate databases is an increasingly common requirement for many applications in data processing, analysis and mining. A medical research project, for example, may need to determine which individuals, whose identities are recorded in a population-based register of people suffering from hepatitis C infection, also appear in a separate population-based register of cases of liver cancer. Traditionally the linkage of records in these two databases would require that identified data on every registered individual be copied from one of the databases to the other, or from both databases to a third party (often the researchers or their proxy) [5]. This clearly invades the privacy of the individuals concerned. It is typically infeasible to obtain consent for this invasion of privacy from the individuals identified in each of the register databases – instead one or more ethics

* Corresponding author

H. Dai, R. Srikant, and C. Zhang (Eds.): PAKDD 2004, LNAI 3056, pp. 121–126, 2004.

committees or institutional review boards must consent for the linkage of the two databases on the behalf of the individuals involved.

However, the invasion of privacy could be avoided, or at least mitigated, if there were some method of determining which records in the two databases matched, or were likely to match on more detailed comparison, without either database having to reveal any identifying information to each other or to a third party. Researchers can then use de-identified (or anonymised) versions of the linked records. If the use of anonymised data is not feasible, then at worst only a small subset of records from each of the databases need be given to the researchers, in which case it may be feasible to obtain direct consent from the individuals concerned.

Anonymous data linkage based on SHA [8] hashed identifiers are used in Switzerland [2] and France [7]. In the French system spelling transformations are performed before the identifiers are hashed (with an added *pad* to prevent dictionary attacks), and probabilistic linkage techniques are then applied based on exact matches between the transformed strings.

2 Methods

Alice holds a database, **A**, which contains one or more attributes (columns, variables), denoted **A.a**, **A.b** and so on, containing confidential strings (like names and addresses) or other character or symbol sequences. The need for confidentiality may arise from the fact that the values in **A.a** identify individuals, or because the information has some commercial value. Bob holds a similar but quite separate database, **B**, also containing one or more confidential columns, **B.a**, **B.b** and so on.

Alice and Bob wish to determine whether any of the values in **A.a** match any of the values in **B.a** without revealing to each other or to any other party what the actual values in **A.a** and **B.a** are. The problem is simple when "matching" is defined as exact equality of the pair of strings or sequences being compared – comparisons can be made between secure one-way message authentication digests of the strings, created using algorithms such as the NIST HMAC [1] which in turn uses a secure one-way hashing function such as SHA [8]. However, the problem becomes more complicated if the strings contain errors (for example typographical variations in names), because even a single character difference between the strings will result in very different message digest values in which a majority of bits will be different.

One method of overcoming this problem is to reduce the dimensionality of the secret strings in **A.a** and **B.a** before they are converted into a secure digest, using, for example, a phonetic encoding function such as Soundex [6]. However, Soundex and other phonetic transformations are not perfect – in particular they are not robust to errors in the initial character, and to truncation differences.

Ideally, a protocol is required which permits the blind calculation by a trusted third party (Carol), of a more general and robust measure of similarity between the pairs of secret strings.

2.1 n-gram Similarity Comparison of Secret Strings or Sequences

The protocol assumes that Carol is trusted by Alice and Bob to (a) adhere to the protocol, (b) not reveal information to other parties except where permitted by the protocol, and (c) not try to determine the values of Alice's or Bob's source strings using cryptologic techniques. There is no assumption that Alice trusts Bob or vice versa. Note that Alice and Bob do need to share meta-data about the nature of the information contained in their databases with each other – in order to decide which columns/attributes can be validly compared – but they do not need to share the actual values of those columns, nor summary measures (such as frequency counts) derived from those values. The protocol is as follows.

1. Alice and Bob mutually agree on a secret random key, K_{AB}, which they share only with each other. This can be done using the Diffie-Hellman key agreement protocol [3]. They also agree on a secure one-way message authentication algorithm, k_{hmac}, such as the NIST HMAC [1] which in turn uses a one-way hashing function k_{owh} (e.g. MD5 or SHA [8]). The shared key K_{AB} is used to protect against "dictionary" attacks. Alice and Bob also need to agree on a protocol for preprocessing strings to render them into a standard form (such as converting all characters to lower case, removal or substitution of punctuation and whitespace, and so on).
2. Alice computes a sorted list of bigrams[1] for each preprocessed (as described in step 1 above) value in the column **A.a**. For example, if a value of **A.a** is "Peter" then the sorted list of bigrams is ("er","et","pe","te"). Note that duplicate bigrams are removed, so each bigram is unique in each list. Alice next calculates all possible sub-lists of all lengths greater than zero for each bigram list – in other words, the power-set of bigrams minus the empty set. For the example given above, Alice computes bigram sub-lists ranging from length 1 to 4.

   ("er"), ("et"), ("pe"), ("te"),
   ("er","et"), ("er","pe"), ("er","te"), ("et","pe"), ("et","te"), ("pe","te"),
   ("er","et","pe"), ("er","et","te"), ("er","pe","te"), ("et","pe","te"),
   ("er","et","pe","te")

   If a bigram list contains b bigrams, the resulting number of sub-lists is $2^b - 1$. Alice then transforms each of the calculated bigram sub-lists into a secure hash digest using K_{AB} as the key. These hashes are stored in column **A.a_hash_bigr_comb**. Alice also creates an encrypted version of the record identifier (key) for the string from which each value in **A.a_hash_bigr_comb** was derived – she stores this in **A.encrypt_rec_key**. She also places the length (that is, number of bigrams) of each **A.a_hash_bigr_comb** in a column called **A.a_hash_bigr_comb_len**, and the length (that is, the number of bigrams) of each original secret string in **A.a**, in a column **A.a_len**. Alice then sends the set of quadruplets (**A.a_hash_bigr_comb**,

[1] In this example, bigrams (2-grams, $n = 2$) are used, but the extension to trigrams ($n = 3$) and other n-grams is direct.

A.a_hash_bigr_comb_len, **A.encrypt_rec_key**, **A.a_len**) to Carol. Note that the number of quadruplets is much larger than the number of original records in **A.a**.
3. Bob carries out the same as in step 2 with his column **B.a**, and also sends the resulting set of quadruplets to Carol.
4. Carol determines the set intersection of the values of **A.a_hash_bigr_comb** and **B.a_hash_bigr_comb** which she has been sent by Alice and Bob respectively. For each value of **a_hash_bigr_comb** shared by **A** and **B**, for each unique pairing of (**A.encrypt_rec_key**, **B.encrypt_rec_key**), Carol calculates a bigram score

$$bigr\_score = \frac{2 \cdot \mathbf{A.a\_hash\_bigr\_comb\_len}}{(\mathbf{A.a\_len} + \mathbf{B.a\_len})}$$

and selects the maximum bigram score value for each possible unique pairing of (**A.encrypt_rec_key**, **B.encrypt_rec_key**) – that is, the highest score for each pair of strings from **A.a** and **B.a**. Note that a bigram score of 1.0 corresponds to an exact match between two values.

What happens next depends on the context. Carol may report the number of strings with a bigram score above an agreed threshold to Alice and Bob, who may then negotiate further steps, or Carol may simply report the similarity scores and the encrypted record keys back to Alice and Bob. Alternatively, Carol may send this information to another third party, *David*, who oversees an overarching *blind data linkage* protocol involving a number of different columns from Alice's and Bob's databases (that is, not just **A.a** and **B.a**, but also **A.b** and **B.b**, **A.c** and **B.c** and so on).

Another strategy which would further reduce the risk of misuse of the information sent to Carol would be to have many Carols available, all functionally equivalent, and for Alice and Bob to decide on which of these Carols to use only at the very last moment. This would mean that a potential attacker would need to suborn or compromise a large number of the Carols in order to have a reasonable chance of gaining access to the information provided to one particular Carol by Alice and Bob.

2.2 Blind Data Linkage

The essence of modern data (or record) linkage techniques [4,9] is the independent comparison of a number of partial identifiers (attributes) between pairs of records, and the combination of the results of these comparisons into a compound or summary score (called *matching weight*) which is then judged against some criteria to classify that pair of records as a match (link), a non-match, or as undecided (potential match). Usually the result of the comparison between individual attributes is weighted in some way – in *deterministic* systems these weights are determined heuristically, whereas in *probabilistic* systems the weights are determined statistically based on the relative reliability of that attribute in deciding matches and non-matches [4], and on the relative frequency of the values of that attribute [9]. The classification criteria for the summary scores (matching weights) are often determined heuristically or statistically, using expectation maximisation (EM) techniques [10].

Thus, the first task is to compare each of the partially-identifying attributes and return a similarity score for each pair. We have demonstrated how this can be done blindly in the previous section.

For each of the partially-identifying (or partially-discriminating) attributes, **a**, **b**, ..., $\boldsymbol{i}$, in their databases **A** and **B**, Alice and Bob dispatch the similarity comparison task to different instances of Carol, which we will term Carol_a, Carol_b, ..., Carol_i and so on. Each of these tasks is independent of the others, and should use a different shared secret key K_{AB}. Each instance of Carol sends the results back to another third party (we will call this party *David*) which oversees the entire data linkage task between **A** and **B**. Thus, David accumulates a series of data sets containing comparison values (or similarity scores) from the Carols.

David joins these data sets and forms a sparse matrix where each entry contains all the comparison values for a record pair. It is now a simple matter for David to multiply this matrix by a vector of weights (for each attribute), and then sum across each row to create a summary *matching weight*, which is compared to some criteria (thresholds) which have been determined through EM [10]. By these methods it is possible for David to arrive at a set of blindly linked records – that is, pairs of (**A.encrypt_rec_key**, **B.encrypt_rec_key**).

3 Conclusions and Future Work

In this paper we have presented some methods for blind "fuzzy" linkage of records using hash encoding, public key encryption and n-gram similarity comparison techniques. Proof-of-concept implementations have demonstrated the feasibility of our approach, albeit at the expense of very high data transmission overheads.

References

1. Bellare M., Canetti R., Krawczyk H.: Message authentication using hash functions – the HMAC construction. RSA Laboratories, CryptoBytes 1996, 2:15.
2. Borst, F., Allaert, F.A. and Quantin, C.: The Swiss Solution for Anonymous Chaining Patient Files. MEDINFO 2001.
3. Diffie W., Hellman M.E.: New directions in cryptography. IEEE Trans. Inform. Theory IT22 1976, 6:644654
4. Fellegi, I. and Sunter, A.: A Theory for Record Linkage. Journal of the American Statistical Society, 1969.
5. Kelman, C.W., Bass, A.J. and Holman, C.D.J.: Research use of linked health data – A best practice protocol. ANZ Journal of Public Health, 26:3, 2002.
6. Lait, A.J. and Randell, B.: An Assessment of Name Matching Algorithms, Technical Report, Dept. of Computing Science, University of Newcastle upon Tyne, UK 1993.

7. Quantin, C., Bouzelat, H., Allaert, F.A.A., Benhamiche, A.M., Faivre, J. and Dusserre, L.: How to ensure data quality of an epidemiological follow-up: Quality assessment of an anonymous record linkage procedure. Intl. Journal of Medical Informatics, vol. 49, pp. 117-122, 1998.
8. Schneider, B.: Applied Cryptography. John Wiley & Sons, second edition, 1996.
9. Winkler, W.E.: The State of Record Linkage and Current Research Problems. RR99/03, US Bureau of the Census, 1999.
10. Winkler, W.E.: Using the EM algorithm for weight computation in the Fellegi-Sunter model of record linkage. RR00/05, US Bureau of the Census, 2000.

Condensed Representation of Emerging Patterns

Arnaud Soulet, Bruno Crémilleux, and François Rioult

GREYC, CNRS - UMR 6072, Université de Caen
Campus Côte de Nacre
F-14032 Caen Cédex France
{Forename.Surname}@info.unicaen.fr

Abstract. Emerging patterns (EPs) are associations of features whose frequencies increase significantly from one class to another. They have been proven useful to build powerful classifiers and to help establishing diagnosis. Because of the huge search space, mining and representing EPs is a hard task for large datasets. Thanks to the use of recent results on condensed representations of frequent closed patterns, we propose here an exact condensed representation of EPs. We also give a method to provide EPs with the highest growth rates, we call them strong emerging patterns (SEPs). In collaboration with the Philips company, experiments show the interests of SEPs.

1 Introduction

The characterization of classes and classification are significant fields of research in data mining and machine learning. Initially introduced in [5], emerging patterns (EPs) are patterns whose frequency strongly varies between two data sets (i.e., two classes). EPs characterize the classes in a quantitative and qualitative way. Thanks to their capacity to emphasize the distinctions between classes, EPs enable to build classifiers [6] or to propose a help for diagnosis. Nevertheless, mining EPs in large datasets remains a challenge because of the very high number of candidate patterns.

In this paper, we propose two contributions for the efficient extraction of emerging patterns. One originality of our approach is to take advantage of recent progresses on the condensed representations of closed patterns [9]. Firstly, we propose an exact condensed representation of the emerging patterns for a dataset. Contrary to the borders approach (Section 2) which provides the EPs with a lower bound of their growth rate, this condensed representation easily enables to know the *exact* growth rate for each emerging pattern. Secondly, we propose a method to easily provide the emerging patterns having the best growth rates (we call them " strong emerging patterns "). This work is also justified by requests from providers of data. In our collaboration with the Philips company, one notices in practice a high number of EPs and the strong emerging patterns are particularly useful to bring more synthetic and more exploitable results.

The paper is organized in the following way. Section 2 introduces the context and the required notations. Section 3 provides an exact condensed representation

H. Dai, R. Srikant, and C. Zhang (Eds.): PAKDD 2004, LNAI 3056, pp. 127–132, 2004.

of the emerging patterns and proposes the strong emerging patterns. Finally, Section 4 presents the experimental evaluations.

2 Context and Related Works

Notations and definitions. Let $\mathcal{D}$ be a dataset (Table 1), which is an excerpt of the data used for the search for failures in a production chain (cf. Section 4).

Table 1. Example of a transactional data set

$\mathcal{D}$

Batch	Items						
B_1	C_1		A	B	C	D	
B_2	C_1		A	B	C		
B_3	C_1		A			D	E
B_4		C_2	A	B	C		
B_5		C_2		B	C	D	E
B_6		C_2		B			E

Each line (or *transaction*) of Table 1 represents a batch (noted $B_1, \ldots, B_6$) described by features (or *items*) : $A, \ldots, E$ denote the advance of the batch within the production chain and C_1, C_2 the class values. $\mathcal{D}$ is partitioned here into two datasets $\mathcal{D}_1$ (the right batches) and $\mathcal{D}_2$ (the defective batches). The transactions having item C_1 (resp. C_2) belong to $\mathcal{D}_1$ (resp. $\mathcal{D}_2$). A *pattern* is a set of items (e.g., $\{A, B, C\}$) noted by the string ABC. A transaction t contains the pattern X if and only if $X \subseteq t$.

The concept of emerging patterns is related to the notion of frequency. The frequency of a pattern X in a dataset $\mathcal{D}$ (noted $\mathcal{F}(X, \mathcal{D})$) is the number of transactions of $\mathcal{D}$ which contain X. X is *frequent* if its frequency is at least the frequency threshold fixed by the user. Intuitively, an emerging pattern is a pattern whose frequency increases significantly from one class to another. The capture of contrast between classes brought by a pattern is measured by its growth rate. The *growth rate* of a pattern X from $\mathcal{D} \backslash \mathcal{D}_i$ to $\mathcal{D}_i$ is defined as:

$$GR_i(X) = \frac{|\mathcal{D}| - |\mathcal{D}_i|}{|\mathcal{D}_i|} \times \frac{\mathcal{F}(X, \mathcal{D}_i)}{\mathcal{F}(X, \mathcal{D}) - \mathcal{F}(X, \mathcal{D}_i)} \tag{1}$$

Definition 1 (emerging pattern or EP). *Given threshold $\rho > 1$, a pattern X is said to be an* emerging pattern *from $\mathcal{D} \backslash \mathcal{D}_i$ to $\mathcal{D}_i$ if $GR_i(X) \geq \rho$.*

Let us give some examples from Table 1. With $\rho = 3$, A and $ABCD$ are EPs from $\mathcal{D}_2$ to $\mathcal{D}_1$. Indeed, $GR_1(A) = 3$ and $GR_1(ABCD) = \infty$.

Related Works. Efficient computation of all EPs in high dimensional datasets remains a challenge because the number of candidate patterns is exponential according to the number of items. The naive enumeration of all patterns with their frequencies fails quickly. In addition, the definition of EPs does not provide anti-monotonous constraints to apply a powerful pruning of the search space for methods stemming from the framework of level-wise algorithms [8]. The approach of handling borders, introduced by Dong and al. [5], enables to give a concise description of the emerging patterns. On the other hand, it requires to repeat for all the $\mathcal{D}_i$ the computation of the intervals and it does not provide for each EP its growth rate. Some other approaches exist like [10,7,1].

In the following, we focus on the condensed representation based on closed patterns [4]. Such an approach enables the implementation of powerful pruning criteria during the extraction, which improves the efficiency of algorithms [9,3]. A *closed* pattern in $\mathcal{D}$ is a maximal set of items shared by a set of transactions [2]. The notion of *closure* is linked to the one of closed pattern. The *closure* of a pattern X in $\mathcal{D}$ is $h(X,\mathcal{D}) = \bigcap\{transaction\ t\ in\ \mathcal{D} | X \subseteq t\}$. An important property on the frequency stems from this definition. The closure of X is a closed pattern and $\mathcal{F}(X,\mathcal{D}) = \mathcal{F}(h(X,\mathcal{D}),\mathcal{D})$. In our example, $h(AB,\mathcal{D}) = ABC$ and $\mathcal{F}(AB,\mathcal{D}) = \mathcal{F}(ABC,\mathcal{D})$. Thus, the set of the closed patterns is a condensed representation of all patterns because the frequency of any pattern can be inferred from its closure.

3 Condensed Representation and Strong Emerging Patterns

3.1 Exact Condensed Representation of Emerging Patterns

Let us move now how to get the growth rate of any pattern X. Equation 1 shows that it is enough to compute $\mathcal{F}(X,\mathcal{D})$ and $\mathcal{F}(X,\mathcal{D}_i)$. These frequencies can be obtained from the condensed representation of frequent closed patterns. Indeed, $\mathcal{F}(X,\mathcal{D}) = \mathcal{F}(h(X,\mathcal{D}),\mathcal{D})$ (closure property) and by the definition of the partial bases $\mathcal{D}_i$, $\mathcal{F}(X,\mathcal{D}_i) = \mathcal{F}(XC_i,\mathcal{D}) = \mathcal{F}(h(XC_i,\mathcal{D}),\mathcal{D})$. Unfortunately, these relations require the computation of two closures ($h(X,\mathcal{D})$ and $h(XC_i,\mathcal{D})$), which it is not efficient. The following properties solve this disadvantage:

Property 1. Let X be a pattern and $\mathcal{D}_i$ a dataset, $\mathcal{F}(X,\mathcal{D}_i) = \mathcal{F}(h(X,\mathcal{D}),\mathcal{D}_i)$

Proof. The properties of the closure operator ensure that for any transaction t, $X \subseteq t \Leftrightarrow h(X,\mathcal{D}) \subseteq t$. In particular, the transactions of $\mathcal{D}_i$ containing X are identical to those containing $h(X,\mathcal{D})$ and we have the equality of the frequencies.

It is now simple to show that the growth rate of every pattern X is obtained thanks to the only knowledge of the growth rate of $h(X,\mathcal{D})$:

Property 2. Let X be a pattern, we have $GR_i(X) = GR_i(h(X,\mathcal{D}))$.

Proof. Let X be a pattern. By replacing $\mathcal{F}(X, \mathcal{D})$ with $\mathcal{F}(h(X, \mathcal{D}), \mathcal{D})$ and $\mathcal{F}(X, \mathcal{D}_i)$ with $\mathcal{F}(h(X, \mathcal{D}), \mathcal{D}_i)$ in Equation 1, we immediately recognize the growth rate of $h(X, \mathcal{D})$.

For instance, $h(AB) = ABC$ and $GR_1(AB) = GR_1(ABC) = 2$. This property is significant because the number of closed patterns is lower (and, in general, much lower) than that of all patterns [4]. Thus, the frequent closed patterns with their growth rates are enough to synthesize the whole set of frequent EPs with their growth rates. We obtain an *exact* condensed representation of the EPs (i.e. the growth rate of each emerging pattern is exactly known). Let us recall that the borders technique (cf. Section 2) only gives a lower bound of the growth rate.

3.2 Strong Emerging Patterns

The number of emerging patterns of a dataset can be crippling for their use. In practice, it is judicious to keep only the most frequent EPs having the best growth rates. But thoughtlessly raising these two thresholds may be problematic. On the one hand, if the minimal growth rate threshold is too high, the found EPs tend to be too specific (i.e. too long). On the other hand, if the minimal frequency threshold is too high, EPs have a too low growth rate.

We define here the strong emerging patterns which are the patterns having the best possible growth rates. They are a trade-off between the frequency and the growth rate.

Definition 2 (strong emerging pattern). *A strong emerging pattern X (SEP in summary) for $\mathcal{D}_i$ is the emerging pattern coming from a closed pattern XC_i in $\mathcal{D}_i$ (i.e. C_i does not belong to the SEP).*

SEPs enable to highlight the following important property (due to space limitation the proof is not given here).

Property 3 (SEPs: EPs with maximum growth rate). Let X be a pattern not containing the item C_i. Then the SEP coming from $h(X, \mathcal{D}_i)$ has a better growth rate than X, i.e. one has $GR_i(X) \leq GR_i(h(X, \mathcal{D}_i) \backslash \{C_i\})$.

Let us illustrate Property 3 on the elementary example. The pattern BC is not a SEP for class 1 (because $h(BC, \mathcal{D}_1) \backslash \{C_1\} = ABC$), its growth rate is 1 and one has well $GR_1(BC) \leq GR_1(ABC) = 2$ and $\mathcal{F}(BC, \mathcal{D}_1) = \mathcal{F}(ABC, \mathcal{D}_1)$.

Compared to EPs, SEPs have two meaningful advantages: they are easy to discover from the condensed representation of frequent closed patterns (by simply filtering those containing C_i), and they have the best possible growth rates. Let us notice that the emerging patterns based on X and $h(X, \mathcal{D}_i)$ have the same frequency, thus they have the same quality according to this criterion. However, the strong emerging pattern coming from $h(X, \mathcal{D}_i)$ has a stronger (i.e. higher) growth rate and thus offers a better compromise between frequency and growth rate.

4 Experiments

Experiments were carried out on a real dataset within a collaboration with the Philips company. The industrial aim is to identify mistaken tools in a silicon plate production chain. The quality test leads to three quasi-homogeneous classes corresponding to three quality levels. Figure 1 depicts the distributions of EPs according to the length of patterns for a minimal frequency threshold of 1.2%. The number of closed EPs (which stemmed from closed patterns) measures of the size of the condensed representation. Two threshold values of the minimal growth rate (1 and ∞) are used. This figure shows that the number of EPs is very high compared to the number of closed EPs or SEPs. This disproportion does not decrease in spite of the rise of the minimal growth rate. These too large numbers of EPs cannot be presented to an expert for his analysis task. Other experiments have shown that the number of closed EPs and SEPs increases less quickly than the number of EPs when the frequency decreases. So, it is possible to examine longer and less frequent patterns.

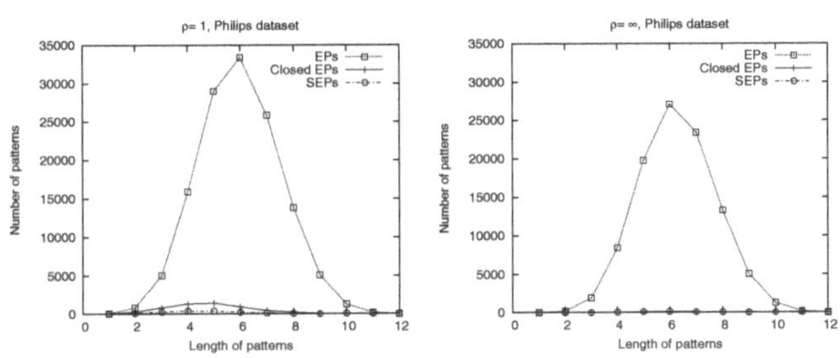

Fig. 1. Comparison between the different kinds of emerging patterns

After talks with the Philips experts, they have confirmed that the stage suspected by the SEPs was the real cause of the failures (an equipment was badly tuned). This experiment shows the practical contribution of SEPs on real-world data.

5 Conclusion

Based on recent results in condensed representations, we have revisited the field of emerging patterns. We have defined an exact condensed representation of emerging patterns for a data base and proposed the strong emerging patterns which are the EPs with the highest growth rates. In addition to the simplicity of their extraction, this approach produces only few SEPs which are particularly useful for helping to diagnosis. So, it is easier to use SEPs than search relevant EPs among a large number of EPs. Dealing with our collaboration with the

Philips company, SEPs enable to successfully identify the failures of a production chain of silicon plates. These promising results encourage the use of strong emerging patterns. Further works concern the use of SEPs for classification tasks.

Acknowledgements. The authors wish to thank Philips company. This work has been partially founded by the AS "Discovery Challenge" supported by CNRS.

References

[1] J. Bailey, T. Manoukian, and K. Ramamohanarao. Fast algorithms for mining emerging patterns. In *PKDD'02*, pages 39–50, 2002.
[2] G. Birkhoff. Lattices theory. *American Mathematical Society, vol. 25*, 1967.
[3] J. F. Boulicaut, A. Bykowski, and C. Rigotti. Free-sets: a condensed representation of boolean data for the approximation of frequency queries. *DMKD journal*, 2003.
[4] T. Calders and B. Goethals. Mining all non-derivable frequent itemsets. In *proceedings of PKDD'02*, 2002.
[5] G. Dong and J. Li. Efficient mining of emerging patterns: Discovering trends and differences. In *Knowledge Discovery and Data Mining*, pages 43–52, 1999.
[6] G. Dong, X. Zhang, W. Wong, and J. Li. CAEP: Classification by aggregating emerging patterns. In *Discovery Science*, pages 30–42, 1999.
[7] J. Li and K. Ramamohanarao. The space of jumping emerging patterns and its incremental maintenance algorithms. In *Proc. ICML*, 2000.
[8] H. Mannila and H. Toivonen. Levelwise search and borders of theories in knowledge discovery. *Data Mining and Knowledge Discovery*, 1(3):241–258, 1997.
[9] N. Pasquier, Y. Bastide, T. Taouil, and L. Lakhal. Discovering frequent closed itemsets for association rules. *Lecture Notes in Computer Science*, 1540, 1999.
[10] X. Zhang, G. Dong, and K. Ramamohanarao. Exploring constraints to efficiently mine emerging patterns from large high-dimensional datasets. In *KDD*, 2000.

Discovery of Maximally Frequent Tag Tree Patterns with Contractible Variables from Semistructured Documents

Tetsuhiro Miyahara[1], Yusuke Suzuki[2], Takayoshi Shoudai[2], Tomoyuki Uchida[1], Kenichi Takahashi[1], and Hiroaki Ueda[1]

[1] Faculty of Information Sciences, Hiroshima City University, Hiroshima 731-3194, Japan
{miyahara@its, uchida@cs, takahasi@its, ueda@its}.hiroshima-cu.ac.jp
[2] Department of Informatics, Kyushu University, Kasuga 816-8580, Japan
{y-suzuki,shoudai}@i.kyushu-u.ac.jp

Abstract. In order to extract meaningful and hidden knowledge from semistructured documents such as HTML or XML files, methods for discovering frequent patterns or common characteristics in semistructured documents have been more and more important. We propose new methods for discovering maximally frequent tree structured patterns in semistructured Web documents by using tag tree patterns as hypotheses. A tag tree pattern is an edge labeled tree which has ordered or unordered children and structured variables. An edge label is a tag or a keyword in such Web documents, and a variable can match an arbitrary subtree, which represents a field of a semistructured document. As a special case, a contractible variable can match an empty subtree, which represents a missing field in a semistructured document. Since semistructured documents have irregularities such as missing fields, a tag tree pattern with contractible variables is suited for representing tree structured patterns in such semistructured documents. First, we present an algorithm for generating all maximally frequent *ordered* tag tree patterns with contractible variables. Second, we give an algorithm for generating all maximally frequent *unordered* tag tree patterns with contractible variables.

1 Introduction

Data model: We use rooted trees as representations of semistructured data such as HTML or XML files, according to Object Exchange Model [1]. In this paper, "ordered" means "with ordered children" and "unordered" means "with unordered children". We consider both ordered trees and unordered trees in order to deal with various semistructured data.

Our approach: To formulate a schema on such tree structured data we have proposed a **tag tree pattern** [5,6,7] (Sec.2). A tag tree pattern is an edge labeled tree which has ordered or unordered children and structured variables. An edge label is a tag or a keyword in Web documents, or a wildcard for any string. A variable can match an arbitrary subtree, which represents a field of a

H. Dai, R. Srikant, and C. Zhang (Eds.): PAKDD 2004, LNAI 3056, pp. 133–144, 2004.

semistructured document. As a special case, a contractible variable can match an empty subtree, which represents a missing field in a semistructured document. Since semistructured data have irregularities such as missing fields, a tag tree pattern with contractible variable is suited for representing tree structured patterns in such semistructured documents. In the examples of tree structured data and tag tree patterns (Fig.1), the variable with label "x" of the tag tree pattern t_2 matches the subtree g_1 and the contractible variable with label "y" of t_2 matches the empty tree g_2.

Novelty of our approach: Graph or tree-based data mining and discovery of frequent structures in graph or tree structured data have been extensively studied [2,3,11,12]. Our target of discovery is neither a simply frequent pattern nor a maximally frequent pattern with respect to syntactic sizes of patterns such as the number of vertices. In order to apply our method to information extraction from heterogeneous semistructured Web documents, our target of discovery is a semantically and *maximally frequent tag tree pattern* (Sec.3) which represents a common characteristic in semistructured documents. "Semantically" means that the maximality is described by the descriptive power (or *language* in Sec.2) of tree structured patterns.

Data mining problems: In this paper, we consider the following data mining problems. **MFOTTP** (resp. **MFUTTP**) (Sec.3) is a problem to generate all Maximally Frequent Ordered (resp. Unordered) Tag Tree Patterns with frequency above a user-specified threshold from a given set of ordered (resp. unordered) semistructured data. Consider the examples in Fig. 1. For a set of semistructured data $\{T_1, T_2, T_3\}$, the tag tree pattern t_2 is a maximally $\frac{2}{3}$-frequent ordered tag tree pattern. In fact, t_2 explains T_2 and T_3, but t_2 does not explain T_1. The tag tree pattern t_1 also explains T_2 and T_3. But t_1 explains any tree with two or more vertices and t_1 is overgeneralized and meaningless. So semantic maximality of desired tag tree patterns is important.

Main results: In Sec.4, we present an algorithm **GEN-MFOTTP** for generating all maximally frequent *ordered* tag tree patterns with contractible variables. In Sec.5, we give an algorithm **GEN-MFUTTP** for generating all maximally frequent *unordered* tag tree patterns with contractible variables.

Related works: A tag tree pattern is different from other representations of tree structured patterns such as in [2,3,11] in that a tag tree pattern has structured variables which can match arbitrary trees and a tag tree pattern represents not a substructure but a whole tree structure. As for our previous works, we proposed a method for generating all maximally frequent *unordered* (resp. *ordered*) tag tree patterns *without* contractible variables [5] (resp. [7]) by using an algorithm for generating unordered (resp. ordered) trees [4] (resp. [9]) and our algorithm for maximality test. Also we gave a polynomial time algorithm for finding *one* of the least generalized *ordered* tag tree patterns with contractible variables [6]. Our algorithms in this paper use polynomial time matching algorithms for ordered and unordered term trees with contractible variables [10] to compute the frequency of tag tree patterns.

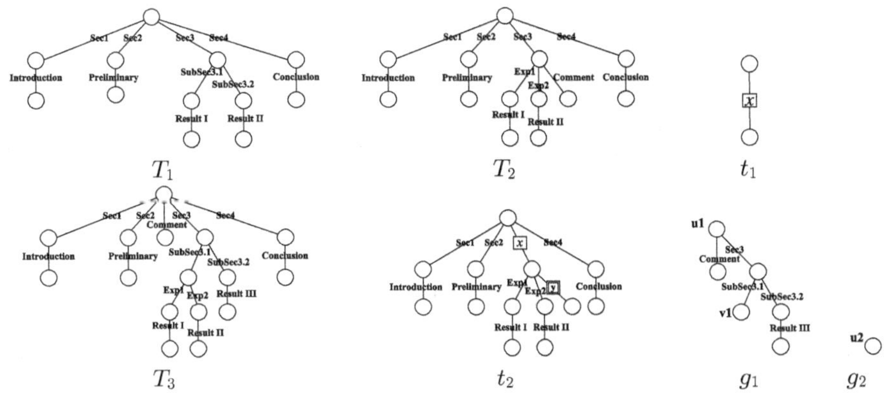

Fig. 1. Tag tree patterns t_1, t_2, and trees T_1, T_2, T_3, g_1, g_2. An uncontractible (resp. contractible) variable is represented by a single (resp. double) lined box with lines to its elements. The label inside a box is the variable label of the variable.

2 Preliminaries

Definition 1. (Ordered term trees and unordered term trees) Let $T = (V_T, E_T)$ be a rooted tree with ordered children or unordered children, which has a set V_T of vertices and a set E_T of edges. We call a rooted tree with ordered (resp. unordered) children an **ordered tree** (resp. an **unordered tree**). Let E_g and H_g be a partition of E_T, i.e., $E_g \cup H_g = E_T$ and $E_g \cap H_g = \emptyset$. And let $V_g = V_T$. A triplet $g = (V_g, E_g, H_g)$ is called an **ordered term tree** if T is an ordered tree, and called an **unordered term tree** if T is an unordered tree. We call an element in V_g, E_g and H_g a *vertex*, an *edge* and a *variable*, respectively.

Below we say a **term tree** or a **tag tree pattern** if we do not have to distinguish between "ordered" and "unordered" ones. We assume that all variable labels in a term tree are different. Λ and X denote a set of edge labels and a set of variable labels, respectively, where $\Lambda \cap X = \phi$. We use a notation $[v, v']$ to represent a variable $\{v, v'\} \in H_g$ such that v is the parent of v'. Then we call v the *parent port* of $[v, v']$ and v' the *child port* of $[v, v']$.

Let X^c be a distinguished subset of X. We call variable labels in X^c *contractible variable labels*. A contractible variable label can be attached to a variable whose child port is a leaf. We call a variable with a contractible variable label a **contractible variable**, which is allowed to substitute a tree with a singleton vertex. We state the formal definition later. We call a variable which is not a contractible variable an **uncontractible variable**. In order to distinguish contractible variables from uncontractible variables, we denote by $[v, v']^c$ (resp. $[v, v']^u$) a contractible variable (resp. an uncontractible variable).

For an ordered term tree g, all children of every internal vertex u in g have a total ordering on all children of u. The ordering on the children of u of an ordered term tree g is denoted by $<_u^g$. Let $f = (V_f, E_f, H_f)$ and $g = (V_g, E_g, H_g)$ be two ordered (resp. unordered) term trees. We say that f and g are *isomorphic*, if there is a bijection φ from V_f to V_g which satisfies the following conditions (1)–(4) (resp. (1)–(3)): (1) The root of f is mapped to the root of g by φ. (2) $\{u,v\} \in E_f$ if and only if $\{\varphi(u),\varphi(v)\} \in E_g$ and the two edges have the same edge label. (3) $[u,v] \in H_f$ if and only if $[\varphi(u),\varphi(v)] \in H_g$, in particular, $[u,v]^c \in H_f$ if and only if $[\varphi(u),\varphi(v)]^c \in H_g$. (4) If f and g are ordered term trees, for any internal vertex u in f which has more than one child, and for any two children u' and u'' of u, $u' <_u^f u''$ if and only if $\varphi(u') <_{\varphi(u)}^g \varphi(u'')$.

Let g be a term tree and x be a variable label in X. Let $\sigma = [u,u']$ be a list of two vertices in g where u is the root of g and u' is a leaf of g. The form $x := [g,\sigma]$ is called a *binding* for x. If x is a contractible variable label in X^c, g may be a tree with a singleton vertex u and thus $\sigma = [u,u]$. It is the only case that a tree with a singleton vertex is allowed for a binding. Let f and g be two ordered (resp. unordered) term trees. A new ordered (resp. unordered) term tree $f\{x := [g,\sigma]\}$ is obtained by applying the binding $x := [g,\sigma]$ to f in the following way. Let $e = [v,v']$ be a variable in f with the variable label x. Let g' be one copy of g and w,w' the vertices of g' corresponding to u,u' of g, respectively. For the variable $e = [v,v']$, we attach g' to f by removing the variable e from H_f and by identifying the vertices v,v' with the vertices w,w' of g', respectively. If g is a tree with a singleton vertex, i.e., $u = u'$, then v becomes identical to v' after the binding. A *substitution* θ is a finite collection of bindings $\{x_1 := [g_1,\sigma_1],\cdots,x_n := [g_n,\sigma_n]\}$, where x_i's are mutually distinct variable labels in X and g_i's are term trees. The term tree $f\theta$, called the *instance* of f by θ, is obtained by applying the all bindings $x_i := [g_i,\sigma_i]$ on f simultaneously. We define the root of the resulting term tree $f\theta$ as the root of f. Further we have to give a new total ordering $<_v^{f\theta}$ on every vertex v of $f\theta$. These orderings are defined in a natural way. Consider the examples g_1,g_2,t_2 and T_3 in Fig. 1. Let $\theta = \{x := [g_1,[u_1,v_1]], y := [g_2,[u_2,u_2]]\}$ be a substitution. Then the instance $t_2\theta$ of the term tree t_2 by θ is the tree T_3.

Definition 2. Let Λ_{Tag} and Λ_{KW} be two languages which consist of infinitely or finitely many words where $\Lambda_{Tag} \cap \Lambda_{KW} = \emptyset$. Let $\Lambda = \Lambda_{Tag} \cup \Lambda_{KW}$. We call a word in Λ_{Tag} a **tag** and a word in Λ_{KW} a **keyword**. An **ordered** (resp. **unordered**) **tag tree pattern** is an ordered (resp. unordered) term tree such that each edge label on it is any of a tag, a keyword, and a special symbol "?". Let $\Lambda_?$ be a subset of Λ. The symbol "?" is a wildcard for any word in $\Lambda_?$. A tag tree pattern with no variable is called a **ground tag tree pattern**.

For an edge $\{v,v'\}$ of a tag tree pattern and an edge $\{u,u'\}$ of a tree, we say that $\{v,v'\}$ *matches* $\{u,u'\}$ if the following conditions (1)-(3) hold: (1) If the edge label of $\{v,v'\}$ is a tag, then the edge label of $\{u,u'\}$ is the same tag or another tag which is considered to be identical with the tag on $\{u,u'\}$. (2) If the edge label of $\{v,v'\}$ is a keyword, then the edge label of $\{u,u'\}$ is a keyword and the label of $\{v,v'\}$ appears as a substring in the edge label of $\{u,u'\}$. (3) If the edge label of $\{v,v'\}$ is "?", then the edge label of $\{u,u'\}$ is in $\Lambda_?$. A ground ordered (resp. unordered) tag tree pattern $\pi = (V_\pi,E_\pi,\emptyset)$ *matches* an ordered (resp. unordered) tree $T = (V_T,E_T)$ if there exists a bijection φ from V_π to V_T

which satisfies the following conditions (1)–(4) (resp. (1)–(3)): (1) The root of π is mapped to the root of T by φ. (2) $\{v, v'\} \in E_\pi$ if and only if $\{\varphi(v), \varphi(v')\} \in E_T$. (3) For all $\{v, v'\} \in E_\pi$, $\{v, v'\}$ matches $\{\varphi(v), \varphi(v')\}$. (4) If π and T are ordered, for any internal vertex u in π which has more than one child, and for any two children u' and u'' of u, $u' <_u^\pi u''$ if and only if $\varphi(u') <_{\varphi(u)}^T \varphi(u'')$. A tag tree pattern π **matches** a tree T if there exists a substitution θ such that $\pi\theta$ is a ground tag tree pattern and $\pi\theta$ matches T.

$\mathcal{OT}_\Lambda$ (resp. $\mathcal{UT}_\Lambda$) denotes the set of all ordered (resp. unordered) trees whose edge labels are in Λ. $\mathcal{OTTP}_\Lambda^c$ (resp. $\mathcal{UTTP}_\Lambda^c$) denotes the set of all ordered (resp. unordered) tag tree patterns with contractible and uncontractible variables whose tags and keywords are in Λ. For π in $\mathcal{OTTP}_\Lambda^c$ (resp. $\mathcal{UTTP}_\Lambda^c$), the *language* $L_\Lambda(\pi)$ is defined as $\{$a tree T in $\mathcal{OT}_\Lambda$(resp. $\mathcal{UT}_\Lambda$) $\mid \pi$ matches $T\}$.

3 Data Mining Problems

Data mining setting: A *set of ordered (*resp. *unordered) semistructured data* $\mathcal{D} = \{T_1, T_2, \ldots, T_m\}$ is a set of ordered (resp. unordered) trees. Let $\Lambda_\mathcal{D}$ be the set of all edge labels of trees in $\mathcal{D}$. The *matching count* of an ordered (resp. unordered) tag tree pattern π w.r.t. $\mathcal{D}$, denoted by $match_\mathcal{D}(\pi)$, is the number of ordered (resp. unordered) trees $T_i \in \mathcal{D}$ $(1 \le i \le m)$ such that π matches T_i. Then the *frequency* of π w.r.t. $\mathcal{D}$ is defined by $supp_\mathcal{D}(\pi) = match_\mathcal{D}(\pi)/m$. Let σ be a real number where $0 < \sigma \le 1$. A tag tree pattern π is **σ-frequent** w.r.t. $\mathcal{D}$ if $supp_\mathcal{D}(\pi) \ge \sigma$. Let Π denotes $\mathcal{OTTP}_\Lambda^c$ or $\mathcal{UTTP}_\Lambda^c$, and $\Lambda' \subseteq \Lambda_{Tag} \cup \Lambda_{KW} \cup \{?\}$. We denote by $\Pi(\Lambda')$ the set of all tag tree patterns $\pi \in \Pi$ such that all edge labels of π are in Λ'. Let Tag be a finite subset of Λ_{Tag} and KW a finite subset of Λ_{KW}. An ordered (resp. unordered) tag tree pattern π in $\mathcal{OTTP}_\Lambda^c(Tag \cup KW \cup \{?\})$ (resp. $\mathcal{UTTP}_\Lambda^c(Tag \cup KW \cup \{?\})$) is **maximally σ-frequent** w.r.t. $\mathcal{D}$ if (1) π is σ-frequent, and (2) if $L_\Lambda(\pi') \subsetneq L_\Lambda(\pi)$ then π' is not σ-frequent for any tag tree pattern π' in $\mathcal{OTTP}_\Lambda^c(Tag \cup KW \cup \{?\})$ (resp. $\mathcal{UTTP}_\Lambda^c(Tag \cup KW \cup \{?\})$).

All Maximally Frequent Ordered Tag Tree Patterns (MFOTTP)
Input: A set of ordered semistructured data $\mathcal{D}$, a threshold $0 < \sigma \le 1$, and two finite sets of edge labels Tag and KW.
Assumption: $\Lambda_\mathcal{D} \subseteq \Lambda_? \subsetneq \Lambda$.
Problem: Generate all maximally σ-frequent ordered tag tree patterns w.r.t. $\mathcal{D}$ in $\mathcal{OTTP}_\Lambda^c(Tag \cup KW \cup \{?\})$.

All Maximally Frequent Unordered Tag Tree Patterns (MFUTTP)
Input: A set of unordered semistructured data $\mathcal{D}$, a threshold $0 < \sigma \le 1$, and two finite sets of edge labels Tag and KW.
Assumption: $\Lambda_\mathcal{D} \subsetneq \Lambda_? \subsetneq \Lambda$, and the cardinality of both $\Lambda - \Lambda_?$ and $\Lambda_? - \Lambda_\mathcal{D}$ is infinite.
Problem: Generate all maximally σ-frequent unordered tag tree patterns w.r.t. $\mathcal{D}$ in $\mathcal{UTTP}_\Lambda^c(Tag \cup KW \cup \{?\})$.

4 Generating All Maximally Frequent Ordered Tag Tree Patterns

We give an algorithm GEN-MFOTTP which generates all maximally σ-frequent *ordered* tag tree patterns. Let $\mathcal{D}$ be an input set of ordered semistructured data. In the following procedure SUB-MFOTTP, we use the algorithm for generating all ordered trees [2]. A *tag tree tree pattern with only uncontractible variables* is a tag tree pattern consisting of only vertices and uncontractible variables. We regard an ordered tag tree pattern with only uncontractible variables as an ordered tree with the same tree structure. By using the same parent-child relation as in [2], we can enumerate without any duplicate all ordered tag tree patterns with only uncontractible variables in a way of depth first search from general to specific and backtracking. Although the semantics of matching of tree structured patterns and tree structured data is different from that in [2], a parent pattern is more general than its child patterns in the generating process of ordered tag tree tree patterns with only uncontractible variables.

Algorithm GEN-MFOTTP;
begin
  $\Pi(\sigma) := \emptyset$; $\pi := (\{v, v'\}, \emptyset, [v, v']^u)$; SUB-MFOTTP($\pi$); **return** $\Pi(\sigma)$;
end.

Procedure SUB-MFOTTP(π);
begin
  if π is not σ-frequent w.r.t. $\mathcal{D}$ **then return else** BASIC-MFOTTP(π);
  foreach child tag tree pattern π' of π **do** SUB-MFOTTP(π');
end;

Procedure BASIC-MFOTTP(π)
begin

Step 1. **Generate σ-frequent tag tree patterns**:
Let $H_\pi = \{h_1, \ldots, h_k\}$ be the variable set of π. We perform procedure SUBSTITUTION-OT(π, h_1, k).

Step 2. **Eliminate redundancy**:
For each $\pi \in \Pi(\sigma)$, if there exists a pair of contractible variables $[u, v]^c$ and $[u, v']^c$ such that v' is the immediately right sibling of v, then we remove π from $\Pi(\sigma)$.

Step 3. **Maximality test1**:

$\theta_X(x) = \{x := [T_X, [R_X, L_X]]\}$
$X \in \{A, B, C, D, E, F\}$

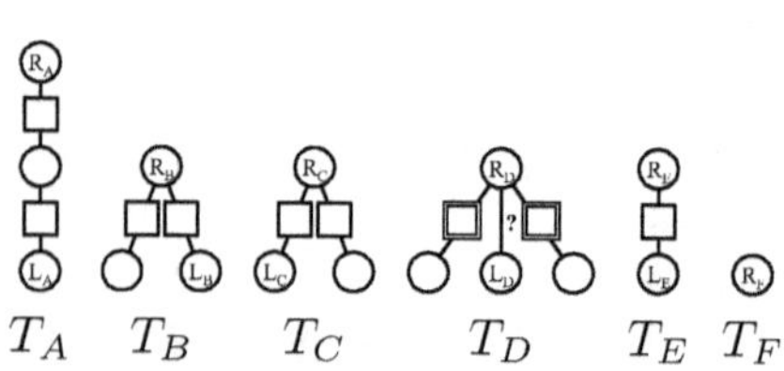

For each $\pi \in \Pi(\sigma)$, if there exists an uncontractible (resp. contractible) variable x in π such that $\pi\theta_X(x)$ is σ-frequent w.r.t. $\mathcal{D}$ for any $X \in \{A, B, C, D\}$

(resp. $X \in \{E, F\}$), then π is not maximally σ-frequent w.r.t. $\mathcal{D}$, and we remove π from $\Pi(\sigma)$.

Step 4. **Maximality test2:**
If there exists an edge with "? " in π such that a tag tree pattern obtained from π by replacing the edge with an edge which has a label in $Tag \cup KW$ is σ-frequent w.r.t $\mathcal{D}$, then π is not maximally σ-frequent w.r.t. $\mathcal{D}$, and we remove π from $\Pi(\sigma)$.

end;

Procedure SUBSTITUTION-OT(π, h_i, k);
begin
  if $i = k + 1$ **then begin** $\Pi(\sigma) := \Pi(\sigma) \cup \{\pi\}$; **return end**;
  If the child port of h_i is not a leaf **then** SUBSTITUTION-OT(π, h_{i+1}, k);
  VARIABLE-REPLACING-OT(π, h_i, k) (Fig. 2);
  return;
end;

For ordered (resp. unordered) tag tree patterns $g = (V, E, H)$ and $g' = (V', E', H')$, we say that g' is an *ordered* (resp. *unordered*) *tag subtree pattern* of g if $V' \subseteq V$, $E' \subseteq E$, and $H' \subseteq H$. For an ordered (resp. unordered) tag tree pattern π'_i in Fig. 3 (resp. Fig. 5), an *occurrence* of π'_i in g is an ordered (resp. unordered) tag subtree pattern of g which is isomorphic to π'_i. The digit in a box $\boxed{k}$ (resp. $\boxed{\geq k}$) near u shows that the number of children of u, which connect to u with edges or uncontractible variables, is equal to k (resp. is more than or equal to k) (Fig. 3, 5).

Lemma 1. *Let π_i and π'_i $(1 \leq i \leq 4)$ be tag tree patterns described in Fig. 3. Let π' be an ordered tag tree pattern which has at least one occurrence of π'_i $(1 \leq i \leq 4)$. For one of occurrences of π'_i, we make a new ordered tag tree pattern π by replacing the occurrence of π'_i with π_i. Then $L_\Lambda(\pi) = L_\Lambda(\pi')$.*

An ordered tag tree pattern π is said to be a *canonical ordered tag tree pattern* if π has no occurrence of π'_i $(1 \leq i \leq 4)$ (Fig. 3). Any ordered tag tree pattern π is transformed into the canonical ordered tag tree pattern by replacing all occurrences of π'_i with π_i $(1 \leq i \leq 4)$ repeatedly. We denote by $c(\pi)$ the canonical ordered tag tree pattern transformed from π. We note that $L_\Lambda(c(\pi)) = L_\Lambda(\pi)$.

Lemma 2. *Let $\Lambda = \Lambda_{Tag} \cup \Lambda_{KW}$. Let $\pi = (V_\pi, E_\pi, H_\pi)$ be an input tag tree pattern which is decided to be maximally σ-frequent w.r.t. $\mathcal{D}$ by* GEN-MFOTTP. *If there is a tag tree pattern $\pi' = (V_{\pi'}, E_{\pi'}, H_{\pi'})$ which is σ-frequent w.r.t. $\mathcal{D}$ and $L_\Lambda(\pi') \subseteq L_\Lambda(\pi)$, then $c(\pi) = c(\pi')$.*

Theorem 1. *Algorithm* GEN-MFOTTP *generates all maximally σ-frequent ordered tag tree patterns in canonical form.*

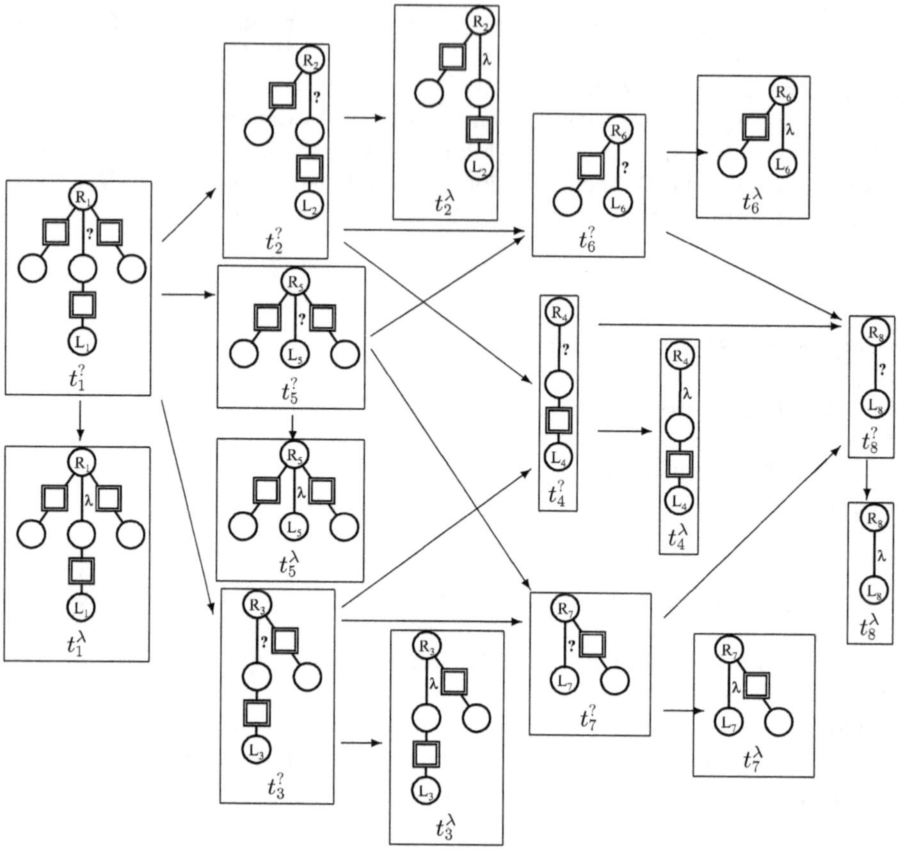

/* Let $\theta_i^?(x) = \{x := [t_i^?, [R_i, L_i]]\}$ $(1 \le i \le 8)$, $\theta_j^\lambda(x) = \{x := [t_j^\lambda, [R_j, L_j]]\}$ $(1 \le j \le 8, \lambda \in Tag \cup KW)$. The tag tree pattern at the end of each arrow is more specific than that at the origin of the arrow. For example, if $\pi\theta_1^?(x)$ is not σ-frequent w.r.t. $\mathcal{D}$, then none of $\pi\theta_1^\lambda(x)$, $\pi\theta_2^?(x)$, $\pi\theta_3^?(x)$, and $\pi\theta_5^?(x)$ are σ-frequent w.r.t. $\mathcal{D}$. */

```
Procedure VARIABLE-REPLACING-OT(π, h_i, k);
begin
  Let x be the variable label of h_i;
  If the child port of h_i is a leaf then Q := {πθ_1^?(x)} else Q := {πθ_5^?(x)};
  while Q ≠ ∅ do begin
    Choose one tag tree pattern πθ_i^a(x) (a ∈ Tag ∪ KW ∪ {?}) from Q;
    Q := Q − {πθ_i^a(x)};
    If πθ_i^a(x) is σ-frequent w.r.t. D then begin
      SUBSTITUTION-OT(πθ_i^a(x), h_{i+1}, k);
      Add all tag tree patterns πθ_j^b(x) (b ∈ Tag ∪ KW ∪ {?}) to Q
      s.t. t_j^b are at the ends of the arrows from t_i^a described in the upper figure.
    end
  end
end;
```

Fig. 2. Procedure VARIABLE-REPLACING-OT(π, h_i, k).

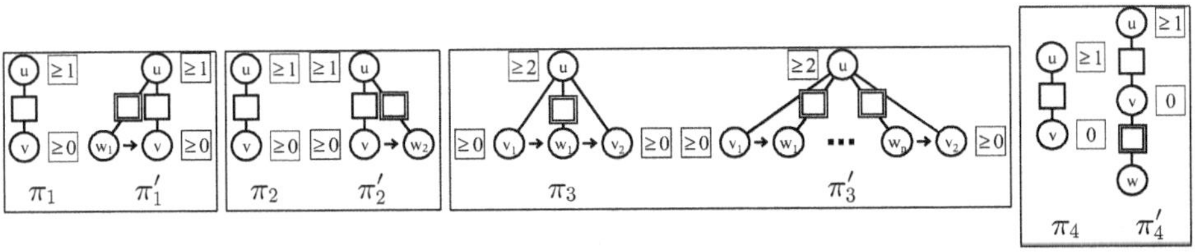

Fig. 3. For $i = 1, 2, 3, 4$, $L_\Lambda(\pi_i) = L_\Lambda(\pi'_i)$. An arrow shows that the right vertex of it is the immediately right sibling of the left vertex.

5 Generating All Maximally Frequent Unordered Tag Tree Patterns

We give an algorithm GEN-MFUTTP which generates all maximally σ-frequent *unordered* tag tree patterns. Let $\mathcal{D}$ be an input set of unordered semistructured data. The descriptions of GEN-MFUTTP, SUB-MFUTTP and SUBSTITUTION-UT are obtained from GEN-MFOTTP,SUB-MFOTTP and SUBSTITUTION-OT, respectively, by replacing "MFOTTP" and "OT" with "MFUTTP" and "UT", respectively. In SUB-MFUTTP, we use the algorithm for generating all unordered trees and the parent-child relation in [3,8] in order to implement the depth first search from general to specific and backtracking.

Procedure BASIC-MFUTTP(π)
begin

Step 1. **Generate σ-frequent tag tree patterns:**
Let $H_\pi = \{h_1, \ldots, h_k\}$ be the variable set of π. We perform procedure SUBSTITUTION-UT(π, h_1, k).

Step 2. **Eliminate redundancy**:
For each $\pi \in \Pi(\sigma)$, if there exists a vertex u having two or more contractible variables such that the parent port of it is u, then we remove π from $\Pi(\sigma)$.

Step 3. **Maximality test1**:

$\theta_X(x) = \{x := [T_X, [R_X, L_X]]\}$
$X \in \{A, B, C, D, E\}$

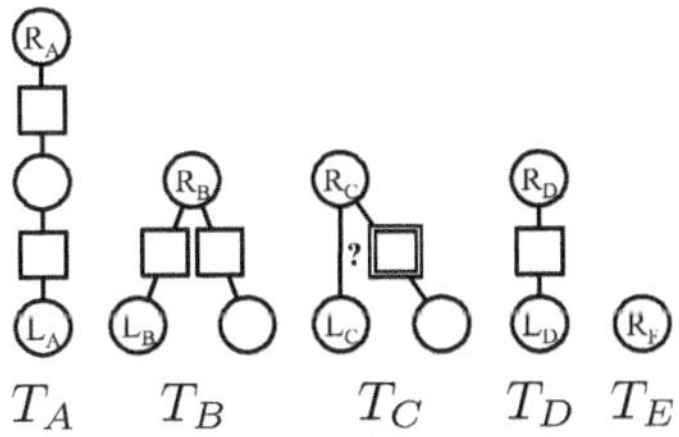

For each $\pi \in \Pi(\sigma)$, if there exists an uncontractible (resp. contractible) variable x in π such that $\pi\theta_X(x)$ is σ-frequent w.r.t. $\mathcal{D}$ for any $X \in \{A, B, C\}$ (resp. $X \in \{D, E\}$), then π is not maximally σ-frequent w.r.t. $\mathcal{D}$, and we remove π from $\Pi(\sigma)$.

Step 4. **Maximality test2**:
If there exists an edge with "? " in π such that a tag tree pattern obtained from π by replacing the edge with an edge which has a label in $Tag \cup KW$ is σ-frequent w.r.t $\mathcal{D}$, then π is not maximally σ-frequent w.r.t. $\mathcal{D}$, and we remove π from $\Pi(\sigma)$.

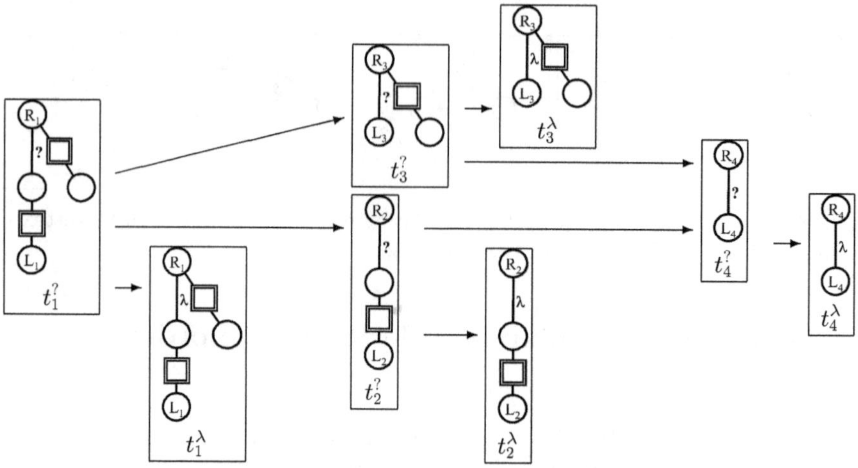

/* Let $\theta_i^?(x) = \{x := [t_i^?, [R_i, L_i]]\}$ $(1 \le i \le 4)$, $\theta_j^\lambda(x) = \{x := [t_j^\lambda, [R_j, L_j]]\}$ $(1 \le j \le 4, \lambda \in Tag \cup KW)$. */

Procedure VARIABLE-REPLACING-UT(π, h_i, k);
begin
  Let x be the variable label of h_i;
  If the child port of h_i is a leaf **then** $Q := \{\pi\theta_1^?(x)\}$ **else** $Q := \{\pi\theta_3^?(x)\}$;
  while $Q \neq \emptyset$ **do begin**
    Choose one tag tree pattern $\pi\theta_i^a(x)$ $(a \in Tag \cup KW \cup \{?\})$ from Q;
    $Q := Q - \{\pi\theta_i^a(x)\}$;
    If $\pi\theta_i^a(x)$ is σ-frequent w.r.t. $\mathcal{D}$ **then begin**
      SUBSTITUTION-UT($\pi\theta_i^a(x), h_{i+1}, k$);
      Add all tag tree patterns $\pi\theta_j^b(x)$ $(b \in Tag \cup KW \cup \{?\})$ to Q
      s.t. t_j^b are at the ends of the arrows from t_i^a described in the upper figure.
    end
  end
end;

Fig. 4. Procedure VARIABLE-REPLACING-UT(π, h_i, k).

end;

Lemma 3. *Let π_i and π'_i $(1 \le i \le 3)$ be unordered tag tree patterns in Fig. 5. Let π' be an unordered tag tree pattern which has at least one occurrence of π'_i $(1 \le i \le 3)$. For one of occurrences of π'_i, we make a new unordered tag tree pattern π by replacing the occurrence of π'_i with π_i. Then $L_\Lambda(\pi) = L_\Lambda(\pi')$.*

An unordered tag tree pattern π is said to be a *canonical unordered tag tree pattern* if π has no occurrence of π'_i $(1 \le i \le 3)$ (Fig. 5). Any unordered tag tree pattern π is transformed into the canonical unordered tag tree pattern by replacing all occurrences of π'_i with π_i $(1 \le i \le 3)$ repeatedly. We denote by $c(\pi)$ the canonical unordered tag tree pattern transformed from π.

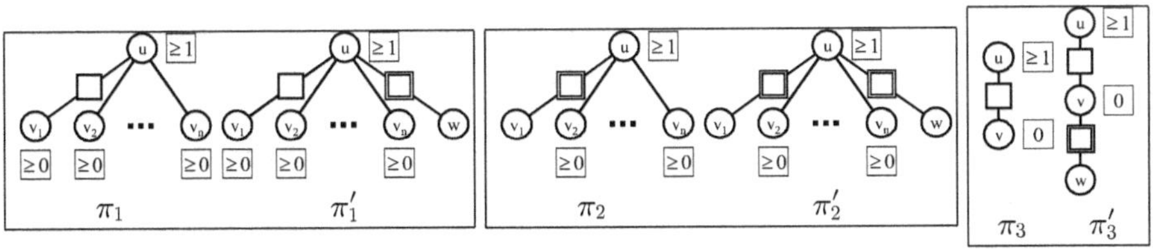

Fig. 5. For $i = 1, 2, 3$ $L_\Lambda(\pi_i) = L_\Lambda(\pi'_i)$. Let u be a vertex of π_i and $v_1, \ldots, v_k$ children of u. We suppose that at least one child among $v_1, \ldots, v_k$ is connected to u by a contractible variable or an uncontractible variable.

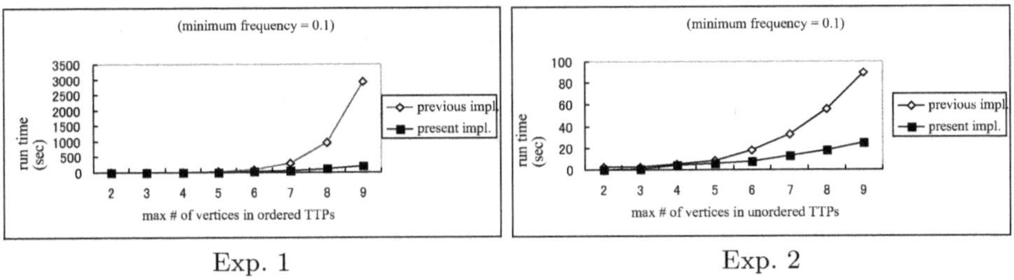

Fig. 6. Experimental results.

Lemma 4. *Let* $\Lambda = \Lambda_{Tag} \cup \Lambda_{KW}$. *Let* $\pi = (V_\pi, E_\pi, H_\pi)$ *be an input tag tree pattern which is decided to be maximally* σ*-frequent w.r.t.* $\mathcal{D}$ *by* GEN-MFUTTP. *If there is a tag tree pattern* $\pi' = (V_{\pi'}, E_{\pi'}, H_{\pi'})$ *which is* σ*-frequent w.r.t.* $\mathcal{D}$ *and* $L_\Lambda(\pi') \subseteq L_\Lambda(\pi)$, *then* $c(\pi) = c(\pi')$.

Theorem 2. *Algorithm* GEN-MFUTTP *generates all maximally* σ*-frequent unordered tag tree patterns in canonical form.*

6 Implementation and Experimental Results

In order to evaluate the performance of the process of searching σ-frequent ordered (resp. unordered) tag tree tree patterns with only uncontractible variables in our algorithms, we have two types of experiments of generating all such patterns by the previous implementation ("previous impl.") and the present implementation ("present impl."). The previous implementation [7] (resp. [5]) cannot prune the search space in the process. The present implementation prunes the search space by modifying GEN-MFOTTP (resp. GEN-MFUTTP). The implementation is by GCL2.2 and on a Sun workstation Ultra-10 clock 333MHz. The sample file is converted from a sample XML file about garment sales data. The sample file consists of 172 tree structured data. The maximum number of vertices over all trees in the file is 11. We can set the maximum number ("max # of vertices in ordered (resp. unordered) TTPs") of vertices of ordered (resp. unordered) tag tree patterns in the hypothesis space. Exp.1 (resp. Exp.2) gives the consumed run time (sec) by the two implementations in case of ordered (resp. unordered) tag tree patterns for the specified minimum frequency =0.1 and varied maximum numbers of vertices of ordered (resp. unordered) tag tree patterns in the hypothesis spaces. These experiments show that the pruning of the search space in the above process is effective.

7 Conclusions

We have studied knowledge discovery from semistructured Web documents such as HTML/XML files. We have presented algorithms for generating all maximally frequent ordered and unordered tag tree patterns with contractible variables. This work is partly supported by Grant-in-Aid for Scientific Research (C) No.13680459 from Japan Society for the Promotion of Science and Grant for Special Academic Research No.2101 from Hiroshima City University.

References

1. S. Abiteboul, P. Buneman, and D. Suciu. *Data on the Web: From Relations to Semistructured Data and XML.* Morgan Kaufmann, 2000.
2. T. Asai, K. Abe, S. Kawasoe, H. Arimura, H. Sakamoto, and S. Arikawa. Efficient substructure discovery from large semi-structured data. *Proc. 2nd SIAM Int. Conf. Data Mining (SDM-2002)*, pages 158–174, 2002.
3. T. Asai, H. Arimura, T. Uno, and S. Nakano. Discovery of frequent substructures in large unordered trees. *Proc. DS-2003, Springer-Verlag, LNAI 2843*, pages 47–61, 2003.
4. T. Beyer and S. Hedetniemi. Constant time generation of rooted trees. *SIAM J. Comput.*, 9:706–712, 1980.
5. T. Miyahara, T. Shoudai, T. Uchida, K. Takahashi, and H. Ueda. Discovery of frequent tree structured patterns in semistructured web documents. *Proc. PAKDD-2001, Springer-Verlag, LNAI 2035*, pages 47–52, 2001.
6. T. Miyahara, Y. Suzuki, T. Shoudai, T. Uchida, S. Hirokawa, K. Takahashi, and H. Ueda. Extraction of tag tree patterns with contractible variables from irregular semistructured data. *Proc. PAKDD-2003, Springer-Verlag, LNAI 2637*, pages 430–436, 2003.
7. T. Miyahara, Y. Suzuki, T. Shoudai, T. Uchida, K. Takahashi, and H. Ueda. Discovery of frequent tag tree patterns in semistructured web documents. *Proc. PAKDD-2002, Springer-Verlag, LNAI 2336*, pages 341–355, 2002.
8. S. Nakano and T. Uno. Efficient generation of rooted trees. *NII Technical Report, NII-2003-005E, National Institute of Infomatics, Japan*, 2003.
9. W. Skarbek. Generating ordered trees. *Theoretical Computer Science*, 57:153–159, 1988.
10. Y. Suzuki, T. Shoudai, S. Matsumoto, T. Uchida, and T. Miyahara. Efficient learning of ordered and unordered tree patterns with contractible variables. *Proc. ALT-2003, Springer-Verlag, LNAI 2842*, pages 114–128, 2003.
11. K. Wang and H. Liu. Discovering structural association of semistructured data. *IEEE Trans. Knowledge and Data Engineering*, 12:353–371, 2000.
12. T. Washio and H. Motoda. State of the art of graph-based data mining. *SIGKDD Explorations*, 5:59–68, 2003.

Mining Term Association Rules for Heuristic Query Construction

Zhenxing Qin, Li Liu, and Shichao Zhang

Faculty of Information Technology, University of Technology, Sydney
PO Box 123, Broadway, Sydney NSW 2007, Australia
{zqin, liliu, zhangsc}@it.uts.edu.au

Abstract. As the Web has become an important channel of information floods, users have had difficulty on identifying what they really want from huge amounts of rubbish-like information provided by the Web search engines when utilizing the Web's low-cost information. This is because most users can only give inadequate (or incomplete) expressions for representing their requirements when querying the Web. In this paper, a heuristic model is proposed for tackling the inadequate query problem. Our approach is based on the potentially useful relationships among terms, called term association rules, in text corpus. For identifying quality information, a constraint is designed for capturing the goodness of queries. The heuristic information in our model assists users in expressing their queries desired.

1 Introduction

User queries to the Web or other information systems are commonly described by using one or more terms as keywords to retrieve information. Some queries might be appropriately given by experienced and knowledgeable users, while others might not be good enough to ensure that those returned results are what the users want. Some users consider that Boolean logic statements are too complicated to be used. Usually, users are not experts in the area in which the information is searched. Therefore, they might lack of the domain-specific vocabulary and the author's preferences of terms used to build the information system. They consequently start searching with generic words to describe the information to be searched for. Sometimes, users are even unsure of what they exactly need in the retrieval. All of these reasons then often lead to uses of incomplete and inaccurate terms for searching. Thus, an information retrieval system should provide tools to automatically help users to develop their search descriptions that match both the need of the user and the writing style of the authors.

One of the approaches to provide the service is the automatic expansion of the queries with some additional terms [1, 2]. These expanded terms for a given query should be semantically or statistically associated with terms in the original query. An expansion method may be using local or global knowledge, i.e. look at the answers to a query or at the entire corpus. An expansion method may be user-interactive or be fully automatic.

H. Dai, R. Srikant, and C. Zhang (Eds.): PAKDD 2004, LNAI 3056, pp. 145–154, 2004.

Global methods use the entire corpus or external knowledge to expand the query. The knowledge used comes in the form of a thesaurus. A thesaurus is a data structure recording associations between terms. Man-made Thesauri such as WordNet [13], whether they are general purpose of domain specific, records linguistic or conceptual associations between the terms as identified by human experts. The expansion is automatic but applies to both the query and the documents.

Some researchers have recently attempted to automatically mine the associations for query expansion from the corpus. Most approaches are based on clustering of terms in the document space. Intuitively, clustering captures synonymous associations. In [Lin 1998] the authors present a global analysis approach based on association rules mining. However, their objective is not to find associations for query expansion but rather to construct a classification of the documents. Moreover, techniques of association rule mining [9, 10] are frequently used for text mining [3, 5, 6, 8] and global query expansion [4, 7].

Local methods are also known as relevance feedback methods. The local set of documents is the set retrieved with an initial unexpanded query. Local methods use the local set to discover the additional candidate terms to be added to the query. In practice local set is kept arbitrarily small compared to the total size of the corpus. The user can indicate manually which documents in the local set are relevant. This step is called relevance feedback. The frequent words in the relevant documents can be used to expand the initial query and/or re-weight the query terms by means of a formula or algorithm, for example the *Rocchio algorithm.* The quality of relevance feedback is the key point of local methods. But not all users like the fussy interactive procedure, and the current algorithms are relatively complicated to common users. Pseudo relevance feedback method just assumes top-ranked documents as the relevance feedback, but it didn't guarantee the quality of relevance feedback.

In this paper, an association analysis approach is proposed for combining the Global and Local information by maintaining a dynamic association rule set for query expansion. It aims to better understanding user queries so as to automatically generate query constructions. In Section 2, the term association analysis and the rule maintenance model are introduced. In Section 3, the structure of our system and different heuristic information used in system are depicted. In Section 4, performance of our method is evaluated based on our experiments.

2 Related Works

2.1 Term Association Rules

Constructing queries with term association rules is to add words to queries from the rules that have query terms in one side with qualified confidence and support. Generally, term association rules are of the form:

$t_1 \rightarrow t_2$, support = s, confidence = c

$s = s(t_1, t_2)$, $c = s(t_1, t_2)/s(t_1)$

where t_1 and t_2 are terms, $s(t_1, t_2)$ is the support of t_1 and t_2 and $s(t_1)$ is the support of t_1.

A rule with a high confidence indicates that term t_2 often occurs in a document where term t_1 occurs. A rule with high support indicates that many examples were found to suggest the rule.

The scope of our investigation is defined by the variation of the range of the two parameters and by the options in using rules of one or more of the forms below given for the query term "nuclear":

"nuclear" → t s, c ;
t → "nuclear" s, c;
t ↔ "nuclear", i.e. t → "nuclear" and "nuclear" → t, s, c1,c2

Of course such rules exist as soon as a term t appears in a document where the term "nuclear" appears. We use the confidence and the support of the rules, indicator of their quality, to select some of them only. We take word "nuclear" in Topic 202 for example to see how it is used.

Example 1:
Nuclear → soviet, (supp = 0.016, conf = 0.4872)
Nuclear → U.S., (supp = 0.0187, conf = 0.5688)
Plutonium → nuclear, (supp = 0.0017,conf = 0.8993)
Reactor → nuclear, (supp = 0.0039,conf = 0.9711)
Weapon ↔ nuclear, (supp = 0.0171, conf1 = 0.2825, conf2 = 0.5194)

Three kinds of rules are as above. We only consider type 1 rules in this paper because user in the interactive procedure confirms the condition words of a query.

2.2 Is There Natural Semantics behind the Rules?

A high confidence rule of the form $t_1 \rightarrow t_2$ indicates that (often) t_2 appears in a document if t_1 appears. This suggests hyper/hyponym or holo/meronym types of relations between the terms. t_2 is equally or more general than t_1 and conversely. Examples mined from our corpus are "Kohl" is a "Chancelor", and "soybean", "corn", and "wheat" are kinds of "grain". The relations found characterize what we could call a contextual holonony: if $t_1 \rightarrow t_2$, then t_1 is part of the vocabulary in the topical context suggested by the concept denoted by t_2. For example we found such associations between the names of "Mandela" and "De Klerk" with the term "Apartheid".
A rule t1 ⇔ t2, i.e. t1 → t2 and t2 → t1 with high and similar respective confidence as well as a high support indicates that t1 and t2 tend to appear together. This suggests a contextual synonymy. An example is the mined association between "conviction" and "sentence".
Many associations were also mined between nouns and adjectives not handled by the stemming algorithm such as "Jew" and "Jewish", "Japan" and "Japanese". There are also many such associations associating the first and last names of personalities: "Saddam" and "Hussein", "Mikail" and "Gorbachev" (provided the first and last name or not ambiguous in the corpus). Of course the synonymy is not proper. The association indicates the similarity of the terms in the topical context such as the association found between the two terms "crisis" and "Gulf" in the 1990 corpus.

Although we have not conducted a formal evaluation of the results obtained, we notice that our subjective evaluation leads to similar conclusion than those presented and substantiated in [4]: it is possible to obtain a topically meaningful thesaurus from the statistical analysis of a corpus. In our case the meaning of the association discovered can be related to the form and parameters of the association rules. A formal evaluation of the semantic quality of the rules could be conducted for instance with a protocol to ask users familiarized with the corpus or the corpus subject matter to assess the rules.

2.3 Maintenance of Association Rules by Weighting

In many applications, the databases are dynamic, i.e., transactions are continuously being added. In particular, we observe that recently added transactions can be more `interesting' than those inserted long ago in the sense that they reflect more closely the current buying trends of customers.

For example, in a toy departmental store, the store will be flooded with purchases of Tarzan and his friends in the period just before, during and immediately after the movie Tarzan was aired in the cinema. From the departmental store's point of view, it will want to capture the sales of these toys accurately to facilitate ordering and sales. On the other hand, some previously popular items in the database may become less popular with very few transactions in the recently added dataset. For example, if the movie Mulan was aired before Tarzan, it is expected that its sales will drop once the movie ended and Tarzan was aired.

Our query construction model is based on the association rules. The association in feedback must be involved in the rule set. We will give simple introduce on the maintenance model which is based on our previous work. Following Figure 1 is the illustration graph of the rule maintenance model. You can check more detail on [10]

3 Methodology and System Architecture

3.1 System Structure

Figure 2 illustrates the system structure that we design for query expansion.

The system consists of four main parts, Interactive Interface, association rule Maintenance module, query constructor, and query processor.

he Interactive Construction Interface shows the heuristic data and last round results. Users can view the heuristic data and take part in the query construction interactively. On the other hand, the results of every round retrieve are also shown here. Users can click on the relevant results or modified the queries. The Interface can gather those feedback information and pass it to rule maintenance module. We will discuss the heuristic data later. The association rule Maintenance module extracts the term associations from feedback data and combines them in the association rule set for query construction. Users can adjust the weight of those new rules.

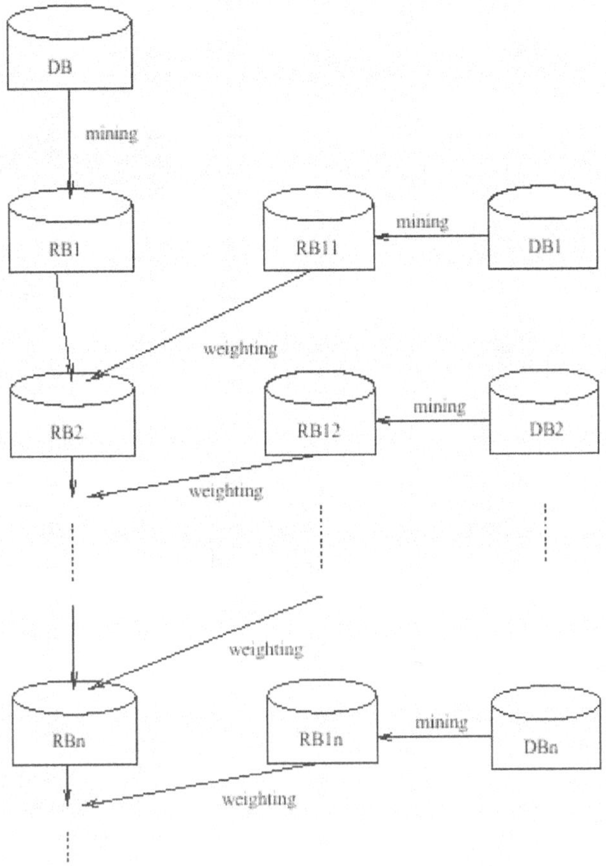

Fig. 1. Maintenance of association rules.

Here, DB: Original Database. DBi (i=1,2,...): incremental data sets to DB;
RBli: Rule base in DBi; RBj: results rule base after maintenance.

The query constructor uses the term association rules and user-input information to construct queries.
The query processor calculates a weighted-cosine similarity between a query and a document.

3.2 Heuristic Data

. Core Query Words
All the words in original query are in the set at first. Users can change the words in the set and see changes in the graph of term association. It is useful for users to find related words and add them to query.

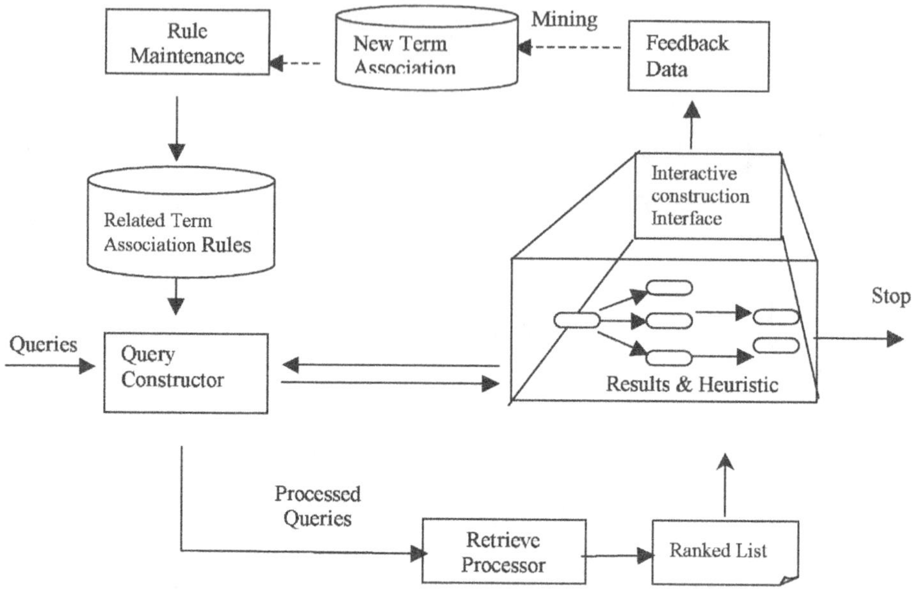

Fig. 2. System structure for heuristic query construction.

. Most Frequent Words in Feedback Data
In the interactive interface, users can click relevant documents as feedback for next round retrieve. Most frequent words in the feedback data are very useful for new query because those documents just be in what the users want to find.

. Exploration of Graph of Term Associations
A high confidence rule of the form $t_1 \rightarrow t_2$ indicates that t_2 often appears in a document if t1 appears. This suggests a certain type of relations between the terms, such as hypernym/hyponym or holonym/meronym, indicating a narrower/broader meaning between the terms. These relations characterize what we could call a contextual holonony, that is, if $t_1 \rightarrow t_2$, then t_1 is part of the vocabulary in the topical context, which is suggested by the concept denoted by t_2. It can help us to understand the latent semantics behind those rules and their effects on the later query expansion. Users can check all words related to the core query words.

3.3 Heuristic Construction Process

In this section, we will show you how the query constructor works. Preprocessed original query is a small set of words Q={Wi | i =1, …, n }. This set is noted as *core set*. Users can operate (add or remove words) the set directly in the interactive interface according to heuristic information. Users also can click the relevant documents of last round retrieve to provide feedback. Maintenance module mines the new rules in the feedback documents and combines them with the old rules. Users can explore the term relation in the rules and operate the core set according to the relations. Association rule set R = {word1 → word2, support, confidence}

Association rule expansion algorithm
Q ← Original query
Q1 = Q
for each rule word1 → word2 in R
  if word1 in Q then Q1=Q1 +{word2}
output Q1

4 The Performance of Query Construction with Association Rules

We used AP90 from TREC4 as our benchmark. The AP90 corpus contains more than 78,000 documents. After stemming and the filtering of stop words, we found more than 133,000 different terms. The average number of different terms in a document is 217. The largest documents contain approximately 900 terms and the smallest 5 terms.

There are 50 queries or topics, called topics in the TREC terminology, associated with AP90. Each topic consists of a natural language proposition or question. For each of these topics, a list of relevant documents has been constructed manually and endorsed by human experts. This list is provided together with the topic. In average, a topic for AP90 calls for 32 documents. The largest topic calls for 145 documents. Two topics call no relevant document at all. Topic 202 is "Status of nuclear proliferation treaties -- violations and monitoring". It calls for 52 documents.

4.1 Rocchio Algorithm

We will compare our association rule expansion method with the Rocchio algorithm [14] that is a classic algorithm of extracting information from relevance feedback. We will briefly describe the Rocchio algorithm here.

The basic idea of the Rocchio algorithm is the combining of document vectors in relevance feedback and original query vectors. It uses both of positive and negative documents to expand the original query. Terms will be assigned positive or negative weights base on whether the terms are in relevant or non-relevant documents on whether the terms are in feedback. Exactly, centroid of relevant feedback is added to the expanded query. The formula of the Rocchio algorithm is as follows [15]:

$$Q_1 = Q_0 + \frac{1}{n_1}\sum_{i=1}^{n_1} R_i - \frac{1}{n_2}\sum_{i=1}^{n_2} S_i$$

where Q_0 is the original query , Q_1 is the modified query , R is the set with relevant documents. And the S is the set with non-relevant documents.

4.2 Performance on Topic 202

In this experiment, we study the effectiveness of feedback mechanisms topic 202. First, we perform the retrieve with original query. We assume all the relevant documents in the top 100 documents are marked as relevant feedback. According to the feedback, the Recchio algorithm is performed.

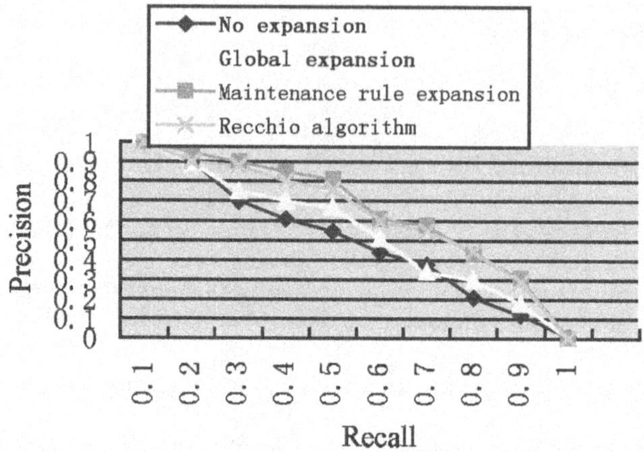

Fig. 3. Precision on topic 202

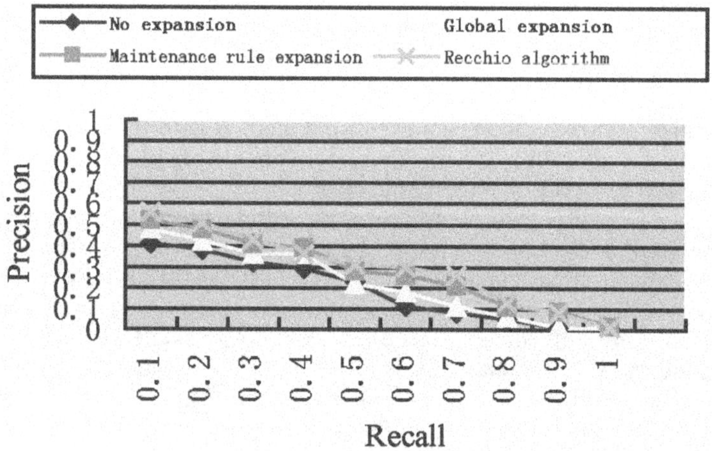

Fig. 4. Precision on all topics

Then we perform the query expansion with global association rules. We also get a ranked list and a relevant feedback from the list. A set of association rules is obtained by mining the relevant documents. After combining with the old global rules using

the maintenance module, we can perform the query expansion again. The retrieve result is shown in Figure 3.
From the Figure 3, both of the lines with relevant feedback have improved the precision significantly, especially the maintenance rule method when recall <0.6.

4.3 Performance on All Topics

Figure 4 is the corresponding graph of average result of all topics. The result is no so exciting as topic 202. The reason is that some queries have very low precisions at the first 100 ranked documents. It means that users cannot catch enough relevant feedback for the query construction.

5 Conclusions and Future Works

In this paper, a new method to handle relevant feedback by association rules maintenance. We present kinds of heuristic information to help users understand and construct their queries interactively. Especially, Users can explore the term associations effectively by changing the words in core query set. Our experimental study showed that the approach could be used as an effective means to represent semantics. We also proposed a new model to integrate the feedback information. The weight rule maintenance model combines different term relations obtained from the feedback chains. Our experiments showed that it is effective for tackling the inadequate query problem.
We plan to extend this work in the following ways. First, since we are mainly concerned with the rules type 1, showing the words can be referred from query word. But the other two kinds of rules should be useful too thought they are difficult to handle currently. We are currently looking into some of these techniques. Second, the efficiency of proposed approach is essentially depended on the relevant document number in the top N ranked list. So how to improve the precision on low recall is the bottleneck of further works. We plan to perform the system on classified text set. More original information is available for users. It will benefit the precision on low recall.

References

[1] E. Efthimiadis. Query expansion. In: Martha E. Williams edited, *Annual Review of Information Scienc and Technology*, Vol. 31, 1996: 121-187.

[2] M. Géry and M. Haddad. Knowledge discovery for automatic query expansion on the World-Wide Web. In: *Proc. of Advances in Conceptual Modeling: ER '99 Workshops*, LNCS 1727, Springer, 1999: 334-347.

[3] H. Haddad, J. Chevallet, M. Bruandet. Relations between Terms Discovered by Association Rules. In: *Proc. of PKDD'2000*, Workshop on Machine Learning and Textual Information Access, Lyon France, September 12, 2000.

[4] J. Wei, S. Bressan, B. Ooi. Mining Term Association Rules for Automatic Global Query Expansion: Methodology and Preliminary Results. *Proc. of WISE'00*. Vol. 1, 2000: 366-373.

[5] R. Feldman, et al. Text Mining at the Term Level. In *Proc. of PKDD'98*, 1998: 65-73.
[6] M. Antonie, O. Zaïane. Text Document Categorization by Term Association. In *Proceedings of ICDM'02*, 2002: 19-26.
[7] J. Wei, Z. Qin, S. Bressan, B. Ooi. Mining Term Association Rules for Automatic Global Query Expansion: A Case Study with Topic 202 from TREC4. In *Proc. of Americas Conference on Information Systems 2000*.
[8] X. Yan, C. Zhang and S. Zhang, Indexing by Conditional Association Semantic. In: *Proceedings of the 14th IRMA International Conference*, Las Vegas, Nevada, USA, April 28-30, 2003.
[9] C. Zhang and S. Zhang, *Association Rules Mining: Models and Algorithms*. Springer-Verlag Publishers in LNCS, Volume 2307, p. 243, 2002.
[10] S. Zhang, C. Zhang and X. Yan, PostMining: Maintenance of Association Rules by Weighting. *Information Systems*. Vol. 28, 7(2003): 691-707.
[11] WordNet - a Lexical Database for English. www.cogsci.princeton.edu/~wn/
[12] http://www.creadersnet.com/newsPool/.
[13] E. Voorhees, Query Expansion using Lexical-Semantic Relations. *ACM-SIGIR*, 1994.
[14] G. Salton, Relevance feedback in information retrieval. In: G.Salton (Ed.), The Smart retrieval system – experiments in automatic document processing. Englewood Cliffs, NJ: Prentice-Hall, 1971.
[15] J. Geertzen, O. Eupen, B. Ariaans, Comparing Rocchino Algorithm and Genetic Algorithm in Relevance Feedback.

FP-Bonsai: The Art of Growing and Pruning Small FP-Trees

Francesco Bonchi[1] and Bart Goethals[2]

[1] Pisa KDD Laboratory, ISTI – CNR, Area della Ricerca di Pisa, Italy
[2] Helsinki Institute for Information Technology (HIIT), Finland

Abstract. In the context of mining frequent itemsets, numerous strategies have been proposed to push several types of constraints within the most well known algorithms. In this paper, we integrate the recently proposed ExAnte data reduction technique within the FP-growth algorithm. Together, they result in a very efficient frequent itemset mining algorithm that effectively exploits monotone constraints.

1 Introduction

The problem of how to push different types of constraints into the frequent itemsets computation has been extensively studied [5,6,3]. However, while pushing anti-monotone constraints deep into the mining algorithm is easy and effective, the case is different for monotone constraints. Indeed, anti-monotone constraints can be used to effectively prune the search space to a small downward closed collection, while the upward closed collection of the search space satisfying the monotone constraints cannot be pruned at the same time. Recently, it has has been shown that a real synergy of these two opposite types of constraints exists and can be exploited by reasoning on both the itemset search space and the input database *together*, using the ExAnte data-reduction technique [2]. This way, pushing monotone constraints does not reduce anti-monotone pruning opportunities, but on the contrary, such opportunities are boosted. Dually, pushing anti-monotone constraints boosts monotone pruning opportunities: the two components strengthen each other recursively. This idea has been generalized in an Apriori-like computation in ExAMiner [1].

In this paper we show how this synergy can be exploited even better within the well known FP-growth algorithm [4]. Thanks to the recursive projecting approach of FP-growth, the ExAnte data-reduction is pervasive all over the computation. All the FP-trees built recursively during the FP-growth computation can be pruned extensively by using the ExAnte property, obtaining a computation with a smaller number of smaller trees. We call such a tiny FP-tree, obtained by growing and pruning, an *FP-bonsai*.

The resulting method overcomes on one hand the main drawback of FP-growth, which is its memory requirements, and on the other hand, the main drawback of ExAMiner which is the I/O cost of iteratively rewriting the reduced datasets to disk.

H. Dai, R. Srikant, and C. Zhang (Eds.): PAKDD 2004, LNAI 3056, pp. 155–160, 2004.

2 Preliminaries

Let $\mathcal{I} = \{x_1, ..., x_n\}$ be a set of *items*. An *itemset* X is a non-empty subset of $\mathcal{I}$. A *transaction* is a couple $\langle tid, X \rangle$ where tid is the transaction identifier and X is an itemset. A *transaction database* $\mathcal{D}$ is a set of transactions. An itemset X is contained in a transaction $\langle tid, Y \rangle$ if $X \subseteq Y$. The *support* of an itemset X in database $\mathcal{D}$, denoted by $supp_{\mathcal{D}}(X)$ is the number of transactions in $\mathcal{D}$ that contain X. Given a user-defined *minimum support* σ, an itemset X is called *frequent* in $\mathcal{D}$ if $supp_{\mathcal{D}}(X) \geq \sigma$. The *frequent itemset mining problem* requires to compute the set of all frequent itemsets.A constraint on itemsets is a function $\mathcal{C} : 2^{\mathcal{I}} \rightarrow \{true, false\}$. We say that an itemset I satisfies a constraint if and only if $\mathcal{C}(I) = true$. Let $Th(\mathcal{C}) = \{X \mid \mathcal{C}(X) = true\}$ denote the set of all itemsets X that satisfy constraint $\mathcal{C}$. In general given a conjunction of constraints $\mathcal{C}$ the the *constrained frequent itemsets mining problem* requires to compute $Th(\mathcal{C}_{freq}) \cap Th(\mathcal{C})$, where $\mathcal{C}_{freq}$ is the frequency constraint.

In particular we focus on two kinds of constraint: a constraint $\mathcal{C}_{AM}$ is *anti-monotone* if $\mathcal{C}_{AM}(X) \Rightarrow \mathcal{C}_{AM}(Y)$ for all $Y \subseteq X$; a constraint $\mathcal{C}_M$ is *monotone* if: $\mathcal{C}_M(X) \Rightarrow \mathcal{C}_M(Y)$ for all $Y \supseteq X$. Since any conjunction of anti-monotone constraints is an anti-monotone constraint, and any conjunction of monotone constraints is a monotone constraint, we consider without loss of generality the conjunction $\mathcal{C}_{freq} \cap Th(\mathcal{C}_M)$ where $\mathcal{C}_M$ is a simple monotone constraint such as $sum(X.prices) \geq n$.

The recently introduced ExAnte method [2] exploits monotone constraints in order to to reduce the input database and thus to prune the search space. This method is based on the synergy of the following two data-reduction operations: (1) *μ-reduction*, which deletes transactions in $\mathcal{D}$ which do not satisfy $\mathcal{C}_M$; and (2) *α-reduction*, which deletes from all transactions in $\mathcal{D}$ singleton items which do not satisfy $\mathcal{C}_{freq}$. The ExAnte property states that a transaction which does not satisfy the given monotone constraint can be deleted from the input database (μ-reduction) since it will never contribute to the support of any itemset satisfying the constraint. A major consequence of reducing the input database in this way is that it implicitly reduces the support of a large amount of itemsets. As a consequence, some singleton items can become infrequent and can not only be removed from the computation, but they can be deleted from all transactions in the input database (α-reduction). This removal also has another positive effect. That is, the reduced transaction might violate the monotone constraint. Obviously, we are inside a loop where two different kinds of pruning (α and μ) cooperate to reduce the search space and the input dataset, strengthening each other step by step until no more pruning is possible (a fix-point has been reached). This is the key idea of the ExAnte preprocessing method [2]. In the end, the reduced dataset resulting from this fix-point computation is usually much smaller than the initial dataset

Given a transaction database $\mathcal{D}$, a conjunction of monotone constraints $\mathcal{C}_M$, and a conjunction of anti-monotone constraints $\mathcal{C}_{AM}$, we define the reduced dataset obtained by the fix-point application of μ and α pruning as: $\mu^+_{\mathcal{C}_{AM},\mathcal{C}_M}(\mathcal{D})$.

3 FP-Bonsai

The FP-growth algorithm [4] stores the actual transactions from the database in a trie structure (prefix tree), and additionally stores a header table containing all items with their support and the start of a linked list going through all transactions that contain that item. This data structure is denoted by *FP-tree* (Frequent-Pattern tree) [4]. For example, consider the transaction database in Figure 2(a) and a minimal support threshold of 4. First, all infrequent items are removed from the database, all transactions are reordered in support descending order and inserted into the FP-tree, resulting in the tree in Figure 2(b).

Given a transaction database $\mathcal{D}$ and a minimal support threshold σ, we denote the set of all frequent itemsets with the same prefix $I \subseteq \mathcal{I}$ by $\mathcal{F}[I](\mathcal{D},\sigma)$. FP-growth recursively generates for every singleton item $\{i\} \in Th(\mathcal{C}_{freq})$ the set $\mathcal{F}[\{i\}](\mathcal{D},\sigma)$ by creating the so called *i-projected* database of $\mathcal{D}$ This database, denoted $\mathcal{D}^i$, is made of all transactions in $\mathcal{D}$ containing i, from which i and all items which come before i, w.r.t. the support descending order, are deleted. This *i-projected* database, which is again stored as an FP-tree, is recursively mined by FP-growth. The FP-growth algorithm is shown in Figure 1.

Algorithm FP-growth
Input: $\mathcal{D},\sigma, I \subseteq \mathcal{I}$
Output: $\mathcal{F}[I](\mathcal{D},\sigma)$
  $\mathcal{F}[I] := \{\}$
  for all $i \in \mathcal{I}$ occurring in $\mathcal{D}$ **do**
    $\mathcal{F}[I] := \mathcal{F}[I] \cup \{I \cup \{i\}\}$
    $H := \{\}; \mathcal{D}^i := \{\}$
    for all $j \in \mathcal{I}$ occurring in $\mathcal{D}$ such that $j > i$ **do**
      if $supp_{\mathcal{D}}(I \cup \{i,j\}) \geq \sigma$ **then**
        $H := H \cup \{j\}$
    for all $(tid, X) \in \mathcal{D}$ with $i \in X$ **do**
      $\mathcal{D}^i := \mathcal{D}^i \cup \{(tid, X \cap H)\}$
    Compute $\mathcal{F}[I \cup \{i\}](\mathcal{D}^i,\sigma)$
    $\mathcal{F}[I] := \mathcal{F}[I] \cup \mathcal{F}[I \cup \{i\}]$

Algorithm FP-pruning
Input: $\mathcal{D}, \mathcal{C}_{AM}, \mathcal{C}_M, I$
Output: $\mu^{+}_{\mathcal{C}_M,\mathcal{C}_{AM}}[\mathcal{D}]$
  repeat
    // μ-pruning of $\mathcal{D}$
    for all transactions t occurring in $\mathcal{D}$ **do**
      if $\mathcal{C}_M(I \cup t) = false$ **then**
        Remove t from $\mathcal{D}$
    // α-pruning of $\mathcal{D}$
    for all items i occurring in $\mathcal{D}$ **do**
      if $\mathcal{C}_{AM}(I \cup \{i\}) = false$ **then**
        Remove i from $\mathcal{D}$
  until nothing changed

Fig. 1. The FP-growth and FP-pruning algorithms.

The main trick exploited in FP-growth is that it only needs to find all singleton frequent itemsets in the given database. Then, for every such item, it creates the corresponding projected database in which again, only the (local) singleton frequent itemsets have to be found. This process goes on until no more (local) items exist. The FP-tree structure guarantees that all this can be done efficiently. In this way, FP-growth implicitly creates a lot of databases, represented by FP-trees. The good news is that all these datasets (trees) can be reduced (pruned) using the ExAnte technique. We call such a pruned FP-tree an *FP-bonsai.*

(a)

tID	Itemset
1	b,c,d,g
2	a,b,d,e
3	b,c,d,g,h
4	a,e,g
5	c,d,f,g
6	a,b,c,d,e
7	a,b,d,f,g,h
8	b,c,d
9	b,e,f,g

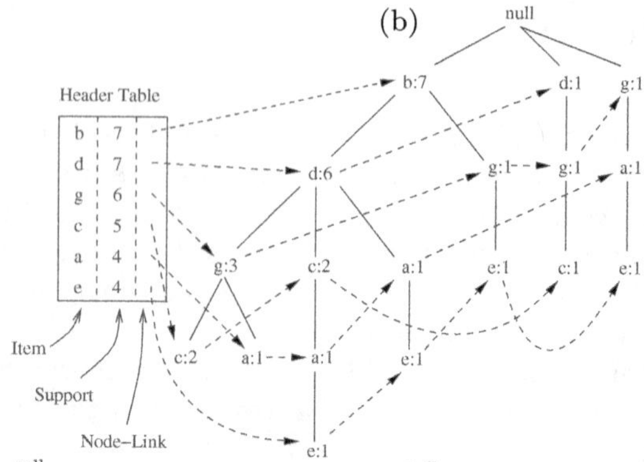

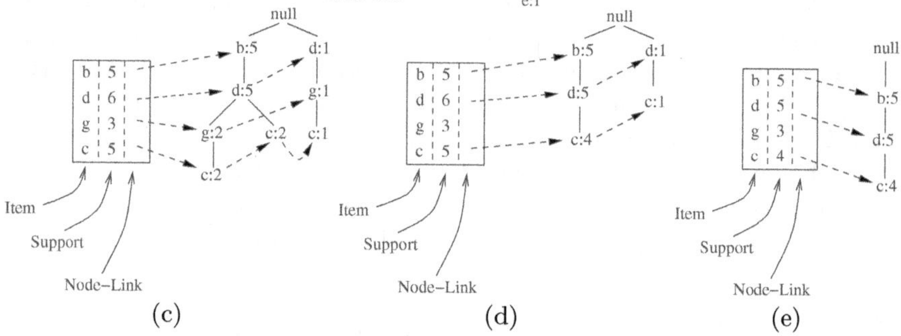

Fig. 2. A transaction database (a), the corresponding FP-tree for minimal support threshold of 4 (b), the initial FP-bonsai for Example 1 (c), the FP-bonsai after the removal of item g (d), and (e) the final FP-bonsai.

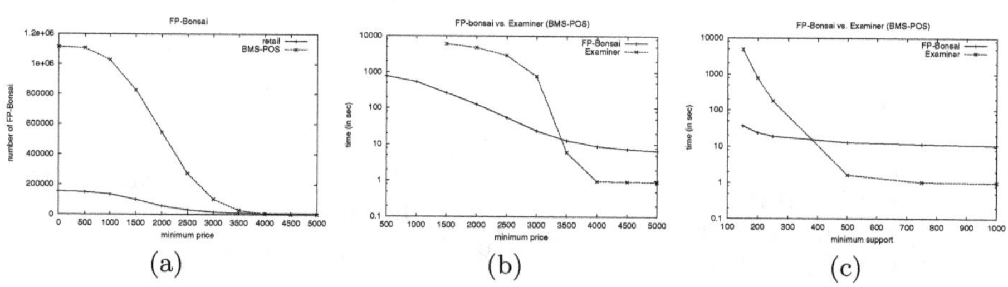

Fig. 3. Number of FP-bonsai built with fixed *minimum_support* and moving monotone threshold (a); run time of ExAMiner and FP-bonsai on dataset *BMS-POS* with *minimum_support* = 200 (b); and minimum sum of prices = 3000 (c).

The FP-pruning procedure is shown in Figure 1. In order to obtain the complete algorithm that finds all itemsets satisfying the given constraints, the FP-pruning algorithm should be called before the first line of the FP-growth algorithm. The fact that the database is stored as an FP-tree is not specifically mentioned. That is because this is actually not necessary, but the FP-tree is simply the most effective data structure for these algorithms to use. How the pruning mechanisms can be effectively applied on the FP-tree structure is described by the following Example.

Example 1. Consider again the transactional database in Figure 2(a) and the price-table: $\{a : 5, b : 8, c : 14, d : 30, e : 20, f : 15, g : 6, h : 12\}$. Suppose that we want to compute itemsets with support no less than 4 and sum of prices no less than 45. The FP-bonsai construction starts with a first scan of $\mathcal{D}$ to count the support of all singleton items. Note that transaction 4 is not used since it does not satisfy the monotone constraint. This causes item a and e to be infrequent and are not included in the header table. Frequent items are ordered in support descending order and the tree is built as seen in Figure 2(c). At this point we find that the item g is no longer frequent than we remove all its occurrences in the tree using the link-node structure. The resulting pruned tree is in Figure 2(d). This α-pruning has created a new opportunity for μ-pruning. in fact, the path on the right edge of the tree does no longer satisfy the monotone constraint and hence it can be removed from the tree. In Figure 2(e) we have the final FP-bonsai (the fix-point has been reached) for the given problem. Note the the final size of the FP-bonsai is 3 nodes (which represents the unique solution to the given problem: itemset bcd with $support = 4$ and sum of prices $= 52$, while the size of the usual FP-tree for the same problem (Figure 2(b)) is 18 nodes!

Once the FP-bonsai has been built (i.e. once the fix-point of α and μ pruning has been reached) we can efficiently mine all frequent itemsets satisfying the given constraints using FP-growth. Thanks to the recursive structure of the FP-growth based algorithm, the ExAnte property is deeply amalgamated with the frequent itemset computation: not only the initial tree is a pruned tree (an FP-bonsai), but also all the other projected trees, built during the recursive growing phase will be much more smaller in number and in size.

The reduction of number of trees built w.r.t. FP-growth (which is the computation with minimum sum of prices = 0) is dramatic, as illustrated in Figure 3(a).

Our experimental study confirms that FP-bonsai outperforms ExAMiner, which is the state-of-the-art algorithm for the computational problem addressed in many occasions.

From those pictures we can see that ExAMiner is faster only when the selectivity of one of the two constraints is very strong, and hence the set of solutions very small. In particular, ExAMiner is faster in recognizing when the user-defined constraints are so selective that the problem has an empty set of solutions. But in all the other cases FP-bonsai performs much better. In particular, when one of the two constraints is not-so-selective, FP-bonsai exhibits a much more stable behaviour, while ExAMiner's computation time increases quickly. Consider, for instance Figure 3(c): at an absolute minimum support of 150, FP-bonsai takes 36 seconds against the 4841 seconds (1 hour and 20 minutes) taken by ExAMiner.

References

1. F. Bonchi, F. Giannotti, A. Mazzanti, and D. Pedreschi. ExAMiner: Optimized level-wise frequent pattern mining with monotone constraints. In *Proc. of ICDM'03.*

2. F. Bonchi, F. Giannotti, A. Mazzanti, and D. Pedreschi. Exante: Anticipated data reduction in constrained pattern mining. In *Proc. of PKDD03.*
3. C. Bucila, J. Gehrke, D. Kifer, and W. White. DualMiner: A dual-pruning algorithm for itemsets with constraints. In *Proc. of ACM SIGKDD'02.*
4. J. Han, J. Pei, and Y. Yin. Mining frequent patterns without candidate generation. In *Proc of ACM SIGMOD'00.*
5. R. T. Ng, L. V. S. Lakshmanan, J. Han, and A. Pang. Exploratory mining and pruning optimizations of constrained associations rules. In *Proc. of ACM SIGMOD'98.*
6. J. Pei, J. Han, and L. V. S. Lakshmanan. Mining frequent item sets with convertible constraints. In *Proc. of ICDE'01.*

Mining Negative Rules Using GRD

Dhananjay R. Thiruvady and Geoff I. Webb

School of Computer Science and Software Engineering,
Monash University, Wellington Road, Clayton, Victoria 3800 Australia,
Dhananjay_Thiruvady@hotmail.com, Geoff.Webb@csse.monash.edu.au

Abstract. GRD is an algorithm for k-most interesting rule discovery. In contrast to association rule discovery, GRD does not require the use of a minimum support constraint. Rather, the user must specify a measure of interestingness and the number of rules sought (k). This paper reports efficient techniques to extend GRD to support mining of negative rules. We demonstrate that the new approach provides tractable discovery of both negative and positive rules.

Keywords: Rule Discovery, Negative Rules.

1 Introduction

Rule discovery involves searching through a space of rules to determine rules of interest to a user. Association rule discovery [1] seek rules between frequent items (literals which satisfy a minimum support constraint). A rule set is developed which can be pruned by using further user defined constraints.

Generalized rule discovery is an alternative rule discovery approach. Rules in GRD are developed based on user defined constraints. There is no need to apply a minimum support constraint. Rather, the user must specify a number of rules to be generated, k. This avoids the inherent limitations of the minimum support methodology.

Mining negative rules from databases has been approached using association rule discovery [3,6,12]. We seek to extend GRD to mining negative rules so as to enable negative rules to be discovered without the need to specify minimum support constraints.

2 Association Rule Discovery

Association rule discovery aims to find rules describing associations between items [1]. A rule has the form $A \Rightarrow B$, where A is the antecedent and B is the consequent. Both A and B are *itemsets* from the database. The rule implies that if an itemset A occurs in a record then itemset B is likely to occur in the same record of the database.

Constraints are defined to limit the space of rules to be searched [8]. For example, 1000 items define 2^{1000} possible combinations of itemsets which results

H. Dai, R. Srikant, and C. Zhang (Eds.): PAKDD 2004, LNAI 3056, pp. 161–165, 2004.

in a large number of rules to explore. The *minimum support* constraint is used to limit the number of itemsets that need be considered.

The *support* of an itemset is the frequency with which the itemset occurs in the database. The itemsets which satisfy the minimum support constraint are *frequent itemsets*. From these itemsets the rules are developed. Further constraints can be applied to prune the set of rules discovered [7].

3 Negative Rules

Mining negative rules has been given some attention and has proved to be useful. Initial approaches [3] considered mining negative associations between two itemsets. Savasere, Omiecinski and Navathe [6] use the method of generating positive rules from which negative rules are mined. The result is that there are fewer but more interesting negative rules that are mined.

Negative association rules are associations rules in which either the antecedent or consequent or both are negated. For example, for the rule $A \Rightarrow B$ the negative rules are $A \Rightarrow \neg B$ (A implies not B), $\neg A \Rightarrow B$, $\neg A \Rightarrow \neg B$ [12].

The rules above specify concrete relationships between each itemset compared to [6] who look at the rule $A \not\Rightarrow B$. Another possibility is to consider itemsets within the antecedent or the consequent being negated (e.g. $(\neg A \,\&\, B) \Rightarrow C$).

4 Generalized Rule Discovery

In some applications minimum support may not be a relevant criterion to select rules. For example, often high-value rules relate to infrequent associations, a problem known as the *vodka and caviar* problem [4].

The GRD algorithm [10,11] implements *k*-most interesting rule discovery. This approach avoids the need to specify a minimum support constraint, replacing it by a constraint on the number of rules to be found together with the specification of an interestingness measure. Further constraints may be specified, including a minimum support constraint if desired, but these are not required.

GRD performs the OPUS search [9] through the space of potential antecedents and for each antecedent the set of consequent conditions are explored. The consequent conditions are limited to single condition to simplify the search.

Space constraints preclude us from presenting the algorithm here. The extensions to the base GRD algorithm [11] are however, straightforward. Based on the idea of a diffset [13], many of the statistics for negative rules can be derived using much statistics already derived for positive rules. Specifically,

- support($A \,\&\, \neg x \Rightarrow B$) = support($A \Rightarrow B$) − support($A \,\&\, x \Rightarrow B$).
- support($A \Rightarrow \neg B$) = support(A) − support($A \Rightarrow B$).

However, the search space is nonetheless considerably larger, as an increase in the number of conditions considered results in an exponential increase in the size of the search space that must be explored.

5 Experiments

The modified GRD program is referred to as *GRDI (GRD new Implementation)*. Experiments were carried out on ten datasets with GRDI. Most of the datasets used were the same datasets used for the comparison of the GRD system with Apriori in [11].

Table 1. Execution times of GRD and GRDI

Data Files	*Records*	*GRD*	*GRDI*	*Ratio*
connect4	67,557	20	106	5.30
covtype	581,012	835	1976	2.37
ipums.la.99	88,443	7	1634	233.43
letter-recognition	20,000	1	34	34.00
mush	8,124	1	8	8.00
pendigits	10,992	1	28	28.00
shuttle	58,000	1	11	11.00
soybean-large	307	1	4	4.00
splice junction	3,177	6	1872	312.00
ticdata2000	5,822	7	647	92.43

Nine out of the ten datasets are taken from the UCI Machine Learning and KDD repositories [2,5]. The other dataset, ticdata2000 is a market-basket dataset used in research in association rule discovery [14]. Three sub ranges were created for numeric attributes. Each sub range approximately contained one third of the records. The experiments were carried out on a Linux server, with a processor speed of 1.20 GHz and main memory of 256 MB RAM.

In all experiments GRD and GRDI search for the 1000 rules with the highest value of the search measure, Leverage. The maximum number of conditions available on the left hand side was 4 and both systems assume that only a single condition was available for the right-hand-side. This will simplify the search task. The executions time for GRD and GRDI are presented in Table 1. Two observations from the results are:

1. GRD: Execution times for some large datasets (large number of records) are very short and some are very long. e.g. connect4 has 67,557 records and requires 20 seconds to develop rules, whereas ipums.la.99 has 88,443 records takes only 7 seconds.
2. GRDI: for most datasets GRDI's execution time is slightly greater than GRD, e.g. mush. However, some datasets require greater execution times for GRDI than GRD, e.g. ticdata2002.

The reason for large increase in the computational time for some datasets (e.g. ipums.la.99) is primarily due to the increase in the size of the search space. If few negative rules are generated then the execution time is only a little greater. If a majority of rules are negative (sometimes all) then the execution times are a lot

Table 2. Comparison of Minimum and Maximum Leverage values

	GRD			*GRDI*		
Data Files	*min. lev.*	*max. lev.*	*mean*	*min. lev.*	*max. lev.*	*mean*
connect4	0.1224	0.1227	0.1225	0.1688	0.1707	0.1698
covtype	0.1083	0.1743	0.1413	0.2459	0.2474	0.2467
ipums.la.99	0.2080	0.2484	0.2282	0.2499	0.2500	0.2500
letter-recognition	0.0455	0.1459	0.0957	0.1020	0.1499	0.1395
mush	0.1558	0.2109	0.1833	0.1994	0.4930	0.3390
pendigits	0.0615	0.1757	0.1186	0.1050	0.1832	0.1441
shuttle	0.0409	0.1599	0.1004	0.0911	0.2040	0.1766
soybean-large	0.2137	0.2359	0.2248	0.2286	0.6182	0.4324
splice junction	0.0404	0.1523	0.0963	0.1244	0.1733	0.1489
ticdata2000	0.1899	0.1922	0.1910	0.2184	0.5341	0.3763

greater. The increase in execution time is directly proportionate to the increase in size of the search space.

The comparison of minimum leverage values of the rules generated by both systems shows that GRDI always contains negative rules in its solution. For the datasets in which GRDI's execution times were greater than GRD, rules with much higher leverage were also generated. This is also true of the maximum leverage values. An important observation is that for several datasets the minimum leverage value of GRDI is greater than the maximum leverage value of GRD. The information contained within the observation is that all the rules generated for those datasets are negative rules.

6 Conclusions

GRD has been extended to discover negative rules, providing negative rule discovery without the need to specify a minimum support constraint. This is useful because such a constraint is not appropriate for some domains and can prevent potentially interesting rules from being discovered. An additional advantage of GRD is that users can generate a specific number of rules that maximize a particular search measure. Incorporating the diffsets technique results in low additional computational time.

The GRD algorithm is modified to iterate through two antecedent sets. One for positive items and the other for negative items. Within each iteration a second consequent set of negative items is explored in addition to the positive consequent set.

A comparison of GRD and GRDI shows that for several datasets GRDI took substantially longer to execute than GRD. The reason for this increase in execution time is because these particular datasets contained many more negative rules than positive rules. With the increased size of the search space, the execution times are bound to be longer for GRDI. For some datasets only negative rules were generated. In conclusion, developing negative and positive rules using GRD is tractable.

References

1. R. Agrawal, T. Imielinski, and A.N. Swami. Mining association rules between sets of items in large databases. In Peter Buneman and Sushil Jajodia, editors, *SIGMOD*, pages 207–216, Washington, D.C., 26–28, 1993.
2. C.L. Blake and C.J. Merz. UCI repository of machine learning databases. http://www.ics.uci.edu/~mlearn/mlrepository.html, University of California, Irvine, Dept. of Information and Computer Sciences, 1998.
3. S. Brin, R. Motwani, and C. Silverstein. Beyond market baskets: generalizing association rules to correlations. *SIGMOD*, pages 265–276, 1997.
4. E. Cohen, M. Datar, S. Fujiwara A. Gionis, R. Indyk, P. Motwani, J. Ullman, and C. Yang. Finding interesting associations without supporting pruning. In *Proc. ICDE*, 2000.
5. S. Hettich and S. D. Bay. The UCI KDD archive. http://www.ics.uci.edu/, University of California, Irvine, Dept. of Information and Computer Sciences, 1999.
6. As. Savasere, E. Omiecinski, and S. B. Navathe. Mining for strong negative associations in a large database of customer transactions. In *ICDE*, pages 494–502, 1998.
7. R. Srikant and R. Agrawal. Mining generalized association rules. *Future Generation Computer Systems*, 13(2–3):161–180, 1997.
8. R. Srikant, Q. Vu, and R. Agrawal. Mining association rules with item constraints. In D. Heckerman, H. Mannila, D. Pregibon, and R. Uthurusamy, editors, *Proc. KDD*, pages 67–73. AAAI Press, 14–17 1997.
9. G. I. Webb. OPUS: An efficient admissible algorithm for unordered search. *Journal of Artificial Intelligence Research*, 3:431–465, 1995.
10. G. I Webb. Efficient search for association rules. *KDD*, pages 99–107, 2000.
11. G. I Webb and S. Zhang. Beyond association rules: Generalized rule discovery. In *Knowledge Discovery and Data Mining.* Kluwer Academic Publishers, 14–17, unpublished manuscript.
12. X. Wu, C. Zhang, and S. Zhang. Mining both positive and negative rules. *ICML*, pages 658–665, 2002.
13. M. Zaki and K. Gouda. Fast vertical mining using diffsets. Technical Report 01-1, Rensselaer Polytechnic Institute, 2001.
14. Z. Zheng, R. Kohavi, and L. Mason. Real world performance of association rule algorithms. *KDD*, pages 401–406, 2001.

Applying Association Rules for Interesting Recommendations Using Rule Templates

Jiye Li[1], Bin Tang[2], and Nick Cercone[2]

[1] School of Computer Science, University of Waterloo
200 University Avenue West, Waterloo, Ontario, Canada N2L 3G1
j27li@uwaterloo.ca
[2] Faculty of Computer Science, Dalhousie University
6050 University Avenue, Halifax, Nova Scotia, Canada B3H 1W5
{btang, nick}@cs.dal.ca

Abstract. In this paper, we propose a new method of applying association rules for recommendation systems. Association rule algorithms are used to discover associations among items in transaction datasets. However, applying association rule algorithms directly to make recommendations usually generates too many rules; thus, it is difficult to find interesting recommendations for users among so many rules. Rule templates define certain types of rules; therefore, they are one of the interestingness measures that reduce the number of rules that do not interest users. We describe a new method. By defining more appropriate rule templates, we are able to extract interesting rules for users in a recommendation system. Experimental results show that our method increases the accuracy of recommendations.

Keywords: Association Rule, Rule Templates, Recommendation Systems

1 Introduction

Generally speaking, there are two types of recommendation systems, content-based recommendation systems and collaborative recommendation systems [1]. Content-based recommendation systems, such as NewsWeeder [2] and InfoFinder [3], require representative properties from the data, which are hard to extract. Collaborative recommendation systems, such as GroupLens [4] and Ringo [5], observe the behaviors, patterns of the current users, and make recommendations based on the similarities between the current users and the new users.

Not many research efforts are found on applying association rule algorithms for collaborative recommendation systems. The association rule algorithm was first introduced in [6]. An association rule [6] is a rule of the form $\alpha \rightarrow \beta$, where both the antecedent α and the consequent β represent itemsets. The quality of the rule is traditionally measured by support and confidence [6]. The disadvantage of using association rule algorithms for collaborative recommendation systems is that there are usually too many rules generated, and it is difficult

H. Dai, R. Srikant, and C. Zhang (Eds.): PAKDD 2004, LNAI 3056, pp. 166–170, 2004.

to make recommendations to the users effectively and efficiently. To solve this problem, Lin et al. [7] designed a new algorithm to adjust the value of support and to limit the number of rules generated. Klemettinen et al. [8] proposed rule templates to find interesting rules from a large set of association rules. Rule templates describe patterns for those items that appear both in the antecedent and in the consequent of the association rules. Defining rule templates more appropriately will result in extracting only interesting rules that match the templates. In this paper, we explore the usage of rule templates on apriori association rule algorithm for collaborative recommendations systems to generate interesting rules. The EachMovie dataset [9] is used in our experiments, in which we consider the merit of movie genres for the generation of rule templates. Our experimental results indicate that, with rule templates, association rule algorithms can efficiently extract interesting rules for collaborative recommendations.

The rest of the paper is organized as follows. Section 2 discusses our methods of making recommendations. The experiments and evaluations are discussed in section 3. The final section makes concluding remarks and describes our future work.

2 Our Method

We would like to apply the association rule algorithm for recommendation systems. In addition to using support and confidence, we examine the role of rule templates to predict the items in which users are most likely interested.

Since we are interested in predicting movies and making recommendations efficiently, we define rule templates to reduce the number of rules.

Template 1, $\langle Movie_1, \ldots, Movie_m \rangle \rightarrow \langle Movie_n \rangle$, specifies that there is only one consequent in the generated rules.

Template 2, $\langle Genre_1 \cap \ldots \cap Genre_m \cap Genre_n \rangle \neq \phi$, specifies that only rules whose antecedents and consequents all belonging to the same movie genre will be generated. In our method, no score or vote information related to the item in the recommendation system is considered.

3 Experiment

3.1 EachMovie Dataset

For this work, we use the EachMovie dataset, the most commonly used test bed for collaborative recommendation task, provided by Compaq's Systems Research Center [9]. This data set is a collection of users' votes on $1,628$ different movies from $72,916$ users over an 18 month period. Each movie is assigned to no less than one of the 10 possible movie genres. Each movie is rated based on a five star evaluation scheme. By removing movies that have no votes and users that have never voted, we are left with $61,265$ valid users, $1,623$ valid movies and $2,811,983$ votes.

For our experiments, only a subset of the data is considered. By counting the frequency of the number of movies that each user voted for, we know that about 40% users voted for no less than 2% (roughly 32) of all the movies. In the following experiments, we consider only users who have voted for no less than 32 movies.

According to the kind of rules we wish to generate, we organize the transaction dataset such that each transaction represents all the movies voted for by a user.

3.2 Performance Evaluation

We use accuracy, defined as $a = \frac{c}{t}$, to evaluate the performance of our method. This function gives the accuracy a computed as a function of c and t, where c stands for the number of times the antecedent and the consequent of a rule belong to the same transaction, and t stands for the number of times the antecedent belongs to a transaction.

3.3 Experiment Results

In our experiment, we applied Borgelt's apriori algorithm [10] to generate frequent itemsets.[1] Our rule templates and rule generation procedure were implemented by C++, and the target compiler and platform are g++ and Unix respectively. We performed 4-fold cross validation for the following experiments. All the experiments were performed on Sun Fire V880, four 900Mhz UltraSPARC III processors, with 8GB of main memory. We applied the two templates to two subsets of data which are commonly used [7], [11].

First Trial. The first subset we tried is from [11]. Training data represent the first 1,000 users who have rated more than 100 movies. Testing data come from the first 100 users whose user ID is larger than 70,000, and who also rated more than 100 movies.

Table 1 a) shows the performance of this experiment when confidence is 80%. The first column shows the support value, the second column shows the accuracy from applying the first template to our algorithm, and the third column shows the accuracy of adding genre information (applying Template 2). Table 1 b) shows the performance when confidence is 90%. From Table 1, we can see that when applying movie genre information, extracting only the rules that all the movies belong to the same genre, we obtain a higher accuracy.

In order to show that the computing overhead is also reduced by applying Template 2, we generated the number of itemsets and rules, as shown in Table 2.

Table 2 show that using Template 2 reduces the number of rules generated. As support value gets lower, there are more frequent itemsets generated, thus more rules are generated. The accuracy increases as the support value decreases.

[1] Downloaded from http://fuzzy.cs.uni-magdeburg.de/~borgelt/software.html#assoc

Table 1. a) Accuracy when confidence = 80%, b) Accuracy when Confidence = 90%

Support	Accuracy_Rules	Accuracy_Genre
70%	73.30%	75.28%
60%	75.83%	77.42%
50%	78.33%	79.52%
40%	80.78%	82.43%

(a)

Support	Accuracy_Rules	Accuracy_Genre
70%	79.97%	80.25%
60%	81.20%	82.93%
50%	84.62%	85.87%
40%	87.25%	88.35%

(b)

Table 2. a) Itemset Size and Rule Size when confidence = 80%, b) Itemset Size and Rule Size when confidence = 90%

Min support	Frequent Itemsets	Assoc. Rules	With Genre Rules
70%	272	608	198
60%	2,773	8,303	1,439
50%	35,276	139,796	10,385
40%	690,382	3,525,426	88,298

(a)

Min support	Frequent Itemsets	Assoc. Rules	With Genre Rules
70%	272	392	127
60%	2,773	5,074	805
50%	35,276	79,353	5,404
40%	690,382	1,994,580	49,278

(b)

Table 3. a) Accuracy when confidence = 80%, b) Accuracy when confidence = 90%

Support	Accuracy_Rules	Accuracy_Genre
20%	78.30%	87.04%
10%	75.79%	91.83%
5%	78.65%	93.86%
4%	81.27%	94.72%

(a)

Support	Accuracy_Rules	Accuracy_Genre
20%	75%	100%
10%	81.44%	98.67%
5%	82.87%	97.64%
4%	83.37%	97.64%

(b)

Table 4. a) Itemset Size and Rule Size when confidence = 80%, b) Itemset Size and Rule Size when confidence = 90%

Min support	Frequent Itemsets	Assoc. Rules	With Genre Rules
20%	171	86	30
10%	9,023	15,671	1,360
5%	579,291	1,926,017	37,855
4%	2,326,891	9,154,962	104,589

(a)

Min support	Frequent Itemsets	Assoc. Rules	With Genre Rules
20%	171	6	4
10%	9,023	3,788	297
5%	579,291	745,971	12,954
4%	2,326,891	3,900,287	41,626

(b)

By increasing the confidence, the accuracy will also be increased; thus, better quality rules will be extracted.

Second Trial. The second subset we tried is from [7]. We used training data for the first 2,000 users. Testing data comes from users whose like ratios are less than 0.75, from which we randomly selected 20 users as one test set. We repeated this choice of test set 4 times, from which we obtained the average accuracy. The accuracy is shown by Table 3, which shows an average of more than 15% increase in accuracy using Template 2.

We list the number of itemsets and rules generated using different support values for different confidence levels in Table 4. As we can see, when confidence gets higher, there are fewer rules generated; when support gets higher, fewer rules are generated.

4 Conclusions and Future Work

We have proposed a new method of applying the association rule algorithms for recommendation systems. Unlike most current recommendation systems, our method does not consider score or vote information associated with every recommended item. By applying appropriate rule templates, we achieved interesting rules. Experimental results show that rule templates increase the accuracy of recommendations. As future work, we would like to explore the performance of different interestingness measures on recommendation systems.

References

1. Balabanovic, M., Shoham,Y.: Fab: Content-based, collaborative recommendation. Communications of the ACM, 40(3):66-72, March (1997)
2. Lang, K.: Newsweeder: Learning to filter netnews. Proceedings of the 12th International Conference on Machine Learning, Tahoe City, Calif (1995)
3. Krulwich, B., Burkey, C.: Learning user information interests through extraction of semantically significant phrases. Proceedings of the AAAI Spring Symposium on Machine Learning in Information Access, Stanford, Calif. March (1996)
4. Resnick, P., Iacovou, N., Suchak, M., Bergstrom, P., Riedl, J.: GroupLens: An open architecture for collaborative filtering of netnews. Proceedings of the ACM Conference on Computer-Supported Cooperative Work, Chapel Hill, NC(1994)
5. Shardanand, U., Maes, P.: Social information filtering: Algorithms for automating "word of mouth". Conf. on Human Factors in Computing Systems-CHI'95 (1995)
6. Agrawal, R., Srikant, R.: Fast Algorithms for Mining Association Rules. Proceedings of 20th International Conference Very Large Data Bases(VLDB), Santiago de Chile, Chile, Morgan Kaufmann (1994) 487–499
7. Lin, W., Alvarez, S., Ruiz, C.: Efficient adaptive-support association rule mining for recommender systems, Data Mining and Knowledge Discovery (2002) 6:83–105
8. Klemettinen, M., Mannila, H., Ronkainen, R., Toivonen, H., Verkamo, A.I.: Finding interesting rules from large sets of discovered association rules. Third International Conference on Information and Knowledge Management, ACM Press (1994)
9. EachMovie Collaborative Filtering data set (1997) http://research.compaq.com/SRC/eachmovie/
10. Borgelt, C.: Efficient Implementations of Apriori and Eclat. Proceedings of the FIMI'03 Workshop on Frequent Itemset Mining Implementations, Melbourne, Florida, USA, November, CEUR Workshop Proceedings (2003) 1613-0073
11. Billsus, D., Pazzani, M.J.: Learning Collaborative Information Filters. Proc. of the Fifteenth International Conference on Machine Learning, Morgan Kaufmann Publishers (1998)

Feature Extraction and Classification System for Nonlinear and Online Data*

Byung Joo Kim[1], Il Kon Kim[2], and Kwang Baek Kim[3]

[1] Youngsan University School of Network and Information Engineering, Korea, bjkim@ysu.ac.kr,
[2] Kyungpook National University Department of Computer Science, Korea, ikkim@knu.ac.kr,
[3] Silla University Department of Computer Engineering, Korea, gbkim@silla.ac.kr

Abstract. A novel incremental feature extraction and classification system is proposed. Kernel PCA is famous nonlinear feature extraction method. The problem of Kernel PCA is that the computation becomes prohibitive when the data set is large. Another problem is that, in order to update the eigenvectors with another data, the whole eigenspace should be recomputed. Proposed feature extraction method overcomes these problems by incrementally eigenspace update and using empirical kernel map as kernel function. Proposed feature extraction method is more efficient in memory requirement than a Kernel PCA and can be easily improved by re-learning the data. For classification extracted features are used as input for Least Squares SVM. In our experiments we show that proposed feature extraction method is comparable in performance to a Kernel PCA and proposed classification system shows a high classification performance on UCI benchmarking data and NIST handwritten data set.

Keywords: Incremental PCA, Kernel PCA, Emperical kernel map, LS-SVM

1 Introduction

In many pattern recognition problem it relies critically on efficient data representation. It is therefore desirable to extract measurements that are invariant or insensitive to the variations within each class. The process of extracting such measurements is called *feature extraction.* Principal Component Analysis(PCA)[1] is a powerful technique for extracting features from possibly high-dimensional data sets. For reviews of the existing literature is described in [2][3][4]. Traditional PCA, however, has several problems. First PCA requires a batch computation step and it causes a serious problem when the data set is large i.e., the PCA computation becomes very expensive. Second problem is that, in order to update

* This study was supported by a grant of the Korea Health 21 R&D Project, Ministry of Health & Welfare, Republic of Korea (02-PJ1-PG6-HI03-0004)

H. Dai, R. Srikant, and C. Zhang (Eds.): PAKDD 2004, LNAI 3056, pp. 171–180, 2004.

the subspace of eigenvectors with another data, we have to recompute the whole eigenspace. Finial problem is that PCA only defines a linear projection of the data, the scope of its application is necessarily somewhat limited. It has been shown that most of the data in the real world are inherently non-symmetric and therefore contain higher-order correlation information that could be useful[5]. PCA is incapable of representing such data. For such cases, nonlinear transforms is necessary. Recently kernel trick has been applied to PCA and is based on a formulation of PCA in terms of the dot product matrix instead of the covariance matrix[8]. Kernel PCA(KPCA), however, requires storing and finding the eigenvectors of a $N \times N$ kernel matrix where N is a number of patterns. It is infeasible method when N is large. This fact has motivated the development of incremental way of KPCA method which does not store the kernel matrix. It is hoped that the distribution of the extracted features in the feature space has a simple distribution so that a classifier could do a proper task. But it is point out that extracted features by KPCA are global features for all input data and thus may not be optimal for discriminating one class from others[6]. This has naturally motivated to combine the feature extraction method with classifier for classification purpose. In this paper we propose a new classifier for on-line and nonlinear data. Proposed classifier is composed of two parts. First part is used for feature extraction. To extract nonlinear features, we propose a new feature extraction method which overcomes the problem of memory requirement of KPCA by incremental eigenspace update method incorporating with an adaptation of kernel function. Second part is used for classificaion. Extracted features are used as input for classification. We take Least Squares Support Vector Machines(LS-SVM)[7] as a classifier. LS-SVM is reformulations to the standard Support Vector Machines(SVM)[8]. SVM typically solving problems by quadratic programming(QP). Solving QP problem requires complicated computational effort and needs more memory requirement. LS-SVM overcomes this problem by solving a set of linear equations in the problem formulation.
Paper is composed of as follows. In Section 2 we will briefly explain the incremental eigenspace update method. In Section 3 KPCA is introduced and to make KPCA incrementally empirical kernel map method is is explained. Proposed classifier combining LS-SVM with proposed feature extraction method is described in Section 4. Experimental results to evaluate the performance of proposed classifier is shown in Section 5. Discussion of proposed classifier and future work is described in Section 6.

2 Incremental Eigenspace Update Method

In this section, we will give a brief introduction to the method of incremental PCA algorithm which overcomes the computational complexity and memory requirement of standard PCA. Before continuing, a note on notation is in order. Vectors are columns, and the size of a vector, or matrix, where it is important, is denoted with subscripts. Particular column vectors within a matrix are denoted with a superscript, while a superscript on a vector denotes a particular observa-

tion from a set of observations, so we treat observations as column vectors of a matrix. As an example, A^i_{mn} is the ith column vector in an $m \times n$ matrix. We denote a column extension to a matrix using square brackets. Thus $[A_{mn}b]$ is an $(m \times (n+1))$ matrix, with vector b appended to A_{mn} as a last column.
To explain the incremental PCA, we assume that we have already built a set of eigenvectors $U = [u_j], j = 1, \cdots, k$ after having trained the input data $\mathbf{x}_i, i = 1, \cdots, N$. The corresponding eigenvalues are Λ and $\bar{\mathbf{x}}$ is the mean of input vector. Incremental building of Eigenspace requires to update these eigenspace to take into account of a new input data. Here we give a brief summarization of the method which is described in [9]. First, we update the mean:

$$\overline{x}' = \frac{1}{N+1}(N\overline{x} + x_{N+1}) \tag{1}$$

We then update the set of Eigenvectors to reflect the new input vector and to apply a rotational transformation to U. For doing this, it is necessary to compute the orthogonal residual vector $\hat{h} = (Ua_{N+1} + \overline{x}) - x_{N+1}$ and normalize it to obtain $h_{N+1} = \frac{h_{N+1}}{\|h_{N+1}\|_2}$ for $\| h_{N+1} \|_2 > 0$ and $h_{N+1} = 0$ otherwise. We obtain the new matrix of Eigenvectors $U^{'}$ by appending h_{N+1} to the eigenvectors U and rotating them :

$$U' = [U, h_{N+1}]R \tag{2}$$

where $R \in \mathbf{R}_{(\mathbf{k+1})\times(\mathbf{k+1})}$ is a rotation matrix. R is the solution of the eigenproblem of the following form:

$$DR = R\Lambda' \tag{3}$$

where Λ' is a diagonal matrix of new Eigenvalues. We compose $D \in \mathbf{R}_{(\mathbf{k+1})\times(\mathbf{k+1})}$ as:

$$D = \frac{N}{N+1}\begin{bmatrix} \Lambda & 0 \\ 0^T & 0 \end{bmatrix} + \frac{N}{(N+1)^2}\begin{bmatrix} aa^T & \gamma a \\ \gamma a^T & \gamma^2 \end{bmatrix} \tag{4}$$

where $\gamma = h^T_{N+1}(x_{N+1} - \bar{x})$ and $a = U^T(x_{N+1} - \bar{x})$. Though there are other ways to construct matrix D[10,11], the only method ,however, described in [9] allows for the updating of mean.

2.1 Eigenspace Updating Criterion

The incremental PCA represents the input data with principal components $a_{i(N)}$ and it can be approximated as follows:

$$\widehat{x}_{i(N)} = Ua_{i(N)} + \bar{x} \tag{5}$$

To update the principal components $a_{i(N)}$ for a new input x_{N+1} , computing an auxiliary vector η is necessary. η is calculated as follows:

$$\eta = \left[U\widehat{h}_{N+1}\right]^T (\overline{x} - \overline{x}') \tag{6}$$

then the computation of all principal components is

$$a_{i(N+1)} = (R')^T \begin{bmatrix} a_{i(N)} \\ 0 \end{bmatrix} + \eta, \qquad i = 1, \cdots, N+1 \tag{7}$$

The above transformation produces a representation with $k+1$ dimensions. Due to the increase of the dimensionality by one, however, more storage is required to represent the data. If we try to keep a k-dimensional eigenspace, we lose a certain amount of information. It is needed for us to set the criterion on retaining the number of eigenvectors. There is no explicit guideline for retaining a number of eigenvectors. Here we introduce some general criteria to deal with the model's dimensionality:

- Adding a new vector whenever the size of the residual vector exceeds an absolute threshold;
- Adding a new vector when the percentage of energy carried by the last Eigenvalue in the total energy of the system exceeds an absolute threshold, or equivalently, defining a percentage of the total energy of the system that will be kept in each update;
- Discarding Eigenvectors whose Eigenvalues are smaller than a percentage of the first Eigenvalue;
- Keeping the dimensionality constant.

In this paper we take a rule described in b). We set our criterion on adding an Eigenvector as $\lambda'_{k+1} > 0.7\bar{\lambda}$ where $\bar{\lambda}$ is a mean of the λ. Based on this rule, we decide whether adding u'_{k+1} or not.

3 Incremental KPCA

A prerequisite of the incremental Eigenspace update method is that it has to be applied on the data set. Furthermore incremental PCA builds the subspace of Eigenvectors incrementally, it is restricted to apply the linear data. But in the case of KPCA this data set $\Phi(x^N)$ is high dimensional and can most of the time not even be calculated explicitly. For the case of nonlinear data set, applying feature mapping function method to incremental PCA may be one of the solutions. This is performed by so-called *kernel-trick*, which means an implicit embedding to an infinite dimensional Hilbert space[8](i.e. feature space) F.

$$K(x, y) = \Phi(x) \cdot \Phi(y) \tag{8}$$

Where K is a given kernel function in an input space. When K is semi positive definite, the existence of Φ is proven[8]. Most of the case ,however, the mapping Φ is high-dimensional and cannot be obtained explicitly. The vector in the feature space is not observable and only the inner product between vectors can be observed via a kernel function. However, for a given data set, it is possible to approximate Φ by empirical kernel map proposed by Scholkopf[12] and Tsuda[13] which is defined as $\Psi_N : \mathbf{R}^d \to \mathbf{R}^N$

$$\begin{aligned} \Psi_N(x) &= [\Phi(x_1) \cdot \Phi(x), \cdots, \Phi(x_N) \cdot \Phi(x)]^T \\ &= [K(x_1, x), \cdots, K(x_N, x)]^T \end{aligned} \tag{9}$$

A performance evaluation of empirical kernel map was shown by Tsuda. He shows that support vector machine with an empirical kernel map is identical with the conventional kernel map[13]. The empirical kernel map $\Psi_N(x_N)$,however, do not form an orthonormal basis in $\mathbf{R}^N$, the dot product in this space is not the ordinary dot product. In the case of KPCA ,however, we can be ignored as the following argument. The idea is that we have to perform linear PCA on the $\Psi_N(x_N)$ from the empirical kernel map and thus diagonalize its covariance matrix. Let the $N \times N$ matrix $\Psi = [\Psi_N(x_1), \Psi_N(x_2), \ldots, \Psi_N(x_N)]$, then from equation (9) and definition of the kernel matrix we can construct $\Psi = NK$. The covariance matrix of the empirically mapped data is:

$$C_\Psi = \frac{1}{N}\Psi\Psi^T = NKK^T = NK^2 \tag{10}$$

In case of empirical kernel map, we diagonalize NK^2 instead of K as in KPCA. Mika shows that the two matrices have the same eigenvectors $\{u_k\}$[14]. The eigenvalues $\{\lambda_k\}$ of K are related to the eigenvalues $\{k_k\}$ of NK^2 by

$$\lambda_k = \sqrt{\frac{k_k}{N}} \tag{11}$$

and as before we can normalize the eigenvectors $\{v_k\}$ for the covariance matrix C of the data by dividing each $\{u_k\}$ by $\sqrt{\lambda_k N}$. Instead of actually diagonalize the covariance matrix C_Ψ, the IKPCA is applied directly on the mapped data $\Psi = NK$. This makes it easy for us to adapt the incremental eigenspace update method to KPCA such that it is also correctly takes into account the centering of the mapped data in an incremental way. By this result, we only need to apply the empirical map to one data point at a time and do not need to store the $N \times N$ kernel matrix.

4 Proposed Classification System

In earlier Section 3 we proposed an incremental KPCA method for nonlinear feature extraction. Feature extraction by incremental KPCA effectively acts a nonlinear mapping from the input space to an implicit high dimensional feature space. It is hoped that the distribution of the mapped data in the feature space has a simple distribution so that a classifier can classify them properly. But it is point out that extracted features by KPCA are global features for all input data and thus may not be optimal for discriminating one class from others. For classification purpose, after global features are extracted using they must be used as input data for classification. There are many famous classifier in machine learning field. Among them neural network is popular method for classification and prediction purpose. Traditional neural network approaches, however have suffered difficulties with generalization, producing models that can overfit the data. To overcome the problem of classical neural network technique, support vector machines(SVM) have been introduced. The foundations of SVM have been developed by Vapnik and it is a powerful methodology for solving problems in nonlinear classification. Originally, it has been introduced within the context of statistical learning theory and structural risk minimization. In the methods one solves convex optimization problems, typically by quadratic programming(QP). Solving QP problem requires complicated computational effort and need more memory requirement. LS-SVM overcomes this problem by solving a set of linear equations in the problem formulation. LS-SVM method is computationally attractive and easier to extend than SVM.

5 Experiment

To evaluate the performance of proposed classification system, experiment is performed by following step. First we evaluate the feature extraction ability of incremental KPCA(IKPCA). The disadvantage of incremental method is their accuracy compared to batch method even though it has the advantage of memory efficiency. So we shall apply proposed method to a simple toy data and image data set which will show the accuracy and memory efficiency of incremental KPCA compared to APEX model proposed by Kung[15] and batch KPCA. Next we will evaluate the training and generalization ability of proposed classifier on UCI benchmarking data and NIST handwritten data set. To do this, extracted features by IKPCA will be used as input for LS-SVM.

5.1 Toy Data

To evaluate the feature extraction accuracy and memory efficiency of IKPCA compared to APEX and KPCA we take nonlinear data used by Scholkoff[5]. Totally 41 training data set is generated by:

$$y = x^2 + 0.2\varepsilon : \ \varepsilon \ from \ N(0,1), x = [-1,1] \tag{12}$$

First we compare feature extraction ability of IKPCA to APEX model. APEX model is famous principal component extractor based on Hebbian learning rule. Applying toy data to IKPCA we finally obtain 2 eigenvectors. To evaluate the performance of two methods on same condition, we set 2 output nodes to standard APEX model.

In table 1 we experimented APEX method on various conditions. Generally neural network based learning model has difficulty in determining the parameters; for example learning rate, initial weight value and optimal hidden layer node. This makes us to conduct experiments on various conditions. $\| w \|$ is norm of weight vector in APEX and $\| w \| = 1$ means that it converges stable minimum. $cos\theta$ is angle between Eigenvector of KPCA and APEX, IKPCA respectively. $cos\theta$ of Eigenvector can be a factor of evaluating accuracy how much IKPCA and APEX is close to accuracy of KPCA. Table 1 nicely shows the two advantages of IKPCA compared to APEX: first, performance of IKPCA is better than APEX; second, the performance of IKPCA is easily improved by re-learning. Another

Table 1. Performance evaluation of IKPCA and APEX

Method	Iteration	Learning Rate	$\| w_1 \|$	$\| w_2 \|$	$cos\theta_1$	$cos\theta_2$	MSE
APEX	50	0.01	0.6827	1.4346	0.9993	0.7084	14.8589
APEX	50	0.05				do not converge	
APEX	500	0.01	1.0068	1.0014	0.9995	0.9970	4.4403
APEX	500	0.05	1.0152	1.0470	0.9861	0.9432	4.6340
APEX	1000	0.01	1.0068	1.0014	0.9995	0.9970	4.4403
APEX	1000	0.05	1.0152	1.0470	0.9861	0.9432	4.6340
IKPCA	100		1	1	1	1	0.0223

factor of evaluating accuracy is reconstruction error. Reconstruction error is defined as the squared distance between the image of x_N and reconstruction when projected onto the first i principal components.

$$\delta = |\Psi(x_N) - P_l\Psi(x_N)|^2 \tag{13}$$

In here P_l is the first i principal component. The MSE (Mean Square Error) value of reconstruction error in APEX is 4.4403 whereas IKPCA is 0.0223. This means that the accuracy of IKPCA is superior to standard APEX and similar to that of batch KPCA. Above results of simple toy problem indicate that IKPCA is comparable to the batch way KPCA and superior in terms of accuracy.

Next we will compare the memory efficiency of IKPCA compared to KPCA. To extract nonlinear features, IKPCA only needs D matrix and R matrix whereas KPCA needs kernel matrix. Table 2 shows the memory requirement of each method. Memory requirement of standard KPCA is 93 times more than IKPCA. We can see that IKPCA is more efficient in memory requirement than KPCA and has similar ability in extracting nonlinear features. By this simple toy problem we can show that IKPCA has similar ability in extracting nonlinear features compare to KPCA and more efficient in memory requirement than KPCA.

Table 2. Memory efficiency of IKPCA compared to KPCA on toy data

	KPCA	IKPCA
Kernel matrix	41 X 41	none
R matrix	none	3 X 3
D matrix	none	3 X 3
Efficiency ratio	93.3889	1

5.2 Reconstruction Ability

To compare the reconstruction ability of incremental eigenspace update method proposed by Hall to APEX model we conducted experiment on US National

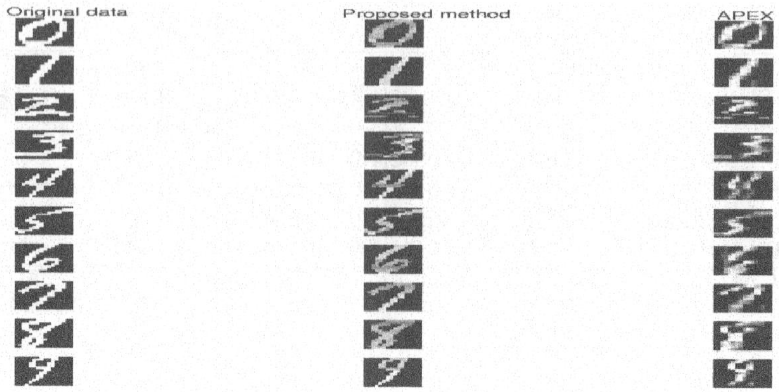

Fig. 1. Reconstructed image by IKPCA and APEX

Institute of Standards and Technology(NIST) handwritten data set. Data has been size-normalized and 16 X 16 images with their values scaled to the interval [0,1]. Applying this data to incremental eigenspace update method we finally obtain 6 Eigenvectors. As earlier experiment we set 6 output nodes to standard APEX method. Figure 1 shows the original data and their reconstructed images by incremental eigenspace update method and APEX respectively. We can see that reconstructed features by incremental eigenspace update method is more clear and similar to original image compared to APEX method.

5.3 UCI Machine Learning Repository

To test the performance of proposed classifier for real world data, we enlarge our experiment to the Cleveland heart disease data and wine data obtained from the UCI Machine Learning Repository. Detailed description of data is available from web site(http://www.ics.uci.edu/ mlearn/MLSummary.html). In this problem we randomly split training data as 80% and remaining as test data. A RBF kernel has been taken with and obtained by 10-fold cross-validation procedure to select the optimal hyperparameter. Table 3 shows the learning and generalization ability by proposed classifier.

Table 3. Training and generalization result by proposed classifier on UCI Machine Learning Repository

	Training	Generalization	Eigenvalue update criterion
Cleveland heart-disease	100%	98.69%	$\lambda' > 0.7\bar{\lambda}$
Wine data	100%	98.62%	$\lambda' > 0.7\bar{\lambda}$

By this result we can see that proposed classification system classifies well on specific data.

5.4 NIST Handwritten Data Set

To validate the above results on a widely used pattern recognition benchmark database, we conducted classification experiment on the NIST data set. This database originally contains 15,025 digit images. For computational reasons, we decided to use a subset of 2000 data set, 1000 for training and 1000 for testing. In this problem we use multiclass LS-SVM classifier proposed by Suykens[16]. An important issue for SVM is model selection. In [17] it is shown that the use of 10-fold cross-validation for hyperparameter selection of LS-SVMs consistently leads to very good results. In this problem RBF kernel has been taken and hyperparameter $\gamma_1 = 1.5198$, $\gamma_2 = 179.731$, $\gamma_3 = 10.51$, $\gamma_4 = 12.81$ and $\sigma_1 = 67.416$, $\sigma_2 = 656.351$, $\sigma_3 = 54.349$, $\sigma_4 = 57.909$ are obtained by 10-fold cross-validation technique.

Table 4. Training and generalization result on NIST handwritten data

	Training	Generalization	Eigenvalue update criterion
Proposed Classifier	100%	98.8%	$\lambda' > 0.7\overline{\lambda}$

Table 5. Misclassification frequency by proposed classification system on test data

Pattern	0	1	2	3	4	5	6	7	8	9	Total
Frequency	0	0	0	3	4	0	0	5	0	0	12

The results on the NIST data are given in Table 4 and 5. For this widely used pattern recognition problem, we can see that proposed classification system classifies well on given data.

6 Conclusion and Remarks

This paper is devoted to the exposition of a new technique on extracting nonlinear features and classification system from the incremental data. To develop this technique, we apply an incremental eigenspace update method to KPCA with an empirical kernel map approach. Proposed IKPCA has following advantages. Firstly, IKPCA has similar feature extracting performance for incremental and nonlinear data comparable to batch KPCA. Secondly, IKPCA is more efficient in memory requirement than batch KPCA. In batch KPCA the $N \times N$ kernel matrix has to be stored, while for IKPCA requirements are $O((k+1)^2)$. Here $k(1 \le k \le N)$ is the number of Eigenvectors stored in each eigenspace updating step, which usually takes a number much smaller than N. Thirdly, IKPCA allows for complete incremental learning using the eigenspace approach, whereas batch KPCA recomputes whole decomposition for updating the subspace of eigenvectors with another data. Finally, experimental results show that extracted features from IKPCA lead to good performance when used as a pre-preprocess data for a LS-SVM.

References

1. Tipping, M.E. and Bishop, C.M. :Mixtures of probabilistic principal component analysers. Neural Computation 11(2), (1998) 443-482
2. Kramer, M.A.:Nonlinear principal component analysis using autoassociative neural networks. AICHE Journal 37(2),(1991) 233-243
3. Diamantaras, K.I. and Kung, S.Y.:Principal Component Neural Networks: Theory and Applications. New York John Wiley & Sons, Inc.(1996)
4. Kim, Byung Joo. Shim, Joo Yong. Hwang, Chang Ha. Kim, Il Kon, "Incremental Feature Extraction Based on Emperical Feature Map," Foundations of Intelligent Systems, volume 2871 of Lecture Notes in Artificial Intelligence, pp 440-444, 2003
5. Softky, W.S and Kammen, D.M, "Correlation in high dimensional or asymmetric data set: Hebbian neuronal processing," Neural Networks vol. 4, pp.337-348, Nov. 1991.
6. Gupta, H., Agrawal, A.K., Pruthi, T., Shekhar, C., and Chellappa., R., "An Experimental Evaluation of Linear and Kernel-Based Methods for Face Recognition," accessible at http://citeseer.nj.nec.com.
7. Suykens, J.A.K. and Vandewalle, J.:Least squares support vector machine classifiers. Neural Processing Letters, vol.9, (1999) 293-300
8. Vapnik, V. N.:Statistical learning theory. John Wiley & Sons, New York (1998)
9. Hall, P. Marshall, D. and Martin, R.: Incremental eigenalysis for classification. In British Machine Vision Conference, volume 1, September (1998)286-295
10. Winkeler, J. Manjunath, B.S. and Chandrasekaran, S.:Subset selection for active object recognition. In CVPR, volume 2, IEEE Computer Society Press, June (1999) 511-516
11. Murakami, H. Kumar.,B.V.K.V.:Efficient calculation of primary images from a set of images. IEEE PAMI, 4(5), (1982) 511-515
12. Scholkopf, B. Smola, A. and Muller, K.R.:Nonlinear component analysis as a kernel eigenvalue problem. Neural Computation 10(5), (1998) 1299-1319
13. Tsuda, K., "Support vector classifier based on asymmetric kernel function," Proc. ESANN, 1999.
14. Mika, S.:Kernel algorithms for nonlinear signal processing in feature spaces. Master's thesis, Technical University of Berlin, November (1998)
15. Diamantaras, K.I. and Kung, S.Y, Principal Component Neural Networks: Theory and Applications, New York, John Wiley & Sons, Inc. 1996.
16. Suykens, J.A.K. and Vandewalle, J.: Multiclass Least Squares Support Vector Machines, In: Proc. International Joint Conference on Neural Networks (IJCNN'99), Washington DC (1999)
17. Gestel, V. Suykens, T. J.A.K. Lanckriet, G. Lambrechts, De Moor, A. B. and Vandewalle, J., "A Bayesian Framework for Least Squares Support Vector Machine Classifiers," Internal Report 00-65, ESAT-SISTA, K.U. Leuven.

A Metric Approach to Building Decision Trees Based on Goodman-Kruskal Association Index

Dan A. Simovici and Szymon Jaroszewicz

University of Massachusetts at Boston,
Department of Computer Science,
Boston, Massachusetts 02125, USA
{dsim,sj}@cs.umb.edu

Abstract. We introduce a numerical measure on sets of partitions of finite sets that is linked to the Goodman-Kruskal association index commonly used in statistics. This measure allows us to define a metric on such partions used for constructing decision trees. Experimental results suggest that by replacing the usual splitting criterion used in C4.5 by a metric criterion based on the Goodman-Kruskal coefficient it is possible, in most cases, to obtain smaller decision trees without sacrificing accuracy.

Keywords: Goodman-Kruskal association index, metric, partition, decision tree

1 Introduction

The construction of decision trees is centered around the selection algorithm of an attribute that generates a partition of the subset of the training data set that is located in the node about to be split. Over the years, several greedy techniques for choosing the splitting attribute have been proposed including the entropy gain and the gain ratio [1], the Gini index [2], the Kolmogorov-Smirnov metric [3, 4], or a metric derived from Shannon entropy [5]. In our previous work [6] we extended the metric splitting criterion introduced by L. de Mántaras by introducing metrics on the set of partitions of a finite set constructed by using generalized conditional entropy (which correspond to a generalization of entropy introduced by Daroczy [7]). This paper introduces a different type of metric on partitions of finite sets that is generated by a coefficient derived from the Goodman-Kruskal association index and shows that this metric can be applied succesfully to the construction of decision trees.

The purpose of this note is to define a metric on the set of partitions of a finite set that is derived from to the Goodman-Kruskal association index. A general framework of classification can be formulated starting with two finite random variables

$$X : \begin{pmatrix} a_1 & \cdots & a_l \\ p_1 & \cdots & p_l \end{pmatrix} \text{ and } Y : \begin{pmatrix} b_1 & \cdots & b_k \\ q_1 & \cdots & q_k \end{pmatrix}$$

H. Dai, R. Srikant, and C. Zhang (Eds.): PAKDD 2004, LNAI 3056, pp. 181–190, 2004.

We assume that we deal with a finite probability space where the elementary events are pairs of values (a_i, b_j), where a is a value of X and b_j is a value of Y. The classification rule adopted here is that an elementary event is classified in the class that has the maximal probability. Thus, in the absence of any knowledge about X, an elementary event will be classified in the Y-class b_j if b_j corresponds to the highest value among the probabilities $P(Y = b_j)$ for $1 \leq j \leq k$. If $P(Y = b_j|X = a_i)$ is the probability of predicting the value b_j for Y when $X = a_i$, then an event that has the component $X = a_i$ will be classified in the Y-class b_j if j is the number for which $P(Y = b_j|X = a_i)$ has the largest value. The probability of misclassification committed by applying this rule is $1 - \max_{1\leq j\leq k} P(Y = b_j|X = a_i)$.

The original Goodman-Kruskal association index $\lambda_{Y|X}$ (see [8,9]) is the relative reduction in the probability of prediction error:

$$\begin{aligned}
&\lambda_{Y|X} \\
&= 1 - \frac{\mathsf{GK}(X,Y)}{1-\max_{1\leq j\leq k} P(Y=b_j)} \\
&= \frac{\sum_{i=1}^{l} P(X=a_i)\max_{1\leq j\leq k} P(Y=b_j|X=a_i) - \max_{1\leq j\leq k} P(Y=b_j)}{1-\max_{1\leq j\leq k} P(Y=b_j)}.
\end{aligned}$$

In other words, $\lambda_{Y|X}$ is the proportion of the relative error in predicting the value of Y that can be eliminated by knowledge of the X-value.

The Goodman-Kruskal coefficient of X and Y that we use is defined by:

$$\begin{aligned}
\mathsf{GK}(X,Y) &= \sum_{i=1}^{l} P(X = a_i)\left(1 - \max_{1\leq j\leq k} P(Y = b_j|X = a_i)\right) \\
&= 1 - \sum_{i=1}^{l} P(X = a_i) \max_{1\leq j\leq k} P(Y = b_j|X = a_i) \\
&= 1 - \sum_{i=1}^{l} \max_{1\leq j\leq k} P(Y = b_j \wedge X = a_i).
\end{aligned}$$

Thus, $\mathsf{GK}(X,Y)$ is the expected value of the probability of misclassification. This coefficient is related to $\lambda_{Y|X}$ by:

$$G(X,Y) = (1 - \lambda_{Y|X})\left(1 - \max_{1\leq j\leq k} P(Y = b_j)\right)$$

Next, we formulate a definition of the Goodman-Kruskal coefficient GK within an algebraic setting, using partitions of finite sets. The advantage of this formulation is the possibility of using lattices of partitions of finite sets and various operations on partitions.

A partition of a set S is a collection of nonempty subsets of S, $\pi = \{B_i \mid i \in I\}$ such that $B_i \cap B_j = \emptyset$ for every $i, j \in I$ such that $i \neq j$ and $\bigcup_{i\in I} B_i = S$. Note that a partition $\pi = \{B_1, \ldots, B_l\}$ of a finite set S generates a finite random variable:

$$X : \begin{pmatrix} 1 & \cdots & l \\ p_1 & \cdots & p_l \end{pmatrix},$$

where $p_i = \frac{|B_i|}{|S|}$ for $1 \leq i \leq l$, and thus, the Goodman-Kruskal coefficient can be formulated in terms of partitions of finite sets.

If $\pi, \sigma \in \mathsf{PART}(S)$ we write $\pi \leq \sigma$ if each block of π is a subset of a block of σ. This is equivalent to saying that every block of σ is a union of blocks of π. We obtain a partial ordered set $(\mathsf{PART}(S), \leq)$. The least partition of S is the unit partition $\iota_S = \{\{a\} \mid a \in S\}$; the largest partition is the one-block partition $\omega_S = \{S\}$. The partial ordered set $(\mathsf{PART}(S), \leq)$ is a semi-modular lattice (see [10]), where $\inf\{\pi, \sigma\}$ is the partition $\pi \wedge \sigma$ whose blocks consist of intersections of blocks $B \cap C$, where $B \in \pi$ and $C \in \sigma$. Note that π is covered by σ (that is, $\pi < \sigma$ and there is no $\theta \in \mathsf{PART}(S)$ such that $\pi < \theta < \sigma$) if and only if σ is obtained from π by fusing together two blocks of π.

The *trace of a partition* $\pi = \{B_1, \ldots, B_k\}$ from $\mathsf{PART}(S)$ on a subset R of S is the partition $\pi_R \in \mathsf{PART}(R)$ given by $\pi_R = \{B_1 \cap R, \ldots, B_l \cap R\}$.

If S, T are two disjoint sets and $\pi \in \mathsf{PART}(S), \sigma \in \mathsf{PART}(T)$, then we denote by $\pi + \sigma$ the partition of $S \cup T$ that consists of all the blocks of π and σ. It is easy to see that "+" is an associative partial operation.

Definition 1. *Let* $\pi = \{B_1, \ldots, B_k\}$ *and* $\sigma = \{C_1, \ldots, C_l\}$ *be two partitions of a set* S*. The Goodman-Kruskal coefficient of* π *and* σ *is the number:*

$$\mathsf{GK}(\pi, \sigma) = 1 - \frac{1}{|S|} \sum_{i=1}^{k} \max_{1 \leq j \leq l} |B_i \cap C_j|.$$

Decision trees are built from data that has a tabular structure common in relational databases. As it is common in the relational terminology (see [11], for example), we regard a table as a triple $\tau = (T, H, \rho)$, where T is a string that gives the name of the table, $H = \{A_1, \ldots, A_n\}$ is a finite set of symbols (called the attributes of τ), and ρ is a relation, $\rho \subseteq \mathrm{Dom}(A_1) \times \cdots \times \mathrm{Dom}(A_n)$. Here $\mathrm{Dom}(A_i)$ is the domain of the attribute A_i for $1 \leq i \leq n$.

A set of attributes $L \subseteq H$ determines a partion π^L on the relation ρ, that is, on the set of tuples of the table τ, where two tuples belong to the same block if they have equal projections on L. It is easy to see that if L, K are two sets of attributes, then $\pi^{LK} = \pi^L \wedge \pi^K$.

The classical technique for building decision trees is using the entropy gain ratio as a criterion for choosing for every internal node of the tree the splitting attribute that maximizes this ratio (see [1]). The construction has an inductive character. If $\tau = (T, H, \rho)$ is the data set used to build the decision tree $\mathcal{T}$, let v be a node of $\mathcal{T}$ that is about to be split and let ρ_v be the set of tuples that corresponds to v. Suppose that the target partition of the data set ρ is θ. Then, the trace of this partition on ρ_v is θ_{ρ_v}.

Choosing the splitting attribute for a node v of a decision tree $\mathcal{T}$ for τ based on the minimal value of $\mathsf{GK}(\pi_v^A, \theta_{\rho_v})$ alone does not yield decision trees with good accuracy. A lucid discussion of those issues can be found in [12,4]. However, we will show that the GK coefficient can be used to define a metric on the set of partitions of a finite set that can be succesfully used for choosing splitting attributes. The decision trees that result are smaller, have fewer leaves (and therefore, less fragmentation) compared with trees built by using the standard gain ratio criterion; also, they have comparable accuracy.

2 The Goodman-Kruskal Metric Space

The main result of this section is a construction of a metric d_{GK} on the set of partitions of a finite set that is related to the Goodman-Kruskal coefficient and can be used for constructing decision trees. To introduce this metric we need to establish several properties of GK. Unless we state otherwise, all sets considered here are finite.

Theorem 1. *Let S be a set and let $\pi, \sigma \in \mathsf{PART}(S)$. We have $\mathsf{GK}(\pi, \sigma) = 0$ if and only if $\pi \leq \sigma$.*

Proof. It is immediate that $\pi \leq \sigma$ implies $\mathsf{GK}(\pi, \sigma) = 0$. Conversely, if $\mathsf{GK}(\pi, \sigma) = 0$, then $\sum_{i=1}^{k} \max_{1 \leq j \leq l} |B_i \cap C_j| = |S|$, which means that for each block B_i of π, there is a block C_j such that $|B_i \cap C_j| = |B_i|$. This is possible only if $B_i \subseteq C_j$, that is, if $\pi \leq \sigma$, which gives the desired conclusion.

Theorem 2. *The function GK is monotonic in the first argument and dually monotonic in the second argument.*

Proof. To prove the first part of the statement let $\pi = \{B_1, \ldots, B_k\}$, $\pi' = \{B'_1, \ldots, B'_m\}$, and $\sigma = \{C_1, \ldots, C_l\}$ be three partitions of S such that $\pi \leq \pi'$. Then, for every block B'_r of π' there is a collection of blocks of π: $B_{i_1}, \ldots, B_{i_s}$ such that $B'_r = B_{i_1} \cup \cdots \cup B_{i_s}$. Consequently for every m, $1 \leq m \leq l$ we can write:

$$\begin{aligned} |B'_r \cap C_m| &= |B_{i_1} \cap C_m| + \cdots + |B_{i_s} \cap C_m| \\ &\leq \max_{1 \leq j \leq l} |B_{i_1} \cap C_j| + \cdots + \max_{1 \leq j \leq l} |B_{i_s} \cap C_j|. \end{aligned}$$

Thus, we obtain:

$$\max_{1 \leq j \leq l} |B'_r \cap C_j| \leq \max_{1 \leq j \leq l} |B_{i_1} \cap C_j| + \cdots + \max_{1 \leq j \leq l} |B_{i_s} \cap C_j|,$$

which implies $\mathsf{GK}(\pi, \sigma) \geq \mathsf{GK}(\pi', \sigma)$.

To prove the second part, let σ, σ' be two partitions such that $\sigma \leq \sigma'$. We show that $\mathsf{GK}(\pi, \sigma) \geq \mathsf{GK}(\pi, \sigma')$. It suffices to show that σ' covers σ, that is, $\sigma = \{C_1, \ldots, C_{l-2}, C_{l-1}, C_l\}$ and $\sigma' = \{C_1, \ldots, C_{l-2}, C_{l-1} \cup C_l\}$. In other words, the blocks of σ' coincide with the blocks of σ with the exception of one block that is obtained by fusing two blocks of σ. Note that for a given block B_i of π we have:

$$\max_{1 \leq j \leq l} |B_i \cap C_j| \leq \max\{\max_{1 \leq j \leq l-2} |B_i \cap C_j|, |B_i \cap (C_{l-1} \cup C_l)|\},$$

which implies $\mathsf{GK}(\pi, \sigma) \geq \mathsf{GK}(\pi, \sigma')$. □

The next result has a technical character:

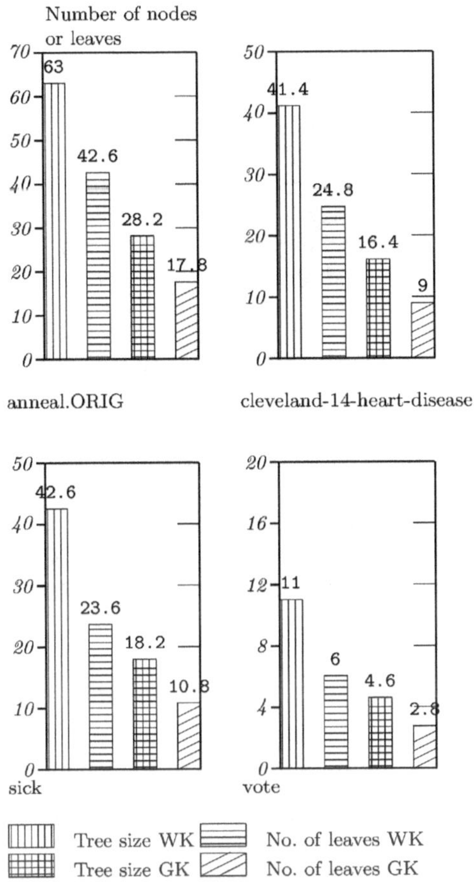

Fig. 1. Comparative Experimental Results

Theorem 3. *For every three partitions θ, π, σ of a finite set S we have:*

$$\mathit{GK}(\pi \wedge \theta, \sigma) + \mathit{GK}(\theta, \pi) \geq \mathit{GK}(\theta, \pi \wedge \sigma).$$

Proof. See Appendix A.

Theorem 4. *Let θ, π, σ be partitions of a set S. We have*

$$\mathit{GK}(\theta, \pi) + \mathit{GK}(\pi, \sigma) \geq \mathit{GK}(\theta, \sigma).$$

Proof. Note that

$$\mathsf{GK}(\theta, \pi) + \mathsf{GK}(\pi, \sigma) \geq \mathsf{GK}(\theta, \pi) + \mathsf{GK}(\pi \wedge \theta, \sigma)$$

due to the monotonicity of GK in its first argument. By Theorem 3

$$\mathsf{GK}(\theta, \pi) + \mathsf{GK}(\pi, \sigma) \geq \mathsf{GK}(\theta, \pi \wedge \sigma) \geq \mathsf{GK}(\theta, \sigma),$$

because of the dual monotonicity of GK in its second argument. □

Corollary 1. *The mapping* $d_{GK} : \mathsf{PART}(S) \times \mathsf{PART}(S) \longrightarrow \mathbb{R}$ *given by:*

$$d_{GK}(\pi, \sigma) = \mathsf{GK}(\pi, \sigma) + \mathsf{GK}(\sigma, \pi)$$

for $\pi, \sigma \in \mathsf{PART}(S)$, *is a metric on the set* $\mathsf{PART}(S)$.

Proof. By Theorem 1 we have $d_{GK}(\pi, \sigma) = 0$ if and only if $\pi = \sigma$. Also, the definition of d_{GK} implies $d_{GK}(\pi, \sigma) = d_{GK}(\sigma, \pi)$ for every $\pi, \sigma \in \mathsf{PART}(S)$.

Finally, the triangular inequality $d_{GK}(\pi, \sigma) + d_{GK}(\sigma, \theta) \geq d_{GK}(\pi, \theta)$ for $\pi, \sigma, \theta \in \mathsf{PART}(S)$ follows immediately from Theorem 4. □

3 The Goodman-Kruskal Splitting Criterion for Decision Trees

Let $\tau = (T, H, \rho)$ be the table that contains the training data set that is used to build a decision tree $\mathcal{T}$. Assume that we are about to expand the node v of the tree $\mathcal{T}$. Using the notations introduced in Section 1, we choose to split the note v using an attribute A_i that minimizes the distance $d_{GK}(\pi^{A_i}_{\rho_v}, \theta_{\rho_v})$.

The d_{GK} metric does not favor attributes with large domains as splitting attributes, an issue that is important for building decision trees.

Theorem 5. *Let* S *be a finite set and let* $\pi, \pi', \sigma \in \mathsf{PART}(S)$ *be such that* $\pi' \leq \pi$. *If there exists a block* C *of* σ *and a block* B *of* π *such that* $B \subseteq C$, *then* $d_{GK}(\pi, \sigma) \leq d_{GK}(\pi', \sigma)$.

Proof. We can assume, without restricting generality, that π' is covered by π, that is, $\pi = \{B_1, \ldots, B_k\}$, $B = B_k$, $\pi' = \{B_1, \ldots, B'_k, B''_k\}$, where $B_k = B'_k \cup B''_k$. Also, let $\sigma = \{C_1, \ldots, C_l\}$, where $C_l = C$.

Theorem 2 implies that $\mathsf{GK}(\sigma, \pi') \leq GK(\sigma, \pi)$ (due to the dual monotonicity in the second argument of GK). We prove that, under the assumptions made in the theorem, we have $\mathsf{GK}(\pi', \sigma) = \mathsf{GK}(\pi, \sigma)$, which implies the desired inequality. Indeed, note that:

$$\begin{aligned}
&\mathsf{GK}(\pi', \sigma)\\
&= 1 - \frac{1}{|S|}\left(\sum_{i=1}^{k-1} \max_{1 \leq j \leq l} |B_i \cap C_j| + \max_{1 \leq j \leq l} |B'_k \cap C_j| + \max_{1 \leq j \leq l} |B''_k \cap C_j|\right)\\
&= 1 - \frac{1}{|S|}\left(\sum_{i=1}^{k-1} \max_{1 \leq j \leq l} |B_i \cap C_j| + |B'_k| + |B''_k|\right)\\
&= 1 - \frac{1}{|S|}\sum_{i=1}^{k} \max_{1 \leq j \leq l} |B_i \cap C_j| = \mathsf{GK}(\pi, \sigma)
\end{aligned}$$

because $B'_k, B''_k \subseteq B_k \subseteq C$. □

We note that the Theorem 5 is similar to the property of the metric generated by the Shannon entropy obtained by L. de Mántaras in [5] and generalized by us in [6].

Table 1. Experimental Results: Entropy Gain Ratio vs. d_{GK}

	Entropy Gain Ratio			d_{GK}		
Dataset	acc	tree size	no. of leafs	acc	tree size	no. of leafs
anneal	98.55	46.4	36.2	98.55	37.2	26.4
anneal.ORIG	90.20	63	42.6	86.30	28.2	17.8
audiology	78.76	46	29	77.41	37.4	24
autos	80	64.6	48.2	67.80	49.6	27.6
balance-scale	78.4	73.8	37.4	77.76	57	29
breast-cancer	73.09	21.2	17.2	73.78	18	13.4
wisc-breast-cancer	94.12	17.4	7.2	94.85	17	9
horse-colic	85.85	8.4	5.8	81.78	7.6	4.4
credit-rating	86.23	29.2	20.8	83.91	20.4	11.6
german-credit	72.9	108	77.6	69.5	63.4	36.8
pima-diabetes	75.65	42.6	21.8	70.96	88.6	44.8
Glass	67.26	39.4	20.2	70.09	33.4	17.2
clev.-14-heart-disease	77.53	41.4	10.4	75.89	16.4	9
hung.-14-heart-disease	78.57	9.8	6.4	80.28	10	6.2
heart-statlog	75.55	26.6	13.8	71.85	17.4	9.2
hepatitis	78.06	13.4	7.2	82.58	9	5
hypothyroid	99.46	25.8	13.4	99.39	21	11
ionosphere	89.73	25.8	13.4	88.89	16.2	8.6
iris	95.33	8.2	4.6	95.33	6.6	3.8
kr-vs-kp	99.15	51.8	27.4	98.46	76.4	39.8
labor	78.63	6.8	4	84.09	3	2
lymphography	80.41	24.4	14.8	79.01	14.8	8.8
mushroom	100	29.4	24.4	100	31.8	25
primary-tumor	40.99	77	41.2	43.64	38.8	21.4
segment	97.09	81.8	41.4	94.02	67	34
sick	98.75	42.6	23.6	98.35	18.2	10.8
sonar	74.03	23.8	12.4	69.16	29.4	15.2
soybean	91.21	89.4	58.4	90.19	105.2	71.2
splice	94.04	199.6	160.8	93.51	194.4	156.6
vehicle	72.10	117.8	59.4	65.60	128.2	64.6
vote	96.55	11	6	94.71	4.6	2.8
vowel	78.18	200.4	120.2	63.43	235	125.8
zoo	93.09	14.6	7.8	93.09	14.6	7.8
average	83.92	50.95	31.84	82.25	45.93	27.29

Next, we compare parameters of decision trees constructed on UCI machine learning datasets [13] by using Entropy Gain Ratio and the Goodman-Kruskal distance d_{GK}. The experiments have been conducted using the J48 (a variant of C4.5) algorithm from the Weka Package [14], modified to use different splitting criteria. The pruning steps of decision tree construction are left unchanged. To verify the accuracy, we used 5-fold cross-validation. For each splitting criterion we present three characteristics of the generated trees: accuracy (percentage of correctly predicted cases), size of the tree (total number of nodes) and the number of leaves in the tree. All are averaged over the 5-fold of cross-validation.

Overall d_{GK} produced smaller trees for 24 out of 33 datasets considered. In 4 cases (`anneal.ORIG, clev.-14-heart-disease, sick, vote`) over 50% reduction was achieved. In one case (`pima-diabetes`) a sharp increase was observed. On average the trees obtained were 10% smaller.

The accuracy of trees constructed using d_{GK} was on average 1.67% worse than that of trees constructed using standard Weka version. In one case (`autos`) the decrease was significant but for all other cases it was rather moderate, and in a few cases d_{GK} produced more accurate trees. Small tree size is an advantage since, in general, small trees are much easier to understand. The total number of nodes and the number of leaves in the tress were highly correlated so we can talk simply about size of the tree.

The best results obtained from experiments are also shown in Figure 1.

Splitting nodes by using an attribute A that minimizes $\mathsf{GK}(\pi^A_{\rho_v}, \theta_{\rho_v})$ instead of $d_{GK}(\pi^A_{\rho_v}, \theta_{\rho_v})$ may result in a substantial loss of accuracy. For example, in the case of the `hungarian-14-heart-disease` dataset, the accuracy obtained using GK, under comparable conditions (averaging over 5-fold cross validation) is just 70.05% compared to 78.57% obtained by using the entropy gain ratio, or 80.28% obtained in the case of d_{GK}. This confirms the claim in the literature of the unsuitability of using $\mathsf{GK}(\pi^A_{\rho_v}, \theta_{\rho_v})$ alone as a splitting criterion.

References

1. Quinlan, J.R.: C4.5: Programs for Machine Learning. Morgan Kaufmann, San Mateo, CA (1993)
2. Breiman, L., Friedman, J.H., Ohlsen, R.A., Stone, C.J.: Classification and Regression Trees. Chapman & Hall/CRC, Boca Raton (1984) Republished 1993.
3. Utgoff, P.E.: Decision tree induction based on efficient tree restructuring. Technical Report 95-18, University of Massachusetts, Amherst (1995)
4. Utgoff, P.E., Clouse, J.A.: A Kolmogorov-Smirnoff metric for decision tree induction. Technical Report 96-3, University of Massachusetts, Amherst (1996)
5. de Mántaras, R.L.: A distance-based attribute selection measure for decision tree induction. Machine Learning **6** (1991) 81–92
6. Simovici, D.A., Jaroszewicz, S.: Generalized conditional entropy and decision trees. In: Proceedings of EGC 2003, Lyon, France (2003) 369–380
7. Daróczy, Z.: Generalized information functions. Information and Control **16** (1970) 36–51
8. Goodman, L.A., Kruskal, W.H.: Measures of Association for Cross-Classification. Volume 1. New York, Springer-Verlag (1980)
9. Liebtrau, A.M.: Measures of Association. SAGE, Beverly Hills, CA (1983)
10. Grätzer, G.: General Lattice Theory. Second edn. Birkhäuser, Basel (1998)
11. Simovici, D.A., Tenney, R.L.: Relational Database Systems. Academic Press, New York (1995)
12. Breiman, L., Friedman, J.H., Olshen, R.A., Stone, C.J.: Classification and Regression Trees. Chapman and Hall, Boca Raton (1998)
13. Blake, C.L., Merz, C.J.: UCI Repository of machine learning databases. University of California, Irvine, Dept. of Information and Computer Sciences, http://www.ics.uci.edu/~mlearn/MLRepository.html (1998)
14. Witten, I.H., Frank, E.: Data Mining. Morgan-Kaufmann, San Francisco (2000)

A Proof of Theorem 3

We begin by showing that if $S_1, \ldots, S_n$ are pairwise disjoint sets, and $\pi_r, \sigma_r \in \mathsf{PART}(S_r)$ for $1 \leq r \leq n$, then

$$\mathsf{GK}(\pi_1 + \cdots + \pi_n, \sigma_1, \ldots, \sigma_n) = \sum_{r=1}^{n} \frac{|S_r|}{|S|} \mathsf{GK}(\pi_r, \sigma_r). \tag{1}$$

Let $\pi_p = \{B_1^p, \ldots, B_{l_p}^p\}$ and $\sigma_q = \{C_1^q, \ldots, C_{k_q}^q\}$ for $1 \leq p, q \leq n$. Then, we can write:

$$\begin{aligned}
&\mathsf{GK}(\pi_1 + \cdots + \pi_n, \sigma_1 + \cdots + \sigma_n)\\
&= 1 - \frac{1}{|S|} \sum_{p,i} \max_{q,j} |B_i^p \cap C_j^q|\\
&= 1 - \frac{1}{|S|} \sum_{p,i} \max_{p,j} |B_i^p \cap C_j^p|\\
&\quad \text{(because } p \neq q \text{ implies } B_i^p \cap C_j^q = \emptyset)\\
&= \sum_{p=1}^{n} \frac{|S_p|}{|S|} \left(1 - \sum_{p=1}^{n} \frac{1}{|S_p|} \sum_{i=1}^{l_p} \max_{1 \leq j \leq k_p} |B_i^p \cap C_j^p|\right)\\
&= \sum_{p=1}^{n} \frac{|S_p|}{|S|} \mathsf{GK}(\pi_p, \sigma_p),
\end{aligned}$$

which is the desired equality.

Let now $\mathcal{K}(\sigma)$ be the number:

$$\mathcal{K}(\sigma) = \mathsf{GK}(\omega_S, \sigma) = 1 - \frac{1}{|S|} \max_{1 \leq j \leq k} |C_j|.$$

We claim that if $\pi, \sigma \in \mathsf{PART}(S)$, then:

$$\mathsf{GK}(\pi, \sigma) \geq \mathcal{K}(\pi \wedge \sigma) - \mathcal{K}(\pi). \tag{2}$$

Let $\pi = \{B_1, \ldots, B_k\}$ and $\sigma = \{C_1, \ldots, C_l\}$. We can write:

$$\begin{aligned}
\mathsf{GK}(\pi, \sigma) &= |S| - \sum_{i=1}^{k} \max_{1 \leq j \leq l} |B_i \cap C_k|\\
&= \sum_{i=1}^{k} (|B_i| - \max_{1 \leq j \leq l} |B_i \cap C_j|\\
&\geq \max_{1 \leq i \leq k} (|B_i| - \max_{1 \leq j \leq l} |B_i \cap C_j|)\\
&\geq \max_{1 \leq i \leq k} |B_i| - \max_{1 \leq i \leq k, 1 \leq j \leq l} |B_i \cap C_j|\\
&= \mathcal{K}(\pi \wedge \sigma) - \mathcal{K}(\pi),
\end{aligned}$$

which proves the inequality (2).

Let $\pi = \{B_1, \ldots, B_k\}$, $\theta = \{D_1, \ldots, D_m\}$ and $\sigma = \{C_1, \ldots, C_l\}$. We have:

$$\pi \wedge \theta = \pi_{D_1} + \cdots + \pi_{D_m} = \theta_{B_1} + \cdots + \theta_{B_k}.$$

Consequently, by Equality (1), we have:

$$\begin{aligned}
\mathsf{GK}(\pi \wedge \theta, \sigma) &= \mathsf{GK}(\pi_{D_1} + \cdots + \pi_{D_m}, \sigma)\\
&= \sum_{h=1}^{m} \frac{|D_h|}{|S|} \mathsf{GK}(\pi_{D_h}, \sigma_{D_h}).
\end{aligned}$$

Also, we have

$$\mathsf{GK}(\theta, \pi) = \sum_{h=1}^{m} \frac{|D_h|}{|S|} \mathcal{K}(\pi_{D_h}),$$

which implies

$$\mathsf{GK}(\pi \wedge \theta, \sigma) + \mathsf{GK}(\theta, \pi) = \sum_{h=1}^{m} \frac{|D_h|}{|S|} \left(\mathsf{GK}(\pi_{D_h}, \sigma_{D_h}) + \mathcal{K}(\pi_{D_h})\right).$$

The Inequality (2) implies:

$$\mathsf{GK}(\pi_{D_h}, \sigma_{D_h}) + \mathcal{K}(\pi_{D_h}) \geq \mathcal{K}(\pi_{D_h} \wedge \sigma_{D_h}) = \mathcal{K}((\pi \wedge \sigma)_{D_h}),$$

so we may conclude that:

$$\mathsf{GK}(\pi \wedge \theta, \sigma) + \mathsf{GK}(\theta, \pi) \geq \sum_{h=1}^{m} \frac{|D_h|}{|S|} \mathcal{K}((\pi \wedge \sigma)_{D_h}) = \mathsf{GK}(\theta, \pi \wedge \sigma).$$

□

DRC-BK: Mining Classification Rules with Help of SVM

Yang Zhang, Zhanhuai Li, Yan Tang, and Kebin Cui

Dept. Computer Science & Engineering, Northwestern Polytechnical University, P.R. China
{zhangy, lzh, tangyan}@co-think.com

Abstract. Currently, the accuracy of SVM classifier is very high, but the classification model of SVM classifier is not understandable by human experts. In this paper, we use SVM, which is applied with a Boolean kernel, to construct a hyper-plan for classification, and mine classification rules from this hyper-plane. In this way, we build *DRC-BK*, a decision rule classifier. Experiment results show that *DRC-BK* has a higher accuracy than some state-of-art decision rule (decision tree) classifiers, such as *C4.5*, *CBA*, *CMAR*, *CAEP* and so on.

Keywords: Decision Rule Classifier, SVM, Boolean Kernel

1 Introduction

Currently, the classification accuracy of SVM classifier is very high. However, its classification model is non-understandable, which is very helpful in some applications, such as diagnoses information classification. In this paper, we use SVM, which is applied with a Boolean kernel, as a learning engine, and mine decision rules from the hyper-plane constructed by SVM, so as to build *DRC-BK* (Decision Rule Classifier based on Boolean Kernel), a decision rule classifier.

To our knowledge, the research to make the classification model of SVM classifiers understandable is not seen from the literature. The classifier-building algorithm and the classification algorithm for some decision tree (rule) classifiers, such as *C4.5*, *CBA*, *CMAR*, and *CAEP* are heuristic, lacking strong mathematical background; while *DRC-BK* mines knowledge from the hyper-plane constructed by SVM, which means that *DRC-BK* is based on the structural risk minimization theory.

2 *DRC-BK* Classifier

When applied with Boolean kernel, SVM could construct an optimal hyper-plane for classification by learning Boolean functions in the high dimensional feature space [1].

Proposition 1. Suppose $U \in \{0,1\}^n, V \in \{0,1\}^n, \sigma > 0$,

$$K_{MDNF}(U,V) = -1 + \prod_{i=1}^{n}(\sigma U_i V_i + 1)$$ is a Boolean kernel.

H. Dai, R. Srikant, and C. Zhang (Eds.): PAKDD 2004, LNAI 3056, pp. 191–195, 2004.

Please refer to [1] for the prove and model selection of K_{MDNF}. In our experiment, we simply set C to 0.5, set σ to 0.01, 0.1 or 0.5, and choose the best σ for mining classification rules.

After training, the classification function learned by SVM is $f(X)=\text{sgn}(\sum_{i=1}^{l}\alpha_i y_i K_{MDNF}(X_i,X)+b)$. Here, $\alpha_i, i=1,2,..l$ (l is the count of support vectors) and b are knowledge learned by SVM, which is non-understandable by human experts. Here, we present our *DRC-BK* classifier, which use SVM as its learning engine, and mine classification rules from the knowledge learned by the SVM. The detailed steps for *DRC-BK* are as following: 1, Construct a hyper-plane for classification by SVM, which is applied with K_{MDNF}. 2, Mining classification rules from this hyper-plane. 3, Classify the testing data with the rules learned in step 2.

Suppose $g(X)=\sum_{i=1}^{l}\alpha_i y_i K_{MDNF}(X_i,X)+b$, then, we have

$$g(X)=\sum(\sum_{i=1}^{l}\alpha_i y_i X_{i,s1}X_{s1})+\sum(\sum_{i=1}^{l}\alpha_i y_i X_{i,s1}X_{i,s2}X_{s1}X_{s2})+...+\sum(\sum_{i=1}^{l}\alpha_i y_i X_{i,s1}X_{i,s2}..X_{i,sn}X_{s1}X_{s2}..X_{sn})+b \cdot$$

If we look each dimension of the input space as a Boolean literal, then each dimension of feature space can be looked as a conjunction of several Boolean literals in the input space. Hence, $g(x)$ could be looked as a weighted linear sum of all these conjunctions. If the input space has n dimensions, then the feature space has 2^n-1 dimensions. However, these dimensions don't make the same contribution to classification. Most of them could be ignored because their contribution is too small.

Let a non-linear mapping ϕ maps X, a vector in input space, to Z, a vector in the feature space, then, $g(X)$ could be written as $g(X)=\sum_{i=1}^{2^n-1}W_{Zi}Z_i+b$. Here, Wz_i is the weight of dimension Z_i. Then, the contribution to classification made by Z_i could be measured as $\frac{\partial g}{\partial Z_i}=W_{Zi}$. This means that the bigger $|Wz_i|$ is, the more contribution to classification Z_i can make; and the smaller $|Wz_i|$ is, the less contribution to classification Z_i can make. This conclusion could be used for feature selection for linear SVM.

For a conjunction z of j Boolean literals, $z=X_{s1}X_{s2}...X_{sj}$, its weight could be calculated as: $w_z=\sum_{i=1}\alpha_i y_i X_{i,s1}X_{i,s2}...X_{i,sj}\sigma^j$. We make the following definition:

Definition 1. Classification Rule: For a classification rule $r=<z, w_z>$, z is a conjunction of j Boolean literals (j>0), and w_z is the weight of this rule calculated by the above formula.

Definition 2. *j*-length Rule: For a rule $r=<z, w_z>$, if z is a Boolean function made up of j Boolean literals, then we say that r is a j-length rule (or, a rule with length j).

In the knowledge learned by SVM, positive support vectors are the sample data which satisfy y_i=+1, and negative support vectors are the sample data which satisfy y_i=□ 1. Here, we write *SVP* and *SVN* for the set of positive support vectors and the set of negative support vectors, respectively. Then, $w_{z,SVP}=\sum_{i\in SVP}\alpha_i y_i X_{i,s1}X_{i,s2}..X_{i,sj}\sigma^j$ and

$w_{z,SVN}=\sum_{i\in SVN}\alpha_i y_i X_{i,s1} X_{i,s2} \ldots X_{i,sj}\sigma^j$ represents the ability of Boolean function $z=x_{s1}x_{s2}\ldots x_{sj}$ to distinguish positive sample data and negative sample data, respectively.

Definition 3. Interesting Rule: An interesting rule $<z, w_z>$ is a rule which satisfies $|w_{z,SVP}| \geq minWeight$ or $|w_{z,SVN}| \geq minWeight$. Here, $minWeight$ is a user defined parameter, representing the minimal weight.

Please note that $\alpha_i > 0 \quad i = 1,2,..l$, so, we have $|w_z| \leq |w_{z,SVP}|$ and $|w_z| \leq |w_{z,SVN}|$. Therefore, if $|w_{z,SVP}| < minWeight$ and $|w_{z,SVN}| < minWeight$, then $|w_z| < minWeight$.

Let's consider the conjunction $z=x_{s1}x_{s2}\ldots x_{sj}$ and $z'=x_{s1}x_{s2}\ldots x_{sj}\ x_{sj+1}$. The length of z is j; the length of z' is $j+1$; and z is contained in z'. The ability for z' to distinguish positive sample data and negative sample data is $w_{z',SVP} = \sum_{i\in SVP}\alpha_i y_i X_{i,s1} X_{i,s2} \ldots X_{i,sj} X_{i,sj+1}\sigma^{j+1}$ and

$w_{z',SVN} = \sum_{i\in SVN}\alpha_i y_i X_{i,s1} X_{i,s2} \ldots X_{i,sj} X_{i,sj+1}\sigma^{j+1}$, respectively. As we have discussed in section 3.3,

the value of σ satisfies $\sigma < 1$. So, we have $w_{z',SVP} \leq w_{z,SVP}$ and $w_{z',SVN} \leq w_{z,SVN}$. This means that for an arbitrary conjunction z', which contains the conjunction $z=x_{s1}x_{s2}\ldots x_{sj}$, z' doesn't have a stronger ability to distinguish positive (negative) sample data than z does.

Please refer to figure 1 for the algorithm for mining interesting classification rules. In this figure, the function *BF*(*R*) is used to calculate the set of conjunctions {z} from the rules in rule set $R=\{<z,w_z>\}$. Figure 2 gives the algorithm for mining classification rules. In our experiment, we simply set the parameter *minWeight* to *b**0.05. Here, *b* is the parameter learned by SVM. Figure 3 gives the classification algorithm for *DCR-BK*. Here, parameter *b* is the knowledge learned by SVM.

Algorithm 1:	Mining interesting rules from positive (negative) support vectors.
Input:	support vector set *SV* (could be *SVP* or *SVN*), weight vector α, minimal weight *minWeight*
Output:	interesting rule set

1, $R_1 = \{\langle z, w_{z,sv}\rangle \ \ |w_{z,sv} \geq minWeight\}$

2, $R_{all} = \Phi$, n=2

3,

$$R_n = \{\langle zz', w_{zz',sv}\rangle \ \ |z \in BF(R_{n-1}) \wedge z' \in BF(R_1) \wedge w_{zz',sv} \geq minWeight\}$$

4, $R_{all} = R_{all} \cup R_n$

5, if $R_n = \Phi$ goto 6, else n=n+1, goto 3

6, Output R_{all}

Fig. 1. Algorithm for Mining Interesting Rules

Algorithm 2: Mining classification rules.
Input: support vector set *SVP* and *SVN*, weight vector α, minimal weight *minWeight*
Output: classification rule set *R*
1, Following algorithm 1, mine interesting rules set R_P from *SVP*
2, Following algorithm 1, mine interesting rules set R_N from *SVN*
3, $R = \{ < r, w_r > / r \in BF(R_P \cup R_N) \}$
4, Output *R*

Fig. 2. Algorithm for Mining Classification Rules

Algorithm 3: Classification Algorithm
Input: rule set *R*, testing data *sample*, parameter *b*
Output: the class type of testing data *sample*
1, *f*=*b*
2, for each *rule* in *R*
    if *rule.z* matches *sample* then *f*=*f*+*rule.w*
3, Output *sgn(f)*

Fig. 3. Classification Algorithm

Table 1. The comparision of classificaion accuracy of *DRC-BK* and 9 other classifiers.

NAME	#INS	#ATTR	#BOOL	C4.5	CBA	CMAR	DEep	CAEP	SVM	DRC-BK	#RULE	TIME
AUSTRA	690	14	50	84.7	84.9	86.1	84.78	86.21	84.49	85.36	1667	3.461
DIABETES	768	8	15	74.2	74.5	75.8	76.82	x	77.73	79.04	1072	1.032
GERMAN	1000	200	60	72.3	73.4	74.9	74.4	72.5	74.9	75.2	93	2.284
HEART	270	13	18	80.8	81.9	82.2	81.11	83.7	81.48	83.33	815	0.811
IONO	351	34	143	90	92.3	91.5	86.23	90.04	90.03	90.6	115	0.252
PIMA	768	8	15	75.5	72.9	75.1	76.82	75	77.21	77.86	96	0.148
SONAR	208	60	42	70.2	77.5	79.4	84.16	x	87.02	85.58	591	0.758
TIC-TAC	958	9	27	99.4	99.6	99.2	99.06	99.06	98.33	99.79	9623	42.63
BREAST	699	10	30	95	96.3	96.4	96.42	97.28	96.42	96.85	321	0.332
CLEVE	303	13	29	78.2	82.8	82.2	87.17	83.25	83.17	83.17	1495	3.189
CRX	690	15	61	84.9	84.7	84.9	84.18	x	86.24	85.51	993	1.852
HEPATIC	155	19	46	80.6	81.8	80.5	81.18	83.03	85.81	86.45	293	0.334
HORSE	368	22	78	82.6	82.1	82.6	84.21	x	82.34	84.51	956	1.87
HYPO	3163	25	57	99.2	98.9	98.4	97.19	x	99.34	97.5	289	0.534
LABOR	57	16	41	79.3	86.3	89.7	87.67	x	92.98	92.98	1984	2.545
SICK	2800	29	63	98.5	97	97.5	94.03	x	97.29	97.32	1740	4.548
AVERAGE	828	30.9	48.4	84.1	85.4	86.03	85.96	85.56	87.17	**87.57**	1384	4.161

3 Experiment Results

In order to compare the classification accuracy of *DRC-BK* with other classifiers, we made experiment on 16 binary datasets from UCI dataset. We made our experiment in the 10-fold cross validation way, and report the average classification accuracy. Table 1 gives the experiment result. In table 1, column 1 lists the name of 16 datasets used in our experiment. Column 2, 3, and 4 gives the number of samples, the number of attributes in the original dataset, and the number of attributes after pre-processing, respectively. Column 5, 6, and 7 give the classification accuracy of *C4.5*, *CBA*, and *CMAR*, respectively. These experiment results are copied from [2]. Column 8, 9, and 10 gives the classification accuracy of *DEep*, *CAEP*, and linear SVM, respectively. Column 11, 12, and 13 gives the classification accuracy, the number of rules, and the executing time for *DRC-BK* to mine classification rules from the hyper plane. From table 1, we can see that *DRC-BK* has the best average classification accuracy among the 7 classifiers.

4 Conclusion and Future Work

In this paper, we present a novel decision rule classifier, *DRC-BK*, which has high classification accuracy and makes it possible to study the generalization error of decision rule classifiers quantitatively in the future. In order to refine the rule set, our future research is to find a better Boolean kernel for mining classification rules.

References

1. Zhang Yang, Li Zhanhuai, Kang Muling, Yan Jianfeng.: Improving the Classification Performance of Boolean Kernels by Applying Occam's Razor. In The 2nd International Conference on Computational Intelligence, Robotics and Autonomous Systems (CIRAS'03), 2003.
2. W. Li, J. Han, and J. Pei.: CMAR: Accurate and Efficient Classification Based on Multiple Class-association Rules. In Proc. the 2001 IEEE International Conference on Data Mining (ICDM'01), 2001.

A New Data Mining Method Using Organizational Coevolutionary Mechanism

Jing Liu, Weicai Zhong, Fang Liu, and Licheng Jiao

Institute of Intelligent Information Processing, Xidian University, Xi'an 710071, China

Abstract. Organizational coevolutionary algorithm for classification (OCEC), is designed with the intrinsic properties of data mining in mind. OCEC makes groups of examples evolved, and then rules are extracted from these groups of examples at the end of evolution. OCEC is first compared with G-NET and JoinGA. All results show that OCEC achieves a higher predictive accuracy. Then, the scalability of OCEC is studied. The results show that the classification time of OCEC increases linearly.

1 Introduction

A new classification algorithm, organizational coevolutionary algorithm for classification (OCEC) is designed to deal with the classification task in data mining. The new approach adopts the coevolutionary model of multiple populations, focusing on extracting rules from examples. The main difference between it and the existed EA-based classification algorithms is its use of a bottom-up search mechanism. This approach makes groups of examples evolved, and then rules are extracted from these groups of examples at the end of evolution.

2 An Organizational Coevolutionary Algorithm for Classification

In order to avoid confusion about terminology, some concepts are explained.

DEFINITION 1. Let $\overline{A_i}$ be a set of attribute values $\{a_{i1}, a_{i2}, \ldots, a_{ik}\}$. An instance space $\mathcal{I}$ is the Cartesian product of sets of attribute values, $\mathcal{I} = \overline{A_1} \times \overline{A_2} \times \ldots \times \overline{A_n}$. An attribute $A_i : \mathcal{I} \rightarrow \overline{A_i}$ is a projection function from the instance space to a set of attribute values. An instance i is an element of $\mathcal{I}$, $\mathcal{E} \subset \mathcal{I} \times \mathcal{C}$ a set of examples and an example e an element of $\mathcal{I} \times \mathcal{C}$, where $\mathcal{C}$ class name set.

DEFINITION 2. An **Organization**, *org*, is a set of examples belonging to the same class and the intersection of different organizations is empty. The examples in an organization are called Members.

DEFINITION 3. If all members of *org* are of the same value for attribute A, then A is a **Fixed-value Attribute**; if A' is a fixed-value attribute and satisfies the condi-

H. Dai, R. Srikant, and C. Zhang (Eds.): PAKDD 2004, LNAI 3056, pp. 196–200, 2004.

tions required for rule extraction, then A' is a **Useful Attribute**. These conditions will be explained later, in ALGORITHM 1. The set of fixed-value attributes of *org* is labeled as F_{org}, and that of useful attributes is labeled as U_{org}.

Organizations are also divided into three classes:

Trivial Organization: An organization whose number of members is 1, and all attributes of such an organization are useful ones;

Abnormal Organization: An organization whose set of useful attributes is empty;

Normal Organization: An organization does not belong to the two classes above.

The sets of the three kinds of organizations are labeled as ORG_T, ORG_A and ORG_N.

Given that two parent organizations, org_{p1} and org_{p2}, are randomly selected from the same population:

Migrating operator: n members randomly selected from org_{p1} are moved to org_{p2}, with two child organizations, org_{c1} and org_{c2}, obtained.

Exchanging operator: n members randomly selected from each parent organization are exchanged, with two child organizations, org_{c1} and org_{c2}, obtained. Here $1 \le n \le \min\{|org_{p1}|, |org_{p2}|\}$, where $|org|$ denotes the number of members in *org*.

Merging operator: The members of the two organizations are merged, with one child organization, org_{c1}, obtained.

Organizational selection mechanism: After an operator creates a pair of new organizations, a tournament will be held between the new pair and the parent pair. The pair containing the organization with the highest fitness survives to the next generation, while the other pair is deleted. If child organizations survive to the next generation, and one of them is an abnormal organization, then it is dismissed and its members are added to the next generation as trivial organizations. If only one organization remains in a population, it will be passed to the next generation directly.

A measure, **Attribute Significance**, is introduced to determine the significance of an attribute. The significance of attribute A is labeled as S_A. All populations use the same S_A. The value of S_A will be updated each time the fitness of an organization is computed, and the method is shown in ALGORITHM 1.

ALGORITHM 1. Attribute significance

t denotes the generation of evolution, the number of attributes is m, N is a predefined parameter, and *org* is the organization under consideration, $org \notin ORG_T$.

Step1: If t=0, then $S_{A_i}^0 \leftarrow 1.0$, $i = 1, 2, \ldots, m$; $U_{org} \leftarrow \varnothing$;

Step2: For each $A_i \in F_{org}$, randomly select a population without *org*, and an organization org_1 from it. If $A_i \in F_{org}$, and the attribute value of A_i in org_1 is different from that of *org*, then $U_{org} \leftarrow U_{org} \cup A_i$; otherwise reduce $S_{A_i}^t$ according to (1) (Case1);

Step3: If $U_{org} = \varnothing$, stop; otherwise randomly select N examples from the classes without *org*. If the combination of the attribute values in U_{org} does not appear in the N examples, then increase the attribute significance of all the attributes in U_{org} according to (1) (Case2); otherwise, $U_{org} \leftarrow \varnothing$.

$$S_A^{t+1} = \begin{cases} 0.9S_A^t + 0.05 & \text{Case1} \\ 0.9S_A^t + 0.2 & \text{Case2} \end{cases} \tag{1}$$

The value of S_A is restricted to the range of [0.5, 2]. The conditions of Case1 and Case2 are the ones required for rule extraction in DEFINITION 3.

The fitness function is defined in (2), where A_i denotes the ith attribute in U_{org}.

$$Fitness(org) = \begin{cases} 0 & org \in ORG_T \\ -1 & org \in ORG_A \\ |org| \prod_{i=1}^{|U_{org}|} S_{A_i} & org \in ORG_N \end{cases} \quad (2)$$

The whole algorithm of OCEC is given in ALGORITHM 2.

ALGORITHM 2. Organizational coevolutionary algorithm for classification

Step1: For each example, if its class name is c_i, $1 \le i \le m$, then add it to population P_i^0 as a trivial organization; $t \leftarrow 0$, $j \leftarrow 1$;

Step2: If $j > m$, go to Step 7;

Step3: If the number of organizations in P_j^t is greater than 1, then go to Step4; otherwise go to Step6;

Step4: Randomly select two parent organizations, org_{p1} and org_{p2}, from P_j^t, and then randomly select an operator from the three evolutionary operators to act on org_{p1} and org_{p2}; update the attribute significance according to ALGORITHM 1 and compute the fitness of child organizations, org_{c1} and org_{c2};

Step5: Perform the selection mechanism on org_{p1}, org_{p2} and org_{c1}, org_{c2}, delete org_{p1}, org_{p2} from P_j^t, and go to Step3;

Step6: Move the organization left in P_j^t to P_j^{t+1}, $j \leftarrow j+1$, and go to Step2;

Step7: If stop conditions are reached, stop; otherwise, $t \leftarrow t+1$, $j \leftarrow 1$, go to Step2.

When the evolutionary process is over, rules will be extracted from organizations. First, all organizations are merged as follows: Merge any two organizations in an identical population into a new organization if one set of useful attributes is a subset of the other set. The members of the new organization are those of the two organizations, and its set of useful attributes is the intersection of the two original sets. Next, a rule will be extracted from each organization based on its set of useful attributes. Each useful attribute forms a condition, and the conclusion is the class of the organization. Then, the ratio of positive examples a rule covers to all examples in the class the rule belongs to is calculated for each rule. Based on the ratio, all rules are ranked. Finally, some rules will be deleted: If the set of examples covered by a rule is a subset of the union of examples covered by some rules before this rule, the rule will be deleted.

3 Experimental Evaluation

11 datasets in the UCI repository[1] are used to test the performance of OCEC. The parameters of OCEC are: the number of generations is 500 for the datasets whose number of examples is less than 1000, and 1000 for the ones whose number of examples is larger than 1000. N is set to 10 percent of the number of examples for each dataset, and n is set to 1 Table 1 shows the comparison between OCEC and G-NET [1]. As can be seen, the predictive accuracies of OCEC on 7 of the 8 datasets are equivalent to or higher than those of G-NET. Table 2 shows the comparison between OCEC and

[1] http://www.ics.uci.edu/~mlearn/MLRepository.html

JoinGA [2]. As can be seen, the predictive accuracies of OCEC are equivalent to or higher than those of JoinGA on all the 4 datasets.

Table 1. Comparison between OCEC and G-NET

	Monk1	Monk2	Monk3	Tictac-toe	Credit	Breast cancer (W)	Vote	Mush-rooms
G-NET	**100.00± 0.00**	**97.20± 3.80**	**100.00± 0.00**	99.03± 0.62	84.20± 4.40	94.71± 2.89	94.90± 3.20	**100.00**
OCEC	**100.00± 0.00**	73.18± 7.31	**100.00± 0.00**	**100.00± 0.00**	**87.97± 4.38**	**96.13± 2.03**	**95.87± 2.61**	**100.00± 0.00**

Table 2. Comparison between OCEC and JoinGA

	Australian	Lymphography	Chess (KR-vs-KP)	Mushrooms
JoinGA	84.9±3.7	82.4±6.3	99.4	**100.0**
OCEC	**87.97±4.04**	**86.38±8.92**	**99.51±0.09**	**100.00±0.00**

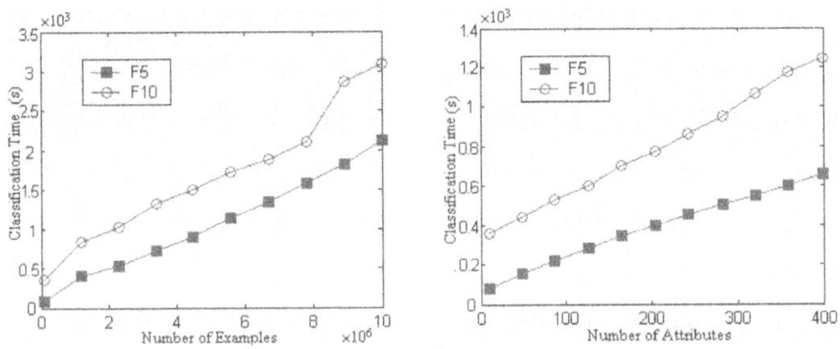

(a) (b)

Fig. 1. (a) The scalability of OCEC on the number of training examples, (b) the scalability of OCEC on the number of attributes

4 The Scalability of OCEC

The evaluation methodology and synthetic datasets proposed in [3] are used. The parameters of OCEC are: the number of generations is 5000 for all datasets, and n is selected from 1~5 randomly. In order to test the predictive accuracy of the obtained rules, another 10,000 instances are generated for each function as the test set.

Fig.4(a) shows the performance of OCEC as the number of training examples is in creased from 100,000 to 10 million in steps of 1100,000. The results show that OCEC can lead to linear classification time. Even when the number of training examples is increased to 10 million, the classification time is still shorter than 3500s.

Fig.4(b) shows the performances OCEC as the number of attributes is increased from 9 to 400 in steps of 39, where the number of training examples is 100,000. The results show that OCEC still leads to linear classification time. When the number of attributes is increased to 400, the classification time is still shorter than 1400s.

References

1. C. Angalano, A. Giordana, G. Lo Bello, and L. Saitta, "An Experimental Evaluation of Coevolutive Concept Learning", in Proceedings of the 15th International Conference on Machine Learning, pp.19-27, Madison, Wisconsin: Morgan Kaufmann, 1998.
2. J. Hekanaho, An Evolutionary Approach to Concept Learning, PhD thesis, Department of Computer Science, Abo Akademi University, 1998.
3. R. Agrawal, T. Imielinski and A. Swami, "Database mining: a performance perspective", IEEE Trans. on Knowledge and Data Engineering, pp.914-925, 5(6), 1993.

Noise Tolerant Classification by Chi Emerging Patterns

Hongjian Fan and Kotagiri Ramamohanarao

Dept. of CSSE, The University of Melbourne, Parkville, Vic 3052, Australia
{hfan, rao}@cs.mu.oz.au

Abstract. Classification is an important data mining problem. A desirable property of a classifier is noise tolerance. Emerging Patterns (EPs) are itemsets whose supports change significantly from one data class to another. In this paper, we first introduce Chi Emerging Patterns (Chi EPs), which are more resistant to noise than other kinds of EPs. We then use Chi EPs in a probabilistic approach for classification. The classifier, Bayesian Classification by Chi Emerging Patterns (BCCEP), can handle noise very well due to the inherent noise tolerance of the Bayesian approach and high quality patterns used in the probability approximation. The empirical study shows that our method is superior to other well-known classification methods such as NB, C4.5, SVM and JEP-C in terms of overall predictive accuracy, on "noisy" as well as "clean" benchmark datasets from the UCI Machine Learning Repository. Out of the 116 cases, BCCEP wins on 70 cases, NB wins on 30, C4.5 wins on 33, SVM wins on 32 and JEP-C wins on 21.

Keywords: Emerging patterns, noise tolerance, classification, Bayesian learning, noise

1 Introduction

Classification is an important data mining problem, and has also been studied substantially in statistics, machine learning, neural networks and expert systems over decades [1]. Data mining is typically concerned with observational retrospective data, i.e., data that has already been collected for some other purpose. For many reasons such as encoding errors, measurement errors, unrecorded causes of recorded features, the information in a database is almost always noisy. The problem of dealing with noisy data is one of the most important research and application challenges in the field of Knowledge Discovery in Databases (KDD). There are three kinds of noise in the training data: attribute noise (wrong attribute values), label noise, also called classification noise (wrong class labels), and mix noise (both classification and attribute noise). The noise in the training data can mislead a learning algorithm to fit it into the classification model. As a result, the classifier finds many meaningless "regularities" in the data. The phenomenon is often referred to as overfitting. In this paper, we address the following question: how can a classifier cope with noisy training data. Our main

H. Dai, R. Srikant, and C. Zhang (Eds.): PAKDD 2004, LNAI 3056, pp. 201–206, 2004.

contributions are the identification of the most discriminating knowledge - Chi Emerging Patterns (Chi EPs), and the probabilistic approach to use Chi EPs for classification to resist noise.

2 Chi Emerging Patterns and Related Work

Emerging Patterns [2] are conjunctions of simple conditions, where each conjunct is a test of the value of one of the attributes. EPs are defined as multivariate features (i.e., itemsets) whose supports (or frequencies) change *significantly* from one class to another. A JEP is a special type of EP, defined as an itemset whose support increases abruptly from zero in one dataset, to non-zero in another dataset – the ratio of support-increase being infinite. Since EPs/JEPs capture the knowledge of sharp differences between data classes, they are very suitable for serving as a classification model. By aggregating the differentiating power of EPs/JEPs, the constructed classification systems [3,4] are usually more accurate than other existing state-of-the-art classifiers.

Recently, the noise tolerance of EP-based classifiers such as CAEP and JEPC has been studied using a number of datasets from the UCI Machine Learning Repository where noise is purposely introduced to the original datasets [5]. The results shows that both CAEP and JEPC do not experience overfitting due to the aggregating approach used in the classification and they are generally noise tolerant. Their comparison of the learning curves of a number of classifiers shows that JEPC and NB are the most noise tolerant, followed by C5.0, CAEP and kNN. However, there are difficulties to apply JEPC on noisy datasets. On one hand, by definition, an itemset which occurs once (or very few times) in one data class while zero times in another class is a JEP. Such JEPs are usually regarded as noise information and are not useful for classification . The number of those useless JEPs can be very large due to the injection of noise, which not only cause lots of difficulties to the mining of JEPs, but also makes JEPC very inefficient or even unusable (although the number of the most expressive JEPs is usually small, we have observed that such number becomes exponential for some noisy datasets). On the other hand, by definition, an itemset with large but finite support growth rate is not a JEP. The number of JEPs for some noisy datasets can be very small or even zero, because of the strict requirement that JEPs must have zero support in one class. Large-growth-rate EPs are also good discriminators to distinguish two classes. The exclusive of using JEPs makes JEPC very unreliable when there are few JEPs to make a decision.

To overcome the problems of JEPC, we propose a new kind of Emerging Patterns, called Chi Emerging Patterns (χEPs).

Definition 1. *An itemset X is called an* Chi emerging pattern *(Chi EP), if all the following conditions are true:*

1. *$supp(X) \geq \xi$, where ξ is a minimum support threshold;*
2. *$GR(X) \geq \rho$, where ρ is a minimum growth rate threshold;*

3. $\neg\exists Y(Y \subset X) \wedge (supp(Y) \geq \xi) \wedge (GR(Y) \geq \rho) \wedge (strength(Y) \geq strength(X))$;
4. $|X| = 1 \vee |X| > 1 \wedge (\forall Y(Y \subset X \wedge |Y| = |X| - 1) \Rightarrow chi(X,Y) \geq \eta)$, *where* $\eta = 3.84$ *is a minimum chi-value threshold and* $chi(X,Y)$ *is computed using the following contingency table [6]*

	X	Y	$\sum row$
D_1	$count_{D_1}(X)$	$count_{D_1}(Y)$	$count_{D_1}(X) + count_{D_1}(Y)$
D_2	$count_{D_2}(X)$	$count_{D_2}(Y)$	$count_{D_2}(X) + count_{D_2}(Y)$
$\sum column$	$count_{D_1+D_2}(X)$	$count_{D_1+D_2}(Y)$	$count_{D_1+D_2}(X) + count_{D_1+D_2}(Y)$

Chi EPs are high quality patterns for classification and are believed to resist noise better because of the following reasons. The first condition ensures a χEP is not noise by imposing a minimum coverage on the training dataset. The second requires that a χEP has reasonably strong discriminating power, because larger growth rate thresholds produces very few EPs and smaller thresholds generates EPs with less discriminating power. The third prefers those short EPs with large strength: subsets of χEPs may satisfy condition (1) and (2), but they will not have strong discriminating power; Super sets of χEPs are not regarded as essential because usually the simplest hypothesis consistent with the data is preferred. Generally speaking, the last condition states that an itemset is a χEP, if the distribution (namely, the supports in two contrasting classes) of its subset is *significantly* different from that of the χEP itself, where the difference is measured by the χ^2-test [6]. In other words, every item in the itemsets contributes *significantly* to the discriminating power of the χEP. Chi EPs can be efficiently discovered by the tree-based pattern fragment growth methods [7].

3 Bayesian Classification by Chi Emerging Patterns

The Naive Bayes (NB) classifier [8] has been shown inherently noise tolerant due to its collection of class and conditional probabilities. The main weakness of NB is the assumption that all attributes are independent given the class. Our previous work [9] shows that extending NB by using EPs can relax the strong attribute independence assumption. In this paper, we propose to use Chi EPs in the probabilistic approach for classification. The classifier is called Bayesian Classification by Chi Emerging Patterns (BCCEP). BCCEP can handle noise very well, because (1) it retains the noise tolerance of the Bayesian approach; (2) it uses high quality patterns (χEPs) of arbitrary size in the probability approximation, which overcomes NB's weakness and provides a better scoring function than previous EP-based classifiers such as CAEP and JEPC. The details about the implementation of BCCEP can be found in [10].

4 Experimental Evaluation

We carry experiments on 29 datasets from the UCI Machine Learning Repository[11]. We regard the original datasets downloaded from UCI as "clean" datasets, although they are not guaranteed to be free from noise. For each dataset, we inject three kinds of noise at the level of 40% to generate three noisy datasets, namely "attribute noise" dataset, "label noise" dataset and "mix noise" dataset. The details about the implementation of noise generation can be found in [10]. Note that when we evaluate the performance under noise, the test datasets do not contain injected noise; only the training datasets are affected by noise. We compare BCCEP against Naive Bayes, decision tree induction C4.5, Support Vector Machines (SVM), and JEP-C [4]. We use WEKA's Java implementation of NB, C4.5 and SVM [12]. All experiments were conducted on a Dell PowerEdge 2500 (Dual P3 1GHz CPU, 2G RAM) running Solaris 8/x86. The accuracy was obtained by using the methodology of *stratified* ten-fold cross-validation (CV-10). We use the Entropy method from the MLC++ machine learning library [13] to discretize datasets containing continuous attributes.

Table 1. Performance Comparison

	# Times as Top Classifier					Average Accuracy				
	NB	C4.5	SVM	JEP-C	BCCEP	NB	C4.5	SVM	JEP-C	BCCEP
Clean	4	10	11	7	19	80.18	84.68	85.99	83.75	87.3
Attribute	9	7	5	11	21	76.94	78.42	76.71	79.98	82.29
Label	5	11	10	2	13	76.78	77.18	81.66	74.94	82.27
Mix	12	5	6	1	17	74.86	70.88	73.02	66.42	76.62
Total	30	33	32	21	70	-	-	-	-	-
Average	7.5	8.25	8	5.25	17.5	77.19	77.79	79.345	76.2725	82.12

A classifier is regard as top classifier when (1) it achieves the highest accuracy; or (2) its accuracy is very close to the highest one (the difference is less than 1%)

For lack of space, we only present a summary of results shown in Table 1. More results can be found in [10]. We highlight some interesting points as follows:

- The average accuracy of NB on clean, attribute noise and label noise datasets are lower than other classifiers. But NB deals with mix noise surprisingly well, when other classifiers are confused by the two-fold noise.
- C4.5 is fast in all cases, clean, attribute noise, label noise or mix noise. There is a big drop of accuracy on mix noise datasets.
- SVM is much slower than C4.5, especially when datasets are large. It deals with noise fairly well: its average accuracy is very close to BCCEP on label noise datasets. But unfortunately we were unable to produce results for some large datasets such as Adult, Shuttle and Splice using the SVM implementation from WEKA [12], as the SVM program does not terminate or exceeds the memory limit.
- JEP-C performs well on clean datasets. When attribute noise, label noise and mix noise is introduced, it is harder and harder to mine JEPs, generating either too many JEPs or too few JEPs, leading to decreased accuracy.
- Our BCCEP deals with all three kinds of noise very well, as evidenced both by the highest accuracy and the number of times being top classifiers. Differ-

ent classifiers are good at handling different kinds of noise. We believe that the success of BCCEP on all three kinds of noise is mainly due to its hybrid nature, combining Bayesian and EP-based classification.
- The accuracy of BCCEP is almost always higher than its ancestor NB. The tree-based EP mining algorithm used in BCCEP mines a relatively small number of χEPs very fast on datasets where finding JEPs is very hard, i.e., JEPC uses up 1GB memory and gives up.

5 Conclusions

In this paper, we have shown that Chi Emerging Patterns resist noise better than other kinds of Emerging Patterns. The classifier, Bayesian Classification by Chi Emerging Patterns (BCCEP), combines the advantages of the Bayesian approach (inherent noise tolerant) and the EP-based approach (high quality patterns with sharp discriminating power). Our extensive experiments on 29 benchmark datasets from the UCI Machine Learning Repository show that BCCEP has good overall predictive accuracy on "clean" datasets and three kinds of noisy datasets; it is superior to other state-of-the-art classification methods such as NB, C4.5, SVM and JEP-C: out of 116 cases (note that there are 29 datasets, each has four versions, namely, clean, attribute noise, label noise and mix noise datasets), BCCEP wins on 70 datasets, which is much higher than any other classifiers. (as comparison, NB wins on 30, C4.5 wins on 33, SVM wins on 32 and JEP-C wins on 21)

References

1. Han, J., Kamber, M.: Data Mining: Concepts and Techniques. Morgan Kaufmann Publishers (2000)
2. Dong, G., Li, J.: Efficient mining of emerging patterns: Discovering trends and differences. In: Proc. 5th ACM SIGKDD, San Diego, CA, USA (1999) 43–52
3. Dong, G., Zhang, X., Wong, L., Li, J.: Classification by aggregating emerging patterns. In: Proc. 2nd Int'l Conf. on Discovery Science (DS'99), Tokyo, Japan (1999) 30–42
4. Li, J., Dong, G., Ramamohanarao, K.: Making use of the most expressive jumping emerging patterns for classification. Knowledge and Information Systems **3** (2001) 131–145
5. Sun, Q., Zhang, X., Ramamohanarao, K.: The noise tolerance of ep-based classifiers. In: Proc. 16th Australian Conf. on Artificial Intelligence, Perth, Australia (2003) 796–806
6. Bethea, R.M., Duran, B.S., Boullion, T.L.: Statistical methods for engineers and scientists. New York : M. Dekker (1995)
7. Fan, H., Ramamohanarao, K.: Efficiently mining interesting emerging patterns. In: Proc. 4th Int'l. Conf. on Web-Age Information Management (WAIM2003), Chengdu, China (2003) 189–201
8. Domingos, P., Pazzani, M.J.: Beyond independence: Conditions for the optimality of the simple bayesian classifier. In: Proc. 13th ICML. (1996) 105–112

9. Fan, H., Ramamohanarao, K.: A bayesian approach to use emerging patterns for classification. In: Proc. 14th Australasian Database Conference (ADC2003), Adelaide, Australia (2003) 39–48
10. Fan, H., Ramamohanarao, K.: Noise tolerant classification by chi emerging patterns. Technical report, Department of Computer Science and Software Engineering, University of Melbourne (2004)
11. Blake, C., Merz, C.: UCI repository of machine learning databases (1998)
12. Witten, I.H., Frank, E.: Data Mining: Practical Machine Learning Tools and Techniques with Java Implementations. Morgan Kaufmann (1999)
13. Kohavi, R., Sommerfield, D., Dougherty, J.: Data mining using MLC++: A machine learning library in C++. International Journal on Artificial Intelligence Tools **6** (1997) 537–566

The Application of Emerging Patterns for Improving the Quality of Rare-Class Classification

Hamad Alhammady and Kotagiri Ramamohanarao

Department of Computer Science and Software Engineering
The University of Melbourne
Parkville, Victoria 3010, Australia
{hhammady, rao}@cs.mu.oz.au

Abstract. The classification of rare cases is a challenging problem in many real life applications. The scarcity of the rare cases makes it difficult for traditional classifiers to classify them correctly. In this paper, we propose a new approach to use emerging patterns (EPs) [3] in rare-class classification (EPRC). Traditional EP-based classifiers [2] fail to achieve accepted results when dealing with rare cases. EPRC overcomes this problem by applying three improving stages: generating new undiscovered EPs for the rare class, pruning low-support EPs, and increasing the supports of the rare-class EPs. An experimental evaluation carried out on a number of rare-class databases shows that EPRC outperforms EP-based classifiers as well as other classification methods such as PNrule [1], Metacost [6], and C4.5 [7].

1 Introduction

Classification of rare cases is an important problem in data mining. This problem is identified as distinguishing rarely-occurring samples from other overwhelming samples in a significantly imbalanced dataset [1]. In this paper, we investigate how to employ emerging patterns (EPs) in rare-case classification. EPs are a new kind of patterns that introduced recently [3]. EPs are defined as itemsets whose supports increase significantly from one class to another. The power of EPs can be used to build high-performance classifiers [2]. Usually, these classifiers achieve higher accuracies than other state-of-the-art classifiers. However, simple EP-based classifiers do not retain their high performance when dealing with datasets which have rare cases. The reason for this failure is that the number of the rare-class EPs is very small, and their supports are very low. Hence, they fail to distinguish rare cases from a vast majority of other cases.

In this paper we propose a new approach to use the advantage of EPs to classify rare cases in imbalanced datasets. The aim of our approach (called EPRC) is to improve the discriminating power of EPs so that they achieve better results when dealing with rare cases. This is achieved through three improving stages; 1) generating new undiscovered EPs for the rare class, 2) pruning the low-support EPs, and 3) increasing the support of rare-class EPs. These stages are detailed in section 3.

H. Dai, R. Srikant, and C. Zhang (Eds.): PAKDD 2004, LNAI 3056, pp. 207–211, 2004.

In this paper we adopt the *weighted accuracy* [8], and the *f-measure* [9] as they are well-known metrics for measuring the performance of rare-case classification. The f-measure depends on the *recall* and *precision* of the rare class.

2 Emerging Patterns

Let $obj = \{a_1, a_2, a_3, \ldots a_n\}$ is a data object following the schema $\{A_1, A_2, A_3, \ldots A_n\}$. $A_1, A_2, A_3, \ldots A_n$ are called *attributes*, and $a_1, a_2, a_3, \ldots a_n$ are *values* related to these attributes. We call each pair (attribute, value) an *item*.

Let *I* denote the set of all items in an encoding dataset *D*. *Itemsets* are subsets of *I*. We say an instance *Y* contains an itemset *X*, if $X \subseteq Y$.

Definition 1. Given a dataset *D*, and an itemset *X*, the support of *X* in *D* is defined as the percentage of the instances in *D* that contain *X*.

Definition 2. Given two different classes of datasets *D1* and *D2*. The growth rate of an itemset *X* from *D1* to *D2* is defined as the ratio between the support of *X* in *D2* and its support in *D1*.

Definition 3. Given a growth rate threshold *p*>1, an itemset *X* is said to be a *p*-emerging pattern (*p*-EP or simply EP) from *D1* to *D2* if $GrowthRate_{D_1 \to D_2}(X) \geq p$.

3 Improving Emerging Patterns

As described earlier, our approach aims at using the discriminating power of EPs in rare-case classification. We introduce the idea of generating new EPs for the rare class. Moreover, we adopt eliminating low-support EPs in both the major and rare classes, and increasing the support of rare-class EPs.

The first step in our approach involves generating new rare-class EPs. Given a training dataset and a set of the discovered EPs, the values that have the highest growth rates from the major class to the rare class are found. The new EPs are generated by replacing different attribute values (in the original rare-class EPs) with the highest-growth-rate values. After that, the new EPs that already exist in the original set of EPs are filtered out. Figure 1 shows an example of this process. The left table shows four rare-class EPs. Suppose that the values that have the highest growth rates for attributes A1 and A3 are V_{11} and V_{33} respectively. Using these two values and EP e4, $\{V_{13}, X, V_{34}, V_{44}, V_{55}\}$, we can generate 2 more EPs (in the right table). The first EP is $\{V_{11}, X, V_{34}, V_{44}, V_{55}\}$ (by replacing V_{13} with V_{11}). The second EP is $\{V_{13}, X, V_{33}, V_{44}, V_{55}\}$ (by replacing V_{34} with V_{33}). However, the first new EP already exists in the original set of EPs (e1). This EP is filtered out. We argue that these new generated EPs have a strong power to discriminate rare-class instances from major-class instances. There are two reasons for this argument. The first reason is that these new EPs are inherited from the original rare-class EPs which themselves have a discriminating power to classify rare cases. The second reason is that they contain the most discriminating attribute values (attribute values with the highest growth rates) obtained from the training dataset.

Fig. 1. Example of generating new rare-class Eps

	A1	A2	A3	A4	A5
e1	V_{11}	X	V_{34}	V_{44}	V_{55}
e2	V_{11}	V_{22}	V_{31}	X	X
e3	V_{12}	V_{22}	V_{33}	V_{43}	X
e4	V_{13}	X	V_{34}	V_{44}	V_{55}

e4-new1	V_{11}	X	V_{34}	V_{44}	V_{55}
e4-new2	V_{13}	X	V_{33}	V_{44}	V_{55}

* V_{ij} = value j for attribute i
* X = undefined value

Based on the above explanation, we have algorithm 1 to generate new rare-class EPs.

```
Algorithm 1 (Generating new rare-class EPs)
Input: the training dataset D, and discovered EPs E.
Output: a set of new rare-class EPs.
Method:
For each attribute i in D
  A_i = value with the highest growth rate of attribute i.
For each rare-class EP e
  For each attribute value k related to i
    If k != A_i
      Generate a new EP  e_new = e
      Replace k by A_i in e_new
      If e_new does not exist in E
```

The second step involves pruning the low-support EPs. This is performed for both the major and rare classes. Given a pruning threshold, the average growth rate of the attribute values in each EP is found. The EPs whose average growth rates are less than the given threshold are eliminated. We argue that these eliminated EPs have the least discriminating power as they contain many least-occurring values in the dataset.

The third step involves increasing the support of rare-class EPs by a given percentage. The postulate behind this point is that this increase compensates the effect of the large number of major-class EPs. That is, the overwhelming major-class EPs make many rare-class instances classified as major-class.

4 Experimental Evaluation

In order to investigate the performance of our approach, we carry out a number of experiments. We use three challenging rare-class databases with different distributions of data between the major and rare classes. These datasets are the insurance dataset [5], the disease dataset [4], and the sick dataset [4]. We compare our approach to other methods such as PNrule, Metacost, C4.5, and CEP. The comparison we present is based on the weighted accuracy, traditional accuracy, major-class accuracy, recall (rare-class accuracy), precision, and F-measure.

4.1 Tuning

Our approach uses three parameters. These parameters are the threshold of pruning the major-class EPs, the threshold of pruning the rare-class EPs, and the percentage of increasing the support of rare-class EPs. The parameters of our approach need to be tuned to achieve the best results. To achieve this aim, we split the training set into 2 partitions. The first partition (70%) is used to train the classifier. The second partition (30%) is used to tune the parameters.

4.2 Comparative Results

After tuning the insurance dataset, the parameters of our approach are fixed to deal with this dataset. We run different methods on the test set of this dataset. These methods include EPRC (our approach), PNrule [1], C4.5 [7], Metacost [6], and CEP (EP-based classifier) [2]. The results of these methods are presented in table 1. Our approach outperforms all other methods in terms of the weighted accuracy and f-measure. This performance is achieved by balancing both the recall and the precision.

Table 1. The results of the insurance dataset

Classifier	Weighted accuracy	Traditional accuracy	Major-class accuracy	Recall (rare-class accuracy)	Precision	F-measure
EPRC	**63.89%**	80.57%	82.82%	44.95%	14.20%	**21.59%**
PNrule	58.91%	87.12%	90.03%	26.89%	15.80%	19.90%
C4.5	52.87%	90.95%	96.09%	9.66%	13.52%	11.27%
Metacost	49.80%	5.95%	0.02%	**99.57%**	5.92%	11.18%
CEP	50.85%	**93.8%**	**99.60%**	2.10%	**25.00%**	3.87%

4.3 The Effects on the EP-Based Classifier

Our basic aim behind the work presented in this paper is to improve the performance of EP-based classifiers in rare-case classification. In this experiment we compare the results obtained for our three datasets using CEP (EP-based classifier) and EPRC. As stated in section 2, EPRC uses CEP as its basic EP-based classifier. The three datasets were tuned using 30% of the training set. Table 2 shows how our approach enhances the performance of the EP-based classifier. There are significant increases in the weighted accuracy and f-measure from CEP to EPRC.

Table 2. The effect on the EP-based classifier

Experiment	Weighted accuracy	F-measure
Insurance dataset (CEP)	50.85%	3.87%
Insurance dataset (EPRC)	63.89%	21.59%
Disease dataset (CEP)	49.94%	Undefined
Disease dataset (EPRC)	65.07%	34.78%
Sick dataset (CEP)	78.89%	70%
Sick dataset (EPRC)	94.57%	79.71%

5 Conclusions and Future Research

In this paper, we propose a new EP-based approach to classify rare classes. Our approach, called EPRC, introduces the idea of generating new rare-class EPs. Moreover, it improves EPs by adopting pruning low-support EPs, and increasing the support of rare-class EPs. We empirically demonstrate how improving EPs enhances the performance of EP-based classifiers in rare-case classification problems. Moreover, our approach helps EP-based classifiers outperform other classifiers in such problems. The proposed approach opens many doors for further research. One possibility is improving the performance of EP-based classifiers further by adding further improving stages to increase the discriminating power of EPs.

References

1. M. V. Joshi, R. Agarwal, and V. Kumar. Mining Needles in a Haystack: Classifying Rare Classes via Two-Phase Rule Induction. In Proceedings of ACM (SIGMOD '01), Santa Barbara, California, USA.
2. J. Bailey, T. Manoukian, and K. Ramamohanarao. Classification Using Constrained Emerging Patterns. In Proceedings of the Fourth International Conference on Web-Age Information Management (WAIM '03), Chengdu, China.
3. G. Dong, and J. Li. Efficient Mining of Emerging Patterns: Discovering Trends and Differences. In Proceedings of 1999 International Conference on Knowledge Discovery and Data Mining (KDD'99), San Diego, CA, USA.
4. C. Blake, E. Keogh, and C. J. Merz. UCI repository of machine learning databases. Department of Information and Computer Science, University of California at Irvine, CA, 1999. http://www.ics.uci.edu/~mlearn/MLRepository.html.
5. P. van der Putten, M. de Ruiter, and M. van Someren. The CoIL Challenge 2000 report. http://www.liacs.nl/~putten/library/cc2000, June 2000.
6. P. Domingos. MetaCost: A General Method for Making Classifiers Cost-Sensitive. In Proceedings of 1999 International Conference on Knowledge Discovery and Data Mining (KDD'99), San Diego, CA, USA.
7. I. H. Witten, E. Frank. Data Mining: Practical Machine Learning Tools and Techniques with Java Implementations. Morgan Kaufmann, San Mateo, CA., 1999.
8. J. Cheng, C. Hatzis, H. Hayashi, M. Krogel, S. Morishita, D. Page, and J. Sese. KDD Cup 2001 report. ACM SIGKDD Explorations, January, 2002.
9. C. J. van Rijsbergan. Information Retrieval. Butterworths, London, 1979.

Finding Negative Event-Oriented Patterns in Long Temporal Sequences

Xingzhi Sun, Maria E. Orlowska, and Xue Li

School of Information Technology and Electrical Engineering
The University of Queensland, QLD, 4072, Australia
{sun, maria, xueli}@itee.uq.edu.au

Abstract. Pattern discovery in a long temporal event sequence is of great importance in many application domains. Most of the previous work focuses on identifying positive associations among time stamped event types. In this paper, we introduce the problem of defining and discovering negative associations that, as positive rules, may also serve as a source of knowledge discovery.
In general, an event-oriented pattern is a pattern that associates with a selected type of event, called a target event. As a counter-part of previous research, we identify patterns that have a negative relationship with the target events. A set of criteria is defined to evaluate the interestingness of patterns associated with such negative relationships. In the process of counting the frequency of a pattern, we propose a new approach, called unique minimal occurrence, which guarantees that the Apriori property holds for all patterns in a long sequence. Based on the interestingness measures, algorithms are proposed to discover potentially interesting patterns for this negative rule problem. Finally, the experiment is made for a real application.

1 Introduction

In reality, a large number of events are recorded with temporal information (e.g., a timestamp). We call a sequence of such events a temporal event sequence. In some data mining domains, there is need to investigate very long temporal event sequences. Substantial work [1,2,3,4,5] has been done for finding frequent patterns in a long temporal event sequence, however, most of them treat every event type equally and do not consider speciality of events.

In many applications, not every type of event has equal impact on overall data analysis goals. One may be particularly interested in a specific type of event and event patterns related to such events. For example, in telecommunication network fault analysis, a variety of signals are recorded to form a long temporal event sequence. In such a sequence, we could be more interested in events representing a fault on the network. Naturally, it is unnecessary to find all frequent patterns in the sequence as done traditionally. We define events of special interest as *target events*, and patterns related to target events as *event-oriented patterns*. The temporal relationship between target events and event-oriented patterns can be specified by *temporal constraints*.

H. Dai, R. Srikant, and C. Zhang (Eds.): PAKDD 2004, LNAI 3056, pp. 212–221, 2004.

Let us consider the application of earthquake predictions. In order to find event patterns that may potentially lead to earthquakes, many types of events that may relate to earthquakes are recorded by independent time controlled devices. A temporal event sequence is then formed by ordering these events based on their timestamps. To predict an earthquake, it would be ideal to find the event pattern (which is a natural phenomenon or a combination of phenomena) that happens frequently *only* in the periods prior to earthquakes. On the contrary, it is also interesting to find patterns that happen frequently all the time *but* before earthquakes. Both types of patterns can contribute to the earthquake prediction and also complement each other.

In the above example, we regard events that represent earthquakes as target events. Considering two types of patterns, we define the former as *positive event-oriented pattern* and the latter as *negative event-oriented pattern*. The temporal constraint would then be the length of periods for the prediction. The problem of finding positive event-oriented patterns has been studied in our previous work [5]. In this paper, we study its counterpart, finding negative event-oriented patterns in a long sequence.

Now, we generalize our problem as follows. Let k types of events be given. A temporal event sequence is a list of events indexed by their timestamps. Let e be a selected event type called target event type. Additionally, let the size of time interval T be given. The problem is to find any frequent pattern P of events such that the occurrences of P are considerably restrained in T-sized intervals before events of type e. The word "restrained" indicates the negative relationship between the pattern and target events. It means that the pattern P either rarely happens before target events or happens far less frequently before target events than in any other periods. Naturally, the mining result of such a problem can be expressed as negative rules in the form of $\neg P \xrightarrow{T} e$. Based on this rule format, we explain further our problem in more details.

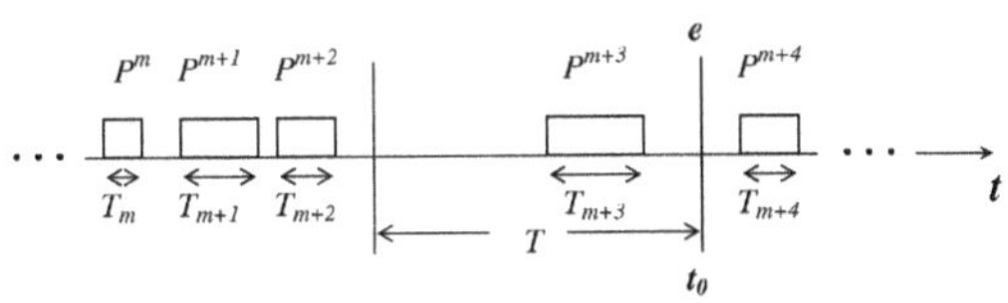

Fig. 1. A fragment of sequence

– The temporal constraints are important to insure the sensibility of negative rules. While T specifies the temporal relationship between patterns and target events, we introduce another size of interval T_P as the temporal constraint of the pattern itself. A pattern P with the temporal constraint T_P regulates that every occurrence of P should happen within an interval of size T_P. Figure 1 illustrates dependencies between all introduced notions in a fragment of a long sequence. A target event occurs at time t_0. T gives

the length of the period before the target event. Let P^i represent the i-th occurrence of pattern P (with temporal constraint T_P) in the sequence. For each P^i, we should have $T_i \leq T_P$ where T_i is the time span of P^i. In the above rule format, T_P is implicitly set with P. Note that both T_P and T are given by domain experts.
- Regarding the order among pattern elements, P could be an existence pattern (i.e., a set of event types) or a sequential pattern (i.e., a list of event types). In this paper, for simplicity of presentation, we only discuss existence patterns. All discussions can be easily extended to sequential patterns.

The challenge of mining negative rules is to define appropriate interestingness measures. Considering the format of negative rules, intuitively, we need find the pattern that 1) occurs frequently in a long sequence, and 2) occurs rarely before target events. To link those two numbers of occurrences, we define a key interestingness measure based on unexpectedness [6,2]. That is, a pattern is interesting because it is unexpected to the prior knowledge. According to the first requirement, all potentially interesting patterns should be frequent in the entire sequence. For any such pattern P, we assume that P is distributed uniformly over the whole period of the sequence S. Based on this important assumption, we can estimate the number of occurrences of P in the periods prior to target events. If the real number of occurrences of P is significantly smaller than our expectation, the pattern P is said to be unexpected and is, therefore, interesting to our problem. According to this idea, we propose the third requirement i.e., 3) the number of occurrences of P before target events is unexpected. Now, the negative event-oriented pattern can be described as: patterns that *should* happen frequently before the target events but actually do *not*.

To evaluate the frequency of a pattern P, we need to count the number of occurrences of P in a long sequence S. Based on the concept of minimal occurrence (MO) [7], we propose a new approach called unique minimal occurrence (UMO), which guarantees that the Apriori property holds for all patterns in a long sequence.

The rest of this paper is organized as follows. Related work is discussed in Section 2. In Section 3, the problem of finding negative event-oriented patterns is formulated and further discussed. In Section 4, we propose algorithms to find a complete set of negative patterns. Section 5 shows the experiment results of a real application. Finally, we conclude this paper in Section 6.

2 Related Work

In [8,7,1], Mannila *et al.* have studied how to find episodes (equivalent to patterns in our paper) in a long temporal event sequence. There are a few differences between our work and theirs. First, we introduce speciality of events and only find the patterns related to a selected type of event. Secondly, while their work focuses on finding positive relationships between events, we target on identifying negative relationships. Last but not least, based on their definition of minimal occurrence, we propose the UMO approach to count the number of occurrences of a pattern in a long sequence, as will be discussed in detail in Section 3.

Mining negative association rules has been studied in [9,10]. The fundamental difference between our work and theirs is that we discover negative rules from a long temporal event sequence rather than a set of transactions. In [10], Wu *et al.* have defined the interestingness measures using probability theory and provided an algorithm to mine both positive and negative association rules simultaneously. However, their interestingness measures cannot deal with our problem because in a long sequence, it is impractical to count how many times a pattern does not happen. In [9], Savasere *et al.* have defined interestingness measures using unexpectedness. They estimated the expected support of rules by applying domain knowledge, which is a taxonomy on items. A rule is interesting if its actual support is significantly smaller the than expected value. In our problem, we estimate the expected support value based on the assumption that frequent patterns are uniformly distributed in the sequence. Hence, our work is domain knowledge independent.

Dong and Li [11] have worked on the problem of discovering emerging patterns (EPs) from two datasets. A pattern is defined as an EP if its support increases significantly from one dataset to another. Formally, P is an EP if $\frac{Supp(P,D_1)}{Supp(P,D_2)} \geq g$ where D_1 and D_2 are two different datasets and g is a threshold. This is similar to one of our interestingness measures, whereas, in our problem, one dataset is a long sequence and the other is a set of sequence fragments. According to this nature, we define different interestingness measures for the pattern in each dataset and define the interestingness based on unexpectedness.

3 Problem Statement

3.1 Negative Event-Oriented Patterns

Let us consider a finite set E of *event types*. An *event* is a pair (a, t) where $a \in E$ and t is the *timestamp* of the event. We define one special event type e as the *target event type*. Any event of type e is called a *target event*. A *temporal event sequence*, or *sequence* in short, is a list of events totally ordered by their timestamps. It is formally represented as $S = \langle (a_1, t_1), (a_2, t_2), \ldots, (a_n, t_n) \rangle$ where $a_i \in E$ for $1 \leq i \leq n$ and $t_i < t_{i+1}$ for $1 \leq i \leq n-1$. The *duration* of sequence S is the time span of S, namely, $Dur(S) = t_n - t_1$.

An *existence pattern* (in short, *pattern*) P with *temporal constraint* T_P is defined as 1) $P \subseteq E$ and 2) events matching pattern P must occur within a time interval of size T_P. The *length* of P is defined as the number of elements in P. A pattern P of length l is called l-pattern.

A *negative rule* is in the form of $r = \left\{ \neg P \xrightarrow{T} e \right\}$, where e is the target event type and P is a *negative event-oriented pattern* implicitly with the temporal constraint T_P. T is the length of interval that specifies the temporal relationship between P and e. Generally, r can be interpreted as: the occurrence of P is unexpectedly rare in T-sized intervals before target events.

3.2 Interestingness Measures

According to the rule format, the right hand side of a rule is always the target event type e. So the key problem is to find negative event-oriented pattern P. Remember that we want P satisfies the following three **requirements**, i.e., P is 1) frequent in S, 2) infrequent before target events, and 3) occurs far less frequently than expected before target events. In the following part, we will discuss how to define a set of interestingness measures for these requirements.

First, let us suppose there exists a method to return the number of occurrences of a pattern P in a sequence S, denoted as $Count\,(P, S)$.

Definition 3.1. *Given a pattern P, a sequence S, and the size of time interval T, the* **global support** *of P in S is defined as $Supp\,(P,S) = \frac{Count(P,S)}{N}$, where $N = \left\lceil \frac{Dur(S)}{T} \right\rceil + 1$.*

From above, we know that the global support of pattern P reflects the frequency of P in S. Note that, under the assumption of uniform distribution, it also gives the expected number of occurrences of P in a T-sized interval.

Considering requirement 2), we first formally define the dataset before target events and then give the interestingness measure.

Let a *window* w be $[t_s, t_e)$, where t_s and t_e are the start time and the end time of w respectively. The *window size* of w is the time span of the window, denoted as $Size(w) = t_e - t_s$. A *sequence fragment* $f\,(S, w)$ is the part of sequence S determined by w. Given a target event type e and a sequence S, we can locate all target events by creating a *timestamp set* $T^e = \{t \mid (e,t) \in S\}$. For each $t_i \in T^e$, we create a T-sized window $w_i = [t_i - T, t_i)$ and get the sequence fragment $f_i = f\,(S, w_i)$. The set of these sequence fragments $D = \{f_1, f_2, \cdots, f_m\}$ is called the *local dataset* of target event type e.

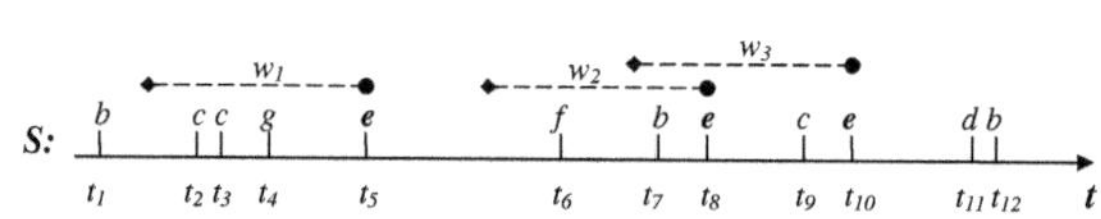

Fig. 2. An Example of local dataset

Example 3.1. Figure 2 presents an event sequence S. Suppose e is the target event type. The timestamp set of e is $\{t_5, t_8, t_{10}\}$. w_1, w_2 and w_3 are three T-sized windows ending at timestamp t_5, t_8 and t_{10} respectively. The sequence fragments f_1, f_2, and f_3 of these three windows are $\langle (c, t_2), (c, t_3), (g, t_4) \rangle$, $\langle (f, t_6), (b, t_7) \rangle$, and $\langle (b, t_7), (e, t_8), (c, t_9) \rangle$. The local dataset D is $\{f_1, f_2, f_3\}$.

Definition 3.2. *Given a pattern P and a local dataset D, the* **local support** *of P in D is defined as $Supp\,(P, D) = \frac{\sum_i Count(P, f_i)}{|D|}$, where $Count\,(P, f_i)$ returns the number of occurrences of P in f_i.*

As every sequence fragment is identified by a window of size T, the local support is the average number of occurrences of P in a T-sized interval before target events. Remember that the global support is the expected number of P in a T-sized interval. Naturally, we can set the ratio between them as the interestingness measure for requirement 3).

Definition 3.3. *Given the global support $Supp(P,S)$ and the local support $Supp(P,D)$ of a pattern P, the* **comparison ratio** *of P is defined as*

$$Cr(P) = \begin{cases} 1, & Supp(P,S) = 0 \wedge Supp(P,D) - 0 \\ \infty, & Supp(P,S) \neq 0 \wedge Supp(P,D) = 0 \\ \frac{Supp(P,S)}{Supp(P,D)}, & otherwise \end{cases}$$

3.3 Problem Definition

The formal definition of finding negative event-orient patterns is: given a sequence S, a target event type e, two sizes of interval T and T_P, and three thresholds s_g, s_l, and cr ($cr > 1$), find the complete set of rule $r = \left\{\neg P \xrightarrow{T} e\right\}$ such that 1) $Supp(P,S) \geq s_g$, 2) $Supp(P,D) < s_l$, and 3) $Cr(P) \geq cr$.

The first condition guarantees that the pattern has statistical significance. While condition 2) requires the *absolute* sparsity of the pattern before target events, condition 3) is a constraint of *relative* sparsity. Note that whether the constraint of absolute sparsity is required or not depends on a domain application.

From Definition 3.3, we can see that the three interestingness measures are not independent. Due to space limitation, we omit the discussion on the relationship among s_g, s_l, and cr.

3.4 Counting the Number of Occurrences

In this section, we discuss how to count the number of occurrences of a pattern in a sequence or a sequence fragment (i.e., how to define the method $Count(P,S)$ or $Count(P,f_i)$). Intuitively, given a pattern P, every occurrence of P corresponds to a window $[t_s, t_e]$ (For brevity, in this section, we do not consider the temporal constraint $[t_s, t_e] \leq T_P$).

Mannila *et al.* [7] have proposed the concept of minimal occurrence (MO) to represent that a pattern occurs in a sequence. Using our model, their definition can be presented as: a window w is a minimal occurrence of a pattern P *iff* 1) $f(S,w)$ contains P, and 2) there does not exist a subwindow $w' \subset w$ (i.e., $t_s \leq t'_s$, $t_e \geq t'_e$, and $Size(w) \neq Size(w')$) such that $f(S,w')$ contains P.

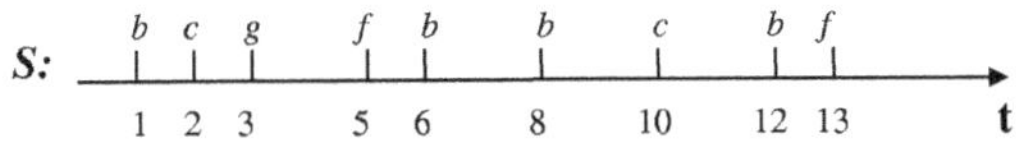

Fig. 3. An example of sequence

Example 3.3. A sequence S is visualized in Figure 3. Let P be an existence pattern $\{b, c\}$. The MOs of P in S are $[1,2], [2,6], [8,10]$, and $[10,12]$.

An important observation is that the Apriori property does not hold in the MO approach. That is, in a sequence S, the number of MOs of a pattern in S may be less than that of its superpattern. For instance, in Example 3.3, the pattern $\{c\}$ happens twice, however, in terms of MO, $\{b, c\}$ happens 4 times.

To rectify this shortcoming, we propose a new approach called unique minimal occurrence (UMO). The *unique* here means that an event can be used to match one pattern at most once. For instance, in Example 3.3, because the event $(c, 2)$ has matched P ($\{b, c\}$) in the window $[1, 2]$, in the UMO approach, we do not consider it any more in the following matching of P. Thus, $[2, 6]$ is not a UMO of P. According to this idea, the UMOs of P in S are $[1, 2]$ and $[8, 10]$. We give the formal definition of the UMO as follows.

Given a sequence S, let $M(S, P)$ be a method to find the first MO of P in S. In addition, let $R(S, t)$ express the part of S after the timestamp t (i.e., the part of S in $(t, t_n]$, where t_n is the timestamp of the last event in S).

Definition 3.5. *Given a sequence, represented as S_0, the* **unique minimal occurrence** *of P is defined as: 1) w_0 is a unique minimal occurrence if $w_0 = M(S_0, P) \neq null$, 2) w_i $(i \geq 1)$ is a unique minimal occurrence if $w_i = M(S_i, P) \neq null$ where $S_i = R(S_{i-1}, w_{i-1}.t_e)$.*

Claim 3.1. *The number of unique minimal occurrences of an existence pattern is no less than that of its supperpattern.*

Proof. Let P and P' be two existence patterns such that $P \subset P'$. For any UMO w' of P', one of following two conditions holds: 1) there exists a subwindow $w \subset w'$ such that w is a UMO of P; 2) there exists a window w'' such that w'' is a UMO of P and $w' \cap w'' \neq \phi$. For any two UMOs of P', denoted as w'_i and w'_j where $i \neq j$, $w'_i \cap w'_j = \phi$ always holds. Therefore, the number of UMOs of P' cannot exceed that of P. □

4 Searching for Interesting Patterns

In this section, we propose two algorithms to find a complete set of interesting patterns. Algorithm 1 is the main algorithm, and Algorithm 2 performs a key subtask for finding UMOs for a set of patterns.

Given a target event type and a set of parameters, Algorithm 1 finds a collection of interesting negative event-oriented patterns from a long sequence. From Figure 4, we can see that the algorithm consists of three phases. In searching phase (phase 1), we discover patterns whose global support is no less than s_g. In phase 2, called testing phase, for every pattern discovered in searching phase, we compute the local support in the dataset before target events. Finally, patterns are pruned based on thresholds of interestingness measures in pruning phase.

Algorithm 2 (shown in Figure 5) finds the number of UMOs for a set of patterns in a sequence (or a sequence fragment). To avoid unnecessary check, we first group patterns according to their event types. For example, *group*(a) is a pattern set that includes all patterns containing event type a. For each pattern P, we use *count* to record the number of UMOs and *event_count* to record how many event types in P have been matched. *P.timestamp*(a) is the timestamp

```
Algorithm 1
Input: A sequence S, a target event type e, two sizes of
interval T and T_P, and four thresholds s_g, s_l, lr and cr.
Output: A complete set of interesting patterns.
Method:
  /* Searching phase */
  F_1 = {frequent 1-patterns};
  for(k = 2; F_{k-1} ≠ φ; k++) do
    C_k =Candidate k-pattern generated from F_{k-1};
    for each P_i ∈ C_k do /* Algorithm 2 */
      Compute Supp(P_i, S);
    F_k = {P_i ∈ C_k | Supp(P_i, S) ≥ s_g};
  Frequent pattern set F = ∪_k F_k;
  /* Testing phase */
  for each (e, t_i) in S do
    Create w_i = [t_i − T, t_i) and f(S, w_i);
    for each P_j ∈ F do
      Compute the number of UMOs of P_j in f_i;
      Update Supp(P_j, D) and Lr(P_j, D);
  /* Pruning phase */
  for each P_i ∈ F do
    if ((Supp(P_i, D) < s_l and Cr(P_i) ≥ cr)
    or Lr(P_i, D) ≥ lr) then
      Output P_i;
```

Fig. 4. Main algorithm

```
Algorithm 2
Input: A set C of patterns, a sequence (or sequence
fragment) S and the size of interval T_P.
Output: the number of UMOs for each pattern.
Method:
  /* Initialization */
  for each P_i ∈ C do
    P_i.event_count = 0; P_i.count = 0;
    for each a_j ∈ P_i do
      P_i.timestamp(a_j) = Null;
      group(a_j) = group(a_j) ∪ {P_i};
  /* Data pass */
  for each (a_i, t_i) in S such that i from 1 to n do
    for each P_j ∈ group(a_i) do
      if (P_j.timestamp(a_i) == Null) then
        P_j.timestamp(a_i) = t_i;
        if (++P_j.event_count == |P_j|) then
          /* Check temporal constraint */
          for each a_k ∈ P_j do
            if (t_i − P_j.timestamp(a_k) > T_P) then
              P_j.timestamp(a_k) = Null;
              P_j.event_count −−;
          if (P_j.event_count == |P_j|) then
            /* a UMO found */
            P_j.count++; P_j.event_count = 0;
            for each a_k ∈ P_j do
              P_j.timestamp(a_k) = Null;
      else /* Update timestamp */
        P_j.timestamp(a_i) = t_i;
```

Fig. 5. Counting the number of UMOs

of the event which matches the event type a in pattern P. When $P.event\_count$ is equal to $|P|$ (i.e., all event types in P have been matched), we check whether the temporal constraint (specified by T_P) is satisfied or not. For any event that matches P but violates the constraint, we eliminate it by setting its timestamp as $Null$ and decrease $P.event\_count$ by 1. If no such events exist (i.e., a UMO has been found), $P.count$ is increased by 1. After finding a UMO of P, we need to set $P.event\_count = 0$ and clear all $P.timestamp\,(a)$.

5 Experiment Results

In this section, we show experiment results of the application of telecommunication network fault analysis. In a telecommunication network, various devices are used to monitor the network continuously, generating a large number of different types of events. One type of event, trouble report (TR), is of particular importance because it indicates the occurrence of errors on the network. To investigate the dependency between TRs and other types of events, all events collected by different devices from the network are integrated into a long sequence in terms of their timestamps. In such a sequence, we regard TR as the target event and try to find patterns that have negative relationship with TR.

The telecommunication event database contains 119,814 events, covering 190 event types. The population of target events is 2,234. In the experiment, we do not consider the constraint on the local support (i.e., disregard the condition $Supp\,(P, D) < s_l$) and always set threshold cr as 2.0. Negative patterns are

Table 1. An example rule

Rule	Interestingness measures
P: $\{E1, F5\}$	$Supp\,(P, S)$: 5.73
e: NTN	$Supp\,(P, D)$: 1.86
T_P (minute): 10	$Cr\,(P)$: 3.08
T (minute): 180	

discovered under different values of other parameters. An example rule is shown in Table 1. The rule indicates that the occurrences that both signal $E1$ and $F5$ happen within 10 minutes are unexpected rare in a 180-minute-interval before trouble report NTN. According to this rule, domain experts believe that the signal $E1$ and $F5$ have negative relationship with NTN.

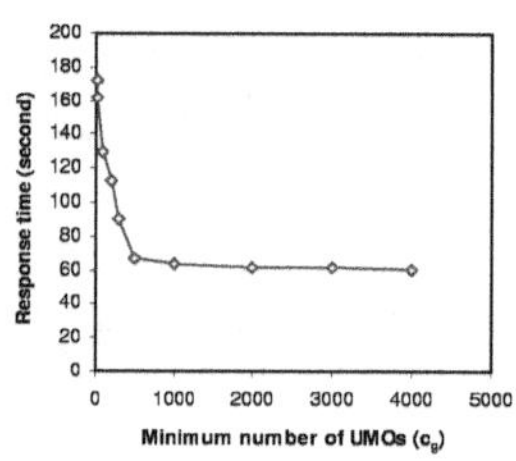

(a) Performance of phase 1 ($\mathbf{c_g}$) $T_P = 20m$

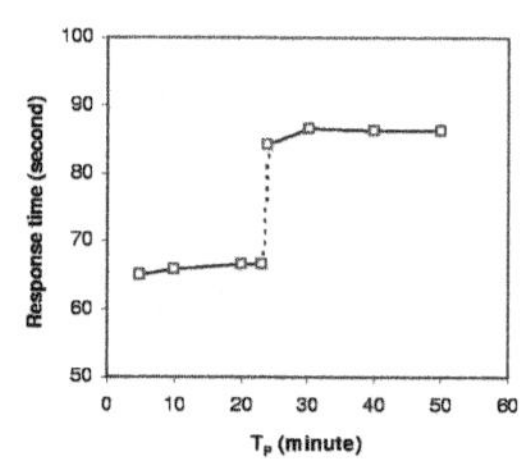

(b) Performance of phase 1 ($\mathbf{T_P}$) $c_g = 500$

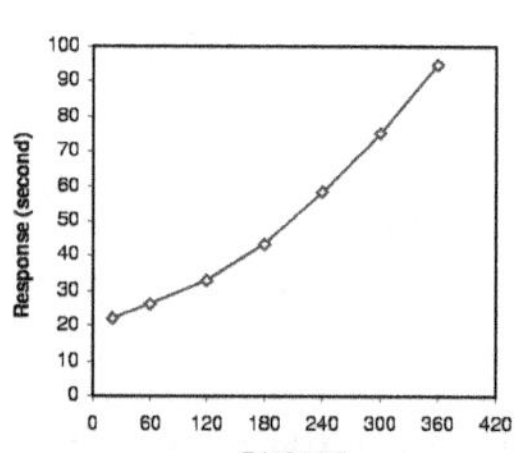

(c) Performance of phase 2 (**T**) $T_P = 20m, c_g = 2000$

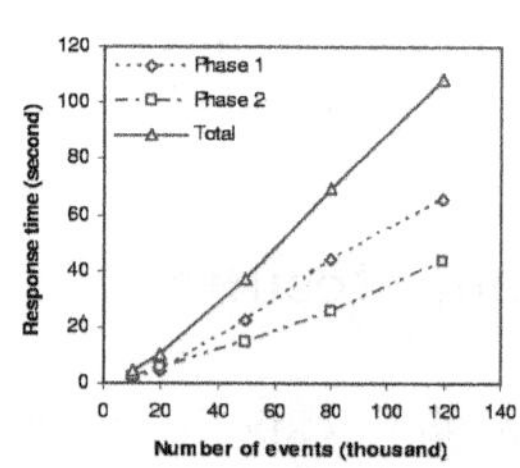

(d) Scalability of Algorithm 1 $T_P = 20m, T = 180m, s_g = 1.0$

Fig. 6. Performance evaluation

Performance evaluation can be seen in Figure 6. Particularly, Figure 6(a) gives the performance curve of phase 1 with respect to c_g, where c_g is a threshold on the number of UMOs, equivalent to $s_g * \left(\left[\frac{Dur(S)}{T}\right] + 1\right)$. Figure 6(b) shows the effect of parameter T_P on the performance of phase 1. In phase 2, a reverse scan on the database is performed. During the scan, we dynamically create the sequence fragment and update local support for patterns discovered in phase 1. The performance of phase 2 is shown in Figure 6(c) in terms of T. Finally, Figure 6(d) illustrates the overall performance and the scalability of Algorithm 1.

6 Conclusions

Mining negative associations should be treated as important as mining positive associations. However, as far as we know, it has been ignored in the research of pattern discovery in a long sequence. In this paper, the problem of finding negative event-oriented patterns has been identified. After proposing a set of interestingness measures, we have designed algorithms to discover a complete set of interesting patterns. To count the frequency of a pattern, we have proposed the UMO approach, which guarantees that the Apriori property holds for all patterns in the sequence. Finally, the experiment is made for a real application, which justifies the applicability of this research problem.

In a long sequence, the relationship between a pattern and target events could be positive, negative, or independent. Using the comparison ratio, we can roughly classify the pattern space into three categories according to the relationship with target events. In future work, we will investigate mining both positive and negative patterns in one process. In such a problem, the performance issue attracts most of our attention.

References

1. Mannila, H., Toivonen, H., Verkamo, A.I.: Discovery of frequent episodes in event sequences. Data Mining and Knowledge Discovery **1** (1997) 259–289
2. Berger, G., Tuzhilin, A.: Discovering unexpected patterns in temporal data using temporal logic. In: Temporal Databases:research and practice. (1998) 281–309
3. Han, J., Dong, G., Yin, Y.: Efficient mining of partial periodic patterns in time series database. In: Proc. 15th ICDE. (1999) 106–115
4. Yang, J., Wang, W., Yu, P.S.: Infominer: mining surprising periodic patterns. In: Proc. 7th KDD. (2001) 395–400
5. Sun, X., Orlowska, M.E., Zhou, X.: Finding event-oriented patterns in long temporal sequences. In: Proc. 7th PAKDD. (2003) 15–26
6. Padmanabhan, B., Tuzhilin, A.: Unexpectedness as a measure of interestingness in knowledge discovery. Decision Support Systems **27** (1999) 303–318
7. Mannila, H., Toivonen, H.: Discovering generalized episodes using minimal occurrences. In: Proc. 2nd KDD. (1996) 146–151
8. Mannila, H., Toivonen, H., Verkamo, A.I.: Discovering Frequent Episodes in Sequences. In: Proc. 1st KDD. (1995) 210–215
9. Savasere, A., Omiecinski, E., Navathe, S.B.: Mining for strong negative associations in a large database of customer transactions. In: Proc. 14th ICDE. (1998) 494–502
10. Wu, X., Zhang, C., Zhang, S.: Mining both positive and negative association rules. In: Proc. 19th ICML. (2002) 658–665
11. Dong, G., Li, J.: Efficient mining of emerging patterns: Discovering trends and differences. In: Proc. 5th KDD. (1999) 43–52

OBE: Outlier by Example

Cui Zhu[1], Hiroyuki Kitagawa[2], Spiros Papadimitriou[3], and Christos Faloutsos[3]

[1] Graduate School of Systems and Information Engineering, University of Tsukuba
[2] Institute of Information Sciences and Electronics, University of Tsukuba
{zhucui, kitagawa}@kde.is.tsukuba.ac.jp
[3] School of Computer Science, Carnegie Mellon University
{spapadim+, christos}@cs.cmu.edu

Abstract. Outlier detection in large datasets is an important problem. There are several recent approaches that employ very reasonable definitions of an outlier. However, a fundamental issue is that the notion of which objects are outliers typically varies between users or, even, datasets. In this paper, we present a novel solution to this problem, by bringing users into the loop. Our *OBE (Outlier By Example)* system is, to the best of our knowledge, the first that allows users to give some examples of what they consider as outliers. Then, it can directly incorporate a *small* number of such examples to successfully discover the hidden concept and spot further objects that exhibit the same "outlier-ness" as the examples. We describe the key design decisions and algorithms in building such a system and demonstrate on both real and synthetic datasets that OBE can indeed discover outliers that match the users' intentions.

1 Introduction

In many applications (e.g., fraud detection, financial analysis and health monitoring), rare events and exceptions among large collections of objects are often more interesting than the common cases. Consequently, there is increasing attention on methods for discovering such "exceptional" objects in large datasets and several approaches have been proposed.

However, the notion of what is an *outlier* (or, exceptional/abnormal object) varies among users, problem domains and even datasets (problem instances): (i) different users may have different ideas of what constitutes an outlier, (ii) the same user may want to view a dataset from different "viewpoints" and, (iii) different datasets do not conform to specific, hard "rules" (if any).

We consider objects that can be represented as multi-dimensional, numerical tuples. Such datasets are prevalent in several applications. From a general perspective [4,7,8,2], an object is, intuitively, an *outlier* if it is in some way "significantly different" from its "neighbors." Different answers to what constitutes a "neighborhood," how to determine "difference" and whether it is "significant," would provide different sets of outliers.

Typically, users are experts in their problem domain, not in outlier detection. However, they often have a few example outliers in hand, which may "describe"

H. Dai, R. Srikant, and C. Zhang (Eds.): PAKDD 2004, LNAI 3056, pp. 222–234, 2004.

their intentions and they want to find more objects that exhibit "outlier-ness" characteristics similar to those examples. Existing systems do not provide a direct way to incorporate such examples in the discovery process.

Example. We give a concrete example to help clarify the problem. The example is on a 2-d vector space which is easy to visualize, but ideally our method should work on arbitrary dimensionality or, even, metric datasets[1].

Consider the dataset in Figure 1. In this dataset, there are a large sparse cluster, a small dense cluster and some clearly isolated objects. Only the isolated objects (circle dots) are outliers from a "bird's eye" view. In other words, when we examine wide-scale neighborhoods (i.e., with large radius—e.g., covering most of the dataset), only the isolated objects have very low neighbor densities, compared with objects in either the large or the small cluster. However, consider the objects on the fringe of the large cluster (diamond dots). These can also be regarded as outliers, if we look closer at mid-scale (i.e., radius) neighborhoods. Also, objects on the fringe of the small cluster (cross dots) become outliers, if we further focus into small-scale neighborhoods. As exemplified here, different objects may be regarded as outliers, depending on neighborhood scale (or, size).

This scenario is intuitive from the users' perspective. However, to the best of our knowledge, none of the existing methods can directly incorporate user examples in the discovery process to find out the "hidden" outlier concept that users may have in mind.

In this paper, we propose *Outlier By Example (OBE)*, an outlier detection method that can do precisely that: discover the desired "outlier-ness" at the appropriate scales, based on a small number of examples. There are several challenges in making this approach practical; we briefly list the most important: **(1)** What are the appropriate features that can capture "outlier-ness?" These should ideally capture the important characteristics *concisely* and be efficient to compute. However, feature selection is only the tip of the iceberg. **(2)** Furthermore, we have to carefully choose exactly what features to extract. **(3)** The method should clearly require minimal user input and effectively use a *small* number of positive examples in order to be practical. Furthermore, it should ideally not need negative examples. **(4)** Given these requirements, can we train a classifier using only the handful of positive examples and unlabeled data? In the paper we describe the key algorithmic challenges and design decisions in detail.

In summary, the main contributions of this paper are: **(1)** We introduce example-based outlier detection. **(2)** We demonstrate its intuitiveness and feasibility. **(3)** We propose OBE, which, to the best of our knowledge, is the first method to provide a solution to this problem. **(4)** We evaluate OBE on both real and synthetic data, with several small sets of user examples. Our experiments demonstrate that OBE can succesfully incorporate these examples in the discov-

[1] A metric dataset consists of objects for which we only know the pairwise distances (or, "similarity"), without any further assumptions.

ery process and detect outliers with "outlier-ness" characteristics very similar to the given examples.

The remainder of the paper is organized as follows: In section 2, we discuss related work on outlier detection. In section 3, we discuss the measurement of "outlier-ness" and the different properties of outliers. Section 4 presents the proposed method in detail. Section 5 reports the extensive experimental evaluation on both synthetic and real datasets. Finally, Section 6 concludes the paper.

2 Related Work

In essence, outlier detection techniques traditionally employ unsupervised learning processes. The several existing approaches can be broadly classified into the following categories: *(1) Distribution-based approach.* These are the "classical" methods in statistics [1,11]. *(2) Depth-based approach.* This computes different layers of k-d convex hulls and flags objects in the outer layer as outliers [5]. *(3) Clustering approach.* Many clustering algorithms detect outliers as by-products [6]. *(4) Distance-based approach.* Distance-based outliers [7,8,9,10,3] use a definition based on a single, global criterion. All of the above approaches regard being an outlier as a binary property. They do not take into account both the degree of "outlier-ness" and where the "outlier-ness" is presented. *(5) Density-based approach,* proposed by M. Breunig, et al. [2]. They introduced a local outlier factor (LOF) for each object, indicating its degree of "outlier-ness." LOF depends on the local density of its neighborhood. The neighborhood is defined by the distance to the MinPts-th nearest neighbor. When we change the value of the parameter MinPts, the degree of "outlier-ness" can be estimated in different scopes. However, LOF is very sensitive to the selection of MinPts values, and it has been proven that LOF cannot cope with the multi-granularity problem. *(6) LOCI.* We proposed the multi-granularity deviation factor (MDEF) and LOCI in [12]. MDEF measures the "outlier-ness" of objects in neighborhoods of different scales. LOCI examines the MDEF values of objects in all ranges and flags as outliers those objects whose MDEF values deviate significantly from the local average in neighborhoods of *some* scales. So, even though the definition of MDEF can capture "outlier-ness" in different scales, these differences are up to the user to examine manually.

Another outlier detection method was developed in [15], which focuses on the discovery of rules that characterize outliers, for the purposes of filtering new points in a security monitoring setting. This is a largely orthogonal problem. Outlier scores from SmartSifter are used to create labeled data, which are then used to find the outlier filtering rules.

In summary, all the existing methods are designed to detect outliers based on some prescribed criteria for outliers. To the best of our knowledge, this is the first proposal for outlier detection using user-provided examples.

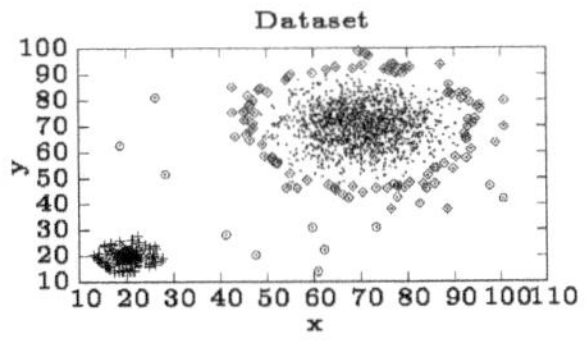

Fig. 1. Illustration of different kinds of outliers in a dataset.

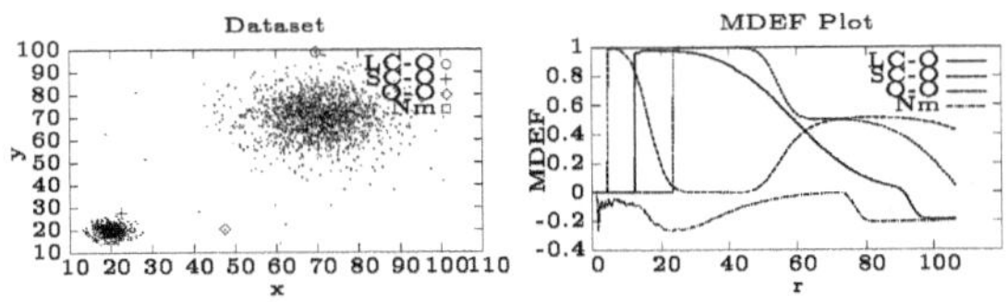

Fig. 2. Illustrative dataset and MDEF plots.

3 Measuring Outlier-ness

In order to understand the users' intentions and the "outlier-ness" they are interested in, a first, necessary step is measuring the "outlier-ness." It is crucial to select features that capture the important characteristics concisely. However, feature selection is only the initial step. In OBE, we employ MDEF for this purpose, which measures "outlier-ness" of objects in the neighborhoods of different scales (i.e., radii).

Detailed definition of the multi-granularity deviation factor (MDEF) is given in [12]. Here we describe some basic terms and notation. Let the r-neighborhood of an object p_i be the set of objects within distance r of p_i. Let $n(p_i, \alpha r)$ and $n(p_i, r)$ be the numbers of objects in the αr-neighborhood (*counting neighborhood*) and r-neighborhood (*sampling neighborhood*) of p_i respectively.[2] Let $\hat{n}(p_i, r, \alpha)$ be the average, over all objects p in the r-neighborhood of p_i, of $n(p, \alpha, r)$.

Definition (MDEF). For any p_i, r and α, the *multi-granularity deviation factor* (*MDEF*) at radius (or scale) r is defined as follows:

$$MDEF(p_i, r, \alpha) = \frac{\hat{n}(p_i, r, \alpha) - n(p_i, \alpha r)}{\hat{n}(p_i, \alpha, r)} \tag{1}$$

Intuitively, the MDEF at radius r for a point p_i is the relative deviation of its local neighborhood density from the average local neighborhood density in its r-neighborhood. Thus, an object whose neighborhood density matches the average local neighborhood density will have an MDEF of 0. In contrast, outliers will have MDEFs far from 0. In our paper, the MDEF values are examined (or, sampled) at a wide range of sampling radii r, $r_{min} \leq r \leq r_{max}$, where r_{max} is the maximum distance of all object pairs in the given dataset and r_{min} is determined based on the number of objects in the r-neighborhood of p_i. In our experiments, for each p_i in the dataset, r_{min} for p_i (denoted by $r_{min,i}$) is the distance to its 20-th nearest neighbor. In other words, we do not examine the MDEF value of an object until the number of objects in its sampling neighborhood reaches 20. This is a reasonable choice which effectively avoids introduction of statistical errors in MDEF estimates in practice.

[2] In all experiments, $\alpha = 0.5$ as in [12].

Next we give some examples to better illustrate MDEF. Figure 2 shows a dataset which has mainly two groups: a large, sparse cluster and a small, dense one, both following a Gaussian distribution. There are also a few isolated points. We show MDEF plots for four objects in the dataset.

- Consider the point in the middle of the large cluster, Nm, (at about $x = 70$, $y = 68$). The MDEF value is low at *all* scales: compared with its neighborhood, whatever the scale is, the local neighborhood density is always similar to the average local density in its sampling neighborhood. So, the object can be always regarded as a normal object in the dataset.
- In contrast, for the other three objects, there exist situations where the MDEFs are very large, some times even approaching 1. This shows that they differ significantly from their neighbors in *some* scales. The greater the MDEF value is, the stronger the degree of "outlier-ness".

Even though all three objects in Figure 2 can be regarded as outliers, they are still different, in that they exhibit "outlier-ness" at different scales.

- The MDEF value of the outlier in the small cluster, SC-O, (at about $x = 22$, $y = 27$), reaches its maximum at radius $r \approx 5$, then it starts to decrease rapidly until it becomes 0 and remains there for a while (in the range of $r \approx 23--45$). Then the MDEF value increases again but only to the degree of 0.6. The change of MDEF values indicates that the object is extremely abnormal compared with objects in the very small local neighborhood (objects in the small cluster).
- On the other hand, the outlier of the large cluster, LC-O, (at about $x = 70$, $y = 98$), exhibits strong "outlier-ness" in the range from $r = 10$ to $r = 30$, then becomes more and more ordinary as we look at a larger scale.
- For the isolated outlier, O-O, (at about $x = 47$, $y = 20$), its MDEF value stays at 0 up to almost $r = 22$, indicating that it is an isolated object. Then, it immediately displays a high degree of "outlier-ness."

4 Proposed Method (OBE)

4.1 Overview

OBE detects outliers based on user-provided examples and a user-specified fraction of objects to be detected as outliers in the dataset. OBE performs outlier detection in three stages: feature extraction step, example augmentation step and classification step. Figure 3 shows the overall OBE framework.

4.2 Feature Extraction Step

The purpose of this step is to map all objects into the MDEF-based feature space, where the MDEF plots of objects capturing the degree of "outlier-ness," as well as the scales at which the "outlier-ness" appears, are represented by vectors. Let D be the set of objects in the feature space. In this space, each object is represented by a vector: $O_i = (m_{i0}, m_{i1}, \ldots, m_{in})$, $O_i \in D$, where $m_{ij} = MDEF(p_i, r_j, \alpha r)$, $0 \le j \le n$, $r_0 = min_k(r_{min,k})$, $r_n = r_{max}$, $r_j = \frac{r_n - r_0}{n} j + r_0$.[3]

4.3 Example Augmentation Step

In the context of outlier detection, outliers are usually few, and the number of examples that users could offer is even less. If we only learn from the given examples, the information is very little to be used to construct an accurate classifier. However, example-based outlier detection is practical only if the number of required examples is small. OBE effectively solves this problem by augmenting the user-provided examples.

In particular, the examples are augmented by adding outstanding outliers and artificial positive examples, based on the original examples.

Outstanding Outliers. After all objects are projected into the feature space, we can detect outstanding outliers. The set of outstanding outliers is defined by $\{O_i \mid max\_M(O_i) > K,\ O_i \in D\}$, where $max\_M(O_i) = max_j(m_{ij})$ and K is a threshold.

Artificial Examples. The examples are further augmented by creating "artificial" data. This is inspired by the fact that an object is sure to be an outlier if *all* of its feature values (i.e., MDEF values) are greater than those of the given outlier examples. Figure 4 shows the created artificial data and the original example.

Artificial data are generated in the following way: (1) Take the difference between the $max\_M(O_i)$ and the threshold K, $Diff\_M(i) = K - max\_M(O_i)$. (2) Divide the difference, $Diff\_M(i)$, into x intervals, where x is the number of artificial examples generated from an original outlier example plus 1. For instance, if the intended augmentation ratio is 200%, two artificial examples are generated from each original example. Then we divide $Diff\_M(i)$ into 3 intervals ($x = 3$), $Intv\_M(i) = Diff\_M(i)/x$. (3) Then, create artificial examples as: $O\_A(i, j) = (m_{i0} + j * Intv\_M(i),\ m_{i1} + j * Intv\_M(i),\ \ldots,\ m_{in} + j * Intv\_M(i))$ for $1 \le j \le x - 1$. Here, $O\_A(i, j)$ is the j-th artificial example generated from object O_i.

In this way, the "outlier-ness strength" of the user's examples is amplified, in a way consistent with these examples.

Putting together the original examples, outstanding outliers and artificial examples, we get the positive training data.

4.4 Classification Step

So far, the (augmented) positive examples, as well as the entire, unlabeled dataset are available to us. The next crucial step is finding an efficient and

[3] More precisely, if $r_j \ge r_{min,i}$, $m_{ij} = MDEF(p_i, r_j, \alpha r)$, otherwise, $m_{ij} = 0$.

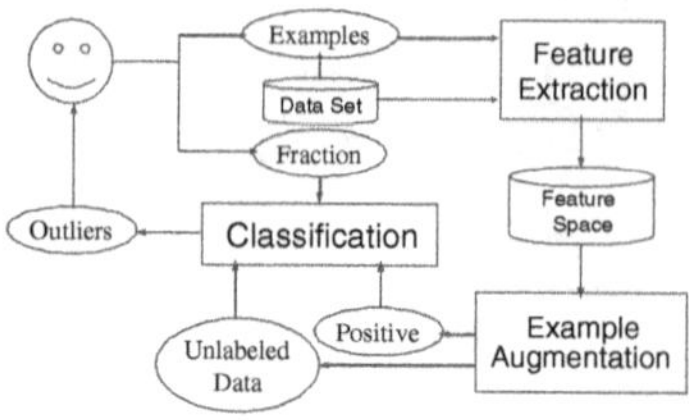

Fig. 3. The Framework of OBE

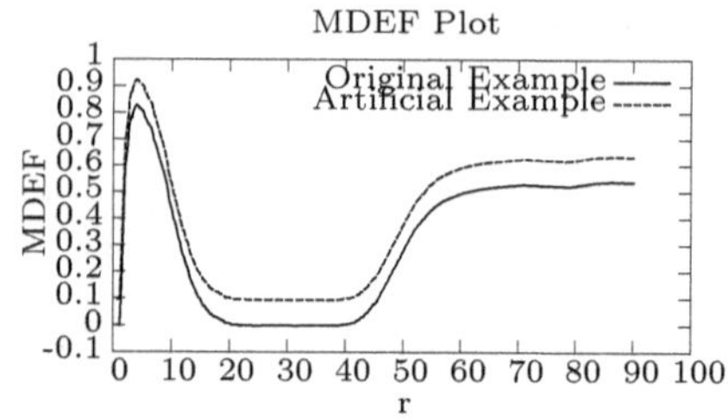

Fig. 4. The Artificial and the Original Examples

effective algorithm to discover the "hidden" outlier concept that the user has in mind.

We use an SVM (Support Vector Machine) classifier to learn the "outlierness" of interest to the user and then detect outliers which match this. Traditional classifier construction needs both positive and negative training data. However, it is too difficult and also a burden for users to provide negative data. Most objects fall in this category and it is unreasonable to expect users to examine them.

However, OBE addresses this problem and can learn only from the positive examples obtained in the augmentation step and the unlabeled data (i.e., the rest of the objects in the dataset). The algorithm shown here uses the marginal property of SVMs. In this sense, it bears some general resemblance to PEBL [13], which was also proposed for learning from positive and unlabeled data. However, in PEBL, the hyperplane for separating positive and negative data is set as close as possible to the set of given positive examples. In the context of OBE, the positive examples are just examples of outliers, and it is not desirable to set the hyperplane as in PEBL. The algorithm here decides the final separating hyperplane based on the fraction of outliers to be detected. Another difference between OBE and PEBL is that strong negative data are determined taking the characteristics of MDEF into consideration.

The classification step consists of the following five sub-steps.

Negative training data extraction sub-step. All objects are sorted in descending order of max_M. Then, from the objects at the bottom of the list, we select a number of (strong) negative training data equal to the number of positive training data. Let the set of strong negative training data be NEG. Also, let the set of positive training data obtained in the example augmentation step be POS.

Training sub-step. Train a SVM classifier using POS and NEG.

Testing sub-step. Use the SVM to divide the dataset into the positive set P and negative set N.

Update sub-step. Replace NEG with N, the negative data obtained in the testing sub-step.

Iteration sub-step. Iterate from the training sub-step to the updating sub-step until the ratio of the objects in P converges to the fraction specified by the user. The objects in the final P are reported to the user as detected outliers.

```
Input:                                      // Example augmentation step:
  Set of outlier examples: E                For each example in E
  Fraction of outliers: F                     Create artificial examples
  Dataset: D                                POS := E ∪ OO ∪ artificial examples
Output:                                     // Classification step:
  Outliers like examples                    NEG := strongest negatives
Algorithm:                                  P := D
  OO := ∅   // Outstanding outliers         Do {
  // Feature extraction step:                 P' := P
  For each p_i ∈ D                            SVM := train_SVM (POS, NEG)
    For each j (0 ≤ j ≤ n)                    (P, N) := SVM.classify (D)
      Compute MDEF value m_ij                 NEG := N
      If m_ij > K                           } while (|P| ≥ F * |D| and |P| ≠ |P'|)
      Then OO := OO ∪ {p_i}                 return P'
```

Fig. 5. The Overall Procedure of OBE

Table 1. Description of synthetic and real datasets.

Dataset	Description
Uniform	A 6000-point group following an uniform distribution.
Ellipse	A 6000-point ellipse following a Gaussian distribution.
Mixture	A 5000-point sparse Gaussian cluster, a 2000-point dense Gaussian cluster and 10 randomly scattered outliers.
NYWomen	Marathon runner data, 2229 women from the NYC marathon: average pace (in minutes per mile) for each stretch (6.2, 6.9, 6.9, and 6.2 miles).

Figure 5 summarizes the overall procedure of OBE.

5 Experimental Evaluation

In this section, we describe our experimental methodology and the results obtained by applying OBE to both synthetic and real data, which further illustrate the intuition and also demonstrate the effectiveness of our method.

We use three synthetic and one real datasets (see Table 1 for descriptions) to evaluate OBE.

5.1 Experimental Procedure

Our experimental procedure is as follows:

1. To simulate interesting outliers, we start by selecting objects which represent "outlier-ness" at some scales under some conditions, for instance, $\bigwedge_q(min_q, max_q, Cond_q, K_q)$, where $(min_q, max_q, Cond_q, K_q)$ stands for the condition that $(m_{ij}\ Cond_q\ K_q)$ for some j such that $min_q \leq j \leq max_q$, where $Cond_q$ could be either ">" or "<".

Table 2. Interesting Outliers, Discriminants and the Performance of OBE. OO denotes outstanding outliers, IO denotes interesting outliers. Precision(Preci-), recall(Reca-) and the number of iterations(Iter-) for convergence in the classification step are used to show the performance of OBE.

Dataset	OO	Cases				OBE		
		Label	Discription	Condition	IO	Preci-	Reca-	Iter-
Uniform Dataset	0	U-Fringe	Fringe	(0.3, 0.6, >, 0.4)	330	82.76	88.18	8.1
		U-Corner	Corner	(1, 2, >, 0.5)	274	91.90	97.92	4.1
Ellipse Dataset	15	E-Fringe	Fringe	(5, 30, >, 0.85)	214	90.20	93.55	6.1
		E-Long	Long Ends	(15, 25, >, 0.8) (30, 40, >, 0.6)	140	88.67	92.14	5.4
		E-Short	Short Ends	(5, 15, >, 0.8) (35, 40, <, 0.6)	169	76.46	80.00	10.4
Mixture Dataset	29	M-All	All	(1, 35, >, 0.9)	166	86.32	93.80	4.5
		M-Large	Large Cluster	(15, 35, >, 0.9)	123	91.52	95.37	4.6
		M-Small	Small Cluster	(1, 5, >, 0.9)	72	91.30	97.92	5.3
NYWomen Dataset	17	N-FS	Very Fast/Slow	(800, 1400, >, 0.7)	91	81.53	84.95	6.5
		N-PF	Partly Fast	(300, 500, >, 0.8) (1400, 1600, <,0.4)	126	73.07	78.81	6.9
		N-SS	Stable Speed	(100, 300, >, 0.8) (400, 600, <, 0.3)	121	66.55	70.74	9.2

2. Then, we "hide" most of these outliers. In particular, we randomly sample $y\%$ of the outliers to serve as examples that would be picked by a user.
3. Next, we detect outliers using OBE.
4. Finally, we compare the detected outliers to the (known) simulated set of outliers. More specifically, we evaluate the success of OBE in recovering the hidden outlier concept using precision/recall measurements.

OBE reports as interesting outliers the outstanding ones, as well as those returned by the classifier. Table 2 shows all the sets of interesting outliers along with the corresponding discriminants used as the underlying outlier concept in our experiments. In the table, for instance, the discriminant (1, 35, >, 0.9) means that objects are selected as interesting outliers when their MDEF values are greater than 0.9 in the range of radii from 1 to 35. The number of the outstanding outliers and interesting outliers is also shown in Table 2. We always randomly sample 10% ($y = 10$) of the interesting outliers to serve as user-provided examples and "hide" the rest.

To detect outstanding outliers, we use $K = 0.97$ for all the synthetic datasets and $K = 0.99$ for the NYWomen dataset. The discovered outstanding outliers of the synthetic datasets are shown in Figure 6. Also, during the augmentation step, we always generate 5 ($x = 6$) artificial examples from each original example.

We use the LIBSVM [14] implementation for our SVM classifier. We extensively compared the accuracy of both linear and polynomial SVM kernels and found that polynomial perform consistently better. Therefore, in all experiments,

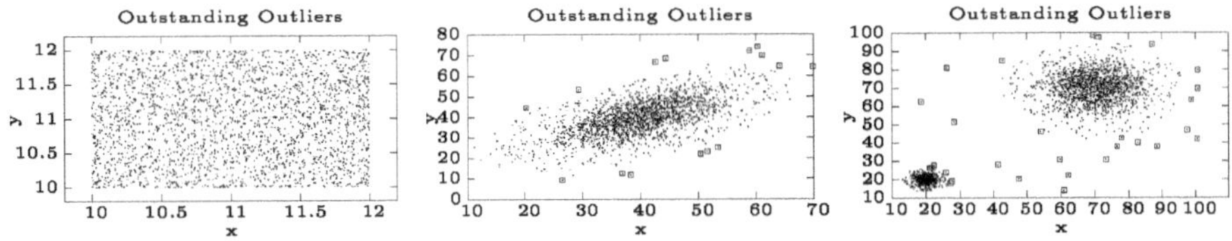

Fig. 6. Outstanding Outliers in the Synthetic Datasets.

we use polynomial kernels and the same SVM parameters[4]. Therefore, the whole processes can be done automatically. We report the effectiveness of OBE in discovering the "hidden" outliers using precision and recall measurements:

$$Precision = \frac{\#\ of\ correct\ positive\ predictions}{\#\ of\ positive\ predictions} \tag{2}$$

$$Recall = \frac{\#\ of\ correct\ positive\ predictions}{\#\ of\ positive\ data} \tag{3}$$

5.2 Results

Uniform dataset. Figure 7 shows the outliers detected by OBE. Although one might argue that no objects from an (infinite!) uniform distribution should be labeled as outliers, the objects at the fringe or corner of the group are clearly "exceptional" in some sense. On the top row, we show the interesting outliers, original examples and the detected results for case U-Fringe. The bottom row shows those for case U-Corner (see Table 2 for a description of the cases). Note that the chosen features can capture the notion of both "edge" and "corner" and, furthermore, OBE can almost perfectly reconstruct these hidden outlier notions!

Ellipse dataset. We simulate three kinds of interesting outliers for the ellipse dataset: (i) the set of fringe outliers whose MDEF values are examined at a wide range of scales, (ii) those mainly spread at the long ends of the ellipse which display outlier-ness in two ranges of scales (from 15 to 25 and from30 to 40), and (iii) mainly in the short ends, which do *not* show strong outlier-ness in the scales from 35 to 40. The output of OBE is shown in Figure 8. Again, the features can capture several different and interesting types of outlying objects and OBE again discovers the underlying outlier notion!

Mixture dataset. We also mimic three categories of interesting outliers: (i) the set of outliers scattered along the fringe of both clusters, (ii) those maily spread along the fringe of the large cluster, and (iii) those mainly in the small cluster. Due to space constraints, the figure is omitted here.

[4] For the parameter C (the penalty imposed on training data that fall on the wrong side of the decision boundary), we use 1000, i.e., a high penalty to mis-classification. For the polynomial kernel, we employ a kernel function of $(u' * v + 1)^2$.

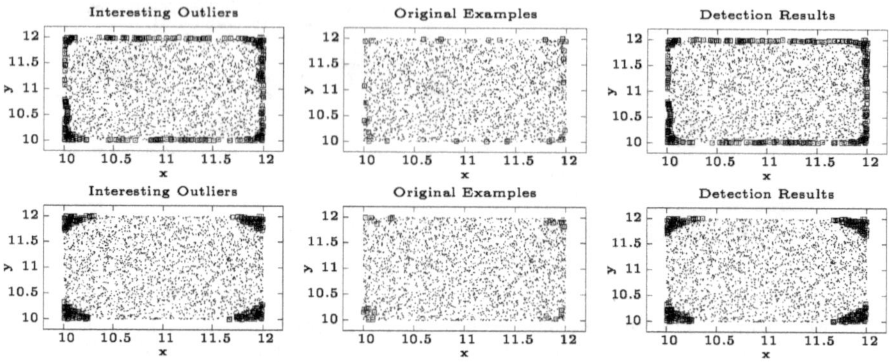

Fig. 7. Detection Results on the Uniform Dataset. Top row: case U-Fringe, bottom row: case U-Corner—see Table 2 for description of each case.

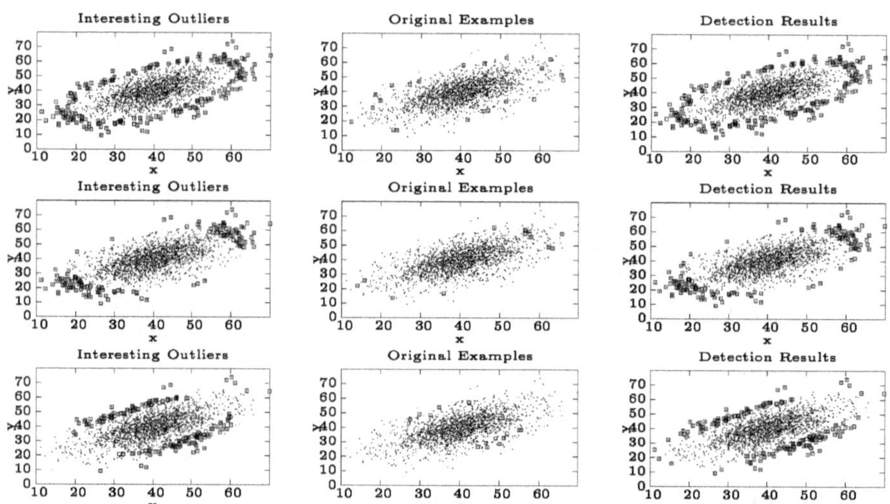

Fig. 8. Detection Results on the Ellipse dataset. From top to bottom, in turn: case E-Fringe, case E-Long, case E-Short—see Table 2 for description of each case.

NYWomen dataset. In the real dataset, we mimic three kinds of intentions for outliers: The first group (case N-FS) is the set of consistently fast or slow runners (i.e., the fastest 7 and almost all of the 70 very slow ones). The second group of outlying runners (case N-PF) are those who are at least partly fast. In this group, we discover both the fastest 23 runners and those runners who were abnormally fast in one or two parts of the four stretches, although they rank middle or last in the whole race. For example, one of them took 47 minutes for the first 6.2 miles, while 91 minutes for the last 6.2 miles. The third set of interesting outliers (case N-SS) is those who run with almost constant speed and rank middle in the whole race. They are very difficult to perceive, but they

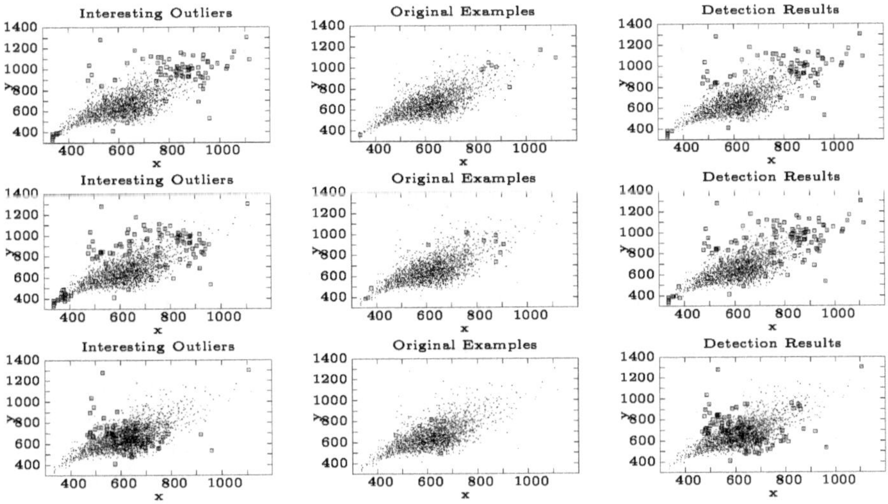

Fig. 9. Detection Results on the NYWomen Dataset. From top to bottom in turn: Case N-FS, Case N-PF, Case N-SS—see Table 2 for description of each case. Only the first and forth dimensions are used for the plots, although NYWomen Dataset is four dimensional.

certainly exhibit "outlier-ness" when we examine them at a small scale. Because of space limits, we only show the result plots in the first and forth dimensions —see Figure 9.

For all datasets, Table 2 shows the precision and recall measurements for OBE, using polynomial kernels (as mentioned, polynomial kernels always performed better than linear kernels in our experiments). It also shows the number of iterations needed to converge in the learning step. In Table 2, all the measurements are averages of ten trials. In almost all cases, OBE detects interesting outliers with both precision and recall reaching 80–90%. In the worst case (case N-SS of NYWomen), it still achieves 66% precision and 70% recall. The number of iterations is always small (less than 10).

6 Conclusion

Detecting outliers is an important, but tricky problem, since the exact notion of an outlier often depends on the user and/or the dataset. We propose to solve this problem with a completely novel approach, namely, by bringing the user in the loop, and allowing him or her to give us some example records that he or she considers as outliers.

The contributions of this paper are the following:

- We propose OBE, which, to the best of our knowledge, is the first method to provide a solution to this problem.

- We build a system, and described our design decisions. Although OBE appears simple to the user ("click on a few outlier-looking records"), there are many technical challenges under the hood. We showed how to approach them, and specifically, how to extract suitable feature vectors out of our data objects, and how to quickly train a classifier to learn from the (few) examples that the user provides.
- We evaluated OBE on both real and synthetic data, with several small sets of user examples. Our experiments demonstrate that OBE can succesfully incorporate these examples in the discovery process and detect outliers with "outlier-ness" characteristics very similar to the given examples.

Acknowledgements. The authors are grateful to Dr. Phillp B. Gibbons for early discussion on example-based outlier detection. This research has been supported in part by Japan-U.S. Cooperative Science Program of JSPS, U.S.-Japan Joint Seminar (NSF grant 0318547) and the Grant-in-Aid for Scientific Research from JSPS and MEXT (#15300027, #15017207).

References

1. V. Barnett and T. Lewis. Outliers in Statistical Data. John Wiley and Sons, 1994.
2. M.M. Breunig, H.P. Kriegel, R.T. Ng, and J. Sander. Lof: Identifying density-based local outliers. In Proc. SIGMOD Conf., pages 93-104,2000.
3. S. D. Bay and M. Schwabacher. Mining Distance-Based Outliers in Near Linear Time with Randomization and a Simple Pruning Rule. In SIGKDD'03, August 24-27, 2003.
4. D.M. Hawkins. Identification of Outliers. Chapman and Hall, 1980.
5. T.Johnson, I. Kwok, and R.T. Ng. Fast computation of 2-dimensional depth contours. In Proc. KDD, pages 224-228, 1998.
6. A.K. Jain, M.N. Murty, and P.J. Flynn. Data clustering: A review. ACM Comp. Surveys, 31(3);264-323,1999.
7. E.M. Knorr and R.T. Ng. A unified notion of outliers: Properties and computation. In Proc. KDD, pages 219-222, 1997.
8. E.M. Knorr and R.T. Ng. Algorithms for mining distance-based outliers in large datasets. In Proc. VLDB 1998, pages 392-403, 1998.
9. E.M. Knorr and R.T. Ng. Finding intentional knowledge of distance-based outliers. In Proc. VLDB, pages 211-222, 1999.
10. E.M. Knorr, R.T. Ng, and V. Tucakov. Distance-based outliers: Algorithms and applications. VLDB Journal, 8:237-253, 2000.
11. P.J. Rousseeuw and A.M. Leroy. Robust Regression and Outlier Detection. John Wiley and Sons, 1987.
12. S. Papadimitriou, H. Kitagawa, P.B. Gibbons and C. Faloutsos. LOCI: Fast Outlier Detection Using the Local Correlation Integral. In Proc. ICDE, pages 315-326, 2003.
13. H. Yu, J. Han and K.Chang. PEBL: Positive Example Based Learning for Web Page Classification Using SVM. In Proc. KDD, 2002.
14. http://www.csie.nut.edu.tw/~cjlin/libsvm.
15. K. Yamanishi and J.Takeuchi. Discovering Outlier Filtering Rules from Unlabeled Data. In Proc. KDD, 2001.

Temporal Sequence Associations for Rare Events*

Jie Chen, Hongxing He, Graham Williams, and Huidong Jin

CSIRO Data Mining
GPO Box 664, Canberra ACT 2601, Australia
{Jie.Chen,Hongxing.He,Graham.Williams,Warren.Jin}@csiro.au

Abstract. In many real world applications, systematic analysis of rare events, such as credit card frauds and adverse drug reactions, is very important. Their low occurrence rate in large databases often makes it difficult to identify the risk factors from straightforward application of associations and sequential pattern discovery. In this paper we introduce a heuristic to guide the search for interesting patterns associated with rare events from large temporal event sequences. Our approach combines association and sequential pattern discovery with a measure of risk borrowed from epidemiology to assess the interestingness of the discovered patterns. In the experiments, we successfully identify a known drug and several new drug combinations with high risk of adverse reactions. The approach is also applicable to other applications where rare events are of primary interest.

1 Introduction

The present work is motivated by the specific domain of temporal data mining in health care, especially adverse drug reactions. Adverse drug reactions occur infrequently but may lead to serious or life threatening conditions requiring hospitalisation. Thus, systematic monitoring of adverse drug reactions is of financial and social importance. The availability of a population-based prescribing data set, such as the Pharmaceutical Benefits Scheme (PBS) data in Australia, linked to hospital admissions data, provides an opportunity to detect common and rare adverse reactions at a much earlier stage. The problem domain has the following characteristics: (1) Primary interest lies in rare events amongst large datasets; (2) Factors leading to rare adverse drug reactions include temporal drug exposure; (3) Rare events are associated with a small proportion of patients yet all data for all patients are required to assess the risk.

For adverse drug reactions, we usually have little prior knowledge of what drug or drug combinations might lead to unexpected outcomes. Our aim is to

* The authors acknowledge the valuable comments from their colleagues, including C. Carter, R. Baxter, R. Sparks, and C. Kelman, as well as the anonymous reviewers. The authors also acknowledge the Commonwealth Department of Health and Ageing, and the Queensland Department of Health for providing data for this research.

H. Dai, R. Srikant, and C. Zhang (Eds.): PAKDD 2004, LNAI 3056, pp. 235–239, 2004.

discover patterns associated with rare events that are then further assessed for their possible relationship with adverse outcomes. Different from some previous work on mining interesting patterns for group difference [2,3] and health data [8, 4], we propose an approach which extends association and sequential pattern discovery with a heuristic measure motivated from epidemiology. We discover patterns that have high local support but generally low overall support and assess their significance using an estimate of a measure of risk ratio. The paper is organised as follows. Formal definitions and our methods are presented in Section 2. Section 3 reports on some encouraging results. The conclusion and discussion are given in Section 4.

2 Mining Temporal Association for Rare Events

Consider a collection of entities ϵ_i $(i = 1, 2, \ldots) \in E$, for example, credit card holders of a bank or patients in hospital. The activities of each entity ϵ_i are recorded as an event sequence $s_i = \langle (e_{i1}, t_{i1}), (e_{i2}, t_{i2}), \ldots, (e_{ij}, t_{ij}), \ldots, (e_{in_i}, t_{in_i}) \rangle$, where n_i is the number of events for the ith entity. For each event (e_{ij}, t_{ij}), e_{ij} indicates an **event type** and the **timestamp** t_{ij} indicates the time of occurrence of the event. For example, the following sequence describes a set of medical services received by a patient:

$$\langle (\text{G03CA}, 1), (\text{J01DA}, 7), (\text{C08CA}, 10), (\text{C09AA}, 10), (\textit{Angioedema}, 30) \rangle \, .$$

On day 1 the patient was dispensed the drug *estrogen*, whose ATC code is G03CA. They then took *cephalosporins and related substances* (J01DA) on the 7th day, and *dihydropyridine derivatives* (C08CA) and *ace inhibitor* (C09AA) on the 10th day. Twenty days later they were hospitalised due to *angioedema*. We refer to this particular event of interest as the **target event**, which can be either within the studied event sequence of an entity or just an external event not included in the event sequence but associated with the entity.

Using the target event as a classification criterion, we partition the entities into two subsets. The first one, denoted by $T \subset E$, contains entities having at least one target event. The second one, $\overline{T} \subset E$, consists of all remaining entities. Note that the time spans can be quite long in any particular sequence. Quite often, only events occurring within a particular lead up period, prior to the target event, are relevant. We first define a time window and its associated segments as follows.

Definition 1. *$[t_s, t_e]$ is a **time window** that starts at time t_s and ends at time t_e, where $w = t_e - t_s$ is constant, and usually specified by a domain expert.*

Definition 2. *$\langle (e_{ip}, t_{ip}), (e_{i,p+1}, t_{i,p+1}), \ldots, (e_{iq}, t_{iq}) \rangle$ is a **windowed segment** of sequence s_i with time window $[t_s, t_e]$ if $t_s \leq t_{ip} \leq t_{i,p+1}, \ldots, t_{iq} < t_e \leq t_{in_i}$, $t_{i,p-1} < t_s$ and $t_{i,q+1} \geq t_e$.*

Definition 3. *For any target entity $\epsilon_i \in T$, a **target segment** of s_i is a windowed segment of s_i with time window $[t_s, t_e]$ where t_e indicates the first occurrence time of the target event.*

Definition 4. *For any entity* $\epsilon_i \in \overline{T}$, *a* ***virtual target segment*** *of* s_i *is a windowed segment of* s_i *with time window* $[t_s, t_e]$.

There may be more than one target event associated with an entity in T. For simplicity, only the first target event is considered. Medical advice suggested that only events occurring within some time window prior to the target event might be considered in this exploration. For example, drug usage within six month prior to the adverse reaction is of main interest in our application. Given the fixed window length w, we have a list of virtual target segments for entity ϵ_i with different starting timestamps, $t_{i1}, t_{i2}, ..., \text{or}, t_{il_i}$, where t_{il_i} is the first element in $\langle t_{i1}, t_{i2}, ..., t_{in_i} \rangle$ such that $t_{il_i} \geq t_{in_i} - w$. We denote all these l_i virtual target segments as $L(i, w)$. Also, we can prove that any non-empty virtual target segment with $t_s \in R^1$ must be in $L(i, w)$.

We introduce a risk ratio, as often used in epidemiological studies, to measure the association between a factor and a disease, i.e., being a ratio of the risk of being disease positive for those with and those without the factor [1, p672]. We use the following estimate of a risk ratio for a candidate pattern p occurring within a fixed sized time window:

$$RR(p, w) = \frac{|T|s_T(p)}{|T|s_T(p) + |\overline{T}|s_{\overline{T}}(p)} / \frac{|T|(1 - s_T(p))}{|T|(1 - s_T(p)) + |\overline{T}|(1 - s_{\overline{T}}(p))}. \quad (1)$$

where $s_T(\text{p})$ is the **support in** T of pattern p, defined as the proportion of entities in T having p in their target segments, and $s_{\overline{T}}(\text{p})$ is the **support in** $\overline{T}$ of p, the proportion of entities in $\overline{T}$ containing p in their virtual target segments, i.e., p is contained in any element of $L(i, w)$. A risk ratio of 1 (i.e., $RR(p, w) = 1$) implies that there is equal risk of the disease with or without pattern p within a fixed sized time window. When $RR(p, w) > 1$, there is a greater risk of the disease in the exposed group.

We employ both the support in T and the estimated risk ratio for identifying interesting patterns. The framework for mining interesting patterns associated with rare events includes the following steps. Firstly, we extract two datasets of entities in the problem domain. Each entity records demographic data and an event sequence. The first dataset contains all entities with at least one target event. The second dataset contains the other entities. Secondly, we partition the entities of the two datasets into sub-populations according to their demographics. Thirdly, we discover candidate association [6] and sequential patterns [5] in T of each sub population. The events within a fixed sized time window prior to the target event are used for pattern discovery. Fourthly, we explore corresponding $\overline{T}$ of each sub-population to identify patterns mined in the above step. Finally, estimated risk ratios of the candidate patterns for each sub-population are calculated according to Equation 1.

3 Experimental Results

The Queensland Linked Data Set [7] links hospital admissions data from Queensland Health with the pharmaceutical prescription data from Commonwealth Department of Health and Ageing, providing a de-identified dataset for analysis.

Table 1. Estimated risk ratio of sample discovered association patterns. $|T|/|\overline{T}|$ for these cohorts are 55/104257 (Male 20-59), 76/194789 (Female 20-59), and 73/128586 (Female 60+)

Gender	Age	Pattern	support %	chi-square	RR
Male	20-59	C09AA N06AA	9.0	8.896	3.684
Male	20-59	N06AA H02AB	9.0	5.613	2.888
Female	20-59	C09AA G03CA	9.2	8.991	3.090
Female	60+	C09AA G03CA	24.6	19.45	3.112
Female	60+	C09AA C08CA	26.0	4.435	1.741

The record for each patient includes demographic variables and a sequence of PBS events for a five year period. Two datasets are extracted. One contains all 299 patients with hospital admissions due to angioedema, e.g., target event. The other contains 683,059 patients who have no angioedema hospitalisations.

It should be noted that the studied population consists of hospital patients rather than the whole Queensland population. We make the assumption that prior to the period of study these patients were not hospitalised for angioedema.

Tables 1 and 2 list experimental results for particular age/gender cohorts. The fourth column lists the support for the pattern in the target dataset. The other columns show the chi-square value and the estimated risk ratio, respectively. Here we use the chi-squares value, which is calculated together with the estimated risk ratio, as a threshold to constrain resulting patterns. Thus, for males aged 20-59 the drugs C09AA and N06AA within six months are over three times more likely to be associated with angioedema patients than with the non-angioedema patients. The following are some of the interesting sequential patterns. Usage of *estrogen* (G03CA) followed by *ace inhibitor* (C09AA), which is a known drug possibly resulting in angioedema, within six months is associated with the estimated risk ratio 2.636 of angioedema for females aged 60+; The sequence consisting of *dihydropyridine derivatives* (C08CA) and *ace inhibitor* (C09AA) within six months is generally associated with a high estimated risk ratio of angioedema for females aged 60+. Interestingly, four of these sequences begin with the pair of C08CA and C09AA, which means the two drugs are supplied to patients on the same day. These proposed hypotheses then form the basis for further statistical study and validation.

Table 2. Estimated risk ratio of sequential patterns. $|T|/|\overline{T}|$ for these cohorts are 76/194789 (Female 20-59), and 73/128586 (Female 60+).

Gender	Age	Pattern	support %	chi-square	RR
Female	20-59	C09AA -1 C09AA	15.7	7.210	2.273
Female	60+	G03CA -1 C09AA	20.5	12.11	2.636
Female	60+	C08CA C09AA -1 C08CA -1 C09AA -1 C08CA	17.8	9.207	2.453
Female	60+	C08CA C09AA -1 C09AA -1 C08CA -1 C08CA	17.8	9.053	2.436
Female	60+	C08CA -1 C09AA -1 C09AA -1 C08CA	20.5	9.395	2.365
Female	60+	C09AA -1 C08CA -1 C09AA -1 C08CA -1 C09AA	19.1	8.751	2.346
Female	60+	C08CA C09AA -1 C09AA -1 C09AA	19.1	8.678	2.339
Female	60+	C09AA -1 G03CA	17.8	7.687	2.280
Female	60+	C08CA C09AA -1 C08CA -1 C09AA	17.8	7.144	2.217
Female	60+	C09AA -1 C08CA -1 C08CA -1 C09AA -1 C09AA	17.8	6.491	2.139

4 Conclusion and Discussion

We have presented a temporal sequence mining framework for rare events and successfully identified interesting associations and sequential patterns of drug exposure which leads to a high risk of certain severe adverse reactions. Our intent is to generate hypotheses identifying potentially interesting patterns, while we realise that further validation and examination are necessitated. We also note that matching of patients has not included matching for time. It is proposed that case matching for time will be a refinement which could improve the approach to reduce the potential for false positives. In estimating risk ratios, it is not clear how to calculate confidence intervals in this case. Confidence intervals are important in identifying the degree of uncertainty in the estimates, and further work is required to investigate this deficiency. Besides adverse drug reactions, our approach may also be applied to a wide range of temporal data mining domains where rare events are of primary interest.

References

1. P. Armitage, G. Berry, and J. N. S. Matthews. *Statistical Methods in Medical Research.* Blackwell Science Inc, 4 edition, 2002.
2. S. D. Bay and M. J. Pazzani. Detecting change in categorical data: Mining contrast sets. In *Proceedings of the 5th ACM SIGKDD International Conference on Knowledge Discovery and Data Mining*, pages 302–306, 1999.
3. G. Dong and J. Li. Efficient mining of emerging patterns: Discovering trends and differences. In *Proceedings of the 5th ACM SIGKDD International Conference on Knowledge Discovery and Data Mining*, pages 43–52, San Diego, August 1999.
4. L. Gu, J. Li, H. He, G. Williams, S. Hawkins, and C. Kelman. Association rule discovery with unbalanced class. In *Proceedings of the 16th Australian Joint Conference on Artificial Intelligence (AI03), Lecture Notes in Artificial Intelligence*, Perth, Western Australia, December 2003.
5. M. Seno and G. Karypis. SLPMiner: An algorithm for finding frequent sequential patterns using length decreasing support constraint. In *Proceedings of the 2nd IEEE International Conference on Data Mining (ICDM)*, pages 418–425, Maebashi City, Japan, Dec 2002. IEEE.
6. G. I. Webb. Efficient search for association rules. In *Proceedings of the 6th ACM SIGKDD International Conference on Knowledge Discovery and Data Mining*, pages 99–107, 2000.
7. G. Williams, D. Vickers, R. Baxter, S. Hawkins, C. Kelman, R. Solon, H. He, and L. Gu. The Queensland Linked Data Set. Technical Report CMIS 02/21, CSIRO Mathematical and Information Sciences, Canberra, 2002.
8. W.-K. Wong, A. Moore, G. Cooper, and M. Wagner. WSARE: What's strange about recent events? *Journal of Urban Health*, 80(2):i66–i75, 2003.

Summarization of Spacecraft Telemetry Data by Extracting Significant Temporal Patterns

Takehisa Yairi[1], Shiro Ogasawara[2], Koichi Hori[1],
Shinichi Nakasuka[1], and Naoki Ishihama[3]

[1] University of Tokyo, 4-6-1 Komaba Meguro-ku, Tokyo, Japan,
yairi@space.rcast.u-tokyo.ac.jp
[2] Nippon Telegraph and Telephone Corporation, Japan
[3] Japan Aerospace Exploration Agency, Japan

Abstract. This paper presents a method to summarize massive spacecraft telemetry data by extracting significant event and change patterns in the low-level time-series data. This method first transforms the numerical time-series into a symbol sequence by a clustering technique using DTW distance measure, then detects event patterns and change points in the sequence. We demonstrate that our method can successfully summarize the large telemetry data of an actual artificial satellite, and help human operators to understand the overall system behavior.

1 Introduction

In recent years, a lot of datamining techniques for time-series data such as similar pattern search[1],[2], pattern clustering[3],[4],[5], event detection[6],[7],[8], change-point detection[9],[10],[11], and temporal association rule mining[12] have been studied actively. These techniques have been successfully applied to various domains dealing with vast time-series data such as finance, medicine, biology, robotics, etc. In the meantime, telemetry data of spacecrafts or artificial satellites is also a huge time-series data set usually containing thousands of sensor outputs from various system components. Though it is known that the telemetry data often contains some symptoms prior to fatal system failures, the limit-checking technique which is ordinarily used in most space systems often fails to detect them.

The purpose of this paper is to propose a data summarization method which helps human experts to find the anomaly symptoms by extracting important temporal patterns from the telemetry data. The summarization process consists of symbolization of the originally numerical time-series and detection of event patterns and change-points. This data-driven approach to the fault detection problem is expected to overcome some limitations of other sophisticated approaches such as expert systems and model-based simulations which require the costly a priori expert knowledge. We also show some results of applying the method to a telemetry data set of an actual artificial satellite ETS-VII (Engineering Test Satellite VII) of NASDA (National Space Development Agency of Japan).

H. Dai, R. Srikant, and C. Zhang (Eds.): PAKDD 2004, LNAI 3056, pp. 240–244, 2004.

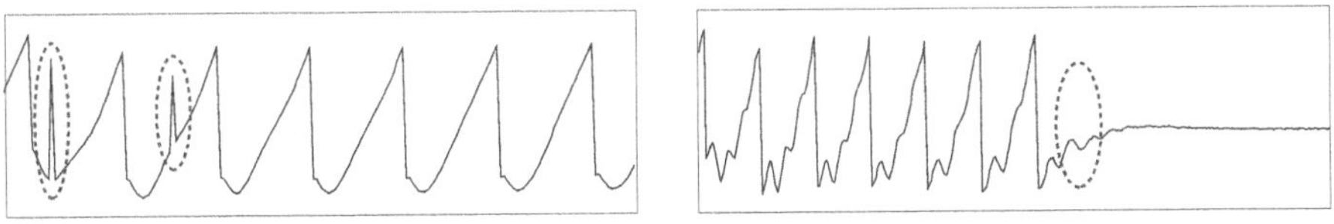

(a) Immediate Events (b) Mode Change

Fig. 1. Examples of event and change patterns in time-series data

2 Proposed Method

2.1 Basic Idea

The purpose of our method is to make a summary of the telemetry (HK) data automatically by preserving only important information while discarding the other. This helps the operators to understand the health status of the systems, and hence increases the chance to find subtle symptoms of anomalies that could not be detected by the conventional techniques.

A non-trivial problem here is that we need to decide "what is important information" in the data beforehand. As to the HK data of spacecrafts, we judged that the following two kinds of information are especially important based on the investigation of past failure cases and interviews with experts.

1. Immediate events ⋯ Patterns that are distinct from other neighboring parts
2. Mode changes ⋯ Points where the characteristics of the time-series change

Fig. 1 shows examples of the event and change patterns. They are considered to have certain important information in that it corresponds to some actual events in the system such as "engine thrustings", "changes of attitude control mode", and so on. In the remainder of this section, we describe the ways of detecting the immediate events and mode changes from the data.

2.2 Detection of Immediate Events

The symbolization of the original HK data and detection of the immediate events are described as follows. First, each time-series in the HK data is divided into a set of subsequences with a fixed length. Then all the subsequences are grouped into clusters based on the DTW (Dynamic Time Warping) distance measure. Finally, each cluster is assigned a unique symbol, and the subsequences contained in "small" clusters are detected as event patterns.

Selecting the number of clusters is a common problem for all clustering methods. In the current implementation, although the system mostly recommends a proper value based on the MDL (Minimum Description Length) criterion. it is sometimes necessary for human operators to adjust the parameter in order to obtain a better result.

2.3 Detection of Mode Changes

In the proposed method, detection of the mode change points in the HK data is realized by finding an optimum segmentation of the symbolized time-series. Suppose a sequence of symbols $[s_1, \ldots, s_t, \ldots, s_N]$ is divided into M segments. Then we model each segment by a 0-order Markov model, and evaluate the goodness of this segmentation by the sum of *modelling losses* for the segments. We search for the best segmentation that minimizes the sum of modelling losses, and define the borders of segments as the mode change points.

This change-detection process also has an open problem of how to decide the number of segments M, which is similar to the decision problem of number of clusters in the symbolization process. In the current implementation, the subtle adjustment is up to the operators.

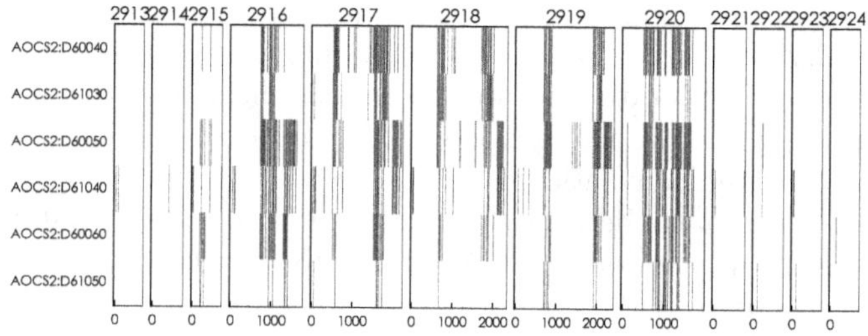

Fig. 2. (Example 1) Co-occurrences of events and changes

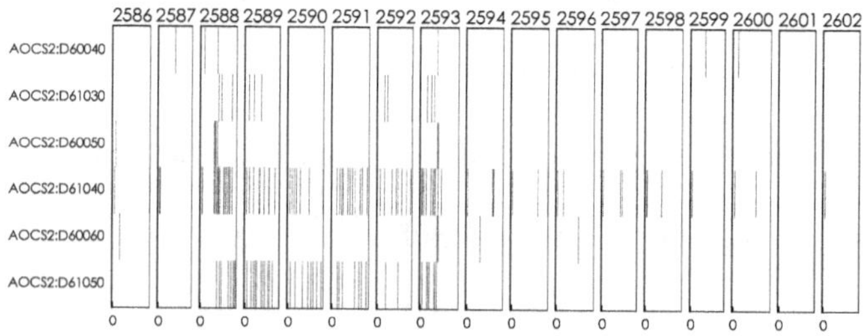

Fig. 3. (Example 2) Association among events and changes of D60050, D61040, and D61050

3 Case Study

We implemented the methods described above and applied it to the HK data of ETS-VII for four years. In this case study, we picked up 6 time-series relating to the AOCS (Attitude & Orbit Control Subsystem). Their brief descriptions are given in Table 1.

Table 1. List of time-series in ETS-VII's HK data chosen for this case study

ID	Explanation
D60040	Drive signal of AOCS reaction wheel (Roll)
D60050	Drive signal of AOCS reaction wheel (Pitch)
D60060	Drive signal of AOCS reaction wheel (Yaw)
D61030	Incremental angle of IRU (Roll)
D61040	Incremental angle of IRU (Pitch)
D61050	Incremental angle of IRU (Yaw)

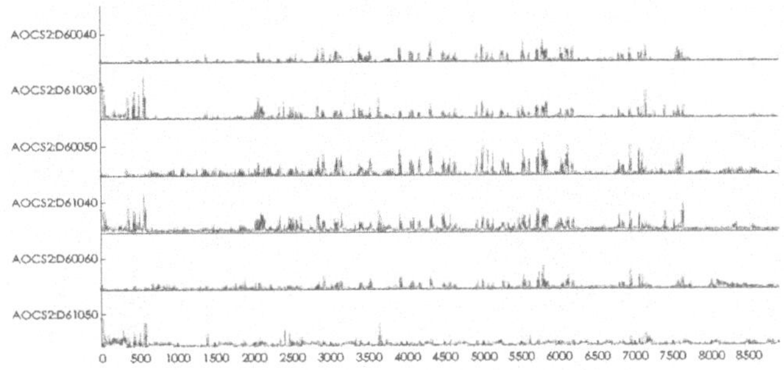

Fig. 4. (Example 3) Transition of relationship among event frequencies

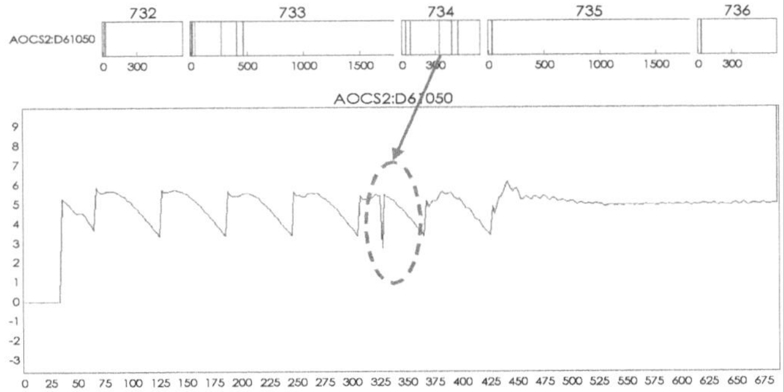

Fig. 5. (Example 4) Detection of an event or distinctive pattern

Fig. 2 shows an example of the occurrence pattern of the events and changes in the 6 time-series for one day. We can easily notice the associations among the series. Fig. 3 also gives a summary of events and changes for another day. In this figure, we can see a characteristic association pattern among the series. That is to say, D61030, D61040 and D61050 become suddenly active right after an event occurs in D60050 (access period 2588), and then become inactive again, responding to the simultaneous events in D60040, D60050 and D60060 (period 2593). Fig. 4 is a summary of the 4 years' data in a more abstract way. It shows the transition of frequencies of the events and changes in each series per day.

We can browse the global trend of the system activities and associations among the series. Fig. 5 shows an example of anomalous patterns detected in D61050 by the proposed method.

4 Conclusion

In this paper, we presented a method to summarize spacecraft telemetry data and to visualize most important information in it for monitoring the health status of the spacecraft systems. It focuses on two kinds of temporal patterns – "event" and "change" in the time-series, and extracts them by combining techniques of pattern clustering and change-point detection.

References

1. Faloutsos, C., Ranganathan, M., Manolopoulos, Y.: Fast subsequence matching in time-series databases. Proc. ACM SIGMOD (1994) 419–429
2. Keogh, E., Smyt,h P.: A probabilistic approach to fast pattern matching in time series databases. Proc. of KDD (1997) 24–30
3. Oates, T., Firoiu, L., Cohen, P.: Using dynamic time warping to bootstrap HMM-based clustering of time series. Sequence Learning (2001) 35–52
4. Hebrail G., Hungueney, B.: Symbolic representation of long time-series. Proc. Xth Int. Symp. Applied Stochastic Models and Data Analysis (2001) 537–542
5. Wijk, J., Selow, E.: Cluster and calendar based visualization of time series data. Proc. IEEE Symposium on Information Visualization (1999) 4–9
6. Keogh, E., Lonardi, S., Chiu, W.: Finding surprising patterns in a time series database in linear time and space. Proc. of KDD (2002) 550–556
7. Shahabi, C., Tian, X., Zhao, W.: Tsa-tree: A wavelet-based approach to improve the efficiency of multi-level surprise and trend queries on time-series data. Proc. of Int. Conf. on Scientific and Statistical Database Management (2000) 55–68
8. Eskin, E.: Anomaly detection over noisy data using learned probability distributions. Proc. 17th ICML (2000) 255–262
9. Guralnik, V., Srivastava, J.: Event detection from time series data. Proc. of KDD (1999) 33–42
10. Oliver, J., Baxter, R., Wallace, C.: Minimum message length segmentation. Proc. of PAKDD (1998) 222–233
11. Fitzgibbon, L., Dowe, D., Allison, L.: Change-point estimation using new minimum message length approximations. Proc. PRICAI (2002) 244–254
12. Das, G., Lin, K., Mannila, H., Renganathan, G., Smyth, P.: Rule discovery for time series. Proc. of KDD (1998) 16–22

An Extended Negative Selection Algorithm for Anomaly Detection

Xiaoshu Hang and Honghua Dai

School of Information Technology, Deakin University, Australia
{xhan,hdai}@deakin.edu.au

Abstract. This paper proposes an extended negative selection algorithm for anomaly detection. Unlike previously proposed negative selection algorithms which do not make use of non-self data, the extended negative selection algorithm first acquires prior knowledge about the characteristics of the problem space from the historical sample data by using machine learning techniques. Such data consists of both self data and non-self data. The acquired prior knowledge is represented in the form of production rules and thus viewed as common schemata which characterise the two subspaces: self-subspace and non-self-subspace, and provide important information to the generation of detection rules. One advantage of our approach is that it does not rely on the structured representation of the data and can be applied to general anomaly detection. To test the effectiveness, we test our approach through experiments with the public data set *iris* and KDD'99 published data set.

1 Introduction

The natural immune system has inspired scientists a great research interest because of its powerful information processing capability. It protects biologic bodies from disease-causing pathogens by pattern recognition, reinforcement learning and adaptive response. The idea of applying computational immunology to solving the problem of computer/network security derives from the inspiration that virus and network intrusion are analogous to pathogens to human bodies. Negative selection algorithm, proposed by Stephanie Forest and her research group in 1994, has been considered to be a highly feasible technique to anomaly detection and has been successfully applied to computer viruses/network intrusion detection[4], tool breakage detection[5], times-series anomaly detection[6], Web document classification[7], etc. The most striking features of the algorithm are that it does not require any prior knowledge of anomalies and can implement distributed anomaly detections in a network environment. For a given data set S to be protected, a set of detectors R is generated in a way that each detector d (a binary string) does not match any string s in S. The negative selection algorithms work in a straightforward way to generate the repertories R. It randomly generates a string d and matches d against each string in S. If d does not match any string in S then store d in R. This process is repeated until we have enough detectors in R to ensure the desired protection. This generate-and-test

H. Dai, R. Srikant, and C. Zhang (Eds.): PAKDD 2004, LNAI 3056, pp. 245–254, 2004.

method requires sampling a quite large number of candidate detectors and the computational complexity is exponential to the size of self data set *S*[8].

The originalnegative selection algorithm can be simply described as follows:

- Define self *S* as the data set a collection of strings of length *l* over a finite alphabet, a collection that needs to be protected ;
- Generate a set *R* of detectors, each of which fails to match any string in *S*;
- Monitor *S* for changes by continually matching the detectors in *R* against *S*. if any detector ever matches, then a change is known to have occurred, as the detector are designed to not match any of the original string in *S*.

Variations of negative selection algorithm have also been investigated mainly focusing on representation scheme, detector generation algorithm and matching rule. Representation scheme has been explored including hyper-rectangle-rule detectors[9], fuzzy-rule detectors[10], and hyper-sphere-rule detectors[11]. Their corresponding detector generation algorithms are negative selection with detection rules, negative selection algorithm with fuzzy rules and randomised real-valued negative selection. And genetic algorithm or other approaches are employed to generate detection rules. Matching rule in negative selection algorithm was analysed and compared in [14] which discussed r-contiguous matching, r-chunk matching, Hamming distance matching, and its variation Rogers and Tanimoto matching.

The drawbacks of previous work on negative selection algorithms can be grossly summarised as follows:

- Non-self data in historical data sample are completely ignored and thus make no contribution to the generation of detector set.
- The distribution feature of self data in the problem space is not investigated which lead to a large number of detectors to be produced in order to implement an effective protection.
- The distribution feature of randomly generated detectors in the non-self space is not analysed. Actually most of the detectors do not take part in any detecting task at all.

In this paper, we propose an extended negative selection algorithm which combines computational immunology and machine learning techniques. It first learns the prior knowledge about both self and non-self by applying machine learning approaches to historical sample data which contains partial non-self data. The prior knowledge is represented in the form of production rules and can be viewed as common schemata which characterise the two subspaces: self-subspace and non-self-subspace. These schemata provide much information for guiding the generation of detection rules which are used to monitor a system for anomaly detection.

The rest of the paper is organized as follows. Section 2 presents the problem description. Section 3 introduces the extended negative algorithm. Section 4 demonstrates some experimental results, and Section 5 comes the conclusions.

2 Problem Formulation

Anomaly detection, from the viewpoint of immunology, aims at identifying whether or not a new element in a problem space is non-self when self is known.

Suppose that a problem space $S=x_1\times x_2\times\cdots\times x_n$ is a n-dimensional vector space, where x_j is either a symbolic feature or a numeric feature. An element $e^* \in S$, where $* \in$ {self, non-self} is a feature vector. Those elements that belong to self form a subspace donated as *SS*, and the non-self subspace *NS* is defined as the complementaryspac e of *SS*. The two subspaces are described as follows:

Self subspace: $SS=\{e_i^* \mid *= self,\ i=1,2,\dots,n\}$
Non-self subspace: $NS=\{ e_j^* \mid *= non\text{-}self,\ j=1,2,\dots,m,\ m< n\}$

where $SS \cup NS=S$ and $SS \cap NS=\varnothing$.

Given a data set *S* for detecting anomaly, the characteristic function χ_{self} for differentiating self and non-self is defined as follows:

$$\chi_{self}(e_j) = \begin{cases} 1, & if\ e_j \in SS \\ 0, & if\ e_j \in NS \end{cases}$$

A detector is usually represented as a detection rule, the structure of which is described as:

$$x_1 \in [val_1^1, val_2^1] \wedge \dots \wedge x_k = d_k \wedge x_m \in [val_1^m, val_2^m] \rightarrow abnormal$$

where $x_i \in [val_1^i, val_2^i]$ means that x_i is a real-value feature and $x_j = d_j$ indicates that feature x_j is symbolic. A detection rule defines a hypercube in the complementary space of self space.

3 The Proposed Approach

3.1 Learning Schemata from Historical Data

Finding common schemata is also inspired from the natural immune system [19]. Bacteria are inherently different from human cells, and many bacteria have cell walls made from polymers that do not occur in humans. The immune system recognizes bacteria partially on the basis of the existence of these unusual molecules. Common schemata represent generic properties of the antigen population and can be obtained by some intelligent computational methods. In GA, a schema is viewed as a template specifying groups of chromosomes with a common characteristic, or a description of hyperplanes through genome space and, generally represented as a string of symbols such as “ ##1101## ”. The character # is explained as “don’t care”. In this paper, we use the well-known classification algorithm C4.5 to learn common schemata which are represented as the conjunctions of attribute-value pairs as follows:

$$x_1 \in [val_1^1, val_2^1] \wedge \ldots \wedge x_k = d_k \wedge x_k \in [val_3^k, val_4^k]$$

Essentially a schema is also a hypercube in the state space(see Figure 1).

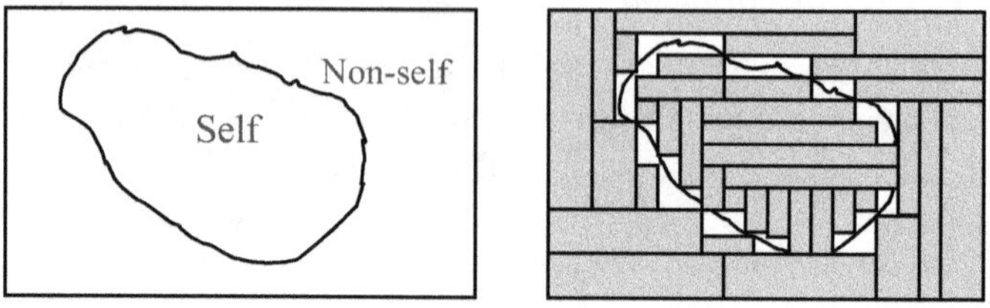

Fig. 1. Self and non-self space characterised by schemata

In this paper, we exploit classification algorithm C4.5 to learn the schemata from historical data. Once a decision tree is built, a path from the root to any leaf forms a production rule, and the premise of which can be viewed as a schema. Generally, a shorter schema coves more examples than a longer schema. Those schemata which identify self data cover the whole self space, whereas the schemata characterising non-self data cover only part of non-self space because the non-self data is not complete. Both self schemata and non-self schemata provide much information for the construction of detection rules.

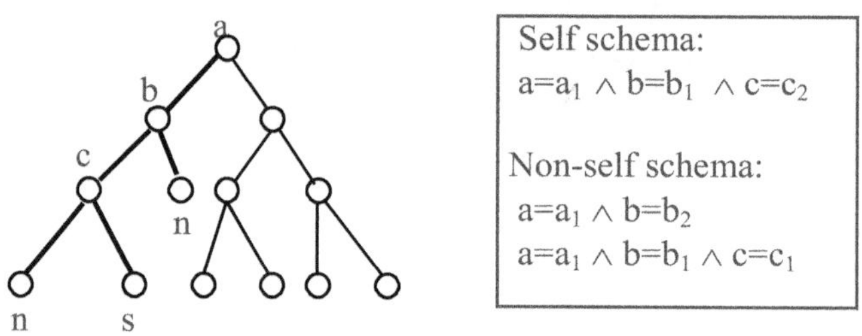

Fig. 2. Some schemata in a decision tree

3.2 The Extended Negative Selection Algorithm

In our approach, detection rules are generated in two ways. Some detection rules are obtained by learning from examples and the others are semi-randomly generated. Thus we have:

Detection rules = learnt from historical sample + semi-randomly generated

The algorithm diagram for producing detector set is described in Figure 3. The detection rules that are semi-randomly generated which will be explained further, match against the common schemata learnt from historical sample. If a detection rule does not matches any common schema, it is considered as a detection rule and stored in the detection rule set, otherwise it is rejected. This process is repeated until an appropriate number of detection rules are obtained.

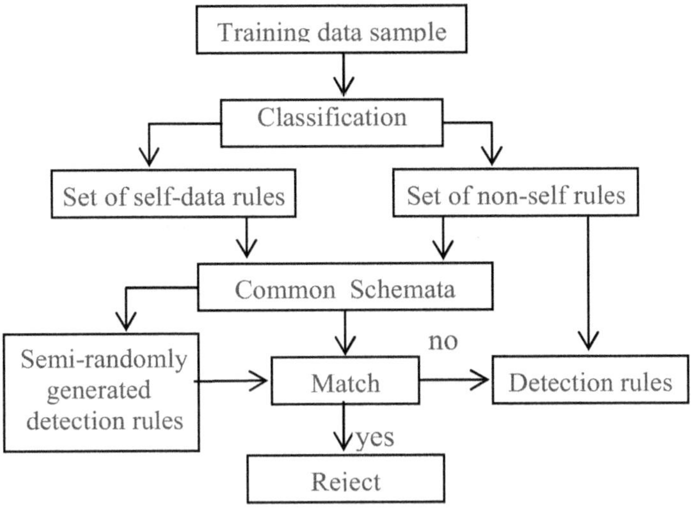

Fig. 3. Diagram of generating detection rules

The two kinds of detection rules perform two different detection tasks respectively:

- The detection rules that are directly obtained from the set of non-self rules detect the non-self data that the same schemata have happened in history.
- The detection rules that are semi-randomly generated and do not match any common schema detect those non-self data that their schemata have never met.

In the monitoring phase, any new data matches against each detection rule in the detector set, an anomaly is detected if a match occurs (see Figure 4).

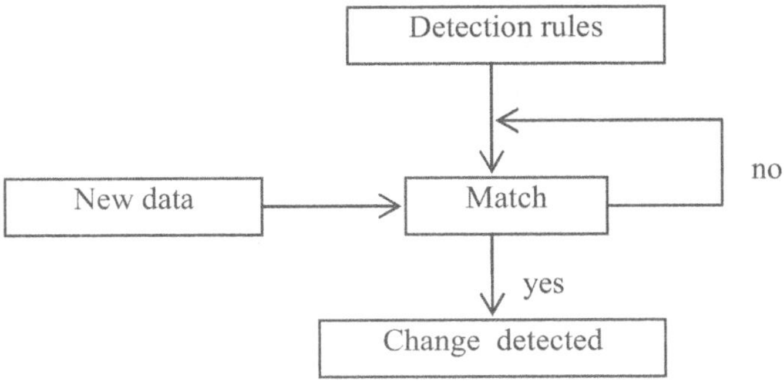

Fig. 4. Monitoring phase

Now we discuss the strategy of semi-randomly generating new detection rules. In the original negative selection algorithm, a detection rule is randomly generated and then match against self-data to testify its validity. The drawback of this method is that it does not use any information that the self-data and non-self data provide and thus creates an exponential computational complexity. We call our method an semi-random generation of detection rules which means it makes use of the information that the common schemata provide, based on that a new detection rule is randomly generated.

After a decision tree is built, the features near the root of the decision tree are more frequently included in common schemata than those features near leaves, which implies that these root-nearby features are more important and thus have more opportunity to be selected into the detection rules. This inspiration is very valuable, especially when the number of features is large. In a huge problem space, it is quite difficult to generate a detection rule locating in the right place. As we know, C4.5 uses info-gain as criterion when selects a feature to classify a data set into subgroups. The feature selected as the root appears in each common schemata and thus has 100% opportunity to be selected into a new detection rule, other attributes has a lower opportunity to be selected.

The possibility for a feature x_i to be selected into a new detection rule is calculated as follows:

$$poss(x_i) = \frac{\text{the number of feature } x_i \text{ appears in schemata}}{\text{the number of the feature at root appears in schemata}}$$

Once a feature is selected into a new detection rule, the next needs be considerd is to choose its one appropriate interval. As shown in figure 5, the light-gray boxes represent that these intervals have occurred in the schemata learnt from historical data set, the white ones denote those idle intervals, and dark-gray boxes refer to the intervals that are selected into detection rules. Our strategy is that a feature x_i with a high $poss(x_i)$ is more likely to be selected with its previously occurred intervals and a feature with a low $poss(xi)$ is more likely to be selected with its idle intervals.

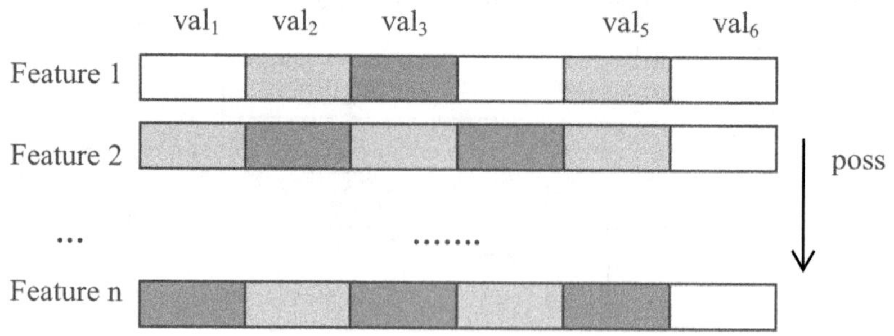

Fig. 5. Semi-randomly generation of detection rules

4 Experiment and Result Analysis

Experiment 1. This experiment is done with the published data set *iris* which has 4 numeric features and 150 examples of three classes. The 100 examples of class 1 and class 3 are used as self data, and the 50 examples of class 2 are viewed as non-self data. 30 examples of class 2, 10 examples of class 1 and 10 examples of class 3 are taken away from the sample data for detection test, thereby the training data set consists of 100 examples for the classification algorithm. Each numeric feature is discretized into 5 equal-length intervals. The schemata and detection rules are learnt from the examples shown in Table 1.

Table1. Schemata and detection rules learnt from training examples

No	Schemata covering self space
1	PL∈[5.72, 6.90]
2	PL∈[1.10, 2.26]
3	PL∈[4.58, 5.72] and PW∈[1.54, 2.02]
4	PL∈[4.58, 5.72] and PW∈[2.02, 2.50]
5	PL∈[3.42, 4.58] and SL∈[4.30, 5.02]
6	PL∈[4.58, 5.72]and PW∈[1.06, 1.54]and SW∈[2.00, 2.48]
7	PL∈[4.58,5.72] and PW∈[1.06, 1.54]and SW∈[2.48, 2.96]and SL∈[5.74,6.46]
	Detection rules
1	If PL∈[2.26, 3.42] then non-self
2	If PL∈[3.42, 4.58] and SL∈([5.02,5.74]or SL∈[5.74, 6.46]or SL∈[6.46, 7.18]) then non-self
3	If PL∈[4.58, 5.72]and PW∈[1.06, 1.54]and SW∈[2.00, 2.48] then non-self
4	If PL∈[4.58,5.72]and PW∈[1.06,1.54]and SW∈[2.48, 2.96]and SL∈[6.46,7.18] then non-self

Four detection rules are produced by the classifier and marked by light-gray squares in Figure3. Another 5 detection rules (dark-gray squares) are semi-randomly generated. Common schemata covering self-data are marked by transparent squares. All the detection rules are produced around the self data and they separate the data of class 1 from the data of class 3. In the testing phase, all the 30 examples of class 2 are detected by the nine detectors, and one example of class 3 is mis-recognized as non-self data.

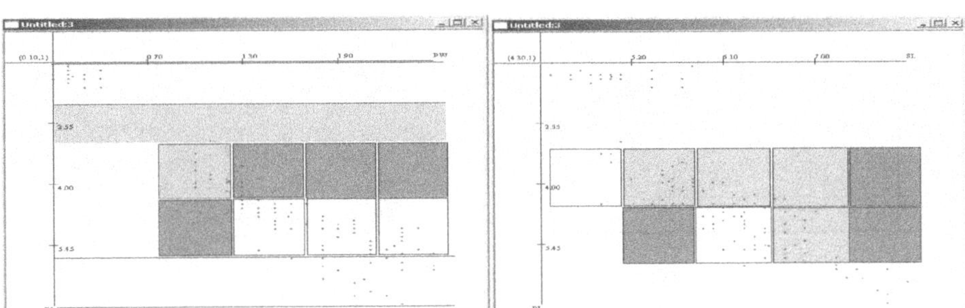

Fig. 6. The distribution of detection rules in two dimensional space

Experiment 2. To compare with other methods, we carry out this second experiment with the network intrusion data set published in KDD Cup 1999. It contains a wide variety of intrusions simulated in a military network environment. The data represent both normal and abnormal information with 42 attributes and approximately 4,900,000 instances, 10% of which are for training and contains only 10 attack types. Other 14 attack types are contained in the test data. This makes the detection task more realistic. The 10% data set is labelled and thus suitable for classification learningIn our experiments, the 10% data set is still large and is split into two parts for training and testing, respectively. The first part contains 60% normal records and the records of 7 attack types, while the second part for testing contains 40% normal data and the records of 10 attack types.

97 classification rules are learnt from the examples, of which 36 rules identify the self-data, and the rest 61 rules characterise the non-self space. The shortest rule consists of 6 attribute-value pairs and the longest rule contains 11 attribute-value pairs. The rest 80 detection rules are semi-randomly generated. The main experimental results are shown in Table2.

Table 2. Experimental results

Self common schema	36
Directors from non-self rules	61
Semi-randomly generated	80
Total detection rules	141
Detection rate	98.67%
False alarm rate	3.52%

In the testing phase, a record is considered to be an anomaly if it is an attack of any type. The detection rate is the ratio of the number of abnormal records that are correctly recognised to the total number of abnormal records and, the false alarm rate is the ratio of the number of normal records that are incorrectly recognised as abnormal to the total number of normal records.

To illustrate the effect of the number of detection rules on the detection rate and the false alarm rate, we set up different total number of detection rules at each run. The best number of detection rules is thus determined as 141(see Figure 7).

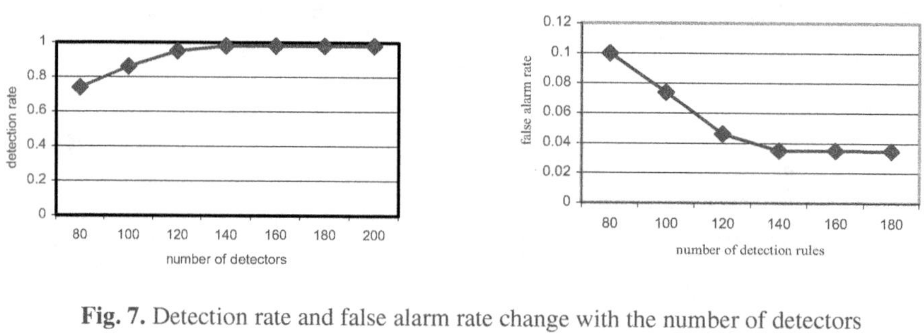

Fig. 7. Detection rate and false alarm rate change with the number of detectors

As we mentioned above, the KDD Cup 1999 Data set has been used in experiments for anomaly detection by other researchers. A brief comparison has been given in [14], and is also copied to table 3, from which we can see that algorithm EFR (Evolving Fuzzy Rules) is the best approach, especially the number of detector rules in EFR is the least. We call our approach CSDR, standing for Common Schema-based Detector Rules. It also exhibits good performance with a high detection rate(DR), a low false alarm rate(FA) and a medium-sized set of detector rules. And the most striking characteristic that our approach possesses is that the detector rules are much shorter than those in other approaches. The right reason for this is that the common schemata in our approach provide much information for the construction of detection rules. We also would like to draw readers' attention to the fact that other approaches which employ genetic algorithms or evolutionary algorithms to generate the detection rules do not make use of any prior knowledge.

Table 3. Comparison with other approaches

Algorithm	DR%	FA%	# Detectors
EFR	98.30	2.0	15
PHC	93.09	2.0	47.4
ERD	60.90	2.0	331.35
EFRID	98.95	7.0	–
RIPPER-AA	94.26	2.02	–
CSDR	**98.67**	**3.52**	**141**

5 Conclusions

In this paper, we propose an extended negative selection algorithm for anomaly detection. It learns prior knowledge about both the normal and abnormal features from historic sample data. The prior knowledge is viewed as common schemata characterising the two subspaces: self and non-self, and used to guide the generation of detection rules. We use two published data sets in our experiments to test the effectiveness of our approach and compare our approaches with other approaches. We conclude that:

- The prior knowledge learnt from examples describes what kinds of schemata existing in the two subspaces and thus provides valuable information for the construction of detection rules that are semi-randomly generated
- The proposed approach is effective and efficient. Among all the 6 different approaches we compared, our extended negative approach achieved the second highest detection rate. The false alarm rate is very competitive.
- The approach does not rely on the structured representation of the data and is applied to the problem of general anomaly detection.

References

[1] Steven A. Hofmeyr, and S. Forrest, "Architecture for an artificial immune system", Evolutionary Computation, 8(4) (2000) 443-473.

[2] D. Dasgupta and S. Forrest, "Novelty detection in time series data using ideas from immunology," in *Proceedings of the International Conference on Intelligent Systems*, pp. 82–87, June 1996.

[3] Dipankar Dasgupta and Fabio Gonzalez "An immunity-based Technique to Characterize Intrusions in Computer Networks", IEEE transaction on evolutionary computation 6(3),pages 1081-1088 June 2002.

[4] Paul K. harmer, Paul D. Williams, Gregg H.Gunch and Gary B.Lamont, "An Artificial Immune System Architure for Computer Security Application" IEEE transaction on evolutionary computer, vol.6. No.3 June 2002.

[5] Dipankar Dasgupta and Stephanie Forrest, "Artificial immune system in industrial application", In the proceeding of *International conference on Intelligent Processing and Manufacturing Material (IPMM). Honolulu, HI (July 10-14, 1999).*

[6] Dipankar Dasgupta and Stephanie Forrest, "Novelty Detection in Time Series data using ideas from Immunology", *In the proceedings of the 5th International Conference on Intelligent Systems, Reno, June 19-21, 1996.*

[7] Jamie Twycross and Steve Cayzer, "An Immune-based approach to document classification", http://citeseer.nj.nec.com/558965.html.

[8] S.Forrest, A.Oerelson, L.Allen, and R.cherukuri. "Slef-nonself discrimination in a computer", *In the proceedings of IEEE symposium on research in security and privacy, 1994.*

[9] Fabio A.Gonzalez and Dipankar Dasgupta, "An Immunogenetic Technique to detect animalies in network traffic", *In the proceeding of GECCO 2002: 1081-1088.*

[10] Jonatan Gomez, Fabio Gonzalez and Dipankar Dasgupta, "An Immune-Fuzzy Approach to Anomaly detection", *In Proceedings of The IEEE International Conference on Fuzzy Systems, St. Louis, MO, May 2003.*

[11] Fabio Gonzalez, Dipankar Dasgupta and Luis Fernando Nino, "A Randomized Real-Value Negative Selection Algorithm", ICARIS-2003.

[12] M. Ayara, J. Timmis, R. de Lemos, L. deCastro and R. Duncan, "Negative Selection: How to Generate Detectors", 1st ICARIS, 2002.

[13] D. Dasgupta, Z.Ji and F.Gonzalez, "Artificial Immune System Research in the last five years". *In the proceedings of the international conference on Evolutionary Computation Conference (CEC), Canbara, Australia, December 8-12, 2003.*]

[14] De Castro, L. N. & Von Zuben, F. J., "Learning and Optimization Using the Clonal Selection Principle", IEEE Transactions on Evolutionary Computation, Special Issue on Artificial Immune Systems, **6**(3), pp. 239-251, 2002.

[15] Jungwon Kim and Peter Bentley, "Negative selection and Niching by an artificial immune system for network intrusion detection", *In the proceeding of Genetic and Evolutionary Computation Conference (GECCO '99), Orlando, Florida, July 13-17.*

Adaptive Clustering for Network Intrusion Detection

Joshua Oldmeadow[1], Siddarth Ravinutala[1], and Christopher Leckie[2]

ARC Special Research Centre for Ultra-Broadband Information Networks
[1] Department of Electrical and Electronic Engineering
[2] Department of Computer Science and Software Engineering
The University of Melbourne
Parkville, Victoria 3010 Australia
`http://www.cs.mu.oz.au/~caleckie`

Abstract. A major challenge in network intrusion detection is how to perform anomaly detection. In practice, the characteristics of network traffic are typically non-stationary, and can vary over time. In this paper, we present a solution to this problem by developing a time-varying modification of a standard clustering technique, which means we can automatically accommodate non-stationary traffic distributions. In addition, we demonstrate how feature weighting can improve the classification accuracy of our anomaly detection system for certain types of attacks.

1 Introduction

Network intrusion detection is the problem of identifying misuse of computer networks [1]. A major challenge for any Intrusion Detection System (IDS) is how to identify new attacks when the normal background traffic is itself changing with time. In this paper, we present a novel approach to detecting network intrusions, based on an adaptive clustering technique. In comparison to earlier approaches that assume a static model of network traffic (e.g. [2, 3, 4]), we demonstrate that this approach has significant advantages in terms of accuracy under changing traffic conditions.

An important approach for intrusion detection research is anomaly detection using unsupervised learning techniques, which do not require labeled data and are able to detect previously "unseen" attacks. A critical issue for the success of unsupervised anomaly detection is how to cope with changing traffic conditions [5]. Our main contribution to this problem has been to develop and evaluate a time varying approach to a fixed-width clustering algorithm. A secondary contribution of this paper is the investigation of the impact of feature-weighting on the performance of an IDS. The performance of these two enhancements on a standard dataset is then compared to three other benchmark algorithms presented in [2].

H. Dai, R. Srikant, and C. Zhang (Eds.): PAKDD 2004, LNAI 3056, pp. 255–259, 2004.

2 Anomaly Detection for Network Security

The process of anomaly detection comprises two phases: training and testing. The *training phase* involves characterizing a set of given network connections. Each connection c is represented by a set of d features, i.e., each connection c is represented as a point in the feature space $\Re^d$. Note that this training data may contain both normal and attack data. In order to build a model that discriminates between normal and attack data points, we need to assume that attack data occurs far less frequently than normal traffic. As a guide we need to assume that less than x % of the data consists of attack connections. The second phase of anomaly detection, the *test phase*, analyses new network connections based upon the information gathered in the training phase. New connections are labeled as anomalous or normal based upon the model developed in the training phase.

The problem of anomaly detection for network intrusion has been an active area of research. Our approach is based on the model of *fixed-width clustering* in [2], which builds a set of clusters that each has a fixed radius in the feature space. Each cluster is represented by a *centroid*, which is the mean of all the points in the cluster. In the training phase of the fixed-width clustering technique, a threshold w is chosen as the maximum radius of a cluster. At the end of training, the clusters that contain less than a threshold τ% of the total set of points are labeled as anomalous. All other clusters are labeled as normal. The test phase operates by calculating the distance between a new point c and each cluster centroid. If the distance from c to the nearest cluster is greater than w, then c lies in a sparse region, and is labeled as anomalous.

The data set used to test these algorithms was from the 1999 KDD Cup Data Mining competition [7]. The 1999 KDD Cup was a supervised learning competition, and contained labeled training and test datasets. The training data contains 24 attack types. The test data contains 14 additional attacks. However, in order to use this dataset for unsupervised learning, we needed to remove two easily detected DOS bandwidth attacks – *smurf* and *neptune*. These two attacks constitute just over 79% of the training data set, and are very easy to detect by other means. We also removed the *snmpgetattack*, which is almost identical to a normal connection. The new attacks in the test set are used to investigate how the system performs when attacks are included that the IDS has not previously seen.

3 Extending Fixed-Width Clustering

Our anomaly detection technique is based on the method of fixed-width clustering. In this section, we first describe the implementation details of how we used the fixed-width clustering algorithm described in [3]. We then describe how we adapted this algorithm for time-varying clustering, in order to improve its detection accuracy on the KDD Cup dataset.

3.1 Implementation of Fixed-Width Clustering

The algorithm for fixed-width clustering is based on the outline in [3]. We are given a set of network connections C_{Tr} for training, where each connection c_i in this set is represented by a d-dimensional vector of features. We then proceed through the following stages of (1) normalization, (2) cluster formation, and (3) cluster labeling.

Normalization – To ensure that all features have the same influence when calculating distance between connections, we normalised each continuous feature x_j in terms of the number of standard deviations from the mean of the feature.

Cluster Formation – After normalization, we measure the distance of each connection c_i in the training set C_{Tr} to the centroid of each cluster that has been generated so far in the cluster set. If the distance to the closest cluster is less than the threshold w, then the centroid of the closest cluster is updated, and the total number of points in the cluster is incremented. Otherwise, a new cluster is formed.

Cluster Labelling – The assumption that the ratio of attack to normal traffic is extremely small is used as a classification criterion in this algorithm. If a cluster contains more than the classification threshold fraction τ of the total points in the data set, it is labeled as a normal cluster, else it is labeled as anomalous. We found the value $\tau = 0.02$ was most effective in our evaluation.

Test Phase – In the testing or real-time phase, each new connection is compared to the each cluster to determine whether it is anomalous. The distance from the connection to each of the clusters is calculated. If the distance to the nearest cluster is less than the cluster width parameter w, then the connection shares the label (normal or anomalous) of its nearest cluster. Otherwise, the connection is labeled as anomalous.

3.2 Our Contributions

We have investigated two techniques for improving the fixed-width clustering for network intrusion detection: feature weighting, and time-varying clusters.

Feature Weighting – We found that some key features have more of an impact in differentiating certain types of attacks from normal traffic. Consequently, we wanted to investigate whether feature weighting could improve the performance of fixed-width clustering on the test data. Based on a manual inspection of the classification accuracy on the training set, we found that the accuracy could be improved by giving greater weight to the feature that encodes the average packet size of a connection.

Time-varying Clusters – Traffic in a network is never stagnant. For example, new services can be activated, work patterns can change as projects start or finish, and so on. Consequently, an IDS needs to be able to adapt to these changing traffic patterns while still maintaining a high level of detection accuracy.

Since the test-data is of a different distribution to the training dataset, we expect a significant improvement in performance when the clusters are allowed to move during real-time operation. In our modified algorithm for time-varying clusters, we allow the clusters to be updated during the on-line testing phase. However, if the formula used in the training phase were to be used in the testing phase, connections added to large clusters would have a lesser influence on the cluster when compared with connections added to smaller clusters. Hence an *influence factor* γ is needed to regulate the effect of new points on a cluster. When assigning a new connection c_i to a cluster ϕ_n during the on-line test phase, the mean of feature x_j of the cluster is updated as follows:

$$\text{mean}_j(\phi_n) = \frac{\text{mean}(\phi_n)\times\gamma + \text{feature}_j(c_i)}{\gamma+1}$$

Table 1 shows how the modified algorithm, with time-varying clusters, performs against the four types of attacks. There is a significant improvement in comparison to the results obtained when only the feature-weighting enhancement implemented. The table shows the percentage improvement resulting from using time-varying clusters by comparing the area under the ROC graphs.

Table 1. Performance of modified fixed-width clustering

Attack Types	Area (original fixed-width clustering)	Area (weighting)	Area (weighting and time-varying)	Improvement over weighted (%)
DOS	0.546	0.982	0.979	-0.31%
Probe	0.962	0.895	0.952	5.98%
U2R	0.961	0.897	0.967	7.24%
R2L	0.979	0.940	0.960	2.08%

As expected, the performance of our algorithm improved against almost all attacks with the implementation of time-varying clusters and feature weighting. In order to evaluate our approach, we have compared our results to those quoted in [2] for k-nearest neighbour, fixed-width clustering, and support vector machines [6]. They evaluated all these techniques on the 1999 KDD Cup dataset [7].

Allowing the clusters to move in real-time and implementing feature weighting allowed us to outperform all the algorithms presented in [2]. Using the area under the ROC graph as a measure, Table 2 lists the percentage improvement in performance our modified clustering algorithm achieved over the results reported in [2].

Table 2. Relative comparison of our time-varying clustering with results from [2]

Algorithm	% Improvement using Modified Time-Varying Clustering
K-NN	8.0%
Fixed-width Clustering	3.4%
SVM	2.5%
Our modified TV-Clustering	-

As mentioned previously, not only did our test data set contain all the attacks present in the training dataset, but it also included attacks which were not present in the training set. This demonstrates the advantages of our modified time-varying clustering algorithm in terms of its ability to detect both new and known attacks.

4 Conclusion

In this paper, we have presented two enhancements to using clustering for anomaly detection. First, we have demonstrated the benefits of feature weighting to improve detection accuracy. Although our feature weighting was based on a manual analysis of the detection accuracy on the training data, our results provide a clear motivation for further research into automated approaches for feature weighting in this context. Second, we developed a time-varying clustering algorithm, which can adapt to changes in normal traffic conditions. We demonstrated that our time-varying approach was able to achieve significant improvement in detection and false positive rates compared to earlier approaches.

Acknowledgements. We thank the organizers of the 1999 KDD Cup for the use of their datasets. This work was supported by the Australian Research Council.

References

1. S. R. Snapp, J. Brentano, G. V. Dias, T. L. Goan, L. T. Heberlein, C. Ho, K. N. Levitt, B. Mukherjee, S. E. Smaha, T. Grance, D. M. Teal, and D. Mansur, "DIDS (Distributed Intrusion Detection System) Motivation, Architecture, and An Early Prototype." Proc. 14th National Computer Security Conference, 1991.
2. E. Eskin, A. Arnold, M. Prerau, L. Portnoy, S. Stolfo, "A Geometric Framework for Unsupervised Anomaly Detection: Detecting Intrusions in Unlabeled Data." Applications of Data Mining in Computer Security, Kluwer, 2002.
3. L. Portnoy, E. Eskin, S. Stolfo, "Intrusion Detection with Unlabeled Data using Clustering," Proc. Workshop on Data Mining for Security Applications, 2001.
4. S. Hofmeyr and S. Forrest, "Architecture for an Artificial Immune System." In Evolutionary Computation 7(1), 1999, pp. 1289-1296.
5. M. Mahoney and P. Chan, "Learning Nonstationary Models of Normal Network Traffic for Detecting Novel Attacks." Proc. 8th ACM KDD, 2002.
6. B. Schölkopf, J. C. Platt, J. Shawe-Taylor, A. J. Smola, R. C. Williamson, "Estimating the Support of a High-Dimensional Distribution." Neural Computation, 13(7), 2001, pp. 1443-1472.
7. S. J. Stolfo, W. Fan, W. Lee, A. Prodromidis P. K. Chan, "Cost-based Modeling and Evaluation for Data Mining With Application to Fraud and Intrusion Detection: Results from the JAM Project." Proc. DARPA Information Survivability Conf, 2000.

Ensembling MML Causal Discovery

Honghua Dai[1], Gang Li[1], and Zhi-Hua Zhou[2]

[1] School of Information Technology, Deakin University, Australia
[2] National Laboratory for Novel Software Technology, Nanjing University, China
{hdai,gangli}@deakin.edu.au zhouzh@nju.edu.cn

Abstract. This paper presents an ensemble MML approach for the discovery of causal models. The component learners are formed based on the MML causal induction methods. Six different ensemble causal induction algorithms are proposed. Our experiential results reveal that (1) the ensemble MML causal induction approach has achieved an improved result compared with any single learner in terms of learning accuracy and correctness; (2) Among all the ensemble causal induction algorithms examined, the weighted voting without seeding algorithm outperforms all the rest; (3) It seems that the ensembled CI algorithms could alleviate the local minimum problem. The only drawback of this method is that the time complexity is increased by δ times, where δ is the ensemble size.

1 Introduction

Discovering causal relationships from a given data with m instances and n variables is one of the major challenges in the areas of machine learning and data mining. In the last ten years , several projects have been done [1,2,3,4] using different techniques. These techniques can be classified into three major categories: one featured in learning Bayesian networks containing compact representations for the conditional probability distributions (CPDs) stored at each node. Such as the work done by Heckerman[5]. The second category is featured with Constraint based methods. The representative of this category is the TETRAD systems done by Peter Spirtes, Clark Glymour and Richard Scheines of the CMU group from the Department of Philosophy, Carnegie Mellon University [6]. The third category is featured with applying informatic (or asymptotically Bayesian) scoring functions such as MML/MDL measures to evaluate the goodness-of-fit of the model discovered to the given data. The representative of this category was originally done in Monash University headed by Chris Wallace[2,3], and then moved to Deakin University directed by Honghua Dai[1,7,4].

All of these three categories of work done so far cannot overcome the local minimum problem. Due to this, they are usually limited to find the best of a local minimum. To further enhance the accuracy and to try to overcome the local minimum problem in discovering causal models, this paper investigates an ensembling approach that aims to achieve a better result in causal induction.

H. Dai, R. Srikant, and C. Zhang (Eds.): PAKDD 2004, LNAI 3056, pp. 260–271, 2004.

2 Causal Discovery

Our proposed method integrates the informatic-MML causal induction method developed so far and the ensemble learning approaches.

A *causal model*, normally in a form of a directed acyclic graph (DAG), is a representation of known or inferred causal relations among a set of variables. Different flavors of causal network can be defined, depending on the nature of the variables concerned (e.g., real or discrete), the form assumed for the effect of one variable on another (e.g., linear or non-linear) and the model assumed for the variation of variable values which is not accounted for by the variable's imputed causes. Here we are concerned with simple linear effects among real variables, and we assume Gaussian unexplained variation. Specifically, if we have a set of T real variables, and N independent instances of data, i.e., if we have the data $\{x_{n,t} : n = 1 \dots N, t = 1 \dots T\}$, a causal model of the data specifies, for each variable v_t, a possibly-empty "parent set" of variables $\{v_{p[t,m]} : m = 1, \dots, d_t\}$ where d_t is the number of "parents" of v_t, and a probabilistic relation among data values

$$x_{n,t} = \sum_m (a_{t,m} * x_{n,p[t,m]}) + r_{n,t} \tag{1}$$

where the coefficients $\{a_{t,m} : m = 1 \dots d_t\}$ give the linear effects of the parents on v_t, and the residual $r_{n,t}$ is assumed to be a random variate from the Gaussian density $N(0, \sigma_t^2)$. Note that we have assumed for simplicity that the mean of each variable is zero.

The model is restricted to ones in which the inter-variable relations are open to a physically-tenable causal interpretation. Thus, if we define the "ancestors" of a variable as its parents, its parents' parents, and so on, we require that no variable has itself as an ancestor. With this restriction, the topology of the causal models among the variables may be represented by a directed acyclic graph (DAG) in which each node or vertex represents a variable, and there is a directed arc from v_i to v_j iff v_i is a parent of v_j.

Our task is: for a given sample data with m instances and T variables, to induce the graph structured knowledge with the highest posterior probability given such a data set.

This discovery involves two major processes, encoding and searching as described in [2,7]: The encoding of a causal model includes describing the causal model and describing the data as in [2].

$$L = L_{Model} + L_{(Data|Model)} \tag{2}$$

The search strategy used in both [2] and [7] is the greedy search method.

3 Ensemble Learning

Ensemble learning is a learning paradigm where multiple component learners are trained for the same task, and the outcomes of these component learners are combined for dealing with future instances.

Since an ensemble is often more accurate than its component learners[8,9,10], such a paradigm has become a hot topic of supervised learning and has already been successfully applied to optical character recognition[11,12], face recognition[13,14], scientific image analysis[15,16], medical diagnosis [17,18], seismic signals classification[19], etc. Building a good ensemble involves increasing both the accuracy of and the diversity among the component learners[20], which is not an easy task. Nevertheless, many practical ensemble methods have been developed.

In general, an ensemble is built in two steps, that is, obtaining multiple component learners and then combining what they learnt. As for obtaining component learners, the most prevailing methods are Bagging[8] and Boosting[21]. Bagging is introduced by Breiman[8] based on bootstrap sampling[22]. It generates several training sets from the original training set and then trains a component learner from each of those training sets. Boosting is proposed by Schapire[21] and improved by Freund et al [23,24]. It generates a series of component learners where the training sets of the successors are determined by the performance of the predecessors in the way that training instances that are wrongly predicted by the predecessors will play more important roles in the training of the successors. There are also many other methods for obtaining the component learners. Some examples are as follows. Breiman's Arc-x4 generates component learners by considering the performance of all the previous learners. Bauer and Kohavi's Wagging[25] use Poisson distribution to sample training sets for component learners from the original training set. Webb's MultiBoost[26] embeds Boosting into Wagging to create component learners. Zhou et al.'s GASEN[18] employs genetic algorithm to select component learners from a set of trained learners. As for combining the predictions of component learners, the methods used for classification and regression are quite different because those two kinds of tasks have distinct characteristics, i.e. the outputs of classification are class labels while those of regression are numeric values. At present, the most prevailing methods for classification tasks are plurality voting or majority voting[27], and those for regression tasks are simple averaging[28] or weighted averaging [29]. There are also many other methods, such as employing learning systems to combine component predictions[10], exploiting principal component regression to determine constrained weights for combining component predictions[30], using dynamic weights determined by the confidence of the component learners to combine their predictions[31], etc. Up to the knowledge of the authors, the research of ensemble learning are mainly focused on supervised learning, yet no work has addressed the issue of building ensemble of causal inducers although this may not only generate strong causal inducers but also extend the usability of ensemble learning methods.

4 Ensemble MML Causal Induction

The overall structure of our ensemble MML causal induction process is shown in Figure 1. For a given original data set, we generate S data sets with the same sample size as the original data set. For each generated data set D_i, the component learner induces a causal model M_i. Then these S models are ensembled with some ensemble learning strategy to come up with a final causal model.

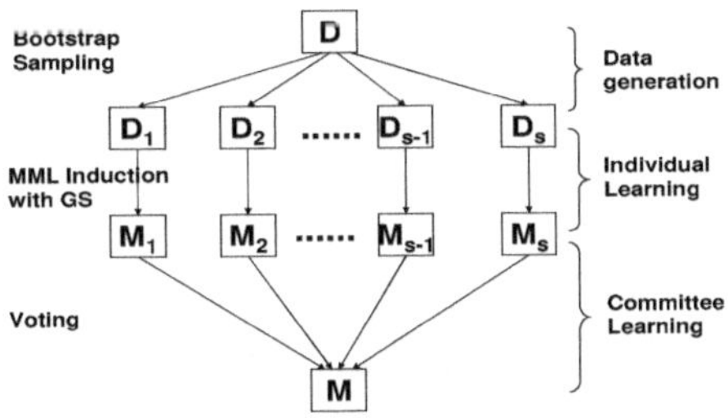

Fig. 1. Processing diagram of Bagging

4.1 The Building Block

The component learners are the building block used in an ensembling learning method. In this research, we use improved MML causal induction method proposed in [7] as the component learners.

4.2 Ensemble Algorithms

In total, we proposed six ensemble causal induction algorithms. The differences among them lie on two choices:

Seeding Graph. The component discovery algorithms in BSV and BWV start with null graph, while the component discovery algorithms in BMSV and BMWV start with seeding graph generated by a Markov Chain Monte Carlo algorithm.

Voting Strategy. Simple voting is used in both BSV and BMSV, where every component has the same voting weight. Howver, in BWV and BMWV, weighted voting is used, where the less the message length, the higher the voting weight.

BSV Causal Inducer. BSV stands for Bagging with simple voting which is a variant of Bagging[8]. It employs bootstrap sampling[22] to generate several training sets from the original data, and induces a linear causal model from each of the generated data sets by the MML causal Inducer[7] which uses the improved encoding scheme and the Greedy search method. The final structure is obtained by majority voting from the individual model structures learnt by individual learners. The path coefficients of the final causal model is obtained using MML based optimization approach. In this method, no seed model is provided.

Algorithm 1 BSV-CI Algorithm

Input: dataset D, inducer I, ensemble size S
Output: ensemble M.
Seed: Empty.

```
for s = 1 to S do
    D_s = bootstrap(D)
    M_s = I(D_s)
end for
M.stru = majorityvoting(M_s.stru)
M.param = MML_Optim(D, M.stru)
```

Algorithm 2 BWV-CI Algorithm

Input: dataset D, inducer I, ensemble size S
Output: ensemble M.
Seed: Empty.

```
for s = 1 to S do
    D_s = bootstrap(D)
    M_s = I(D_s)
    w_s = MML(M_s, D_s)
end for
M.stru = weightedvoting(M_s.stru)
M.param = MML_Optim(D, M.stru, M_s.param)
```

The BSV-CI algorithm is shown in Algorithm 1, where T bootstrap samples $D_1, D_2, \ldots, D_S$ are generated from the original data set D and a component causal model M_s is obtained from D_s, an ensembled causal model M is built from $M_1, M_2, \ldots, M_S$. The structure of M is determined through majority voting, where a connection exists if and only if such a connection existing in majority component causal models. The path coefficients of M are determined by MML based optimization approach[2].

BWV Causal Inducer. BWV stands for Bagging with weighted voting as shown in Algorithm 2. It is almost the same as BSV, but during the voting of final structure, individual structures be assigned different weights. Weights are decided using the message length of the individual causal model. The less the message length is, the higher the weight of the structure.

In BSV-CI, all the component causal models is regarded equally important as only simple majority voting is used. In general, the quality of the component causal models may be quite different. So, it might be better to attach a weight to each component model, which measures the quality of the model to some extent, and then perform weighted voting at ensembling. In order to measure the quality of the causal models, a possible choice is to use MML measurement[2,7].

BMSV and BMWV Causal Inducers. These two methods are similar to BSV-CI and BWV-CI, except that in BSV-CI the seeding graphs are empty. In this method, a seed is provided to the causal inducer. The Seeding structures are generated from MCMC.

Note that in both BSV-CI and BWV-CI, all the component causal models are obtained with a null seed. In fact, start from different seeds may result in different causal models. In building ensemble of neural networks, Maclin and Shavlik[32] has obtained good results through initializing different component neural networks at different points in the weight space. So, another possible ensemble strategy is to obtain many seeds, generate a component causal model from each seed, and then combine the component models.

Thus, we obtained BMSV-CI. The algorithm is shown in Algorithm 3. It is obvious that seeds sampled from MCMC could also be used to BWV-CI to generalize BMWV-CI, as that has been done to BSV-CI. The result is the fourth algorithm, which is shown in Algorithm 4.

Algorithm 3 BMSV-CI Algorithm

Input: dataset D, inducer I, ensemble size S
Output: ensemble M.

 for $s = 1$ to S **do**
  $Seed_s = MCMC(D)$
  $M_s = I(D, Seed_s)$
 end for
 $M.Stru = majorityvoting(M_s.Stru), s = 1, \ldots, S$
 $M.weight = weightsearch(M.Stru, D)$

Algorithm 4 BMWV-CI Algorithm

Input: dataset D, inducer I, ensemble size S
Output: ensemble M.

 for $s = 1$ to S **do**
  $Seed_s = MCMC(D)$
  $M_s = I(D, Seed_s)$
  $w_s = MML(M_s, D)$
 end for
 $M.Stru = weightedvoting(M_s.Stru), s = 1, \ldots, S$
 $M.weight = weightsearch(M.Stru, D)$

BMSV-S and BMWV-S Causal Inducers. Note that in both **BMSV-CI** and **BMWV-CI**, the component causal models are directly generated from the original data set start from a generated seed model. If bootstrap sampling is introduced so that the component causal models are generated from sampled data sets with different seeds, another two algorithms, i.e. **BMSV-S-CI** and **BMWV-S-CI** are obtained, as shown in Algorithm 5 and Algorithm 6.

In applying all the ensembling algorithms, if in case the ensembled model is not a DAG, a further repairing strategy will apply.

Algorithm 5 BMSV-S-CI Algorithm

Input: dataset D, inducer I, ensemble size S
Output: ensemble M.
  for $s = 1$ to S **do**
    $D_s = bootstrap(D)$
    $Seed_s = MCMC(D_s)$
    $M_s = I(D_s, Seed_s)$
  end for
  $M.Stru = majorityvoting(M_s.Stru), s = 1, \ldots, S$
  $M.weight = weightsearch(M.Stru, D)$

Algorithm 6 BMWV-S-CI Algorithm

Input: dataset D, inducer I, ensemble size S
Output: ensemble M.
  for $s = 1$ to S **do**
    $D_s = bootstrap(D)$
    $Seed_s = MCMC(D_s)$
    $M_s = I(D_s, Seed_s)$
    $W_s = MML(M_s, D)$
  end for
  $M.Stru = weightedvoting(M_s.Stru), s = 1, \ldots, S$
  $M.weight = weightsearch(M.Stru, D)$

5 Empirical Results and Analysis

In our experiments, we compare six learning algorithms tested on 7 models as shown in Table 1(a). For each one of the 7 models, we generate a data set with 1000 instances.

Table 1. Experimental Data sets and related Results

(a) Information of Data Set

Data Set	Nodes	Sample Size
Fiji	4	1000
Evans	5	1000
Blau	6	1000
Rodgers	7	1000
Case9	9	1000
Case10	10	1000
Case12	12	1000

(b) Performance Comparison of Algorithms

Alg	Fiji	Blau	Rodgers	Evans	Case12
MML-CI	[2,0,1]	[0,1,2]	[2,1,2]	[2,1,3]	[0,0,0]
MCMC	[2,0,1]	[0,0,1]	[0,0,0]	[3,1,1]	[0,0,1]
BMSV	[2,0,1]	[0,0,1]	[0,0,0]	[3,1,1]	[0,0,0]
BMWV	[2,0,1]	[0,0,1]	[0,0,0]	[3,1,1]	[0,0,0]
BSV	[2,0,1]	[0,0,1]	[2,1,2]	[2,1,3]	[0,0,0]
BWV	[2,0,1]	[0,0,1]	[0,0,0]	[1,0,2]	[0,0,0]

The experimental results in Table 1(b) reports the performance comparison of the following algorithms: MML-CI, MCMC, BSV, BWV, BMSV and BMWV.

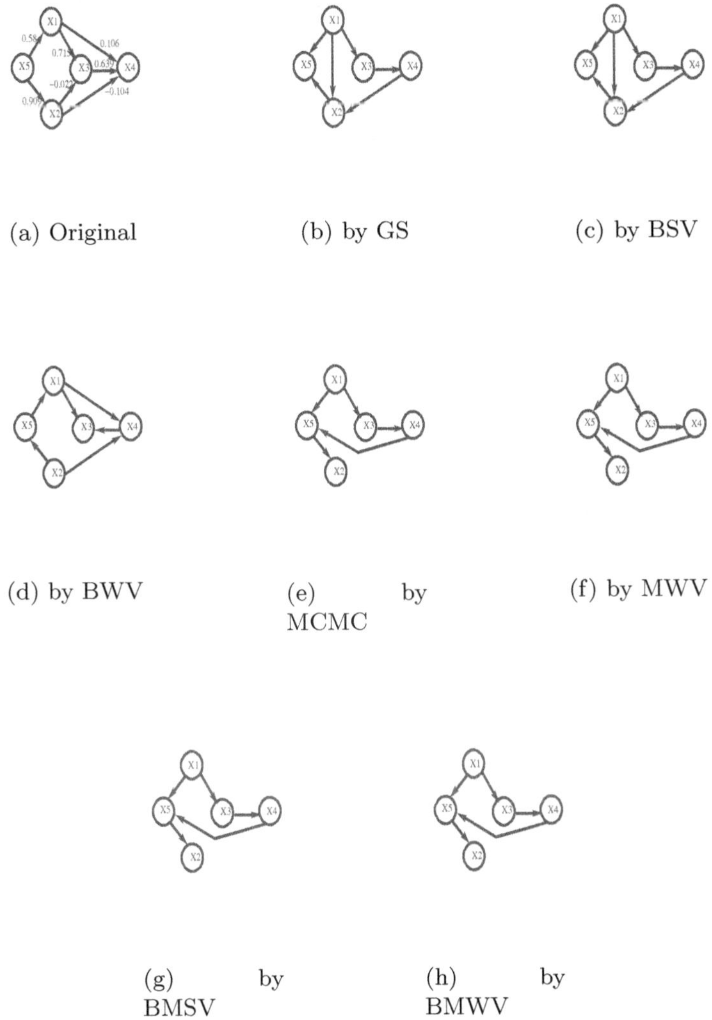

(a) Original (b) by GS (c) by BSV

(d) by BWV (e) by MCMC (f) by MWV

(g) by BMSV (h) by BMWV

Fig. 2. Comparison of Discovery Result on ***Evans***

In Table 1(b), a triple-tuple $[m, a, r]$ is used to represent the results in which m is the number of missing edges, a is the number of added edges, and r is the number of reversed edges.

From Table 1(b) and Figures 2 and 3, we can see that:

1. All of the four ensemble algorithms works better than the individual causal induction algorithm MML-CI, and the MCMC.
2. Among all the ensemble CI algorithms, the BWV achieved the best result.

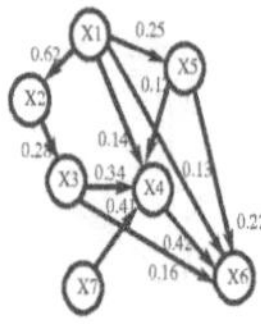

(a) Original

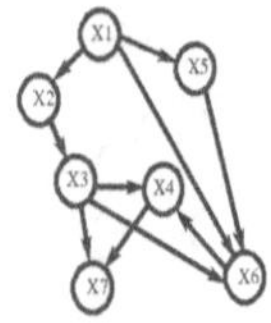

(b) by GS

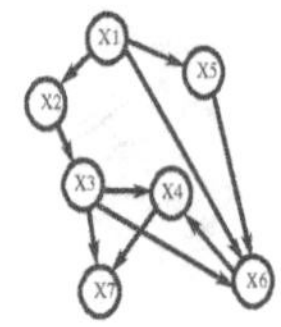

(c) by BSV

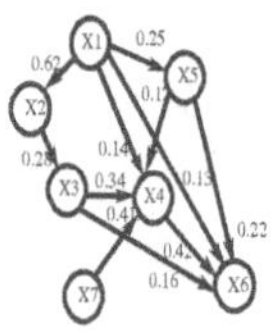

(d) by BWV

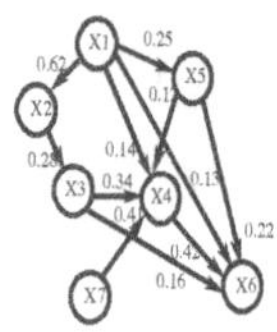

(e) by MCMC

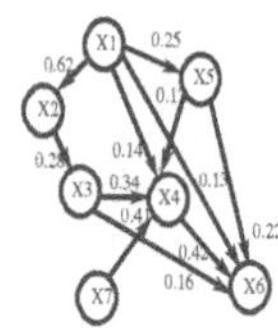

(f) by MWV

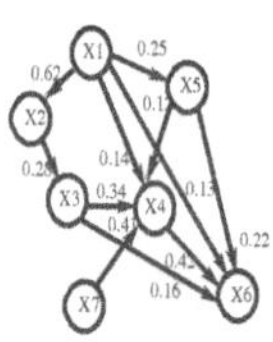

(g) by BMSV

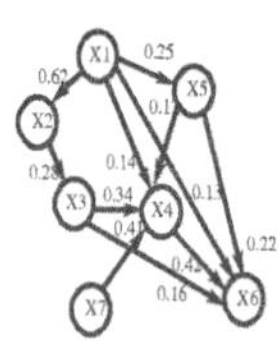

(h) by BMWV

Fig. 3. Comparison of Discovery Result on ***Rodgers***

Our experiments also show that:

1. It is interesting to see that weighted voting without seed achieved a better result than with seeding. This can be seen from the result of BWV and BMWV tested on Evans model.
2. Bagging with seedings from MCMC sampling doesn't improve the performance of the causal discovery algorithm.

3. For MCMC algorithm, even though the number of samples needed by CMC is theoretically polynomial (not exponential) in the dimensionality of the search space, in practice it has been found that MCMC does not converge in reasonable time, especially when a model has more than about 10 nodes.
4. As ensemble learning strategy could form the best from a number of discovered models, and the different seeding model could provide different start point in the model space, it is most likely the case that this method could overcome the local minimum problem. The performance results seem to support this. But it needs to be further proved.
5. In terms of precision of resulting structure, Bagging with weighted voting outperforms all the other algorithms examined, including *Message Length-based Greedy Causal Discovery*, MCMC algorithm, etc.

Table 2 and 3 list the results of the 6 algorithms with various sample size from 3 to 19, there is no indication showing that a larger size achieves a better result or vice versa.

Table 2. MML on *Rodgers* under different ensemble size

Alg	3	7	11	15	19
MCMC	52.42	57.63	52.42	53.23	63.30
MWV	52.42	52.42	52.42	52.42	61.96
BMSV	52.42	52.42	52.42	52.42	61.96
BMWV	52.42	52.42	52.42	52.42	52.42
BSV	62.28	62.28	62.28	62.28	62.28
BWV	52.06	52.06	52.06	52.06	52.06

(93 in front of each number is ommitted. So 53.2 should be "9353.2)

Table 3. Number of Errors on *Rodgers* under different ensemble size

Alg	3	5	7	9	11	13	15	17	19
MCMC	5	3	4	5	4	0	5	1	7
MWV	3	3	3	3	3	0	3	0	7
BMSV	3	3	3	3	3	0	3	0	7
BMWV	3	3	3	3	3	0	3	0	3
BSV	5	5	5	5	5	5	5	5	5
BWV	0	0	0	0	0	0	0	0	0

6 Conclusions

In this paper, we proposed an ensemble MML causal Induction method. Several ensembling strategies are examined. Compared with this new Causal induction algorithm based on Bagging with the previous algorithms [2,7], the following conclusions appear to be supported by our experimental results:

1. In terms of correctness and accuracy, Bagging ensembling approach with weighted voting without seeding outperforms all the other algorithms examined.
2. Large ensemble size does not seem to be able to improve performance.
3. The model induced by ensembling algorithms is almost always better than that by any single causal learner.

Acknowledgements. The authors would like to thank Geoff Webb for his support, both financially and intellectually, to the project described in this paper. This work contributed by the author, Zhi-Hua Zhou, was partially supported by the National Outstanding Youth Foundation of China under the Grant No. 60325207, the Excellent Young Teachers Program of MOE of China, the Fok Ying Tung Education Foundation under the Grant No. 91067, and the National 973 Fundamental Research Program of China under the Grant No. 2002CB312002.

References

1. Dai, H., Li, G.: An improved approach for the discovery of causal models via MML. In: Proceedings of The 6th Pacific-Asia Conference on Knowledge Discovery and Data Mining (PAKDD-2002), Taiwan (2002) 304–315
2. Wallace, C., Korb, K., Dai, H.: Causal discovery via MML. In: Proceedings of the 13th International Conference on Machine Learning (ICML'96). (1996) 516–524
3. Dai, H., Korb, K., Wallace, C., Wu, X.: A study of causal discovery with small samples and weak links. In: Proceedings of the 15th International Joint Conference On Artificial Intelligence IJCAI'97, Morgan Kaufmann Publishers, Inc. (1997) 1304–1309
4. Honghua Dai: A minimal causal model learner. In: Proceedings of The Third Pacific-Asia Conference on Knowledge Discovery and Data Mining (PAKDD-99), Beijing (1999) 400–408
5. Heckerman, D., Geiger, D., Chickering, D.M.: Learning Bayesian networks: The combination of knowledge and statistical data. Machine Learning **20** (1995) 197–243
6. Spirtes, P., Glymour, C., Scheines, R., Meek, C.: TETRAD II: tools for causal modeling. Lawrence Erlbaum, Hillsdale, New Jersey (1994)
7. Dai, H., Li, G., Tu, Y.: An empirical study of encoding schemes and search strategies in discovering causal networks. In: Proceedings of 13th European Conference on Machine Learning (Machine Learning: ECML 2002), Helsinki, Finland, Springer (2002) 48–59
8. Breiman, L.: Bagging predictors. Machine Learning **24** (1996) 123–140
9. Quinlan, J.: Bagging, boosting, and c4.5. In: Proceedings of the AAAI96. (1996) 725–730
10. Wolpert, D.: Stacked generalization. Neural Networks **5** (1992) 241–259
11. Drucker, H., Schapire, R.: Improving performance in neural networks using a boosting algorithm. In: Advances in Neural Information Processing Systems 5. (1993) 42–49
12. Mao, J.: A case study on bagging, boosting and basic ensembles of neural networks for ocr. In: Proceedings of IJCNN-98. (1998) 1827–1833

13. Gutta, S., Wechsler, H.: Face recognition using hybrid classifier systems. In: Proceedings of ICNN-96. (1996) 1017–1022
14. Huang, F., Zhou, Z., Zhang, H., Chen, T.: Pose invariant face recognition. In: Proceedings of the 4th IEEE International Conference on Automatic Face and Gesture Recognition. (2000) 245–250
15. Asker, L., Maclin, R.: Ensembles as a sequence of classifiers. In: Proceedings of IJCAI1997. (1997) 860–865
16. Cherkauer, K.: Human expert level performance on a scientific image analysis task by a system using combined artificia neural networks. In: Proceedings of AAAI96 Workshop on Integrating Multiple Learned Models for Improving and Scaling Machine Learning Algorithms. (1996) 15–21
17. Cunningham, P., J., C., Jacob, S.: Stability problems with artificial neural networks and the ensemble solution. Artificial Intelligence in Medicine **20** (2000) 217–225
18. Zhou, Z.H., Wu, J., Tang, W.: Enselbling neural networks: Many could be better than all. Artificial Intelligence **137** (2002) 993–1001
19. Shimshoni, Y., Intrator, N.: Classification of seismic signals by integrating ensembles of neural networks. IEEE Transaction on Signal Processing **46** (1998) 1194–1201
20. Krogh, A., Vedelsby, J.: Neural network ensembles, cross validation, and active learning. In: Advances in Neural Information Processing Systems 7. (1995) 231–238
21. Schapire, R.: The strength of weak learnability. Machine Learning **5** (1990) 197–227
22. Efron, B., Tibshirani, R.: An introduction to the bootstrap. Chapman and Hall, New York (1993)
23. Freund, Y.: Boosting a weak algorithm by majority. Information and Computation **121** (1995) 256–285
24. Freund, Y., Schapire, R.: A decision-theoretic generalization of on-line learning and an application to boosting. In: Proceedings of EuroCOLT-94. (1995) 23–37
25. Bauer, E., Kohavi, R.: An empirical comparison of voting classification algorithms: bagging, boosting, and variants. machine learning **36** (1999) 105–139
26. Webb, G.: Multiboosting: a technique for combining boosting and wagging. Machine Learning **40** (2000) 159–196
27. Hansen, L., Salamon, P.: Neural network ensembles,. IEEE Transaction on Pattern Analysis and Machine Intelligence **12** (1990) 993–1001
28. Opitz, D., Shavlik, J.: Actively searching for an effective neural network ensemble. Connection Science **8** (1996) 337–353
29. M.P. Perrone, L.C.: When networks disagree: ensemble method for neural networks. In: Artificial Neural Networks for Speech and Vision. Chapman and Hall, New York (1993) 126–142
30. Merz, C., Pazzani, M.: Combining neural network regression estimates with regularized linear weights. In: Advances in Neural Information Processing Systems 9. (1997) 564–570
31. Jimenez, D.: Dynamically weighted ensemble neural networks for classification. In: Proceedings of IJCNN-98. (1998) 753–756
32. R. Maclin, J.S.: Combining the predictions of multiple classifiers: using competitive learning to initialize neural networks. In: Proceedings of IJCAI-95. (1995) 524–530

Logistic Regression and Boosting for Labeled Bags of Instances

Xin Xu and Eibe Frank

Department of Computer Science
University of Waikato
Hamilton, New Zealand
{xx5, eibe}@cs.waikato.ac.nz

Abstract. In this paper we upgrade linear logistic regression and boosting to multi-instance data, where each example consists of a labeled bag of instances. This is done by connecting predictions for individual instances to a bag-level probability estimate by simple averaging and maximizing the likelihood at the bag level—in other words, by assuming that all instances contribute equally and independently to a bag's label. We present empirical results for artificial data generated according to the underlying generative model that we assume, and also show that the two algorithms produce competitive results on the Musk benchmark datasets.

1 Introduction

Multi-instance (MI) learning differs from standard supervised learning in that each example is not just a single instance: examples are collections of instances, called "bags", containing an arbitrary number of instances. However, as in standard supervised learning there is only one label for each example (i.e. bag). This makes MI learning inherently more difficult because it is not clear how the bag label relates to the individual instances in a bag. The standard way to approach the problem is to assume that there is one "key" instance in a bag that triggers whether the bag's class label will be positive or negative.[1] This approach has been proposed for predicting the activity of a molecule, where the instances correspond to different shapes of the molecule, and the assumption is that the presence of a particular shape determines whether a molecule is chemically active or not [1]. The task is then to identify these key instances. A different approach is to assume that all instances contribute equally and independently to a bag's class label. In the above context this corresponds to assuming that the different shapes of a molecule have a certain probability of being active and the average of these probabilities determines the probability that the molecule will be active. In this paper, we use the latter approach to upgrade logistic regression and boosting to MI problems.

The paper is organized as follows. In Section 2 we explain the underlying generative model that we assume and illustrate it using artificial MI data based on

[1] Throughout this paper we assume a classification task with two possible classes.

H. Dai, R. Srikant, and C. Zhang (Eds.): PAKDD 2004, LNAI 3056, pp. 272–281, 2004.

this model. In Section 3 we describe how we use the generative model to upgrade linear logistic regression and boosting to MI learning. In Section 4 we evaluate the resulting algorithms on both the artificial data and the Musk benchmark problems [1]. Section 5 summarizes our results.

2 A Generative Model

We assume that the class label of a bag is generated in a two-stage process. The first stage determines class probabilities for the individual instances in a bag, and the second stage combines these probability estimates in some fashion to assign a class label to the bag. The difficulty lies in the fact that we cannot observe class labels for individual instances and therefore cannot estimate the instance-level class probability function directly.

This general two-stage framework has previously been used in the Diverse Density (DD) algorithm [2] to learn a classifier for MI problems. In the first step, DD assumes a radial (or "Gaussian-like") formulation for the instance-level class probability function $Pr(y|x)$ (where y is a class label—either 0 or 1—and x an instance). In the second step, it assumes that the values of $Pr(y|x)$ for the instances in a bag are combined using either a multi-stage (based on the noisy-or model) or a one-stage (based on the most-likely-cause model) Bernoulli process to determine the bag's class label. In both cases DD essentially attempts to identify ellipsoid areas of the instance space for which the probability of observing a positive instance is high, i.e, it attempts to identify areas that have a high probability of containing "key" instances. The parameters involved are estimated by maximizing the bag-level likelihood.

In this paper we use the same two-stage framework to upgrade linear logistic regression and boosting to MI data. However, we assume a different process for combining the instance-level class probability estimates, namely that all instances contribute equally to a bag's class probability. Consider an instance-level model that estimates $Pr(y|x)$ or the log-odds function $\log \frac{Pr(y=1|x)}{Pr(y=0|x)}$, and a bag b that corresponds to a certain area in the instance space. Given this bag b with n instances $x_i \in b$, we assume that the bag-level class probability is either given by

$$Pr(Y|b) = \frac{1}{n}\sum_{i=1}^{n} Pr(y|x_i) \tag{1}$$

or by

$$\log \frac{Pr(y=1|b)}{Pr(y=0|b)} = \frac{1}{n}\sum_{i=1}^{n} \log \frac{Pr(y=1|x_i)}{Pr(y=0|x_i)}$$
$$\Rightarrow \begin{cases} Pr(y=1|b) = \frac{[\prod_i^n Pr(y=1|x_i)]^{1/n}}{[\prod_i^n Pr(y=1|x_i)]^{1/n}+[\prod_i^n Pr(y=0|x_i)]^{1/n}} \\ Pr(y=0|b) = \frac{[\prod_i^n Pr(y=0|x_i)]^{1/n}}{[\prod_i^n Pr(y=1|x_i)]^{1/n}+[\prod_i^n Pr(y=0|x_i)]^{1/n}} \end{cases} \tag{2}$$

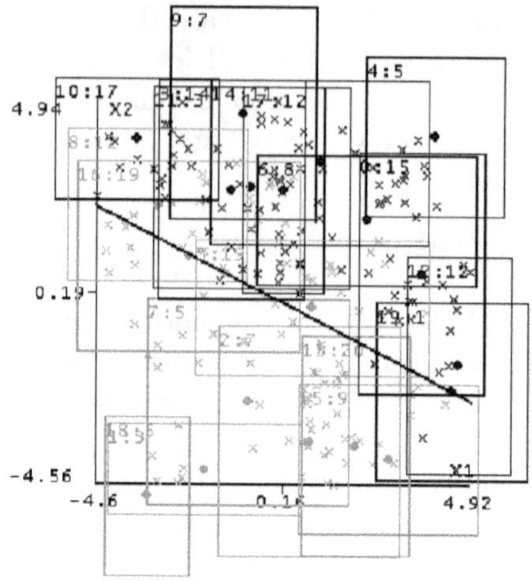

Fig. 1. An artificial dataset with 20 bags.

Equation 1 is based on the arithmetic mean of the instance-level class probability estimates while Equation 2 involves the geometric mean.

The above generative models are best illustrated based on an artificial dataset. Consider an artificial domain with two attributes. We create bags of instances by defining rectangular regions in this two-dimensional instance space and sampling instances from within each region. First, we generate coordinates for the centroids of the rectangles according to a uniform distribution with a range of $[-5, 5]$ for each of the two dimensions. The size of a rectangle in each dimension is chosen from 2 to 6 with equal probability. Each rectangle is used to create a bag of instances. For sampling the instances, we assume a symmetric triangle distribution with ranges $[-5, 5]$ in each dimension (and density function $f(x) = 0.2 - 0.04|x|$). From this distribution we sample n instances from within a rectangle.

The number of instances n for each bag is chosen from 1 to 20 with equal probability. We define the instance-level class probability by the linear logistic model $Pr(y = 1|x_1, x_2) = \frac{1}{1+e^{-x_1-2x_2}}$, where x_1 and x_2 are the two attribute values of an instance. Thus the log-odds function is a simple linear function, which means the instance-level decision boundary is a line. Then we take Equation 2 to calculate $Pr(y|b)$. Finally we label each bag by flipping a coin according to this probability.

Figure 1 shows a dataset with 20 bags that was generated according to this model. The black line in the middle is the instance-level decision boundary (i.e. where $Pr(y = 1|x_1, x_2) = 0.5$) and the sub-space on the right side has instances with higher probability to be positive. A rectangle indicates the region used to sample points for the corresponding bag (and a dot indicates its centroid). The top-left corner of each rectangle shows the bag index, followed by the number of instances in the bag. Bags in gray belong to class "negative" and bags in black to class "positive". Note that bags can be on the "wrong" side of the instance-level decision boundary because each bag was labeled by flipping a coin based on its class probability.

3 Upgrading Linear Logistic Regression and Boosting

In this section we first show how to implement linear logistic regression based on the above generative model. Then we do the same for additive logistic regression (i.e. boosting)—more specifically, AdaBoost.M1 [3].

3.1 Linear Logistic Regression

The standard logistic regression model does not apply to multi-instance data because the instances' class labels are masked by the "collective" class label of a bag. However, suppose we know how the instance-level class probabilities are combined to form a bag-level probability, say, by Equation 2. Then we can apply a standard optimization method (e.g. gradient descent) to search for a maximum of the (bag-level) binomial likelihood based on the training data. This gives us an indirect estimate of the instance-level logistic model.

The instance-level class probabilities are given by $Pr(y = 1|x) = 1/(1 + \exp(-\beta x))$ and $Pr(y = 0|x) = 1/(1 + \exp(\beta x))$ respectively, where β is the parameter vector to be estimated. If we assume bag-level probabilities are given by Equation 2, this results in the bag-level model

$$\begin{aligned} Pr(y=1|b) &= \frac{[\prod_i^n Pr(y=1|x_i)]^{1/n}}{[\prod_i^n Pr(y=1|x_i)]^{1/n}+[\prod_i^n Pr(y=0|x_i)]^{1/n}} = \frac{\exp(\frac{1}{n}\beta\sum_i \mathbf{x_i})}{1+\exp(\frac{1}{n}\beta\sum_i \mathbf{x_i})} \\ Pr(y=0|b) &= \frac{[\prod_i^n Pr(y=0|x_i)]^{1/n}}{[\prod_i^n Pr(y=1|x_i)]^{1/n}+[\prod_i^n Pr(y=0|x_i)]^{1/n}} = \frac{1}{1+\exp(\frac{1}{n}\beta\sum_i \mathbf{x_i})} \end{aligned} . \quad (3)$$

Based on this we can estimate the parameter vector β by maximizing the bag-level binomial log-likelihood function:

$$LL = \sum_i^N [y_i \log Pr(y = 1|b) + (1 - y_i) \log Pr(y = 0|b)] \quad (4)$$

where N is the number of bags.

Note that this problem can actually be reduced to standard single-instance logistic regression by converting each bag of instances into a single instance representing the bag's mean (because the instances only enter Equation 3 in the sum $\sum_i x_i$). The reason for this is the simple linear structure of the model and the fact that the bag-level probabilities are generated based on the geometric mean (Equation 2). In practice, it is usually not possible to say whether the geometric mean is more appropriate than the arithmetic one, so we may want to use Equation 1 as an alternative. In that case, we change the formulation of $Pr(y|b)$ based on Equation 1 and use this in conjunction with the same log-likelihood function (Equation 4). This problem can no longer be solved by

1. Initialize weight of each bag to $W_i = 1/N$, $i = 1, 2, \ldots, N$.
2. Repeat for $m = 1, 2, \ldots, M$:
   a) Set $W_{ij} \leftarrow W_i/n_i$, assign the bag's class label to each of its instances, and build an instance-level model $h_m(x_{ij}) \in \{-1, 1\}$.
   b) Within the ith bag (with n_i instances), compute the error rate $e_i \in [0, 1]$ by counting the number of misclassified instances within that bag, i.e. $e_i = \sum_j 1_{(h_m(x_{ij}) \neq y_i)}/n_i$.
   c) If $e_i < 0.5$ for all i's, go to Step 3.
   d) Compute $c_m = argmin \sum_i W_i \exp[(2e_i - 1)c_m]$ using numeric optimization.
   e) If($c_m \leq 0$), go to Step 3.
   f) Set $W_i \leftarrow W_i \exp[(2e_i - 1)c_m]$ and renormalize so that $\sum_i W_i = 1$.
3. return $sign[\sum_j \sum_m c_m h_m(x_j)]$.

Fig. 2. The MIBoosting Algorithm.

applying the above transformation in conjunction with standard single-instance logistic regression. In the remainder of this paper, we will call the former method MILogisticRegressionGEOM and the latter one MILogisticRegressionARITH.

As usual, the maximization of the log-likelihood function is carried out via numeric optimization because there is no direct analytical solution. The optimization problem can be solved very efficiently because we are working with a linear model. (Note that the radial formulation in the DD [2] algorithm, for example, poses a difficult global optimization problem that is expensive to solve.) In our implementation we use a quasi-Newton optimization procedure with BFGS updates suggested in [4], searching for parameter values around zero.

3.2 Boosting

Both linear logistic regression and the model used by the DD algorithm assume a limited family of patterns. Boosting is a popular algorithm for modeling more complex relationships. It constructs an ensemble of so-called "weak" classifiers. In the following we explain how to upgrade the AdaBoost.M1 algorithm to MI problems, assuming the same generative model as in the case of linear logistic regression. Note that AdaBoost.M1 can be directly applied to MI problems (without any modifications) if the "weak" learner is a full-blown MI learner, in the same way as ensemble classifiers for MI learning have been built using bagging in [5]. In the following we consider the case where the weak learner is a standard single-instance learner (e.g. C4.5 [6]).

Boosting, more specifically, AdaBoost.M1, originates from work in computational learning theory [3], but received a statistical explanation as additive logistic regression in [7]. It can be shown that it minimizes an exponential loss function in a forward stage-wise manner, ultimately estimating the log-odds function $\frac{1}{2}\log\frac{Pr(y=1|X)}{Pr(y=0|X)}$ based on an additive model [7]. To upgrade AdaBoost.M1 we use the generative model described in Section 2 in conjunction with Equation 2 (i.e. the geometric average). The pseudo code for the algorithm, called "MIBoosting" is shown in Figure 2. Here, N is the number of bags, and we use the subscript i to denote the ith bag, where $i = 1, 2, \ldots, N$. There are n_i instances in the ith bag, and we use the subscript j to refer to the jth instance, where $j = 1, 2, \ldots, n_i$. The jth instance in the ith bag is x_{ij}. Note that, for convenience, we assume that the class label of a bag is either 1 or -1 (i.e. $y \in \{1, -1\}$), rather than 1 or 0.

The derivation of the algorithm is analogous to that in [7]. We regard the expectation sign E as the sample average instead of the population expectation. We are looking for a function (i.e. classifier) $F(b)$ that minimizes the exponential loss $E_B E_{Y|B}[\exp(-yF(b))]$. In each iteration of boosting, the aim is to expand $F(b)$ into $F(b)+cf(b)$ (i.e. adding a new "weak" classifier) so that the exponential loss is minimized. In the following we will just write $E[.]$ to denote $E_B E_{Y|B}[.]$, and use $E_w[.]$ to denote the weighted expectation, as in [7].

In each iteration of the algorithm, we search for the best $f(b)$ to add to the model. Second order expansion of $\exp(-ycf(b))$ about $f(b) = 0$ shows that we can achieve this by searching for the maximum of $E_w[yf(b)]$, given bag-level weights $W_B = \exp(-yF(b))$ [7]. If we had an MI-capable weak learner at hand that could deal with bag weights, we could estimate $f(b)$ directly. However we are interested in wrapping our boosting algorithm around a single-instance weak learner. Thus we expand $f(b)$ into $f(b) = \sum_n h(x_j)/n$, where $h(x_j) \in \{-1, 1\}$ is the prediction of the weak classifier $h(.)$ for the jth instance in b. We are seeking a weak classifier $h(.)$ that maximizes

$$E_w[yh(x_b)/n] = \sum_{i=1}^{N} \sum_{j=1}^{n_i} [\frac{1}{n_i} W_i y_i h(x_{ij})].$$

This is maximized if $h(x_{ij}) = y_i$, which means that we are seeking a classifier $h(.)$ that attempts to correctly predict the label of the bag that each instance pertains to. More precisely, $h(.)$ should minimize the weighted instance-level classification error given instance weights $\frac{W_i}{n_i}$ (assuming we give every instance its bag's label). This is exactly what standard single-instance learners are designed to do (provided they can deal with instance weights). Hence we can use any (weak) single-instance learner to generate the function $h(.)$ by assigning each instance the label of its bag and the corresponding weight $\frac{W_i}{n_i}$. This constitutes Step 2a of the algorithm in Figure 2.

Now that we have found $f(b)$, we have to look for the best multiplier $c > 0$. To do this we can directly optimize the objective function

$$\begin{aligned} E_B E_{Y|B}[\exp(-yF(b) + c(-yf(b)))] &= \sum_i W_i \exp[c_m(-\frac{y\sum_j h(x_{ij})}{n_i})] \\ &= \sum_i W_i \exp[(2e_i - 1)c_m], \end{aligned}$$

where $e_i = \sum_j 1_{(h_m(x_{ij}) \neq y_i)}/n_i$ (computed in Step 2b).

Minimization of this expectation constitutes Step 2d. Note that this function will not have a global minimum if all $e_i < 0.5$. In that case all bags will be correctly classified by $f(b)$, and no further boosting iterations can be performed. Therefore this is checked in Step 2c. This is analogous to what happens in standard AdaBoost.M1 if the current weak learner has zero error on the training data.

Note that the solution of the optimization problem involving c may not have an analytical form. Therefore we simply search for c using a quasi-Newton method in Step 2d. The computational cost for this is negligible compared to the time it takes to learn the weak classifier. The resulting value for c is not necessarily positive. If it is negative we can simply reverse the prediction of the weak classifier and get a positive c. However, we use the AdaBoost.M1 convention and stop the learning process if c becomes negative. This is checked in Step 2e.

Finally, we update the bag-level weights in Step 2f according to the additive structure of $F(b)$, in the same way as in standard AdaBoost.M1. Note that, the more misclassified instances a bag has, the greater its weight in the next iteration. This is analogous to what happens in standard AdaBoost.M1 at the instance level.

To classify a test bag, we can simply regard $F(b)$ as the bag-level log-odds function and take Equation 2 to make a prediction (Step 3). An appealing property of this algorithm is that, if there is only one instance per bag, i.e. for single-instance data, the algorithm naturally degenerates to normal AdaBoost.M1. To see why, note that the solution for c in Step 2d will be exactly $\frac{1}{2}\log\frac{1-err_W}{err_W}$, where err_W is the weighted error. Hence the weight update will be the same as in standard AdaBoost.M1.

4 Experimental Results

In this section we first discuss empirical results for the artificial data from Section 2. Then we evaluate our techniques on the Musk benchmark datasets.

4.1 Results for the Artificial Data

Because we know the artificial data from Section 2 is generated using a linear logistic model based on the geometric mean formulation in Equation 2, MILogisticRegressionGEOM is the natural candidate to test on this specific data. Figure 3 shows the parameters estimated by this method when the number of bags increases. As expected, the estimated parameters converge to the true parameters asymptotically. However, more than 1000 bags are necessary to produce accurate estimates.

To evaluate the performance of MIBoosting we created an independent test set with 10,000 bags based on the same generative model. As the "weak" learner we used decision stumps, performing 30 iterations of the boosting algorithm. The test error for different numbers of training bags is plotted in Figure 4, and compared to that of MI linear logistic regression. Not surprisingly, the latter outperforms boosting because it matches the underlying generative model exactly. However, both methods approach the optimum error rate on the test data.

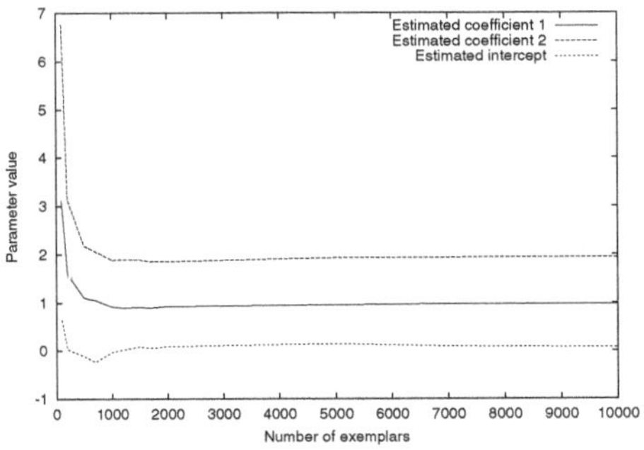

Fig. 3. Parameters estimated by MI Logistic Regression on the artificial data.

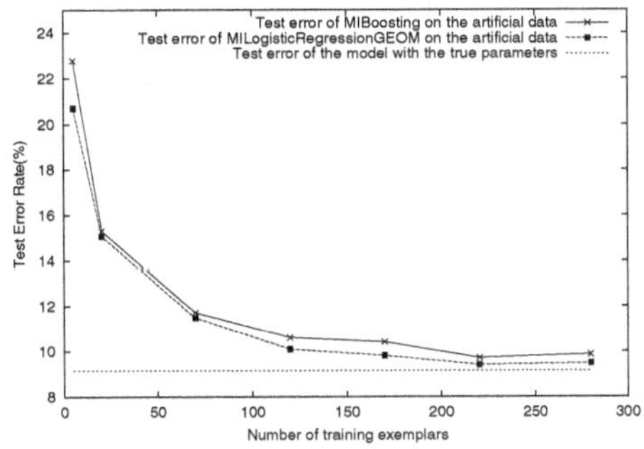

Fig. 4. Test error rates for boosting and logistic regression on the artificial data.

4.2 Results for the Musk Problems

In this section we present results for the Musk drug activity datasets [1]. The Musk 1 data has 92 bags and 476 instances, the Musk 2 data 102 bags and 6598 instances. Both are two-class problems with 166 attributes. Because of the large number of attributes, some regularization proved necessary to achieve good results with logistic regression. To this end we added an L_2 penalty term $\lambda||\beta||^2$ to the likelihood function in Equation 4, where λ is the ridge parameter.[2] In our experiments we set $\lambda = 2$. For boosting, we used C4.5 trees as the "weak" classifiers and 50 boosting iterations. We introduced some regularization by setting the minimum number of instances per leaf to twice the average bag size (10 instances for Musk 1, and 120 instances for Musk 2). The post-pruning mechanism in C4.5 was turned off.

The upper part of Table 1 shows error estimates for the two variants of MILogisticRegression and for MIBoosting. These were obtained by 10 runs of stratified 10-fold cross-validation (CV) (at the bag level). The standard deviation of the 10 estimates from the 10 runs is also shown. The lower part summarizes some results for closely related statistical methods that can be found in the literature. A summary of results for other multi-instance learning algorithms can be found in [8]. Note also that the performance of MI learning algorithms can be improved further using bagging, as shown in [5].

The first result in the lower part of Table 1 is for the Diverse Density algorithm applied in conjunction with the noisy-or model [2]. This result was also obtained by 10-fold cross-validation. The second result was produced by a neural network. The MI Neural Network is a multi-layer perceptron adapted to multi-instance problems by using a soft-max function to hone in on the "key" instances in a bag [9].[3] The last entry in Table 1 refers to a support vector machine (SVM) used in conjunction with a Gaussian multi-instance kernel [8]. This method re-

---
[2] We standardized the training data using the weighted mean and variance, each instance being weighted by the inverse of the number of instances in its bag.

[3] It is unclear how the error estimates for the neural network were generated.

Table 1. Error rate estimates for the Musk datasets and standard deviations (if available).

Method	Musk 1	Musk 2
MILogisticRegressionGEOM	14.13±2.23	17.74±1.17
MILogisticRegressionARITH	13.26±1.83	15.88±1.29
MIBoosting with 50 iterations	12.07±1.95	15.98±1.31
Diverse Density	11.1	17.5
MI Neural Network	12.0	18.0
SVM with Gaussian MI kernel	13.6±1.1	12.0±1.0

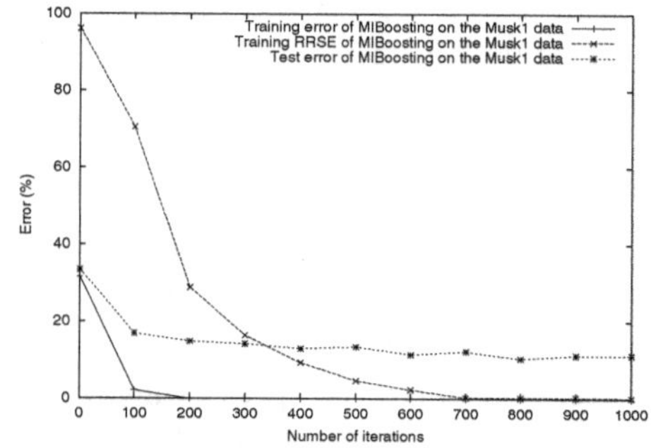

Fig. 5. The effect of varying the number of boosting iterations on the Musk1 data.

places the dot product between two instances (used in standard SVMs) by the sum of the dot products over all possible pairs of instances from two bags. This method resembles ours in the sense that it also implicitly assumes all instances in a bag are equally relevant to the classification of the bag. The error-estimates for the SVM were generated using leave-10-out, which produces similar results as 10-fold cross-validation on these datasets.

The results for logistic regression indicate that the arithmetic mean (Equation 1) is more appropriate than the geometric one (Equation 2) on the Musk data. Furthermore, boosting improves only slightly on logistic regression. For both methods the estimated error rates are slightly higher than for DD and the neural network on Musk 1, and slightly lower on Musk 2. Compared to the SVM, the results are similar for Musk 1, but the SVM appears to have an edge on the Musk 2 data. However, given that logistic regression finds a simple linear model, it is quite surprising that it is so competitive.

Figure 5 shows the effect of different numbers of iterations in the boosting algorithm. When applying boosting to single-instance data, it is often observed that the classification error on the test data continues to drop after the error on the training data has reached zero. Figure 5 shows that this effect also occurs in the multi-instance version of boosting—in this case applied to the Musk 1 data and using decision stumps as the weak classifiers.

Figure 5 also plots the Root Relative Squared Error (RRSE) of the probability estimates generated by boosting. It shows that the classification error on the training data is reduced to zero after about 100 iterations but the RRSE keeps decreasing until around 800 iterations. The generalization error, estimated using 10 runs of 10-fold CV, also reaches a minimum around 800 iterations, at 10.44% (standard deviation: 2.62%). Note that this error rate is lower than the one for boosted decision trees (Table 1). However, boosting decision stumps is too slow for the Musk 2 data, where even 8000 iterations were not sufficient to minimize the RRSE.

5 Conclusions

We have introduced multi-instance versions of logistic regression and boosting, and shown that they produce results comparable to other statistical multi-instance techniques on the Musk benchmark datasets. Our multi-instance algorithms were derived by assuming that every instance in a bag contributes equally to the bag's class label—a departure from the standard multi-instance assumption that relates the likelihood of a particular class label to the presence (or absence) of a certain "key" instance in the bag.

Acknowledgment. This research was supported by Marsden Grant 01-UOW-019.

References

1. T.G. Dietterich, R.H. Lathrop, and T. Lozano-Pérez. Solving the multiple-instance problem with the axis-parallel rectangles. *Artificial Intelligence*, 89(1-2):31–71, 1997.
2. O. Maron. *Learning from Ambiguity*. PhD thesis, MIT, United States, 1998.
3. Y. Freund and R.E. Schapire. Experiments with a new boosting algorithm. In *Proc of the 13th Int Conf on Machine Learning*, pages 148–156. Morgan Kauffman, 1996.
4. P.E. Gill, W. Murray, and M.H. Wright. *Practical Optimization*. Academic Press, 1981.
5. Z.-H. Zhou and M.-L. Zhang. Ensembles of multi-instance learners. In *Proc of the 14th European Conf on Machine Learning*, pages 492–501. Springer, 2003.
6. J.R. Quinlan. *C4.5: Programs for Machine Learning*. Morgan Kaufmann, 1993.
7. J. Friedman, T. Hastie, and R. Tibshirani. Additive logistic regression, a statistical view of boosting (with discussion). *Annals of Statistics*, 28:307–337, 2000.
8. T. Gärtner, P.A. Flach, A. Kowalczyk, and A.J. Smola. Multi-instance kernels. In *Proc of the 19th Int Conf on Machine Learning*, pages 179–186. Morgan Kaufmann, 2002.
9. J. Ramon and L. De Raedt. Multi instance neural networks. In *Attribute-Value and Relational Learning: Crossing the Boundaries*, 2000. Workshop at the 17th Int Conf on Machine Learning.

Fast and Light Boosting for Adaptive Mining of Data Streams

Fang Chu and Carlo Zaniolo

University of California, Los Angeles, CA 90095, USA
{fchu, zaniolo}@cs.ucla.edu

Abstract. Supporting continuous mining queries on data streams requires algorithms that (i) are fast, (ii) make light demands on memory resources, and (iii) are easily to adapt to concept drift. We propose a novel boosting ensemble method that achieves these objectives. The technique is based on a dynamic sample-weight assignment scheme that achieves the accuracy of traditional boosting without requiring multiple passes through the data. The technique assures faster learning and competitive accuracy using simpler base models. The scheme is then extended to handle concept drift via change detection. The change detection approach aims at significant data changes that could cause serious deterioration of the ensemble performance, and replaces the obsolete ensemble with one built from scratch. Experimental results confirm the advantages of our adaptive boosting scheme over previous approaches.

Keywords: Stream data mining, adaptive boosting ensembles, change detection

1 Introduction

A substantial amount of recent work has focused on continuous mining of data streams [4, 10,11,15,16]. Typical applications include network traffic monitoring, credit card fraud detection and sensor network management systems. Challenges are posed by data ever increasing in amount and in speed, as well as the constantly evolving concepts underlying the data. Two fundamental issues have to be addressed by any continuous mining attempt.

Performance Issue. Constrained by the requirement of on-line response and by limited computation and memory resources, continuous data stream mining should conform to the following criteria: (1) Learning should be done very *fast*, preferably in one pass of the data; (2) Algorithms should make very *light* demands on memory resources, for the storage of either the intermediate results or the final decision models. These *fast and light* requirements exclude high-cost algorithms, such as support vector machines; also decision trees with many nodes should preferably be replaced by those with fewer nodes as base decision models.

Adaptation Issue. For traditional learning tasks, the data is stationary. That is, the underlying concept that maps the features to class labels is unchanging [17]. In the context of data streams, however, the concept may drift due to gradual or sudden changes of the external environment, such as increases of network traffic or failures in sensors. In fact, mining changes is considered to be one of the core issues of data stream mining [5].

H. Dai, R. Srikant, and C. Zhang (Eds.): PAKDD 2004, LNAI 3056, pp. 282–292, 2004.

In this paper we focus on continuous learning tasks, and propose a novel *Adaptive Boosting Ensemble* method to solve the above problems. In general, ensemble methods combine the predictions of multiple base models, each learned using a learning algorithm called the base learner [2]. In our method, we propose to use very simple base models, such as decision trees with a few nodes, to achieve fast and light learning. Since simple models are often weak predictive models by themselves, we exploit boosting technique to improve the ensemble performance. The traditional boosting is modified to handle data streams, retaining the essential idea of dynamic sample-weight assignment yet eliminating the requirement of multiple passes through the data. This is then extended to handle concept drift via change detection. Change detection aims at significant changes that would cause serious deterioration of the ensemble performance. The awareness of changes makes it possible to build an active learning system that adapts to changes promptly.

Related Work. Ensemble methods are hardly the only approach used for continuous learning. Domingos et al. [4] devised a novel decision tree algorithm, the Hoeffding tree, that performs asymptotically the same as or better than its batched version. This was extended to CVFDT in an attempt to handle concept drift [11]. But, Hoeffding-tree like algorithms need a large training set in order to reach a fair performance, which makes them unsuitable to situations featuring frequent changes. Domeniconi et al. [3] designed an incremental support vector machine algorithm for continuous learning.

There has been work related to boosting ensembles on data streams. Fern et al. [6] proposed online boosting ensembles, and Oza et al. [12] studied both online bagging and online boosting. Frank et al. [7] used a boosting scheme similar to our boosting scheme. But none of these work took concept drift into consideration.

Previous ensemble methods for drifting data streams have primarily relied on bagging-style techniques [15,16]. Street et al. [15] gave an ensemble algorithm that builds one classifier per data block independently. Adaptability relies solely on retiring old classifiers one at a time. Wang et al. [16] used a similar ensemble building method. But their algorithm tries to adapt to changes by assigning weights to classifiers proportional to their accuracy on the most recent data block. As these two algorithms are the most related, we call them *Bagging* and *Weighted Bagging*, respectively, for later references in our experimental comparison. [1]

This paper is organized as follows. Our adaptive boosting ensemble method is presented in section 2, followed by a change detection technique in section 3. Section 4 contains experimental design and evaluation results, and we conclude in section 5.

2 Adaptive Boosting Ensembles

We use the boosting ensemble method since this learning procedure provides a number of formal guarantees. Freund and Schapire proved a number of positive results about its generalization performance [13]. More importantly, Friedman et al. showed that boosting is particularly effective when the base models are simple [9]. This is most desirable for *fast and light* ensemble learning on steam data.

[1] The name "*bagging*" derives from their analogy to traditional bagging ensembles [1].

Algorithm 1 Adaptive boosting ensemble algorithm

Ensure: Maintaining a boosting ensemble E_b with classifiers $\{C_1, \cdots, C_m\}$, $m \leq M$.
1: **while** (1) **do**
2: Given a new block $B_j = \{(x_1, y_1), \cdots, (x_n, y_n)\}$, where $y_i \in \{0, 1\}$,
3: Compute ensemble prediction for sample i: $E_b(x_i) = \text{round}(\frac{1}{m}\sum_{k=1}^{m} C_k(x_i))$,
4: *Change Detection:* $E_b \Leftarrow \emptyset$ *if a change detected!*
5: **if** ($E_b \neq \emptyset$) **then**
6: Compute error rate of E_b on B_j: $e_j = E[1_{E_b(x_i) \neq y_i}]$,
7: Set new sample weight $w_i = (1 - e_j)/e_j$ if $E_b(x_i) \neq y_i$; $w_i = 1$ otherwise
8: **else**
9: set $w_i = 1$, for all i.
10: **end if**
11: Learn a new classifier C_{m+1} from weighted block B_j with weights $\{w_i\}$,
12: Update E_b: add C_{m+1}, retire C_1 if $m = M$.
13: **end while**

In its original form, the boosting algorithm assumes a static training set. Earlier classifiers increase the weights of misclassified samples, so that the later classifiers will focus on them. A typical boosting ensemble usually contains hundreds of classifiers. However, this lengthy learning procedure does not apply to data streams, where we have limited storage but continuous incoming data. Past data can not stay long before making place for new data. In light of this, our boosting algorithm requires only two passes of the data. At the same time, it is designed to retain the essential idea of boosting—the dynamic sample weights modification.

Algorithm 1 is a summary of our boosting process. As data continuously flows in, it is broken into blocks of equal size. A block B_j is scanned twice. The first pass is to assign sample weights, in a way corresponding to AdaBoost.M1 [8]. That is, if the ensemble error rate is e_j, the weight of a misclassified sample x_i is adjusted to be $w_i = (1 - e_j)/e_j$. The weight of a correctly classified sample is left unchanged. The weights are normalized to be a valid distribution. In the second pass, a classifier is constructed from this weighted training block.

The system keeps only the most recent classifiers, up to *M*. We use a traditional scheme to combine the predictions of these base models, that is, by averaging the probability predictions and selecting the class with the highest probability. Algorithm 1 is for binary classification, but can easily be extended to multi-class problems.

Adaptability Note that there is a step called "*Change Detection*" (line 4) in Algorithm 1. This is a distinguished feature of our boosting ensemble, which guarantees that the ensemble can adapt promptly to changes. Change detection is conducted at every block. The details of how to detect changes are presented in the next section.

Our ensemble scheme achieves adaptability by actively detecting changes and discarding the old ensemble when an alarm of change is raised. No previous learning algorithm has used such a scheme. One argument is that old classifiers can be tuned to the new concept by assigning them different weights. Our hypothesis, which is borne out by experiment, is that obsolete classifiers have bad effects on overall ensemble performance even they are weighed down. Therefore, we propose to learn a new ensemble from scratch when changes occur. Slow learning is not a concern here, as our base learner is fast and light, and boosting ensures high accuracy. The main challenge is to detect changes with a low false alarm rate.

3 Change Detection

In this section we propose a technique for change detection based on the framework of statistical decision theory. The objective is to detect changes that cause significant deterioration in ensemble performance, while tolerating minor changes due to random noise. Here, we view ensemble performance θ as a random variable. If data is stationary and fairly uniform, the ensemble performance fluctuations are caused only by random noise, hence θ is normally assumed to follow a Gaussian distribution. When data changes, yet most of the obsolete classifiers are kept, the overall ensemble performance will undergo two types of decreases. In case of an abrupt change, the distribution of θ will change from one Gaussian to another. This is shown in Figure 1(a). Another situation is when the underlying concept has constant but small shifts. This will cause the ensemble performance to deteriorate gradually, as shown in Figure 1(b). Our goal is to detect both types of significant changes.

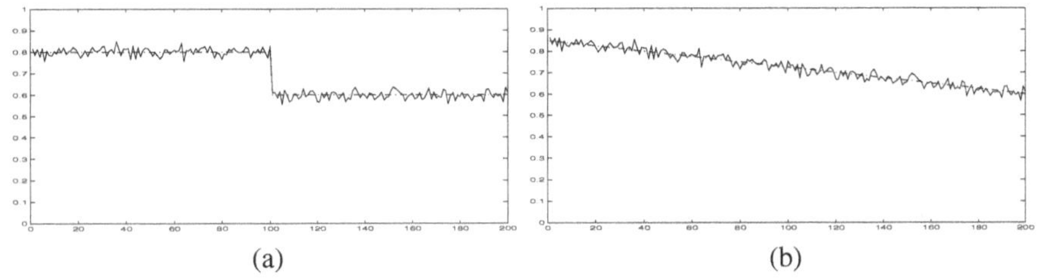

Fig. 1. Two types of significant changes. Type I: abrupt changes; Type II: gradual changes over a period of time. These are the changes we aim to detect.

Every change detection algorithm is a certain form of *hypothesis test*. To make a decision whether or not a change has occurred is to choose between two competing hypotheses: the *null hypothesis* $\mathcal{H}_0$ or the *alternative hypothesis* $\mathcal{H}_1$, corresponding to a decision of *no-change* or *change*, respectively. Suppose the ensemble has an accuracy θ_j on block j. If the conditional probability density function (*pdf*) of θ under the null hypothesis $p(\theta|\mathcal{H}_0)$ and that under the alternative hypothesis $p(\theta|\mathcal{H}_1)$ are both known, we can make a decision using a *likelihood ratio test*:

$$L(\theta_j) = \frac{p(\theta_j|\mathcal{H}_1)}{p(\theta_j|\mathcal{H}_0)} \underset{\mathcal{H}_0}{\overset{\mathcal{H}_1}{\gtrless}} \tau. \tag{1}$$

The ratio is compared against a threshold τ. $\mathcal{H}_1$ is accepted if $L(\theta_j) \geq \tau$, and rejected otherwise. τ is chosen so as to ensure an upper bound of false alarm rate.

Now consider how to detect a possible type I change. When the null hypothesis $\mathcal{H}_0$ (no change) is true, the conditional *pdf* is assumed to be a Gaussian, given by

$$p(\theta|\mathcal{H}_0) = \frac{1}{\sqrt{2\pi\sigma_0^2}} exp\left\{-\frac{(\theta-\mu_0)^2}{2\sigma_0^2}\right\}, \tag{2}$$

where the mean μ_0 and the variance σ_0^2 can be easily estimated if we just remember a sequence of most recent θ's. But if the alternative hypothesis $\mathcal{H}_1$ is true, it is not possible to estimate $P(\theta|\mathcal{H}_1)$ before sufficient information is collected. This means a long delay before the change could be detected. In order to do it in time fashion, we perform a *significance test* that uses $\mathcal{H}_0$ alone. A significant test is to assess how well the null hypothesis $\mathcal{H}_0$ explains the observed θ. Then the general likelihood ratio test in Equation 1 is reduced to:

$$p(\theta_j|\mathcal{H}_0) \underset{\mathcal{H}_1}{\overset{\mathcal{H}_0}{\gtrless}} \tau. \tag{3}$$

When the likelihood $p(\theta_j|\mathcal{H}_0) \geq \tau$, the null hypothesis is accepted; otherwise it is rejected. Significant tests are effective in capturing large, abrupt changes.

For type II changes, we perform a typical hypothesis test as follows. First, we split the history sequence of θ's into two halves. A Gaussian *pdf* can be estimated from each half, denoted as G_0 and G_1. Then a likelihood ratio test in Equation 1 is conducted.

So far we have described two techniques aiming at two types of changes. They are integrated into a two-stage method as follows. As a first step, a significant test is performed. If no change is detected, then a hypothesis test is performed as a second step. This two-stage detection method is shown to be very effective experimentally.

4 Experimental Evaluation

In this section, we first perform a controlled study on a synthetic data set, then apply the method to a real-life application.

In the synthetic data set, a sample x is a vector of three independent features $< x_i >$, $x_i \in [0, 1], i = 0, 1, 2$. Geometrically, samples are points in a 3-dimension unit cube. The class boundary is a sphere defined as: $B(x) = \sum_{i=0}^{2}(x_i - c_i)^2 - r^2 = 0$, where c is the center of the sphere, r the radius. x is labelled class 1 if $B(x) \leq 0$, class 0 otherwise. This learning task is not easy, because the feature space is continuous and the class boundary is non-linear.

We evaluate our boosting scheme extended with change detection, named as *Adaptive Boosting*, and compare it with *Weighted Bagging* and *Bagging*.

In the following experiments, we use decision trees as our base model, but the boosting technique can, in principle, be used with any other traditional learning model. The standard C4.5 algorithm is modified to generate small decision trees as base models, with the number of terminal nodes ranging from 2 to 32. Full-grown decision trees generated by C4.5 are also used for comparison, marked as *fullsize* in Figure 2-4 and Table 1-2.

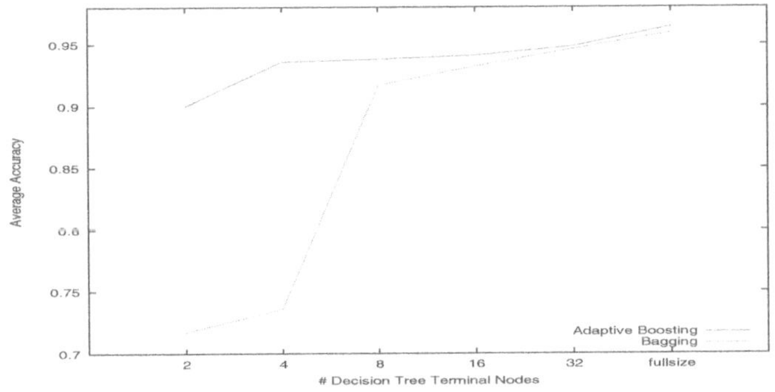

Fig. 2. Performance comparison of the adaptive boosting vs the bagging on stationary data. The weighted bagging is omitted as it performs almost the same as the bagging.

4.1 Evaluation of Boosting Scheme

The boosting scheme is first compared against two bagging ensembles on stationary data. Samples are randomly generated in the unit cube. Noise is introduced in the training data by randomly flipping the class labels with a probability of p. Each data block has n samples and there are 100 blocks in total. The testing data set contains 50k noiseless samples uniformly distributed in the unit cube. An ensemble of M classifiers is maintained. It is updated after each block and evaluated on the test data set. Performance is measured using the generalization accuracy averaged over 100 ensembles.

Figure 2 shows the generalization performance when p=*5%*, n=2k and M=*30*. Weighted bagging is omitted from the figure because it makes almost the same predictions as bagging, a not surprising result for stationary data. Figure 2 shows that the boosting scheme clearly outperforms bagging. Most importantly, boosting ensembles with very simple trees performs well. In fact, the boosted two-level trees(2 terminal nodes) have a performance comparable to bagging using the full size trees. This supports the theoretical study that boosting improves weak learners.

Higher accuracy of boosted weak learners is also observed for (1) block size n of 500, 1k, 2k and 4k, (2) ensemble size M of 10, 20, 30, 40, 50, and (3) noise level of 5%, 10% and 20%.

4.2 Learning with Gradual Shifts

Gradual concept shifts are introduced by moving the center of the class boundary between adjacent blocks. The movement is along each dimension with a step of $\pm\delta$. The value of δ controls the level of shifts from small to moderate, and the sign of δ is randomly assigned. The percentage of positive samples in these blocks ranges from 16% to 25%. Noise level p is set to be *5%, 10%* and *20%* across multiple runs.

The average accuracies are shown in Figure 3 for small shifts ($\delta = 0.01$), and in Figure 4 for moderate shifts ($\delta = 0.03$). Results of other settings are shown in Table 1. These experiments are conducted where the block size is 2k. Similar results are obtained for other block sizes. The results are summarized below:

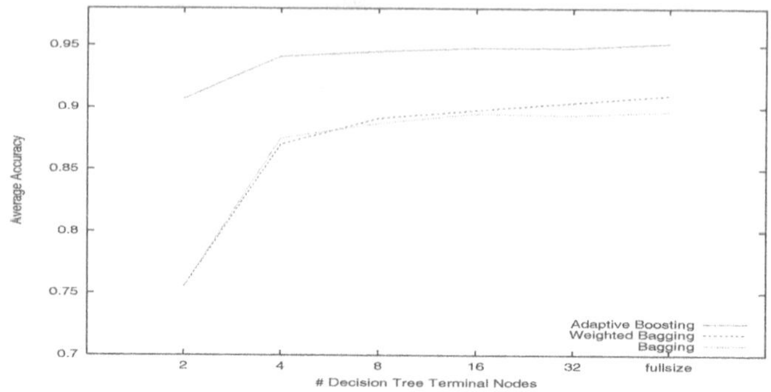

Fig. 3. Performance comparison of the three ensembles on data with small gradual concept shifts.

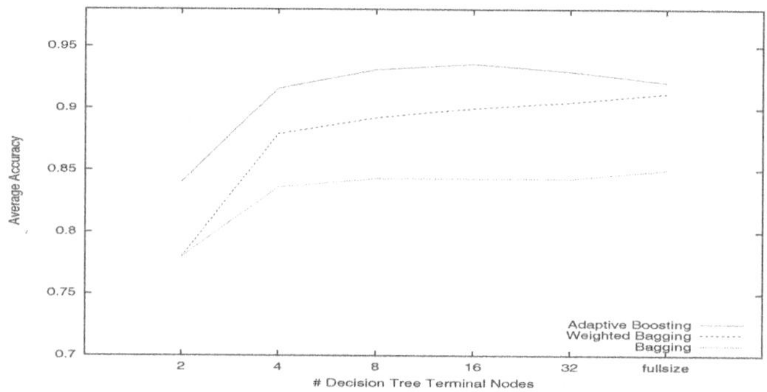

Fig. 4. Performance comparison of the ensembles on data with moderate gradual concept shifts.

Table 1. Performance comparison of the ensembles on data with varying levels of concept shifts. Top accuracies shown in bold fonts.

	$\delta = .005$				$\delta = .02$			
	2	4	8	fullsize	2	4	8	fullsize
Adaptive Boosting	**89.2%**	**93.2%**	**93.9%**	**94.9%**	**92.2%**	**94.5%**	**95.7%**	**95.8%**
Weighted Bagging	71.8%	84.2%	89.6%	91.8%	83.7%	92.0%	93.2%	94.2%
Bagging	71.8%	84.4%	90.0%	92.5%	83.7%	91.4%	92.4%	90.7%

- Adaptive boosting outperforms two bagging methods at all time, demonstrating the benefits of the change detection technique; and
- Boosting is especially effective with simple trees (terminal nodes ≤ 8), achieving a performance compatible with, or even better than, the bagging ensembles with large trees.

4.3 Learning with Abrupt Shifts

We study learning with abrupt shifts with two sets of experiments. Abrupt concept shifts are introduced every 40 blocks; three abrupt shifts occur at block 40, 80 and 120. In one

Table 2. Performance comparison of three ensembles on data with abrupt shifts or mixed shifts. Top accuracies are shown in bold fonts.

$\delta_1 = \pm 0.1$	$\delta_2 = 0.00$		$\delta_2 = \pm 0.01$	
	4	fullsize	4	fullsize
Adaptive Boosting	**93.2%**	**95.1%**	**93.1%**	**94.1%**
Weighted Bagging	86.3%	92.5%	86.6%	91.3%
Bagging	86.3%	92.7%	85.0%	88.1%

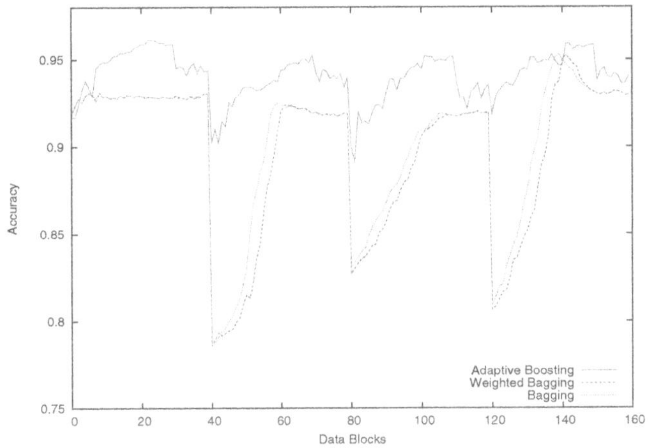

Fig. 5. Performance comparison of the three ensembles on data with abrupt shifts. Base decision trees have no more than 8 terminal nodes.

set of experiments, data stays stationary between these blocks. In the other set, small shifts are mixed between adjacent blocks. The concept drift parameters are set to be $\delta_1 = \pm 0.1$ for abrupt shifts , and $\delta_2 = \pm 0.01$ for small shifts.

Figure 5 and Figure 6 show the experiments when base decision trees have no more than 8 terminal nodes. Clearly the bagging ensembles, even with an empirical weighting scheme, are seriously impaired at changing points. Our hypothesis, that obsolete classifiers are detrimental to overall performance even if they are weighed down, are proved experimentally. Adaptive boosting ensemble, on the other hand, is able to respond promptly to abrupt changes by explicit change detection efforts. For base models of different sizes, we show some of the results in Table 2. The accuracy is averaged over 160 blocks for each run.

4.4 Experiments on Real Life Data

In this subsection we further verify our algorithm on a real life data containing 100k credit card transactions. The data has 20 features including the transaction amount, the time of the transaction, etc. The task is to predict fraudulent transactions. Detailed data description is given in [14]. The part of the data we use contains 100k transaction each

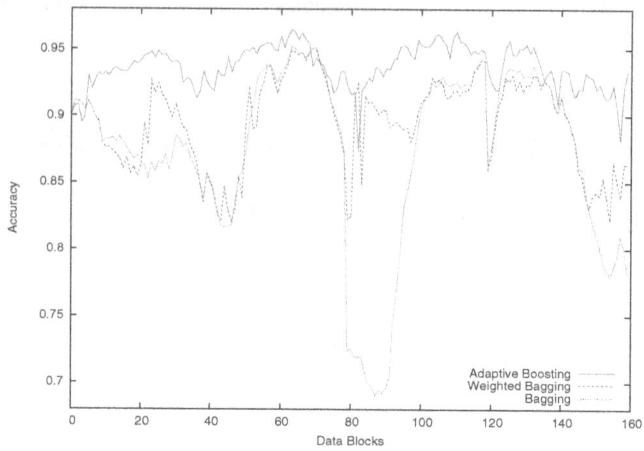

Fig. 6. Performance comparison of the three ensembles on data with both abrupt and small shifts. Base decision trees have no more than 8 terminal nodes.

with a transaction amount between \$0 and \$21. Concept drift is simulated by sorting transactions by changes by the transaction amount.

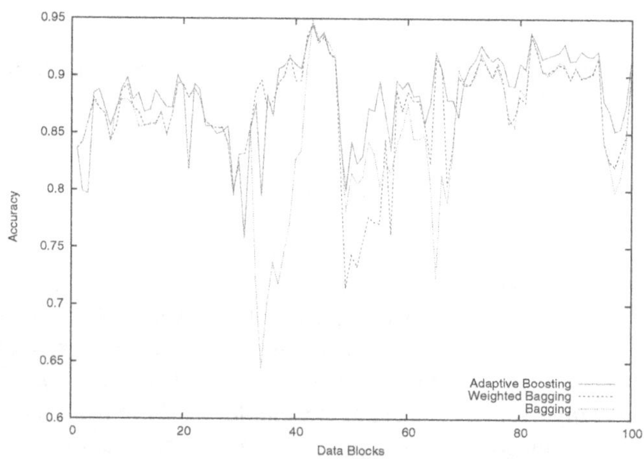

Fig. 7. Performance comparison of the three ensembles on credit card data. Concept shifts are simulated by sorting the transactions by the transaction amount.

We study the ensemble performance using varying block sizes (1k, 2k, 3k and 4k), and different base models (decision trees with terminal nodes no more than 2, 4, 8 and full-size trees). We show one experiment in Figure 7, where the block size is 1k, and the base models have at most 8 terminal nodes. The curve shows three dramatic drops in accuracy for bagging, two for weighted bagging, but only a small one for adaptive boosting. These drops occur when the transaction amount jumps. Overall, the boosting ensemble is much better than the two baggings. This is also true for the other experiments, whose details are omitted here due to space limit.

The boosting scheme is also the fastest. Moreover, the training time is almost not affected by the size of base models. This is due to the fact that the later base models tend to have very simple structures; many of them are just decision stumps (one level decision trees). On the other hand, training time of the bagging methods increases dramatically as the base decision trees grow larger. For example, when the base decision tree is full-grown, the weighted bagging takes 5 times longer to do the training and produces a tree 7 times larger on average. The comparison is conducted on a 2.26MHz Pentium 4 Processor. Details are shown in Figure 8.

To summarize, the real application experiment confirms the advantages of our boosting ensemble methods: it is fast and light, with good adaptability.

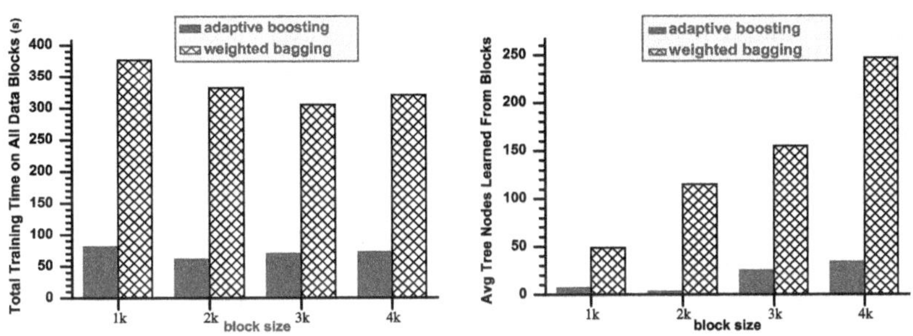

(a) (b)

Fig. 8. Comparison of the adaptive boosting and the weighted bagging, in terms of (a) building time, and (b) average decision tree size. In (a), the total amount of data is fixed for different block sizes.

5 Summary and Future Work

In this paper, we propose an adaptive boosting ensemble method that is different from previous work in two aspects: (1) We boost very simple base models to build effective ensembles with competitive accuracy; and (2) We propose a change detection technique to actively adapt to changes in the underlying concept. We compare adaptive boosting ensemble methods with two bagging ensemble-based methods through extensive experiments. Results on both synthetic and real-life data set show that our method is much faster, demands less memory, more adaptive and accurate.

The current method can be improved in several aspects. For example, our study of the trend of the underlying concept is limited to the detection of significant changes. If changes can be detected on a finer scale, new classifiers need not be built when changes are trivial, thus training time can be further saved without loss on accuracy. We also plan to study a classifier weighting scheme to improve ensemble accuracy.

References

1. L. Breiman. Bagging predictors. In *ICML*, 1996.
2. T. Dietterich. Ensemble methods in machine learning. In *Multiple Classifier Systems*, 2000.
3. C. Domeniconi and D. Gunopulos. Incremental support vector machine construction. In *ICDM*, 2001.
4. P. Domingos and G. Hulten. Mining high-speed data streams. In *ACM SIGKDD*, 2000.
5. Guozhu Dong, Jiawei Han, Laks V.S. Lakshmanan, Jian Pei, Haixun Wang, and Philip S. Yu. Online mining of changes from data streams: Research problems and preliminary results. In *ACM SIGMOD MPDS*, 2003.
6. A. Fern and R. Givan. Online ensemble learning: An empirical study. In *ICML*, 2000.
7. E. Frank, G. Holmes, R. Kirkby, and M. Hall. Racing committees for large datasets. In *Discovery Science*, 2002.
8. Y. Freund and R. Schapire. Experiments with a new boosting algorithm. In *ICML*, 1996.
9. J. Friedman, T. Hastie, and R. Tibshirani. Additive logistic regression: A statistical view of boosting. In *The Annals of Statistics, 28(2):337–407*, 1998.
10. V. Ganti, J. Gehrke, R. Ramakrishnan, and W. Loh. Mining data streams under block evolution. In *SIGKDD Explorations 3(2):1-10*, 2002.
11. G. Hulten, L. Spencer, and P. Domingos. Mining time-changing data streams. In *ACM SIGKDD*, 2001.
12. N. Oza and S. Russell. Experimental comparisons of online and batch versions of bagging and boosting. In *ACM SIGKDD*, 2001.
13. R. Schapire, Y. Freund, and P. Bartlett. Boosting the margin: A new explanation for the effectiveness of voting methods. In *ICML*, 1997.
14. S. Stolfo, W. Fan, W. Lee, A. Prodromidis, and P. Chan. Credit card fraud detection using meta-learning: Issues and initial results. In *AAAI-97 Workshop on Fraud Detection and Risk Management*, 1997.
15. W. Street and Y. Kim. A streaming ensemble algorithm (sea) for large-scale classification. In *ACM SIGKDD*, 2001.
16. H. Wang, W. Fan, P. Yu, and J. Han. Mining concept-drifting data streams using ensemble classifiers. In *ACM SIGKDD*, 2003.
17. G. Widmer and M. Kubat. Learning in the presence of concept drift and hidden contexts. In *Machine Learning*, 1996.

Compact Dual Ensembles for Active Learning

Amit Mandvikar[1], Huan Liu[1], and Hiroshi Motoda[2]

[1] Arizona State University, Arizona, USA.
[2] Osaka University, Osaka, Japan.
{huanliu,amitm}@asu.edu and motoda@sanken.osaka-u.ac.jp

Abstract. Generic ensemble methods can achieve excellent learning performance, but are not good candidates for active learning because of their different design purposes. We investigate how to use diversity of the member classifiers of an ensemble for efficient active learning. We empirically show, using benchmark data sets, that (1) to achieve a good (stable) ensemble, the number of classifiers needed in the ensemble varies for different data sets; (2) feature selection can be applied for classifier selection from ensembles to construct compact ensembles with high performance. Benchmark data sets and a real-world application are used to demonstrate the effectiveness of the proposed approach.

1 Introduction

Active learning is a framework in which the learner has the freedom to select which data points are added to its training set [11]. An active learner may begin with a small number of labeled instances, carefully select a few additional instances for which it requests labels, learn from the result of those requests, and then using its newly-gained knowledge, carefully choose which instances to request next. More often than not, data in forms of text (including emails), image, multi-media are unlabeled, yet many supervised learning tasks need to be performed [2,10] in real-world applications. Active learning can significantly decrease the number of required labeled instances, thus greatly reduce expert involvement. Ensemble methods are learning algorithms that construct a set of classifiers and then classify new instances by taking a weighted or unweighted vote of their predictions. An ensemble often has smaller expected loss or error rate than any of the n individual (member) classifiers. A good ensemble is one whose members are both *accurate* and *diverse* [4]. This work explores the relationship between the two learning frameworks, attempts to take advantage of the learning performance of ensemble methods for active learning in a real-world application, and studies how to construct ensembles for effective active learning.

2 Our Approach

2.1 Ensembles and Active Learning

Active learning can be very useful where there are limited resources for labeling data, and obtaining these labels is time-consuming or difficult [11]. There

H. Dai, R. Srikant, and C. Zhang (Eds.): PAKDD 2004, LNAI 3056, pp. 293–297, 2004.

exist widely used active learning methods. Some examples are: Uncertainty sampling [7] selects the instance on which the current learner has lowest certainty; Pool-based sampling [9] selects the best instances from the entire pool of unlabeled instances; and Query-by-Committee [6,12] selects instances that have high classification variance themselves.

Constructing good ensembles of classifiers has been one of the most active areas of research in supervised learning [4]. The main discovery is that ensembles are often much more accurate than the member classifiers that make them up. A necessary and sufficient condition for an ensemble to be more accurate than any of its members is that the member classifiers are accurate and diverse. Two classifiers are diverse if they make different (or uncorrelated) errors on new data points. Many methods for constructing ensembles have been developed such as Bagging [3] and Boosting [5]. We consider Bagging in this work as it is the most straightforward way of manipulating the training data to form ensembles [4].

Disagreement or diversity of classifiers are used for different purposes for the two learning frameworks: in generic ensemble learning, diversity of classifiers is used to ensure high accuracy by voting; in active learning, disagreement of classifiers is used to identify critical instances for labeling. In order for active learning to work effectively, we need a *small* number of *highly* accurate classifiers so that they seldom disagree with each other. Since ensemble methods have shown their *robustness* in producing *highly accurate* classifiers, we have investigated the use of class-specific ensembles (dual ensembles), and shown their effectiveness in our previous work [8]. Next, we empirically investigate whether it is necessary to find compact dual ensembles and then we present a method to find them while maintaining good performance.

2.2 Observations from Experiments on Benchmark Data Sets

Ensemble's goodness can be measured by accuracy and diversity. Let $\hat{Y}(x) = \hat{y}_1(x), ...\hat{y}_n(x)$ be the set of the predictions made by member classifiers $C_1, ..., C_n$ of ensemble E on instance $\langle x, y \rangle$ where x is input, and y is the true class. The **ensemble prediction** of a uniform voting ensemble for input x under loss function l is, $\hat{y}(x) = argmin_{y \in Y} E_{c \in C}[l(\hat{y}_c(x), y]$. The **loss** of an ensemble on instance $\langle x, y \rangle$ under loss function l is given by $L(\langle x, y \rangle) = l(\hat{y}(x), y)$. The **diversity** of an ensemble on input x under loss function l is given by $\overline{D} = E_{c \in C}[l(\hat{y}_c(x), \hat{y}(x))]$. The error rate for a data set with N instances can be calculated as $e = \frac{1}{N} \sum_1^N L_i$, where L_i is the loss for instance x_i. **Accuracy** of ensemble E is $1-e$. **Diversity** is the expected loss incurred by the predictions of the member classifiers relative to the ensemble prediction. Commonly used loss functions include square loss, absolute loss, and zero-one loss. We use zero-one loss in this work.

The purpose of these experiments is to observe how diversity and error rate change as ensemble size increases. We use benchmark data sets from the UCI repository [1] in these experiments. We use Weka [13] implementation of Bagging [3] as the ensemble generation method and J4.8 (without pruning) as the base learning algorithm. For each data set, we run Bagging with increasing ensemble sizes from 5 to 151 and record each ensemble's error rate e and diversity

D. We run 10-fold cross validation and calculate the average values, $\overline{e}$ and $\overline{D}$. We observed that as the ensemble sizes increase, diversity values increase and approach to the maximum, and error rates decrease and become stable. The results show that smaller ensembles (with 30-70 classifiers) can achieve accuracy and diversity values similar to those of larger ensembles. We will now show a procedure for selecting compact dual ensembles from larger ensembles.

2.3 Selecting Compact Dual Ensembles via Feature Selection

The experiments with the benchmark data sets show that there exist smaller ensembles that can have similar accuracy and diversity as that of large ensembles. We need to select classifiers with these two criteria. We build our initial ensemble, E_{max} by setting $max = 100$ member classifiers. We now need to *efficiently find* a compact ensemble E_M (with M classifiers) that can have similar error rate and diversity of E_{max}. We use all the learned classifiers (C_k) to generate predictions for instances $\langle x_i, y_i \rangle : \hat{y}_i^k = C_k(x_i)$. The resulting dataset consists of instances of the form $((\hat{y}_i^1, ..., \hat{y}_i^K), y_i)$. After this data set is constructed, the problem of selecting member classifiers becomes one of feature selection. Here features actually represent member classifiers, therefore we also need to consider this special nature for the feature selection algorithm.

DualE: selecting compact dual ensembles

input: Tr: Training data, $FSet$: All classifiers in E_{max}, N: max;
output: E_1: Optimal ensemble for class=1, E_0: Optimal ensemble for class=0;

01 Generate N classifiers from T_r with Bagging;
02 $Tr_1 \leftarrow$ Instances(Tr) with class label= 1;
03 $Tr_0 \leftarrow$ Instances(Tr) with class label= 0;
04 Calculate diversity, D_0 and error rate, e_0 for E_{max} on Tr_1;
05 $U \leftarrow N$; $L \leftarrow 0$; $M \leftarrow \frac{U-L}{2}$;
06 **while** $|U - M| > 1$
07   Pick M classifiers from $FSet$ to form E';
08   Calculate diversity, D' and error rate, e' for E' on Tr_1;
09   **if** ($\frac{D_0 - D'}{D_0} < 1\%$) and ($\frac{e' - e_0}{e_0} < 1\%$)
10         $U \leftarrow M$; $M \leftarrow M$ - $\frac{M-L}{2}$;
11   **else**
12         $L \leftarrow M$; $M \leftarrow M + \frac{U-M}{2}$;
13 $E_1 \leftarrow E'$;
14 Repeat steps 5 to 12 for Tr_0; $E_0 \leftarrow E'$;

Fig. 1. Algorithm for Selecting Dual Ensembles

We design an algorithm **DualE** that takes $O(\log max)$ to determine M where max is the size of the starting ensemble (e.g., 100). In words, we test an ensemble E_M with size M which is between upper and lower bounds U and L (initialized as max and 0 respectively). If E_M's performance is similar to that of E_{max}, we set $U = M$ and $M = (L+M)/2$; otherwise, set $L = M$ and $M = (M+U)/2$. The details are given in Fig. 1. Ensemble performance is defined by error rate e and diversity D. The diversity values of the two ensembles are similar if $\frac{D_0 - D'}{D_0} \leq p$ where p is user defined ($0 < p < 1$) and D_0 is the reference ensemble's (E_{max}'s) diversity. Similarly, error for the two ensembles is similar if $\frac{e' - e_0}{e_0} \leq p$ where e_0 is the reference ensemble's error rate.

3 Experiments

Two sets of experiments are conducted with **DualE**: one is on a image data set and other on a benchmark data set (breast). The purpose is to examine if the compact dual ensembles selected by **DualE** can work as well as the entire ensemble, E_{max}. When dual ensembles are used, it is possible that they disagree. These instances are called *uncertain* instances (UC). In context of active learning, the uncertain instances will be labeled by an expert. Prediction of E_{max} is by majority and there is no disagreement. So for E_{max} only the accuracy is reported.

The image mining problem, that we study here, is to classify Egeria Densa in aerial images. To automate Egeria classification, we ask experts to label images, but want to minimize the task. Active learning is employed to reduce this expert involvement. The idea is to let experts label some instances, learn from these labeled instances, and then apply the active learner to unseen images. We have 17 images with 5329 instances, represented by 13 attributes of color, texture and edge. One image is used for training and the rest for testing. We first train an initial ensemble E_{max} with $max = 100$ on the training image, then obtain accuracy of E_{max} for the 17 testing images. Dual E_s are the ensembles selected by using **DualE**. Table 3 clearly shows that accuracy of E_s is similar to that of E_{max}. The number of uncertain regions is also relatively small. This demonstrates the effectiveness of using the dual ensembles, E_s to reduce expert involvement for labeling.

Table 1. E_s vs. E_{max} for Image data

Image	Dual E_s		E_{max}	
	Acc%	#UC	Acc%	Acc Gain%
1	81.91	1	81.90	-0.0122
2	90.00	0	90.00	0.0000
3	78.28	38	79.28	1.2775
4	87.09	34	86.47	-0.7119
5	79.41	0	79.73	0.4029
6	84.51	88	84.77	0.3076
7	85.00	3	85.41	0.4823
8	85.95	18	86.6	0.7562
9	71.46	0	72.32	1.2035
10	91.08	2	90.8	-0.3074
11	89.15	31	88.82	-0.3702
12	75.91	0	76.02	0.1449
13	66.84	0	67.38	0.8079
14	73.06	49	73.73	0.9170
15	83.1	1	83.24	0.1684
16	76.57	14	76.82	0.3265
17	87.67	31	88.42	0.8555
Average	81.58	18.24	81.86	0.3676

For the breast dataset, we design a new 3-fold cross validation scheme, which uses 1-fold for training, the remaining 2 folds for testing. This is repeated for all the 3 folds of the training data. The results for the breast data set are shown in Table 2. We also randomly select member classifiers to form random dual ensembles, Dual E_r. We do so 10 times and report the average accuracy and number of uncertain instances. In Table 2, Dual E_s are the selected ensembles (using **DualE**), and E_{max} is the initial ensemble. Accuracy gains for Dual E_r and E_{max} (and UC Incr for Dual E_r) against E_s are reported. Comparing dual

Table 2. Comparing E_s with E_r and E_{max} for Breast data

	Dual E_s		Dual E_r				E_{max}	
	Acc%	#UC	Acc%	#UC	Acc Gain%	UC Incr%	Acc%	Acc Gain%
Fold 1	95.9227	3	94.0773	13.6	-1.9238	353.33	96.1373	0.2237
Fold 2	97.2103	5	94.4206	15.2	-2.8698	204.00	96.9957	-0.2208
Fold 3	94.8498	12	93.5193	8.7	-1.4027	-27.50	94.4206	-0.4525
Average	95.9943	6.67	94.0057	12.5	-2.0655	176.61	95.8512	-0.1498

E_s and dual E_r, dual E_r exhibit lower accuracy and more uncertain instances. Comparing dual E_s and E_{max}, we observe no significant change in accuracy. This is consistent with what we want (maintain both accuracy and diversity).

4 Conclusions

In this work, we point out that (1) generic ensemble methods are not suitable for active learning (2) dual ensembles are very good for active learning if we can build compact dual ensembles. Our empirical study suggests that there exist such compact ensembles. We propose **DualE** that can find compact ensembles with good performance via feature selection. Experiments on a benchmark and an image data set exhibit the effectiveness of dual ensembles for active learning.

References

1. C.L. Blake and C.J. Merz. UCI repository of machine learning databases, 1998. http://www.ics.uci.edu/~mlearn/MLRepository.html.
2. A. Blum and T. Mitchell. Combining labeled and unlabeled data with co-training. In *COLT: Proc. of Workshop on Computational Learning Theory*, Morgan Kaufmann Publishers, 1998.
3. L. Breiman. Bagging predictors. Machine Learning, 24:123–140, 1996.
4. T.G. Dietterich. Ensemble methods in machine learning. In *Proc. of Intl. Workshop on Multiple Classifier Systems*, pp. 1–15. Springer-Verlag, 2000.
5. Y. Freund and R.E. Schapire. Experiments with a new boosting algorithm. In *Proc. of Intl. Conf. on Machine Learning*, pp. 148–156. Morgan Kaufmann, 1996.
6. Y. Freund, H. Seung, E. Shamir, and N. Tishby. Selective sampling using the query by committee algorithm. *Machine Learning*, 28:133–168, 1997.
7. D. Lewis and W. Gale. A sequential algorithm for training text classifiers. In *Proc. of ACM-SIGR Conference on Research and Development in Information Retrieval*, pp. 3–12, 1994.
8. A. Mandvikar and H Liu. Class-Specific Ensembles for Active Learning in Digital Imagery. In *Proc. of SIAM Intl. Conf. on Data Mining*, in press, 2004.
9. A. McCallum and K. Nigam. Employing EM in pool-based active learning for text classification. In *Proc. of Intl. Conf. on Machine Learning*, pp. 350–358, 1998.
10. K. Nigam, A. K. Mccallum, S. Thrun, and T. Mitchell. Text classification from labeled and unlabeled documents usingEM. *Machine Learning*, 39:103–134, 2000.
11. N. Roy and A. McCallum. Toward optimal active learning through sampling estimation of error reduction. In *Proc. of Intl. Conf. On Machine Learning*, 2001.
12. H.S. Seung, M. Opper, and H. Sompolinsky. Query by committee. In *Proc. of Workshop on Computational Learning Theory*, pp. 287–294, 1992. ACM Press.
13. I.H. Witten and E. Frank. *Data Mining – Practical Machine Learning Tools and Techniques with JAVA Implementations.* Morgan Kaufmann Publishers, 2000.

On the Size of Training Set and the Benefit from Ensemble

Zhi-Hua Zhou[1], Dan Wei[1], Gang Li[2], and Honghua Dai[2]

[1] National Laboratory for Novel Software Technology
Nanjing University, Nanjing 210093, China
zhouzh@nju.edu.cn dwei@ai.nju.edu.cn
[2] School of Information Technology
Deakin University, Burwood, Vic3125, Australia
{gangli, hdai}@deakin.edu.au

Abstract. In this paper, the impact of the size of the training set on the benefit from ensemble, i.e. the gains obtained by employing ensemble learning paradigms, is empirically studied. Experiments on Bagged/Boosted J4.8 decision trees with/without pruning show that enlarging the training set tends to improve the benefit from Boosting but does not significantly impact the benefit from Bagging. This phenomenon is then explained from the view of bias-variance reduction. Moreover, it is shown that even for Boosting, the benefit does not always increase consistently along with the increase of the training set size since single learners sometimes may learn relatively more from additional training data that are randomly provided than ensembles do. Furthermore, it is observed that the benefit from ensemble of unpruned decision trees is usually bigger than that from ensemble of pruned decision trees. This phenomenon is then explained from the view of error-ambiguity balance.

1 Introduction

Ensemble learning paradigms train a collection of learners to solve a problem. Since the generalization ability of an ensemble is usually better than that of a single learner, one of the most active areas of research in supervised learning has been to study paradigms for constructing good ensembles [5].

This paper does not attempt to propose any new ensemble algorithm. Instead, it tries to explore how the change of the training set size impacts the benefit from ensemble, i.e. the gains obtained by employing ensemble learning paradigms. Having an insight into this may be helpful to better exerting the potential of ensemble learning paradigms. This goal is pursued in this paper with an empirical study on ensembles of pruned or unpruned J4.8 decision trees [9] generated by two popular ensemble algorithms, i.e. Bagging [3] and Boosting (In fact, *Boosting* is a family of ensemble algorithms, but here the term is used to refer the most famous member of this family, i.e. AdaBoost [6]). Experimental results show that enlarging training set does not necessarily enlarges the benefit from ensemble. Moreover, interesting issues on the benefit from ensemble, which is related to the

H. Dai, R. Srikant, and C. Zhang (Eds.): PAKDD 2004, LNAI 3056, pp. 298–307, 2004.

characteristics of Bagging and Boosting and the effect of decision tree pruning, have been disclosed and discussed.

The rest of this paper is organized as follows. Section 2 describes the empirical study. Section 3 analyzes the experimental results. Section 4 summarizes the observations and derivations.

2 The Empirical Study

Twelve data sets with 2,000 to 7,200 examples, 10 to 36 attributes, and 2 to 10 classes from the UCI Machine Learning Repository [2] are used in the empirical study. Information on the experimental data sets are tabulated in Table 1.

Table 1. Experimental data sets

Data set	Size	Attribute		Class
		Categorical	Continuous	
allbp	2,800	22	7	3
ann	7,200	15	6	3
block	5,473	0	10	5
hypothyroid	3,772	22	7	2
kr-vs-kp	3,196	36	0	2
led7	2,000	7	0	10
led24	2,000	24	0	10
sat	6,435	0	36	6
segment	2,310	0	19	7
sick	3,772	22	7	2
sick-euthyroid	3,156	22	7	2
waveform	5,000	0	21	3

Each original data set is partitioned into ten subsets with similar distributions. At the first time, only one subset is used; at the second time, two subsets are used; and so on. The earlier generated data sets are proper subsets of the later ones. In this way, the increase of the size of the data set is simulated.

On each generated data set, 10-fold cross validation is performed. In each fold, Bagging and Boosting are respectively employed to train an ensemble comprising 20 pruned or unpruned J4.8 decision trees. For comparison, a single J4.8 decision tree is also trained from the training set of the ensembles. The whole process is repeated for ten times, and the average error ratios of the ensembles generated by Bagging and Boosting against the single decision trees are recorded, as shown in Tables 2 to 5, respectively. The predictive error rates of the single decision trees are shown in Tables 6 and 7. In these tables the first row indicates the percentage of data in the original data sets that are used, and the numbers following '±' are the standard deviations.

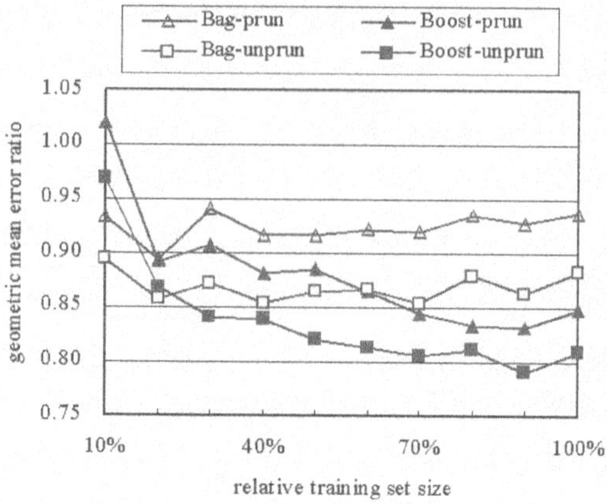

Fig. 1. The geometrical mean error ratio of Bagged/Boosted J4.8 decision trees with/without pruning against single J4.8 decision trees with/without pruning

Here the error ratio is defined as the result of dividing the predictive error rate of an ensemble by that of a single decision tree. A smaller error ratio means relatively bigger benefit from ensemble, while a bigger error ratio means relative smaller benefit from ensemble. If an ensemble is worse than a single decision tree, then its error ratio is bigger than 1.0.

In order to exhibit the overall tendency, the geometric mean error ratio, i.e. average ratio across all data sets, are also provided in Tables 2 to 5, which is then explicitly depicted in Fig. 1.

3 Discussion

3.1 Bagging and Boosting

An interesting phenomenon exposed by Fig. 1 and Tables 2 to 5 is that the benefit from Bagging and Boosting exhibit quite different behaviors on the change of the training set size. In detail, although there are some fluctuations, the benefit from Bagging remains relatively unvaried while that from Boosting tends to be enlarged when the training set size increases. In order to explain this phenomenon, it may be helpful to consider the different characteristics of Bagging and Boosting from the view of bias-variance reduction.

Given a learning target and the size of training set, the expected error of a learning algorithm can be broken into the sum of three non-negative quantities, i.e. the intrinsic noise, the bias, and the variance [7]. The intrinsic noise is a lower bound on the expected error of any learning algorithm on the target. The bias measures how closely the average estimate of the learning algorithm is able to approximate the target. The variance measures how much the estimate of the learning algorithm fluctuates for the different training sets of the same size.

Table 2. Error ratios of Bagged J4.8 decision trees against single J4.8 decision trees. All the trees are pruned.

Data set	10%	20%	30%	40%	50%
allbp	.909±.193	.914±.111	.952±.070	1.01±.125	.961±.149
ann	1.15±.203	.899±.135	1.03±.180	.952±.209	.931±.086
block	.911±.186	.908±.061	.906±.119	.886±.067	.885±.046
hypothyroid	1.10±.236	1.01±.223	1.14±.243	1.04±.135	1.19±.247
kr-vs-kp	.938±.104	.897±.180	1.06±.376	.991±.138	.892±.119
led7	.976±.054	.976±.049	.978±.017	.986±.025	.988±.019
led24	.902±.063	.920±.042	.941±.048	.961±.027	.958±.035
sat	.786±.052	.751±.031	.736±.043	.722±.040	.729±.021
segment	.800±.142	.898±.128	.913±.106	.851±.105	.816±.102
sick	1.22±.352	.938±.201	.999±.121	.975±.150	1.08±.143
sick-euthyroid	.826±.257	.952±.212	.955±.159	.945±.139	.911±.080
waveform	.672±.138	.663±.068	.664±.084	.671±.073	.655±.056
geometric-mean	.933	.894	.940	.916	.916
Data set	**60%**	**70%**	**80%**	**90%**	**100%**
allbp	.958±.078	1.02±.085	.940±.080	1.01±.116	1.01±.038
ann	1.08±.170	.985±.152	1.04±.140	1.09±.161	1.15±.162
block	.876±.043	.907±.057	.873±.055	.864±.044	.862±.048
hypothyroid	1.14±.255	.908±.132	1.04±.194	.955±.076	.973±.089
kr-vs-kp	.810±.169	.957±.136	1.02±.142	1.02±.130	1.07±.082
led7	.996±.014	1.01±.014	1.00±.012	1.01±.011	1.01±.013
led24	.960±.036	.967±.023	.969±.028	.972±.028	.979±.021
sat	.703±.029	.719±.027	.724±.014	.715±.019	.704±.026
segment	.849±.121	.859±.085	.845±.086	.826±.095	.857±.092
sick	1.07±.096	1.10±.250	1.14±.365	.978±.104	.954±.081
sick-euthyroid	.950±.129	.956±.119	.941±.085	.992±.068	1.00±.108
waveform	.670±.057	.650±.037	.690±.051	.699±.028	.679±.030
geometric-mean	.922	.920	.935	.928	.937

Since the intrinsic noise is an inherent property of the given target, usually only the bias and variance are concerned.

Previous research shows that Bagging works mainly through reducing the variance [1][4]. It is evident that such a reduction is realized by utilizing bootstrap sampling to capture the variance among the possible training sets under the given size and then smoothing the variance through combining the trained component learners. Suppose the original data set is D, a new data set D' is bootstrap sampled from D, and the size of D' is the same as that of D, i.e. $|D|$. Then, the size of the shared part between D and D' can be estimated according to Eq. 1, which shows that the average overlap ratio is a constant, roughly 63.2%.

$$\left(1 - (1 - 1/|D|)^{|D|}\right) |D| \approx (1 - 0.368)\, |D| = 0.632\, |D| \tag{1}$$

Table 3. Error ratios of Boosted J4.8 decision trees against single J4.8 decision trees. All the trees are pruned.

Data set	10%	20%	30%	40%	50%
allbp	1.00±.190	.930±.202	.895±.096	1.02±.163	.883±.157
ann	1.26±.349	1.01±.238	1.07±.138	.926±.116	.924±.145
block	.858±.101	.934±.110	.963±.149	.951±.073	1.01±.044
hypothyroid	1.63±.506	.983±.197	1.07±.236	.924±.394	1.12±.394
kr-vs-kp	.707±.244	.669±.151	.823±.261	.827±.179	.751±.113
led7	.998±.008	1.00±.000	1.00±.000	1.00±.000	1.00±.000
led24	1.06±.104	1.09±.079	1.16±.069	1.15±.054	1.17±.059
sat	.717±.041	.679±.047	.653±.040	.658±.038	.657±.025
segment	.783±.138	.702±.226	.654±.084	.605±.139	.559±.141
sick	1.47±.749	1.05±.251	1.02±.130	.898±.113	.956±.071
sick-euthyroid	1.09±.329	1.02±.272	.971±.221	.980±.209	.969±.162
waveform	.675±.165	.644±.075	.604±.057	.628±.061	.626±.056
geometric-mean	1.02	.893	.907	.881	.885
Data set	60%	70%	80%	90%	100%
allbp	.896±.077	.931±.092	.817±.059	.844±.058	.854±.041
ann	1.04±.135	.981±.089	.894±.073	.977±.118	1.09±.146
block	.986±.050	.979±.061	.986±.074	.997±.039	.968±.033
hypothyroid	.984±.314	.797±.214	.886±.255	.720±.184	.771±.130
kr-vs-kp	.549±.220	.522±.160	.559±.153	.623±.243	.652±.177
led7	1.00±.000	1.00±.000	1.00±.000	1.00±.000	1.00±.000
led24	1.16±.033	1.19±.042	1.20±.030	1.18±.028	1.18±.032
sat	.621±.031	.656±.037	.645±.028	.641±.015	.636±.018
segment	.589±.149	.519±.086	.497±.079	.491±.110	.523±.092
sick	.967±.149	.934±.179	.896±.167	.838±.165	.820±.078
sick-euthyroid	.952±.224	.990±.193	.975±.170	1.02±.217	1.02±.229
waveform	.624±.047	.629±.043	.655±.042	.650±.032	.669±.029
geometric-mean	.864	.844	.834	.832	.849

This means that the variance among the possible samples with the same size that could be captured by a given number of trials of bootstrap sampling might not significantly change when the training set size changes. Therefore, when the training set size increases, the improvement of the ensemble owes much to the improvement of the component learners caused by the additional training data instead of the capturing of more variance through bootstrap sampling. Since the single learner also improves on the additional training data in the same way as the component learners in the ensemble do, the benefit from Bagging might not be significantly changed when the training set is enlarged.

As for Boosting, previous research shows that it works through reducing both the bias and variance but primarily through reducing the bias [1][4]. It is evident that such a reduction on bias is realized mainly by utilizing adaptive sampling,

Table 4. Error ratios of Bagged J4.8 decision trees against single J4.8 decision trees. All the trees are unpruned.

Data set	10%	20%	30%	40%	50%
allbp	1.03±.047	.733±.029	.965±.005	.890±.129	.898±.107
ann	.995±.111	.980±.109	.985±.146	1.00±.178	.879±.153
block	.856±.205	.902±.058	.862±.120	.847±.059	.830±.049
hypothyroid	1.08±.149	.867±.283	1.02±.236	.864±.078	1.07±.292
kr-vs-kp	.865±.151	.998±.325	.891±.200	.899±.141	.887±.252
led7	.979±.041	.968±.045	.978±.021	.999±.026	.988±.019
led24	.823±.082	.803±.046	.803±.043	.827±.030	.806±.043
sat	.750±.061	.735±.036	.716±.051	.693±.039	.702±.017
segment	.795±.149	.871±.127	.898±.121	.812±.085	.810±.110
sick	1.08±.337	.927±.188	.866±.091	.883±.148	1.00±.112
sick-euthyroid	.818±.203	.848±.100	.827±.079	.858±.113	.836±.074
waveform	.669±.134	.664±.067	.668±.090	.665±.070	.656±.062
geometric-mean	.895	.858	.873	.853	.864
Data set	**60%**	**70%**	**80%**	**90%**	**100%**
allbp	.906±.089	.885±.035	.876±.077	.858±.030	.901±.025
ann	.991±.149	1.01±.095	.953±.111	1.11±.245	1.23±.273
block	.836±.092	.841±.028	.872±.065	.844±.030	.804±.044
hypothyroid	1.11±.453	.788±.108	.993±.355	.870±.160	.858±.066
kr-vs-kp	.864±.158	.869±.144	1.02±.181	.893±.133	1.01±.135
led7	.997±.015	1.01±.012	.996±.016	1.01±.009	1.01±.013
led24	.796±.029	.785±.022	.800±.018	.799±.017	.807±.013
sat	.680±.026	.687±.027	.699±.014	.694±.023	.685±.023
segment	.792±.138	.822±.105	.808±.071	.806±.078	.806±.096
sick	.976±.145	1.07±.170	1.03±.242	.940±.121	.909±.064
sick-euthyroid	.795±.097	.824±.091	.832±.077	.843±.062	.924±.062
waveform	.663±.051	.647±.028	.682±.053	.693±.025	.669±.028
geometric-mean	.867	.853	.880	.863	.884

i.e. adaptively changing the data distribution to enable a component learner focus on hard examples for its predecessor. When the training set is enlarged, the adaptive sampling process becomes more effective since more hard examples for a component learner could be effectively identified and then passed on to the successive learner, some of which might not be identified when the training set is a relatively smaller one. Therefore, the reduction on bias may be enhanced along with the increase of the size of training set, which leads to that the benefit from Boosting tends to be enlarged.

It is worth noting that Fig. 1 and Tables 2 to 5 also show that the benefit from ensemble, even for Boosting, does not always increase consistently when the training set size increases. This is not difficult to understand because the chances for an ensemble to get improved from the additional training data that

Table 5. Error ratios of Boosted J4.8 decision trees against single J4.8 decision trees. All the trees are unpruned.

Data set	10%	20%	30%	40%	50%
allbp	.953±.187	.746±.090	.984±.128	.883±.175	.851±.131
ann	1.47±.300	1.07±.123	.939±.146	1.06±.366	.832±.168
block	.860±.210	.898±.122	.911±.075	.890±.032	.943±.048
hypothyroid	1.44±.453	.949±.345	.951±.350	.753±.316	1.00±.262
kr-vs-kp	.646±.243	.757±.236	.762±.281	.865±.315	.656±.197
led7	1.00±.000	1.00±.000	1.00±.000	1.00±.000	1.00±.000
led24	.968±.107	.974±.061	.956±.052	.980±.022	.965±.036
sat	.691±.071	.672±.070	.640±.038	.631±.029	.624±.015
segment	.723±.087	.646±.103	.549±.064	.594±.097	.539±.065
sick	1.23±.886	1.11±.259	.917±.130	.876±.123	.960±.152
sick-euthyroid	.985±.242	.948±.171	.860±.162	.906±.097	.877±.095
waveform	.665±.114	.643±.101	.621±.066	.631±.063	.600±.066
geometric-mean	.969	.868	.841	.839	.821
Data set	60%	70%	80%	90%	100%
allbp	.875±.138	.875±.054	.751±.105	.755±.069	.727±.050
ann	.989±.221	1.00±.108	.948±.159	1.02±.150	1.17±.103
block	.928±.048	.938±.075	.975±.048	.930±.011	.923±.051
hypothyroid	.869±.366	.742±.291	.811±.118	.696±.121	.726±.103
kr-vs-kp	.673±.282	.585±.075	.719±.175	.619±.066	.613±.143
led7	1.00±.000	1.00±.000	1.00±.000	1.00±.000	1.00±.000
led24	.953±.026	.958±.020	.976±.024	.959±.020	.972±.016
sat	.614±.038	.615±.027	.611±.012	.620±.020	.624±.017
segment	.541±.064	.506±.051	.501±.047	.497±.081	.511±.057
sick	.892±.103	.936±.147	.931±.209	.860±.122	.864±.074
sick-euthyroid	.823±.077	.870±.079	.870±.089	.895±.064	.936±.078
waveform	.601±.038	.638±.028	.637±.045	.636±.050	.659±.040
geometric-mean	.813	.805	.811	.791	.810

are randomly provided might be less than that of a single learner since the ensemble is usually far stronger than the single learner. It is analogous to the fact that improving a poor learner is more easier than improving a strong learner. Therefore, the benefit of ensemble shrinks if the improvement of the ensemble is not so big as that of the single learner on additional training data. However, if the additional training data have been adequately selected so that most of them can benefit the ensemble, then both the ensemble and the single learner could be significantly improved while the benefit from ensemble won't be decreased.

3.2 Pruned and Unpruned Trees

Another interesting phenomenon exposed by Fig. 1 and Tables 2 to 5 is that the benefit from ensemble comprising unpruned decision trees is always bigger than

Table 6. Predictive error rate (%) of pruned single C4.5 decision trees.

Data set	10%	20%	30%	40%	50%
allbp	4.40±1.36	4.00±0.58	3.57±0.52	3.25±0.55	3.35±0.38
ann	0.98±0.41	0.73±0.25	0.52±0.16	0.49±0.14	0.45±0.08
block	4.39±1.00	3.97±0.27	3.59±0.53	3.60±0.32	3.19±0.29
hypothyroid	1.45±0.57	1.24±0.39	0.97±0.34	0.69±0.14	0.57±0.16
kr-vs-kp	4.62±1.37	2.75±0.58	1.75±0.57	1.36±0.40	1.14±0.26
led7	35.21±2.92	31.83±3.07	30.08±1.87	28.81±1.67	28.01±1.48
led24	36.19±6.13	33.17±3.35	30.67±2.83	30.93±1.49	29.93±1.41
sat	19.22±1.68	17.02±1.04	15.87±0.59	15.33±0.52	14.63±0.54
segment	9.16±3.07	6.94±2.00	5.85±1.41	5.59±1.37	4.94±1.21
sick	2.21±0.97	2.27±0.97	1.86±0.36	1.79±0.27	1.63±0.33
sick-euthyroid	4.07±1.97	3.39±1.68	3.20±1.56	2.95±1.00	2.67±0.80
waveform	11.27±1.87	10.43±1.59	10.44±1.00	9.70±0.93	9.96±0.58
Data set	60%	70%	80%	90%	100%
allbp	3.10±0.32	2.88±0.28	2.94±0.30	2.87±0.29	2.76±0.17
ann	0.39±0.07	0.41±0.06	0.39±0.05	0.33±0.03	0.30±0.03
block	3.20±0.18	3.11±0.21	3.15±0.21	3.08±0.15	3.03±0.11
hypothyroid	0.53±0.16	0.54±0.12	0.53±0.11	0.49±0.06	0.45±0.04
kr-vs-kp	1.01±0.20	0.93±0.16	0.80±0.13	0.72±0.13	0.57±0.08
led7	27.82±1.41	27.39±0.68	27.02±0.56	26.90±0.50	26.73±0.27
led24	29.02±1.28	28.56±1.03	28.30±0.72	28.20±0.34	27.78±0.54
sat	14.83±0.59	14.11±0.46	14.11±0.53	13.70±0.49	13.54±0.30
segment	4.11±1.06	3.82±0.86	3.46±0.82	3.21±0.63	2.93±0.59
sick	1.50±0.30	1.42±0.28	1.33±0.35	1.39±0.25	1.38±0.29
sick-euthyroid	2.69±1.08	2.51±0.62	2.42±0.52	2.23±0.49	2.21±0.49
waveform	9.88±0.48	9.75±0.56	9.38±0.45	9.13±0.46	8.95±0.31

that comprising pruned decision trees, despite whether Bagging or Boosting is employed. In order to explain this phenomenon, it may be helpful to consider the effect of decision tree pruning from the view of error-ambiguity balance.

It has been shown that the generalization error of an ensemble can be decomposed into two terms, i.e. $E = \bar{E} - \bar{A}$, where $\bar{E}$ is the average generalization error of the component learners while $\bar{A}$ is the average ambiguity [8]. The smaller the $\bar{E}$ and the bigger the $\bar{A}$, the better the ensemble.

In general, the purpose of decision tree pruning is to avoid overfitting. With the help of pruning, the generalization ability of a decision tree is usually improved. Thus, the $\bar{E}$ of an ensemble comprising pruned decision trees may be smaller than that of an ensemble comprising unpruned decision trees. But on the other hand, pruning usually causes the decrease of the ambiguity among the decision trees. This is because some trees may become more similar after pruning. Thus, the $\bar{A}$ of an ensemble comprising pruned decision trees may be smaller than that of an ensemble comprising unpruned decision trees.

Table 7. Predictive error rate (%) of unpruned single C4.5 decision trees.

Data set	10%	20%	30%	40%	50%
allbp	4.44±1.27	4.62±0.59	3.91±0.66	3.56±0.72	3.65±0.44
ann	1.00±0.40	0.73±0.26	0.52±0.16	0.51±0.14	0.47±0.10
block	4.50±1.04	4.00±0.26	3.69±0.54	3.77±0.27	3.34±0.20
hypothyroid	1.74±0.88	1.40±0.43	1.11±0.37	0.75±0.16	0.61±0.18
kr-vs-kp	4.41±1.16	2.51±0.56	1.65±0.42	1.34±0.37	1.16±0.19
led7	35.36±2.85	32.04±3.15	30.06±1.72	28.61±1.63	28.09±1.66
led24	40.29±7.11	38.42±2.67	36.95±2.76	36.96±1.32	36.58±2.23
sat	19.92±1.69	17.40±0.99	16.34±0.63	15.93±0.56	15.09±0.52
segment	9.92±1.89	7.60±1.17	6.43±0.76	6.02±0.43	5.23±0.52
sick	2.38±1.10	2.14±0.67	2.03±0.43	1.89±0.31	1.59±0.27
sick-euthyroid	3.84±1.09	3.42±0.97	3.22±0.87	2.96±0.68	2.80±0.20
waveform	11.27±1.71	10.37±1.58	10.36±1.02	9.77±0.88	9.99±0.61
Data set	**60%**	**70%**	**80%**	**90%**	**100%**
allbp	3.36±0.35	3.16±0.37	3.17±0.34	3.09±0.42	3.02±0.14
ann	0.40±0.09	0.40±0.07	0.38±0.05	0.33±0.05	0.27±0.04
block	3.33±0.24	3.24±0.20	3.26±0.24	3.23±0.13	3.24±0.09
hypothyroid	0.56±0.16	0.59±0.15	0.55±0.15	0.51±0.09	0.48±0.05
kr-vs-kp	1.00±0.21	0.91±0.14	0.77±0.12	0.71±0.10	0.60±0.08
led7	27.81±1.48	27.34±0.69	27.08±0.71	27.06±0.49	26.92±0.26
led24	35.88±1.52	35.89±1.04	35.19±0.75	35.23±0.76	34.41±0.63
sat	15.27±0.62	14.66±0.46	14.54±0.53	14.06±0.55	13.84±0.30
segment	4.55±0.62	4.11±0.31	3.78±0.34	3.44±0.31	3.17±0.17
sick	1.52±0.23	1.34±0.19	1.26±0.29	1.26±0.20	1.22±0.09
sick-euthyroid	2.90±0.21	2.78±0.44	2.66±0.15	2.53±0.15	2.39±0.14
waveform	9.99±0.54	9.80±0.49	9.47±0.44	9.21±0.44	9.02±0.32

In other words, in constituting an ensemble, the advantage of stronger generalization ability of pruned decision trees may be killed to some degree by its disadvantage of smaller ambiguity. Thus, although an ensemble comprising pruned decision trees may be stronger than that comprising unpruned decision trees, the gap between the former and the pruned single decision trees may not be so big as that between the latter and the unpruned single decision trees. Therefore, the benefit from ensemble of unpruned decision trees is usually bigger than that from ensemble of pruned decision trees.

4 Conclusion

In summary, the empirical study described in this paper discloses:

- Enlarging the training set tends to enlarge the benefit from Boosting but does not significantly impact the benefit from Bagging. This is because the

increase of the training set size may enhance the bias reduction effect of adaptive sampling but may not significantly benefit the variance reduction effect of bootstrap sampling.
- The benefit from ensemble does not always increase along with the increase of the size of training set. This is because single learners sometimes may learn relatively more from randomly provided additional training data than ensembles do.
- The benefit from ensemble of unpruned decision trees is usually bigger than that from ensemble of pruned decision trees. This is because in constituting an ensemble, the relatively big ambiguity among the unpruned decision trees counteracts their relatively weak generalization ability to some degree.

These findings suggest that when dealing with huge volume of data, ensemble learning paradigms employing adaptive sampling are more promising, adequately selected training data are more helpful, and the generalization ability of the component learners could be sacrificed to some extent if this leads to a very significant increase of the ambiguity.

Acknowledgements. This work was partially supported by the National Outstanding Youth Foundation of China under the Grant No. 60325207, the Excellent Young Teachers Program of MOE of China, the Fok Ying Tung Education Foundation under the Grant No. 91067, and the National 973 Fundamental Research Program of China under the Grant No. 2002CB312002.

References

1. Bauer, E., Kohavi, R.: An empirical comparison of voting classification algorithms: Bagging, Boosting, and variants. Machine Learning **36** (1999) 105–139
2. Blake, C., Keogh, E., Merz, C.J.: UCI repository of machine learning databases [http://www.ics.uci.edu/~mlearn/MLRepository.html], Department of Information and Computer Science, University of California, Irvine, CA (1998)
3. Breiman, L.: Bagging predictors. Machine Learning **24** (1996) 123–140
4. Breiman, L.: Bias, variance, and arcing classifiers. Technical Report 460, Statistics Department, University of California, Berkeley, CA (1996)
5. Dietterich, T.G.: Machine learning research: four current directions. AI Magazine **18** (1997) 97–136
6. Freund, Y., Schapire, R.E.: A decision-theoretic generalization of on-line learning and an application to boosting. In: Proceedings of the 2nd European Conference on Computational Learning Theory, Barcelona, Spain (1995) 23–37
7. German, S., Bienenstock, E., Doursat, R.: Neural networks and the bias/variance dilemma. Neural Computation **4** (1992) 1–58
8. Krogh, A., Vedelsby, J.: Neural network ensembles, cross validation, and active learning. In: Tesauro, G., Touretzky, D.S., Leen, T.K. (eds.): Advances in Neural Information Processing Systems, Vol. 7. MIT Press, Cambridge, MA (1995) 231–238
9. Witten, I.H., Frank, E.: Data Mining: Practical Machine Learning Tools and Techniques with Java Implementations. Morgan Kaufmann, San Francisco, CA (2000)

Identifying Markov Blankets Using Lasso Estimation

Gang Li, Honghua Dai, and Yiqing Tu

School of Information Technology, Deakin University,
221 Burwood Highway, Vic 3125, Australia
{gangli,hdai,ytu}@deakin.edu.au

Abstract. Determining the causal relation among attributes in a domain is a key task in data mining and knowledge discovery. The Minimum Message Length (MML) principle has demonstrated its ability in discovering linear causal models from training data. To explore the ways to improve efficiency, this paper proposes a novel Markov Blanket identification algorithm based on the Lasso estimator. For each variable, this algorithm first generates a Lasso tree, which represents a pruned candidate set of possible feature sets. The Minimum Message Length principle is then employed to evaluate all those candidate feature sets, and the feature set with minimum message length is chosen as the Markov Blanket. Our experiment results show the ability of this algorithm. In addition, this algorithm can be used to prune the search space of causal discovery, and further reduce the computational cost of those score-based causal discovery algorithms.

1 Introduction

Graphical models carrying probabilistic or causal information have a long and rich tradition, which began with the geneticist Sewall Wright [1,2]. There appeared many variants in different domains, such as *Structural Equations Models* [3] and *Path Diagrams* [2] within social science, and *Bayesian Networks* in artificial intelligence. The year of 1988 marked the publication of Pearl's influential book [4] on graphical models, and since then, many research papers that address different aspects and applications of graphical models in various areas have been published.

In social sciences, there is a class of limited *Graphical Models*, usually referred as *Linear Causal Models*, including *Path Diagrams* [2], and *Structural Equation Models* [3]. In *Linear Causal Models*, effect variables are strictly linear functions of exogenous variables. Although this is a significant limitation, its adoption allows for a comparatively easy environment in which to develop causal discovery algorithms.

In 1996, Wallace et al. successfully introduced an information theoretic approach to the discovery of *Linear Causal Models*. This algorithm uses Wallace's *Minimum Message Length* (MML) criterion [5] to evaluate and guide the search of *Linear Causal Models*, and their experiments indicated that MML criterion

H. Dai, R. Srikant, and C. Zhang (Eds.): PAKDD 2004, LNAI 3056, pp. 308–318, 2004.

is capable of recovering a *Linear Causal Model* which is very close to the original model [6]. In 1997, Dai et al. further studied the reliability and robustness issues in causal discovery [7], they closely examined the relationships among the complexity of the causal model to be discovered, the strength of the causal links, the sample size of the given data set and the discovery ability of individual causal discovery algorithms. In 2002, Dai and Li proposed a new encoding scheme for model structure, and *Stirling's approximation* is further applied to simplify the computational complexity of the discovery process [8,9]. Different encoding schema and search strategies have been compared in [10] and empirical results revealed that greedy search works very well when compared with other more sophisticated search strategies. To further enhance the accuracy of causal discovery, Dai and Li introduced *Bagging* into causal discovery to improve the learning accuracy. The theoretical analysis and experimental results confirm that this method can also be used to alleviate the local minimum problem in discovering causal models [11].

Despite many successful applications, people who intend to use graphical models still encounter the problem of extensive computational cost. Indeed, it is a NP-hard problem to find the best model structure among the model space, except a special case that each node has no more than one parent. Clearly, increasing the amount of pruning and reducing the model space is the basis of solving this problem.

The *Markov Blanket* is the minimal set of variables conditioned on which all other variables are independent of the particular variable. It is an important concept in graphical models, and a number of researchers have recently applied it to address the issue of feature selection [12,13]. It has been shown that the Markov Blanket of a variable consists of strongly relevant variables related to this variable. Therefore, from a learned graphical model, it is easy to identify the set of relevant variables employing the concept of Markov Blanket. That is, first, we need to learn the graphical model, then identify the Markov blanket of the target variable from the model structure, finally use those variables in the Markov blanket as the selected features. The relationship between Markov Blankets and the optimal feature set motivates our novel method for the identification of Markov blankets: first, we can use some method to find out the optimal feature set; then, the optimal feature set can be used as an approximation to the Markov Blanket.

This paper presents a novel algorithm to identify the Markov blanket for each variable using the lasso estimator [14]. The rest of this paper is organized as follows. In Section 2 we briefly review the basic concepts of linear causal models and lasso estimation. In Section 3 the Minimum Message Length (MML) principle is then employed to chose the optimal feature set. In Section 4 the experimental results are given to evaluate the performance of this algorithm. Finally, we conclude this paper in Section 5.

2 Preliminaries

2.1 Linear Causal Models and Markov Blanket

A *Linear Causal Model* is a *Directed Graphical Model* in which every variable concerned is continuous. Informally speaking, it consists of two parts: *Structure*, which qualitatively describes the relation among different variables; and *Parameters*, which quantitatively describe the relation between a variable and its parents.

The model *Structure* is restricted to ones in which the inter-variable relations are open to a physically-tenable causal interpretation. Thus, if we define the "ancestors" of a variable as its parents, its parents' parents, and so on, we require that no variable has itself as an ancestor. With this restriction, the *Structure* of a *Linear Causal Model* among those continuous variables may be represented by a directed acyclic graph (DAG) in which each node represent a variable, and there is a directed edge from V_i to V_j if and only if V_i is a parent of V_j. The local relation between a variable and its parents is captured by a linear function. Specifically, if we have a set of continuous variables, a *Linear Causal Model* specifies, for each variable V_i, a possibly-empty "parent set" of it, and a probabilistic relation among data values,

$$V_i = \sum_{j=1}^{K_i} \alpha_{ij} \times Pa_{ij} + R_i \tag{1}$$

Where K_i is the number of parents for node V_i, α_{ij} $(j = 1, \ldots, K_i)$ is the linear coefficient reflecting the strength of the relationship between V_i and its j-th parent Pa_{ij}, and R_i is assumed to be identically distributed following a *Gaussian* distribution with zero mean and a standard deviation, that is $R_i \sim N(0, \sigma_i^2)$, so the set of local *Parameter* for a node V_i with parents is $\{\sigma_i^2, \alpha_{i1}, \ldots, \alpha_{iK_i}\}$. On the other hand, for a node V_i with an empty set of parents, we assume it as a random sample from a *Gaussian* distribution, $V_i \sim N(\mu_i, \sigma_i^2)$, where μ_i is the expected value of node V_i, so the local *Parameter* at node V_i is $\{\mu_i, \sigma_i^2\}$.

In a linear causal model, the *Markov Blanket* $MB(V_i)$ of a particular variable V_i is the minimal set of variables such that conditioned on $MB(V_i)$, any other variable is independent of V_i. From the model structure, the Markov Blanket of a variable V_i can be easily identified: it is the union of parents and children of V_i, and parents of children of V_i. An example of linear causal model Markov Blanket is shown in Fig. 1, in which the Markov Blanket of variable V_4 is $\{V_3, V_5, V_6\}$, and this means that the variables V_1 and V_2 are independent of V_4 conditioned on $\{V_3, V_5, V_6\}$.

Because of the relationship between Markov Blankets and the optimal feature set, this paper focuses on the task of identifying Markov Blanket. Before we delve into the specific algorithm, we introduce the concept of the Lasso Estimator.

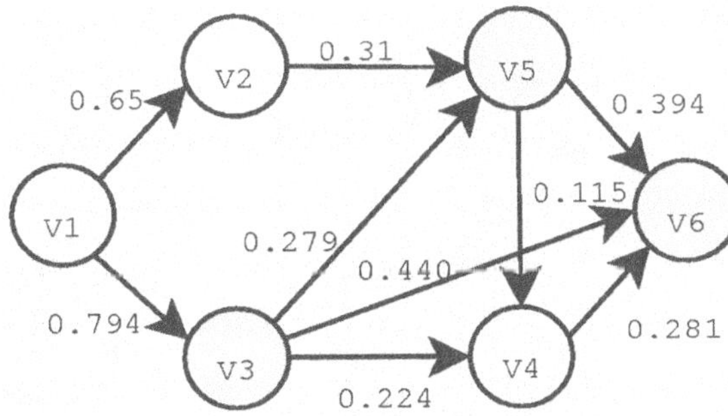

Fig. 1. An example of a Markov Blanket within a Linear causal model

2.2 The Lasso Estimator

Consider a regular linear regression problem with m predictor variables $\{X_1, \ldots, X_m\}$ and one response variable Y, and the training data is $(x_{k1}, \ldots, x_{km}, y_k), k = 1, \ldots, n$. In this situation, the ordinary least squares (OLS) regression finds the linear combination of all those m predictor variables that minimizes the residual sum of squares. A closely related optimization problem is

$$\hat{\alpha^0} = arg\min \sum_{k=1}^{n} (y_k - \sum_{j=1}^{m} \alpha_j^0 x_{kj})^2$$

where $\hat{\alpha^0} = \{\alpha_1^0, \ldots, \alpha_m^0\}$ is the OLS regression coefficient vector.

However, when the predictor variables are highly correlated or when m is large, the variances of the least squares coefficient may be unacceptably high. Standard solutions to this difficulty include ridge regression and feature selection. As an alternative to standard ridge regression and subset selection techniques, Tibshirani proposed the *Least Absolute Shrinkage and Selection Operator* (Lasso) [14], which is a constrained version of the ordinary least squares. The Lasso estimator solves the following optimization problem

$$\hat{\alpha}(t) = arg\min\{\sum_{k=1}^{n} (y_k - \sum_{j=1}^{m} \alpha_j x_{kj})^2 : \sum_{j=1}^{m} |\alpha_j| \leq t\}$$

where t is a tuning parameter, if t is greater than or equal to $\sum_{j=1}^{m} |\alpha_j^0|$, the lasso regression coefficients are the same as the OLS coefficients. For smaller values of t, the Lasso estimator shrinks the regression coefficient vector $\hat{\alpha}(t)$ towards the origin, and usually puts some of the regression coefficients to be zero.

Fig. 2(a) shows all Lasso solutions $\hat{\alpha}(t)$ for the variable V_4 in the *Blau* model, as t increases from 0, where $\hat{\alpha} = \hat{0}$, to $t = 0.86558$, where $\hat{\alpha}$ equals the OLS regression coefficient vector. Fig. 2(b) shows all Lasso solutions for the variable V_1. We see that the Lasso estimator tends to shrink the regression coefficients toward 0, more so for small values of t.

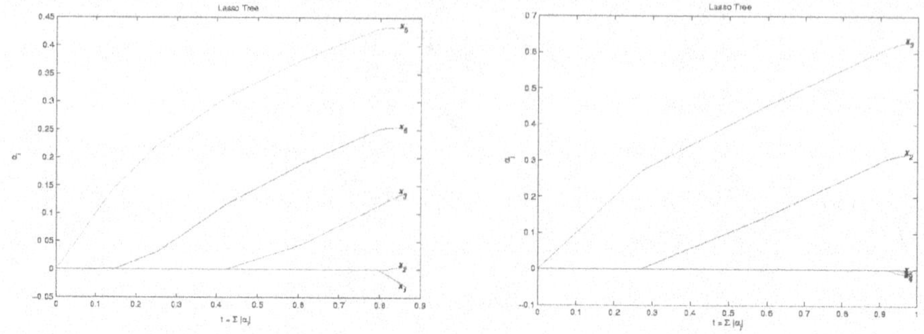

(a) Lasso Tree for Variable V_4 (b) Lasso Tree for Variable V_1

Fig. 2. Lasso Trees for variables in *Blau* model

Thus the Lasso estimator has a parsimony property: for any given constraint parameter t, only a subset of predictor variables have non-zero coefficients, and this property makes the Lasso a useful tool for feature selection. One difficulty faced by many regular feature selection algorithms is the exponential growth of the size of the power set of the predictor variables. However, this difficulty has been gracefully avoided by Lasso estimation: Along with the shrinkage of t, normally the regression coefficients will become 0 one by one. Thus, the number of candidate feature sets is usually equal to the number of predictor variables, rather than the size of the power set of the predictor variables. This is evident from Fig. 2, for example, in Fig. 2(a), the possible feature sets for variable V_4 are $\{\}$, $\{V_5\}$,$\{V_5, V_6\}$,$\{V_5, V_6, V_3\}$,$\{V_5, V_6, V_3, V_1\}$ and $\{V_5, V_6, V_3, V_1, V_2\}$.

Now, the problem is *how to find an optimal feature set from those candidate sets.*

3 Choosing the Optimal Feature Set Using the MML Principle

A fundamental problem in model selection is the evaluation of different models for fitting data. In 1987, Wallace and Freeman [15] extended their earlier method [5] for model selection based on *Minimum Message Length* (MML) coding. The basic idea of the MML principle is to find the hypothesis H which leads to the shortest message M. Its motivation is that this method can automatically embrace selection of an hypothesis of appropriate complexity as well as leading to good estimates of any free parameters of the hypothesis.

In this section we apply the MML principle to choosing the optimal feature set. This requires the estimate of the optimal code for describing the model and the data set, and it can be considered as a specific case of the general MML application to discover the linear causal model [9]. To ease exposition, suppose the candidate feature set with m predictor variables $\{X_1, \dots, X_m\}$, and Y is the response variable, the data set D consists of n instances, i.e.,

$D = \{(x_{k1}, \ldots, x_{km}, y_k), k = 1, \ldots, n\}$. Then the relation between Y and its m predictor variables can be captured by the following linear function

$$Y = \sum_{i=1}^{m} \alpha_i X_i + R \tag{2}$$

Where R represents the regression residuals, and they are assumed to be i.i.d. and follow some *Gaussian* distribution $\mathbf{N}(0, \sigma^2)$. All those local parameters $\theta = \{\alpha_1, \cdots, \alpha_m, \sigma\}$ can be estimated by the ordinary least squares (OLS) regression which attempts to minimize the residual sum of squares.

Therefore, for this linear model with local parameter $\theta = \{\alpha_1, \cdots, \alpha_m, \sigma\}$, the encoding length for this model becomes

$$msgLen(\theta_i) = -\log \frac{h(\theta)}{\sqrt{|F(\theta)|}} + \frac{m+1}{2} \log \kappa_{m+1} + \frac{m+1}{2} \tag{3}$$

where the term $h(\theta)$ is the prior probability of parameter θ, the term $\kappa_{(m+1)}$ is the $(m+1)$-dimensional optimal quantizing lattice constant [16], and the term $|F(\theta)|$ is the determinant of the empirical Fisher information matrix [17](page 96), which can be calculated by

$$F(\theta) = \begin{pmatrix} \frac{n}{2\sigma^4} & 0 \\ 0 & \frac{X'X}{\sigma^2} \end{pmatrix} \tag{4}$$

where $\theta = \{\alpha_1, \ldots, \alpha_m, \sigma\}$, X is the $n \times m$ matrix consisting of n observations of m predictor variables, and we can use a $m \times m$ matrix A to represent $X'X$. So the determinant of $F(\theta)$ becomes $\frac{n}{2}|A|(\sigma^2)^{-(m+2)}$.

In this paper, we use a uniform prior over the finite parameter space, $h(\theta) = \frac{1}{\sigma^2}$. Thus, the encoding length for this linear model is

$$msgLen(\theta) = \frac{1}{2}\log(\frac{n}{2}) + \frac{1}{2}\log|A| - \frac{m}{2}\log\sigma^2 + \frac{m+1}{2}\log\kappa_{(m+1)} + \frac{m+1}{2} \tag{5}$$

Given the local parameter θ, the encoding length for the training data D is

$$\begin{aligned} msgLen(D|\theta) &= msgLen(R|\theta) \\ &= -\log Prob(R|\theta) \\ &= \frac{n}{2}\log 2\pi + \frac{n}{2}\log\sigma^2 + \sum_{j=1}^{n} \frac{(y_j - \sum_i \alpha_i x_{ij})^2}{2\sigma^2} - N\log\epsilon_r \end{aligned} \tag{6}$$

where ϵ_r is the measurement accuracy of regression residual R, and it can be ignored when comparing two models.

Thus, for each candidate feature set, we can estimate its minimum message length by $msgLen(\theta) + msgLen(D|\theta)$, where θ is the set of local parameters related to this feature set. Because the number of candidate feature set returned by lasso estimator is usually equal to the number of predictor variables, we can compare each candidate set by simply calculating their corresponding encoding message length. As such, the candidate with the minimum message length will be chosen as the optimal feature set.

As pointed out in Section 1, the close relation between Markov Blanket and the optimal feature set leads to a novel Markov Blanket identification algorithm, that is, the chosen optimal feature set can be used as an approximation to the Markov Blanket. The pseudo code of this algorithm is shown in Alg. 1.

Algorithm 1 Lasso-based Markov Blanket Identification

```
Require: A lasso estimator Lasso, training set D, response variable Y.
Ensure: MB = Markov Blanket of Y
  CFS = Lasso(D, Y) {Generate Candidate Feature sets using Lasso estimation}
  OFS = full set of predictor variables
  for each CFS_i ∈ CFS do
    if msgLen(D, CFS_i) < msgLen(D, OFS) then
      OFS = CFS_i
    end if
  end for{Iterate each candidate feature set to choose the optimal one}
  MB = OFS {The Markov Blanket can be approximated by the optimal feature set.}
```

4 Experimental Result

In this section, the performance of the Lasso-based Markov Blanket identification (LMB) algorithm is evaluated. First, a full set of MB identification result for the *Verbal&Mechanical Ability* model is given to show the ability of our LMB algorithm. Then, different data sets are tested to establish a general performance measure for this algorithm.

4.1 A Case Study: Verbal and Mechanical Ability Model

Fig. 3(a) illustrates the *Verbal&Mechanical Ability* model described by Loehlin [18]. In this model, variable V_1, V_2 and V_3 are three hypothetical variables introduced by factor analysis, while V_4 and V_5 are two verbal tests, and V_6 and V_7 are two mechanical tests.

For this case, the Lasso-based Markov Blanket identification algorithm is applied to find the Markov Blanket for each variable. Fig. 3 shows all those results: Fig. 3(b) gives the lasso tree for variable V_1, the MML selection chooses the optimal feature set as $\{V_2, V_3\}$; Fig. 3(c) gives the lasso tree for variable V_2, the LMB algorithm finds the optimal feature set as $\{V_1, V_4, V_5\}$, as indicated by the vertical line at $t = 0.83$.

From these figures, it is clear that the algorithm closely recovered all those Markov Blankets of variables in the Verbal&Mechanical Ability model.

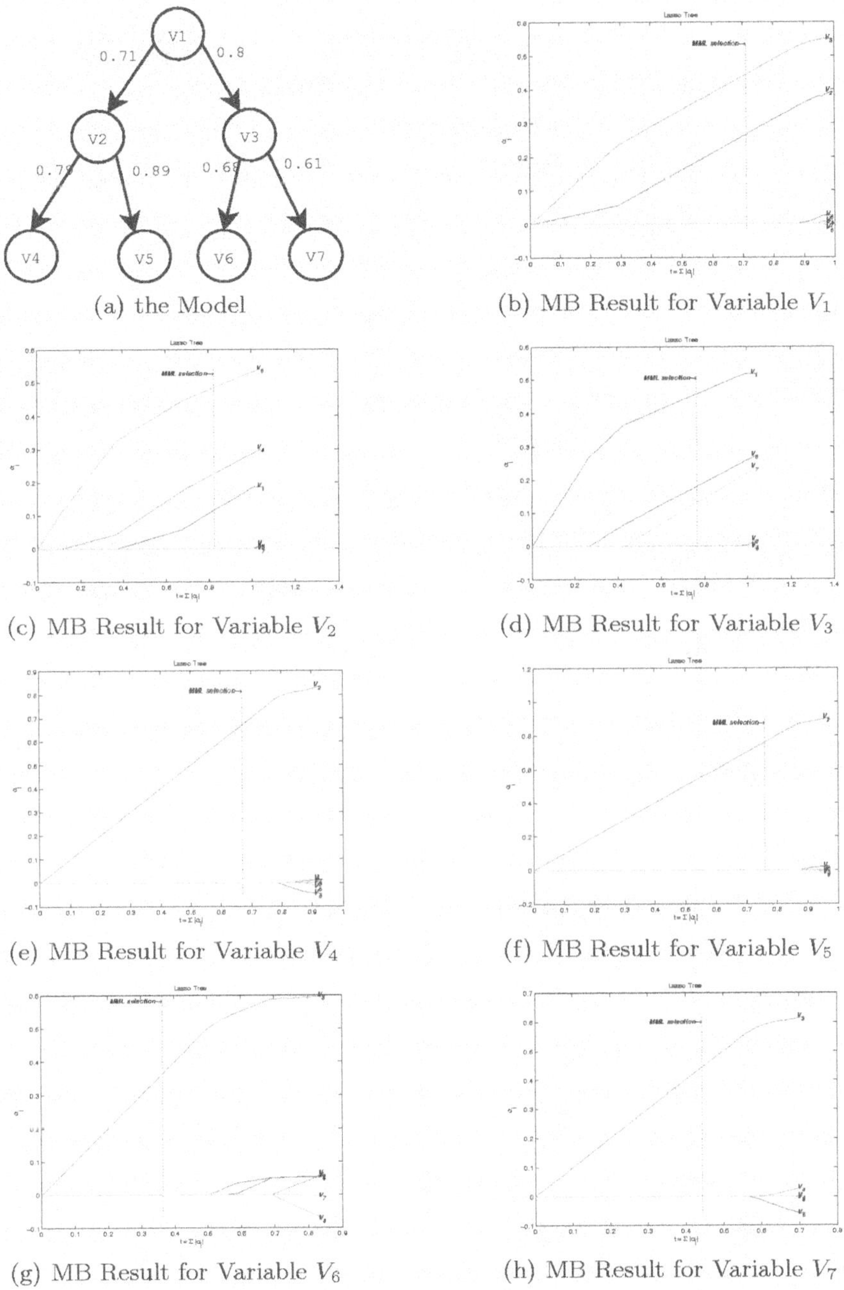

(a) the Model

(b) MB Result for Variable V_1

(c) MB Result for Variable V_2

(d) MB Result for Variable V_3

(e) MB Result for Variable V_4

(f) MB Result for Variable V_5

(g) MB Result for Variable V_6

(h) MB Result for Variable V_7

Fig. 3. A Case Study for Learning MB using Lasso Estimator

4.2 Performance Evaluation

In order to get a general idea of this algorithm's performance, eight linear causal models reported in related literature [6,8] are re-examined: *Fiji, Evans, Blau, Goldberg, V&M Ability, case9, case10 and case12.* The details of these models are described in Table 1.

Table 1. Information of Examined Data Set

Data Set	Number of Variables
Fiji	4
Evans	5
Blau	6
Goldberg	6
V&M Ability	7
Case9	9
Case10	10
Case12	12

The identified Markov Blankets can be compared against the actual Markov Blankets in terms of *recall rate* and *exclude rate*, where the *recall rate* is the ratio of correctly identified Markov Blanket variables over the true Markov Blanket variables, and the *exclude rate* is the ratio of identified non-Markov Blanket variables over the true non-Markov Blanket variables.

Table 2. Result of LMB algorithm over examined Data set

Data Set	Recall Rate	Exclude Rate
Fiji	1	
Evans	0.88	1
Blau	0.94	0.92
Goldberg	1	1
V&M Ability	1	1
Case9	0.91	0.93
Case10	1	1
Case12	1	1

Table 2 gives the results of LMB algorithm over examined data sets: Out of these eight models, the structure of the *Fiji* model is a complete DAG, and any variable belonging to the Markov blanket of any other variable, so it is impossible to calculate the exclude rate. Five out of eight models can be correctly identified by LMB algorithm. For the *Evans* model, all those non-Markov blanket variables can be excluded, however, some of the Markov blanket variables were also excluded by mistake, such that the recall rate is less than 1. For model *Blau* and *Case9*, their recall rates and exclude rates are all less than 1, and this indicates that fact that some non-Markov blanket variables were selected while some Markov blanket variables were excluded.

5 Conclusion and Future Work

This paper presented a novel algorithm (LMB) to identify Markov Blanket using Lasso estimator. The experimental results reported in this paper show that in general this algorithm can correctly distinguish those Markov blanket variables from those non-Markov Blanket variables.

We take these results to be significant confirmation that Lasso can be used as a useful assisting tool for the discovery of linear causal model. Future work can be carried out on the following aspects:

- First, the identification result can be further refined using some characters of Markov Blanket, such as, if a variable X is in the Markov Blanket of variable Y, then the variable Y is also in the Markov Blanket of variable X.
- Second, the identified Markov Blanket can be used to prune the search space of model selection. This is a potential method to further increase the efficiency of linear causal model discovery.
- Third, how to distinguish parent variables from the other variables in Markov blanket can be another interesting topic.
- Finally, the reason that the MML principle plays very well for the identification of Markov Blankets has not been studied, and its comparative performance with the cross-validation selection strategy also calls for further experimental work.

References

[1] Wright, S.: Correlated and causation. Journal of Agricultural Research **20** (1921) 557–585

[2] Wright, S.: The method of path coefficients. Annals of Mathematical Statistics **5** (1934) 161–215

[3] Bollen, K.: Structural Equations with Latent Variables. Wiley, New York (1989)

[4] Pearl, J.: Probabilistic Reasoning in Intelligent Systems. Revised second printing edn. Morgan Kauffmann Publishers, San Mateo, California (1988)

[5] Wallace, C., Boulton, D.: An information measure for classification. Computer Journal **11** (1968) 185–194

[6] Wallace, C., Korb, K.B., Dai, H.: Causal discovery via MML. In: Proceedings of the 13th International Conference on Machine learning (ICML'96), San Francisco, Morgan Kauffmann Publishers (1996) 516–524

[7] Dai, H., Korb, K., Wallace, C., Wu, X.: A study of causal discovery with small samples and weak links. In: Proceedings of the 15th International Joint Conference On Artificial Intelligence **IJCAI'97**, Morgan Kaufmann Publishers, Inc. (1997) 1304–1309

[8] Dai, H., Li, G.: An improved approach for the discovery of causal models via MML. In: Proceedings of The 6th Pacific-Asia Conference on Knowledge Discovery and Data Mining (PAKDD-2002), Taiwan (2002) 304–315
[9] Li, G., Dai, H., Tu, Y.: Linear causal model discovery using MML criterion. In: Proceedings of 2002 IEEE International Conference on Data Mining, Maebashi City, Japan, IEEE Computer Society (2002) 274–281
[10] Dai, H., Li, G., Tu, Y.: An empirical study of encoding schemes and search strategies in discovering causal networks. In: Proceedings of 13th European Conference on Machine Learning (Machine Learning: ECML 2002), Helsinki, Finland, Springer (2002) 48–59
[11] Dai, H., Li, G., Zhou, Z.H., Webb, G.: Ensembling MML causal induction. In: Technical Report, Deakin University. (2003)
[12] Koller, D., Sahami, M.: Toward optimal feature selection. In: Proceedings of the 13th International Conference in Machine Learning (ICML'96), San Francisco, Morgan Kauffmann Publishers (1996) 284–292
[13] Tsamardinos, I., Aliferis, C.: Towards principled feature selection: Relevancy, filters and wrappers. In: Proceedings of the ninth International Workshop on Artificial Intelligence and Statistics, IEEE Computer Society (2003) ??–??
[14] Tibshirani, R.: Regression shrinkage and selection via the lasso. Journal of the Royal Statistical Society, Series B **58** (1996) 267–288
[15] Wallace, C., Freeman, P.: Estimation and inference by compact coding. Journal of the Royal Statistical Society **B,49** (1987) 240–252
[16] Conway, J., Sloane, N.: Sphere Packings, Lattices and Groups. Springer-Verlag, London (1988)
[17] Harvey, A.: The Econometric Analysis of Time Series. 2 edn. The MIT Press, Cambridge, Massachusetts (1990)
[18] Loehlin, J.C.: Latent Variable Models: An Introduction to Factor, Path and Structural Analysis. second edn. Lawrence Erlbaum Associates, Hillsdale, New Jersey (1992)

Selective Augmented Bayesian Network Classifiers Based on Rough Set Theory

Zhihai Wang[1], Geoffrey I. Webb[2], and Fei Zheng[2]

[1] School of Computer and Information Technology, Beijing Jiaotong University, Beijing, 100044, China
zhhwang@center.njtu.edu.cn

[2] School of Computer Science and Software Engineering, Monash University, Clayton, Victoria, 3800, Australia
{webb, feizheng}@mail.csse.monash.edu.au

Abstract. The naive Bayes classifier is widely used in interactive applications due to its computational efficiency, direct theoretical base, and competitive accuracy. However, its attribute independence assumption can result in sub-optimal accuracy. A number of techniques have explored simple relaxations of the attribute independence assumption in order to increase accuracy. TAN is a state-of-the-art extension of naive Bayes, that can express limited forms of inter-dependence among attributes. Rough sets theory provides tools for expressing inexact or partial dependencies within dataset. In this paper, we present a variant of TAN using rough sets theory and compare their tree classifier structures, which can be thought of as a selective restricted trees Bayesian classifier. It delivers lower error than both pre-existing TAN-based classifiers, with substantially less computation than is required by the $SuperParent$ approach.

Keywords: Naive Bayes, Bayesian Network, Machine Learning,

1 Introduction

A classification task in data mining is to build a classifier which can assign a suitable class label to an unlabelled instance described by a set of attributes. Many approaches and techniques have been developed to create a classification model. The naive Bayesian classifier is one of the most widely used in interactive applications due to its computational efficiency, competitive accuracy, direct theoretical base, and its ability to integrate the prior information with data sample information [1,7,3,5,18,15,14]. It is based on Bayes' theorem and an assumption that all attributes are mutually independent within each class. Assume X is a finite set of instances, and $A = \{A_1, A_2, \cdots, A_n\}$ is a finite set of n attributes. An instance $x \in X$ is described by a vector $< a_1, a_2, \cdots, a_n >$, where a_i is a value of attribute A_i. C is called the class attribute. Prediction accuracy will be maximized if the predicted class

$$L(x) = argmax_c(P(c \mid < a_1, a_2, \cdots, a_n >). \tag{1}$$

H. Dai, R. Srikant, and C. Zhang (Eds.): PAKDD 2004, LNAI 3056, pp. 319–328, 2004.

Unfortunately, unless the vector occurs many times within X, it will not be possible to directly estimate $P(c \mid < a_1, a_2, \cdots, a_n >)$ from the frequency with which each class $c \in C$ co-occurs with $< a_1, a_2, \cdots, a_n >$ within the training instances. Bayes' theorem provides an equality that might be used to help estimate the posterior probability $P(c_i \mid x)$ in such a circumstance:

$$P(c_i \mid x) = \alpha \cdot P(c_i) \cdot P(< a_1, a_2, \cdots, a_n > \mid c_i) \tag{2}$$

where $P(c_i)$ is the prior probability of class $c_i \in C$, $P(< a_1, a_2, \cdots, a_n > \mid c_i)$ is the conditional probability of $x \in T$ given the class c_i, and α is a normalization factor. According to the chain rule, equation 2 can be written as:

$$P(c_i \mid x) = \alpha \cdot P(c_i) \cdot \prod_{k=1}^{n} P(a_k \mid a_1, a_2, \cdots, a_{k-1}, c_i) \tag{3}$$

Therefore, an approach to Bayesian estimation is to seek to estimate each $P(a_k \mid a_1, a_2, \cdots, a_{k-1}, c_i)$.

If the n attributes are mutually independent within each class value, then the conditional probability can be calculated in the following way:

$$P(< a_1, a_2, \cdots, a_n > \mid c_i) = \prod_{k=1}^{n} P(a_k \mid c_i). \tag{4}$$

Classification selecting the most probable class as estimated using formulae 2 and 4 is the well-known naive Bayesian classifier.

2 Approaches of Improving Naive Bayesian Method

In real world problems, the performance of a naive Bayesian classifier is dominated by two explicit assumptions: the attribute independence assumption and the probability estimation assumption. Of numerous proposals to improve the accuracy of naive Bayesian classifiers by weakening its attribute independence assumption, both Tree Augmented Naive Bayes(TAN) [4,3,5] and Lazy Bayesian Rules(LBR) [18] have demonstrated remarkable error performance [14]. Friedman, Geiger and Goldszmidt presented a compromise representation, called tree-augmented naive Bayes (TAN, simply called the basic TAN), in which the class node directly points to all attributes' nodes and an attribute node can have only at most one additional parent to the class node. Keogh & Pazzani took a different approach to constructing tree-augmented Bayesian networks [5](simply called $SuperParent$ or SP). The two methods mainly differ in two aspects. One is the criterion of attribute selection used to select dependence relations among the attributes while building a tree-augmented Bayesian network. Another is the structure of the classifiers. The first one always tends to construct a tree including all attributes, the second one always tends to construct a tree with fewer dependence relations among attributes and better classification accuracy. Zheng and Webb proposed a lazy Bayesian rule (LBR) learning technique [18].

LBR can be thought of as applying lazy learning techniques to naive Bayesian rule induction. Both *LBR* and *TAN* can be viewed as variants of naive Bayes that relax the attribute independence assumption [14].

In this paper, however, we concentrate on the eager strategy, which holds a computational advantage when a single model can be applied to classify many instances. First of all, we mainly analyze the implementations of two different *TAN* classifiers and their tree classifier structures, and experimentally show how different dependence relations impact on the accuracy of *TAN* classifiers. Second, based on the definition of dependence in the basic rough set theory, we propose a definition of dependence measurement given the class variable, and use it to build a new dependence relation matrix. We believe that the directions of dependence relations are very important for performance of a classifier. Using this kind of definition, we can actually gain a directed-graph description. Third, we present a new algorithm for building selective augmented Bayesian network classifiers, which reduce error relative to the *TAN* classifiers, and has similar computational overheads. Experimental results also show that can deliver some improvements on performance, while requiring substantially less computation.

3 Some Issues in the Implementations

Now, we discuss some extended issues in the implementations of the *TAN* classifiers. First of all, the problem is related to the probability estimation assumption. In the basic *TAN*, for each attribute we assess the conditional probability given the class variable and another attribute. This means that the number of instances used to estimate the conditional probability is reduced as it is estimated from the instances that share three specific values (the class value, the parent value and the child value). Thus it is not surprising to encounter unreliable estimates, especially in small datasets. Friedman, Geiger and Goldszmidt dealt with this problem by introducing a smoothing operation [3]. There is a problem with this strategy when attribute value a does not occur in the training data (this situation can occur during cross validation testing), the value of the estimate will be zero. In our implementation, we use both these smoothing adjustments to estimate any conditional probability with $|\pi(a)| = 2$, i.e.,

$$\widehat{P}(a \mid \pi(a)) = \frac{counts(a, \pi(a)) + N^0 \cdot \frac{counts(a)+1}{|T|+|A|}}{counts(\pi(a)) + N^0}. \tag{5}$$

where $|A|$ is the number of values for attribute A. We use Laplace estimation to estimate any other probability. In Keogh and Pazzani's *SuperParent* algorithm, they replace zero probabilities with a small epsilon (0.0001). Kohavi, Becker and Sommerfield [8] have shown that different methods for estimating the base probabilities in naive Bayes can greatly impact upon its performance. Similarly, different estimation methods will gain different effects on the same *TAN* classification model. We think that estimation methods should depend on the distribution of variables, as well as the number of training instances to support these kinds of estimation methods. Estimation methods should be independent of classification models for the same probability. In order to compare the performances of classification models, we use the same method to estimate probabilities. In all the algorithms mentioned in our experiments, we always use formular 5 to estimate any conditional probability with $|\pi(a)| = 2$ and only standard Laplace estimation applied to any other probability.

Secondly, regarding the problem of missing values, in both the basic TAN classifiers and $SuperParent$ classifiers, instances with missing values were deleted from the set of training instances. We keep all the instances, but ignore missing values from the counts for missing attribute values. Also, when we estimate a conditional probability $P(a_k \mid c_i)$, for a prior probability of class value c_i we exclude the occurrences of class value c_i with missing values on attribute A_k. Obviously, this makes the estimation of the condition more reliable while estimating any conditional probability.

Thirdly, although the choice of root variable does not change the log-likelihood of the basic TAN network, we have to set the direction of all edges for classification. When each edge (Ai, Aj) is added to the current tree structure, we always set the direction from Ai to Aj $(i < j)$ at once. In Keogh and Pazzani's $SuperParent$ classifiers, the direction of an arc is always from the super parent node to the favorite child node. That means that when an dependence relation is singled out, it always has the specific direction.

4 Selective Augmented Bayesian Classifiers

There are further factors that influence the performance of an augmented naive Bayesian classifier. The first one is the criterion for selecting dependence relations among the attributes. The second one is the criterion of terminating the selection. In this section, we describe our new algorithm for selective augmented Bayesian classifiers, explain how it works and is different from the basic TAN classifiers and $SuperParent$ classifiers, and experimentally show its better performance and preferable computational profile.

4.1 Dependence Relation Matrix Based on Rough Set Theory

Friedman, Geiger and Goldszmidt explain why they use the conditional mutual information as the criterion of selecting dependence relations among the attributes [3,4]. One problem with this criterion, as mentioned above, is how to decide the directions of dependence relations. Keogh and Pazzani use leave-one-out cross-validation to handle this problem in the process of building a classifier [5,6]. When the best super parent and its favorite child are found, the dependence relation between them is naturally from the best super parent to its favorite child, i.e., the favorite child is depends on the corresponding best super parent. For each arc to be added into the structure, $SuperParent$ needs many times to execute leave-one-out cross-validations on the whole set of training instances. Although they proposed some shortcuts to speed up the process of evaluating many classifiers, the algorithm is still very time consuming. In our algorithm, we will use a dependence measurement based on rough sets theory.

Rough set [11] provides, in particular, tools for expressing inexact or partial dependencies within dataset. Given a dataset of some description or measurements concerning available instances, rough set methods enable to extract dependence relations between corresponding attributes or variables. These dependence relations can be applied to inductive reasoning about new, so far unseen cases, in a way well understandable for the user. Above advantages, as well as very effective computational framework for extraction of the most interesting dependencies from real-life data, cause a rapid development of applications of rough sets to more and more scientific fields and practical tasks [13].

According to the values of class variable, we define a new dependence relation matrix, in which each item of conditional dependence relation, $D(A_i, A_j \mid C)$, can be described as follows:

$$D(A_i, A_j \mid C) = \sum_{A_i, A_j, Y_c} \frac{\mid POS_{A_i}^{Y_c}(A_j) \mid}{counts(Y_c)} \tag{6}$$

where $POS_{A_i}^{Y_c}(A_j)$ represents the positive region of attribute A_j relative to attribute A_i within the class value Y_c [11]. Using this kind of definition, we can actually gain a directed-graph description. Each item not only reflects the degree of the dependence between two attributes, but also tells us the direction of the dependence relation.

4.2 A New Selective Augmented Bayesian Algorithm

In an augmented Bayesian classifier, the second important issue is how to decide the candidate dependence set and when terminate the the selection. There are $n(n-1)$ different dependence relations among n attributes. When there are $n-1$ or no any edge with the conditional mutual information more than 0, a basic TAN structure will have $n-1$ arcs. We also try to add $n-1$ arcs to our augmented Bayesian classifier, but the candidate set is different from the basic TAN. Because the way of weighting a candidate arc and the way of setting the direction of an arc are different from the basic TAN structure, the candidate arc set is different.

Based on above discussions, we can describe our selective augmented Bayesian algorithm, simply called *Select* in all tables, as follows.

1. Compute the dependence relation matrix conditional mutual information using formula 6.

2. Select an arc using a near maximum branching directed arborescence alogorithm [10], based on the dependence relation matrix.

3. Use leave-one-out cross-validation to evaluate the current set of arborescences to decide whether this arc should be added into the current structure or not, adding it only if doing so reduces cross-validation error.

4. Repeat the previous iteration $n-1$ times, or until no more arcs can be tested.

An important difference from the basic *TAN* algorithm is that this algorithm tries to build a maximum directed branch or arborescence [10], not a maximum undirected spanning tree. We believe that the direction of dependence relations is a critical issue to minimizing error. In the procedure of building an augmented Bayesian classifier the network is always a directed structure.

5 Experimental Results

There are thirty-two natural domains used in our experiments, shown in Table 1. Twenty-nine of these are totally drawn from the previous research paper [18]. The other three (*Satellite*, *segment*, and *Shuttle*) are larger datasets. "$S\sharp$" means the number of instances. "$C\sharp$" means the number of values of a class attribute. "$A\sharp$" means the number of attributes, not including the class attribute. All the experiments were performed in the Weka system [17]. The error rate of each classification model on each domain is determined by running 10-fold cross-validation on a dual-processor 1.7Ghz Pentium 4 Linux computer with 2Gb RAM. We use the default discretization method "weka.filters.DiscretizeFilter" as the discretization method for continuous values, which is based on Fayyad and Irani's method [2].

Table 1 also shows the error rates of naive Bayes classifier (*NB*), the basic *TAN* classifier, the *SuperParent* clssifier (*SP*), and our selective augmented Bayesian classifier (*Select*) on each domain, respectively. The last row contains the mean error rates for each column. The best one for a given dataset is shown in bold text. It shows the selective augmented Bayesian classifier has the lowest mean error rate. Table 2 presents the WIN/LOSS/DRAW records for comparing with all other classifiers. This is a record of the number of data sets for which the nominated algorithm achieves lower, higher, and equal error to the comparison algorithm, measured to two decimal places. The table also includes the outcome of a two-tailed binomial sign test. This indicates the probability that the observed outcome or more extreme should occur by chance if wins and losses were equiprobable. The selective augmented Bayesian classifier demonstrates the best performance. In the thirty-two databases, there are nineteen databases which the selective augmented Bayesian classifier has higher classification accuracy than Naive Bayes, twenty-one databases than the basic *TAN* classifier, sixteen databases than the *SuperParent* classifier. It is remarkable that there are eighteen databases which the basic *TAN* classifier has higher classification accuracy than Naive Bayes, and twelve datasets worse than Naive Bayes.

This suggests that selective augmented Bayesian classifier has similar error to the *SuperParent* classifier, but the selective augmented Bayesian classifier is much more efficient than the *SuperParent* classifier. Table 3 shows the time of building each classifier. On the most of domains, the selective augmented Bayesian classifier is much faster than the *SuperParent* classifier.

Table 4 shows the number of arcs in the Bayesian network built by the *SuperParent* classifier (*SP*), the basic *TAN* classifier, and our (*Select*) on each domain, respectively. These experimental results show that the basic *TAN* al-

Table 1. Descriptions of Data and Average Error Rates

	Domain	S♯	C♯	A♯	M	*NB*	*SP*	*TAN*	*Select*
1	Anealing	898	6	38	Y	5.46	**3.3**	4.34	4.12
2	Audiology	226	24	69	Y	29.20	27.88	27.43	**26.55**
3	Breast Cancer	699	2	9	Y	**2.58**	**2.58**	**5.01**	**2.58**
4	Chess(kr-vs-kp)	3169	2	39	N	12.36	**5.19**	6.54	8.01
5	Credit	69	2	15	Y	15.07	15.22	**14.93**	15.23
6	Echocardiogram	131	2	6	Y	**27.48**	29.01	35.88	28.24
7	Glass	214	7	9	N	41.12	41.59	37.85	**35.98**
8	Heart	270	2	13	N	**15.19**	16.30	20.74	17.04
9	Hepatitis	155	2	19	Y	16.13	16.13	**11.61**	14.84
10	Horse Colic	368	2	21	Y	20.11	19.29	19.84	**19.02**
11	House Votes 84	435	2	16	N	9.89	**6.67**	7.59	9.20
12	Hypothyroid	3163	2	25	Y	2.94	2.81	**2.66**	2.75
13	Iris	150	3	4	N	6.00	6.67	**5.33**	8.00
14	Labors	57	2	16	Y	**3.51**	**3.51**	12.28	**3.51**
15	LED	1000	10	7	N	26.20	26.60	**25.90**	26.70
16	Bupa	345	2	6	N	**36.81**	38.84	37.68	39.13
17	Lung Cancer	32	3	56	Y	46.88	50.00	46.88	**43.75**
18	Lymphography	148	4	18	N	14.19	14.86	18.92	**12.84**
19	PID	768	2	8	N	25.00	25.52	25.00	**24.74**
20	Post Operative	90	3	8	Y	**28.89**	31.11	35.56	30.00
21	Primary Tumor	339	22	17	Y	**48.97**	51.15	54.28	50.15
22	Promoters	106	2	57	N	**8.49**	**8.48**	16.04	11.32
23	Satellite	6435	6	36	N	18.90	12.18	12.29	**12.15**
24	Segment	2310	7	19	N	11.08	7.01	6.15	**5.54**
25	Shuttle	58000	7	9	N	10.07	**5.09**	8.06	6.95
26	Solarflare	1389	3	10	N	3.89	**1.08**	1.08	1.87
27	Sonar	208	2	60	N	25.48	25.96	29.33	**23.08**
28	Soybean	683	19	35	Y	7.17	6.59	10.98	**6.44**
29	Splice	3177	3	60	N	4.66	4.50	4.60	**4.41**
30	TTT	958	2	9	N	29.54	27.45	**26.10**	29.54
31	Wine	178	3	13	N	3.37	3.37	3.93	**2.81**
32	Zoology	101	7	16	N	5.94	6.93	**4.95**	6.93
	Error Mean					17.58	16.93	18.11	**16.67**

ways tends to construct a tree including all attributes, the SP always tends to construct a tree with fewer dependence relations among attributes and better classification accuracy. In order to show the candidate arc set and the structure are different from the basic TAN classifier, we give an example produced by the basic TAN algorithm, the *SuperParent* algorithm, and our selective augmented Bayesian algorithm on dataset *Soybean* respectively. The basic TAN algorithm produced a tree with $n - 1$ arcs ($n = 35$), shown in figure 1, where a node which parent is node 0 and has no any child is omitted for simplicity. This tree includes all the attribute nodes. The selective augmented Bayesian algorithm produced eight arcs in four branches, shown in figure 2. The structure of this

Table 2. Comparison of *Select* to others

	WIN	*LOSS*	*DRAW*	*P*
NaiveBayes	19	10	3	0.1360
SuperParent	16	11	5	0.4420
TAN	21	11	0	0.1102

Table 3. Time in CPU Seconds

Domain	*SP*	*TAN*	*Select*
1 Anealing	110.31	0.17	13.95
2 Audiology	530.96	1.79	126.38
3 Breast Cancer	1.20	0.17	0.40
4 Chess	662.98	1.14	61.81
5 Credit	5.02	0.14	1.35
6 Echocardiogram	0.07	0.03	0.10
7 Glass	0.37	0.11	0.31
8 Heart	0.31	0.11	0.25
9 Hepatitis	2.13	0.11	1.59
10 Horse Colic	6.78	0.13	1.42
11 House Votes	1.51	0.10	1.50
12 Hypothyroid	130.58	0.25	28.8
13 Iris	0.07	0.02	0.07
14 Labor	0.09	0.09	0.17
15 LED	1.48	0.07	1.54
16 Bupa	0.24	0.06	0.13
17 Lung Cancer	7.23	0.14	5.56
18 Lymphography	0.36	0.10	1.52
19 PID	1.37	0.12	1.32
20 Post Operative	0.10	0.02	0.11
21 Ptn	2.63	0.11	5.22
22 Promoters	11.75	0.34	10.77
23 Satellite	6247.56	4.9	269.66
24 Segment	202.69	1.84	24.36
25 Shuttle	137.68	5.14	79.82
26 Solarflare	8.85	0.07	2.56
27 Sonar	165.06	1.88	22.62
28 Soybean	51.59	0.43	51.63
29 Splice	1090.46	3.71	313.44
30 TTT	1.07	0.06	0.53
31 Wine	0.17	0.13	0.48
32 Zoo	1.80	0.09	0.61

Table 4. Arcs for each algorithm

Domain	*SP*	*TAN*	*Select*
1 Anealing	9	27	12
2 Audiology	6	61	11
3 Breast Cancer	0	8	1
4 Chess	8	35	17
5 Credit	4	13	7
6 Echocardiogram	0	5	17
7 Glass	3	5	7
8 Heart	0	10	3
9 Hepatitis	2	18	3
10 Horse Colic	4	18	7
11 House Votes	1	15	3
12 Hypothyroid	7	24	14
13 Iris	0	3	1
14 Labor	0	10	3
15 LED	1	6	4
16 Bupa	1	2	1
17 Lung Cancer	2	19	6
18 Lymphography	1	17	4
19 PID	1	5	1
20 Post Operative	2	7	3
21 Ptn	0	16	5
22 Promoters	0	56	4
23 Satellite	35	35	35
24 Segment	13	15	17
25 Shuttle	2	6	8
26 Solarflare	6	9	8
27 Sonar	3	56	7
28 Soybean	2	34	8
29 Splice	2	59	7
30 TTT	1	8	0
31 Wine	0	12	2
31 Zoo	2	15	3

Bayesian classifier is a forest, where some of arcs belong to the *TAN* structure, but some others do not belong to the *TAN* structure. This example also shows our selective augmented Bayesian algorithm produces different results from the *sTAN* method described by Keogh and Pazzani [6]. In the *sTAN* model, they only select from those edges that appear in the basic *TAN* tree structure. The

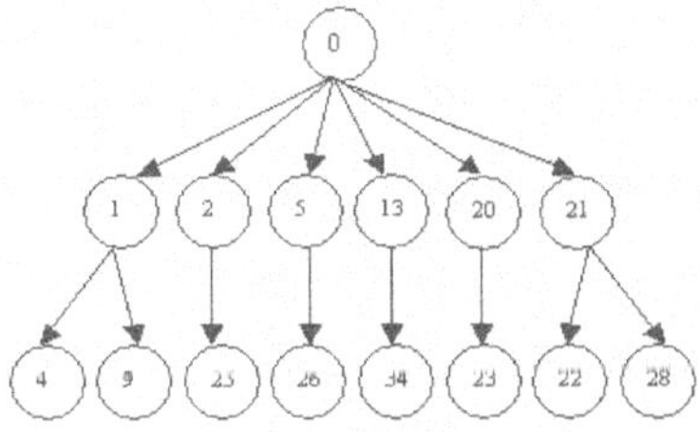

Fig. 1. The Bayesian network of TAN on dataset Soybean

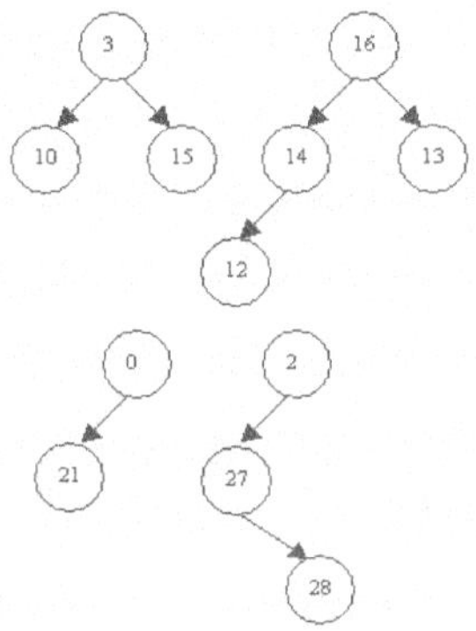

Fig. 2. The Bayesian network of selective augmented TAN on dataset Soybean

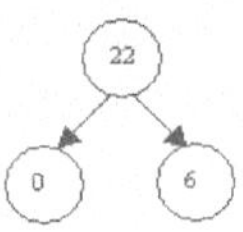

Fig. 3. The Bayesian network of super parent TAN on dataset Soybean

result of the *SuperParent* algorithm is shown in figure 3. The *SuperParent* algorithm only uses leave-one-out cross validation to determine the directions of arcs, but it can not always obtain better performance. For example, on dataset PID, the Bayesian networks built by both the *SuperParent* algorithm and the selective augmented Bayesian algorithm have only one arc, but the directions are different. The tree structure built by the *SuperParent* algorithm is only the arc: $\langle 4, 1 \rangle$, but the selective augmented Bayesian algorithm produces one arc with reverse direction.

6 Conclusion

In this paper we have investigated three issues that affect the quality of TAN classifiers learned from data. The first issue is how to estimate the base probabilities. We conclude that it is important to use a consistent estimation to compare with each other. The second is how to measure dependence relations between two attributes with directions. Based on the definition of dependence degree in the basic rough set theory, we propose a definition of dependence measurement given class variable for building a classification model. Thirdly, we mainly present a selective augmented Bayesian network classifier that reduces error relative to the original TAN, and with similar computational overheads, but much lower computational overheads than the $SuperParent$, which is a state-of-the-art variant of the basic TAN classifier. Experimental results show that it can deliver some improvements on performance, while requiring substantially less computation.

References

1. P. Domingos and M. Pazzani.: Beyond Independence: Conditions for the Optimality of the Simple Bayesian Classifier. In: Proceedings of the Thirteenth ICML, Morgan Kaufmann Publishers, San Francisco, CA(1996)105–112,
2. U. Fayyad and K. Irani.: Multi-Interval Discretization of Continuous-valued Attributes for Classification Learning. In: Proceedings of the 13th IJCAI, Seattle, WA: Morgan Kaufmann Publishers, (1993), 1022-1027
3. N. Friedman, D. Geiger and M. Goldszmidt.: Bayesian Network Classifiers. Machine Learning, Kluwer Academic Publishers, Boston, 29 (1997) 131–163
4. N. Friedman and M. Goldszmidt: Building Classifiers Using Bayesian Networks. In: Proc. of the 13h National Conf. on AI, Menlo Park, 1996 (1277-1284)
5. E. J. Keogh and M. J. Pazzani.: Learning Augmented Bayesian Classifiers: A Comparison of Distribution-Based and Classification-Based Approaches. In: Proceedings of the 7th International Workshop on AI and Statistics. (1999) 225-230
6. E. J. Keogh and M. J. Pazzani.: Learning the Structure Augmented Bayesian Classifiers. Int. J. on AI Tools, World Scientific Publishing Company, 11 (2002), 587-601
7. R. Kohavi.: Scaling up the Accuracy of Naive-Bayes Classifiers: A Decision-Tree Hybird, In: Simoudis, E.,Han, J. W., Fayyad, U. M.(eds.): Proceedings of the 2nd International Conference on KDD, The AAAI Press, Menlo Park, (1996)202-207
8. R. Kohavi, B. Becker and D. Sommerfield.: Improving Simple Bayes. In: Proceedings of the ECML, 1997
9. P. Langley and S. Sage.: Induction of Selective Bayesian Classifiers. In: Proceedings of the 10th Conf. on UAI, Morgan Kaufmann Publishers, Seattle, 1994 (339–406)
10. E. Minieka.: Optimization Algorithms for Networks and Graphs. New York: Marcel Dekker, Inc., (1978) 26-36
11. Z. Pawlak.: Rough Sets: Theoretical Aspects of Reasoning about Data, Dordrecht: Kluweer Academic Publishers, 1991
12. M. Sahami.: Learning Limited Dependence Bayesian Classifiers. In: Proceedings of the 2nd Int. Conf.on KDD, AAAI Press, Portland, OR, 1996, 335–338
13. D. Slezak.: Searching for Frequential Reduts in Decision Tables with Uncertain Objects. In: Proc.of the First Int.Conf.on RSCTC, Springer, 1998, 52-59
14. Wang. Z., Webb, G. I.: Comparison of Lazy Bayesian Rule and Tree-Augmented Bayesian Learning. In: Proc. of ICDM2002, IEEE Computer Society, (2002) 490-497
15. Webb, G. I., Boughton, J., Wang, Z: Averaged One-Dependence Estimators: Preliminary Results. In: Proc. of ADM 2002, Canberra, Australia: UTS, (2002) 65-73
16. Webb, G. I., Pazzani, M. J.: Adjusted Probability Naive Bayesian Induction. In: Proceedings of the 11th AJCAI, Springer-verlag, Brisbane, 1998 (285–295)
17. Witten, I. H., Frank, E.: Data Mining: Practical Machine Learning Tools and Techniques with Java Implementations. Seattle, WA: Morgan Kaufmann. (2000)
18. Zheng, Z., Webb, G. I.: Lazy Learning of Bayesian Rules. Machine Learning, Kluwer Academic Publishers, Boston, 41(2000) 53-84

Using Self-Consistent Naive-Bayes to Detect Masquerades

Kwong H. Yung

Stanford University Statistics Department
390 Serra Mall; Stanford CA 94305-4020; USA
khyung@stat.stanford.edu

Abstract. To gain access to account privileges, an intruder masquerades as the proper account user. This paper proposes a new strategy for detecting masquerades in a multiuser system. To detect masquerading sessions, one profile of command usage is built from the sessions of the proper user, and a second profile is built from the sessions of the remaining known users. The sequence of the commands in the sessions is reduced to a histogram of commands, and the naive-Bayes classifier is used to decide the identity of new incoming sessions. The standard naive-Bayes classifier is extended to take advantage of information from new unidentified sessions. On the basis of the current profiles, a newly presented session is first assigned a probability of being a masquerading session, and then the profiles are updated to reflect the new session. As prescribed by the expectation-maximization algorithm, this procedure is iterated until both the probabilities and the profiles are self-consistent. Experiments on a standard artificial dataset demonstrate that this self-consistent naive-Bayes classifier beats the previous best-performing detector and reduces the missing-alarm rate by 40%.

Keywords: self-consistent naive-Bayes, expectation-maximization algorithm, semisupervised learning; masquerade detection, anomaly detection, intrusion detection.

1 Introduction

This paper presents a simple technique for identifying masquerading sessions in a multiuser system. Profiles of proper and intruding behavior are built on the sessions of known users. As new, unidentified sessions are presented, the profiles are adjusted to reflect the credibility of these new sessions. Based on the expectation-maximization algorithm, the proposed *self-consistent* naive-Bayes classifier markedly improves on earlier detection rates.

1.1 Motivation

Unauthorized users behave differently from proper users. To detect intruders, behaviors of proper users are recorded and deviant behaviors are flagged. Typically, each user account is assigned to a single proper user with a specified role.

H. Dai, R. Srikant, and C. Zhang (Eds.): PAKDD 2004, LNAI 3056, pp. 329–340, 2004.

Hence, future sessions under that user account can be compared against the recorded profile of the proper user. This approach to detecting masquerades is called anomaly detection.

More generally, intrusion detection is amenable to two complementary approaches, misuse detection and anomaly detection. Misuse detection trains on examples of illicit behavior; anomaly detection trains on examples of proper behavior. Because examples of intrusions are generally rare and novel attacks cannot be anticipated, anomaly detection is typically more flexible. In the context of masquerade detection, anomaly detection is especially appropriate because each user account is created with an assigned proper user, whose behavior must not deviate far from the designated role.

1.2 Previous Approaches

The artificial dataset used in this study has been studied by many researchers [3,4,7,9,11,12]. Subsection 3.1 presents a summary of the dataset, and full details are available in [11]. Before this paper, the best result was achieved by using a naive-Bayes classifier with updating [9]. Yung [14] markedly improved detection rates but relied on confirmed updates. This paper also substantially improves over [9] and is competitive with [14].

There are at least two limitations to the updating scheme used in [9] and [11]. Principally, the detector must make a concrete decision whether to update the proper profile with the new session. In many cases, the judgment of a session is not always clear. Yet the detector is forced to make a binary decision before the detector can continue, because future decisions depend on the decision for the current case.

Furthermore, all future decisions depend on the past decisions, but the converse does not hold. In principle, the detector can defer the decision on the current session until additional new sessions from a later stage are studied. Yet in both [9] and [11], the detector is forced to make an immediate, greedy decision for each new session, without considering the information from future cases. Scores for previous sessions cannot be revised as new sessions are encountered. Hindsight cannot improve the detector's performance on earlier sessions, because backtracking is not allowed.

1.3 New Strategy

Without a concrete optimality criterion, there is no basis on which to judge the performance of the greedy sequential strategy. In general, however, the best greedy strategy is not always the best strategy overall. By ignoring strategies that involve backtracking, the greedy search may fail to make decisions most consistent with the entire dataset.

This paper presents an alternative strategy for identifying masquerading sessions. To start, an optimality criterion is defined by specifying a concrete objective function. Moreover, each new session is assigned a score indicating how likely that session is a masquerading session. Instead of a binary decision, the detector assigns to the new session a probability between 0 and 1. This new session is then used to update the detector in a continuous fashion. The technique presented in this paper is an extension of the naive Bayes classifier used in [9] and will be called the *self-consistent naive-Bayes* classifier.

Although the self-consistent naive-Bayes classifier is a significant step beyond the simple naive-Bayes classifier used by Maxion and Townsend [9], three elements introduced there still apply here: the classification formulation, the bag-of-words model, and the naive-Bayes classifier. These three strategies are not unique to detecting masquerading sessions, but rather they appear in the far more general context of intrusion detection and text classification. Yung [14] provides a brief review of these three elements, within the context of identifying masquerading sessions.

2 Theory

The simple naive-Bayes classifier is typically built on a training set of labeled documents. Nigam and his colleagues [10] demonstrated that classification accuracy can be improved significantly by incorporating labeled as well as unlabeled documents in the training set. The algorithm for training this new *self-consistent* naive-Bayes classifier can be derived from the formalism of the EM-algorithm. Only the simplest version of this extended naive-Bayes classifier is presented here.

2.1 Likelihood from Test Sessions

Only two distinct classes of sessions are considered, the proper class and the masquerading class. The indicator *variable* $1_s = 1$ exactly when session s is a masquerading session. Let $1-\epsilon$ and ϵ be the prior probabilities that a session is proper and masquerading, respectively. Moreover, let p_c and p'_c be the probability of command c in a proper and masquerading session, respectively.

The log-likelihood L_s of a test session s is simply, up to an additive constant

$$L_s = (1-1_s)(\log(1-\epsilon) + \sum_{c=1}^{C} n_{sc} \log p_c) + 1_s(\log \epsilon + \sum_{c=1}^{C} n_{sc} \log p'_c), \quad (1)$$

where n_{sc} is the total count of command c in session s.

Assuming that all test sessions are generated independently of each other, the cumulative log-likelihood L^t after t test sessions is, up to an additive constant

$$L^t_+ = \sum_{s=1}^{t} L_s = w^t_+ \log(1-\epsilon) + \sum_{c=1}^{C} n^t_{+c} \log p_c + w'^t_+ \log \epsilon + \sum_{c=1}^{C} n'^t_{+c} \log p'_c, \quad (2)$$

where

$$w^t_+ = \sum_{s=1}^{t}(1-1_s), w'^t_+ = \sum_{s=1}^{t} 1_s, n^t_{+c} = \sum_{s=1}^{t}(1-1_s)n_{sc}, n'^t_{+c} = \sum_{s=1}^{t} 1_s n_{sc}. \quad (3)$$

Here w_+^t and $w_+'^t = t - w_+^t$ are the cumulative numbers of proper and masquerading sessions, respectively; n_{+c}^t and $n_{+c}'^t$ are the cumulative counts of command c amongst proper sessions and masquerading sessions, respectively, in the t total observed test sessions.

2.2 Likelihood from Training Sessions

Now let n_{+c}^0 denote the total count of command c among proper sessions in the training set. Likewise, let $n_{+c}'^0$ denote the total count of command c among masquerading sessions in the training set. Letting r denote a session in the training set,

$$n_{+c}^0 = \sum_r (1 - 1_r) n_{rc}, n_{+c}'^0 = \sum_r 1_r n_{rc}. \tag{4}$$

Assuming that the sessions in the training set are generated independently, the log-likelihood L_+^0 of the proper and masquerading sessions in the training set is

$$L_+^0 = \sum_{c=1}^{C} n_{+c}^0 \log p_c + \sum_{c=1}^{C} n_{+c}'^0 \log p_c'. \tag{5}$$

This log-likelihood L_+^0 is useful for providing initial estimates of p_c and p_c' but provides no information about ϵ.

2.3 Posterior Likelihood

Rare classes and rare commands may not be properly reflected in the training set. To avoid zero estimates, smoothing is typically applied to the maximum-likelihood estimators. This smoothing can also be motivated by shrinkage estimation under Dirichlet priors on the parameters ϵ, p_c, and p_c'.

Here the parameters ϵ, p_c, and p_c' are drawn from known prior distributions with fixed parameters, and a simple standard Bayesian analysis is applied. Suppose that

$$(1 - \epsilon, \epsilon) \sim \text{Beta}(\beta, \beta'), \tag{6}$$
$$p \sim \text{Dirichlet}(\alpha), \tag{7}$$
$$p' \sim \text{Dirichlet}(\alpha'), \tag{8}$$

where α, α', β, and β' are taken to be known fixed constants specified in advance. Then the cumulative posterior log-likelihood $\tilde{L}^t$ is, up to an additive constant

$$\tilde{L}^t = (\beta - 1 + w_+^t) \log(1 - \epsilon) + \sum_{c=1}^{C} (\alpha_c - 1 + n_{+c}^0 + n_{+c}^t) \log p_c +$$
$$(\beta' - 1 + w_+'^t) \log \epsilon + \sum_{c=1}^{C} (\alpha_c' - 1 + n_{+c}'^0 + n_{+c}'^t) \log p_c'. \tag{9}$$

Here w_+^t, $w_+'^t$, n_{+c}^t, and $n_{+c}'^t$, defined in Equation 3, are cumulative quantities determined by the t available test sessions; n_{+c}^0 and $n_{+c}'^0$, defined in Equation 4, are the fixed quantities determined by the training sessions.

2.4 Shrinkage Estimators

Equation 9 gives the cumulative posterior log-likelihood $\tilde{L}^t$ after observing t test sessions, in addition to the training sessions. Shrinkage estimators are just the maximum-likelihood estimators calculated from the posterior log-likelihood $\tilde{L}^t$. So cumulative shrinkage estimators $\hat{\epsilon}^t$, $\hat{p}_c^t$ and $\hat{p}_c'^t$ for ϵ, p_c and p_c' after $t > 0$ sessions are

$$\hat{\epsilon}^t = \frac{\beta' - 1 + w_+'^t}{\beta - 1 + \beta' - 1 + t}, \tag{10}$$

$$\hat{p}_c^t = \frac{\alpha_c - 1 + n_{+c}^0 + n_{+c}^t}{\sum_{v=1}^C (\alpha_v - 1 + n_{+v}^0 + n_{+v}^t)}, \tag{11}$$

$$\hat{p}_c'^t = \frac{\alpha_c' - 1 + n_{+c}'^0 + n_{+c}'^t}{\sum_{v=1}^C (\alpha_v' - 1 + n_{+v}'^0 + n_{+v}'^t)}. \tag{12}$$

2.5 Complete Log-Likelihood

Initially, the training set is used to build the naive-Bayes classifier. As a new unidentified session is presented to the classifier, that new session is scored by the classifier and used to update the classifier. This scoring and updating procedure is repeated until convergence. As each new session is presented, the classifier is updated in a self-consistent manner.

For a session s in the training set, the identity is known. Specifically, $1_s = 0$ for a proper session, and $1_s = 1$ for a masquerading session in the training set. For a new unidentified session, 1_s is a missing variable that must be estimated from the available data.

2.6 EM-Algorithm for Naive-Bayes Model

The EM-algorithm [1] is an iterative procedure used to calculate the maximum-likelihood estimator from the complete log-likelihood. Each iteration of the EM-algorithm includes two steps, the expectation step and the maximization step.

In the expectation step, the indicators 1_s are replaced by their expectations, calculated from the current estimates of model parameters. In the maximization step, the new estimates of the model parameters are calculated by maximizing the expected log-likelihood. For each stage t, these two-steps are repeated through multiple iterations until convergence. Below the iterations of the EM-algorithm during a fixed stage t are described.

Let $\theta^t = (\epsilon^t, p_c^t, p_c'^t)$ denote the estimate of $\theta = (\epsilon, p_c, p_c')$ at stage t, after observing the first t test sessions. For each known session r in the training set, 1_r is a known constant. So n_{+c}^0 and $n_{+c}'^0$ of Equation 4 remain fixed even as θ^t changes.

For each *unidentified* test session $s = 1, 2, \ldots, t$, the expectation step estimates $E[1_s|c_s;\theta^t] = P(1_s = 1|c_s;\theta^t)$ via Bayes rule as

$$P(1_s = 1|c_s;\theta^t) = \frac{P(1_s = 1;\theta^t)P(c_s|1_s = 1;\theta^t)}{P(c_s;\theta^t)}, \tag{13}$$

where

$$P(c_s;\theta^t) = P(1_s = 0;\theta^t)P(c_s|1_s = 0;\theta^t) \\ +P(1_s = 1;\theta^t)P(c_s|1_s = 1;\theta^t). \quad (14)$$

For the expectation step of the current iteration, the estimate $\hat{\theta}^t$ from the maximization step of the previous iteration is used for θ^t.

The maximization step calculates updated estimate $\hat{\theta}^t$ of the parameter θ^t. The usual cumulative shrinkage estimators of Equations 10–12 are used. However, the counts w^t_+, w'^t_+, n^t_{+c}, and n'^t_{+c} are now no longer known constants but rather are random variables. These random variables are replaced by their estimates from the expectation step above. In other words, the maximization step of the current iteration uses the estimates from the expectation step of the current iteration.

3 Results

The technique proposed in this paper is tested on a standard dataset previously analyzed by other authors using different techniques. The dataset contains command logs from 50 users on a UNIX multiuser system. To create artificial intrusion sessions, sessions from another 20 users are injected at random into the command logs of the 50 known users. The dataset contains a training portion, without masquerading sessions. The remaining test portion contains possible masquerading sessions from the artificially created intruders. A summary of the dataset is provided below, but full details are provided in [11].

3.1 Reference Dataset

For privacy reasons, the command logs were stripped of the command options and arguments, leaving only the commands themselves. For each of 70 users, the first 15000 recorded commands were kept. Then 50 users were chosen at random to serve as the dataset, and the remaining 20 users were excluded, except to provide artificial intrusion data. For each user, the sequence of 15000 commands was divided into 150 sequential blocks, each containing 100 commands in sequence. Each artificially created block is treated as a session. Therefore, each of the 50 users had 150 sessions of 100 commands.

The first 50 sessions of the 50 users are assumed to be free of intruders. These 50 sessions for 50 users constitute the training set. The remaining 100 sessions for each of the 50 users form the test set. Sessions from the 20 excluded users are injected into the test session of the 50 known users. The extra sessions in the test sets are then removed, leaving just 100 test sessions for each of the 50 known users.

To prevent over-training on the data creation process, the exact details of how the test set is created cannot be used in the intrusion detection procedure. Essentially, once an intruder has broken into a user account, he is likely to return on the very next session. A proper session is followed by an intruder session with 0.01 probability. A masquerading session is followed by a masquerading session from the same intruder with 0.80 probability. A chain of masquerading sessions is taken from the intruder in a contiguous sequence. Out of the total 5000 test sessions, there were 231 total masquerading sessions, and the remaining 4769 were proper sessions. These aspects of the data creation are not used as part of the classification procedure.

3.2 Experimental Design

The details for one single experiment are described below. The example shows how masquerading sessions on User 12 were detected. For other users, the same analysis applies.

The full training set was divided into two classes. The proper class consisted of 50 sessions for User 12. The masquerade class consisted of the 50 sessions for each of the other 49 users. So the masquerade class was built on a total of 2450 sessions.

The test set for user 12 had 100 sessions, all purportedly proper sessions. The classifier for user 12 was then run against each of the 100 sessions in a sequential fashion. As a new session is added, the scores of the previous sessions are also updated. At the end of the 100 sessions, the final set of all 100 scores were recorded.

The ROC curve for user 12 was generated by thresholding against the *final* set of scores, after all test sessions were presented sequentially. In this case, a session's score is the probability that the session was generated from the proper profile. Therefore, sessions with lower scores were flagged before sessions with higher scores.

3.3 Classifier Settings

A self-consistent naive-Bayes classifier was built for each individual user. The algorithm is outlined in Subsection 2.6. The initial naive-Bayes was built on the training set. As each new session is presented, that new session is first scored by the previous classifier. Afterwards, all the unidentified sessions are used to construct the self-consistent naive-Bayes. This procedure is repeated for each new session.

In the full self-consistent naive-Bayes classifier presented in Section 2, the fraction ϵ of masquerading sessions was adjusted in each iteration. In the results presented here, the fraction $\epsilon = 0.50$ was held fixed and was not estimated as part of the EM-algorithm. Keeping the prior weight $(1-\epsilon, \epsilon)$ fixed at $(0.50, 0.50)$ allowed for faster convergence and did not adversely affect the final results.

For each new session, the EM-algorithm was used to adjust the fractions p_c and p'_c. The Dirichlet parameters $\alpha_c = 1.01$ and $\alpha'_c = 1.01$ for all $c \in \mathbf{C}$ were chosen to match the parameters used in [9]. The EM-algorithm was iterated until the change between iteration i and $i+1$, measured by the quantity $||p^{(i+1)} - p^{(i)}||^2 + ||p'^{(i+1)} - p'^{(i)}||^2$ averaged over the total number of estimated parameters, was less than some tolerance, set to be 2.2×10^{-16}. In practice, the algorithm was not sensitive to the precise tolerance.

3.4 Composite ROC Curves

A separate classifier was created for each user and used to score that user's test sessions. The scores of test sessions from one user can be compared because those scores were assigned by that user's classifier. However, scores from different users cannot easily be compared because those scores are assigned from different classifiers.

To evaluate a strategy's success over the entire dataset of 50 users, a composite ROC curve can be useful. There are several common ways to integrate the individual 50 ROC curves into one single curve, but all methods are at best arbitrary.

In this paper, the scores from the 50 different classifiers were taken at face value. The 5000 total test sessions from the 50 user were sorted, and sessions with the lowest scores were flagged first. Because the scores were probability values, this global thresholding strategy has some merit. This method for constructing the composite ROC curve was also used in [9] and [11]. Because the same methodology was used in these previous papers, the composite ROC curves can be meaningfully compared.

3.5 Experimental Results

Two ROC curves are used to compare the self-consistent naive-Bayes classifier to the one-step discrete adaptive naive-Bayes classifier of [9]. These curves together demonstrate that the self-consistent naive-Bayes classifier offers significant improvements. All results reported here are based on offline evaluation, after all the sessions have been presented sequentially. Online evaluation and deferral policies are discussed in a separate paper.

Figure 1 shows in fine detail the composite ROC curves of all 50 users, for false-alarm rate 0.00–0.10. Indeed, the self-consistent naive-Bayes outperforms the adaptive naive-Bayes uniformly for all but the small portion of false-alarm rate 0.00–0.01. In particular, for false-alarm rate 0.013, the self-consistent naive-Bayes reduces the missing-alarm rate of the adaptive naive-Bayes by 40%.

Figure 2 shows the ROC curve for user 12 alone. As noted by the other authors [11,9], user 12 is a challenging case because the masquerading sessions in the test set appear similar to the proper sessions. On user 12, the adaptive naive-Bayes performs worse than random guessing, but self-consistent naive-Bayes encounters little difficulty. In fact, the 50 ROC curves over each individual user show that self-consistent naive-Bayes typically outperforms adaptive naive-Bayes.

4 Discussion

The self-consistent naive-Bayes classifier outperforms the adaptive naive-Bayes classifier in all practical circumstances, where only low false-alarm rates can be tolerated. For false-alarm rate 0.013, the self-consistent naive-Bayes classifier lowers the missing-alarm rate by 40%.

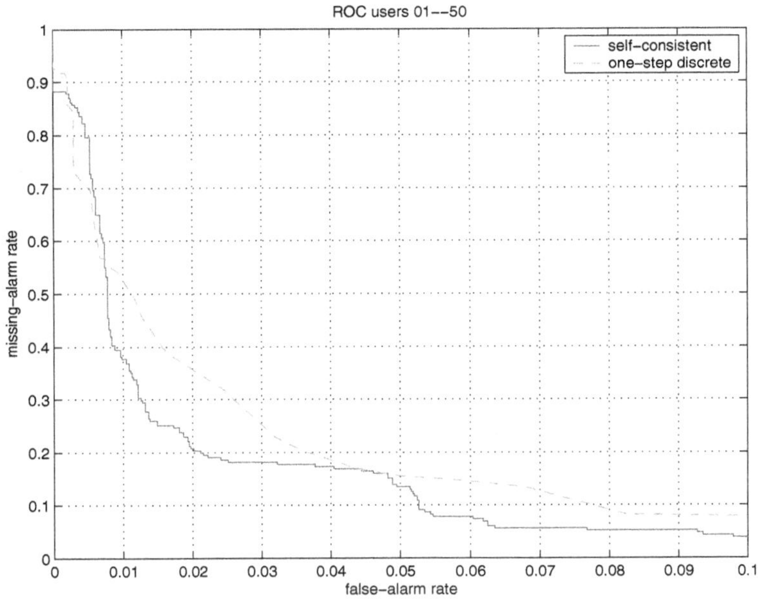

Fig. 1. ROC curve over users 01–50, less than 0.10 false-alarm rate.

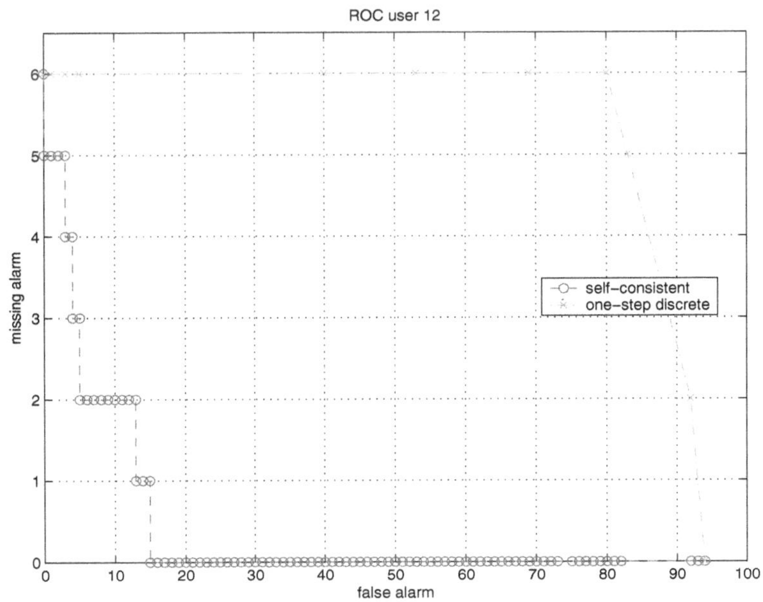

Fig. 2. ROC curve user 12

4.1 Long-Term Scenario

Unlike the adaptive naive-Bayes, the self-consistent naive-Bayes is not forced to make a binary decision as each new session is presented. This advantage will become more dramatic as the number of users and the number of sessions increase. As the classifier learns from more cases, mistakes from earlier decisions are propagated onwards and magnified. Because the adaptive naive-Bayes is forced to make a binary decision immediately, it is also more likely to make errors. In a long-term scenario, the improvements of the self-consistent naive-Bayes classifier are expected to become far more pronounced.

4.2 Computation Time

Although the self-consistent naive-Bayes classifier offers a great number of advantages, more computation is required. The iterative procedure used to calculate the probabilities and to update the profile must be run until convergence. This additional effort allows the self-consistent naive-Bayes classifier to assign scores in a non-greedy fashion. Naturally, searching through the space of models requires more effort than the simple greedy approach.

In practice, convergence is often quick because most sessions have probability scores at the extremes, near 0 or 1. In the context of sequential presentation, only one single session is presented at a time. Computing the score of a single session requires little additional effort. Moreover, the set of scores from one stage can be used as the starting point to calculate scores for the next stage. This incremental updating approach relieves the potential computational burden.

4.3 Nonstationary User Behavior

For some users, the sessions in the test set differ significantly from the sessions in the training set. Because user behavior changes unpredictably with time, a detector based on the old behavior uncovered in the training set can raise false alarms. This high rate of false alarms can render any detector impractical, because there are not enough resources to investigate these cases.

In certain real world applications, however, identifying changed behavior may well be useful because a proper user can misuse his own account privileges for deviant and illicit purposes [8]. If the detector is asked to detect all large deviations in behavior, then flagging unusual sessions from the proper user is acceptable. This redefinition is especially useful to prevent a user from abusing his own privileges.

4.4 Future Directions

Only the simplest model of user sessions has been presented. In [10], EM-algorithm was applied to more elaborate and realistic models. These extensions offer much potential, and future work will explore these possibilities.

The iterative nature of the EM-algorithm is quite intuitive. In fact, many instantiations [2,5,6,13] of the EM-algorithm existed well before the general EM-algorithm was formulated in [1]. Moreover, the self-consistency paradigm can be applied even to classifiers not based on a likelihood model. A self-consistent version can be constructed for potentially any classifier, in the same way that a classifier can be updated by including the tested instance as part of the modified training set.

The naive-Bayes classifier relies only on command frequencies. As a user's behavior changes over time, the profile built from past sessions become outdated. An exponential-weighing process can be applied to counts from past sessions. For the naive-Bayes classifier, this reweighting of counts is especially simple. Such an extension becomes even more valuable in realistic scenarios, where a sequential classifier is used in an online context for a long time.

5 Summary

In previous approaches to detecting masquerading sessions, the detector was forced to make a binary decision about the identity of a new session. The self-consistent naive-Bayes does not make a binary decision but rather estimates the probability of a session being a masquerading session. Moreover, past decisions can be adjusted to accommodate newer sessions. Experiments prove that this sensible extension markedly improves over more restrictive adaptive approaches, by reducing the missing-alarm rate by 40%.

The self-consistent naive-Bayes classifier extends the usual naive-Bayes classifier by taking advantage of information from unlabeled instances. New instance are assigned probabilistic labels, and then the profiles are updated in accordance with the assigned probabilities of the new instances. This procedure is iterated until convergence to a final set of probabilities which are consistent with the updated profile.

By its very nature, the self-consistent naive-Bayes classifier is adaptive to the new sessions. Moreover, information from new sessions is also used to adjust scores of previous new sessions. In this way, the scores of sessions are assigned in a self-consistent manner. As a specific instance of the EM-algorithm, the self-consistent naive-Bayes classifier finds a model optimal with respect to its likelihood given the available data.

Acknowledgments. This research project was funded in part by the US Department of Justice grant 2000-DT-CX-K001. Jerome H. Friedman of the Stanford University Statistics Department provided invaluable advice through many helpful discussions. Jeffrey D. Ullman of the Stanford University Computer Science Department introduced the author to the field of intrusion detection and offered insightful critiques throughout the past three years. The author is grateful for their guidance and support.

References

1. Arthur P. Dempster, Nam M. Laird, and Donald B. Rubin. "Maximum likelihood from incomplete data via the EM algorithm." *Journal of the Royal Statistical Society, Series B*, 39(1), 1–38, 1977.
2. N. E. Day. "Estimating the components of a mixture of two normal distributions." *Biometrika*, **56**, 463–474, 1969.
3. William DuMouchel. "Computer intrusion detection based on Bayes factors for comparing command transition probabilities." Technical Report 91, National Institute of Statistical Sciences, Research Triangle Park, North Carolina 27709-4006, 1999.
4. William DuMouchel and Matthias Schonlau. "A comparison of test statistics for computer intrusion detection based on principal components regression of transition probabilities." *Proceedings of the 30th Symposium on the Interface: Computing Science and Statistics*, 30:404–413, 1999.
5. Victor Hasselblad. "Estimation of parameters for a mixture of normal distributions." *Technometrics*, **8**, 431–444, 1966.
6. Victor Hasselblad. "Estimation of finite mixtures of distributions from the exponential family." *Journal of American Statistical Association*, **64**, 1459–1471, 1969.
7. Wen-Hua Ju and Yehuda Vardi. "A hybrid high-order Markov chain model for computer intrusion detection." Technical Report 92, National Institute for Statistical Sciences, Research Triangle Park, North Carolina 27709–4006, 1999.
8. Vernon Loeb. "Spy case prompts computer search." *Washington Post*, 05 March 2001, page A01.
9. Roy A. Maxion and Tahlia N. Townsend. "Masquerade Detection Using Truncated Command Lines." *International Conference on Dependable Systems and Networks (DSN-02)*, pp. 219–228, Washington, DC, 23–26 June 2002. IEEE Computer Society Press, Los Alamitos, California, 2002.
10. Kamal Nigam, Andrew K. McCallum, Sebastian Thrun, Tom Mitchell. "Text classification from labeled and unlabeled documents using EM." *Machine Learning*, 39:2:103–134, 2000.
11. Matthias Schonlau, William DuMouchel, Wen-Hua Ju, Alan F. Karr, Martin Theus, and Yehuda Vardi. "Computer intrusion: detecting masquerades." *Statistical Science*, 16(1):58–74, February 2001.
12. Matthias Schonlau and Martin Theus. "Detecting masquerades in intrusion detection based on unpopular commands." *Information Processing Letters*, 76(1–2):33–38, November 2000.
13. John H. Wolfe. "Pattern clustering by multivariate mixture analysis." *Multivariate Behavioral Research*, **5**, 329–350, 1970.
14. Kwong H. Yung. "Using feedback to improve masquerade detection." *Lecture Notes in Computer Science: Applied Crytopgraphy and Network Security.* Proceedings of the ACNS 2003 Conference. LNCS **2846**, 48–62. Springer-Verlag, October 2003.

DB-Subdue: Database Approach to Graph Mining*

Sharma Chakravarthy, Ramji Beera, and Ramanathan Balachandran

Information Technology Laboratory and CSE Department
The University of Texas at Arlington
{sharma, beera, balachan} @cse.uta.edu

Abstract. In contrast to mining over transactional data, graph mining is done over structured data represented in the form of a graph. Data having structural relationships lends itself to graph mining. Subdue is one of the early main memory graph mining algorithms that detects the best substructure that compresses a graph using the minimum description length principle. Database approach to graph mining presented in this paper overcomes the problems – performance and scalability – inherent to main memory algorithms. The focus of this paper is the development of graph mining algorithms (specifically Subdue) using SQL and stored procedures in a Relational database environment. We have not only shown how the Subdue class of algorithms can be translated to SQL-based algorithms, but also demonstrated that scalability can be achieved without sacrificing performance.

1 Introduction

Database mining has been a topic of research for quite some time [1-5]. Most of the work in database mining has concentrated on discovering association rules from transactional data represented as (binary) relations. The ability to mine over graphs is important, as graphs are capable of representing complex relationships. Graph mining uses the natural structure of the application domain and mines directly over that structure (unlike others where the problem has to be mapped to transactions or other representations). Graphs can be used to represent structural relationships in many domains (web topology, protein structures, chemical compounds, relationship of related transactions for detecting fraud or money laundering, etc). Subdue is a mining approach that works on a graph representation. Subdue uses the principle of Minimum description length (or MDL) to evaluate the substructures. The MDL principle has been used for decision tree induction [8], pattern discovery in bio-sequences [9], image processing [6], and concept learning from relational data.

Main memory data mining algorithms typically face two problems with respect to scalability. Graphs could be larger than available main memory and hence cannot be loaded into main memory or the computation space required at runtime exceeds the available main memory. Although Subdue provides us with a tool for mining

* This work was supported, in part, by the Office of Naval Research, the SPAWAR System Center-San Diego & by the Rome Laboratory (grant F30602-02-2-0134), and by NSF grant IIS-0097517

H. Dai, R. Srikant, and C. Zhang (Eds.): PAKDD 2004, LNAI 3056, pp. 341–350, 2004.

interesting and repetitive substructures within the data, it is a main memory algorithm. The algorithm constructs the entire graph as an adjacency matrix in main memory and then mines by iteratively expanding each vertex into larger subgraphs.

The focus of this paper is on the development of graph mining algorithms (specifically Subdue class of algorithms) using SQL and stored procedures using a Relational database management system (RDBMS). Representation of a graph, and its manipulation – generation of larger subgraphs, checking for exact and inexact matches of subgraphs – are not straightforward in SQL. They have to be cast into join-based operations and at the same time avoid manipulations that are inefficient (correlated subqueries, cursors on large relations, in-place inserts and deletes from a relation). This paper elaborates on a suite of algorithms that have been carefully designed to achieve functionality as well as scalability. We have not only shown how graph mining can be mapped to relations and operations on them, but we have also demonstrated that scalability can be achieved without sacrificing performance.

Section 2 provides an overview of the main memory algorithm Subdue and includes related work. Section 3 describes graph representation and how the substructures are represented and generated. Section 4 presents the relational approach to graph mining. Section 5 presents performance evaluation including comparison with the main memory algorithm. Conclusions and future work are presented in section 6.

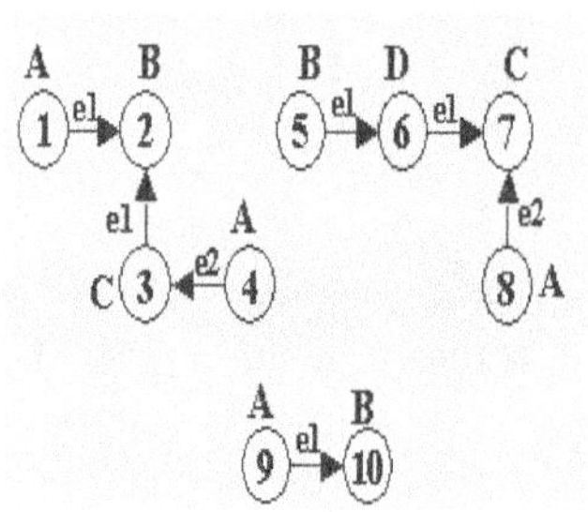

Fig. 1. Input Graph

V	1	A	
V	2	B	
V	3	C	
V	4	A	
V	5	D	
V	6	E	
V	7	F	
V	8	A	
V	9	A	
V	10	B	
D	1	2	e1
D	3	2	e1
D	4	3	e2
D	5	6	e1
D	7	6	e1
D	8	7	e2
D	9	10	e1

Fig. 2. Input file for Subdue

2 Related Work

In this section we briefly describe the working of the subdue algorithm and the parameters associated with it. Subdue represents the data as a labeled graph. Objects in the data are represented as either vertices or as small subgraphs and the relationships between the objects are represented as edges. For the graph shown in Fig. 1, the input to Subdue is a file as shown in Fig. 2, which describes the graph. Each vertex has a unique number, and a label (not necessarily unique). Each edge has a label (not necessarily unique) and the numbers of the vertices that it connects – from source to destination. The edge can be undirected or directed. Each substructure is expanded in all possible ways, by adding an edge and a vertex to the instance or just an edge if both the vertices are already present in the instance. Some of the instances, which were expanded in a different way but match the substructure within a threshold using the inexact graph match, are also included in the instances of that substructure. Each of these substructures is evaluated using the MDL heuristic and only a limited number of substructures (specified as **beam**) are selected for future expansion. **Beam** represents the maximum number of substructures kept in the substructure list to be expanded. The algorithm's run time is bounded by the user specified beam width, and the **Limit**, which is the total number of substructures considered by the algorithm.

The output for Subdue for the input file in Fig. 2 is a list of substructures. The best substructure returned by Subdue is A → B with two instances. The description length as calculated by MDL and the compression achieved by this substructure are also output. Since Subdue executes in main memory and keeps the entire graph in main memory, its performance degrades for larger data sets.

Frequent Subgraphs or FSG is another approach [10] to graph mining. FSG is used to discover subgraphs that occur frequently in a data set of graphs. FSG uses a mechanism called canonical labeling [12] for graph comparison. The heuristic used by FSG for evaluating the substructures is count, which is the total number of substructures available in the whole dataset of graphs. In the candidate generation stage, joining two frequent k size substructures generates the k+1 size substructure candidates. For two k size frequent substructures to be joined they must have the same *core,* which means they must have a common k - 1 size subgraph.

Vertex1	Vertex2	EdgeName
1	2	e1
3	2	e1
4	3	e2
5	6	e1
6	7	e1
8	7	e2
9	10	e1

Fig. 3. Tuples in the edges relation

VertexNo	VertexLabel
1	A
2	B
3	C
4	A
5	B
6	D
7	C
8	A
9	A
10	B

Fig. 4. Tuples in the vertices relation

3 Graph Representation and Generation of Substructures

This section describes how graphs are represented in a database. Since databases have only relations, we need to convert the graph into tuples in a relation. The vertices in the graph are inserted into a relation called Vertices and the edges are inserted into a relation called Edges. The input is read from a delimited ASCII file and loaded into the relations. For the graph shown in Fig. 1, the corresponding vertices and the edges relations are shown in Fig. 3 and Fig. 4. The joined_1 relation will consist of all the substructures of size one – size representing the number of edges. The new relation joined_1 is created because the edges relation does not contain information about the vertex labels. So the edges relation and the vertices relation are joined to get the joined_1 relation. The resultant joined_1 relation is shown in Fig. 5. For a single edge substructure, the edge direction is always from the first vertex to the second vertex.

For a higher edge substructure, we need to know the directionality of the edges between the vertices of the substructure. In case of a two-edge substructure, the third vertex can either be expanded from the first vertex or the second vertex. Therefore a new attribute called the extension attribute (extn) has been introduced for each expansion of the substructure. For example, if the 5 → 6 substructure in Fig. 1 is extended to get the 5 → 6 → 7 substructure then the extension attribute will have value 2, indicating that the third vertex is extended from the second vertex. The edge direction can be either into the extended vertex or away from the extended vertex. When the edge comes into the extended vertex, it is represented by a negative value in the extension attribute. For example, if 1 → 2 substructure in Fig. 1 is expanded to 1 → 2 ← 3 then the extension attribute will have value –2, indicating that the third vertex is expanded from the second vertex and the edge is going towards the second vertex. In general if the extension *i (attribute ext)* is –j, then the edge i+1 is from vertex i+2 to j. If the extension i is j, then the edge i+1 is from vertex j to i+2. For a substructure of size n, we need n-1 extension attributes.

vertex1	vertex2	vertex1name	vertex2name	edgename
1	2	A	B	e1
3	2	C	B	e1
4	3	A	C	e2
5	6	B	D	e1
6	7	D	C	e1
8	7	A	C	e2
9	10	A	B	e1

Fig. 5. Joined_1 relation

4 Approach

The subgraph discovery process using relations uses the above representation for the graph. We start with a one-edge substructure and compute a count for each tuple (or substructure) to indicate the number of substructures it is isomorphic to (either exact match or an inexact match using a threshold). This is done using the Group by clause in DB2 [11] to update the count (number of instances) of each substructure. Based on count as a heuristic, the substructures are ranked. Substructures are then expanded to three edge substructures and the process is repeated until we find the best n sized substructures.

4.1 Algorithm

The pseudocode for this algorithm is given below:

```
Subdue-DB (input file, size)
Load vertices into vertices relation;
Load edges into edges relation;
Load joined_base relation and joined_1 relation by joining
vertices and edges relation
```

```
i = 2
WHILE (i < size)
        Load joined_i relation
          (substructures of size i)
        from  beam_joined_i-1, joined_base
        create frequent_i relation
        DECLARE  Cursor c1 on frequent_i order by count
        DECLARE  Cursor c2 on frequent_i
        WHILE (c1.count < beam)
             FETCH c1 into g1
             WHILE (c2)
                  FETCH c2 into g2
                  If (! Isomorphic (g1, g2)=0)
                    Insert c1 into frequent_beam_i
        Insert into beam_joined_i
          from frequent_beam_i,joined_i
i++
```

4.2 Substructure Discovery

The substructure discovery algorithm starts with one-edge substructures unlike in Subdue, which starts with all the unique vertex labels. In the database version, each instance of the substructure is represented as a tuple in the relation. The algorithm starts with initializing the vertices and the edges relation. The joined_base relation is then constructed by joining the vertices and edges relation. The joined_base relation will be used for future substructure expansion. The joined_1 relation is a copy of the joined_base relation. The frequent_1 relation is created to keep track of substructures of size 1 and their counts. Here, size refers to the number of edges as opposed to Subdue main memory algorithm in which size refers to the number of vertices. Projecting the vertex labels and edge labels attributes on the joined_1 relation and then grouping them on the same attributes, produces the count for the frequent_1 relation as shown in Fig. 6.

vertex1name	vertex2name	edge1	count1
A	B	e1	2
A	C	e2	2
B	D	e1	1
C	B	e1	1
D	C	e1	1

Fig. 6. Frequent_1 relation

Therefore the frequent_1 relation does not have vertex numbers. Group by clause is used so that all the exact instances of the substructure are grouped as one tuple with their count updated.

The pruning of substructures with a single instance corresponds to deleting the tuples with count value 1 from the frequent_1 relation. Since these substructures do not contribute to larger repeated substructures, they are deleted from the joined_base relation as well, so that further expansions using that relation do not produce substructures with single instances (there may be 1 instance substructures produced later when repeated substructures of length i do not grow to repeated substructures of length i + 1). *This is based on the property that the number of instances of a larger substructure cannot be more than the number of instances of a smaller structure embedded in it. This allows for pruning of substructures that will not produce larger substructures of significance. This property is similar to the subset property used in association rule mining.*

vertex1	vertex2	vertex1name	vertex2name	edgename
1	2	A	B	e1
4	3	A	C	e2
8	7	A	C	e2
9	10	A	B	e1

Fig. 7. Updated joined_base relation

The updated frequent_1 relation contains first two rows of frequent_1 relation shown in Fig. 6. and the resultant joined_base relation is shown in Fig. 7. The substructures are then sorted in decreasing order of the count attribute and the beam number of substructures is inserted into another relation called the frequent_beam_1 relation. Only the substructures present in this relation are expanded to larger substructures. Since all instances of the substructure are not present in the frequent_beam_1 relation, this relation is joined with the joined_1 relation to construct the joined_beam_1 relation. The joined_beam_1 is in turn joined with the joined_base relation to generate the two edge substructures.

4.3 Extending to Higher-Edge Substructures

In the main memory approach, every substructure, which is necessarily a subgraph , is defined as a *structure* in the programming language. Extensions to two or more edges are generated by growing the substructure appropriately. In the database representation, as there are no *structures,* the only information to be used will be the single edge substructures, which are basically tuples in the joined_base relation. The number of attributes of the relation needs to be increased to capture substructures of increased size. The joined_beam_1 relation contains all the instances of the single edge substructure. Each single edge substructure can be expanded to a two-edge substructure on any of the two vertices in the edge. In general, an n edge substructure can be expanded on n + 1 vertices in the substructure. So by making a join with the joined_base relation we can always extend a given substructure by one edge. In order to make an extension, one of the vertices in the substructure has to match a vertex in the joined_base relation. The resultant tuples are stored in the joined_2 relation. The following query extends a single edge substructure in all possible ways,

```
Insert into joined_2(vertex1,vertex2,vertex3,
                  vertex1name,vertex2name,vertex3name,
                       edge1name,edge2name,ext1)
 (select j1.vertex1,j1.vertex2,j2.vertex2,
        j1.vertex1name, j1.dge1name,j2.edge1name,1
```

```
from   joined_1 j1, joined_base j2
where j1.vertex1=j2.vertex1 and
       j1.vertex2!=j2.vertex2
Union
Select j1.vertex1,j1.vertex2,j2.vertex2,
       j1.vertex1name,j1.vertex2name,
       j2.vertex2name, j1.edge1name, j2.edge1name,2
from   joined_1 j1, joined_base j2
where j1.vertex2 = j2.vertex1
Union
Select j.vertex1,  j.vertex2,j1.vertex1,
       j.vertex1name,j.vertex2name, j1.vertex1name,
       j.edge1,j1.edge1,-2
from   joined_1 j, joined_base j1
where j.vertex2 = j1.vertex2  and
       j.vertex1 != j1.vertex1)
```

In the above there are 3 queries (and 2 unions). Since there are 2 nodes (in a one-edge substructure), the first query corresponds to the positive extension of the first node, and the other two correspond to the positive and negative extensions of the second node. In general, for a substructure with n nodes, there will be (n-1)*2 +1 queries and (n-1)*2 unions. The reason for not performing the negative extension on the first node is that it will be covered by the positive expansion of the node whose edge is coming into it. Also, every node is a starting point for expanding it to a larger size substructure. Hence the incoming edges of the starting node need not be considered. The rest of the steps are the same as the single edge substructure expansion, except for the addition of a new extension attribute. The n-1 edge substructures are stored in the joined_beam_n-1 relation. In order to expand to n edge substructures, an edge is added to the existing n-1 edge substructure. Therefore, the joined_beam_n-1 relation is joined with the joined_base relation to add an edge to the substructure. The frequent_n relation is in turn generated from the joined_n relation. The frequent_beam_n relation contains the beam tuples. The joined_beam_n is then generated which has only the instances of the beam tuples. The main halting condition for the algorithm is the user-specified parameter – the max size (limit). Once the algorithm discovers all the substructures of the max size the program terminates.

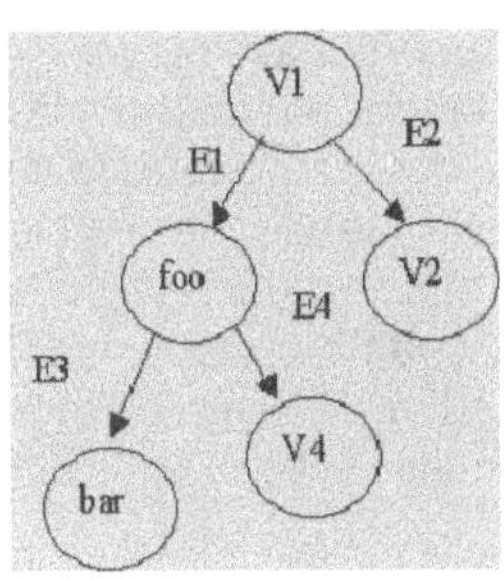

Fig. 8. Embedded Substructure

5 Performance Evaluation of Implementation Alternatives

This section discusses the performance comparison between DB-Subdue and Sudbue main memory algorithm for various datasets. Several optimizations were done before the final approach was implemented. This process was very useful in understanding the effect of various types of query formulations (correlated, using the except or minus operator, use of indexes, etc.). We use ***DB2 6.1, running on a SUNW, Ultra-5_10 5.6 with 348MB of RAM***. The embedded substructure in the graph on which the performance comparison will be made is shown in Fig. 8. Each experiment was performed 4 times and the fist result was discarded (so that cold start is avoided). The numbers presented show the average of the next 3 runs. The data set is encoded, as TnVmE where n is the number of vertices in the graph and m is the number of edges in the graph.

The graphical comparisons between subdue and DB-Subdue is shown in Fig. 9 and Fig. 10 for beam value of 4 and 10, respectively. The timings are shown on the logarithmic scale, as the range is very large. The final timings and their comparison with Subdue are shown. The numbers of instances of the substructures discovered by all the algorithms are the same. DB-Subdue approach was applied to graphs that had 800K vertices and 1600K edges. The main memory algorithm would not go beyond 50K vertices and 100K edges. The performance crossover point for the algorithm takes place at as low as 100 edges in the graph (which was a surprise!). The main memory algorithm took more than 60,000 seconds to initialize the T400KV1600KE graph and took more than 20 hours to initialize the T800KV1600KE data set. Testing was not done beyond the T800KV1600KE data set for the database approach because the graph generator could not produce the required graph. We also ran the same data set with beam size set to 10. The run time for the data set T800KV1600KE using the database approach with a beam size of 10 is 12,703 seconds.

The time taken by the Subdue algorithm for a data set of T50KV100KE and a beam size of 10 is 71,192 seconds. One of the main reasons for such improvement has been using pure SQL statements to achieve the functionality. The database approach is also insensitive to the beam size as compared to Subdue. For the largest data set, with beam=4, it takes 34,259 seconds and with beam=10 it takes 71,192 seconds. The time taken when the beam is increased does not increase with the same rate with which the Subdue algorithm increases. This is a significant improvement from scalability viewpoint. Also, the absolute numbers are not important here, as the experiments have been done on an old machine.

It took us several iterations before arriving at the final version of the algorithm. Initial naïve versions of DB-subdue performed worse than the main memory algorithm. We eliminated correlated subqueries and used the minus operator, changed the approach from deleting the tuples (which were large in number) to retaining the tuples that we wanted to expand (small in number). The performance using the above optimizations have resulted in orders of magnitude of difference in computation time allowing us to scale to very large data sets.

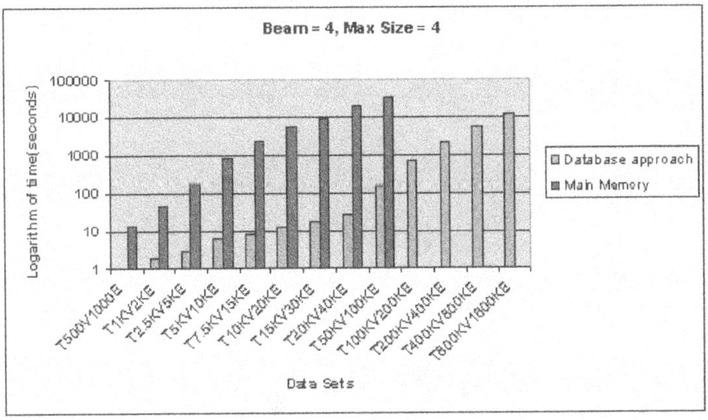

Fig. 9. Graphical comparison of the approaches

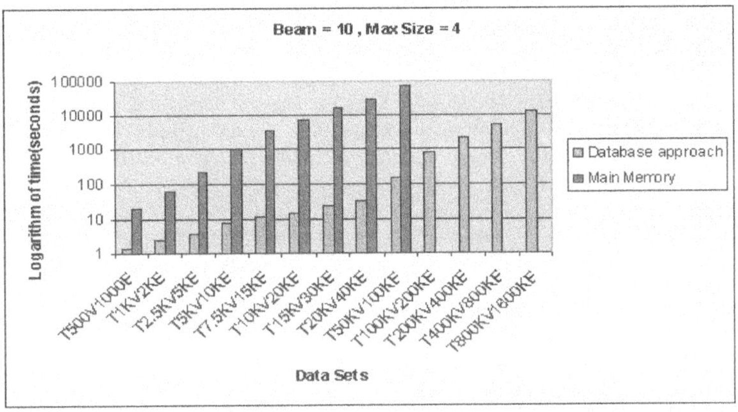

Fig. 10. Graphical comparison of the approaches

6 Conclusion and Future Work

We believe that this is the first attempt to mine substructures over graphs using the database approach. The idea behind graph-based data mining is to find interesting and repetitive substructures in a graph. One of the main challenges was the representation and generation of substructures in a database. The next challenge was to optimize the approach to achieve performance superiority over the main memory version. Finally, we were able to run graph mining on data sets that have 800K vertices and 1600K edges to achieve the desired functionality and scalability.

Currently, DB-Subdue is being extended in several ways: handle cycles in a graph, to include the concept of overlap among substructures (where a smaller subgraph is common between two subgraphs), inexact graph match and develop the equivalent of MDL for the database approach.

References

[1] Agrawal, R. and R. Srikant, *Fast Algorithms for Mining Association Rules*. in *Proceedings 20th International Conference Very Large Databases, VLDB*. Chile: p. 487-499, 1994.

[2] Han, J., J. Pei, and Y. Yin, *Mining Frequent Patterns without Candidate Generation*. in *ACM SIGMOD International Conference on Management of Data*, 2000.

[3] Thomas, S. *Architectures and Optimizations for integrating Data Mining algorithms with Database Systems*, Ph.d Thesis CISE Dept, University of Florida. 1998.

[4] Sarawagi, S., S. Thomas, and R. Agrawal, *Integrating Mining with Relational Database Systems: Alternatives and Implications*. in *SIGMOD*. Seattle: p. 343-354, 1998.

[5] Mishra, P. and S. Chakravarthy, *Performance Evaluation and Analysis of SQL-92 Approaches for Association Rule Mining*. in *BNCOD Proceedings*: p. 95 - 114, 2003.

[6] Leclerc, Y.G., *Constructing simple stable descriptions for image partitioning*. International journal of Computer Vision, 1989. **3(1)**: p. 73 -102.

[7] Rissanen, J., *Stochastic Complexity in statistical inquiry*. in *World Scientific Publishing Company*, 1989.

[8] Quinlan, J.R. and R.L. Rivest, *Inferring decision trees using the minimum description length principle*. Information and Computation, vol. **80** p. 227-248, 1989.

[9] Brazma, A., et al., *Discovering patterns and subfamilies in biosequences*. in *Proceedings of the Fourth International Conference on Intelligent Systems for Molecular Biology*: p. 34-93, 1996.

[10] Kuramochi, M. and G. Karypis. *An Efficient Algorithm for Discovering Frequent Subgraphs*, in Technical Report. Department of Computer Science/Army HPC Research Center, University of Minnesota. 2002.

[11] Chamberlin, D., *A Complete Guide to DB2 Universal Database*. 1998: Morgan Kaufmann Publishers, Inc.

[12] Read, R.C. and D.G. Corneil, *The graph isomorph disease*. Journal of Graph Theory, 1977. **1**: p. 339–363.

Finding Frequent Structural Features among Words in Tree-Structured Documents

Tomoyuki Uchida[1], Tomonori Mogawa[2], and Yasuaki Nakamura[1]

[1] Faculty of Information Sciences,
Hiroshima City University, Hiroshima 731-3194, Japan
{uchida@cs, nakamura@cs}.hiroshima-cu.ac.jp

[2] Department of Computer and Media Technologies, Hiroshima City University,
Hiroshima 731-3194, Japan
t_mogawa@toc.cs.hiroshima-cu.ac.jp

Abstract. Many electronic documents such as SGML/HTML/XML files and LaTeX files have tree structures. Such documents are called tree-structured documents. Many tree-structured documents contain large plain texts. In order to extract structural features among words from tree-structured documents, we consider the problem of finding frequent structured patterns among words in tree-structured documents. Let $k \geq 2$ be an integer and $(W_1, W_2, \ldots, W_k)$ a list of words which are sorted in lexicographical order. A *consecutive path pattern on* $(W_1, W_2, \ldots, W_k)$ is a sequence $\langle t_1; t_2; \ldots; t_{k-1} \rangle$ of labeled rooted ordered trees such that, for $i = 1, 2, \ldots, k-1$, (1) t_i consists of only one node having the pair (W_i, W_{i+1}) as its label, or (2) t_i has just two nodes whose degrees are one and which are labeled with W_i and W_{i+1}, respectively. We present a data mining algorithm for finding all frequent consecutive path patterns in tree-structured documents. Then, by reporting experimental results on our algorithm, we show that our algorithm is efficient for extracting structural features from tree-structured documents.

1 Introduction

Background: Many electronic documents such as SGML/HTML/XML files and LaTeX files have tree structures which are no rigid structure. Such documents are called tree-structured documents. Since many tree-structured documents contain large plain texts, we focus on the characteristics such as the usage of words and the structural relations among words in tree-structured documents. The aim of this paper is to present an efficient data mining technique for extracting interesting structures among words in tree-structured documents.

Data model and consecutive path pattern: As a data model for tree-structured documents, we use an Object Exchange Model (OEM, for short) presented by Abiteboul et al. in [1]. As usual, a tree-structured document is represented by a labeled ordered rooted tree, which is called an *OEM tree*. As a value, each node of an OEM tree has a string such as a tag in HTML/XML files, or a text such as a text written in the field of PCDATA in XML files. Moreover,

H. Dai, R. Srikant, and C. Zhang (Eds.): PAKDD 2004, LNAI 3056, pp. 351–360, 2004.

```
⟨REUTERS⟩
   ⟨DATE⟩ 26-FEB-1987 ⟨/DATE⟩
   ⟨TOPICS⟩ ⟨D⟩ cocoa ⟨/D⟩ ⟨/TOPICS⟩
   ⟨TITLE⟩ BAHIA COCOA REVIEW ⟨/TITLE⟩
   ⟨DATELINE⟩ SALVADOR, Feb 26 - ⟨/DATELINE⟩
   ⟨BODY⟩
      Showers continued throughout the week
      in the Behia cocoa zone, ...
   ⟨/BODY⟩
⟨/REUTERS⟩
```

xml_sample

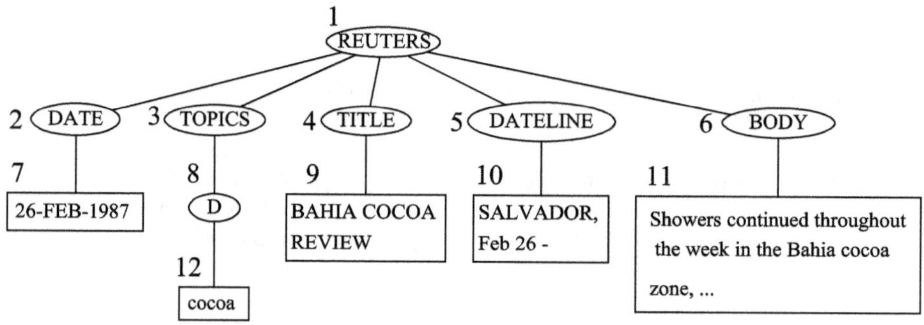

T

Fig. 1. An XML file *xml_sample* and the OEM tree T of *xml_sample*.

an internal node has children which are ordered. For example, we give an XML file *xml_sample* and its OEM tree T in Fig. 1. In T, the node 3 has "TOPICS" as a value and is the 2nd child of the node 1.

Many tree-structured documents have no absolute schema fixed in advance, and their structures may be irregular or incomplete. The formalization of representing knowledge is important for finding useful knowledge. For an integer $k \geq 2$, let $(W_1, W_2, \ldots, W_k)$ be a list of k words appearing in a given set of tree-structured documents such that words are sorted in ASCII-based lexicographical order. Then, a **consecutive path pattern** on $(W_1, W_2, \ldots, W_k)$ (a **CPP**, for short) is a sequence $\langle t_1; t_2; \ldots; t_{k-1} \rangle$ of labeled ordered rooted trees such that, for $i = 1, 2, \ldots, k-1$, (1) t_i consists of only one node having the pair (W_i, W_{i+1}) as its label, or (2) t_i has just two nodes whose degrees are one and which are labeled with W_i and W_{i+1}, respectively. A CPP $\alpha = \langle t_1; \ldots; t_k \rangle$ is said to appear in the OEM tree T_d of a given tree-structured document d, if there exists a sequence $(s_d^1, s_d^2, \ldots, s_d^k)$ of subtrees $s_d^1, s_d^2, \ldots, s_d^k$ of T_d satisfying the following conditions (1) and (2). (1) For each $1 \leq i \leq k$, s_d^i is isomorphic to 2-tree t_i in α. (2) For each $1 \leq j \leq k-1$, the righthand leaf of s_d^j and the

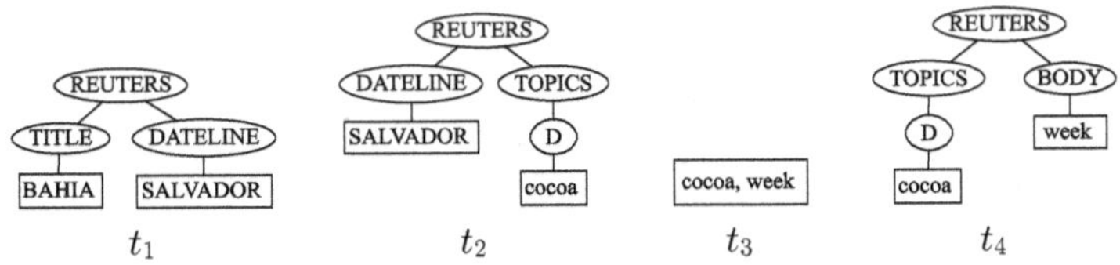

Fig. 2. 2-trees t_1, t_2, t_3 and t_4

lefthand leaf of s_d^{j+1} are same in T_d. For example, the sequences $\alpha = \langle t_1; t_2; t_4 \rangle$ and $\beta = \langle t_1; t_2; t_3 \rangle$ consisting of 2-trees t_1, t_2, t_3 and t_4 given in Fig. 2 are CPPs on (BAHIA, SALVADOR, cocoa, week). We can see that α appears in T but β does not appear in T, where T is the labeled rooted tree given in Fig. 1.

Main results: We consider the problem for finding all CPPs which appear in a given set of tree-structured documents in a frequency of more than a user-specified threshold, which is called a minimum support. For this problem, we present an efficient algorithm which is based on level-wise search strategy with respect to length of a CPP such as Apriori algorithm [2] in making new CPPs from founded CPPs. Next, in order to show the efficiency of our algorithm, we apply our algorithm to Reuters news-wires [9], which contain 21,578 SGML documents and whose total size is 28MB. By reporting experimental results on our algorithm, we show that our algorithm have good performance for a set of a large number of tree-structured documents.

Related works: In [6], we presented the algorithm for extracting all frequent CPPs from a given set of tree-structured documents. This algorithm was designed under the following idea. If a set f of words is not a σ-frequent itemset w.r.t D, then any CPP, which contains all words in f, is not frequent. This idea leads us to reduce the size of search space of possible representative CPPs. However, when there are many high frequent and large itemsets in tree-structured documents, this idea does not work well for pruning a search space. Hence, in this paper, we present a new efficient algorithm, which is based on levelwise search strategy, for solving Frequent CPP Problem.

As a data mining method for unstructured texts, Fujino et al. [5] presented an efficient algorithm for finding all frequent phrase association patterns in a large collection of unstructured texts, where a phrase association pattern is a set of consecutive sequences of keywords which appear together in a document. In order to discover frequent schema from semistructured data, several data mining methods were proposed. For example, Wang and Liu [11] presented an algorithm for finding frequent tree-expression patterns and Fernandez and Suciu [4] presented an algorithm for finding optimal regular path expressions from semistructured data. In [3], Asai et al. presented an efficient algorithm for discovering frequent substructures from a large collection of semistructured data. In [10], Miyahara et al. presented an algorithm for extracting all frequent maximally structural relations between substructures from tree-structured documents.

In order to discover characteristic schema, many researches including above results focused on tags and their structured relations appearing in semistructured data. But in this paper, we are more interested in words and their structural relations rather than tags in tree-structured documents. The algorithm given in this paper is useful for avoiding inappropriate documents and searching interesting documents for users.

2 Preliminaries

In this section, in order to represent structural features among words in tree-structured documents, we formally define a consecutive path pattern as consecutive paths between words. In this paper, we deal with only tree-structured documents. Therefore, we simply call a tree-structured document a document.

An *alphabet* is a set of finite symbols and is denoted by Σ. We assume that Σ includes the space symbol " ". A finite sequence $(a_1, a_2, \ldots, a_n)$ of symbols in Σ is called a *string* and it is denoted by $a_1a_2\cdots a_n$ for short. A *word* is a substring $a_2a_3\cdots a_{n-1}$ of $a_1a_2\cdots a_n$ over Σ such that both a_1 and a_n are space symbols and each a_i $(i = 2, 3, \ldots, n-1)$ is a symbol in Σ which is not the space symbol. Let T_d be the OEM tree of a document d. For a word w, $\mathcal{JS}_d(w)$ denotes the set of all nodes of T_d whose value contains w as a word. Moreover, let $D = \{d_1, d_2, \ldots, d_m\}$ be a set of documents and T_i $(i = 1, 2, \ldots, m)$ the OEM tree of the document d_i in D. For a word w, $\mathcal{JS}_D(w) = \bigcup_{1 \le i \le m} \mathcal{JS}_{d_i}(w)$ is called a *appearance node set* of w in D. For a set or a list S, the number of elements of S is denoted by $\#S$.

Let W and W' be two words over Σ. We call an ordered rooted tree t a **2-tree** over the pair (W, W') if (1) t consists of only one node labeled with (W, W') or (2) t has just two nodes whose degrees are one and which are labeled with W and W', respectively. For a 2-tree t, the lefthand node and the righthand node of t whose degrees are one, are denoted by $ln(t)$ and $rn(t)$, respectively. If t consists of only one node, we remark that $ln(t) = rn(t)$. For a 2-tree t and a document d, a *matching function of t for T_d* is any function $\pi : V_t \to V_{T_d}$ that satisfies the following conditions (1)-(3), where T_d is the OEM tree of d, and V_t and V_{T_d} are the node sets of t and T_d, respectively.

(1) π is a one-to-one mapping. That is, for any $v_1, v_2 \in V_t$, if $v_1 \neq v_2$ then $\pi(v_1) \neq \pi(v_2)$.
(2) π preserves the parent-child relation. That is, for the edge sets E_t and E_{T_d} of t and T_d, respectively, $(v_1, v_2) \in E_t$ if and only if $(\pi(v_1), \pi(v_2)) \in E_{T_d}$.
(3) If t has two leaves, the label of each node $v \in V_t$ is contained in the value of the node $\pi(v)$ in V_{T_d} as a word. If t consists of only one node v, two words in the label of v appear individually in the value of $\pi(v)$ in V_{T_d}.

Moreover, a *pseudo-matching function from t to T_d* is any function $\psi : V_t \to V_{T_d}$ that satisfies the above conditions (1) and (2) and the following condition (3').

(3') If t has two leaves, the label of each leaf $v \in V_t$ is contained in the value of $\psi(v)$ in V_{T_d} as a word. If t consists of only one node v, two words in the label of v are appeared individually in the value of $\psi(v)$ in V_{T_d}.

That is, in the definition of a pseudo-matching function, we ignore a label of any inner node in t. A 2-tree t is said to be *representative* if any inner node of t has a special symbol, which is not a word, as a label.

In this paper, we assume a total order over words such as an ASCII-based lexicographical order. For two words W and W', if W is less than W' in this order, we denote $W \leq W'$. Let k be an integer greater than 1 and $(W_1, W_2, \ldots, W_k)$ a list of k words $W_1, W_2, \ldots, W_k$ such that $W_i \leq W_{i+1}$ for $1 \leq i \leq k-1$. Then, a list $\langle t_1; t_2; \ldots; t_{k-1} \rangle$ of $k-1$ consecutive 2-trees t_1, t_2, $\ldots$ and t_{k-1} is called a **consecutive path pattern** for $(W_1, W_2, \ldots, W_k)$ if for any $1 \leq i \leq k-1$, the i-th 2-tree t_i is a 2-tree over (W_i, W_{i+1}). We remark that, for each $1 \leq i \leq k-2$, both the righthand node $rn(t_i)$ of t_i and the lefthand node $ln(t_{i+1})$ of t_{i+1} have the same label. A consecutive path pattern $\langle t_1; t_2; \ldots; t_{k-1} \rangle$ is said to be *representative* if for any $1 \leq i \leq k-1$, the i-th 2-tree t_i is a representative 2-tree over (W_i, W_{i+1}). For a CPP α, it is simply called a consecutive path pattern, if k words do not need to specify. A consecutive path pattern is shortly called a **CPP**. The number of 2-trees in a CPP α is called a *length* of α.

In the same way as a 2-tree, for a consecutive path pattern $\alpha = \langle t_1; t_2; \cdots, t_k \rangle$ and a document d, we can define a *matching function* $\pi : \bigcup_{1 \leq i \leq k} V_i \to V_{T_d}$ as follows, where T_d is the OEM tree of d, V_{T_d} is the node set of T_d and V_i is the node set of t_i for $1 \leq i \leq k$.

(1) For any $1 \leq i \leq k$, there exists a matching function $\pi_i : V_i \to V_{T_d}$ such that for a node v in t_i, $\pi(v) = \pi_i(v)$.
(2) For any $1 \leq i \leq k-1$, the node $rn(t_i)$ in t_i and the node $ln(t_{i+1})$ in t_{i+1}, $\pi(rn(t_i)) = \pi(ln(t_{i+1}))$

Moreover, we also define a *pseudo-matching function* ψ from a representative CPP α to T_d by interchanging a matching function π_i and a pseudo-matching function ψ_i.

For a CPP $\alpha = \langle t_1; t_2; \cdots; t_k \rangle$ and a document d, the set $\{(\pi, \pi(rn(t_k))) \mid \pi$ is a matching function of α for $T_d\}$ is called the *junction set* of α in d and is denoted by $\mathcal{JS}_d(\alpha)$, where T_d is the OEM tree of d. If $\#\mathcal{JS}_d(\alpha) \geq 1$, we say that α *appears in* d. In the same way, for a representative CPP α' and a document d, the junction set $\mathcal{JS}_d(\alpha')$ is the set $\{(\pi', \pi'(rn(t_k))) \mid \pi'$ is a pseudo-matching function of α' for $T_d\}$. If $\#\mathcal{JS}_d(\alpha') \geq 1$, we say that α' *appears in* d. For example, four 2-trees t_1, t_2, t_3 and t_4 in Fig. 2 appear in the document *xml_sample* in Fig. 1. The CPP $\alpha = \langle t_1; t_2; t_4 \rangle$ appears in *xml_sample*. Since there is no function π such that $\pi(rn(t_2)) = \pi(ln(t_3)) = \pi(rn(t_3))$, π is a matching function of t_2 for T in Fig. 1 and is also a matching function of t_3 for T in Fig. 1, the CPP $\langle t_1; t_2; t_3 \rangle$ does not appear in *xml_sample* in Fig. 1.

3 Extracting Consecutive Path Patterns from Tree-Structured Documents

In this section, we formally define the data mining problem of finding all frequent consecutive path patterns from tree-structured documents. Next, we give an efficient algorithm for solving this problem.

For a set $D = \{d_1, d_2, \ldots, d_m\}$ of documents, a word w and a CPP α, let $Occ_D(w) = \#\{i \mid 1 \leq i \leq m,\ \mathcal{JS}_{d_i}(w) \neq \emptyset\}$ and let $Occ_D(\alpha) = \#\{i \mid 1 \leq i \leq m,\ \mathcal{JS}_{d_i}(\alpha) \neq \emptyset\}$. For a set D of m documents and a real number σ $(0 < \sigma \leq 1)$, a word w and a CPP α are *σ-frequent* w.r.t. D if $Occ_D(w)/m \geq \sigma$ and $Occ_D(\alpha)/m \geq \sigma$, respectively. In general, a real number σ is given by a user and is called a *minimum support*. For a tree-structured document d, $\mathcal{W}(d)$ denotes the set of all words appearing in d. For a set D of documents, a **CPP w.r.t** D is a CPP on a list of words in $\mathcal{W}(D) = \bigcup_{d \in D} \mathcal{W}(d)$. Then, we consider the following data mining problem.

Frequent (Representative) CPP Problem
Instance: A set D of tree-structured documents and a minimum support $0 \leq \sigma \leq 1$.
Problem: Find all σ-frequent (representative) CPPs w.r.t. D.

In [6], we presented the Apriori based algorithm for solving the Frequent Representative CPP Problem. This algorithm was designed under the following idea. If a set f of words is not a σ-frequent itemset w.r.t D, then any representative CPP, which contains all words in f, is not frequent. This algorithm finds all σ-frequent itemsets w.r.t D by regarding a word as an item and the set $\mathcal{W}(d)$ of all words appearing in each document d as a transaction and using a FP-tree which is a compact data structure given by Han et al. in [8]. This idea leads us to reduce the size of search space of possible representative CPPs. However, when there are many high frequent and large itemsets in documents, this idea does not work well for pruning a search space. Hence, in this paper, we present a new efficient algorithm, which is based on levelwise search strategy, for solving Frequent CPP Problem. Moreover, as heuristics of our algorithm for extracting interesting patterns from given tree-structured documents and reducing search space, a quite frequent word such as "a" or "the" is not used as a word in a CPP. Such a word is called a *Stop Word*.

In Fig. 3, we present an algorithm *Find_Freq_CPP* for solving Frequent CPP Problem. As input for our algorithm, given a set D of tree-structured documents, a minimum support $0 \leq \sigma \leq 1$ and a set of Stop Words, our algorithm outputs the set of all σ-frequent CPPs w.r.t. D. Our algorithm is based on a strategy of level-wise search. A CPP consists of consecutive 2-trees. For any two words w_1, w_2 in a document d, there exists just one path from the node of T_d whose value contains w_1 to the node of T_d whose value contains w_2, where T_d is an OEM tree of d. Hence, by using junction sets of all frequent CPPs whose length is $k \geq 1$ and appearance node sets of all frequent words, we can correctly obtain all frequent CPPs whose length is $k+1$ and their junction sets from all frequent CPPs whose length is k.

In reading given documents, we construct a *trie* (see [7]), which is a compact data structure, for efficiently storing and searching words and all appearance node set of it. By using this structure, the line 1 of our algorithm can be executable in linear time. Let M and N be the numbers of σ-frequent words in D and the extracted σ-frequent CPPs, respectively. Let n be the number of

Algorithm Find_Freq_CPP

Input: A set $D = \{d_1, d_2, \ldots, d_m\}$ of tree-structured documents, a minimum support $0 < \sigma \leq 1$ and a set Q of stop words.

Output: The set $\mathcal{F}$ of all σ-frequent CPPs w.r.t. D.

$Freq\_Word := \{(w, \mathcal{JS}_D(w)) \mid w \in \mathcal{W}(D) - Q,\ w$ is σ-frequent w.r.t. $D\}$;
$\mathcal{F}(1) :=$ *Make_2Tree*$(Freq\_Word, \sigma)$;
$k := 1$;
while $\mathcal{F}(k) \neq \emptyset$ **do**;
    begin
        $\mathcal{F}[k+1] :=$ *Expand_Pattern*$(F(k), Freq\_Word, \sigma)$;
        $k := k + 1$;
    end; /* end of while loop */
return $\mathcal{F} = \bigcup_{1 \leq i \leq k} \{\alpha \mid (\alpha, \mathcal{JS}_D(\alpha)) \in \mathcal{F}[i]\}$;

Fig. 3. The data mining algorithm *Find_Freq_CPP*

Procedure Make_2Tree

Input: A set *Freq_Word* of pairs consisting of words and the lists of nodes, and a minimum support $0 < \sigma \leq 1$.

Output: The set $\mathcal{F} = \{(\langle t \rangle, \mathcal{JS}_D(\langle t \rangle)) \mid \langle t \rangle$ is a σ-frequent CPP w.r.t. $D\}$.

for each $(W, list_W) \in Freq\_Word$ **do**
    for each $(P, list_P) \in Freq\_Word$ such that $W \leq P$ **do**
        $tmp := \emptyset$;
        for each node $x \in list_W$ and each node $y \in list_P$ **do**,
            search a 2-tree t over the pair (W, P) from x to y,
                if both x and y are in the same tree;
            execute the following instruction (1) or (2);
                (1) add y to $list$ if there exists $(t, list)$ in tmp such that $y \notin list$;
                (2) add (t, y) to tmp if $t \notin \{s \mid (s, L) \in tmp\}$;
        end do
        if $((\#Occ_D(\langle t \rangle))/(\#D) \geq \sigma$ then $\mathcal{F} := \mathcal{F} \cup \{(\langle t \rangle, list) \mid (t, list) \in tmp\}$;
    end do; /* end of inner for loop */
end do; /* end of outer for loop */
return $\mathcal{F}$;

Fig. 4. The procedure *Make_2Tree*

nodes of all OEM trees each of which corresponds to a document in D. Then, *Make_2Tree* and *Expand_Pattern* are executable in $O(n^2 M^2 \log n \log M)$ and $O(n^2 NM \log n \log M)$, respectively. Hence, given a set D of tree-structured documents, a minimum support $0 < \sigma \leq 1$ and a set of Stop Words, *Find_Freq_CPP* can find all σ-frequent CPPs in $O(\max(n^2 M^2 \log n \log N,\ kn^2 MN \log n \log N))$

Procedure Expand_Pattern

Input: A set $F[k]$ of pairs of CPPs whose lengths are k and lists of nodes, a set $Freq\_Word$ of pairs consisting of words and lists of nodes, and a minimum support $0 < \sigma \leq 1$.

Output: The set $F[k+1] = \{(\alpha, \mathcal{JS}_D(\alpha) \mid \alpha$ is a σ-frequent CPPs w.r.t.D whose lengths are $k+1.\}$

1. **for** each $(\alpha, \mathcal{JS}_D(\alpha)) \in F[k]$ **do**
    /* let $\alpha = \langle t_1; t_2; \cdots; t_k \rangle$ be a CPP over $(W_1, W_2, \ldots, W_{k+1})$ */
2. **for** each $(W, list_W) \in Freq\_Words$ such that $W_{k+1} \leq W$ **do**
3. $tmp := \emptyset$;
4. **for** each node $x \in \mathcal{JS}_D(\alpha)$ and each node $y \in list_W$ **do**,
5. search a 2-tree t over the pair (W_{k+1}, W) from x to y,
    if both x and y are in the same tree;
6. execute the following instruction (1) or (2);
    (1) add y to $list$ if there exists a pair $(t, list)$ in tmp with $y \notin list$;
    (2) add (t, y) to tmp if $t \notin \{s \mid (s, L) \in tmp\}$;
7. **end do**
8. if $(\#Occ_D(\langle t_1; t_2; \cdots; t_k; t \rangle))/(\#D) \geq \sigma$ then
    $\mathcal{F}[k+1] := \mathcal{F}[k+1] \cup \{(\langle t_1; t_2; \cdots; t_k; t \rangle, list) \mid (t, list) \in tmp\}$;
9. **end do**; /* end of inner for loop */
10. **end do**; /* end of outer for loop */
11. **return** $\mathcal{F}[k+1]$;

Fig. 5. The procedure *Expand_Pattern*

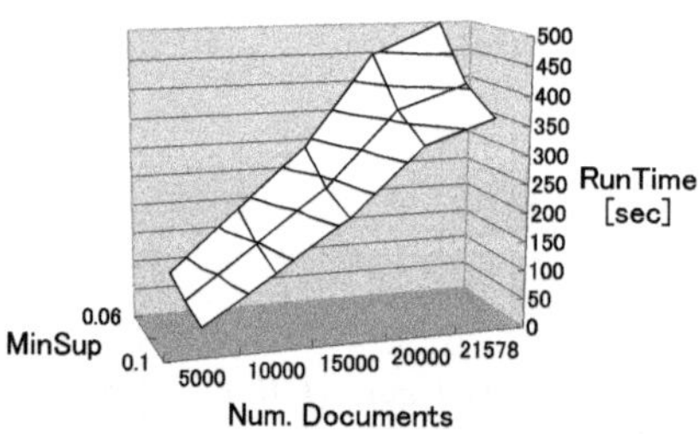

(a) *Find_Freq_CPP*

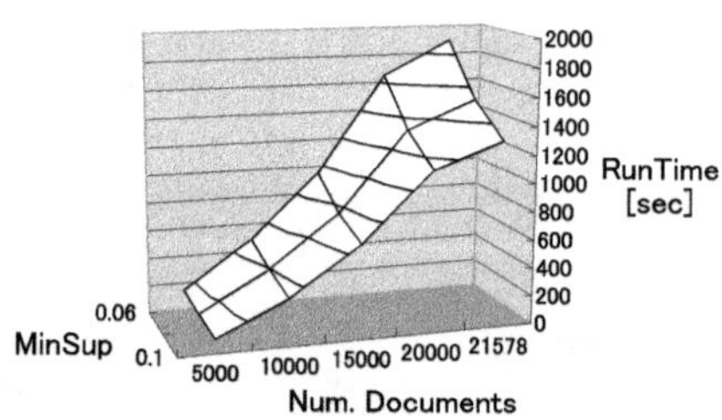

(b) *Prev_Rep_Algorithm*

Fig. 6. The running times of *Find_Freq_CPP* and *Prev_Rep_Algorithm*

time, where k is the maximum length of founded CPPs. This shows that if M is in polynomial of the input size, *Find_Freq_CPP* terminates in a polynomial time. For lower minimum supports, M and N become to be large numbers. So, we need to give an appropriate set of Stop Words as input, when we want to gain all frequent CPPs.

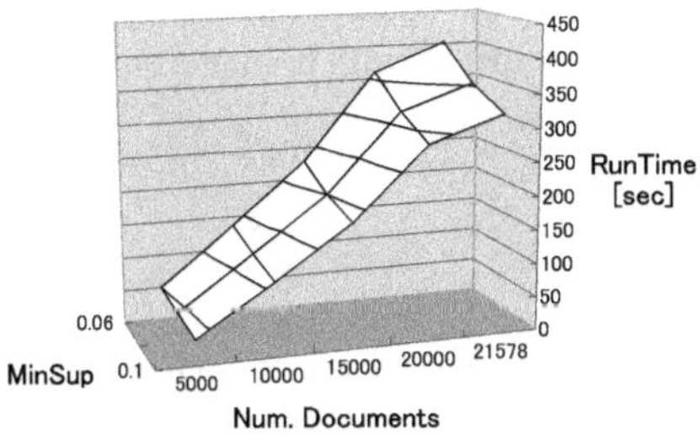

(c) *Find_Freq_Rep_CPP*

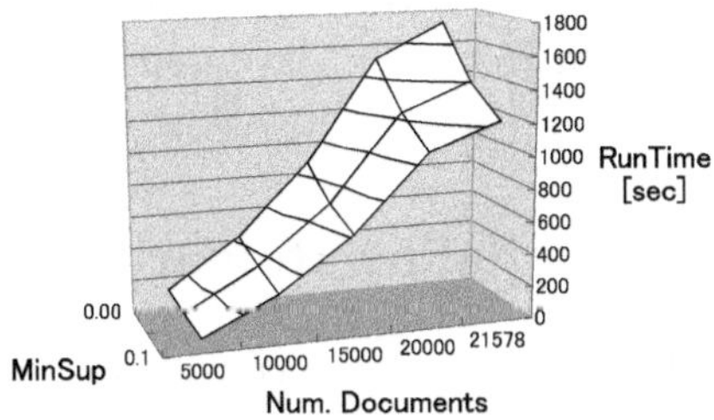

(d) *Prev_Rep_Algorithm*

Fig. 7. The running times of *Find_Freq_Rep_CPP* and *Prev_Rep_Algorithm*

By slightly modifying the above algorithm *Find_Freq_CPP*, we construct an algorithm for solving Frequent Representative CPP Problem. This algorithm is denoted by *Find_Freq_Rep_CPP*.

4 Experimental Results

In this section, we show the efficiency of our new algorithms *Find_Freq_CPP* and *Find_Freq_Rep_CPP* given in previous section. We implemented *Find_Freq_CPP* and *Find_Freq_Rep_CPP*, and our previous algorithms *Prev_Algorithm* and *Prev_ Rep_Algorithm* in [6] which solve Frequent CPP Problem and Frequent Representative CPP Problem, respectively. All algorithms are implemented in C++ on a PC running Red Hat Linux 8.0 with two 3.2 GHz Pentium 4 processors and 1GB of main memory. In the following experiments, as Stop Words, we chose symbols such as "-,+,...", numbers such as "0,1,2,...", pronouns such as "it, this, ...", articles "a, an, the", and auxiliary verbs "can, may, ..." and so on.

We apply these algorithms for Reuters-21578 text categorization collection in [9], which has 21,578 SGML documents and its size is about 28.0MB, in cases of each minimum support in $\{0.06,\ 0.08,\ 0.10\}$ and each number of documents in $\{5,000,\ 10,000,\ 15,000,\ 20,000,\ 21,578\}$.

Fig. 6 (a), (b) and Fig. 7 (c), (d) show the running times of *Find_Freq_CPP*, *Prev_Algorithm*, *Find_Freq_Rep_CPP* and *Prev_Rep_Algorithm* in the experiments, respectively. Each running time is a time needed from reading all input documents up to finding all frequent CPPs or all frequent representative CPPs. Tree structures such as SGML/XML files can be freely defined by using strings given by users as tags. So, in case of analyzing miscellaneous tree-structured documents, it may not be necessary to focus on tag's names. Then, *Find_Freq_Rep_CPP* is suitable for extracting structural features among words. From both Fig. 6 and 7, *Find_Freq_CPP* and *Find_Freq_Rep_CPP* are more efficient than *Prev_Algorithm* and *Prev_Rep_Algorithm* given in [6], respectively.

Moreover, for lower minimum supports, we see that both *Find_Freq_CPP* and *Find_Freq_Rep_CPP* have good performance for analyzing a large set of tree-structured documents such as Reuters news-wires.

5 Concluding Remarks

In this paper, we have considered the problem of extracting structural features among words from a set of tree-structured documents. We have defined a consecutive path pattern on a list of words and have shown efficient data mining algorithms for solving the problem. Moreover, in order to show the efficiency of our algorithms, we have reported some experimental results on applying our algorithms to Reuters news-wires. This work was partially supported by Grant-in-Aid for Young Scientists (B) No.14780303 from the Ministry of Education, Culture, Sports, Science and Technology and Grant for Special Academic Research No.2117 from Hiroshima City University.

References

1. S. Abiteboul, P. Buneman, and D. Suciu. *Data on the Web: From Relations to Semistructured Data and XML.* Morgan Kaufmann, 2000.
2. R. Agrawal and R. Srikant. Fast algorithms for mining association rules. *Proc. of the 20th VLDB Conference*, pages 487–499, 1994.
3. T. Asai, K. Abe, S. Kawasoe, H. Arimura, H. Sakamoto, and S. Arikawa. Efficient substructure discovery from large semi-structured data. *Proc. 2nd SIAM Int. Conf. Data Mining (SDM-2002) (to appear)*, 2002.
4. M. Fernandez and Suciu D. Optimizing regular path expressions using graph schemas. *Proc. Int. Conf. on Data Engineering (ICDE-98)*, pages 14–23, 1998.
5. R. Fujino, H. Arimura, and S. Arikawa. Discovering unordered and ordered phrase association patterns for text mining. *Proc. PAKDD-2000, Springer-Verlag, LNAI 1805*, pages 281–293, 2000.
6. K. Furukawa, T. Uchida, K. Yamada, T. Miyahara, T. Shoudai, and Y. Nakamura. Extracting characteristic structures among words in semistructured documents. *Proceedings of the 6th Pacific-Asia Conference on Knowledge Discovery and Data Mining (PAKDD-2002), Springer-Verlag, LNAI 2336*, pages 356–367, 2002.
7. G. Gonnet and R. Baeza-Yates. *Handbook of Algorithms and Data Structures.* Addison-Wesley, 1991.
8. J. Han and M. Kamber. *Data Mining: Concepts and Techniques.* Morgan Kaufmann Publishers, 2001.
9. D. Lewis. Reuters-21578 text categorization test collection. *UCI KDD Archive, http://kdd.ics.uci.edu/databases/reuters21578/reuters21578.html*, 1997.
10. T. Miyahara, T. Shoudai, T. Uchida, K. Takahashi, and H. Ueda. Discovery of frequent tree structured patterns in semistructured web documents. *Proc. PAKDD-2001, Springer-Verlag, LNAI 2035*, pages 47–52, 2001.
11. K. Wang and H. Liu. Discovering structural association of semistructured data. *IEEE Trans. Knowledge and Data Engineering*, 12:353–371, 2000.

Exploring Potential of Leave-One-Out Estimator for Calibration of SVM in Text Mining

Adam Kowalczyk, Bhavani Raskutti, and Herman Ferrá

Telstra Corporation, 770 Blackburn Road, Clayton, Victoria 3168, Australia
{Adam.Kowalczyk, Bhavani.Raskutti, Herman.Ferra}@team.telstra.com

Abstract. This paper investigates a number of techniques for calibration of the output of a Support Vector Machine in order to provide a posterior probability $P(\text{target class}|\text{instance})$.

Five basic calibration techniques are combined with five ways of correcting the SVM scores on the training set. The calibration techniques used are addition of a simple ramp function, allocation of a Gaussian density, fitting of a sigmoid to the output and two binning techniques. The correction techniques include three methods that are based on recent theoretical advances in leave-one-out estimators and two that are variants of hold-out validation set. This leads us to thirty different settings (including calibration on uncorrected scores). All thirty methods are evaluated for two linear SVMs (one with linear and one with quadratic penalty) and for the ridge regression model (regularisation network) on three categories of the Reuters Newswires benchmark and the WebKB dataset. The performance of these methods are compared to both the probabilities generated by a naive Bayes classifier as well as a calibrated centroid classifier.

The main conclusions of this research are: (i) simple calibrators such as ramp and sigmoids perform remarkably well, (ii) score correctors using leave-one-out techniques can perform better than those using validation sets, however, cross-validation methods allow more reliable estimation of test error from the training data.

1 Introduction

When applying data mining techniques to a particular predictive modelling task, one typically trains a model to produce a score for each instance such that the instances may be ordered based on their likelihood of target class membership. However, in many applications of interest, it is not sufficient to be able to induce an ordering of cases. For instance in a marketing campaign if one needs to action say, 20% of the target class cases, an ordering based on likelihood of membership is not enough. Instead, what is required is an accurate estimate of the posterior probability $P(\text{target class}|\text{instance})$.

The problem of such estimation, particularly for decision tree classifiers, has been an area of active research for some time. It has been noted that decision trees trained to maximise classification accuracy generally behave poorly as

H. Dai, R. Srikant, and C. Zhang (Eds.): PAKDD 2004, LNAI 3056, pp. 361–372, 2004.

conditional probability estimators. This has prompted methods for improving decision tree estimates. In the paper by Zadrozny and Elkan [18], a number of such methods including various smoothing techniques, a method known as curtailment and an alternative decision tree splitting criteria are compared. In addition, this paper also describes a histogram method known as binning and applies this method to naive Bayes classifiers with some success. This success prompted the application of the binning method to calibrate the output of Support Vector Machines (SVM) [4]. Other calibration techniques converting classifier scores to conditional probability estimates include the use of the training set to fit a sigmoid function [14] and other standard statistical density estimation techniques [6,16]. All of these techniques may be applied to any ranking classifier, i.e., a classifier that returns a score indicating likelihood of class membership.

It is known that SVMs tend to overtrain, providing a separation of support vectors which is too optimistic. Hence, for reliable empirical estimates based on the training data, it is advantageous to use corrected scores for support vectors prior to calibration. We investigate correction techniques based on recent theoretical advances in the leave-one-out estimator [7,8,13] and compare them to techniques based on hold-out validation sets.

In this paper, we evaluate the applicability of different calibration (Section 3.2) and score correction (Section 3.1) techniques for allocating posterior probabilities in the text categorisation domain. We also compare these probabilities against those estimated by naive Bayes classifiers (Section 4.2) and evaluate the reliability of test error estimates from training data (Section 4.3).

2 Metrics

In this section we briefly introduce the basic concepts and methodology for quantifying the performance of learning machines which are suitable for the purpose of this paper. We focus on the simplest case of discrimination of two classes.

We assume that there is given a space of labelled observations $(x, y) \in X \times \{\pm 1\}$ with a probability distribution $P(x, y)$. Our ultimate goal is to estimate the posterior probability $P[y|x]$. This will be done via estimates of the posterior $P[y|f]$, where $f : X \to \mathbb{R}$ is a model able to score each observation $x \in X$. These estimates will be of the form

$$\hat{\mathcal{P}}[y|f] = \begin{cases} \phi \circ f(x) & \text{for } y = 1, \\ 1 - \phi \circ f(x) & \text{for } y = -1. \end{cases}$$

where $\phi : \mathbb{R} \to [0, 1] \subset \mathbb{R}$ is a particular *calibration method* , i.e. a method of allocating a class probability to scores.

The main metric chosen in this paper for quantification of quality of these estimators is the *standard error*, which is the square root of the mean squared error (MSE). Other metrics such as average log loss have been investigated. In [18] MSE and average log loss were found to provide similar evaluations of candidate probability estimators. We have used the standard error because the log loss metric can become infinite under certain conditions.

The squared error for a single labelled example $x \in X$ is defined as

$$\sum_{y=\pm 1}\left(P[y|x] - \hat{\mathcal{P}}[y|f(x)]\right)^2.$$

In practice, for a test set of m examples

$$\overrightarrow{xy}^m := \big((x_1, y_1), ..., (x_m, y_m)\big) \in (X \times \{\pm 1\})^m, \tag{1}$$

where true labels y_i are known but not true probabilities $P[y_i|x_i]$, we shall substitute 1 for $P[y_i|x_i]$ and 0 for $P[-y_i|x_i]$ [1,18]. This leads us to the following empirical formula for *the standard error*:

$$SE_\phi(f) = \sqrt{\frac{2}{m}\sum_{i=1}^{m}\left(\frac{y_i+1}{2} - \phi\circ f(x_i)\right)^2}. \tag{2}$$

3 Classifiers and Estimators

In this research, we have used five classifiers: two linear *support vector machines* (SVM), one Ridge Regression model, a simple centroid classifier and a naive Bayes classifier. Naive Bayes have been implemented as described in [11] and class probability was defined accordingly by generated posteriors for two categories of interest. The other four classifiers are implemented as described below.

Support Vector Machine (*SVM*). Given a training m-sample (1), a learning algorithm used by SVM [2,3,17] outputs a model $f_{\overrightarrow{xy}^m} : X \to \mathbb{R}$ defined as the minimiser of *the regularised risk functional*:

$$f \mapsto ||f||^2_{\mathcal{H}} + \sum_{i=1}^{m} L([1 - y_i f(x_i)]_+). \tag{3}$$

Here $\mathcal{H}$ denotes a reproducing kernel Hilbert space (RKHS) [17] of real valued functions $f : X \to \mathbb{R}$, $||.||_{\mathcal{H}}$ the corresponding norm, $L : \mathbb{R} \to \mathbb{R}^+$ is a *non-negative*, convex *cost function* penalising for the deviation $1 - y_i f(x_i)$ of the estimator $f(x_i)$ from target y_i and $[\xi]_+ := \max(0, \xi)$. In this paper we shall consider exclusively the following two types of the loss function: (*i*) *linear* (*SVM^1*): $L(\xi) := C\xi$, and (*ii*) *quadratic* (*SVM^2*): $L(\xi) := C\xi^2$, where $C > 0$ is *a regularisation constant.*

The minimisation of (3) can be solved by quadratic programming [2] with the formal use of the following expansion known to hold for the minimiser (3):

$$f_{\overrightarrow{xy}^m}(x) = \sum_{i=1}^{m} \alpha_i y_i k(x_i, x), \tag{4}$$

$$||f_{\overrightarrow{xy}^m}||^2_{\mathcal{H}} = \sum_{i,j=1}^{m} \alpha_i \alpha_j y_i y_j k(x_i, x_j), \tag{5}$$

where $k : X \times X \to \mathbb{R}$ is the kernel corresponding to the RKHS $\mathcal{H}$ [9,15]. The coefficients α_i are unique and they are the Lagrange multipliers of the quadratic minimisation problem corresponding to the constraints $y_i f_{\overline{xy}^m}(x_i) > 0$.

If $\alpha_i \neq 0$, then x_i is called *a support vector*. It can be shown that $y_i f_{\overline{xy}^m}(x_i) \leq 1$ for each support vector. For experiments in this paper, it is convenient to slightly expand the concept of a support vector to the subset

$$SV := \{i \; ; \; y_i f_{\overline{xy}^m}(x_i) \leq 1 + \epsilon \text{ for } i = 1, ,.., m\}$$

where $\epsilon > 0$ is a constant. This definition attempts to accommodate the fact that typically practical solutions are sub-optimal due to early stopping.

Ridge Regression (RN^2). In addition to the SVMs we also use a Regularisation Network or ridge regression predictor, RN^2 [3,5,9,15]. Formally, this predictor is closely related to SVM^2, the only difference being that it minimises a modified risk (3), with loss $L = C(1 - y_i f(x_i))^2$ rather than $L = C[1 - y_i f(x_i)]_+^2$. As we shall see, both machines may display quite different performance.

Centroid Classifier ($Cntr$). The centroid classifier is a simple linear classifier with the solution,

$$f_{\overline{xy}^m}(x) = \frac{\sum_{i, y_i = +1} k(x_i, x)}{2 \max(1, m_+)} - \frac{\sum_{i, y_i = -1} k(x_i, x)}{2 \max(1, m_-)},$$

and m_+ and m_- denote the numbers of examples with labels $y_i = +1$ and $y_i = -1$, respectively. In terms of the feature space, the centroid classifier implements the projection in the direction of the weighted difference between the centroids of data from each class. It can be shown that the centroid solution approximates the solution obtained by SVMs at very low values of C [10].

3.1 Score Correctors

It is known that support vector machine tends to overtrain providing a separation of support vectors which is too optimistic. Hence, for reliable empirical estimates based on the training data it is advantageous to use corrected scores for support vectors prior to calibration. We investigate three correction methods based on leave-one-out (LOO) estimator for SVMs. The first two corrections are based on a theoretical bound, and are applied to "spread" the scores of support vectors and other training instances that have scores close to ± 1. The Jaakkola-Haussler correction ($f^{JH}_{\overline{xy}^m}(x)$) is the lower bound of the LOO estimator derived in [7,15] and is applied to the support vector set SV only. The Opper-Winther correction ($f^{OW}_{\overline{xy}^m}(x)$) is a simplified approximation of the LOO estimator derived in [13,15] and is applied to the the union of the support vector and the *margin* support vector sets ($SV \cup mSV$). We use the following "practical" definition for the set of margin support vectors:

$$mSV := \{i \; ; \; |y_i f_{\overline{xy}^m}(x_i) - 1| \leq \epsilon \text{ for } i = 1, ,.., m\},$$

where $\epsilon > 0$ is a constant. This set is a "practical" substitute for the theoretical set of the margin support vectors studied in [13] taking into account that a particular practical solution is only an approximation of the true minimiser of the empirical risk.

These two LOO-based corrections are defined as follows:

$$f^{JH}_{\overline{xy}^m}(x_i) := f_{\overline{xy}^m}(x_i) - \alpha_i y_i k(x_i, x_i)\, I_{i \in SV}, \tag{6}$$

$$f^{OW}_{\overline{xy}^m}(x_i) := \begin{cases} f_{\overline{xy}^m}(x_i) - \alpha_i y_i \sum_{j,j'}^{mSV} k_{ij} [\mathbf{k}^{-1}_{mSV}]_{jj'} k_{j'i} & \text{if } i \in SV - mSV, \\ f_{\overline{xy}^m}(x_i) - \alpha_i y_i [\mathbf{k}^{-1}_{mSV}]_{ii}\, I_{i \in mSV} & \text{otherwise.} \end{cases} \tag{7}$$

Here $i = 1, ..., m$ and $I_{i \in SV}$ is the indicator function equal to 1 if $i \in SV$ and 0, otherwise. $I_{i \in mSV}$ is defined analogously. Here $k_{ij} := k(x_i, x_j)$, $\mathbf{k}_{mSV} = (k_{ij})_{i,j \in mSV}$ is a square matrix of dimension $= |mSV|$ and $[K]_{ii}$ denotes the element of the main diagonal of the matrix K corresponding to index i.

The third method introduced in this paper is a heuristic modification to alleviate the fact that the lower bound of Jaakkola and Haussler correction is known to be pessimistic. It is defined as follows:

$$f^{uni}_{\overline{xy}^m}(x_i) := f_{\overline{xy}^m}(x_i) - r\alpha_i y_i k(x_i, x_i)\, I_{i \in SV}, \qquad \text{for } i = 1, ..., m.$$

Here, r is a number drawn from a uniform distribution on the segment $[0, 1]$, hence its working name *uniform corrector*. The intention here is to "spread" the scores for support vectors away from the margin, but not necessarily as much as in the $f^{JH}_{\overline{xy}^m}$ case.

The fourth and fifth methods of score correction evaluated in this paper are based on use of a hold-out set. For the first hold-out method ($Hold^1$) the training data is split into two parts, one is used for training the classifier, the second one is used to derive the parameters of the calibration method. In other words, we apply the classifier to the hold-out set, and use this to adjust the calibration method optimally to the trained classifier. The second hold-out method ($Hold^2$) is the same as $Hold^1$ except that after adjustment of the calibration method the whole training set is used to train a new classifier which is used in conjunction with the previously derived calibration.

We report here the results for a single split: 50% for training and 50% for $Hold^1$ and $Hold^2$. Note that this is different from the split used in [4] and was used to accommodate the larger variations in class sizes that is common in text categorisation.

3.2 Calibration Methods

We have tested five techniques for allocation of class probability to a classifier's score. The first of them is independent of the training data, while the remaining four use the training set.

Ramp Calibrator. The first method is a simple ramp calibrator:

$$\phi^{ramp}(t) := \begin{cases} (1+t)/2 & \text{for } |t| \le 1, \\ (1 + sgn(t))/2 & \text{otherwise} \end{cases} \tag{8}$$

for $t \in \mathbb{R}$. This calibrator is independent of the training data.

Gauss Calibrator. The second method implements a routine statistical technique for density estimation [6,16]:

$$\phi^{Gauss}(t) := \frac{\sum_{i,y_i=1} K_h(f_{\overline{xy}^m}(x_i) - t)}{\sum_i K_h(f_{\overline{xy}^m}(x_i) - t)}, \tag{9}$$

where $K_h(t) := \frac{1}{\sqrt{2\pi}h}\exp(-0.5t^2/h^2)$ is the popular Gauss kernel. We have used $h = 0.5$ in our experiments and applied (9) to $f_{\overline{xy}^m}$ clipped to the range $[-3, 3]$.

Sigmoidal calibrator. The third method fits a sigmoid to the SVM output:

$$\phi^{Sigm}(t) := \left(1 + e^{At+B}\right)^{-1}, \tag{10}$$

where constants A, B are determined from the training data using Platt's method [14]. This finds parameters A, B by minimising the negative log likelihood of the training data, which is a cross-entropy error function:

$$min_{A,B}\left(-\sum_i t_i \log\frac{1}{1+e^{Af(x_i)+B}} + (1-t_i)\log(1 - \frac{1}{1+e^{Af(x_i)+B}})\right),$$

where $t_i = (n_+ + 1)/(n_+ + 2)$ if $y_i = 1$ and $t_i = 1/(n_- + 2)$, otherwise, where n_+ and n_- are the numbers of positive and negative examples in the training set.

Binning Methods. The fourth and fifth methods for calibration are variants of the histogram method [4,18]. This method proceeds by first ranking the training examples according to their scores, then dividing them into b subsets, called *bins*. Each bin is defined by its lower and upper boundary. Given a test example x, it falls into the bin if and only if its score $f(x)$ is within bin boundaries. The estimate $\hat{\mathcal{P}}[y|x]$ of the probability that x belongs to the class y is set to the fraction of the training examples of class y which are in the same bin.

In our experiment we have used a fixed number of bins $b = 10$ as in [4,18]. We have used two different methods for definition of bin boundaries. The first binning method, referred to as Bin^1, is exactly as in the above references: it selects bin boundaries such that each bin contains the same number of training examples (irrespective of example class). A potential problem arises with this method when the data is heavily biased and well separable by a classifier. This is exactly the case often encountered in text mining using SVMs, where the target (minority) class accounts for less than 10% of the data and almost all target examples fall into the top 10% segment of scores, i.e. a single bin. To alleviate this problem, we have used another binning method, referred to as Bin^2. In this method we first set a default bin as the open segment $(-\infty, a)$, where $a = min\{f(x_i)\ ;\ y_i = 1\}$ (we always assume that the target class corresponds to the label $y = +1$). Hence this bin contains no target class examples. Then we set boundaries of the remaining top $b-1$ bins in such a way that the target class cases in the training set are evenly split into $b-1$ equal subsets. In this case we typically encounter some larger and smaller bins, with the largest, the default bin, containing no target class cases. As the experimental results show, this method outperforms Bin^1 in cases where there are few target cases in the training set (c.f. Figure 1).

4 Experiments and Results

In our evaluation, we use two distinct corpora of documents and different categories from each of the corpora. The experimental results presented for each of the settings are averages over 20 models with a one-standard deviation bar to show the variance in the results. Each of the 20 models is created using a fraction (ρ) of the total number of examples in the corpus, and testing on the remaining, where these training examples are selected randomly and the sampling is not proportional, i.e., non-stratified random sampling is performed.

4.1 Data Collections

Our first corpus is the Reuters-21578 news-wires benchmark. Here we used a collection of 12902 documents (combined test and training sets of so called modApte split available from http://www.research.att.com/lewis) which are categorised into 115 overlapping categories. Each document in the collection has been converted to a 20,197 dimensional word-presence vector, using a standard stop-list, and after stemming all of the words using a standard Porter stemmer.

Our second corpus is the WebKB dataset which consists of web pages gathered from university computer science departments [12]. After removing all those pages that specify browser relocation only, this set consists of 7,563 web pages split across 7 categories. In this paper, we use the three most populous categories ($\neq$ 'other'): Course, Student and Faculty. We parse the HTML tags, and use all the words without stemming, which results in a feature set consisting of 87,601 words. We do not use a stoplist since some of the typical stopwords such as "my" are useful for prediction in this case.

4.2 Standard Error Results

Figure 1 shows the standard error for three Reuters categories: category "earn" the largest category that is roughly 30% of the data (first row), category "crude" that is just 4% of the data (second row) and category "corn" which form 2% of the data (third row). For this set of results, we have used a split of 10% training and 90% test ($\rho = 0.1$).

Results are presented for calibration of raw scores (first column) as well as the three LOO-based score correctors (columns 2-4) and the two correctors based on hold-out sets (columns 5-6). Within each column, calibration results are shown for all of the five calibrators described in Section 3.2, although the ramp calibrator is unaffected by correction of training scores and hence has a different result only for $HOLD^1$ which uses a smaller training set for model generation. Standard error is presented for all 30 combinations for three machines: SVM with quadratic penalty (SVM^2, first marker: cross), SVM with linear penalty (SVM^1, second marker: diamond) and ridge regression (RN^2, third marker: square). For all machines, the regularisation constant in the SVM equation in Section 3 is set to $C = 800/m$, where m is the number of instances in the training set. In addition, we present the error for the centroid classifier (fourth marker:

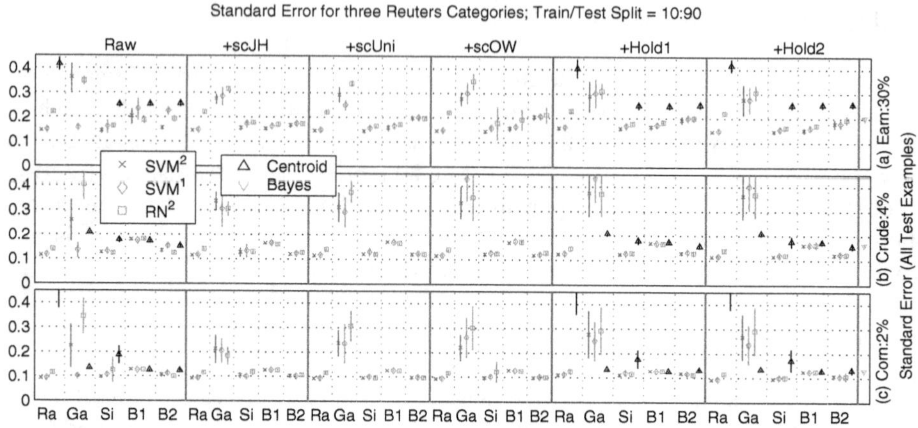

Fig. 1. Standard error for three Reuters categories for different score correctors and calibrators for different machines. We denote the various calibrators as follows: ramp (Ra), Gauss (Ga), sigmoidal (Si), Bin^1 (B1) and Bin^2 (B2). Score correctors are denoted as follows: $f^{JH}(x)$ (+scJH), $f^{OW}(x)$ (+scOW), $f^{uni}(x)$ (+scUni), $Hold^1$ (+Hold1) and $Hold^2$ (+Hold2). Raw refers to calibrated, uncorrected scores. The seventh column shows the probability estimates obtained from naive Bayes Classifier.

triangle) for the raw scores and for score correctors based on hold-out sets. The last column (column 7) shows the probability estimates obtained from naive Bayes Classifier (shown by inverted triangle). Points missing from the graph are those with standard error larger than 0.42.

Comparing different calibration techniques, we note the following:

1. The ramp calibrator performs extremely well across all score correction methods except $f^{JH}_{\overline{xy}^m}(x)$, and across all classifiers except centroid. In general the ramp method applied to the raw scores of either SVM^2 or SVM^1 produces estimates very close (or equal) to the minimum standard error over all the other methods. This is quite impressive given the simplicity of the method. A further point to note is that the method is very consistent over different training set samples showing very little variance (the notable exception being column 2, i.e., after correction using $f^{JH}_{\overline{xy}^m}(x)$).

2. The ramp calibrator performs very poorly with the centroid classifier, however, other calibrators such as sigmoids and gauss as well as the binning ones, return good estimates for the centroid classifier. Thus, more sophisticated calibration methods are useful when the classifier is simple and does not separate with a margin.

3. Regardless of the correction method used, the gauss calibrator estimates are much worse than any other calibrator and also have much higher variance.

4. Binning methods, particularly Bin^1, do not have any advantage over any calibrators other than Gauss. As expected, Bin^2 performs better than Bin^1 for the two smaller categories (rows 2 and 3), and this is true with score correction using hold-out methods, as well as raw scores.

Comparing the performance of different score correction techniques, we note the following:

1. Adjustment of the model scores by use of any of the correction techniques prior to calibration has a very limited positive impact, and this happend only with RN^2 which usually performs worse than the two SVMs on the raw scores. One explanation for this behaviour could be the fact that RN^2 concentrates training case scores at -1 and $+1$ which leads to a significantly larger number of training set points which are subsequently corrected.

2. The performance of all of the LOO-based correction methods is very similar. However, there are significant computational advantages for the $f^{JH}_{\overrightarrow{xy}^m}(x)$ and $f^{uni}_{\overrightarrow{xy}^m}(x)$ methods since these, unlike $f^{OW}_{\overrightarrow{xy}^m}(x)$ do not involve a matrix inversion.

3. Although our $f^{OW}_{\overrightarrow{xy}^m}(x)$ correction is theoretically justified for the linear penalty case only, it is worth noting that the performance in SVM^2 case is at least comparable to that of SVM^1. This is rather unexpected and requires further investigation. Specifically, what is needed is to properly derive the Opper-Winther LOO correction in the quadratic case to compare the difference.

4. Not surprisingly, correction of scores prior to calibration helps to reduce the variance of estimates. However, there is no significant advantage in using cross-validation sets for correction. This suggests that good probability estimates can be obtained if the model used for categorisation is trained on all of the available data.

Finally, the probability estimate obtained from the naive Bayes classifier is reasonably accurate, although always worse than that obtained from ramp calibration on raw scores with the SVMs. It would be interesting to explore the effect of binning on these probabilities [4].

Generalisation to Other Datasets. Figure 2 shows the standard error for three WebKB categories: Course (12% of data – first row), Student (22% second row) and Faculty (14% of data, third row). The notation, the format and the training/test split (10%/90%) is the same as that for Figure 1. These experiments were performed in order to understand how the above observations hold with other datasets,

Comparing Figures 1 and 2, we see that there are two significant differences regarding performance of different machines. Firstly, SVM^1 performs significantly better than SVM^2. However, the observation regarding ramp calibrator with raw scores still holds with SVM^1. Secondly, all score correction methods have a positive impact on both SVM^2 and RN^2 and bring it closer to the performance of SVM^1, however, the impact of score correction on the best performer, SVM^1 is not consistent. This is in line with the observation that correctors improve the performance when the baseline performance is sub-optimal (e.g., improvement for RN^2 for the Reuters dataset).

Impact of Training Set Size. In order to determine the effect of the number of training examples on the different correctors and calibrators, the experiments

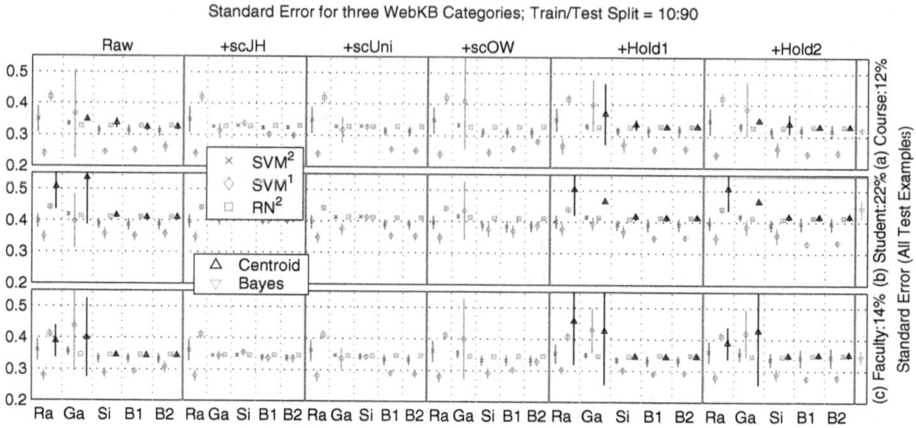

Fig. 2. Standard error for three WebKB categories for different score correctors and calibrators. The settings and notations are the same as that for Figure 1.

described in the previous Section with a training/test split of 10:90 were repeated with a split of 50:50 (results not shown here due to space considerations). We observed that the only impact of the larger training set size is the lower standard error and the lower variance across all categories, calibrators, correctors and classifiers. This is true for both the Reuters and the WebKB datasets.

4.3 Training/Test Error Ratio

In order to select the best performing methods we need to have an estimate of the performance of the method on unseen data. One way to achieve this is to take part of the labelled data and hold it in reserve as an independent test set. Alternatively, the performance on the training set can be used as an estimate of the method's performance on unseen data. As a guide to the accuracy of such an estimate, we use $TrTeErrorRatio$, the training/test error ratio which is defined as the ratio of standard error on the training set to that on the test set. For each method we characterise the accuracy of the training set estimate using this $TrTeErrorRatio$. Methods that provide $TrTeErrorRatio$ close to 1 and good estimates on the training set are preferable to methods that provide good estimates on the training set, but have poor $TrTeErrorRatio$ (i.e., $<< 1$). If we are confident that a particular method has $TrTeErrorRatio$ close to 1, then we can use all of the labelled data for model generation, which as noted above leads to more accurate models.

Figure 3 shows the $TrTeErrorRatio$ for the three Reuters categories for the same settings and in the same format as in Figure 1. As can be seen from Figure 3, the estimates of performance derived from the training set are consistently optimistic for the SVMs ($TrTeErrorRatio << 1$). This is common to all the calibration methods, with the binning techniques, especially Bin^1, appearing a little more immune to this than the other methods. This optimism is consistent with the observation that the SVM raw scores are known to be overtrained.

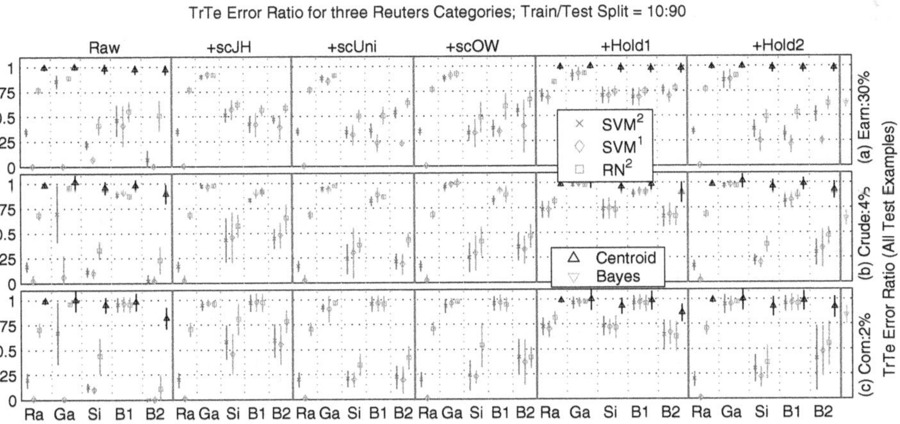

Fig. 3. *TrTeErrorRatio* for three Reuters categories for different score correctors and calibrators. The settings and notations are the same as that for Figure 1.

SVM^1 is the worst performer having *TrTeErrorRatio* close to 0 in most cases, while RN^2, and naive Bayes have a *TrTeErrorRatio* around 0.75. The centroid classifier provides very good estimates (≈ 1), however, the standard error for this classifier is generally quite high (Figures 1 and 2).

All correction methods tend to move the *TrTeErrorRatio* closer to 1, with hold-out sets having the most impact. $Hold^1$, in particular, provides consistently good estimates even with little available data, while the other correction methods tend to become more optimistic as training data becomes scarce. This is expected since overfitting is usually worse when there are few examples to learn from.

5 Conclusion

The paper has demonstrated the feasibility of extracting accurate probability estimates from SVM scores for text categorisation. Our evaluation with multiple text datasets shows that simple calibrators such as ramp calibrator can provide very accurate probabilities. While score correction does not improve the probability estimates, it does indeed provide better estimation of the accuracy of the classifier on unseen data. The best compromise between accuracy of probability estimates and prediction of performance on unseen data is provided by the Bin^1 calibrator on raw scores.

Future work indicated by these results include investigation of other real data sets and carefully designed artificial data. A significant effort should be directed towards developments of practical techniques (heuristics) for training score correction. Results of this paper for our uniform score corrector show that such heuristics exist. Finally, the probability estimates obtained here should be tested against calibrated naive Bayes classifiers, and tested for their utility in multi-class text categorisation.

Acknowledgements. The permission of the Chief Technology Officer, Technology Strategy and Research, Telstra Corporation, to publish this paper is gratefully acknowledged.

References

1. B. Bauer and R. Kohavi. An empirical comparison of voting classification algorithms: Bagging, boosting and variants. *Machine Learning*, 36:105 – 142, 1999.
2. C. Corte and V. Vapnik. Support vector networks. *Machine Learning*, 20:273 – 297, 1995.
3. N. Cristianini and J. Shawe-Taylor. *An Introduction to Support Vector Machines and other kernel-based learning methods.* Cambridge University Press, Cambridge, 2000.
4. J. Drish. Obtaining calibrated probability estimates from support vector machines.
5. F. Girosi, M. Jones, and T. Poggio. Regularization theory and neural networks architectures. *Neural Computation*, 7(2):219–269, 1995.
6. D. J. Hand. *Construction and Assessment of Classification Rules.* John Wiley and Sons, Chichester, 1982.
7. T. Jaakkola and D. Haussler. Probabilistic kernel regression models. In *Proc. of the 1999 Conference on AI and Statistics*, 1999.
8. T. Joachims. Estimating the Generalization Performance of an SVM Efficiently. In P. Langley, editor, *Seventh International Conference on Machine Learning*, pages 431–438, San Francisco, 2000. Morgan Kaufman.
9. G. Kimeldorf and G. Whaba. A correspondence between Bayesian estimation of stochastic processes and smoothing by splines. *Ann. Math. Statist.*, 41:495–502, 1970.
10. A. Kowalczyk and B. Raskutti. Exploring Fringe Settings of SVMs for Classification. In *Proceedings of the Seventh European Conference on Principle and Practice of Knowledge Discovery in Databases (PKDD03)*, 2003.
11. T. Mitchell. *Machine Learning.* McGraw Hill, 1996.
12. Kamal Nigam, Andrew Kachites Mccallum, Sebastian Thrun, and Tom Mitchell. Text classification from labeled and unlabeled documents using EM. *Machine Learning*, 39(2/3):103–134, 2000.
13. M. Opper and O. Winther. Gaussian process classification and svm: Mean field results and leave-one out estimator. In A. Smola, P.Bartlett, B. Schölkopf, and D. Schuurmans, editors, *Advances in Large Margin Classifiers*, pages 301–316, Cambridge, MA, 2000. MIT Press.
14. J. Platt. Probabilities for SV Machines. In A. Smola, P.Bartlett, B. Schölkopf, and D. Schuurmans, editors, *Advances in Large Margin Classifiers*, pages 61–74, Cambridge, MA, 2000. MIT Press.
15. B. Schölkopf and A. J. Smola. *Learning with Kernels: Support Vector Machines, Regularization, Optimization and Beyond.* MIT Press, 2001.
16. P. Smyth, A. Gray, and U. Fayyad. Retrofitting decision tree classifiers usinf kernel density estimation. In *Proceedings of the 12th International Conference on Machine Learning*, pages 506–514, 1995.
17. V. Vapnik. *Statistical Learning Theory.* Wiley, New York, 1998.
18. B. Zadrozny and C. Elkan. Obtaining calibrated probability estimates from decision trees and naive bayesian classiers. In *Proceedings of the Eighteenth International Conference on Machine Learning ICML2001*, 2001.

Classifying Text Streams in the Presence of Concept Drifts

Gabriel Pui Cheong Fung[1], Jeffrey Xu Yu[1], and Hongjun Lu[2]

[1] The Chinese University of Hong Kong
Hong Kong, China
{pcfung,yu}@se.cuhk.edu.hk

[2] The Hong Kong University of Science and Technology
Hong Kong, China
luhj@cs.ust.hk

Abstract. Concept drifting is always an interesting problem. For instance, a user is interested in a set of topics, X, for a period, may switches to a different set of topics, Y, in the next period. In this paper, we focus on two issues of concept drifts, namely, concept drifts detection and model adaptation in a text stream context. We use statistical control to detect concept drifts, and propose a new multi-classifier strategy for model adaptation. We conducted extensive experiments and reported our findings in this paper.

1 Introduction

For many learning tasks where data is collected over an extended period of time, its underlying distribution may change. A changing context will induce changes in the concept of the collection of data, producing what is known as concept drifts.

Knowing concept drifts is important in many applications. For instance, a news agent needs to decide the top stories and headlines every day or even every hour based on the interests of its users. An information dissemination system needs to decide how to broadcast news stories or promotions to its users who will be most interested in. An email filtering system should be able to filter out the spam email. All these require to trace the pattern of the users interests.

In practice, up to some extent, it is possible to trace users interests with historical data. For instance, we can trace the set of articles that the users are interested in by monitoring the time that they spent on browsing the articles. For removing the spam email, it could be done by tracking the email that are kept and deleted by the users. Yet, in different time horizons, users may be interested in different things. For example, one may be interested in receiving materials related to president election during the election campaign period, but this does not mean that the user is always interested in politics.

Broadly speaking, the problem of concept drifts can be divided into two subproblems: drift detection and model adaptation [3,4,5,6,7,10,11,12]. Detecting changes are mostly discussed in information filtering [5,6,7]. On the other hand,

H. Dai, R. Srikant, and C. Zhang (Eds.): PAKDD 2004, LNAI 3056, pp. 373–383, 2004.

model adaptation is often done by maintaining a sliding window, either fixed size or adaptive size, on the training data [4,12]. While these heuristics are intuitive and work very well in many domains, most of them required complicated tuning on their parameters, and the tunings are not transferable to other domains.

[3] describes a roubust window approach for model adaptation using support vectors machine (SVM). However, their window adjustment algorithm requires to compute the $\xi\alpha$-estimator on every batch from scratch which is too expensive in practice. [10] illustrates how to maintain the support vectors for handling concept drifts. However, they did not show any situations where concepts drift. [11] proposes a framework for mining concept-drifting data streams by using a weighted ensemble classifiers. The authors show that using a weighted ensemble classifiers will perform substantial better than just using a single classifier.

In this paper, we re-visit the problem of concept drifts in text stream contexts, and study both concept drifts detection and model adaptation. We focus on a single classifier but not a community of classifiers. For detecting concept drifts, we make use of statistical quality control to detect any changes in a text stream. For model adaptation, we propose an approach that uses different kind of model sketches (short-term and long-term). No complex parameter tuning is necessary. We do not use sliding window for model adaptation.

2 Concept Drifts

A set of text documents is classified into multiple *topics*. Each topic is labeled as either *interesting* (C_+) or *non-interesting* (C_-). A *concept* is the main theme of the interesting topic(s). *Concept drift* means that the main theme of the interesting event is changed. In the following, we follow the assumption that text streams arrive as batches with variable-length:

$$d_{1,1}, ..., d_{1,m_1}, d_{2,1}, ..., d_{2,m_2}, \cdots, d_{n,1}, ..., d_{n,m_n}, ...$$

where $d_{i,j}$ denotes the jth document in the ith batch. The i-th batch consists of m_i documents ($m_i \geq 1$). We use C_+^t and C_-^t to denote the two classes C_+ and C_- at a given time t. A document at time t is labeled as either interesting if it belongs to C_+^t or non-interesting if it belongs to C_-^t.

Figure 1 illustrates a simple concept drifts scenario by using two sets of topics, X and Y. There are 10 time intervals: $t_1, t_2, \cdots, t_{10}$. Documents arrived within $[t_{i-1}, t_i]$ form a batch. From t_0 to t_3, the users are interested in X but not Y. Beginning from t_3, the users decrease their interest in X but gain interest in Y. Finally, from t_7, the users totally lose their interest in X and interest in Y only.

Our problem is how to maintain such interesting and non-interesting concepts in a text stream environment, with two objectives: 1) maximizing the accuracy of classification; and 2) minimizing the cost of re-classification. Two related issues are: concept drifts detection and model adaptation. Concept drifts detection is to figure out whether the concept in the stream is changed whereas model adaptation aims at capturing new concepts immediately.

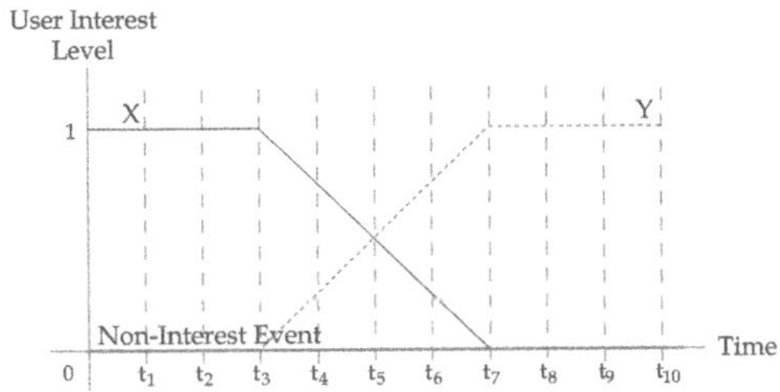

Fig. 1. A simple example of concept drift.

Table 1. The contingency table for measuring the quality of a classifier.

		Expert judgments	
		Yes	No
Classifier	Yes	TP	FP
judgments	No	FN	TN

3 Concept Drifts Detection

In this section, we introduce some measurements for classification quality, and address our concerns. Then, we discuss our solutions.

When a text stream is being classified by a classifier and the true class labels of these documents are given, the relationships between the classifier decisions and their corresponding true class labels can be summarized as shown in Table 1. For measuring the effectiveness of a classifier, a common approach is to use precision (π) recall (ρ) and F_1 [9].

$$\pi = \frac{TP}{TP + FP} \qquad \rho = \frac{TP}{TP + FN} \qquad F_1 = \frac{2 \cdot \pi \cdot \rho}{\pi + \rho} \tag{1}$$

Yet, none of them are suitable for detecting concept drifts. The reason is that π (ρ) does not take FN and TN (FP and TN) into account, meanwhile F_1 does not take TN into consideration. Therefore, we can not notify any drift in the non-interesting events. In this paper, we use error rate, ε, for measuring the effectiveness of a classifier, as it captures all four dimensions.

$$\varepsilon = \frac{FP + FN}{TP + TN + FP + FN} \tag{2}$$

However, detecting concept drifts by simply comparing the error rates of two consecutive batches is insufficient. Consider Figure 2 in which users gradually lose their interest in topic X after time t_n. The bottom of Figure 2 shows a typical result for the error rate of the classification. There always exist small fluctuations on the error rate. Comparing the error rates directly may fire false alarms.

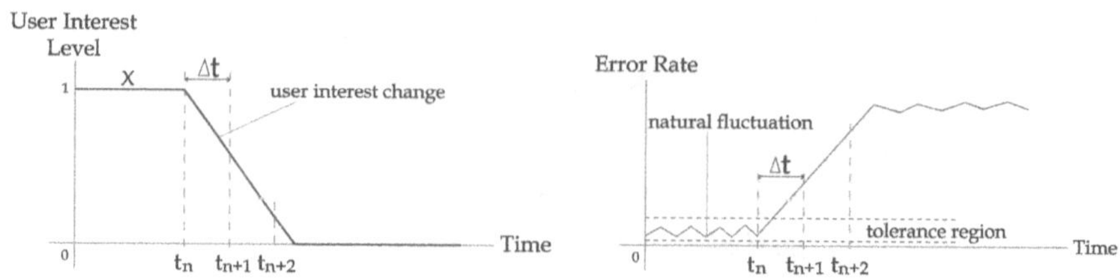

Fig. 2. Concept drift and error rate.

In order to determine whether it is a fluctuation or a drift, we use statistical quality control [8]. The basic idea is to test whether or not a single observation (ε) is within a *tolerance region:*

$$\bar{\varepsilon}_n - N\sigma_n \leq \varepsilon \leq \bar{\varepsilon}_n + N\sigma_n \tag{3}$$

where $\bar{\varepsilon}_n$ is the mean error rate of the n previous observations; σ_n is the corresponding standard deviation and N is the control tolerance. In this paper, we set $N = 2$.

When a change is detected at t, we will adapt the classifier to include or remove any concepts from C^t_+ and C^t_-. Note that detecting possible concept drifts is insufficient for maintaining a good classification model. For instance, consider two scenarios. In Scenario-A, the concepts X and Y interchange slowly, while in Scenario-B X and Y interchange fast. In order words, the *drift rates* are different. The consequence of ignoring such drift rate is that the new model may adopt the drift too quickly (too slowly) and drop the existing concepts too early (too late). Here, we introduce the term *expected drift rate.*

Assume that a drift is being detected at time t_{n+1}. The expected drift rate, ϕ, is measured by the slope of δt:

$$\phi = \frac{\varepsilon_{n+1} - \bar{\varepsilon}_n}{t_{n+1} - t_n} \tag{4}$$

Note that $\bar{\varepsilon}_n$ is used instead of ε_n. This is because taking only one previously observation (ε_n) may yield a very biasing result. By combining ϕ and the current error rate, the expected error for the next time interval is:

$$\hat{\varepsilon}_{n+2} = \phi \cdot \delta t_{n+1} + \varepsilon_{n+1} \tag{5}$$

4 Model Adaptation

The key point of model adaption is to maintain a high quality model such that the newly adapted model is best suitable for the next incoming document.

We argue that model adaptation based on sliding window is not the best choice. Let us consider Figure 1. From t_3 to t_7, the users interest toward X (Y)

decrease (increase). At time t_5, 50% of the documents in X should be classified as $C_+^{t_5}$, whereas another 50% should be classified as $C_-^{t_5}$. In such a case, the classifier will be confused because the number of the positive and negative examples are the same. In this paper, we extend our previous work on a similarity-based classifier, called discriminative category matching (DCM) [1]. Classification is achieved by comparing the unseen document sketch with different class sketches. DCM summarized the information of features in a document, within a class and across different classes.

4.1 Dynamic Sketches Constructing/Removing

We create and maintain two class sketches for C_+ and C_-, denoted $\tilde{k}_+$ and $\tilde{k}_-$, respectively. Let us explain dynamic sketch constructing/removing using Figure 1 as an example.

Initially, two sketches, $\tilde{k}_+$ and $\tilde{k}_-$, are created at t_0, for C_+ and C_-. At t_3, the users interest begins to change. Suppose that this change is being detected at t_4, we create two new sketches, $\tilde{k}'_+$ and $\tilde{k}'_-$ at t_4, which are served as short-term classifiers. From then, we will maintain four sketches: $\tilde{k}'_+$, $\tilde{k}'_-$, $\tilde{k}_+$ and $\tilde{k}_-$. During the period from t_3 to t_7, suppose that the expected drift rate at t_i is ϕ_i for $i \in [4..6]$) (ϕ_i at t_i can be different from ϕ_j at t_j for $t_i \neq t_j$). According to ϕ_i, we can estimate the probability of user interest, γ_i, for Y in next batch as:

$$\gamma_{i+1} = \phi_i \cdot t_i + \gamma_i \tag{6}$$

In other words, γ documents belonging to Y should be classified as C_+ and the remaining should be classified as C_-. Similar applies to X. Finally, after t_7, the error rate for $\tilde{k}'_+$ and $\tilde{k}'_-$ becomes stable and within the tolerance region and the error rate for $\tilde{k}_+$ and $\tilde{k}_-$ increase to maximum. Now, there is no reason to kept $\tilde{k}_+$ and $\tilde{k}_-$, they will be removed and $\tilde{k}'_+$ and $\tilde{k}'_-$ will be kept for classification.

A question remaining unanswered here is: how to create short-term sketches? For every learning algorithm, two types of examples must be provided: positive and negative. Positive examples for $\tilde{k}'_+$ and $\tilde{k}'_-$ can be collected by choosing the mis-classified examples from $\tilde{k}_+$ and $\tilde{k}_-$ as the mis-classified documents should be those with changing concepts. However, we do not know the negative examples.

Our solution is given here. First, we select feature that acts *actively* in the positive examples such that its occurrence is skewed in positive examples. These features are selected according to the feature strength:

$$\hat{\omega}(d, i) = \frac{WC_{i,+}}{WC_{i,\cup}} \cdot (WC_{i,+} - WC_{i,\cup}) \tag{7}$$

where $WC_{i,+}$ is the weight of feature i in the positive examples and $WC_{i,\cup}$ is the weight of the corresponding feature in all of the documents in a batch. Refer to [1] for the calculation of WC. Furthermore, let $\Gamma(d, j)$ be a sorted list of feature strengths for d. We select the top K features in each positive example and form a set of positive features, ζ. K in each document may be different, and

Table 2. 10 selected topics from the Reuters-21578 data set and the number of documents assigned to them.

Topic	Documents	Topic	Documents
Ship	158	Acquisition	2362
Money-fx	307	Earn	3945
Ship	158	Coffee	116
Sugar	143	Crude	408
Trade	361	Interest	285

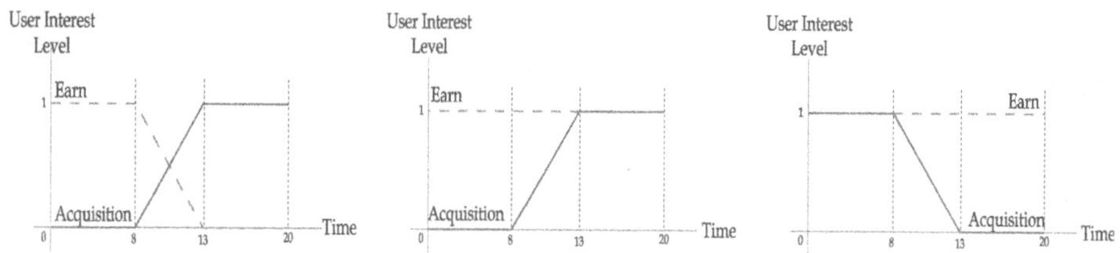

Fig. 3. Different type of concept drift.

is determined using mean square error, such that $E(\Gamma(d,j)) \leq E(\Gamma(d,j+1))$ for $j < K$ and $E(\Gamma(d,K)) > E(\Gamma(d,K+1))$, where

$$E(\Gamma(d,j)) = \frac{1}{2}(\Gamma(d,j)^2 - \Gamma(d,j+1)^2) \quad (8)$$

We repeat the similar procedure, and extract the negative examples from the universe such that $d \cap \zeta = \emptyset$.

5 Evaluations

The data set used for evaluation is Reuters-21578. Table 2 shows the 10 selected topics. All features are stemmed and converted to lower cases. Four measures are used for evaluating the performance of the classifier: 1) Precision (ρ); 2) Recall (π); 3) Error rate (ε); and 4) F_1. Following the existing works [3,4,7], we simulate three drifting scenarios between two topics: `Earn` and `Acquisition`. All other topics are served as noise, and are not relevant to the user interest. In each scenario, documents from each topic are evenly distributed into 21 batches (batch 0 to batch 20). Batch 0 serves as the initial training set, and the remaining 20 batches represent the temporal development:

- **Scenario A** (Figure 3 (a)): From batch 0 to batch 8, documents from `Earn` are relevant with respect to the user interest. This changes gradually from batch 9 to batch 12, and finally only `Acquisition` is relevant beginning from batch 13.

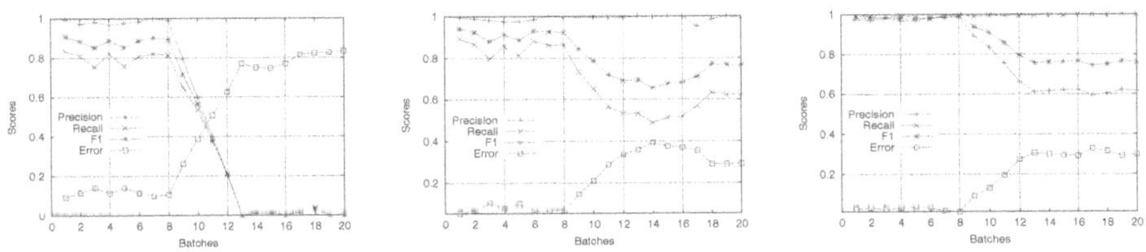

(a) Scenario A (b) Scenario B (c) Scenario C

Fig. 4. Results for Exp 1

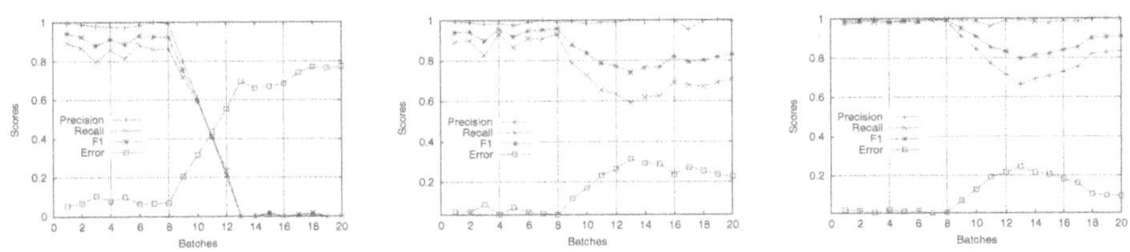

(a) Scenario A (b) Scenario B (c) Scenario C

Fig. 5. Results for Exp 2

- **Scenario B** (Figure 3 (b)): A new relevant topic arises gradually while another relevant topic is always present. From batch 0 to batch 8, only `Earn` is relevant. `Acquisition` is added to the relevant text starting from batch 9 to batch 12.
- **Scenario C** (Figure 3 (c)): The relevant topic is split into two, and one of them gradually becomes non-relevant. Initially, `Earn` and `Acquisition` are considered as relevant. Starting from batch 9, `Acquisition` gradually becomes non-relevant.

5.1 A Comparison of Different Approaches

Four experiments are conducted for comparison:

- **Exp 1 (Only-Once)**: The classifier is learned only from batch 0, i.e. the classifier will not be updated.
- **Exp 2 (Incremental)**: The classifier is learned throughout all of the batches, i.e. at batch i, the classifier is learned from batch 0 to batch $(i-1)$, and batch i is used for testing.
- **Exp 3 (Based on Concept Drifts)**: The classifier is constructed in the same way as discuss in this paper.

In Exp 1 (Figure 4), initially, the performance of only-once classification in all scenarios is good. However, whenever a drift occurs, its performance deteriorates

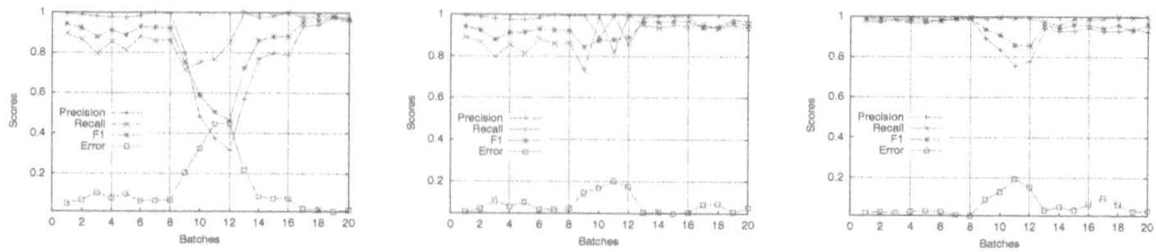

(a) Scenario A (b) Scenario B (c) Scenario C

Fig. 6. Results for Exp 3

significantly. Scenario B shows that the measurement "precision" cannot detect the concept drifts, while Scenario C shows that the measurement "recall" fails to do so. In Exp 2 (Figure 5), the performance of incremental classification in all scenarios is slightly better than that of only-once classification (Exp 1), but still unacceptable. Exp 3 (Figure 6) performs the best.

6 Detailed Analysis

In this section, three situations are test: 1) Varying the amount of noise; 2) Varying the drift rate; and 3) Varying the extend of drift. Due to the space limitation, only Scenario A is demonstrated.

6.1 Varying the Amount of Noise

Two situations are simulated:

- **Case-1**: $\alpha\%$ of the target concepts, `Earn` and `Acquisition`, will be used randomly. For example, in the extreme case, when $\alpha = 100$, a completely random data will be generated. Here, three α are chosen: 10, 20 and 30.
- **Case-2**: Some topics are added and/or removed. Three situations are modeled: 1) Last five topics from Table 2 are removed. 2) Additional topics are added to the non-interesting event on top of the first situation. 3) Additional five topics are added to the relevant event on top of the second situation. The topics are chosen randomly. This case aims at determining the effectiveness of the statistical quality control.

Figure 7 and Figure 8 show the results for Case-1 and Case-2, respectively. In Figure 7, the general shape deteriorates from (a) to (c). This is expected as more noise present. Note that immediate after the drifts, all situations return to their corresponding stable stage quickly. This suggests that the proposed drift rate is useful.

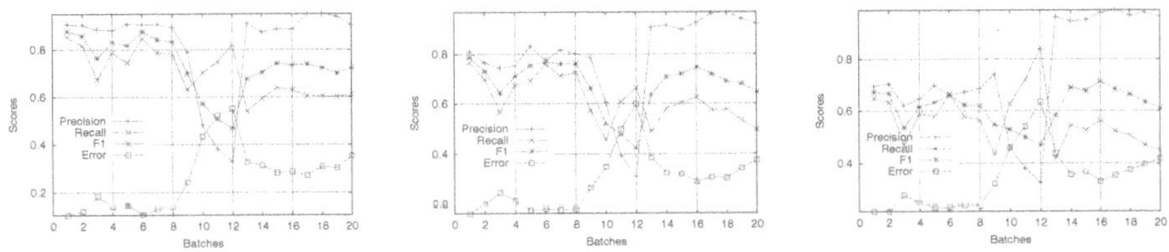

(a) $\alpha = 10\%$ (b) $\alpha = 20\%$ (c) $\alpha = 30\%$

Fig. 7. Results of varying noise for Case-1

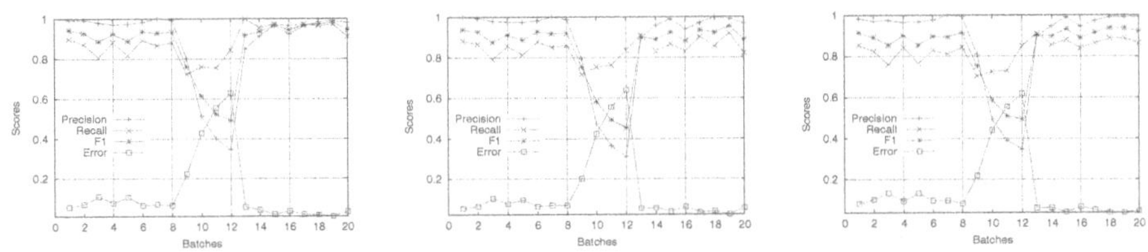

(a) Situation 1 (b) Situation 2 (c) Situation 3

Fig. 8. Results for varying noise for Case-2

6.2 Varying the Drift Rate

Sometimes concepts will change gradually, creating a period of uncertainty between stable states. In other words, the new concept only gradually takes over. Consider Figure 2 again. By varying Δt, we can simulate the situation of varying the drift rate. Three drift rates are chosen: 1) $\Delta t = 0$ (abrupt drift); 2) $\Delta t = 1$ (fast drift); and 3) $\Delta t = 8$ (slow drift). All of these units represent the number of batches. The concept drifts here all begin at batch 8.

Figure 9 shows the results of varying the drift rate. For $\Delta t = 0$, the quality of the classifier drops to minimum in batch 9 when the concept begins to drift. When $\Delta t = 0$, the concept is not *drift* but is *shift*. In this situation, we do not have any previous knowledge to predict when a shift will happen. However, the classifier can adapt the new concept in the next batch quickly. This is due to the drift rate proposed in Section 3.

6.3 Varying the Extend of Drift

Computational learning theory quantifies the extend of drift as the relative error between two concepts, which is the probability that a randomly drawn example which should be classified as A is mis-classified as B. Thus, the extend of drift is the dissimilarity between two successive concepts A and B.

In general, such an experiment is not easy to conduct, because there is no such a benchmark to denote the similarity between two text categories. However,

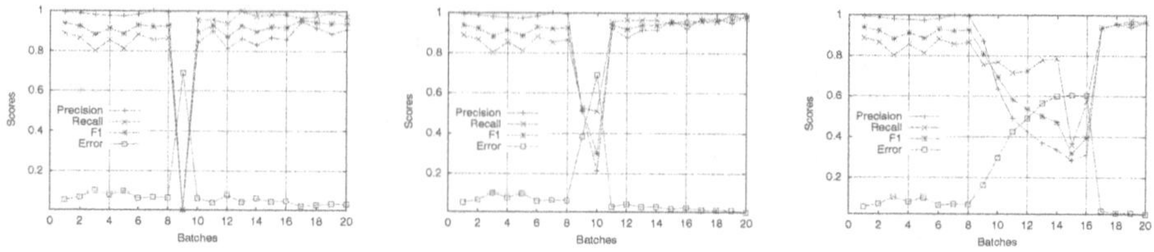

(a) $\Delta t = 0$ (b) $\Delta t = 1$ (c) $\Delta t = 8$

Fig. 9. Results for varying the rate of concept drift

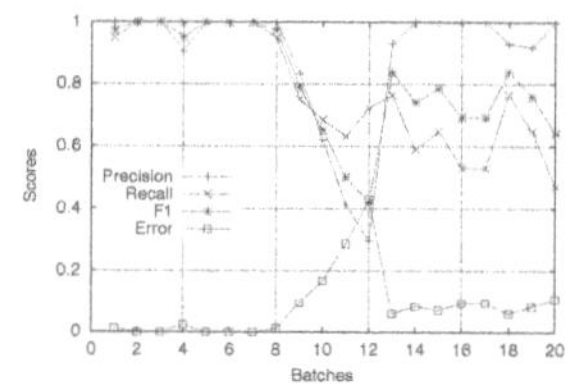

Fig. 10. Results for varying the extent of concept drift

as pointed out by many researchers that some Reuters' categories are "difficult categories" [13,2,9]. [9] shows that `Crude` and `Trade` are more difficult to be classified correctly rather than `Earn` and `Acquisition`. We conduct an experiment using the two "difficult categories", `Crude` and `Trade`. Figure 10 shows the results. As expected, the quality decreases further. However, concerning with the detection of concept drifts, it seems that it does not have difficulties.

7 Conclusion

In this paper, we presented an effective approach for classifying text streams in the presence of concept drifts. We plan to study how to deal with recurring concepts. In the case where a previously existing concept re-appears, it is a waste of time to re-learn the characteristics of this concept from scratch. It is possible to maintain the potentially useful class sketches and re-examine at a later stage for faster convergence.

Acknowledgment. The work described in this paper was supported by grants from the Research Grants Council of the Hong Kong Special Administrative Region (CUHK4229/01E, HKUST6175/03E).

References

1. G. P. C. Fung, J. X. Yu, and H. Lu. Discriminative category matching: Efficient text classification for huge document collection. In *ICDM02*, 2002.

2. T. Joachims. Text categorization with support vector machines: Learning with many relevant features. In *ECML98*, 1998.
3. R. Klinkenberg and T. Joachims. Detecting concept drift with support vector machines. In *ICML00*, 2000.
4. R. Klinkenberg and I. Renz. Adaptive information filtering: Learning in the presence of concept drifts. In *ICML98*, 1998.
5. C. Lanquillon. Information filtering in changing domains. In *Proceedings of SIGIR-95 18th ACM International Conference on Research and Development in Information Retrieval*, 1995.
6. C. Lanquillon. Information filtering in changing domains. In *IJCAI99 Workshop*, 1999.
7. C. Lanquillon and I. Renz. Adaptive information filtering: Detecting changes in text stream. In *CIKM99*, page 538, 1999.
8. D. C. Montgomery. *Introduction to Statistical Quality Control*. Wiley, New York, 3rd edition, 1997.
9. F. Sebastiani. Machine learning in automated text categorization. *ACM Computing Surveys*, 34(1):1–47, 2002.
10. N. A. Syed, H. Liu, and K. K. Sung. Incremental learning with support vector machines. In *KDD99*, 1999.
11. H. Wang, W. Fan, P. S. Yu, and J. Han. Mining concept-drifting data streams using ensemble classifiers. In *KDD03*, 2003.
12. G. Widmer and M. Kubat. Learning in the presence of concept drift and hidden contexts. *Machine Learning*, 23:69–101, 2000.
13. Y. Yang and X. Liu. A re-examination of text categorization methods. In *SIGIR99*, 1999.

Using Cluster-Based Sampling to Select Initial Training Set for Active Learning in Text Classification*

Jaeho Kang, Kwang Ryel Ryu, and Hyuk-Chul Kwon

Department of Computer Engineering, Pusan National University,
San 30 Jangjeon-Dong, Kumjeong-Ku, Busan 609-735, Korea
{jhkang,krryu,hckwon}@pusan.ac.kr

Abstract. We propose a method of selecting initial training examples for active learning so that it can reach high performance faster with fewer further queries. Our method divides the unlabeled examples into clusters of similar ones and then selects from each cluster the most representative example which is the one closest to the cluster's centroid. These representative examples are labeled by the user and become the members of the initial training set. We also promote inclusion of what we call model examples in the initial training set. Although the model examples which are in fact the centroids of the clusters are not real examples, their contribution to enhancement of classification accuracy is significant because they represent a group of similar examples so well. Experiments with various text data sets have shown that the active learner starting from the initial training set selected by our method reaches higher accuracy faster than that starting from randomly generated initial training set.

1 Introduction

For a successful application of machine learning to classification tasks, we usually need to provide the learner with a sufficiently large number of labeled examples. If a large number of examples are available, however, obtaining the labels for these data often takes a lot of effort and time depending on the application area. Active learning [1,2] copes with this problem by carefully selecting examples to be labeled and trying to achieve the best possible performance using as few such labeled examples as possible, with the learning and querying stages repeated alternately. The focus of previous works on active learning was mainly on what and how to select the next example to query for its label, while not much attention has been given to the issue of selecting the initial training set. Given a better initial training set, we expect that the active learner can reach high performance faster with fewer further queries.

It is also worth pointing out that asking for category information for a single example at each querying stage could be practically inefficient. The user might

* This work was supported by National Research Laboratory Program (Contract Number: M10203000028-02J0000-01510) of KISTEP.

H. Dai, R. Srikant, and C. Zhang (Eds.): PAKDD 2004, LNAI 3056, pp. 384–388, 2004.

show better performance at categorizing examples when not just one but multiple examples are given simultaneously because he or she can compare different examples freely and work on them in any flexible order and thus becomes able to assign labels more precisely in a more time efficient manner. Especially when we have available enough manpower to answer the queries we can save time by having multiple examples labeled in parallel. Considering that active learning aims at achieving high performance using limited resources of time and manual efforts, selection of multiple examples for initial training seems desirable.

In this paper, we propose a method of selecting initial training examples for the purpose of enhancing the performance of active learning. When the number of labeled examples to be used for training is severely limited as is the case in the setting of active learning, it is advantageous not to have too similar examples to be included together in the training set. Our method therefore first divides the given unlabeled examples into clusters of similar ones and then selects from each cluster the most representative one to be labeled. In our empirical study, we applied k-means algorithm to cluster text data and selected from each cluster the document closest to the centroid as the cluster's representative. The selected documents were then labeled to form an initial training set. Experiments with this training set showed much better learning curves than that with randomly selected initial training set.

2 Cluster-Based Sampling to Select Initial Training Set

Our method of selecting examples for an initial training set starts by clustering the unlabeled examples by applying a k-means algorithm. Suppose we want to select two examples to form an initial training set. We first select two examples randomly. However, instead of using these examples directly as an initial training set (by querying to the user for their labels), our method takes them as initial seeds to group all the unlabeled examples into two clusters. Once the k-means algorithm converges to two final clusters after some iterations, a *representative example* is selected from each cluster as shown in Fig. 1(a). Since each cluster consists of similar examples which are likely to belong to the same category, it seems to be a good idea to pick a representative example from each cluster to form an initial training set. As the best candidate for a representative of each cluster, one might easily consider the centroid. However, direct labeling of the centroids may be difficult because they are not real existing examples. When working in the domain of text classification, for example, there is no document corresponding to a centroid. We therefore take the example closest to the centroid as the representative and ask the user for its category. In fact, the category thus obtained can also be assigned to the centroid itself because the two are so close to each other. The representatives, after being labeled, optionally together with the correspondingly labeled centroids, constitute an initial training set. The classifier learned from such an initial training set is shown in Fig. 1(b). The centroids, when they are labeled (at no cost of making extra query) and used for training examples, are called *model examples*. We will later show that inclusion of the

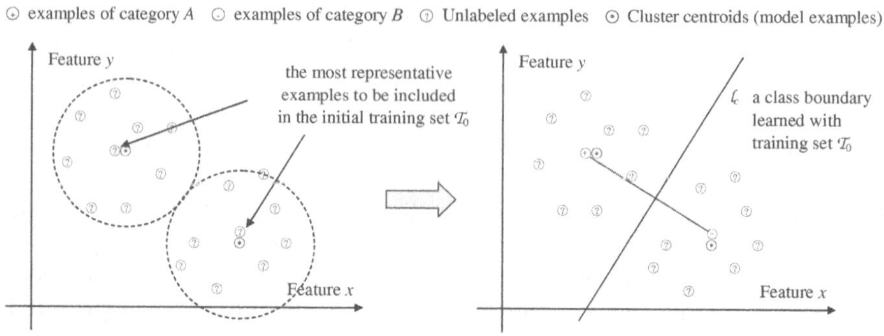

(a) Selecting initial training examples from stabilized clusters

(b) A class boundary learned with an initial training set $\mathcal{T}_0$ (with two cluster-basis examples)

Fig. 1. A result of learning with an initial training set selected by k-means clustering

model examples in the initial training set results in even more enhancement of learning performance.

3 Experimental Results

We have tested our cluster-based sampling method on several text classification tasks to form initial training set for active learning, using the two well-known text data sets `Reuters-21578` and `Newsgroups-20` [3]. From `Reuters-21578`, we selected only those articles which belong to one of the two most common topics (`earn` and `acq`). `Newsgroups-20` consists of about 20,000 articles that had been posted to 20 different newsgroups. To see the effect of our method on tasks of different difficulty levels, we generated from `Newsgroups-20` two data sets: `Different-3` and `Same-3` as also used and described in [4]. `Different-3` consists of articles which were posted to one of the three very different newsgroups, i.e., `alt.atheism`, `rec.sport.baseball`, and `sci.space`. The data set `Same-3` consists of articles from three very similar newsgroups, i.e., `comp.graphics`, `comp.os.ms-windows`, and `comp.windows.x`. Each article was converted to a final form through the process of stemming and removing USENET headers except the subject (for `Different-3` and `Same-3`), SGML tags (for `Reuters`), and stop-words. For classification tasks, we applied k-NN (k-nearest neighbor) algorithm which is known to work very well for text classification.

We implemented and compared three different strategies for selecting an initial training set from a bigger set of unlabeled examples. *Random* selects initial training examples just randomly. The other two strategies both start by selecting random seeds and applying a k-means algorithm to cluster the given data before forming an initial training set. While *KMeans* samples only one representative example from each cluster to form an initial training set, *KMeans+ME* collects the model example of each cluster in addition to the representative example. Ten-fold cross-validation was repeated for ten times to compare the effects of the differently generated initial training sets.

Figure 2 shows accuracies of the classifiers learned by using the initial training sets generated by the three strategies with the sizes of the initial training sets varied from 10 to up to 50. The horizontal axis of each graph indicates the number of training examples used. The vertical axis of each graph is the accuracy of classifier. The value of k for the k-NN algorithm was set to 5. Both the classifiers obtained by using our methods *KMeans* and *KMeans+ME* achieved higher accuracies than that by *Random* for all the three data sets whether they are easy or hard. In Fig. 2, we can see that the ten examples selected by *KMeans+ME* were more valuable for learning than the fifty examples selected by *Random*. These results show how important it is to select good examples for training, and also show that our strategies are really effective for the selection work. Notice that the model examples have positive effect on learning.

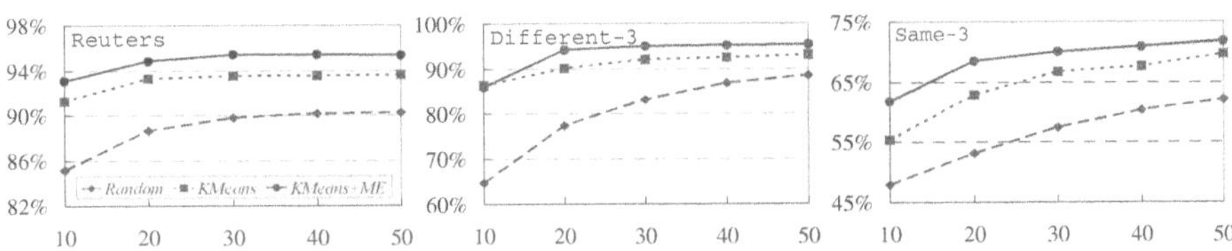

Fig. 2. Accuracies of classifiers with various sizes of initial training set

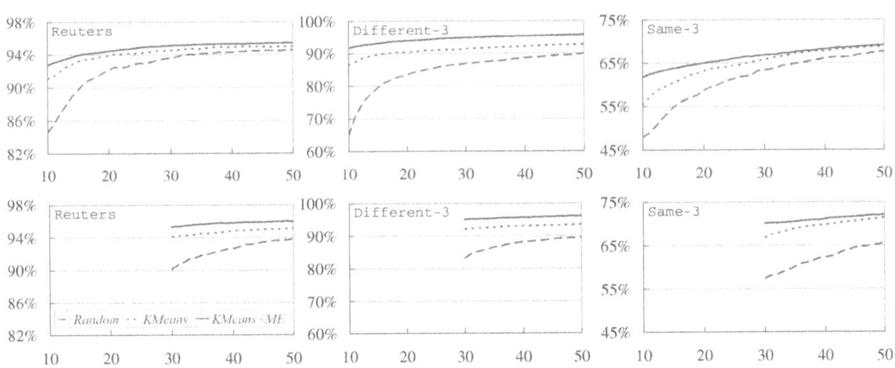

Fig. 3. Accuracies of active learner starting with 10 or 30 initial training examples

Figure 3 shows learning curves of the active learner when started from 10 or 30 initial training examples. The active learner was allowed to query for up to 50 examples including the given initial training set. The active learner used in this experiment selected the most uncertain example for the next query. The parameter k was fixed to 5 as in the previous experiments. We can easily see that our methods consistently outperform *Random*, especially when the size of the initial training set is relatively large. These results reveal that the quality of the initial training set makes significant influence on the subsequent on-going performance of the active learner.

4 Related Works

The main query selection strategy in active learning is uncertainty sampling [1] which selects the most ambiguous example for the next query. Roy and McCallum [2] argued that the best example for query is the one which, if labeled, can be used in the next learning stage to derive a classifier of minimum expected loss. The focus of these works was mainly on what and how to select an example for the next query after the initial training.

The text bundling technique recently proposed by Shih *et. al.* [5] is a new method for condensing the text data. Their method divides texts belonging to the same category into groups of similar ones and then averages each group to generate a bundled text. They showed that text bundling is useful especially when we solve text classification tasks by using a support vector machine which is a highly accurate but time-consuming learning algorithm. The role of bundled texts is very similar to that of our model examples in that they help the learner to derive classifiers of higher classification accuracy. However, text bundling requires examples which are already labeled.

5 Conclusions and Future Work

In this paper, we presented a method of selecting initial training examples for active learning so that it can reach high performance faster with fewer further queries. The core of our method is the selection of representative and model examples from the given unlabeled data after dividing the data into clusters of similar examples. Experiments with various text data sets have shown that the active learner starting from the initial training set selected by our method reaches higher accuracy faster than that starting from randomly selected initial training set.

As one of the future works, we would try to verify our method by applying it to various other classification tasks. We also plan to find a way to generalize our method to incorporate other clustering algorithms such as the EM algorithm.

References

1. Lewis, D. and Gale, W.: A sequential algorithm for training text classifiers. In Proc. of the 17th ACM-SIGIR Conference (1994) 3–12
2. Roy, N. and McCallum, A.: Toward optimal active learning through sampling estimation of error reduction. In Proc. of the 18th ICML (2001) 441–448
3. UCI Knowledge Discovery in Databases Archive: `http://kdd.ics.uci.edu/`
4. Basu, S., Banerjee, A., and Mooney, R.: Semi-supervised clustering by seeding, In Proc. of the 19th ICML (2002) 19–26
5. Shih, L., Rennie, J. D. M., Chang, Y.-H., and Karger, D. R.: Text bundling: statistics-based data reduction, In Proc. of the 20th ICML (2003) 696–703

Spectral Analysis of Text Collection for Similarity-Based Clustering

Wenyuan Li, Wee-Keong Ng, and Ee-Peng Lim

Centre for Advanced Information Systems
School of Computer Engineering
Nanyang Technological University, Singapore
liwy@pmail.ntu.edu.sg, {awkng, aseplim}@ntu.edu.sg

Abstract. Clustering of natural text collections is generally difficult due to the high dimensionality, heterogeneity, and large size of text collections. These characteristics compound the problem of determining the appropriate similarity space for clustering algorithms. In this paper, we propose to use the spectral analysis of the similarity space of a text collection to predict clustering behavior before actual clustering is performed. Spectral analysis is a technique that has been adopted across different domains to analyze the key encoding information of a system. Spectral analysis for prediction is useful in first determining the quality of the similarity space and discovering any possible problems the selected feature set may present. Our experiments showed that such insights can be obtained by analyzing the spectrum of the similarity matrix of a text collection. We showed that spectrum analysis can be used to estimate the number of clusters in advance.

1 Introduction

With the rapid growth of the World Wide Web, there is an abundance of text information to general users. Text collections on the Web are more heterogeneous, exhibit more frequent changes in terms of content, size, and feature than classical text collections. In clustering, one common preprocessing step is to capture or predict the characteristic of the target text collection before the clustering algorithm is applied. In this paper, we refer to this characteristic as the *clustering behavior* of a text collection. For instance, when clustering Web text collections that are heterogeneous and dynamic (frequent updates), prior knowledge of their characteristic is helpful in terms of determining the factors that may affect the clustering algorithm. In general, prediction can be thought as a *snapshot* (or rough view) of a text collection. In this sense, we refer to such a characteristic as the *fingerprint* of the target text collection. In this paper, our work is motivated by the investigation of the following two facts:

- **Function of similarity.** Clustering is described or defined by many researchers as a process based on the similarity among objects. In most cases, when the similarities among documents have been computed, the clustering

H. Dai, R. Srikant, and C. Zhang (Eds.): PAKDD 2004, LNAI 3056, pp. 389–393, 2004.

results are also largely determined. Such pairwise similarity among documents in a text collection can be recorded in a similarity matrix. Hence, the similarity matrix *encodes* the clustering characteristics of a text collection. Based on the function of similarity, it is feasible to establish the fingerprint of a text collection.

- **Function of domain knowledge.** The selection of feature sets, the feature weighting scheme, and the similarity measure can be viewed as the implicit or explicit domain knowledge in a clustering process [5]. For example, based on one's experience, one may select a set of meaningful terms or phrases to construct the feature set for a text collection. This feature set would be more effective than other feature sets that are computed mechanically. The effectiveness of a feature set determines the efficacy of the similarity matrix in terms of revealing the expected relationships of documents [4,7,9].

The above two observations serve as guiding principles for our work in this paper, which aims to perform text clustering from another perspective. In this paper, we proposed the use of the *normalized spectrum* (the set of the eigenvalues) of a similarity matrix as the fingerprint of a text collection. We discovered observations of the normalized spectrum of a target text collection that could establish the relationship between the similarity spectrum and the clustering behavior of the text collection.

2 Spectral Analysis of Similarity Matrix

2.1 Weighted Undirected Graph and Variant of Weighted Laplacian

Given the similarity matrix $\mathbf{S} = (s_{ij})_{n \times n}$, we define $\mathcal{G}(\mathbf{S}) = \langle V, E, \mathbf{S} \rangle$ as its associated graph where V is the set of n vertices and E is the set of weighted edges. In this graph, each vertex v_i corresponds to the i-th column (or row) and the weight of each edge $\widehat{v_i v_j}$ corresponds to the non-diagonal entry s_{ij}. Therefore, we have established a mapping between $\mathbf{S}$ and $\mathcal{G}(\mathbf{S})$. This mapping binds the clustering behaviors of $\mathbf{S}$ with the properties and structure of $\mathcal{G}(\mathbf{S})$.

Due to the different scale of the spectrums of different $\mathbf{S}$'s, it is difficult to analyze and compare them. Thus, we transform $\mathbf{S}$ to a weighted Laplacian $\mathbf{L} = (\ell_{ij})$ which has "normalized" eigenvalues with the same scale [1]. The eigenvalues of $\mathbf{L}$ lie in the interval $[0, 2]$. For ease of analysis and computation, we use L, a variant of the weighted Laplacian, instead of $\mathbf{L}$, which is defined as:

$$\mathsf{L} = \mathbf{D}^{-1/2}(\mathbf{S} - \mathbf{I})\mathbf{D}^{-1/2} \tag{1}$$

where $\mathbf{D} = \text{diag}(d_i)$ denotes the diagonal matrix ($d_i = \sum_j s_{ij}$ is the degree of the vertex v_i in graph $\mathcal{G}(\mathbf{S})$). From the definition of $\mathbf{L}$ [1] and Equations 1, we deduce $\text{eig}(\mathsf{L}) = \{1 - \lambda | \lambda \in \text{eig}(\mathbf{L})\}$ (where $\text{eig}(\cdot)$ represents the set of eigenvalues of a matrix). Hence, the eigenvalues of L are in the interval $[-1, 1]$ and all the conclusions and properties of the weighted Laplacian $\mathbf{L}$ are also applicable to L. In this paper, we use eigenvalues of a variant of the weighted Laplacian, $\text{eig}(\mathsf{L})$, to perform the spectral analysis of the similarity matrix $\mathbf{S}$. Given the similarity matrix **S**, let $\lambda_1 \geqslant \lambda_2 \geqslant \cdots \geqslant \lambda_n$ be the eigenvalues of $\mathcal{G}(\mathbf{S})$ in non-increasing order. There are some basic facts of the $\mathcal{G}(\mathbf{S})$: (1) $\sum \lambda_i = 0$; (2) $-1 \leqslant \lambda_i \leqslant 1$, $i = 1, 2, \cdots, n$; (3) $\lambda_1 = 1$. Proofs of these facts are not given as they can be found in spectral graph theory [1,2].

Suppose h is the number of non-zeros in **S**, then the computational complexity of this transformation is $O(h)$. the complexity of our method is $O(k(h+n))$ using Lanczos method [2].

2.2 Some Observations about the $\mathcal{G}(\mathbf{S})$ Spectrum

Two observations of the $\mathcal{G}(\mathbf{S})$ spectrum for the clustering behavior of **S** are given for the relationships between the clustering behavior of **S** and the principal properties and structure of $\mathcal{G}(\mathbf{S})$. Due to the limitation of space, intuitive and theoretical explanations for these observations are not presented in this paper.

Observation 1. *Suppose the similarity matrix* **S** *has the block structure:*

$$\mathbf{S} = \begin{bmatrix} \mathbf{S}_{11} & \cdots & \mathbf{S}_{1k} \\ \vdots & \ddots & \vdots \\ \mathbf{S}_{k1} & \cdots & \mathbf{S}_{kk} \end{bmatrix} \begin{matrix} n_1 \\ \vdots \\ n_k \end{matrix} \qquad (2)$$
$$\quad n_1 \cdots n_k$$

where $n_1 + \cdots + n_k = n$, *for each diagonal block matrix* $\mathbf{S}_{ii}$, *they satisfy* $0 < n_i - \|\mathbf{S}_{ii}\|_F < \delta(\delta \to 0)$. *For each non-diagonal block matrix* $\mathbf{S}_{ij}$, *they satisfy* $\|\mathbf{S}_{ij}\|_F \to 0$ ($\|\cdot\|_F$ *is the Frobenius norm*). **S** *shows a good clustering behavior with* n *clusters. The spectrum of* $\mathcal{G}(\mathbf{S})$ *is*

$$\begin{aligned} &\lambda_i \to 1 \quad (i = 1, \cdots, k \text{ and } 0 < \lambda_i \leqslant 1) \\ &|\lambda_i| \to 0 \ (i = k+1, \cdots, n) \end{aligned} \qquad (3)$$

In this observation, the above two conditions guarantee the extremely good clustering behavior of **S** with intra-similarities approaching 1 and inter-similarities approaching 0 among clusters. The spectrum of $\mathcal{G}(\mathbf{S})$ shows the typical distribution in (3). However, for text collections in the real world, the spectrum of $\mathcal{G}(\mathbf{S})$ shows complex distribution that is greatly different from that of the extreme case. The following is the observation about the large text collection.

Observation 2. *Suppose* $1 = \lambda_1 \geqslant \lambda_2 \geqslant \cdots \geqslant \lambda_n$ *be the* $\mathcal{G}(\mathbf{S})$ *spectrum we have*

(1) *The distribution of the spectrum exhibits a polynomial of high odd order, and most of eigenvalues are centered on the near neighbor of 0.*
(2) *If* λ_2 *is higher, there exists a better bipartition for* **S**.
(3) *For the sequence* $\alpha_i = \frac{\lambda_i}{\lambda_2}$ $(i \geqslant 2)$, $\exists k \geqslant 2$, *it has* $\alpha_i \to 1$ *and* $\alpha_i - \alpha_{i+1} > \delta$ $(0 < \delta < 1)$, *then* k *indicates the cluster number of the text collection.* (Here, δ is a predefined threshold to measure the first large gap between α_i)
(4) *If the curve of the spectrum is closer to the* x*-axis, the clustering behavior of* **S** *is worse.* (This closeness can be measured by $\sum_i (\lambda_i)^2$)

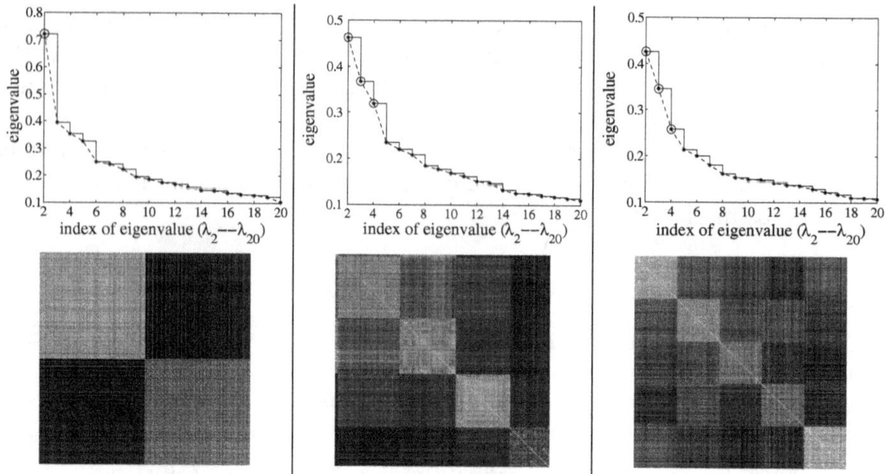

(a) WEB2 (b) CLASSIC4 (c) NEWS5

Fig. 1. Part of Spectrum and Gray Scale Images on Three Text Collections

Table 1. Estimation of Cluster Number by Three Indices (√ means correct estimation)

	Bisecting k-means			Graph-based			Hierarchical		
	WEB2	CLASSIC4	NEWS5	WEB2	CLASSIC4	NEWS5	WEB2	CLASSIC4	NEWS5
CH	5	2	3	3	3	3	7	2	5 (√)
KL	29	17	22	27	21	22	22	21	9
Hart	6	13	4 (√)	6	10	4 (√)	1	2	1

3 Experimental Results

Three text collections were used in experiments: (1) "WEB2" is from two categories in Yahoo with each category containing 600 documents [8]. (2) "CLASSIC4" is from four classes of traditional text collections in IR area (CACM/CISI/CRAN/MED) with each class containing 1,000 documents. (3) "NEWS5" is from five newsgroups of UseNet news postings with each newsgroup containing 500 documents. In this experiment, we investigated the prediction of cluster number to test the viability of spectral analysis on text collections.

Spectral Analysis. Specifically, we estimated the number of clusters in text collection by Observation 2(3). The analysis starts from λ_2, as λ_1 is always 1. From Figure 1, we found and circled the largest eigenvalues with the largest gaps. It can be observed by the stair steps among the eigenvalues in Figure 1. According to Observation 2(3), the number of these large eigenvalues (including λ_1) with wide gaps indicates the number of clusters. We can verify it by analyzing their gray scale images and class topics.

WEB2 has two completely different topics, which can be seen in its gray scale image (Figure 1(a)). Observing the spectrum of its similarity matrix (Figure 1(a)), λ_2 has a higher value than other eigenvalues. The rest of eigenvalues fall along a smooth gentle line rather than a steep line. This phenomenon conforms to Observation 2(3). Hence, in this case, $k = 2$. For CLASSIC4, similar spectral analysis can be made from its spectrum (Figure 1(b)). The result shows that this collection has $k = 4$ clusters. Unlike the previous two collections with disjoint topics, NEWS5 has five overlapping topics: "atheism", "comp.sys", "comp.windows", "misc.forsale" and "rec.sport". They are exhibited by the gray scale image (Figure 1(c)). λ_2, λ_3, and λ_4 have higher values and wider gaps than other eigenvalues. It indicates that there are $k = 4$ clusters in this collection. The topics "comp.sys" and "comp.windows" can be viewed as one topic "comp" which are more different from other three topics. From this point of view, the estimation of four clusters is a reasonable result.

Comparison with Existing Methods. Traditional methods for estimating the number of cluster typically run a clustering algorithm many times with predefined cluster numbers k from 2 to k_{max}. Then the optimum k is obtained by an internal index based on those clustering results. Their difference is the index they used. In this experiment, we computed three widely used statistical indices [3]: Calinski and Harabasz (CH), Krzanowski and Lai (KL) and Hartigan (Hart), based on three clustering algorithms [6]. In Table 1, our method remarkably outperforms these three methods. It may be due to the dependence of the specific clustering algorithms and their ineffectiveness for high dimensional data. In contrary, our method is independent of the specific clustering algorithms, is suitable for high-dimensional data sets, and has low complexity.

References

1. F. R. K. Chung. *Spectral Graph Theory.* Number 92 in CBMS Regional Conference Series in Mathematics. American Mathematical Society, 1997.
2. G. Golub and C. V. Loan. *Matrix Computations.* The Johns Hopkins University Press, 1996.
3. A. Gordon. *Classification.* Chapman and Hall/CRC, 2nd edition, 1999.
4. V. Hatzivassiloglou, L. Gravano, and A. Maganti. An investigation of linguistic features and clustering algorithms for topical document clustering. In *Proc. of the 23rd ACM SIGIR*, pages 224–231, Greece, July 2000.
5. A. K. Jain, M. N. Murty, and P. J. Flynn. Data clustering: A review. *ACM Computing Surveys*, 31(13):264–323, 1999.
6. G. Karypis. Cluto: A clustering toolkit. TR #02 017, Univ. of Minnesota, 2002.
7. D. Modha and S. Spangler. Feature weighting in k-means clustering. *Machine Learning*, 52(3):556–561, 2003.
8. M. P. Sinka and D. W. Corne. *Soft Computing Systems: Design, Management and Applications*, chapter A Large Benchmark Dataset for Web Document Clustering, pages 881–890. IOS Press, 2002.
9. A. Strehl, J. Ghosh, and R. Mooney. Impact of similarity measures on web-page clustering. In *Proc. of AAAI Workshop on AI for Web Search*, pages 58–64, Texas, July 2000.

Clustering Multi-represented Objects with Noise

Karin Kailing, Hans-Peter Kriegel, Alexey Pryakhin, and Matthias Schubert

Institute for Computer Science
University of Munich
Oettingenstr. 67, 80538 Munich, Germany
{kailing | kriegel | pryakhin | schubert}@dbs.informatik.uni-muenchen.de

Abstract. Traditional clustering algorithms are based on one representation space, usually a vector space. However, in a variety of modern applications, multiple representations exist for each object. Molecules for example are characterized by an amino acid sequence, a secondary structure and a 3D representation. In this paper, we present an efficient density-based approach to cluster such multi-represented data, taking all available representations into account. We propose two different techniques to combine the information of all available representations dependent on the application. The evaluation part shows that our approach is superior to existing techniques.

1 Introduction

In recent years, the research community spent a lot of attention to clustering resulting in a large variety of different clustering algorithms [1]. However, all those methods are based on one representation space, usually a vector space of features and a corresponding distance measure. But for a variety of modern applications such as biomolecular data, CAD- parts or multi-media files mined from the internet, it is problematic to find a common feature space that incorporates all given information. Molecules like proteins are characterized by an amino acid sequence, a secondary structure and a 3D representation. Additionally, protein databases such as Swissprot [2] provide meaningful text descriptions of the stored proteins. In CAD-catalogues, the parts are represented by some kind of 3D model like Bezier curves, voxels or polygon meshes and additional textual information like descriptions of technical and economical key data. We call this kind of data *multi-represented data*, since any data object might provide several different representations that may be used to analyze it.

To cluster multi-represented data using the established clustering methods would require to restrict the analysis to a single representation or to construct a feature space comprising all representations. However, the restriction to a single feature space would not consider all available information and the construction of a combined feature space demands great care when constructing a combined distance function.

In this paper, we propose a method to integrate multiple representations directly into the clustering algorithm. Our method is based on the density-based

H. Dai, R. Srikant, and C. Zhang (Eds.): PAKDD 2004, LNAI 3056, pp. 394–403, 2004.

clustering algorithm DBSCAN [3] that provides several advantages over other algorithms, especially when analyzing noisy data. Since our method employs a separated feature space for each representation, it is not necessary to design a new suitable distance measure for each new application. Additionally, the handling of objects that do not provide all possible representations is integrated naturally without defining dummy values to compensate for the missing representations. Last but not least, our method does not require a combined index structure, but benefits from each index that is provided for a single representation. Thus, it is possible to employ highly specialized index structures and filters for each representation.We evaluate our method for two example applications. The first is a data set consisting of protein sequences and text descriptions. Additionally, we applied our method to the clustering of images retrieved from the internet. For this second data set, we employed two different similarity models.

The rest of the paper is organized as follows. After this introduction, we present related work. Section 3 formalizes the problem and introduces our new clustering method. In our experimental evaluation that is given in section 4, we introduce a new quality measure to judge the quality of a clustering with respect to a reference clustering and display the results achieved by our method in comparison with the other mentioned approaches. The last section summarizes the paper and presents some ideas for future research.

2 Related Work

There are several problems that are closely related to the clustering of multi-represented data. Data mining of multi-instance objects [4] is based on the precondition that each data object might be represented by more than one instance in a common data space. However, all instances that are employed are elements of the same data space and multi-instance objects were predominantly treated with respect to classification not to clustering.

A similar setting to the clustering of multi-represented objects is the clustering of heterogenous or multi-typed objects [5,6] in web mining. In this setting, there are also multiple databases each yielding objects in a separated data space. Each object within these data spaces may be related to an arbitrary amount of data objects within the other data spaces. The framework of reinforcement clustering employs an iterative process based on an arbitrary clustering algorithm. It clusters one dedicated data space while employing the other data spaces for additional information. It is also applicable for multi-represented objects. However, due to its dependency on the data space for which the clustering is started, it is not well suited to solve our task. Since to the best of our knowledge reinforcement clustering is the only other clustering algorithm directly applicable to multi-represented objects, we use it for comparison in our evaluation section.

Our approach is based on the formal definitions of density-connected sets underlying the algorithm DBSCAN [3]. Based on two input parameters (ε and k), DBSCAN defines dense regions by means of core objects. An object $o \in DB$ is called *core object*, if its ε-neighborhood contains at least k objects. Usually clusters contain several core objects located inside a cluster and border objects located at the border of the cluster. In addition, objects within a clusters must be “density-connected”. DBSCAN is able to detect arbitrarily shaped clusters by one single pass over the data. To do so, DBSCAN uses the fact, that a density-connected cluster can be detected by finding one of its core-objects o and computing all objects which are density-reachable from o. The correctness of DBSCAN can be formally proven (cf. lemmata 1 and 2 in [3], proofs in [7]).

3 Clustering Multi-represented Objects

Let DB be a database consisting of n objects. Let $R := \{R_1, ..., R_m\}$ be the set of different representations existing for objects in DB. Each object $o \in DB$ is therefore described by maximally m different representations, i.e. $o := \{R_1(o), R_2(o), ..., R_m(o)\}$. If all different representations exist for o, than $|o| = m$, else $|o| < m$. The distance function is denoted by $dist$. We assume that $dist$ is symmetric and reflexive. In the following, we call the ε_i-neighborhood of an object o in one special representation R_i its local ε-neighborhood w.r.t. R_i.

Definition 1 (local ε_i-neighborhood w.r.t R_i).
Let $o \in DB$, $\varepsilon_i \in I\!R^+$, $R_i \in R$, $dist_i$ the distance function of R_i. The local ε_i-neighborhood w.r.t. R_i *of o, denoted by $\mathcal{N}_{\varepsilon_i}^{R_i}(o)$, is defined by*

$$\mathcal{N}_{\varepsilon_i}^{R_i}(o) = \{x \in DB \mid dist_i(R_i(o), R_i(x)) \leq \varepsilon_i\}.$$

Note that ε_i can be chosen optimally for each representation. The simplest way of clustering multi-represented objects, is to select one representation R_i and cluster all objects according to this representation. However, this approach restricts data analysis to a limited part of the available information and does not use the remaining representations to find a meaningful clustering. Another way to handle multi-represented objects is to combine the different representations and use a combined distance function. Then any established clustering algorithm can be applied. However, it is very difficult to construct a suitable combined distance function that is able to fairly weight each representation and handle missing values. Furthermore, a combined feature space, does not profit from specialized data access structures for each representation.

The idea of our approach is to combine the information of all different representations as early as possible, i.e. during the run of the clustering algorithm, and as late as necessary, i.e. after using the different distance functions of each representation. To do so, we adapt the core object property proposed for DBSCAN. To decide whether an object is a core object, we use the local ε-neighborhoods of each representation and combine the results to a global neighborhood. Therefore, we must adapt the predicate direct density-reachability proposed for DBSCAN. In the next two subsections, we will show how we can use the concepts of union and intersection of local neighborhoods to handle multi-represented objects.

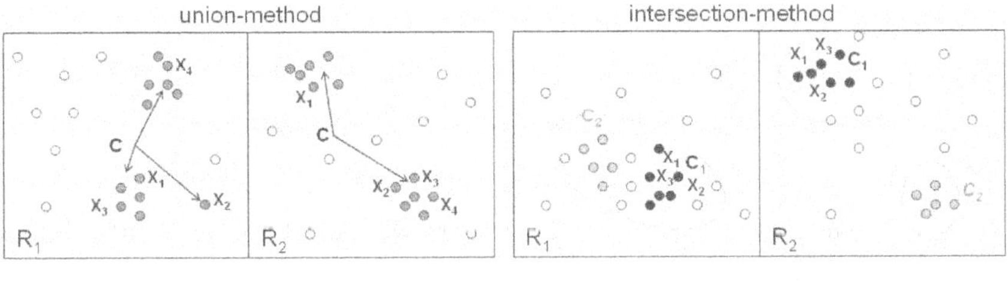

Fig. 1. The left figure displays local clusters and a noise object that are aggregated to a multi-represented cluster **C**. The right figure illustrates, how the intersection-method divides a local clustering into clusters C_1 and C_2.

3.1 Union of Different Representations

This variant is especially useful for sparse data. In this setting, the clusterings in each single representation will provide several small clusters and a large amount of noise. Simply enlarging ε would relief the problem, but on the other hand, the separation of the clusters would suffer. The union-method assigns objects to the same cluster, if they are similar in at least one of the representations. Thus, it keeps up the separation of local clusters, but still overcomes the sparsity. If the object is placed in a dense area of at least one representation, it is still a core object regardless of how many other representations are missing. Thus, we do not need to define dummy values. The left part of figure 1 illustrates the basic idea. We adapt some of the definitions of DBSCAN to capture our new notion of clusters. To decide whether an object o is a union core object, we unite all local ε_i-neighborhoods and check whether there are enough objects in the global neighborhood, i.e. whether the global neighborhood of o is dense.

Definition 2 (union core object).
Let $\varepsilon_1, \varepsilon_2, ..., \varepsilon_m \in \mathbb{R}^+$, $k \in \mathbb{N}$. An object $o \in DB$ is called union core object, *denoted by $\text{CoreU}^k_{\varepsilon_1,..,\varepsilon_m}(o)$, if the union of all local ε-neighborhoods contains at least k objects, formally:*

$$\text{CoreU}^k_{\varepsilon_1,..,\varepsilon_m}(o) \Leftrightarrow \Big| \bigcup_{R_i(o)\in o} \mathcal{N}^{R_i}_{\varepsilon_i}(o) \Big| \geq k.$$

Definition 3 (direct union-reachability).
Let $\varepsilon_1, \varepsilon_2, .., \varepsilon_m \in \mathbb{R}^+$, $k \in \mathbb{N}$. An object $p \in DB$ is directly union-reachable *from $q \in DB$ if q is a union core object and p is an element of at least one local $\mathcal{N}^{R_i}_{\varepsilon_i}(q)$, formally:*

$$\text{DirReachU}^k_{\varepsilon_1,..,\varepsilon_m}(q,p) \Leftrightarrow \text{CoreU}^k_{\varepsilon_1,..,\varepsilon_m}(q) \wedge \exists\, i \in \{1,..,m\} : R_i(p) \in \mathcal{N}^{R_i}_{\varepsilon_i}(q).$$

The predicate direct union-reachability is obviously symmetric for pairs of core objects, because the $dist_i$ are symmetric distance functions. Thus, analogously to DBSCAN reachability and connectivity can be defined.

3.2 Intersection of Different Representations

The intersection method is well suited for data containing unreliable representations, i.e. there is a representation, but it is questionable, whether it is a good description of the object. In those cases, the intersection-method requires that a cluster should contain only objects which are similar according to all representations. Thus, this method is useful, if all different representations exist, but the derived distances do not adequately mirror the intuitive notion of similarity. The intersection-method is used to increase the cluster quality by finding purer clusters.

To decide, whether an object o is an intersection core object, we examine, whether o is a core object in each involved representation. Of course, we use different ε-values for each representation to decide, whether locally there are enough objects in the ε-neighborhood. The parameter k is used to decide, whether globally there are still enough objects in the ε-neighborhood, i.e. the intersection of all local neighborhoods contains at least k objects.

Definition 4 (intersection core object).
Let $\varepsilon_1, \varepsilon_2, ..., \varepsilon_m \in I\!R^+$, $k \in I\!N$. An object $o \in DB$ is called intersection core object, *denoted by* $\text{CoreIS}^k_{\varepsilon_1,..,\varepsilon_m}(o)$, *if the intersection of all its local ε_i-neighborhoods contain at least k objects, formally:*

$$\text{CoreIS}^k_{\varepsilon_1,..,\varepsilon_m}(o) \Leftrightarrow |\bigcap_{i=1,..,m} \mathcal{N}^{R_i}_{\varepsilon_i}(o)| \geq k.$$

Using this new property, we can now define direct intersection-reachability in the following way:

Definition 5 (direct intersection-reachability).
Let $\varepsilon_1, \varepsilon_2, ..., \varepsilon_m \in I\!R^+$, $k \in I\!N$. An object $p \in DB$ is directly intersection-reachable *from $q \in DB$ if q is an intersection core object and p is an element of all local $\mathcal{N}_\varepsilon(q)$, formally:*

$$\text{DirReachIS}^k_{\varepsilon_1,..,\varepsilon_m}(q,p) \Leftrightarrow \text{CoreIS}^k_{\varepsilon_1,..,\varepsilon_m}(q) \wedge \forall i = 1,..,m : R_i(p) \in \mathcal{N}^{R_i}_{\varepsilon_i}(q) \,.$$

Again, reachability and connectivity can be defined analogously to DBSCAN. The right part of figure 1 illustrates the effects of this method.

3.3 Determination of Density Parameters

In [3], a heuristic is presented to determine the ε-value of the "thinnest" cluster in the database. This heuristic is based on a diagram that represents sorted *knn*-distances of all given objects. In the case of multi-represented objects, we have to choose ε for each dimension separately, whereas k can be chosen globally. A user determines a value for global k. The system computes the *knn*-distance diagrams for the given global k (one diagram for every representation). The user has to choose a so-called border object o for each representation. The ε for the

i-th representation is given by the knn-distance of the border object of R_i. Let us note that this method still allows a certain range of ε-values to be chosen. The selection should mirror the different requirements of the proposed methods. For the union method, it is more advisable to chose a lower or conservative value, since its characteristic demands that the elements of the local ε-neighborhood should really be similar. For the intersection-method, the ε-value should be selected progressively, i.e. at the upper rim of the range. This selection reflects that the objects of a cluster need not be too similar for a single representation, because it is required that they are similar with respect to all representations.

4 Performance Evaluation

To demonstrate the capability of our method, we performed a thorough experimental evaluation for two types of applications. We implemented the proposed clustering algorithm in Java 1.4. All experiments were processed on a work station with a 2.6 GHz Pentium IV processor and 2 GB main memory.

4.1 Deriving Meaningful Groupings in Protein Databases

The first set of experiments was performed on protein data that is represented by amino-acid sequences and text descriptions. Therefore, we employed entries of the Swissprot protein database [2] belonging to 5 functional groups (cf. Table 1) and transformed each protein into a pair of feature vectors. Each amino acid sequence was mapped into a 436 dimensional feature space. The first 400 features are 2-grams of successive amino-acids. The last 36 dimensions are 2-grams of 6 exchange groups that the single amino-acids belong to [8]. To compare the derived feature vectors, we employed Euclidian distance. To process text documents, we rely on projecting the documents into the feature space of relevant terms. Documents are described by a vector of term frequencies weighted by the inverse document frequency (TFIDF) [9]. We chose 100 words of medium frequency as relevant terms and employed cosine distance to compare the TFIDF-vectors. Since Swissprot entries provide a unique mapping to the classes of Gene Ontology [10], a reference clustering for the selected proteins was available. Thus, we are able to measure a clustering of Swissprot entries by the degree it reproduces the class structure provided by Gene Ontology.

To have an exact measure for this degree, we employed the class entropy in each cluster. However, there are two effects that have to be considered to obtain a

Table 1. Description of the protein data sets.

	Set 1	Set 2	Set 3	Set 4	Set 5
Name	Isomerase	Lyase	Signal Transducer	Oxidore-ductase	Transferase
Classes	16	35	39	49	62
Objects	501	1640	2208	3399	4086

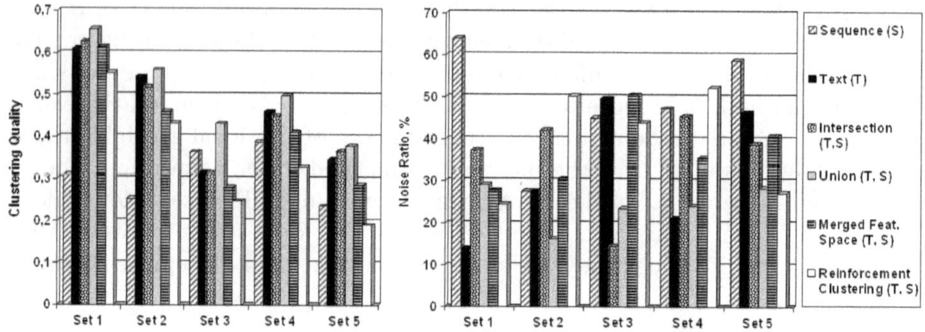

Fig. 2. Clustering quality and noise ratio.

fair measure of a clustering with noise. First, a large cluster of a certain entropy should contribute more to the overall quality of the clustering than a rather small cluster providing the same quality. The second effect is that a clustering having a 5 % noise ratio should be ranked higher than a clustering having the same average entropy for all its clusters, but contains 50 % noise.

To consider both effects we propose the following quality measure for comparing different clusterings with respect to a reference clustering.

Definition 6. *Let O be the set of data objects, let $C = \{C_i | C_i \subset O\}$ be the set of clusters and let $K = \{K_i | K_i \subset O\}$ be the reference clustering of O. Then we define:*

$$Q_K(C) = \sum_{C_i \in C} \frac{|C_i|}{|O|} \cdot (1 + entropy_K(C_i))$$

where $entropy_K(C_i)$ denotes the entropy of cluster C_i with respect to K.

The idea is to weight every cluster by the percentage of the complete data objects being part of it. Thus, smaller clusters are less important than larger ones and a clustering providing an extraordinary amount of noise can contribute only the percentage of clustered objects to the quality. Let us note that we add 1 to the cluster entropies. Therefore, we measure the reference clustering K with the quality score of 1 and a worst case clustering – e.g. no clusters are found at all– with the score of 0. To relate the quality of the clustering achieved by our methods to the results of former methods, we compared it to 4 alternative approaches. First, we clustered text and sequences separately using only one of the representations. A second approach combines the features of both representations into a common feature space and employs the cosine distance to relate the resulting feature vectors. As the only other clustering method that is able to handle multi-represented data, we additionally compared reinforcement clustering using DBSCAN as underlying cluster algorithm. For reinforcement clustering, we ran 10 iterations and tried several values of the weighting parameter α. The local ε-parameters were selected as described above and we chose

Fig. 3. Example of an image cluster. The left rectangle contains images clustered by the intersection-method. The right rectangles display additional images that were grouped with the corresponding cluster when clustering the images with respect to a single representation.

$k = 2$. To consider the different requirements of both methods, for each data set a progressive and a conservative ε-value was determined. All approaches were run for both settings and the best results are displayed.

The left diagram of figure 2 displays the derived quality for those 4 methods and the two variants of our method. In all five test sets, the union-method using conservative ε-values outperformed any of the other algorithms. Furthermore, the noise ratio for each data set was between 16% and 28% (cf. figure 2, right), indicating that the main portion of the data objects belongs to some cluster. The intersection method using progressive ε-parameters performed comparably well, but was to restrictive to overcome the sparseness of the data as good as the union-method.

4.2 Clustering Images by Multiple Representations

Clustering image data is a good example for the usefulness of the intersection-method. A lot of different similarity models exists for image data, each having its own advantages and disadvantages. Using for example text descriptions of images, one is able to cluster all images related to a certain topic, but these images must not look alike. Using color histograms instead, the images are clustered according to the distribution of color in the image. But as only the color information is taken into account a green meadow with some flowers and a green billiard table with some colored shots on it, can of course not be distinguished by this similarity model. On the other hand, a similarity model taking content information into account might not be able to distinguish images of different colors.

Our intersection approach is able to get the best out of all these different types of representations. Since the similarity in one representation is not really sound, the intersection-method is well-suited to find clusters of better quality for this application. For our experiments, we used two different representations. The first representation was a 64-dimensional color histogram. In this case, we used the weighted distance between those color histograms, represented as a quadratic form distance function as described for example in [11]. The second representation were segmentation trees. An image was first divided into segments of similar color by a segmentation algorithm. In a second step, a tree was created from those segments by iteratively applying a region-growing algorithm which merges neighboring segments, if their colors are alike. In [12] an efficient technique is described to compute the similarity between two such trees using filters for the complex edit-distance measure.

As we do not have any class labels to measure the quality of our clustering, we can only describe the results we achieved. In general, the clusters we got using both representations were more accurate than the clusters we got using each representation separately. Of course, the noise ratio increased for the intersection-method. Due to space limitations we only show one sample cluster of images we found with the intersection-method (see Figure 3). Using this method, very similar images are clustered together. When clustering each single representation, a lot of additional images were added to the corresponding cluster. As one can see, using the intersection-method only the most similar images of both representations still belong to the cluster.

5 Conclusions

In this paper, we discussed the problem of clustering multi-represented objects. A multi-represented object is described by a set of representations where each representation belongs to a different data space. Contrary to existing approaches our proposed method is able to cluster this kind of data using all available representations without forcing the user to construct a combined data space. The idea of our approach is to combine the information of all different representations as early as possible and as late as necessary. To do so, we adapted the core object property proposed for DBSCAN. To decide whether an object is a core object, we use the local ε-neighborhoods of each representation and combine the results to a global neighborhood. Based on this idea, we proposed two different methods for varying applications. For sparse data, we introduced the union-method that assumes that an object is a core object, if k objects are found within the union of its local ε-neighborhoods. Respectively, we defined the intersection-method for data where each local representation yields rather big and unspecific clusters. Therefore, the intersection-method requires that at least k objects are within the intersection of all local ε-neighborhoods of a core object. In our experimental evaluation, we introduced an entropy based quality measure that compares a given clustering with noise to a reference clustering. Employing this quality measure, we demonstrated that the union method was most suitable to overcome the sparsity of a given protein data set. To demonstrate the ability of the intersection method to increase the cluster quality, we applied it to a set of images using two different similarity models. For future work, we plan to examine applications providing more than two representations. We are especially interested, in clustering proteins with respect to all of the mentioned representations. Another interesting challenge is to extend our method to an multi-instance and multi-representation clustering. In this setting each object may be represented by several instances in some of the representations.

References

1. Han, J., Kamber, M.: "Data Mining: Concepts and Techniques". Morgan Kaufman (2001)
2. Boeckmann, B., Bairoch, A., Apweiler, R., Blatter, M.C., Estreicher, A., Gasteiger, E., Martin, M., Michoud, K., O'Donovan, C., Phan, I., Pilbout, S., Schneider, M.: "The SWISS-PROT Protein Knowledgebase and its Supplement TrEMBL in 2003". Nucleic Acid Research **31** (2003) 365–370
3. Ester, M., Kriegel, H.P., Sander, J., Xu, X.: "A Density-Based Algorithm for Discovering Clusters in Large Spatial Databases with Noise". In: Proc. KDD'96, Portland, OR, AAAI Press (1996) 291–316
4. Weidmann, N., Frank, E., Pfahringer, B.: "A Two-Level Learning Method for Generalized Multi-instance Problems". In: Proc. ECML 2003, Cavtat-Dubrovnik,Cr. (2003) 468–479
5. Wang, J., Zeng, H., Chen, Z., Lu, H., Tao, L., Ma, W.: "ReCoM: reinforcement clustering of multi-type interrelated data objects. 274-281". In: Proc. SIGIR 2003, July 28 - August 1, Toronto, CA, ACM (2003) 274–281
6. Zeng, H., Chen, Z., Ma, W.: "A Unified Framework for Clustering Heterogeneous Web Objects". In: Proc. 3rd WISE 2002, 12-14 December, Singapore, IEEE Computer Society (2002) 161–172
7. Sander, J., Ester, M., Kriegel, H.P., Xu, X.: "Density-Based Clustering in Spatial Databases: The Algorithm GDBSCAN and its Applications". In: Data Mining and Knowledge Discovery, An International Journal, Kluwer Academic Publishers (1998) 169–194
8. Deshpande, M., Karypis, G.: "Evaluation of Techniques for Classifying Biological Sequences". In: Proc. PAKDD'02. (2002) 417–431
9. Salton, G.: "Automatic Text Processing: The Transformation, Analysis, and Retrieval of Information by Computer". Addison-Wesley (1989)
10. Consortium, T.G.O.: "Gene Ontology: Tool for the Unification of Biology". Nature Genetics **25** (2000) 25–29
11. Hafner, J., Sawhney, H., W., E., Flickner, M., Niblack, W.: Efficient color histogram indexing for quadratic form distance functions. IEEE Transactions on Pattern Analysis and Machine Intelligence (TPAMI) **17 (7)** (1995) 729–736
12. Kailing, K., Kriegel, H.P., Schönauer, S., Seidl, T.: Efficient similarity search for hierachical data in large databases. In: to appear in Proc. EDBT 2004. (2004)

Providing Diversity in K-Nearest Neighbor Query Results

Anoop Jain, Parag Sarda, and Jayant R. Haritsa*

Database Systems Lab, SERC/CSA
Indian Institute of Science, Bangalore 560012, INDIA.

Abstract. Given a point query Q in multi-dimensional space, K-Nearest Neighbor (KNN) queries return the K closest answers in the database with respect to Q. In this scenario, it is possible that a majority of the answers may be very similar to one or more of the other answers, especially when the data has clusters. For a variety of applications, such homogeneous result sets may not add value to the user. In this paper, we consider the problem of providing diversity in the results of KNN queries, that is, to produce the closest result set such that each answer is sufficiently different from the rest. We first propose a user-tunable definition of diversity, and then present an algorithm, called MOTLEY, for producing a diverse result set as per this definition. Through a detailed experimental evaluation we show that MOTLEY can produce diverse result sets by reading only a small fraction of the tuples in the database. Further, it imposes no additional overhead on the evaluation of traditional KNN queries, thereby providing a seamless interface between diversity and distance.

Keywords: Nearest Neighbor, Distance Browsing, Result Diversity

1 Introduction

Over the last few years, there has been considerable interest in the database community with regard to supporting K-Nearest Neighbor (KNN) queries [8]. The general model of a KNN query is that the user gives a point query in multidimensional space and a distance metric for measuring distances between points in this space. The system is then expected to find, with regard to this metric, the K closest answers in the database from the query point. Typical distance metrics include Euclidean distance, Manhattan distance, etc.

It is possible that a majority of the answers to a KNN query may be very similar to one or more of the other answers, especially when the data has clusters. In fact, there may even be *duplicates* w.r.t. the attributes of the multidimensional space. For a variety of applications, such as online restaurant selection [6], providing homogeneous result sets may not add value to the user. It is our contention in this paper that, for such applications, the user would like to have not just the closest set of answers, but the closest *diverse* set of answers.

Based on the above motivation, we consider here the problem of providing diversity in the results of KNN queries, that is, to produce the closest result set such that each

* Contact Author: haritsa@dsl.serc.iisc.ernet.in

H. Dai, R. Srikant, and C. Zhang (Eds.): PAKDD 2004, LNAI 3056, pp. 404–413, 2004.

answer is sufficiently diverse from the rest. We hereafter refer to this problem as the *K-Nearest Diverse Neighbor (KNDN)* problem, which to the best of our knowledge has not been previously investigated in the literature, and cannot be handled by traditional clustering techniques (details in [6]).

An immediate question that arises is how to define diversity. This is obviously a user dependent choice, so we address the issue by providing a *tunable* definition, which can be set with a single parameter, *MinDiv*, by the user. *MinDiv* values range over [0,1] and specify the minimum diversity that should exist between *any pair* of answers in the result set. Setting *MinDiv* to zero results in the traditional KNN query, whereas higher values give more and more importance to diversity at the expense of distance.

Finding the optimal result set for KNDN queries is an *NP-complete problem* in general, and is computationally extremely expensive even for fixed K, making it infeasible in practice. Therefore, we present here an online algorithm, called **MOTLEY**[1], for producing a sufficiently diverse and close result set. MOTLEY adopts a greedy heuristic and leverages the existence of a spatial-containment-based multidimensional index, such as the R-tree, which is natively available in today's commercial database systems [7]. The R-tree index supports a "distance browsing" mechanism proposed in [5] through which database points can be efficiently accessed in increasing order of their distance from the query point. A pruning technique is incorporated in MOTLEY to minimize the R-tree processing and the number of database tuples that are examined.

Through a detailed experimental evaluation on real and synthetic data, we have found that MOTLEY can produce a diverse result set by reading only a small fraction of the tuples in the database. Further, the quality of its result set is very close to that provided by an off-line brute-force optimal algorithm. Finally, it can also evaluate traditional KNN queries without any added cost, thereby providing a *seamless interface between the orthogonal concepts of diversity and distance.*

2 Basic Concepts and Problem Formulation

In the following discussion, for ease of exposition and due to space limitations, we focus on a *restricted* instance of the KNDN problem – a significantly more general formulation is available in the full version of the paper [6].

We model the database as composed of N tuples over a D-dimensional space with each tuple representing a point in this space[2]. The domains of all attributes are numeric and normalized to the range [0,1]. The user specifies a point query Q over an M-sized subset of these attributes ($M \leq D$). We refer to these attributes as "point attributes". The user also specifies K, the number of desired answers, and a L-sized subset of attributes on which she would like to have diversity ($L \leq D$). We refer to these attributes as "diversity attributes" and the space formed by diversity attributes as *diversity-space*. Note that the *choice* of the diversity attributes is *orthogonal* to the *choice* of the point attributes. Finally, the result is a set of K database points.

Given that there are N points in the database and that we need to select K points for the result set, there are ${}^{N}C_{K}$ possible choices. We apply the diversity constraints first

[1] Motley: A collection containing a variety of sorts of things [9].

[2] We use point and tuple interchangeably in the remainder of this paper

to determine the feasible sets and then bring in the notion of distance from the query point to make a selection from these sets. Viewed abstractly, we have a *two-level* scoring function: The first level chooses candidate result sets based on diversity constraints, and the second level selects the result set that is spatially closest to the query point.

Result Diversity. We begin by defining *point diversity* and then, since the result is viewed as a set, extend the definition to *set diversity*. Point diversity is defined with regard to a *pair* of points and is evaluated with respect to the diversity attributes, $V(Q)$, mentioned in the query. Specifically, given points P_1, P_2 and $V(Q)$, the function DIV $(P_1, P_2, V(Q))$ returns true if P_1 and P_2 are diverse with respect to each other on the specified diversity attributes. A sample DIV function is described later in this section.

For a set to be *fully diverse*, all the points in the set should be mutually diverse. That is, given a result set $\mathcal{R}$ with points $R_1, R_2, \ldots, R_K$, we require $DIV(R_i, R_j, V(Q))$ $=$ true $\forall\, i, j$ such that $i \neq j$ and $1 \leq i, j \leq K$. For the restricted scenario considered here, we assume that at least one fully diverse result set is always available for the user query.

Diversity Function. Our computation of the diversity between two points P_1 and P_2, is based on the classical *Gower coefficient* [2], wherein the difference between two points is defined as a weighted average of the respective attribute differences. Specifically, we first compute the differences between the attributed values of these two points in *diversity-space*, sequence these differences in *decreasing* order of their values, and then label them as $(\delta_1, \delta_2, \ldots, \delta_L)$. Now, we calculate $divdist$, the diversity distance between points P_1 and P_2 with respect to diversity attributes $V(Q)$ as

$$divdist(P_1, P_2, V(Q)) = \sum_{j=1}^{L} (W_j \times \delta_j) \tag{1}$$

where the W_j's are weighting factors for the *differences*. Since all δ_j's are in the range [0,1] (recall that the values on all dimensions are normalized to [0,1]), and by virtue of the W_j assignment policy discussed below, diversity distances are also bounded in the range [0,1].

The assignment of the weights is based on the heuristic that *larger* weights should be assigned to the larger differences. That is, in Equation 1, we need to ensure that $W_i \geq W_j$ if $i < j$ (recall that δ_j's are sorted in decreasing order.) The rationale for this assignment is as follows: Consider the case where point P_1 has values (0.2, 0.2, 0.3), point P_2 has values (0.19, 0.19, 0.29) and point P_3 has values (0.2, 0.2, 0.27). Consider the diversity of P_1 with respect to P_2 and P_3. While the aggregate difference is the same in both cases, yet intuitively we can see that the pair (P_1, P_2) is more homogeneous as compared to the pair (P_1, P_3). This is because P_1 and P_3 differ considerably on the third attribute as compared to the corresponding differences between P_1 and P_2.

Now consider the case where P_3 has value (0.2, 0.2, 0.28). Here, although the aggregate δ_j is higher for the pair (P_1, P_2), yet again it is pair (P_1, P_3) that appears more diverse since its difference on the third attribute is larger than any of the individual differences in pair (P_1, P_2).

Based on the above discussion, the weighting function should have the following properties: Firstly, all weights should be positive, since having difference in any dimension should never decrease the diversity. Second, the sum of the weights should add up to 1 (i.e., $\sum_{j=1}^{L} W_j = 1$) to ensure that $divdist$ values are normalized to the [0,1] range. Finally, the weights should be *monotonically decaying* ($W_i \geq W_j$ if $i < j$) to reflect the preference given to larger differences.

Example 1. A candidate weighting function that obeys the above requirements is the following:

$$W_j = \frac{a^{j-1} \times (1-a)}{1-a^L} \quad (1 \leq j \leq L) \tag{2}$$

where a is a tunable parameter over the range (0,1). Note that this function implements a *geometric* decay, with the parameter 'a' determining the rate of decay. Values of a that are close to 0 result in faster decay, whereas values close to 1 result in slow decay. When the value of a is nearly 0, almost all weight is given to maximum difference i.e., $W_1 \simeq 1$, modeling (in the language of vector *p-norms*) the L_∞ (i.e., Max) distance metric, and when a is nearly 1, all attributes are given similar weights, modeling a L_1 (i.e., Manhattan) distance metric.

Minimum Diversity Threshold. We expect that the user provides a quantitative notion of the minimum diversity distance that she expects in the result set through a threshold parameter *MinDiv* that ranges between [0,1][3]. Given this threshold setting, two points are diverse if the diversity distance between them is greater than or equal to *MinDiv*. That is, $DIV(P_1, P_2, V(Q))$ = true iff $divdist(P_1, P_2, V(Q)) \geq MinDiv$.

The *physical* interpretation of the *MinDiv* value is that if a pair of points are deemed to be diverse, then these two points have a difference of *MinDiv* or more on atleast one diversity dimension. For example, a *MinDiv* of 0.1 means that any pair of diverse points differ in atleast one diversity dimension by atleast 10% of the associated domain size. This physical interpretation can guide the user in determining the appropriate setting of *MinDiv*. In practice, we would expect that *MinDiv* settings would be on the low side, typically not more than 0.2. As a final point, note that with the above formulation, the DIV function is *symmetric* with respect to the point pair $\{P_1, P_2\}$. However, it is *not transitive* in that even if $DIV(P_1, P_2, V(Q))$ and $DIV(P_2, P_3, V(Q))$ are both true, it does not imply that $DIV(P_1, P_3, V(Q))$ is true.

Integrating Diversity and Distance. Let function $SpatialDist(P, Q)$ calculate the spatial distance of point P from query point Q (this distance is computed with regard to the point attributes specified in Q.) The choice of *SpatialDist* function is based on the user specification and could be any monotonically increasing distance function such as Euclidean, Manhattan, etc. We combine distances of all points in a set into a single value using an aggregate function Agg which captures the overall distance of the set from Q. While a variety of aggregate functions are possible, the choice is constrained by the fact

[3] This is similar to the user specifying minimum support and minimum confidence in association rule mining to determine what constitutes interesting correlations.

that the aggregate function should ensure that as the points in the set move farther away from the query, the distance of the set should also increase correspondingly. Sample aggregate functions which obey this constraint include the Arithmetic, Geometric, and Harmonic Means.

Finally, we use the reciprocal of the aggregate of the spatial distances of the result points from the query point to determine the score of the (fully diverse) result set. (Note that the *MinDiv* threshold only determines the identities of the fully diverse result sets, but not their scores.) Putting all these formulations together, given a query Q and a candidate fully diverse result set $\mathcal{R}$ with points $R_1, R_2, \ldots, R_K$, the score of $\mathcal{R}$ with respect to Q is computed as

$$Score(\mathcal{R}, Q) = \frac{1}{Agg(SpatialDist(Q, R_1), \ldots, SpatialDist(Q, R_K))} \quad (3)$$

Problem Formulation. In summary, our problem formulation is as follows:

Given a point query Q on a D-dimensional database, a desired result cardinality of K, and a MinDiv threshold, the goal of the K-Nearest Diverse Neighbor (KNDN) problem is to find the set of K mutually diverse tuples in the database, whose score, as per Equation 3, is the maximum, after including the nearest tuple to Q in the result set.

The requirement that the nearest point to the user's query should *always* form part of the result set is because this point, in a sense, *best fits* the user's query. Further, the nearest point R_1 serves to seed the result set since the diversity function is meaningful only for a *pair* of points. Since point R_1 of the result is fixed, the result sets are differentiated based on their remaining $K - 1$ choices.

An important point to note here is that when *MinDiv* is set to zero, all points (including duplicates) are *diverse* with respect to each other and hence the KNDN problem reduces to the traditional KNN problem.

3 The MOTLEY Algorithm

Finding the optimal result set for the KNDN problem is computationally hard. We can establish this (proof in [6]) by mapping KNDN to the well known *independent set problem* [3], which is NP-complete. Therefore, we present an alternative algorithm here called MOTLEY, which employs a greedy selection strategy in combination with a distance-browsing-based accessing of points; our experimental evaluation, presented later in Section 4, shows that the result sets obtained are extremely close to the optimal solution.

3.1 Distance Browsing

In order to process database tuples (i.e., points) incrementally, we adopt the "distance browsing" approach proposed in [5], through which it is possible to efficiently access data points in increasing order of their distance from the query point. This approach is predicated on having a containment-based index structure such as the R-Tree[4], built collectively on all dimensions of the database (more precisely, the index needs to cover only those dimensions on which point predicates may appear in the query workload.)

To implement distance browsing, a priority queue, *pqueue*, is maintained which is initialized with the root node of the R-Tree. The *pqueue* maintains the R-Tree nodes and data tuples in increasing order of their distance from the query point. While the distance between a data point and the query Q is computed in the standard manner, the distance between a R-tree node and Q is computed as the minimum of the distances between Q and all points in the region enclosed by the MBR (Minimum Bounding Rectangle) of the R-tree node. The distance of a node from Q is zero if Q is within the MBR of that node, otherwise it is the distance of the closest point on the MBR periphery. For this, we first need to compute the distances between the MBR and Q along each query dimension – if Q is inside the MBR on a specific dimension, the distance is zero, whereas if Q is outside the MBR on this dimension, it is the distance from Q to either the low end or the high end of the MBR, whichever is nearer. Once the distances along all dimensions are available, they are combined (based on the distance metric in operation) to get the effective distance.

Example 2. Consider an MBR, M, specified by ((1,1,1),(3,3,3)) in a 3-D space. Let $P_1(2,2,2)$ and $P_2(4,2,0)$ be two data points in this space. Then, $SpatialDist(M, P_1) = \sqrt{0^2+0^2+0^2} = 0$ and $SpatialDist(M, P_2) = \sqrt{(4-3)^2+0^2+(0-1)^2} = 1.414$.

To return the next nearest neighbor, we pick up the first element of the *pqueue*. If it is a tuple, it is immediately returned as next nearest neighbor. However, if the element is an R-tree node, all the children of that node are inserted in the *pqueue*. Note that during this insertion process, the *spatial* distance of the object from the query point is calculated and used as the insertion key. The insertion process is repeated until we get a tuple as the first element of the queue, which is then returned.

The above distance browsing process continues until either the diverse result set is found, or until all points in the database are exhausted, signaled by the *pqueue* becoming empty.

3.2 Finding Diverse Results

We first present a simple greedy approach, called *Immediate Greedy*, and then its extension *Buffered Greedy*, for efficiently finding result sets that are both close and diverse.

Immediate Greedy Approach. In the ImmediateGreedy (IG) method, tuples are sent in increasing order of their spatial distance from the query point using distance browsing, as discussed above. The first tuple is always inserted into the result set, $\mathcal{R}$, to satisfy the requirement that the closest tuple to the query point must figure in the result set. Subsequently, each new tuple is added to $\mathcal{R}$ if it is diverse with respect to *all* tuples currently in $\mathcal{R}$; otherwise, it is discarded. This process continues until $\mathcal{R}$ grows to contain K tuples. Note that the result set obtained by this approach has following property: Let $\mathcal{B} = b_1, \ldots, b_K$ be the sequence formed by any other fully diverse set such that elements are listed in increasing order of spatial distance from Q. Now if i is the smallest index such that $b_i \neq R_i (R_i \epsilon \mathcal{R})$, then $SpatialDist(b_i, Q) \geq SpatialDist(R_i, Q)$.

While the IG approach is straight forward and easy to implement, there are cases where it may make poor choices as shown in Figure 1. Here, Q is the query point, and P_1

through P_5 are the tuples in the database. Let us assume that the goal is to report 3 diverse tuples with *MinDiv* of 0.1. Clearly, $\{P_1,\ P_3,\ P_4\}$ satisfies the diversity requirement. Also $DIV(P_1, P_2, V(Q)) = true$. But inclusion of P_2 disqualifies the candidatures of P_3 and P_4 as both $DIV(P_2, P_3, V(Q)) = false$ and $DIV(P_2, P_4, V(Q)) = false$. By inspection, we observe that the overall best choice could be $\{P_1,\ P_3,\ P_4\}$, but Immediate Greedy would give the solution as $\{P_1,\ P_2,\ P_5\}$. Moreover, if point P_5 is not present in the database, then this approach will fail to return a fully diverse set even though such a set, namely $\{P_1,\ P_3,\ P_4\}$, is available.

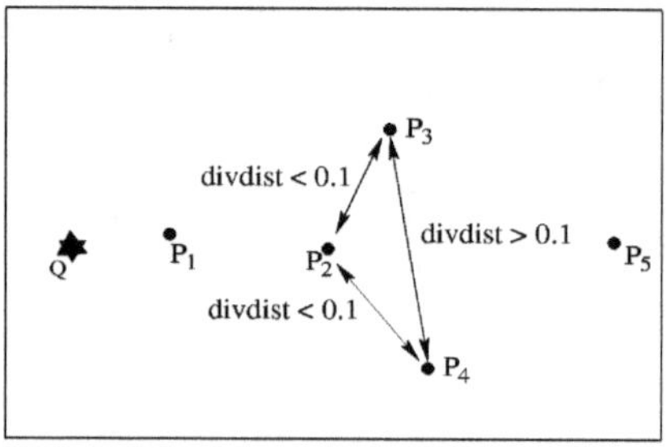

Fig. 1. Poor Choice by Immediate Greedy

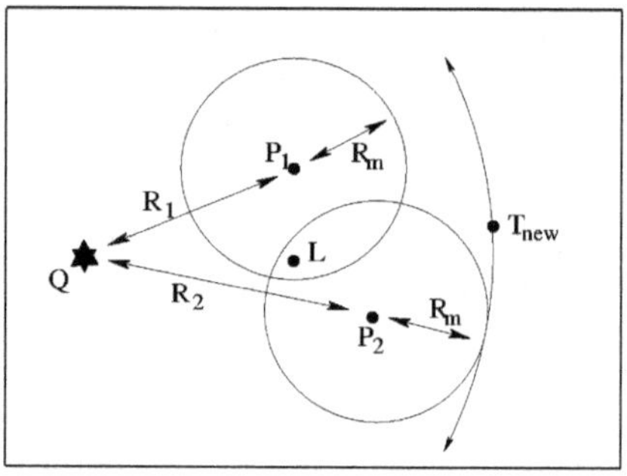

Fig. 2. Heuristic in Buffered Greedy Approach

Buffered Greedy Approach. The above problems are addressed in the BufferedGreedy (BG) method by recognizing that in IG, only the diverse points (hereafter called "leaders") in the result set, are retained at all times. Specifically, BG maintains with each leader a bounded buffered set of "dedicated followers" – a dedicated follower is a point that is not diverse with respect to a specific leader but is diverse with respect to *all remaining* leaders. Our empirical results show that a buffer of capacity K points (where K is the desired result size) for each leader, is sufficient to produce a near-optimal solution. The additional memory requirement for the buffers is small for typical values of K and D (e.g., for K=10 and D=10, and using 8 bytes to store each attribute value, we need only 8K bytes of additional storage).

Given this additional set of dedicated followers, we adopt the heuristic that a current leader, L_i, is *replaced* in the result set by its dedicated followers $F_i^1, F_i^2, \ldots, F_i^j (j > 1)$ as leaders if (a) these dedicated followers are *all* mutually diverse, and (b) incorporation of these followers as leaders does not result in the premature disqualification of future leaders. The first condition is necessary to ensure that the result set contains only diverse points, while the second is necessary to ensure that we do not produce solutions that are worse than Immediate Greedy. For example, if in Figure 1, point P_5 had happened to be only a little farther than point P_4 such that $DIV(P_2, P_5, V(Q)) = true$, then the replacement of P_2 by P_3 and P_4 could be the wrong choice since $\{P_1, P_2, P_5\}$ may turn out to be the best solution.

To implement the second condition, we need to know when it is "safe" to go ahead with a replacement i.e., when it is certain that all future leaders will be diverse from the current set of followers. To achieve this, we take the following approach: For each point, we consider a hypothetical sphere that contains all points in the domain space that may be non-diverse with respect to it. That is, we set the radius R_m of the sphere to be equal to the distance of the farthest non-diverse point in the domain space. Note that this sphere may contain some diverse points as well, but our objective is to take a conservative approach. Now, the replacement of a leader by selected dedicated followers can be done as soon as we have reached a distance greater than R_m with respect to the farthest follower from the query – this is because all future leaders will be diverse with respect to selected dedicated followers and there is no possibility of disqualification beyond this point. To clarify this technique, consider the following example:

Example 3. In Figure 2, the circles around P_1 and P_2 show the areas that contain all points that are not diverse with respect to P_1 and P_2, respectively. Due to distance browsing technique, when we access the point T_{new} (Figure 2), we know that all future points will be diverse from P_1 and P_2. At this time, if P_1 and P_2 are dedicated followers of L and mutually diverse, then we can replace L by $\{P_1, P_2\}$.

The integration of Buffered Greedy with distance browsing as well as pruning optimizations for minimizing the database processing are discussed in [6]. Further, the complexity of Buffered Greedy is shown to be $O(NK^2)$ in [6].

4 Experiments

We conducted a detailed suite of experiments to evaluate the quality and efficiency of the MOTLEY algorithm with regard to producing a diverse set of answers. While a variety of datasets were used in our experiments (see [6] for details), we report on only one dataset here, namely Forest Cover [10], a real dataset containing 581,012 tuples and 4 attributes representing *Elevation*, *Aspect*, *Slope*, and *Distance*.

Our experiments involve uniformly distributed point queries across the whole data space, with the attribute domains normalised to the range [0, 1]. The default value of K, the desired number of answers, was 10, unless mentioned otherwise, and *MinDiv* was varied across [0, 1]. In practice, we expect that *MinDiv* settings would be on the low side, typically not more than 0.2, and we therefore focus on this range in our experiments. The decay rate (a) of the weights (Equation 2) was set to 0.1, Harmonic Mean was used for the Agg function (Equation 3), and spatial distances were computed using the Euclidean metric. The R-tree (specifically, the R* variant [1]) was created with a fill factor of 0.7 and branching factor 64.

Result-set Quality. We begin by characterizing the quality of the result set provided by MOTLEY, which is a greedy online algorithm, against an off-line brute-force optimal algorithm. This performance perspective is shown in Figure 3, which presents the average and worst case ratio of the result set scores. As can be seen in the figure, the average case is almost optimal (note that the Y-axis of the graph begins from 0.8), indicating that MOTLEY typically produces a *close-to-optimal* solution. Moreover, even in the worst-case, the difference is only around 10 percent. Importantly, even in the situations

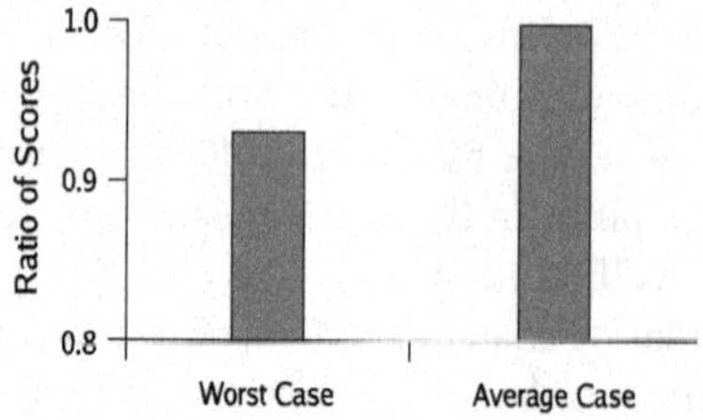

Fig. 3. MOTLEY vs. Optimal

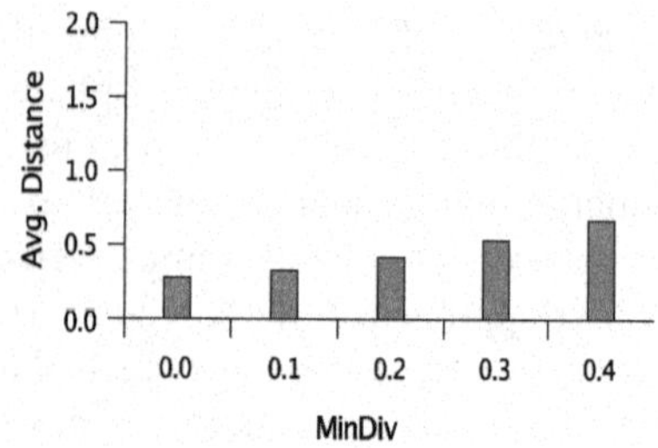

Fig. 4. MOTLEY vs. KNN

where MOTLEY did not provide the complete optimal result set, the errors were mostly restricted to only *one or two* points, out of the total of ten answers.

Figure 4 shows, as a function of *MinDiv*, the average *distance* of the points in MOTLEY's result set – this metric effectively captures the cost to be paid in terms of distance in order to obtain result diversity. The important point to note here is that for values of *MinDiv* up to 0.2, the distance increase is *marginal*, with respect to the traditional KNN query (*MinDiv* = 0). Since, as mentioned earlier, we expect that users will typically use *MinDiv* values between 0 and 0.2, it means that *diversity can be obtained at relatively little cost in terms of distance.*

Execution Efficiency. Having established the high-quality of MOTLEY answers, we now move on to evaluating its execution efficiency. In Figure 5, we show the average fraction of tuples read to produce the result set as a function of *MinDiv*. Note firstly that the tuples scanned are always less than 15% of the complete dataset. Secondly, at lower values of *MinDiv*, the number of tuples read are small because we obtain K diverse tuples after processing only a small number of points, whereas at higher values of *MinDiv*, pruning is more effective and hence the number of tuples processed continues to be small in comparison to the database size.

Effect of K. We also evaluated the effect of K, the number of answers, on the algorithmic performance. Figure 6 shows the percentage of tuples read as a function of *MinDiv* for

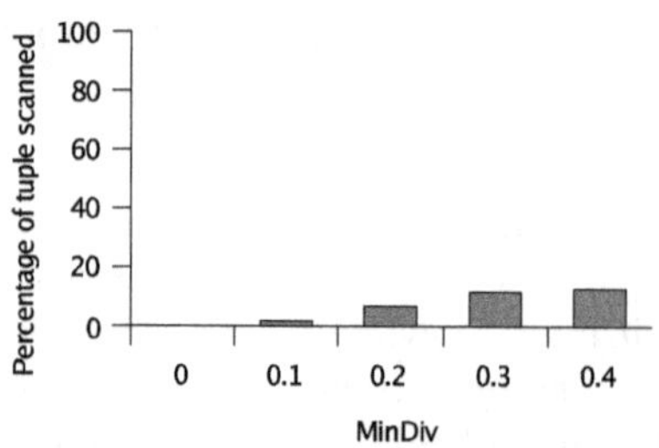

Fig. 5. Execution Efficiency of MOTLEY

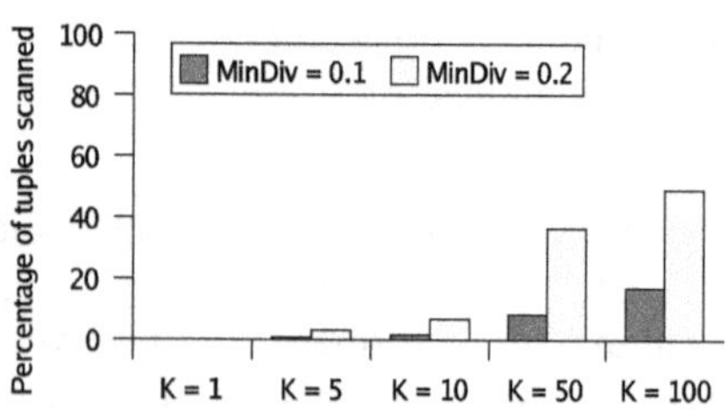

Fig. 6. Effect of K

different values of K ranging from 1 to 100. For $K = 1$, it is equivalent to the traditional NN search, irrespective of *MinDiv*, due to requiring the closest point to form part of the result set. As the value of K increases, the number of tuples read also increases, especially for higher values of *MinDiv*. However, we can expect that users will specify lower values of *MinDiv* for large K settings.

5 Conclusions

In this paper, we introduced the problem of finding the K Nearest Diverse Neighbors (KNDN), where the goal is to find the closest set of answers such that the user will find each answer sufficiently different from the rest, thereby adding value to the result set. We provided a quantitative notion of diversity that ensured that two tuples were diverse if they differed in at least one dimension by a sufficient distance, and presented a two-level scoring function to combine the orthogonal notions of distance and diversity.

We described MOTLEY, an online algorithm for addressing the KNDN problem, based on a buffered greedy approach integrated with a distance browsing technique. Pruning optimizations were incorporated to improve the runtime efficiency. Our experimental results demonstrated that MOTLEY can provide high-quality diverse solutions at a low cost in terms of both result distance and processing time. In fact, MOTLEY's performance was close to the optimal in the average case and only off by around ten percent in the worst case.

Acknowledgements. This work was supported in part by a Swarnajayanti Fellowship from the Dept. of Science & Technology, Govt. of India.

References

1. N. Beckmann, H. Kriegel, R. Schneider and B. Seeger, *The R*-tree: An efficient and robust access method for points and rectangles*, Proc. of ACM SIGMOD Intl. Conf. on Management of Data, 1990.
2. J. Gower, *A general coefficient of similarity and some of its properties*, Biometrics 27, 1971.
3. M. Grohe, *Parameterized Complexity for Database Theorists*, SIGMOD Record 31(4), December 2002.
4. A. Guttman, *R-trees: A dynamic index structure for spatial searching*, Proc. of ACM SIGMOD Intl. Conf. on Management of Data, 1984.
5. G. Hjaltason and H. Samet, *Distance Browsing in Spatial Databases*, ACM Trans. on Database Systems, 24(2), 1999.
6. A. Jain, P. Sarda and J. Haritsa, *Providing Diversity in K-Nearest Neighbor Query Results*, Tech. Report TR-2003-04, DSL/SERC, Indian Institute of Science, 2003.
7. R. Kothuri, S. Ravada and D. Abugov, *Quadtree and R-tree indexes in Oracle Spatial: A comparison using GIS data*, Proc. of ACM SIGMOD Intl. Conf. on Management of Data, 2002.
8. N. Roussopoulos, S. Kelley and F.Vincent, *Nearest Neighbor Queries*, Proc. of ACM SIGMOD Intl. Conf. on Management of Data, 1995.
9. *www.thefreedictionarity.com.*
10. ftp://ftp.ics.uci.edu/pub/machine-learning-databases/covtype

Cluster Structure of K-means Clustering via Principal Component Analysis

Chris Ding and Xiaofeng He

Lawrence Berkeley National Laboratory, Berkeley, CA 94720 USA

Abstract. K-means clustering is a popular data clustering algorithm. Principal component analysis (PCA) is a widely used statistical technique for dimension reduction. Here we prove that principal components are the continuous solutions to the discrete cluster membership indicators for K-means clustering, with a clear simplex cluster strcuture. Our results prove that PCA-based dimension reductions are particularlly effective for for K-means clustering. New lower bounds for K-means objective function are derived, which is the total variance minus the eigenvalues of the data covariance matrix.

Introduction

Data analysis methods are essential for analyzing the ever-growing massive quantity of high dimensional data. On one end, cluster analysis attempts to pass through data quickly to gain first order knowledge by partitioning data points into disjoint groups such that data points belonging to same cluster are similar while data points belonging to different clusters are dissimilar. One of the most popular and efficient clustering methods is the K-means method [2] which uses prototypes to represent clusters minimizing the squared error function.

On the other end, high dimensional data are often transformed into lower dimensional data via the principal component analysis (PCA) [3] (or singular value decomposition) where coherent patterns can be detected more clearly. Such unsupervised dimension reduction is used in broad areas such as meteorology, image processing, genomic analysis and information retriev al.

The main basis of PCA-based dimension reduction is that PCA picks up the directions with the largest variances. Mathematically, this is equivalent to finding the best low rank approximation (in L_2 norm) of the data via the singular value decomposition (SVD). However, this noise reduction property alone is inadequate to explain the effectiveness of PCA.

In this paper, we prove that principal components are actually the continuous solution of the cluster membership indicators in the K-means clustering method, i.e., the PCA dimension reduction automatically performs data clustering according to the K-means objective function. This provides an important justification for PCA-based data reduction.

Our results also provide effective ways to solve the K-means clustering problem. K-means method uses K prototypes, the centroids of clusters, to charac-

H. Dai, R. Srikant, and C. Zhang (Eds.): PAKDD 2004, LNAI 3056, pp. 414–418, 2004.

terize the data. They are determined by minimizing the sum of squared errors,

$$J_K = \sum_{k=1}^{K} \sum_{i \in C_k} (\mathbf{x}_i - \mu_k)^2$$

where $(\mathbf{x}_1, \cdots, \mathbf{x}_n) = X$ is the data matrix, $\mu_k = \sum_{i \in C_k} \mathbf{x}_i / n_k$ is the centroid of cluster C_k and n_k is the number of points in C_k. Standard iterative solution to K-means suffers from a well-known problem: as iteration proceeds, the solutions are trapped in the local minima due to the greedy nature of the update algorithm.

Notations on PCA. X represents the original data matrix ; $Y = (\mathbf{y}_1, \cdots, \mathbf{y}_n)$, $\mathbf{y}_i = \mathbf{x}_i - \bar{\mathbf{x}}$, represents the centered data matrix, where $\bar{\mathbf{x}} = \sum_i \mathbf{x}_i / n$. The covarance matrix (ignoring the factor $1/n$) is $\sum_i (\mathbf{x}_i - \bar{\mathbf{x}})(\mathbf{x}_i - \bar{\mathbf{x}})^T = YY^T$. Principal directions $\mathbf{u}_k$ and principal components $\mathbf{v}_k$ are eigenvectors satisfying: $YY^T \mathbf{u}_k = \lambda_k \mathbf{u}_k$, $Y^T Y \mathbf{v}_k = \lambda_k \mathbf{v}_k$. These are the defining equations for the SVD of Y: $Y = \sum_k \lambda_k \mathbf{u}_k \mathbf{v}_k^T$. Elements of $\mathbf{v}_k$ are the projected values of data points on the principal direction $\mathbf{u}_k$.

The simplex structure of K-means clustering

The main results of the paper to propose the simplex structure in the cluster membership indicator space as a key concept in clustering and prove that this indicator space is an orthogonal transformation of the PCA subspace.

Indicator space and simplex structure. The solution of clustering can be represented by K unsigned cluster membership indicator vectors: $H_K = (\mathbf{h}_1, \cdots, \mathbf{h}_K)$, where

$$\mathbf{h}_k = (0, \cdots, 0, \overbrace{1, \cdots, 1}^{n_k}, 0, \cdots, 0)^T / n_k^{1/2} \tag{1}$$

(Without loss of generality, we index the data such that data objects within each cluster are adjacent.) Consider the simplex consisting of K basis vectors $\mathbf{h}_k$ plus the origin in the K-dimensional *indicator space* spanned by H_K. All objects belonging to a cluster will collapse into the same corner of the simplex. Different clusters collapse into different corners and they become well separated.

Thus solving the clustering problem is reduced to computing the simplex in the indicator space.

Theorem 1 (Simplex Structure Theorem).
(i) The continuous solution of the transformed indicators, $Q_K = H_K T$, are $V_K = (\mathbf{v}_1, \cdots, \mathbf{v}_{K-1}, \sqrt{\frac{1}{n}}\mathbf{e})$. In other words the K-dimensional indicator space is an orthogonal transformation of the principal component subspace.
(ii) The $K \times K$ transformation matrix $T = (\mathbf{t}_1, \cdots, \mathbf{t}_K)$ are K eigenvectors of the following eigenvalue equation

$$\Gamma \mathbf{t}_k = \lambda_k \mathbf{t}_k \tag{2}$$

where

$$\Gamma = \Omega^{-1/2}\, \bar{\Gamma}\, \Omega^{-1/2}, \ \Omega = \mathrm{diag}(\sqrt{n_1}, \cdots, \sqrt{n_K}).$$

and

$$\bar{\Gamma}_{ij} = -\alpha_{ij},\ i \neq j;\ \bar{\Gamma}_{ii} = \sum_j \alpha_{ij}, \tag{3}$$

where $\alpha_{ij} > 0$ $(i \neq j)$, $\alpha_{ii} = 0$, $\alpha_{ij} = \alpha_{ji}$, and $\sum_{ij} \alpha_{ij} = 1$.
(iii) J_{K} satisfies the upper and lower bounds

$$n\overline{\mathbf{y}^2} - \sum_{k=1}^{K-1} \lambda_k < J_{\mathrm{K}} < n\overline{\mathbf{y}^2} \tag{4}$$

where $n\overline{\mathbf{y}^2}$ is the total variance and λ_k are the principal eigenvalues of the covariance matrix YY^T.

The proof of Theorem 1 is outlined in appendix. The spectral relaxation of K-means is first studied in [4]. Now we prove the effectiveness PCA dimension reduction as related to K-means cluster with the following.

Proposition 2. The K-means clustering has two invariant properties: (i) it is invariant under any orthogonal coordinate rotation operation

$$R : \mathbf{x} \to T\mathbf{x},$$

(ii) it is invariant under coordinate translation (shift) operation

$$L : \mathbf{x} \to \mathbf{x} + \ell.$$

Proof. The K-means objective function can be written as

$$J_{\mathrm{K}} = \sum_{k=1}^{K} \sum_{i,j \in C_k} \frac{(\mathbf{x}_i - \mathbf{x}_j)^2}{2n_k}. \tag{5}$$

Now (i) follows from the fact that for orthonormal transformations, T, $TT^T = I$; thus $||T\mathbf{x}_i - T\mathbf{x}_j|| = ||\mathbf{x}_i - \mathbf{x}_j||$. (ii) is a simple consequence of $||(\mathbf{x}_i + \ell) - (\mathbf{x}_j + \ell)|| = ||\mathbf{x}_i - \mathbf{x}_j||$. □

Proposition 3. PCA dimension reduction is effective K-means clustering.
Proof. From Thereom 1, the eigen-space V_K are the relaxed solutions of the transformed indicators Q_K, i.e., K-means clustering in eigen-space V_K are approximately equivalent to that in the transformed indicator space Q_K. Because K-means clustering is invariant w.r.t. the orthogonal transformation T (Proposition 2), K-means clustering in Q_K space is equivalent to K-means clustering in the indicator space H_K. In H_K space, all objects belonging to a cluster (approximately) collapse into the same corner of the simplex. Different clusters become well separated. Hence clustering in H_K space is particularly effective — our results provides a theoretical basis for the use of PCA dimension reduction for K-means clustering.

Table 1. Clustering accuracy as the PCA dimension is reduced from original 1000.

Dim	A5(balanced)	A5(un-bal.)	B5(balanced)	B5(un-bal.)
5	0.81/0.91	0.88/0.86	0.59/0.70	0.64/0.62
6	0.91/0.90	0.87/0.86	0.67/0.72	0.64/0.62
10	0.90/0.90	0.89/0.88	0.74/0.75	0.67/0.71
20	0.89	0.90	0.74	0.72
40	0.86	0.91	0.63	0.68
1000	0.75	0.77	0.56	0.57

Experiments on Internet Newsgroups

A 20-newsgroup dataset is from www.cs.cmu.edu/afs/cs/project/theo-11/www/naive-bayes.html. Word - document matrix is first constructed. 1000 words are selected according to the mutual information. Standard `tf.idf` term weighting is used. Each document is normalized to 1. We focus on two sets of 5-newsgroup combinations:

```
      A5:                          B5:
NG2:  comp.graphics          NG2:  comp.graphics
NG9:  rec.motorcycles        NG3:  comp.os.ms-windows
NG10: rec.sport.baseball     NG8:  rec.autos
NG15: sci.space              NG13: sci.electronics
NG18: talk.politics.mideast  NG19: talk.politics.misc
```

We apply K-means clustering in the PCA subspace. Here we reduce the data from the original 1000 dimensions to 40, 20, 10, 6, 5 dimensions respectively. The clustering accuracy on 10 random samples of each newsgroup combination and size composition are averaged and the results are listed in Table 1. To see the subtle difference between centering data or not at 10, 6, 5 dimensions; results for original uncentered data are list at left and the results for centered data are listed at right.

Two observations. (1) It is clear that as dimensions are reduced, the results systematically and significantly improves. For example, for datasets A5-balanced, the cluster accuracy improves from 75% at 1000-dim to 91% at 5-dim. (2) For very small number of dimensions, PCA based on the centered data seem to lead to better results.

Appendix

Here we outline the proof of Theorem 1. First, J_K can be written as

$$J_K = \sum_i \mathbf{x}_i^2 - \sum_k \frac{1}{n_k} \sum_{i,j \in C_k} \mathbf{x}_i^T \mathbf{x}_j, \tag{6}$$

The first term is a constant. The second term is the sum of the K diagonal block elements of $X^T X$ matrix representing within-cluster (inner-product) similarities.

Using Eq.(1), Eq.(6) becomes

$$J_K = \mathrm{Tr}(X^T X) - (\mathbf{h}_1^T X^T X \mathbf{h}_1 + \cdots + \mathbf{h}_k^T X^T X \mathbf{h}_k). \tag{7}$$

There are redundancies in H_K. For example, $\sum_{k=1}^{K} n_k^{1/2} \mathbf{h}_k = \mathbf{e}$. Thus one of the $\mathbf{h}_k$'s is linear combination of others. We remove this redundancy by (a) performing a linear transformation T into $\mathbf{q}_k$'s: $Q_K = (\mathbf{q}_1, \cdots, \mathbf{q}_K) = H_K T$, or $\mathbf{q}_\ell = \sum_k \mathbf{h}_k t_{k\ell}$, where $T = (t_{ij})$ is a $K \times K$ orthonormal matrix: $T^T T = I$, and (b) requiring that the last column of T is

$$\mathbf{t}_K = (\sqrt{n_1/n}, \cdots, \sqrt{n_k/n})^T, \tag{8}$$

leading to $\mathbf{q}_K = \sqrt{n_1/n}\,\mathbf{h}_1 + \cdots + \sqrt{n_k/n}\,\mathbf{h}_K = \sqrt{1/n}\,\mathbf{e}$.

The mutual orthogonality of $\mathbf{h}_k$, $\mathbf{h}_k^T \mathbf{h}_\ell = \delta_{k\ell}$ implies the mutual orthogonality of $\mathbf{q}_k$, $\mathbf{q}_k^T \mathbf{q}_\ell = \delta_{k\ell}$. Let $Q_{K-1} = (\mathbf{q}_1, \cdots, \mathbf{q}_{K-1})$, the orthogonality relation become

$$Q_{K-1}^T Q_{K-1} = I_{K-1}, \tag{9}$$

$$\mathbf{q}_k^T \mathbf{e} = 0, \text{for } k = 1, \cdots, K-1. \tag{10}$$

which must be maintained. The K-means objective can now be written as

$$J_K = \mathrm{Tr}(X^T X) - \mathbf{e}^T X^T X \mathbf{e}/n - \mathrm{Tr}(Q_{k-1}^T X^T X Q_{k-1}). \tag{11}$$

J_K does not distinguish the original data $\{\mathbf{x}_i\}$ and the centered data $\{\mathbf{y}_i\}$. Repeating the above derivation on $\{\mathbf{y}_i\}$, we have

$$J_K = \mathrm{Tr}(Y^T Y) - \mathrm{Tr}(Q_{k-1}^T Y^T Y Q_{k-1}), \tag{12}$$

noting $Y\mathbf{e} = 0$. The first term is constant. The problem reduces to optimization of the second term subject to the constraints Eqs.(9,10).

Now we relax the restriction that $\mathbf{q}_k$ must take discrete values, and let $\mathbf{q}_k$ take continuous values, while keeping constraint Eq.(9). This maximization problem is well-known via the Ky Fan theorem [1]. The solution are $Q_{K-1} = (\mathbf{v}_1, \cdots, \mathbf{v}_{K-1})$, the principal components. which also automatically satisfy the constraint Eq.(10). From this, (i) and (iii) follow.

To prove (ii), we can show that (a) $\mathbf{t}_K$ of Eq.(8) is an eigenvector of Γ with eigenvalue $\lambda = 0$. (b) The symmetric Γ is also semi-positive definite. Other $K-1$ eigenvectors are mutually orthonormal, $\mathbf{t}_k^T \mathbf{t}_l = \delta_{kl}$. These prove (ii). □

Acknowledgements Work supported by U.S. Department of Energy (Office of Science, through a LBNL LDRD) under contract DE-AC03-76SF00098.

References

1. K. Fan. On a theorem of Weyl concerning eigen values of linear transformations. *Proc. Natl. Acad. Sci. USA*, 35:652–655, 1949.
2. J.A. Hartigan and M.A. Wang. A K-means clustering algorithm. *Applied Statistics*, 28:100–108, 1979.
3. I.T. Jolliffe. *Principal Component Analysis*. Springer Verlag, 1986.
4. H. Zha, C. Ding, M. Gu, X. He, and H.D. Simon. Spectral relaxation for K-means clustering. *Advances in Neural Information Processing Systems 14*, pages 1057–1064, 2002.

Combining Clustering with Moving Sequential Pattern Mining: A Novel and Efficient Technique*

Shuai Ma, Shiwei Tang, Dongqing Yang, Tengjiao Wang, and Jinqiang Han

Department of Computer Science, Peking University, Beijing 100871, China
{mashuai,tjwang,jqhan}@db.pku.edu.cn
{tsw,dqyang}@pku.edu.cn

Abstract. Sequential pattern mining is a well-studied problem. In the context of mobile computing, moving sequential patterns that reflects the moving behavior of mobile users attracted researchers' interests recently. In this paper a novel and efficient technique is proposed to mine moving sequential patterns. Firstly the idea of clustering is introduced to process the original moving histories into moving sequences as a preprocessing step. Then an efficient algorithm called PrefixTree is presented to mine the moving sequences. Performance study shows that PrefixTree outperforms LM algorithm, which is revised to mine moving sequences, in mining large moving sequence databases.

1 Introduction

Mining moving sequential patterns has great significance for effective and efficient location management in wireless communication systems. The problem of mining moving sequential patterns is a special case of mining traditional sequential patterns with the extension of support. There are mainly four differences between mining conventional sequential patterns and moving sequential patterns. Firstly, if two items are consecutive in a moving sequence α, and α is a subsequence of β, those two items must be consecutive in β. This is because we care about what the next move is for a mobile user in mining moving sequential patterns. Secondly, in mining moving sequential patterns the support considers the number of occurrences in a moving sequence that helps a more reasonable pattern discovery, so the support of a moving sequence is the sum of the number of occurrence in all the moving sequences of the whole moving sequence database. Thirdly, the Apriori property plays an important role for efficient candidate pruning in mining traditional sequential patterns. For example, suppose <ABC> is a frequent length-3 sequence, and then all the length-2 subsequences {<AB>, <AC>, <BC>} must be frequent in mining sequential patterns. In mining moving sequential patterns <AC> may not be frequent. This is because a

* Supported by the National High Technology Development 863 Program of China under Grant No. 2002AA4Z3440; the National Grand Fundamental Research 973 Program of China under Grant No. G1999032705.

H. Dai, R. Srikant, and C. Zhang (Eds.): PAKDD 2004, LNAI 3056, pp. 419–423, 2004.

mobile user can only move into a neighboring cell in a wireless system and items must be consecutive in mining moving sequential patterns. In addition, <AC> is not a subsequence of <ABC> any more in mining moving sequential patterns and that any subsequence of a frequent moving sequence must be frequent is still fulfilled from that meaning, which is called Pseudo-Apriori property. The last difference is that a moving sequence is an order list of items, but not an order list of itemsets, where each item is a cell id.

Wen-Chih Peng et al. presented a data-mining algorithm, which involves mining for user moving patterns in a mobile computing environment in [1]. Moving pattern mining is based on a roundtrip model [2], and their LM algorithm selects an initial location S, which is either VLR or HLR whose geography area contains the homes of the mobile users. Suppose a mobile user goes to a strange place for one month or longer, the method in [1] cannot find the proper moving pattern to characterize the mobile user. A more general method should not give any assumption of the start point of a moving pattern. Basically, algorithm LM is a variant one from GSP [3]. The Apriori-based methods can efficiently prune candidate sequence patterns based on Aprior property, but in moving sequential pattern mining we cannot prune candidate sequences efficiently because the moving sequential pattern only preserves Pseudo-Apriori property. In the meanwhile Apriori-based algorithms still encounter problem when a sequence database is large and/or when sequential patterns to be mined are numerous and/or long [4].

Time factor is considered for personal paging area design in [5]. G. Das et al. present a clustering based method to discretize a times series in [6]. Time is also a very important factor in mining moving sequential patterns. In this paper, firstly the idea of clustering is also introduced into the mining of moving sequential patterns to discretize the time attribute of the moving histories, and the moving histories are transformed into moving sequences based on the clustering result. Then based on the idea of projection and Pseudo-Apriori property, an efficient moving sequential pattern mining algorithm called PrefixTree is proposed, which can effectively represent candidate frequent moving sequences with a key tree structure of prefix trees. In addition, the wireless network topology based optimization approach is also presented to improve the efficiency of the PrefixTree algorithm.

The rest of the paper is organized as follows. Data preprocessing and the PrefixTree algorithm are given in section 2. Section 3 gives the experimental results from different viewpoints. Discussion is made in section 4.

2 Mining Moving Sequential Patterns

User moving history is the moving logs of a mobile user, which is an ordered (c, t) list where c is the cell ID and t is the time when the mobile user reaches cell c. Let MH=$<(c_1, t_1), (c_2, t_2), (c_3, t_3), \ldots (c_n, t_n)>$ be a moving history, and MH means the mobile user enters c_1 at t_1, leaves c_1 and enters c_2 at t_2, leaves c_2 and enters c_3 at t_3...and finally enters c_n at t_n. Each $(c_i, t_i) \in$ MH ($1 \le i \le n$) is called an element of MH. The time difference between two consecutive elements in a moving history reflects a mobile users' moving speed (or sojourn time in a cell). If the difference is high, it shows the mobile user moves at a relatively low speed; otherwise, if the difference is low, it shows the mobile user moves at a relatively high speed. The idea of using clustering to discretize time is that if a mobile user possesses regular moving behavior, then he often moves on the same set of paths, and the arrival time to each point of the paths is similar.

The clustering algorithm CURD [7] is used to discretize the time attribute of the moving histories. For each cell c in the cell set, all the elements in the moving history database D is collected, denoted by ES(c)={(c, t)|∃(c, t)∈ MH_i and MH_i∈D}. Then the CURD algorithm is used to cluster the element set ES(c), where Euclidean distance on time t is used as the similarity function between two elements in ES(c). This is a clustering problem in one-dimension space. After the clustering processing we have a clustering result as {(c, T_s, T_e)|T_s, T_e∈T and $T_s \leq T_e$}(T is the time domain), which means the mobile user often enters cell c at the period [T_s, T_e].

The main idea of data transformation is to replace the MH elements with the corresponding clusters that they belong to. The transformed moving history is called a moving sequence, and the transformed moving history database is called moving sequence database. Each moving history can be transformed into one and only one moving sequence because the clustering method guarantees that any MH element belongs to one and only one cluster.

The PrefixTree algorithm only need scan the database three times, and the key idea of PrefixTree is the use of the prefix trees. Prefix tree is a compact representation of candidate moving sequential patterns. The root is the frequent item, and is defined at depth one. Each node contains three attributes: one is the item, one is the count attribute which means the support of the item, and the last one is the flag indicating whether the node is traversed. The items of a node's children are all contained in its candidate consecutive items. In the first two scans PrefixTree generates the frequent itmes, frequent length-2 moving sequential patterns and CCIs of each frequent item, and the prefix trees are constructed in the third scan. For each item C_i, we call the items that may appear just after it candidate consecutive items. It is easy to know that only the items after C_i in a moving sequence may be the consecutive items of C_i, denoted by CCI(C_i). And for any item C_j∈CCI(C_i), length-2 moving sequence $<C_iC_j>$ is frequent. It is easy for us to generate the moving sequential patterns based on the prefix trees. Every moving sequence from the root node to the leaf node is a candidate frequent moving sequences. Scanning all the prefix trees once can generate all the moving sequential patterns. The support of each node decreases with the depth increase, so a new frequent moving sequence is generated when we traverse the prefix trees from the root to the leaves when encountering a node whose count is less than the support threshold.

The extension of support is considered when generating the prefix trees, which can be done by generating projected sequences. Another novelty of the PrefixTree algorithm is that it doesn't generate projected physical files, which is different from traditional projection-based sequential algorithms [4] and is a time-cost work. Due to the room reason, we cannot give too many details here.

As pointed in [8], the initial candidate set generation, especially for the length-2 frequent patterns, is the key issue to improve the performance of data mining. Fortunately, sometimes the apriori of the wireless network topology can be got. From the apriori the neighboring cells for each cell can be known in advance. It has been pointed out that for any two consecutive items in a moving sequence, one item must be the other one's neighboring cell or itself. The number of neighbor cells is often small, for example it is two in one-dimension model, and six in two-dimension hexagonal model and eight in two-dimension mesh model, and is relative small even in graph model [9]. So based on this apriori the number of the candidate length-2 moving sequential patterns is decreased much, and then the frequent length-2 moving sequential patterns can be efficiently generated.

3 Experimental Results and Performance Study

All experiments are performed on a 1.7GHz Pentium 4 PC machine with 512M main memory and 60G hard disk, running Microsoft Windows 2000 Professional. All the methods are implemented using JBuilder 6.0. The synthetic dataset used in our experiments comes from SUMATRA (Stanford University Mobile Activity TRAces, which is available at http://www-db.stanford.edu/sumatra/). BALI-2: Bay Area Location Information (real-time) dataset records the mobile users' moving and calling activities in a day. The mobile user averagely moves 7.2 times in a day in 90 zones, so the number of items is 90 and the average length of the moving sequences is 8.2. BALI-2 contains about 40,000 moving sequences, which are used for our experiments.

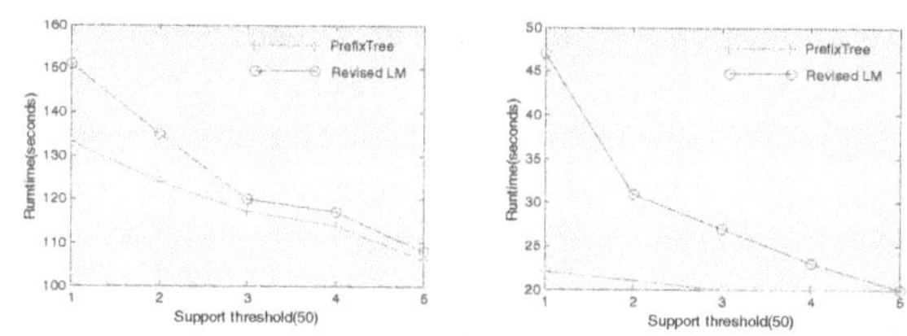

Fig. 1. Performance study with varying support threshold, where the left part and the right part are without and with optimization respectively

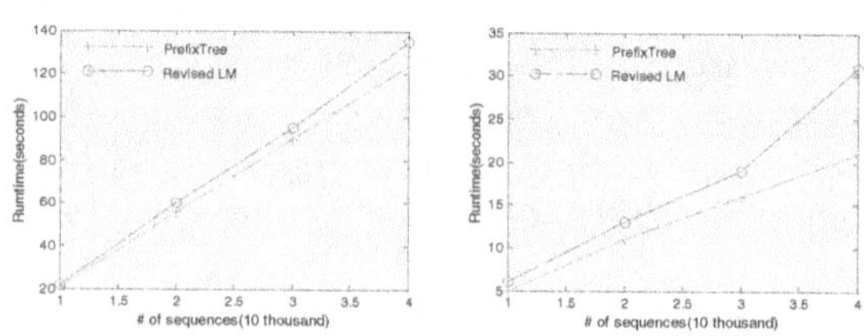

Fig. 2. Performance study with varying number of moving sequences, where the left part and the right part are without and with optimization respectively

Fig.1 and Fig.2 show that the PrefixTree algorithm, which needs only three scans of the moving sequence database, is more efficient and scalable than the Revised LM algorithm, which needs multiple scans of the moving sequence database limited with the length of the longest moving sequential pattern. In addition, the optimization based on the wireless network topology improves much the efficiency of mining processing. Let |D| is the number of moving sequences in database D, and L is the length of the longest moving sequential pattern. Let reading a moving sequence in a data file costs 1 unit of I/O. The I/O cost of the Revised LM algorithm is equal to L(|D|); the I/O cost of PrefixTree is equal to 3(|D|). From the I/O costs analysis we could get a coarse conclusion that the PrefixTree algorithm is more efficient than the Revised LM algorithm if L is bigger than 3. The above simple I/O analysis of the PrefixTree algorithm and the Revised LM algorithm gives an evience of the Prefix-Tree algorithm's efficiency showed in the experimental results.

4 Discussion

In this paper the idea of clustering method is introduced to discretize the time attribute in moving histories. And then a novel and efficient method, called PrefixTree, is proposed to mine the moving sequences. Its main idea is to generate projected sequences and construct the prefix trees based on candidate consecutive items. It is highly desirable because in most cases the user tends to try a few minimum supports before being satisfied with the result. Another valuable function of the PrefixTree algorithm is supporting parameter tuning, which means the prefix trees with a higher support threshold can be generated directly from the prefix trees with a smaller one.

References

1. Wen-Chih Peng, Ming-Syan Chen. Mining User Moving Patterns for Personal Data Allocation in a Mobile Computing System. Proc. of ICPP, pp. 573-580, 2000.
2. N. Shivakumar, J. Jannink, and J. Widom. Per-user Profile Replication in Mobile Environments: Algorithms Analysis and Simulation Result. ACM/Baltzer Journal of Mobile Networks and Applications, vol. 2, no. 2, pp. 129-140, 1997.
3. Ramakrishnan Srikant, Rakesh Agrawal. Mining Sequential Patterns: Generalizations and Performance Improvements. Proc. of EDBT, pp. 3-17, 1996.
4. J. Pei, J. Han, B. Mortazavi-Asl et al. PrefixSpan: Mining Sequential Patterns Efficiently by PrefixProjected Pattern Growth. Proc. of ICDE, pp. 215-224, 2001.
5. Hsiao-Kuang Wu, Ming-Hui Jin, Jorng-Tzong Horng. Personal Paging Area Design Based On Mobiles Moving Behaviors. Proc. of INFOCOM, pp. 21-30, 2001.
6. G. Das, K. I. Lin, H. Mannila, G. Renganathan, and P. Smyth. Rule Discovery from Time Series. Proc. of KDD, pp. 16-22, 1998.
7. Shuai Ma, Tengjiao Wang, Shiwei Tang et al. A New Fast Clustering Algorithm Based on Reference and Density. Proc. of WAIM, pp. 214-225, 2003.
8. Jong Soo Park, Ming-Syan Chen, Philip S. Yu. Using a Hash-Based Method with Transaction Trimming for Mining Association Rules. IEEE TKDE, vol. 9, no. 5, pp. 813-825, 1997.
9. Vincent W.S. Wong, Victor C. M. Leung. Location Management for Next Generation Personal Communication Networks. IEEE Network, vol. 14, no. 5, pp. 18-24, 2000.

An Alternative Methodology for Mining Seasonal Pattern Using Self-Organizing Map

Denny and Vincent C.S. Lee[2*]

[1] Faculty of Computer Science, University of Indonesia; and
School of Computer Science and Software Engineering,
Faculty of Information Technology, Monash University, denny@cs.ui.ac.id
[2] School of Business Systems, Faculty of Information Technology,
Monash University, vincent.lee@infotech.monash.edu.au

Abstract. In retail industry, it is very important to understand seasonal sales pattern, because this knowledge can assist decision makers in managing inventory and formulating marketing strategies. Self-Organizing Map (SOM) is suitable for extracting and illustrating essential structures because SOM has unsupervised learning and topology preserving properties, and prominent visualization techniques. In this experiment, we propose a method for seasonal pattern analysis using Self-Organizing Map. Performance test with real-world data from stationery stores in Indonesia shows that the method is effective for seasonal pattern analysis. The results are used to formulate several marketing and inventory management strategies.

Keywords: Visualization, Clustering, Temporal Data, Self-Organizing Maps.

1 Introduction

In retail industry, it is very important to understand the seasonal sales pattern of groups of similar items, such as different brands of pencils. This knowledge can be used by decision makers in decision making process such as managing inventory and formulating marketing strategy. However, it is difficult to understand the seasonal sales pattern from large sales transaction database. Common ways to extract and illustrate essential structures from data are by visualization and clustering. Visualization is important since humans can absorb information and extract knowledge efficiently from images. Groups of similar items (such as pencils, pens) that have similar seasonal sales pattern are put together in the same cluster.

Self-organizing map (SOM), a special type of artificial neural network that performs unsupervised competitive learning [9], is suitable for extracting and illustrating essential structures from data, because it requires no *a priori* assumptions about the distribution of the data [4], allows visualization and exploration of the high-dimensional data space by projecting onto simple geometric relationships, and follows the distribution of the original data set [9]. Because prototype vectors in SOM are topological ordered, i.e. similar objects are mapped nearby, SOM is suitable for clustering.

---

* Corresponding author: Dr. Vincent C. S. Lee, http://www.bsys.monash.edu.au

H. Dai, R. Srikant, and C. Zhang (Eds.): PAKDD 2004, LNAI 3056, pp. 424–430, 2004.

The aim of this paper is to extract and illustrate essential seasonal pattern structures from groups of similar sales items using Kohonen's Self-Organizing Maps. This kind of approach had not been fully explored in real world application to a retail industry. This method is applied to real-world data from a retail industry in Indonesia that has not been used in other researches.

The structure of the paper is as follows. Section 2 briefly describes the data and proposed method. The experiences in applying the method to a retail industry are discussed in Section 3 and the paper closes with a concluding section.

2 Proposed SOM Method and Data

2.1 Description of the Data

In this experiment, we use real-life data from a major stationery retail chain based in Indonesia. The data used in this experiment is two years sales data, from 1999 to 2000, from seven branches, with about 1.5 millions of transactional data for 17,836 products. The products are divided into 34 groups of similar items (such as school equipment), and further divided into 551 sub-groups of more similar items (such as pencil cases).

2.2 Proposed SOM Method for Seasonal Pattern Analysis

Seasonal pattern analysis process using SOM can be generally divided into data preparation stages, map training, mapping the data set to the map, and lastly analysis based on visualization and clustering of the map.

Data Pre-processing. In comparing time series, some difficulties might arise due to noise, offset translation, amplitude scaling, longitudinal scaling, linear drift, and discontinuities [8]. However, some of these problems can be considered not important in comparing seasonality pattern. Longitudinal scaling problems are not important, since data should be compared for the same time-period. Moreover, time-series with linear drift can be assumed as different time-series.

Discontinuities, amplitude scaling and offset translation problem can be solved by performing pre-processing and normalization of the data set. Discontinuities problem can be alleviated by aggregating daily sales quantity in one month or using moving average. Amplitude scaling and offset translation problems can be solved by performing normalization of the data set. In our experiment, the monthly total sales quantities are divided by the average monthly sales in a year. This measurement is called seasonality index. The seasonality index value '5' for a subgroup in February can be interpreted sales quantity in February is five times larger compared to the average monthly sales of that subgroup. In this experiment, seasonal pattern is defined as a series of seasonality index of 12 months.

To compare similarity between time-series data, several researchers developed measurement based on longest common subsequence [1,2]. However, this kind of similarity measures is not appropriate for measuring similarity for seasonality pattern for several reasons. Firstly, the measurement allows gaps of the sequences that match, which allows comparing seasonality indexes from different time periods, which is illogical. Secondly, this measure also does not consider the degree of differences. For instance, this measurement produces the same results for comparing {3 1 1 1}, {1 1 1 3}, and {1 1 1 10}. As a result, the similarity of time-series data used in this experiment is Euclidian similarity measurement, because it is fast to compute thus allowing using it for clustering tasks.

Map Training. In this experiment, the map size is 16x20 with hexagonal grid topology. The map is initialized based on two principal eigenvectors of the input data vectors, and trained using the batch version of SOM. After the map is trained, each data vector is mapped to the most similar prototype vector in the map (the best matching unit). The software used is SOM Toolbox 2.0 [12].

Analysis of SOM and Its Clustering Results. Generally, it is not easy to capture knowledge from SOM solely by observing the prototype vectors in SOM, especially when the map is large. Visualization of SOM can be used to capture pattern in the data set. Some useful visualization tools for seasonal pattern analysis are unified distance matrix (u-matrix) and component planes. The unified distance matrix (u-matrix) shows the distances between neighbouring nodes in the data space using a grey scale representation on the map grid [7]. Long distances show highly dissimilar features between neighbouring nodes are represented with darker colour, which divides clusters (dense parts of the maps with similar features). Component planes show the values of a certain component of prototype vectors in SOM [10].

Since visualization of the u-matrix can show the rough cluster structure of the data set, to get a finer clustering result, the prototype vectors from the map are clustered using partitive clustering technique (*k*-means) [11]. After that, for each cluster simple descriptive statistical analyses, such as averages, of the cluster members' attributes (such as sales quantity) are calculated.

3 Results and Discussions

Due to business data confidentiality, the complete results cannot be shown in this paper. However, permission to use small amount of examples of the results is granted by the owner of the data.

After the data is pre-processed, the total input vector for SOM training is 572 with 12 dimensions. Once the map is trained, each node of the SOM stores a prototype vector of 12 components each of which represents seasonality index for each month from January to December and a set of subgroups that are mapped to the node. The values of each prototype vectors are visualized using bar plot in Fig. 1.

Since SOM follows the distribution of the original data set, the structure of the original data set can be extracted from the map. In Fig. 1, there are more nodes with roughly stable seasonality indexes throughout the year (the nodes in the centre of the map) than nodes with non-uniform seasonality indexes (the nodes at the edge of the map), it can be inferred that most of the subgroups in the original data set also have roughly same seasonal pattern. This knowledge is difficult to acquire by observing the

Fig. 1. The bar plot of each component values of each prototype vector

original data set or calculating mean and standard deviation of each variable. For that reason, visualization of SOM makes it easier for analysts to acquire knowledge about the structure of the original data set.

Table 1. Clusters' characteristics

No	Total Member	Total Sales Qty in 2 Years			Average of Seasonal Index											
		Average	Sum	%	Jan	Feb	Mar	Apr	May	Jun	Jul	Aug	Sep	Oct	Nov	Dec
1	79	11,862	937,126	12.2%	0.1	0.1	0.1	0.2	0.7	1.2	1.6	1.5	1.9	1.7	1.6	1.2
2	20	22,953	459,055	6.0%	0.2	0.3	0.3	0.4	0.5	1.1	**4.8**	1.7	0.9	0.7	0.6	0.4
3	127	21,668	2,751,805	35.9%	0.6	0.7	0.9	1.0	0.9	1.0	1.2	1.3	1.2	1.1	1.0	0.9
4	56	3,318	185,820	2.4%	0.0	0.0	0.0	0.0	0.1	0.1	0.2	0.2	2.4	**3.8**	**2.9**	**2.4**
5	25	3,223	80,583	1.1%	0.1	0.1	0.1	0.1	0.2	0.5	0.7	**2.2**	**4.0**	1.9	1.3	0.8
6	23	4,732	108,834	1.4%	0.1	0.1	0.1	0.5	1.6	**4.0**	1.9	1.5	0.8	0.6	0.4	0.3
7	145	17,467	2,532,653	33.0%	1.3	1.4	1.5	1.4	1.0	0.9	1.1	0.9	0.7	0.6	0.6	0.5
8	18	3,165	56,968	0.7%	**6.4**	**2.5**	0.8	0.6	0.5	0.2	0.2	0.3	0.3	0.1	0.1	0.1
9	7	5,120	35,839	0.5%	0.2	1.2	**2.7**	**5.4**	0.8	0.4	0.6	0.1	0.2	0.2	0.1	0.1
10	21	2,107	44,242	0.6%	0.0	0.0	0.0	0.0	0.0	0.0	0.0	0.0	0.5	**2.3**	**5.7**	**3.2**
11	17	16,824	286,005	3.7%	0.9	0.7	0.5	0.5	0.5	0.8	1.0	0.7	1.0	1.0	1.8	**2.7**
12	34	5,565	189,200	2.5%	**2.1**	**3.3**	**2.7**	1.5	0.8	0.4	0.4	0.3	0.2	0.2	0.1	0.1

3.1 Clusters of Seasonal Pattern

Based on visualization of the u-matrix in Fig. 2, there are at least two large clusters (cluster A and B) and two relatively medium-sized clusters (cluster C and D). The position of these two largest clusters in Fig. 1 shows that these clusters contains subgroups that have stable or uniform sales pattern throughout the year. Therefore, the visualization of u-matrix and visualization of the value of prototype vectors can give a rough clusters structure in the original data set.

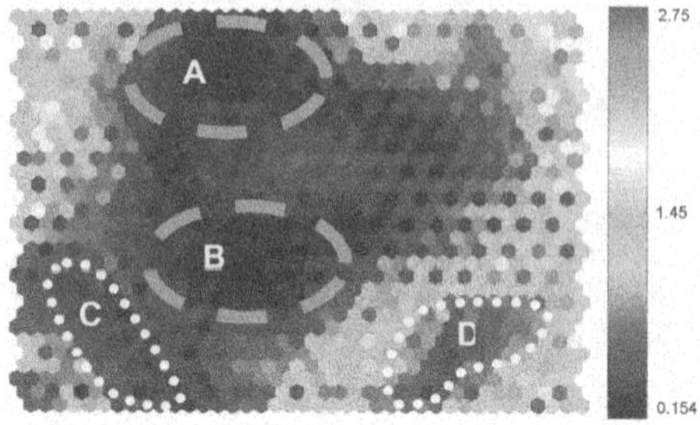

Fig. 2. The u-matrix

Fig. 3. Clustering result of SOM using *k*-means with 12 clusters

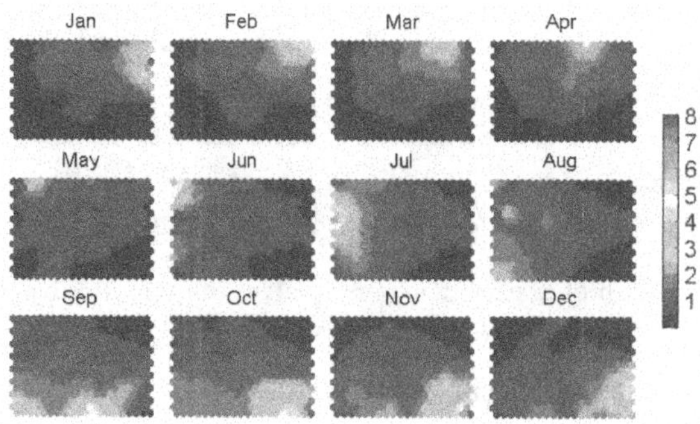

Fig. 4. The component planes

In order to get a finer clustering result, two-level clustering is used. In this experiment, we try to find clustering with total cluster from 2 to 12. Davies-Bouldin index [3] is used to determine the optimal number of cluster. Based on Davies-Bouldin index, the best clustering result from 2 to 12 total clusters is 12 clusters. Although any number of clusters can be chosen based on the purpose of the analysis, this experiment uses the clustering results with 12 clusters as shown in Fig. 3.

Table 1 reveals some interesting clusters. The most interesting clusters are cluster 7 and cluster 3, since they are the two largest clusters that contribute almost 69% of total sales quantity. The subgroups in these clusters have roughly stable seasonal pattern throughout the year with cluster 3 has a slightly higher sales in the second half-year while cluster 7 in the first half-year. This fact confirmed the analysis based on visualization of the u-matrix described earlier. Another interesting cluster is cluster 2, since the average sales quantity of the subgroups in this cluster is the highest and sales quantities of these subgroups in July are 4.8 higher compared to the average monthly sales quantity.

These clustering results can be used to determine marketing strategy for product combination package (cross-selling). The common practice in this company is combining items that are complementary without taking seasonality factor into account, such as combining pens and writing pads. However, selling items with the same seasonality pattern in one package during their sales peak times might not be effective since the cost of the promotion (such as discount value) might be higher than the increase in sales of each item. As a result, we recommend that items from different cluster (different seasonal pattern) are sold as combo-pack to increase the sale of one product with the help of the other product. This strategy can be exercised throughout the year with different combination. For example, correction pens subgroup from cluster 6 that has a peak in May and refills for gel-based pens subgroup from cluster 10 that has a peak in November can be combined in one package. This combination can increase the sales of white pens in its off-peak time as the sales of refill for gel-based pens are high near the end of the year.

3.2 Component Planes of the Map

The component planes, as shown in Fig. 4, show the seasonality indexes for a particular month. The colours in the component planes represent the value seasonality index. For example, the sixth component plane shows the value of seasonality indexes for June. In June, the subgroups in the top left region have the highest seasonality index, which are over 5, while the subgroups in the top right and bottom right region have a low seasonality index. These component planes also show the distribution of seasonal index for each month.

These component planes can be used as a guide for devising strategies for each month, with giving more attention to the subgroups in red, yellow, and green regions. Two strategies are proposed based on the analysis of these component planes.

Firstly, the stores can provide more variety of items for seasonal subgroups, which are the subgroups in red and yellow regions. For example, based on July's component plane, the sale of the pencil cases subgroup has a non-uniformly sales distributed throughout the year with a peak in June. Consequently, the store can provide more variety of pencil cases to its customers during this time. As a result, the customer may view the stores as a store that provides a wide variety of items. Additionally, the sales of this subgroup may increase. This strategy will change the current practice that provides an approximately constant variety of items throughout the year.

Secondly, the component planes can be used as a guide to arrange store layout and space allocation for specific purposes, especially for subgroups that lies in red region in that particular month. For example, pencil cases subgroup can be allocated a larger area and placed at a special place in June. This strategy can be valuable since the space of the store is very limited. This strategy will also change the current practice that allocates approximately constant space for each subgroup throughout the year.

4 Conclusions

The above discussion has highlighted the use of SOM in the seasonal pattern analysis using visualization and clustering. Based on our experiments, SOM is an effective tool for seasonal pattern analysis, since it provides visualization that reveals structure in data in comprehensible form. Based on visualization and clustering of SOM, several strategies based on seasonal pattern are proposed, such as determining product combination package and managing inventory. Therefore, hopefully, sale of these seasonal items can be increased.

The ideal way to measure seasonality is by measuring demand after considering other factors that impact sales, such as discounts, inventory, and random effects. Future work incorporating with multi-attribute measure of seasonality is underway. In addition, future work includes combining the clustering result of seasonal sales pattern with the association rules discovered from the data set for determining promotional packages. Items with different seasonal pattern that frequently sold together can be combined as a promotional package.

Acknowledgement. The authors would like to express gratitude to K. Julianto Mihardja and Liany Susanna for allowing the use of their data in this experiment.

References

1. Agrawal, R., Lin, K.-I., Sawhney, H.S., and Shim, K. (1995). "Fast Similarity Search in the Presence of Noise, Scaling, and Translation in Time-Series Databases". In *Proc. 21th Intl. Conf. on Very Large Databases (VLDB'95).*
2. Das, G., Gunopulos, D., and Mannila, H. (1997). "Finding Similar Time Series". In *Principles of Data Mining and Knowledge Discovery*, pp. 88-100.
3. Davies, D.L., and Bouldin, D.W. (1979). "A cluster separation measure". *IEEE Transactions on Pattern Analysis and Machine Intelligence*, 1(2), pp. 224-227.
4. Deboeck, G. and Kohonen, T. (1998). *Visual Explorations in Finance with Self-Organizing Maps*. Springer-Verlag, Berlin.
5. Faloutsos, C., Ranganathan, M., and Manolopoulos, Y. (1994). "Fast Subsequence Matching in Time-Series Databases". In *Proceedings of the 1994 Annual ACM SIGMOD Conference*, New York, NY: ACM Press, pp. 419-429.
6. Hand, D.J., Mannila, H., and Smyth, P. (2001). *Principles of Data Mining*. The MIT Press, Cambridge.
7. Iivarinen, J., Kohonen, T., Kangas, J., and Kaski, S. (1994). "Visualizing the Clusters on the Self-Organizing Map". In Carlsson, C., Järvi, T., and Reponen, T. (eds.), *Proceedings of the Conference on Artificial Intelligence Research in Finland*, 12, 122-126. Helsinki, Finland. Finnish Artificial Intelligence Society.
8. Keogh, E.J. and Pazzani, M.J. (1998). "An enhanced representation of time series which allows fast and accurate classification, clustering and relevance feedback". In R. Agrawal, P. Stolorz and G. Piatetsky-Shapiro (Eds.), *Fourth International Conference on Knowledge Discovery and Data Mining (KDD'98)*, pp. 239-241. ACM Press, New York City, NY.
9. Kohonen, T. (2001). *Self-Organizing maps* (third edition). Springer-Verlag, Berlin.
10. Tryba, V., Metzen, S., and Goser, K. (1989). "Designing of basic Integrated Circuits Self-Organizing Feature Maps". *Neuro-Nimes'89. Int Workshop on Neural Networks and their applications* (pp. 225-235). Nanterre, France.
11. Vesanto, J. and Alhoniemi, E. (2000). "Clustering of the Self-Organizing Map". *IEEE Transactions on Neural Networks*, 11(3), 586-600.
12. Vesanto, J., Himberg, J., Alhoniemi, E., and Parhankangas, J. (2000). *SOM Toolbox for Matlab 5* (Tech. Rep. No. A57). Espoo, Finland: Neural Networks Research Centre, Helsinki University of Technology.

ISM: Item Selection for Marketing with Cross-Selling Considerations

Raymond Chi-Wing Wong and Ada Wai-Chee Fu

Department of Computer Science and Engineering
The Chinese University of Hong Kong
{cwwong, adafu}@cse.cuhk.edu.hk

Abstract. Many different algorithms are studied on association rules in the literature of data mining. Some researchers are now focusing on the application of association rules. In this paper, we will study one of the application called Item Selection for Marketing (ISM) with cross-selling effect consideration. The problem ISM is to find a subset of items as marketing items in order to boost the sales of the store. We prove a simple version of this problem is NP-hard. We propose an algorithm to deal with this problem. Experiments are conducted to show that the algorithms are effective and efficient.

1 Introduction

In the literature of data mining, there are a lot of studies on association rules [2]. Such studies are particularly useful with a large amount of data in order to understand the customer behavior in their stores. However, it is generally true that the results of association rule mining are not directly useful for the business sector. Therefore there has been research in examining more closely the business requirements and finding solutions that are suitable for particular issues, such as marketing and inventory control. Recently, some researchers [10] studied the utility of data mining such as association and clustering, on decision making for revenue-maximizing enterprises. They have formulated the general problem as an optimization problem where a profit is to be maximized by determining a best strategy. The profit is typically generated from the customer behaviour in such an enterprise. More specific problems for revenue-maximizing are considered in more recent works [6,14,9,5,4,17]. The related problem of mining user behaviour is also of much research interest recently and a number of results can be found in [18].

In this paper we consider the problem of selecting a subset of items in a store for marketing in order to boost the overall profit. The difficulty of the problem is that we need to estimate the cross-selling effect to determine the influence of the marketed items on the sales of the other items. It is known that the records of sales transactions are very useful [3] and we determine the cross-selling effect with such information. We call the problem defined this way *Item Selection for Marketing (ISM)*. We show that a simple version of this problem is NP-hard. We propose a hill climbing approach to tackle this problem. In our experiment,

H. Dai, R. Srikant, and C. Zhang (Eds.): PAKDD 2004, LNAI 3056, pp. 431–440, 2004.

we apply the proposed approach to a set of real data and the approach is found to be effective and efficient.

2 Related Work

One major target of data mining is solving decision making problems for the business sector. A study of the utility of data mining for such problems is investigated in [10], published in 1998. A framework based on optimization is presented for the evaluation of data mining operations. In [10] the general decision making problem is considered as a maximization problem as follows

$$\max_{x \in \mathcal{D}} \sum_{i \in \mathcal{C}} g(x, y_i) \tag{1}$$

where $\mathcal{D}$ is the set of all possible decisions in the domain problem (e.g. inventory control and marketing), $\mathcal{C}$ is the set of customers, y_i is the data we have on customer i, and $g(x, y_i)$ is the utility (benefit) from a decision x and y_i. However, when we examine some such decision problems more closely, we find that we are actually dealing with a maximization problem of the form

$$\max_{x \in \mathcal{D}} g(x, Y) \tag{2}$$

where Y is the set of all y_i, or the set of data collected about all customers. The above is more appropriate when there are correlations among the behaviours of customers (e.g. cross-selling, the purchase of one item is related to the purchase of another item), or when there are interactions among the customers themselves (e.g. viral marketing, or marketing by word-of-month among customers). This is because we cannot determine $g()$ based on each single customer alone.

We illustrate the above in two different problems that have been studied. The first problem is about optimal product selection [5,4,16,17] (in SIGKDD 1999,2000,2002, and ICDM 2003, respectively). The problem is that in a typical retail store, the types of products should be refreshed regularly so that losing products are discarded and new products are introduced. Hence we are interested to find a subset of the products to be discontinued so that the profit can be maximized. The formulation of the problem considers the important factor of cross-selling which is the influence of some products on the sales of other products. The cross-selling factor is embedded into the calculation of the maximum profit gain from a decision. This factor can be obtained from an analysis of the history of transactions kept from previous sales which corresponds to the set Y in formulation (2).[1]

The second such problem is about viral marketing where we need to choose a subset of the customers to be the targets of marketing so that they can influence more of other customers. Some related work can be found in [6,14,9,1] (in

[1] The problem is related to inventory management which has been studied in management science, however, previous works are mostly on the problems of when to order, where to order from, how much to order and the proper logistics [15].

SIGKDD 2001,2002,2003 and WWW2003, respectively). Again the profit gain from any decision relies on an analysis based on the knowledge collected about all customers.

The problem that we tackle here is of a similar nature since we also consider the factor of cross-selling when calculating the utility or benefit of a decision. In our modeling, we adopt concepts of the association rules to model the cross-selling effects among items.

Suppose we are given a set I of items, and a set of transactions. Each transaction is a subset of I. An *association rule* has the form $X \rightarrow I_j$, where $X \subseteq I$ and $I_j \in I - X$; the *support* of such a rule is the fraction of transactions containing all items in X and item I_j; the confidence for the rule is the fraction of the transactions containing all items in set X that also contain item I_j. The problem is to find all rules with sufficient support and confidence given some thresholds. Some of the earlier work include [13,2,12].

3 Problem Definition

In this section we introduce the problem of ISM. To the best of our knowledge, this is the first definition of item selection problem for marketing with the consideration of cross-selling effects. Item Selection for Marketing (ISM) is a problem to select a set of items for marketing, called *marketing items*, so as to maximize the total profit of marketing items and non-marketing items among all choices. In ISM, we assume that the sales of some items are affected by the sales of some other items. Given a data set with m transactions, $t_1, t_2, ..., t_m$, and n items, $I_1, I_2, ..., I_n$. Let $I = \{I_1, I_2, ..., I_n\}$. The profit of item I_a in transaction t_i before marketing is given by $prof(I_a, t_i)$. Let $S \subset I$ be a set of selected items. In each transaction t_i, we define two symbols, t'_i and d_i, for the calculation of the total profit.

$$t'_i = t_i \cap S, \qquad d_i = t_i - t'_i$$

Definition 1 (Profit Before Marketing). *The original profit $Profit_0$ before marketing for all transactions is defined as:*

$$Profit_0 = \textstyle\sum_{i=1}^{m} \sum_{I_a \in t_i} prof(I_a, t_i) \tag{3}$$

Suppose we select a subset S of marketing items. Marketing action such as discounting will be taken on S. Let us consider a transaction t_i containing the marketing items I_a and non-marketing items I_b. If we market item I_a with a cost of $cost(I_a, t_i)$ (e.g. discount of item), the profit of item I_a after marketing in transaction t_i will become $prof(I_a, t_i) - cost(I_a, t_i)$. After the marketing actions are taken, more of the marketing items, says I_a, will be purchased. We define the changes in the sales by $\alpha(T)$, where T is a set of items:

$$\alpha(T) = \frac{\text{sale volume of } T \text{ after marketing}}{\text{sale volume of } T \text{ before marketing}}. \tag{4}$$

In the above the sale volume of T is measured by the total amount of the items in T that are sold in a fixed period of time.[2] If $\alpha(\{I_a\}) = 1$, then there is no increase of the sales of items I_a. If $\alpha(\{I_a\}) = 2$, then the sales of I_a is doubled compared with the sales before marketing.

On the other hand, without the consideration of cross-selling effect due to marketing, the profit of non-marketing items I_b is still $prof(I_b, t_i)$. With the consideration of cross-selling effects, some of the non-marketing items I_b will be purchased more if there is an increase of sales of marketing items I_a. The cross-selling factor is modelled by $csfactor(T, I_b)$, where T is a set of marketing items I_a, and $0 \leq csfactor(T, I_b) \leq 1$. That is, more customers may come to buy item I_b if some other items in T are being marketed. The increase of the sale of item I_b is modelled by $(\alpha(T) - 1)csfactor(T, I_b)$. [3] If $\alpha(T) = 1$, then there is no increase of sales of marketing items in set T. So, there is no increase of sales of non-marketing item I_b. The term $(\alpha(T) - 1)csfactor(T, I_b)$ becomes zero. Similarly, if $\alpha(T) = 2$, the sales of items in set T is doubled. Thus, the increase of sales is modelled by $csfactor(T, I_b)$.

Definition 2 (Profit After Marketing). *The profit after marketing $Profit_1$ is defined as follows.*

$$\begin{aligned} Profit_1 = \textstyle\sum_{i=1}^{m} \Big[\textstyle\sum_{I_a \in t'_i} \alpha(\{I_a\})(prof(I_a, t_i) - cost(I_a, t_i)) \\ + \textstyle\sum_{I_b \in d_i} (1 + (\alpha(t'_i) - 1)csfactor(t'_i, I_b))prof(I_b, t_i)\Big] \end{aligned} \quad (5)$$

Recall that t'_i is the set of items in transaction t_i that are selected to be marketed. For each transaction t_i, we compute the profit from the marketing items (discounted by $cost(I_a, t_i)$), and the profit from the non-marketing items whose sales are influenced by $csfactor()$. $Profit_1$ is the sum of the profits from all tranasctions. The objective of marketing is to increase the profit gain compared with the profit before marketing. The profit gain is defined as follows.

Definition 3 (Profit Gain). *Profit gain is :*

$$\textit{Profit Gain} = Profit_1 - Profit_0 \quad (6)$$

From the above definitions, we can rewrite the profit gain as follows.

$$\begin{aligned} \text{Profit Gain} &= Profit_1 - Profit_0 \\ &= \textstyle\sum_{i=1}^{m} \Big[\textstyle\sum_{I_a \in t'_i} [(\alpha(\{I_a\}) - 1)prof(I_a, t_i) - \alpha(\{I_a\})cost(I_a, t_i)] \\ &\quad + \textstyle\sum_{I_b \in d_i} (\alpha(t'_i) - 1)csfactor(t'_i, I_b)prof(I_b, t_i)\Big] \end{aligned} \quad (7)$$

[2] We note that different items may have their different increase ratio of the sales (i.e. $\alpha(\{I_i\})$). However, it is difficult to predict this parameter $\alpha(\{I_i\})$ for each item I_i. For simplicity, we set all $\alpha(\{I_i\})$ to be the same (e.g. α_0) in this paper, which is the same as [6,14].

[3] If $\alpha(\{I_i\}) = \alpha_0$ for all i, then it is easy to see that $\alpha(T) = \alpha_0$ for any T (a subset of I) .

Next we can formally define the problem of ISM:

ISM: Given a set of transactions with profits assigned to each item in each transaction and the cross-selling factors, $csfactor()$, pick a set S from all given items which gives a maximum profit gain.

This problem is at least as difficult as the following decision problem.

ISM Decision Problem: Given a set of items and a set of transactions with profits assigned to each item in each transaction, a minimum profit gain G, and cross-selling factors, $csfactor()$, can we pick a set S such that $Profit\ Gain \geq G$?

Note that the cross-selling factor can be determined in different ways, one way is by the domain experts. Let us consider the very simple version where $csfactor(t'_i, I_a) = 1$ for any non-empty set of t'_i. That is, any selected items in the transaction will increase the sale of the other items with the same volume. This may be a much simplified version of the problem, but it is still very difficult.

Theorem 1 (NP-hardness). *The item selection for marketing (ISM) decision problem where $csfactor(t'_i, I_a) = 1$ for $t'_i \neq \phi$ and $csfactor(t'_i, I_a) = 0$ for $t'_i = \phi$ is NP-hard.*

Proof: We shall transform the problem of MAX CUT to the ISM problem. MAX CUT [7] is an NP-complete problem defined as follows: *Given a graph - (V, E) with weight $w(e) = 1$ for each $e \in E$ and positive integer K, is there a partition of V into disjoint sets V_1 and V_2 such that the sum of the weights of the edges from E that have one endpoint in V_1 and one endpoint in V_2 is at least K?* The transformation from MAXCUT to ISM problem is described as follows. Let $G = K$, $\alpha(\{I_a\}) = 2$, and $\alpha(t'_i) = 2$. For each vertex $v \in V$, construct an item. For each edge $e \in E$, where $e = (v_1, v_2)$, create a transaction with 2 items $\{v_1, v_2\}$. Set $prof(I_j, t_i) = 1$ and $cost(I_j, t_i) = 0.5$, where t_i is a transaction created in the above, $i = 1, 2, ..., |E|$, and I_j is an item in t_i. It is easy to check that Profit Gain $= \sum_{i=1}^{m} \sum_{I_b \in d_i} csfactor(t'_i, I_b)$. The above transformation can be constructed in polynomial time. When the problem is solved in the transformed ISM, the original MAX CUT problem is also solved. Since MAX CUT is an NP-complete problem, ISM problem is NP-hard. □

4 Association Based Cross-Selling Effect

In the previous section, we see that the cross-selling factor is important in the problem formulation. The factor is indicated by $csfactor(t'_i, I_j)$, where t'_i is a set of items selected for marketing and I_j is another item. This factor can be provided by domain experts if they can estimate the impact of t'_i on I_j. However, in typical application, the amount of items would be large and it would be impractical to expect purely human analysis on these values. We suggest that the factor is to be determined by data mining technique based on the history of transactions collected for the application. We shall adopt the concepts of association rules for this purpose.

Definition 4. *Let $d_i = \{Y_1, Y_2, Y_3, ..., Y_q\}$ where Y_i refers to a single item for $i = 1, 2, .., q$, then $\diamond d_i = Y_1 \vee Y_2 \vee Y_3 \vee .... \vee Y_q$.* □

In our remaining discussion, $csfactor(t'_i, I_j)$ is equal to $conf(\diamond t'_i \to I_j)$, where $conf(\diamond t'_i \to I_j)$ is the confidence of the rule $\diamond t'_i \to I_j$ The definition of confidence here is similar to the definition of association rules. That is,

$$\begin{aligned} &csfactor(t'_i, I_j) = conf(\diamond t'_i \to I_j) \\ &= \frac{\text{number of transactions containing any item in } t'_i \text{ and } I_j}{\text{number of transactions containing any items in } t'_i} \end{aligned} \tag{8}$$

The reason for the above formulation is given as follows. A transaction can be viewed as a customer behavior. In transaction t_i, there are the cross-selling effect between any marketing items I_a in t'_i and non-marketing items in set d_i. Let us consider some cases. If all items in t_i are being marketed, then there are no non-marketing items, and the profit gain is the difference between the profit of marketing items after marketing and that before marketing. If all items in t_i are not marketed, as there are no marketing items, in transaction t_i, there is no cross-selling effect from marketing items in transaction t_i. Thus, the profit gain due to marketing becomes zero. Now, consider the case of a transaction containing both marketing items and non-marketing items. Suppose the customer who purchases any marketing items in set t'_i always purchases non-marketing items I_b. This phenomenon is modelled by a gain rule $\diamond t'_i \to I_b$. The greater the confidence of these rules is, the greater the cross-selling effect is. That is, if this confidence is high, then when more of t'_i are sold, it means that very likely more of I_b will also be sold. [4]

5 Hill Climbing Approach

The ISM problem is likely to be very difficult. We propose here a hill climbing approach to tackle the problem.[5]

Let $f(S)$ be the function of the profit gain of the selection S of marketing items. Initially, we assign $S = \{\}$. Then, we will calculate $f(S \cup \{I_a\})$ for each item I_a. We choose the item I_b with the greatest value of $f(S \cup \{I_b\})$ and insert it into set S. The above process repeats for the remaining items whenever $f(S \cup \{I_b\}) > f(S)$.

5.1 Efficient Calculation of the Profit Gain

As the formula of the profit gain is computationally intensive, an efficient calculation of this formula is required. The hill climbing approach chooses the item

[4] The rule $I \to \diamond d_i$ is called a *loss rule* in [17], because in [17], the problem is to determine a set of items to be discontinued from a store, d_i refers to some items to be removed, and it may cause some loss in profit from other items.

[5] We have also tried to apply the well-known optimization technique of quadratic programming. However, we could only approximate the problem by a quadratic programming problem and the approximation is not very accurate since we need to throw away terms in a Taylor's series which may not be insignificant. The resulting performance is not as good as the hill climbing method and hence are not shown.

with the greatest profit gain for each iteration. Suppose S now contains k items at the k-th iteration. At this iteration, we store the value of $f(S)$ in a variable f_S. At the $(k+1)$-th iteration, we can calculate $f(S \cup \{I_x\})$ from f_S efficiently for all $I_x \notin S$. Let $\mathcal{T}$ be the set of transactions containing item I_x and at least one item in selection set S. We can calculate $f(S \cup \{I_x\})$ as

$$f(S \cup \{I_x\}) \leftarrow f_S + g(I_x) - h(S, \mathcal{T}) + h(S \cup \{I_x\}, \mathcal{T}) \tag{9}$$

$$\text{where } g(I_x) = \sum_{i=1}^{m} [(\alpha(\{I_x\}) - 1)prof(I_x, t_i) - \alpha(\{I_x\})cost(I_x, t_i)]$$

$$h(X, \mathcal{T}) = \sum_{t_i \in \mathcal{T}} \sum_{I_b \in d_i} (\alpha(t'_i) - 1)csfactor(t'_i, I_b)prof(I_b, t_i)$$

For $h(X, \mathcal{T})$ we assume all items in set X are selected for marketing, i.e. $t'_i = t_i \cap X$, and $d_i = t_i - t'_i$. Function $g(I_x)$ is the profit gain of marketing item I_x in all transactions. Function $h(X, \mathcal{T})$ is the profit gain of non-marketing items for the selection X in all transactions in set $\mathcal{T}$.

Let us consider the calculation of $f(S \cup \{I_x\})$. For $g(I_x)$, we need to add the profit gain of the newly added marketing item I_x after marketing (i.e. $g(I_x)$) to f_S. For the remaining parts, we only deal with the transactions in set $\mathcal{T}$. We need to subtract the profit gain of non-marketing items for the selection S in all the transactions in set $\mathcal{T}$ (i.e. $h(S, \mathcal{T})$) and then add the profit gain of non-marketing items for the new selection $S \cup \{I_x\}$ in all the transactions in set $\mathcal{T}$ (i.e. $h(S \cup \{I_x\}, \mathcal{T})$). As the set $\mathcal{T}$ is typically small compared with the whole database, we can save much computation by restricting the scope of search to $\mathcal{T}$. In the actual implementation the scope restriction is realized by a special search procedure of a special FP-tree as described below.

5.2 FP-Tree Implementation

The transactions in the database are examined for computation whenever the confidence term $conf(\diamond t'_i \rightarrow I_j)$ is calculated. So, we need to do this operation effectively. If we actually scan the given database, which typically contains one record for each transaction, the computation will be very costly. Here we make use of the FP-tree structure [8].

We construct an FP-tree $\mathcal{FPT}$ once for all transactions, setting the support threshold to zero, and recording the occurrence count of itemsets at each tree node. With the zero threshold, $\mathcal{FPT}$ retains all information in the given set of transactions. Then we can traverse $\mathcal{FPT}$ instead of scan the original database. The advantage of $\mathcal{FPT}$ is that it forms a single path for transactions with repeated patterns. In many applications, there exist many transactions with the same pattern, especially when the number of transactions is large. These repeated patterns are processed only once with $\mathcal{FPT}$. By traversing $\mathcal{FPT}$ once, we can count the number of transactions containing any items in set t'_i and item I_b and number of transactions containing any items in set t'_i.

The details of the procedure can be found in the description of the function parseFPTree(N,D) in [17]. From our experiments this mechanism can greatly reduce the overall running time.

6 Empirical Study

We have used Pentium IV 2.2GHz PC to conduct our experiments. In our experiments, we study the resulting profit gain of marketing using the proposed algorithm. After the execution of our algorithm, there will be a number of selected marketing items. Let there be J resulting items (or marketing items). Note that J is not an input parameter. We compare our results with the naive approach of marketing by choosing J items with the greatest values of profit gain in Definition 3 in Section 3 as marketing items, assuming no cross-selling effect (i.e. $csfactor(t_i', I_b) = 0$ for any set t_i and item I_b). This naive approach is called *direct marketing*.

6.1 Data Set

We adopted the data set from BMS WebView-1, which contains clickstream and purchase data collected by a web company and is part of the KDD-Cup 2000 data [11]. There are 59,602 transactions and 497 items. The average transaction size is 2.5. The profit of each item is generated similarly as [17].

6.2 Experimental Results

For the data set, we study two types of marketing method - discounted items and free items. For discounted items, the selling price is half of the original price. Free items are free of charge. As remarked in Section 3, we shall assume a uniform change in the sale volume for all marketing items, i.e. $\alpha(I_i) = \alpha$ for all items I_i. This set up is similar to that in [6]. For the real data set, the experimental results of profit gains and execution time against α for the situation of discount items are shown in Figure 1 and Figure 2. Those for the situation of free items are shown in Figure 3 and Figure 4. In the graphs showing the profit gains, we show the number of resulting marketing items next to each data point of the hill climbing method. This number is also the number of iterations in the hill climbing method.

In all the experiments, the profit gain for the hill climbing approach is always greater than that for direct marketing. This is because the proposed algorithm considers the cross-selling effect among items, but the direct marketing does not.

The execution time of direct marketing is roughly constant and is very small in all cases. For the hill climbing approach, the execution time increases significantly with the increase in α. This is explained by the fact that when α is increased, the marketing effect increases, meaning that the increase in sale of marketing items will be greater, which also increases the sale of non-marketing items by cross-selling effect. The combined increase in sale will be able to bring more items to be profitable for marketing since they can now counter the cost of marketing. This means that the hill climbing approach will have more iterations as α increases since the introduction of each marketing item requires one iteration, and this means longer execution time.

Note that in the scenario of free marketing items, direct marketing leads to zero or negative profit gain. This is because the items are free and generate no profit, and hence when compared to the profit before marketing, the profit gain is zero or negative. In the synthetic data, it is found that the gain is zero in most cases, since the marketing items are chosen to be those with no recorded transaction. The gain becomes negative for the real data set. Such results are similar to those for direct marketing in [6].

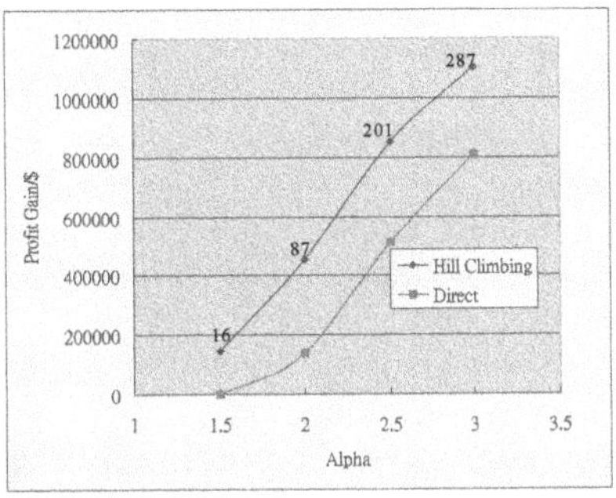

Fig. 1. Profit Gains against α for Real Data Set (Discount)

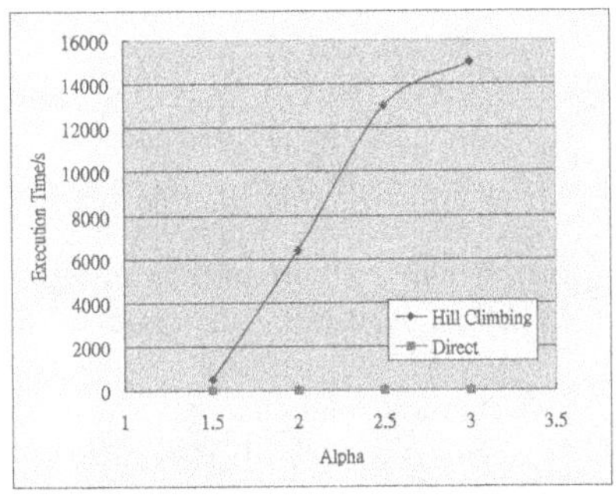

Fig. 2. Execution Time against α for Real Data Set (Discount)

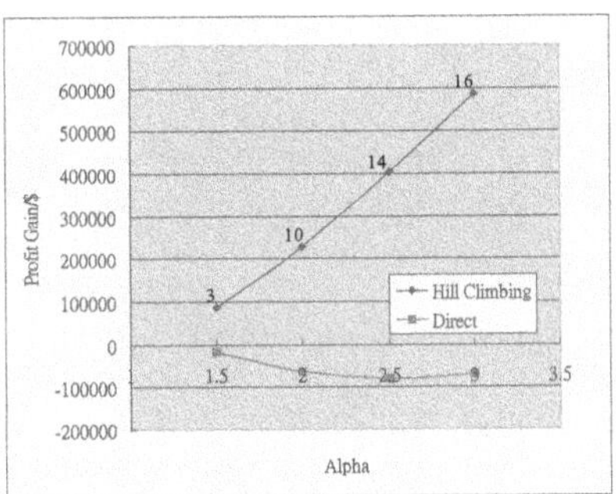

Fig. 3. Profit Gains against α for Real Data Set (Free)

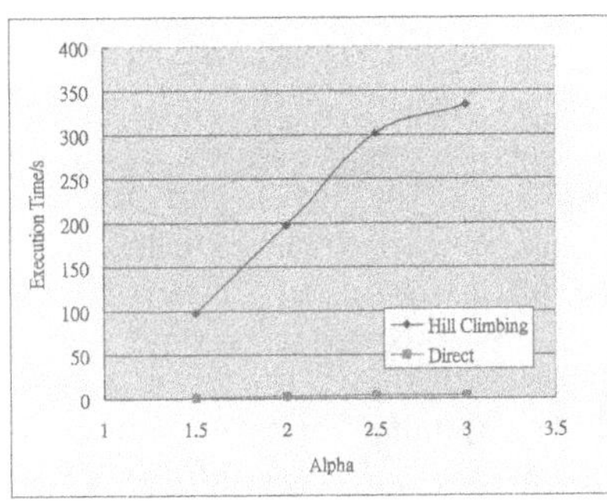

Fig. 4. Execution Time against α for Real Data Set (Free)

7 Conclusion

In this paper, we have formulated the problem Item Selection for Marketing (ISM) with the consideration of cross-selling effect among the items. We proved that a simple version of this problem is NP-hard. We adopt the concepts of association rules to the determination of the cross-selling factor. Then we propose a hill climbing approach to deal with this problem. We have conducted some experiments on both real data and synthetic data to compare our method with the results of a naive marketing method. The results show that our algorithm is highly effective and efficient.

Acknowledgements. This research was supported by the RGC Earmarked Research Grant CUHK 4179/01E.

References

1. R. Agrawal, S. Rajagopalan, R. Srikant, and Y. Xu. Mining newsgroups using networks arising from social behavior. In *Proc. of WWW*, 2003.
2. R. Agrawal and R. Srikant. Fast algorithms for mining association rules. In *Proc. of VLDB*, 1994.
3. T. Blischok. Every transaction tells a story. In *Chain Store Age Executive with Shopping Center Age, 71 (3)*, pages 50–57, 1995.
4. T. Brijs, B. Goethals, G. Swinnen, K. Vanhoof, and G. Wets. A data mining framework for optimal product selection in retail supermarket data: The generalized profset model. In *Proc. of SIGKDD*, 2000.
5. T. Brijs, G. Swinnen, K. Vanhoof, and G. Wets. Using association rules for product assortment decisions: A case study. In *Proc. of SIGKDD*, 1999.
6. P. Domingos and M. Richardson. Mining the network value of customers. In *Proc. of SIGKDD*, 2001.
7. M.R. Garey and D.S. Johnson. Computers and intractability: A guide to the theory of np-completeness. Freeman, 1979.
8. J. Han, J. Pei, and Y. Yin. Mining frequent patterns without candidate generation. In *Proc. of SIGMOD*, 2000.
9. D. Kempe, J. Kleinberg, and E. Tardos. Maximizing the spread of influence through a social network. In *Proc. of SIGKDD*, 2003.
10. J. Kleinberg, C. Papadimitriou, and P. Raghavan. A microeconomic view of data mining. In *Knowledge Discovery and Data Mining*, volume 2:4, pages 254–260. Kluwer Academic Publishers, 1998.
11. Ron Kohavi, Carla Brodley, Brian Frasca, Llew Mason, and Zijian Zheng. KDD-Cup 2000 organizers' report: Peeling the onion. *SIGKDD Explorations*, 2(2):86–98, 2000.
12. H. Mannila. Methods and problems in data mining. In *Proc. of Int. Conf. on Database Theory*, 1997.
13. H. Mannila, H. Toivonen, and A. I. Verkamo. Efficient algorithms for discovering association rules. In *Proc. of KDD*, 1994.
14. M. Richardson and P. Domingos. Mining knowledge-sharing sites for viral marketing. In *Proc. of SIGKDD*, 2002.
15. B.W. Taylor. Chapter 16: Inventory management. In *Introduction to Management Science, 7th Edition.* Prentice Hall, 2001.
16. K. Wang and M.Y. Su. Item selection by "hub-authority" profit ranking. In *Proc. of SIGKDD*, 2002.
17. R.C.W. Wong, A.W.C. Fu, and K.Wang. Mpis: Maximal-profit item selection with cross-selling considerations. In *ICDM*, 2003.
18. X. Wu, A. Tuzhilin, and J. Shavlik (Ed.). Third IEEE international conference on data mining, ICDM. 2003.

Efficient Pattern-Growth Methods for Frequent Tree Pattern Mining*

Chen Wang[1], Mingsheng Hong[1], Jian Pei[2], Haofeng Zhou[1], Wei Wang[1], and Baile Shi[1]

[1] Fudan University, China
{chenwang, 9924013, haofzhou, weiwang1, bshi}@fudan.edu.cn
[2] State University of New York at Buffalo, USA. jianpei@cse.buffalo.edu

Abstract. Mining frequent tree patterns is an important research problems with broad applications in bioinformatics, digital library, e-commerce, and so on. Previous studies highly suggested that pattern-growth methods are efficient in frequent pattern mining. In this paper, we systematically develop the pattern growth methods for mining frequent tree patterns. Two algorithms, Chopper and XSpanner, are devised. An extensive performance study shows that the two newly developed algorithms outperform TreeMinerV [13], one of the fastest methods proposed before, in mining large databases. Furthermore, algorithm XSpanner is substantially faster than Chopper in many cases.

1 Introduction

Recently, many emerging application domains encounter tremendous demands and challenges of discovering knowledge from complex and semi-structural data. For example, one important application is mining semi-structured data [2,4,7,11, 12,13]. In [12], Wang and Liu adopted an Apriori-based technique to mine frequent path sets in ordered trees. In [7], Miyahara et al. used a directly generate-and-test method to mine tree patterns. Recently, Zaki [13] and Asai et al. [2] proposed more efficient algorithms for frequent subtree discovery in a forest, respectively. They adopted the method of rightmost expansion, that is, their methods add nodes only to the rightmost branch of the tree.

Recently, some interesting approaches for frequent tree pattern mining have been proposed. Two typical examples are reported in [6,13]. These methods observe the *Appriori* property among the frequent tree sub-patterns: *every non-empty subtree of a frequent tree pattern is also frequent.* Thus, they smartly extend the candidate-generation-and-test approach to tackle the mining. The method for frequent tree pattern mining is efficient and scalable when the patterns are not too complex. Nevertheless, if there are many complex patterns in

* This research is supported in part by the Key Program of National Natural Science Foundation of China (No. 69933010), China National 863 High-Tech Projects (No. 2002AA4Z3430 and 2002AA231041), and US NSF grant IIS-0308001.

H. Dai, R. Srikant, and C. Zhang (Eds.): PAKDD 2004, LNAI 3056, pp. 441–451, 2004.

the data set, there can be a huge number of candidates need to be generated and tested. That may degrade the performance dramatically.

Some previous studies strongly indicate that the depth-first search based, pattern-growth methods, such as FP-growth [5], TreeProjection [1] and H-mine [8] for frequent itemset mining, and PrefixSpan [9] for sequential pattern mining, can mine long patterns efficiently from large databases. That stimulates our thinking: "*Can we extend the pattern-growth methods for efficient frequent tree mining?*" This is the motivation of our study.

Is it straightforward to extend the pattern-growth methods to mine tree patterns? Unfortunately, the previously developed pattern-growth methods cannot be extended simply to tackle the frequent tree pattern mining problem efficiently. There are two major obstacles. On the one hand, one major cost in frequent tree pattern mining is to test whether a pattern is a subtree of an instance in the database. New techniques must be developed to make the test efficient. On the other hand, there can be many possible "growing points" (i.e., possible ways to extend an existing pattern to more complex ones) in a tree pattern. It is non-trivial to determine the "good" growth strategy and avoid redundance.

In this paper, we systematically study the problem of frequent tree pattern mining and develop two novel and efficient algorithms, *Chopper* and *XSpanner*, to tackle the problem. In algorithm *Chopper*, the mining of sequential patterns and the extraction of frequent tree patterns are separated as two phases. For each sequential pattern, *Chopper* generates and tests all possible tree patterns against the database. In algorithm *XSpanner*, the mining of sequential patterns and the extraction of frequent tree patterns are integrated. Larger frequent tree patterns are "grown" from smaller ones.

Based on the above ideas, we develop effective optimizations to achieve efficient algorithms. We compare both *Chopper* and *XSpanner* with algorithm *TreeMinerV* [13], one of the best algorithms proposed previously, by an extensive performance study. As an Apriori-based algorithm, *TreeMinerV* achieves the best performance for mining frequent tree patterns among all published methods. Our experimental results show that both *Chopper* and *XSpanner* outperform *TreeMinerV* while the mining results are the same. *XSpanner* is more efficient and more scalable than *Chopper*.

The remainder of this paper is organized as follows. In Section 2, we define the problem of frequent tree pattern mining. Algorithm *Chopper* and *XSpanner* are developed in Section 3 and 4 respectively. In Section 5, we present the results on synthetic and real dataset via comparing with *TreeMinerV*. Section 6 concludes the paper.

2 Problem Definition

A *tree* is an acyclic connected graph. In this paper, we focus on *ordered, labelled, rooted trees*. A tree is denoted as $T(v_0, N, L, E)$, where (1) $v_0 \in N$ is the *root node*; (2) N is the set of *nodes*; (3) L is the set of *labels of nodes*, for any node

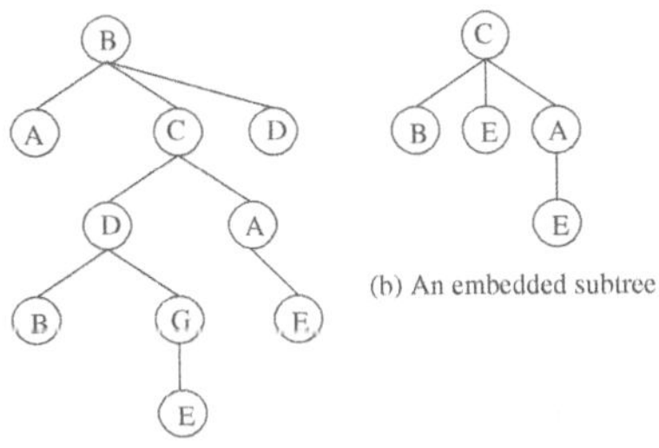

(a) Tree T

(b) An embedded subtree

Fig. 1. Tree and embedded subtree.

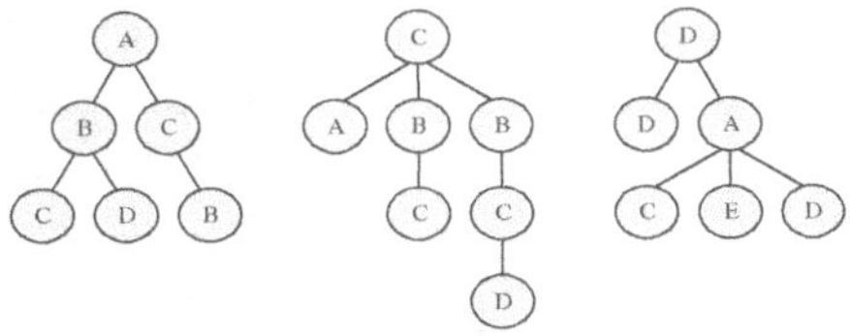

TID	Tree
1	A1B2C3D3C2B3
2	C1A2B2C3B2C3D4
3	D1D2A2C3E3D3

Fig. 2. An example of tree database TDB.

$u \in N$, $L(u)$ is the label of u; and (4) E is the set of *edges* in the tree. Please note that two nodes in a tree may carry the identical label.

Let v be a node in a tree $T(v_0, N, L, E)$. The *level* of v is defined as the length of the shortest path from v_0 to v. The *height* of the tree is the maximum level of all nodes in the tree. For any nodes u and v in tree $T(v_0)$, if there exists a path v_0-...-u-...-v such that every edge in the path is distinct, then u is called an *ancestor* of v and v a *descendant* of u. Particularly, if (u, v) is an edge in the tree, then u is the *parent* of v and v is a *child* of u. For nodes u, v_1 and v_2, if u is the parent of both v_1 and v_2, then v_1 and v_2 are *siblings*. A node without any child is a *leaf node*, otherwise, it is an *internal node*. In general, an internal node may have multiple children. If for each internal node, all the children are ordered, then the tree is an *ordered tree*. We denote the k-th child of node u as $child^k(u)$. In the case that such a child does not exist, $child^k(u) = null$.

Hereafter, without special mention, all trees are labelled, ordered, and rooted. An example is shown in Figure 1(a). The tree is of height 4.

Given a tree $T(v_0, N, L, E)$, tree $T'(v'_0, N', L', E')$ is called an *embedded subtree* of T, denoted as $T' \sqsubseteq T$, if (1) $N' \sqsubseteq N$; (2) for any node $u \in N'$, $L(u) = L'(u)$; and (3) for every edge $(u, v) \in E'$ such that u is the parent of v, u is an ancestor of v in T. Please note that the concept embedded subtree defined here is different from the conventional one.[1] In Figure 1(b), an embedded subtree of tree T in Figure 1(a) is shown.

A *tree database* is a bag of trees. Given a tree database TDB, the *support* of a tree T is the number of trees in TDB such that T is an embedded subtree, i.e., $sup(T) = \|\{T' \in TDB | T \sqsubseteq T'\}\|$. Given a *minimum support threshold min_sup*, a tree T is called as a *frequent tree pattern* if $sup(T) \geq min\_sup$.

Problem statement. Given a tree database TDB and a minimum support threshold *min_sup*, The problem of *frequent tree pattern mining* is to find the complete set of frequent tree patterns from database TDB.

In an ordered tree, by a preorder traversal of all nodes in the tree, a *preorder traversal label sequence* (or *l-sequence* in short) can be made. For example, the

[1] Conventionally, a tree G' whose graph vertices and graph edges form subsets of the graph vertices and graph edges of a given tree G is called a subtree of G.

l-sequence of tree T in Figure 1(a) is $BACDBGEAED$. Preorder traversal sequences are not unique with respect to trees. That is, multiple different trees may result in an identical preorder traversal sequence. To overcome this problem, we can add the levels of nodes into the sequence and make up the *preorder traversal label-level sequence* (or *l^2-sequence* in short). For example, the l^2-sequence of tree T is $B0A1C1D2B3G3E4A2E3D1$. We have the following results.

Theorem 1 (Uniqueness of l^2-sequences). *Given trees $T_1(v_1, N_1, L_1, E_1)$ and $T_2(v_2, N_2, L_2, E_2)$, their l^2-sequences are identical if and only if T_1 and T_2 are isomorphic, i.e., there exists a one-to-one mapping $f : N_1 \to N_2$ such that (1) $f(v_1) = f(v_2)$; (2) for every node $u \in N_1$, $L_1(u) = L_2(f(u))$; and (3) for every edge $(u_1, u_2) \in E_1$, $(f(u_1), f(u_2)) \in E_2$.*

Lemma 1. *Let S be the l^2-sequence of tree $T(v_0, N, L, E)$.*

1. *The first node in S is v_0 whose level number is 0;*
2. *For every immediate neighbors $L(u)iL(v)j$ in S, $j \leq (i + 1)$; and*
3. *For nodes u and v such that u is the parent of v, $L(u)i$ $(i \geq 0)$ is the nearest left neighbor of $L(v)j$ in S such that $j = (i - 1)$.*

Although an l-sequence is not unique with respect to trees, it can serve as an *isomorph* for multiple trees. In other words, multiple l^2-sequences and thus their corresponding trees can be *isomers* of an l-sequence.

Given a sequence $S = s_1 \cdots s_n$. A sequence $S' = s'_1 \cdots s'_m$ is called a *subsequence* of S and S as a *super sequence* of S', denoted as $S' \sqsubseteq S$ if there exist $1 \leq i_1 < \cdots < i_m \leq n$ such that $s_{i_j} = s'_j$ for $(1 \leq j \leq m)$. Given a bag of sequences SDB, the *support* of S in SDB is number of S's super sequences in SDB, i.e., $sup(S) = \|\{S' \in SDB | S \sqsubseteq S'\}\|$.

Given a tree database TDB, the bag of the l-sequences of the trees in TDB form a sequence database SDB. We have the following interesting result.

Theorem 2. *Given a tree database TDB, let SDB be the corresponding l-sequence database. For any tree pattern T, let $l(T)$ be the l-sequence of T. Then, $sup(T) \leq sup(l(T))$; and T is frequent in TDB only if $l(T)$ is frequent in SDB.*

Theorem 2 provides an interesting heuristic for mining frequent tree patterns: we can first mine the sequential patterns in the l-sequence database, and then mine tree patterns accordingly. In particular, a sequential pattern in the l-sequence database (with respect to the same support threshold in both tree database and l-sequence database) is called an *l-pattern*.

Given a tree T, not every subsequence of T's l-sequence corresponds to an embedded subtree of T. For example, consider the tree T in Figure 1(a). The l-sequence is $BACDBGEAED$. CBD is a subsequence. However, there exists no an embedded subtree T' in T such that its l-sequence is CBD.

Fortunately, whether a subsequence corresponds to an embedded subtree can be determined easily from an l^2-sequence. Let S be the l^2-sequence of a tree T. For a node v i in S, the *scope* of v is the longest subsequence S'

Input: a tree database TDB and a support threshold min_sup;
Output: all frequent tree patterns with respect to min_sup;
Method:
(1) scan database TDB once, generate its l-sequence database SDB;
(2) mine l-patterns from SDB using algorithm l-PrefixSpan;
(3) scan TDB, generate candidates according to l-patterns and find frequent tree;

Fig. 3. Algorithm Chopper

starting from v such that the label-level of each node in S', except for v itself, is greater than i. For example, the l^2 sequence of tree T in Figure 1(a) is $B0A1C1D2B3G3E4A2E3D1$. The scope of $B0$ is $B0A1C1D2B3G3E4A2E3D1$ and the scope of $D2$ is $B3G3E4$. We have the following result.

Lemma 2. *Given a tree T. An l-sequence $S = v_1 \ldots v_k$ corresponds to an embedded subtree in T if and only if there exists a node v_1 in the l^2-sequence of T such that $v_2 \ldots v_k$ is a subsequence in the scope of a node v_1.*

3 Algorithm Chopper

Algorithm *Chopper* is shown in Figure 3. The correctness of the algorithm follows Theorem 2. In the first 2 steps, *Chopper* finds the sequential patterns (i.e., l-patterns) from the l-sequence database. Based on Theorem 2, we only need to consider the trees whose l-sequence is an l-pattern. In the last step, *Chopper* scans the database to generate candidate tree patterns and verify the frequent tree patterns.

To make the implementation of *Chopper* as efficient as possible, several techniques are developed. The first step of *Chopper* is straightforward. Since the trees are stored as l^2-sequences in the database, *Chopper* does not need to form the explicit l-sequence database SDB. Instead, it uses the l^2-sequence database and just ignores the level numbers.

In the second step of *Chopper*, we need to mine sequential patterns (i.e., l-patterns) from the l-sequence database. A revision of algorithm PrefixSpan [9], called l-PrefixSpan, is used. Some specific techniques have been developed to enable efficient implementation of PrefixSpan. Interested readers should refer to [9] for a detailed technical discussion.

While PrefixSpan can find the sequential patterns, the tree pattern mining needs only part of the complete set as l-patterns. One key observation here is that *only those l-patterns having a potential frequent embedded tree pattern should be generated.* The idea is illustrated in the following example.

Example 1. Figure 2 shows a tree database as the running example in this paper. Suppose that the minimum support threshold is 2, i.e., every embedded subtree is a frequent tree pattern if it appears at least in two trees in the database.

The l^2-sequences of the trees are also shown in Figure 2. If we ignore the level numbers in the l^2-sequences, we get the l-sequences. Clearly, sequence $\langle BCB \rangle$ appears twice in the l-sequence database, and thus it is considered as a sequential pattern by PrefixSpan. However, we cannot derive any embedded tree in the database whose l-sequence is BCB. Thus, such patterns should not be generated.

We revise PrefixSpan to l-PrefixSpan to mine only the promising l-patterns and prune the unpromising ones. In l-PrefixSpan, when counting $sup(S)$ for a possible sequential pattern S from the (projected) databases, we count only those trees containing an embedded subtree whose l-sequence is S. Following Lemma 2, we can determine whether an l-sequence corresponds to an embedded subtree in a tree easily. For example, $sup(BCB) = 0$ and $sup(ABC) = 1$. Thus, both BCB and ABC will be pruned in l-PrefixSpan.

It can be verified that many patterns returned by PrefixSpan, such as AB, ABC, $ABCD$, etc., will be pruned in l-PrefixSpan. In our running example, only 9 l-patterns are returned, i.e., A, AC, AD, B, BC, BCD, BC, C and D.

In the last step of *Chopper*, a naïve implementation would be as follows. We can generate all possible tree patterns as candidates according to the l-patterns found by l-PrefixSpan, and then scan the tree database once to verify them. Unfortunately, such a naïve method does not scale well for large databases: There can be a huge number of candidate tree patterns!

Chopper adopts a more elegant approach here. It scans the tree database against the l-patterns found by l-PrefixSpan. For each tree in the tree database, *Chopper* firstly verifies whether the tree contains some candidate tree patterns. If so, the counters of the tree patterns will be incremented by one. Then, *Chopper* also verifies whether more tree candidate patterns corresponding to some l-patterns can be generated from the current tree. If so, they will be generated and counters will be set up with an initial value 1. Moreover, to facilitate the matching between a tree in the tree database and the l-patterns as well as candidate tree patterns, all l-patterns and candidate tree patterns are indexed by their prefixes. Please note that a tree is stored as an l^2-sequence in the database.

The major cost of *Chopper* comes from two parts. On the one hand, l-PrefixSpan mines the l-patterns. The pruning of unpromising l-patterns improves the performance of the sequential pattern mining here. On the other hand, *Chopper* has to check every tree against the l-patterns and the candidate tree patterns in the last step. In this step, only one scan needed.

Although l-PrefixSpan can prune many unpromising l-patterns, some unpromising l-patterns still may survive from the pruning, such as BCD in our running example. The reason is that the l-pattern mining process and the tree pattern verification process are separated in *Chopper*. Such unpromising l-patterns may bring unnecessary overhead to the mining. *Can we integrate the l-pattern mining process into the tree pattern verification process so that the efficiency can be improved further?* This observation motivates our design of *XSpanner*.

Input and output: same as Chopper;
Method:

```
(1) scan database TDB once, find frequent length-1 l-patterns;
(2) for each length-1 l-pattern x_i, do
(3)     output a frequent tree pattern x_i0;
(4)     form the <x_i0>-projected database TDB_{x_i0};
(5)     if TDB_{x_i0} has at least min_sup trees then mine the projected database;
```

Fig. 4. Algorithm XSpanner

4 Algorithm XSpanner

Algorithm *XSpanner* is shown in Figure 4. Clearly, by scanning the tree database TDB only once, we can find the frequent items in the database, i.e., the items appearing in at least min_sup trees in the database. They are length-1 l-patterns. This is done in Step (1) in algorithm XSpanner.

Suppose that $x_1, \ldots, x_m$ are the m length-1 l-patterns found. We have the following two claims. On the one hand, for $(1 \leq i \leq m)$, x_i0 is a frequent tree pattern in the database. On the other hand, the complete set of frequent tree patterns can be divided into m exclusive subsets: the i-th subset contains the frequent tree patterns having x_i as the root.

Example 2. Let us mine the frequent tree patterns in the tree database TDB in Figure 2. By scanning TDB once, we can find 4 length-1 l-patterns, i.e., A, B, C and D. Clearly, $A0$, $B0$, $C0$ and $D0$ are the four frequent tree patterns. On the other hand, the complete set of tree patterns in the database can be divided into 4 exclusive subsets: the ones having A, B, C and D as the root, respectively.

The remainder of *XSpanner* (line (2)-(5)) mines the subsets one by one.

For each length-1 l-pattern x_i, *XSpanner* first outputs a frequent tree pattern x_i0 (line (3)). Clearly, to find frequent tree patterns having x_i0 as a root, we need only the trees containing x_i. Moreover, for each tree in the tree database containing x_i, we need only the subtrees having x_i as a root. We collect such subtrees as the $\langle x_i0 \rangle$*-projected database*. This is done in line (4) in the algorithm.

In implementation, constructing a physical projected database can be expensive in both time and space. Instead, we apply the *pseudo-projection techniques* [9,8]. The idea is that, instead of physically constructing a copy of the subtrees, we reuse the trees in the original tree database. For each node, the tree id and a hyperlink are attached. By linking the nodes labelled x_i together using the hyperlinks, we can easily get the $\langle x_i0 \rangle$-projected database. Please note that such hyperlinks can be reused latter to construct other projected databases.

Example 3. Figure 5 shows the $B0$-projected database. Basically, all the subtrees rooted at a node B are linked, except for the leaf nodes labelled B. The leaf node labelled B in the left tree is not linked.

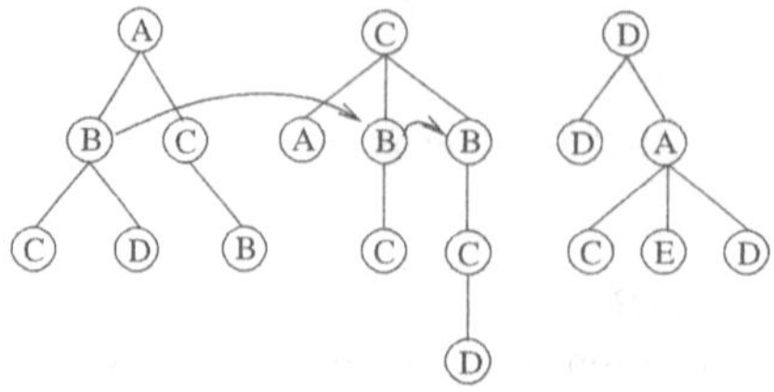

Fig. 5. The $B0$-projected database

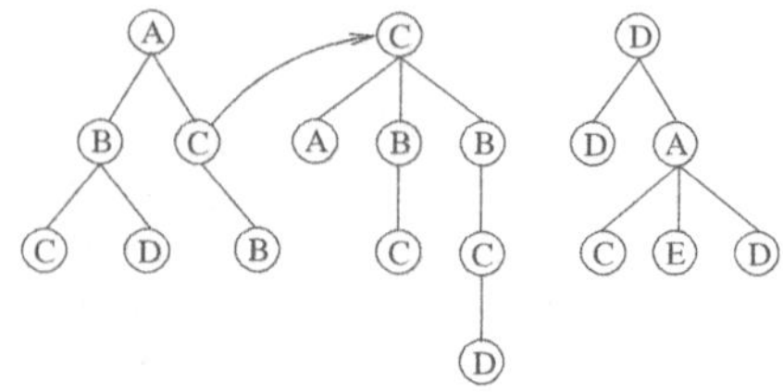

Fig. 6. The $C0$-projected database

Why the leaf nodes should not be linked? The leaf node cannot contain any embedded subtree larger than the tree pattern we have got so far. In other words, not linking such leaf nodes will not affect the result of the mining. Thus, it is safe to prune them in the projected database.

Please note that there can be more than one node with the same label in a tree, such as the tree at the middle of Figure 5. These subtrees are processed as follows: a node labelled with B is linked only if it is not a descendant of some other node that is also labelled with B. As another example, Figure 6 shows the $C0$-projected database. In the tree at the middle, only the root node is linked.

As shown in Figure 5, more than one subtree from a tree in the tree database may be included in the projected database. When counting the support of a tree pattern, such as BC, the pattern should gain support 1 from the same original tree, even it may be matched in multiple subtrees.

The projected databases can be mined recursively. In the $\langle x_i 0 \rangle$-projected database, the length-1 l-patterns should be found. For each length-1 l-pattern x_j, $x_i 0 x_j 1$ is a frequent tree pattern. The set of frequent tree patterns having x_i as a root can be divided into smaller subsets according to x_j's.

In general, suppose that P is a frequent tree pattern and P-projected database is formed. Let S be the l-pattern of P. By scanning the P-projected database once, *XSpanner* finds frequent items in the projected database. Then, for each frequent item x_i, *XSpanner* checks whether Sx_i is an l-pattern in the P-projected database and there exists a frequent tree pattern corresponds to Sx_i. If so, then the new frequent tree pattern is output and the recursive mining continues. Otherwise, the search in the branch is terminated.

The correctness of the algorithm can be proved. Limited by space, we omit the details here.

5 Experiments and Performance Study

In this section, we will evaluate the performance of *XSpanner* and *Chopper* in comparison with *TreeMinerV* [13]. All the experiments are performed on a Pentium IV 1.7GHz PC machine with 512MB RAM. The OS platform is Linux Red Hat 9.0 and the algorithms are implemented in C++.

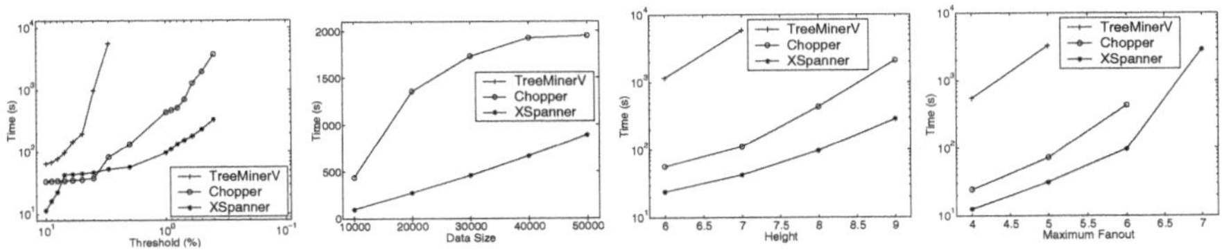

(a) Sup vs. time (b) Size vs. time (c) Height vs. time (d) Fanout vs. time

Fig. 7. Results on synthetic data sets

We wrote a synthetic data generation program to output all the test data. There are 8 parameters for adjustment: the number of the labels $|S|$, the probability threshold of one node in the tree to generate children or not p, the number of the basic pattern trees (BPT) $|L|$, the average height of the BPT $|I|$, the maximum fanout(children) of nodes in the BPT $|C|$, the data size of synthetic trees $|N|$, the average height of synthetic trees $|H|$ and the maximum fanout of nodes $|F|$ in synthetic trees. The actual height of each (basic pattern) tree is determined by the Gaussian distribution having the average of $|H|(|I|)$ and the standard deviation of 1.

At first, we consider the scalability with $minsup$ of the three algorithms, while other parameters are:$S = 100, p = 0.5, L = 10, I = 4, C = 3, N = 10000, H = 8, F = 6$. Figure 7(a) shows the result, where the $minsup$ is set from 0.1 to 0.004. In this figure, both X and Y axes have been processed by $\log_{10} T$ for the convenience of observation.

From the figure 7(a), we can find that when support threshold is larger than 5%, the three algorithms perform approximately the same. With the threshold becoming smaller, the algorithm *XSpanner* and *Chopper* begin to outperform *TreeMinerV*. What should be explained here is the reason when the threshold changes from 2% to 1%, the running time went up suddenly. We choose 10000 trees as our test dataset; however, there are only 100 node labels. So much more frequent structures are generated, which leads to the greater time consuming. During this period, we can find that *TreeMinerV* runs out of memory, while *XSpanner* and *Chopper* retain the ability to finish computing. This is because of the large amount of candidates generated by Apriori-based algorithms. It is also clear that when the threshold continues to decrease, *XSpanner* surpasses *Chopper* in performance. *XSpanner* estimates and differentiates between the isomers during the generation of frequent sequences, which can reduce a large number of frequent sequences but infrequent substructures generated when the threshold is low.

Then, figures 7((b) shows the scalability with data size. The data size N varies from 10000 to 50000, while other parameters are:$S = 100, p = 0.5, L = 10, I = 4, C = 3, H = 8, F = 6, minsup = 0.01$. Here we find the cost of both time and space of *XSpanner* and *Chopper* is extremely smaller than that of *TreeMinerV* which is halted for memory overflow. The reason is that *XSpanner*

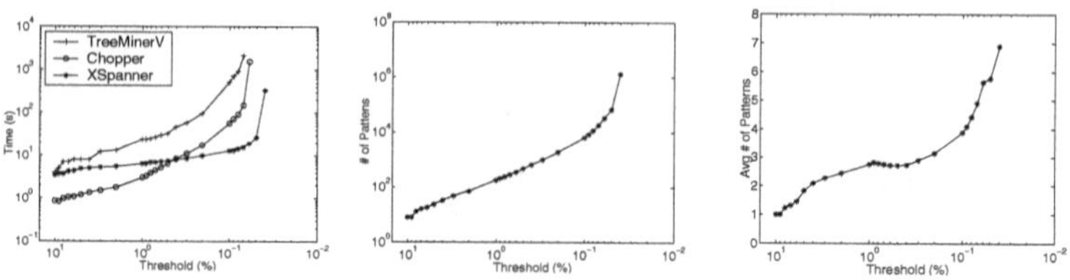

(a) Sup vs. time (b) Sup vs. # patterns (c) Sup vs. avg # nodes

Fig. 8. Results on real data sets

and *Chopper* can save time and space cost by avoiding false candidate subtree generation.

Finally, the scalability with tree size is shown in Figures 7(c) and (d). The Tree size of H and F varies, while other parameters are: $S = 100, p = 0.5, L = 10, I = 4, C = 3, N = 10000, minsup = 0.01$. In figure 7(c), we only vary H from 6 to 9. It is easy to find that, when H equals to 6 or 7, the performance of *XSpanner* and *Chopper* is better than that of *TreeMinerV*. However, when the trees become higher, the superiority grows. In particular, when H equals to 8 or 9, the two algorithms thoroughly defeat *TreeMinerV* for the reason that *TreeMinerV* is halted for memory overflow. In figure 7(d), the performance of the two algorithms and *TreeMinerV* is similar to the case above. *XSpanner* and *Chopper* performs better than *TreeMinerV*, while the fanout continues to increase.

From all of the experiments we did, we can conclude that it is an acceptable and efficient way to put the process of distinguishing isomers in the process of generating frequent sequences.

We also tested *XSpanner* and *Chopper* in Web Usage Mining. We downloaded the Weblog of Hyperreal (http://music.hyperreal.org), chose those dating from Sep. 10 to Oct.9 in 1998 as the input data. and then transformed the Weblog into tree-like data set which includes 12000 more records totally.

Figure 8(a) shows the performance of the three algorithms, where the $minsup$ is set from 0.1 to 0.0003. In this figure, both X and Y axes have been processed by $\log_{10} T$ for the convenience of observation. We can find, that the performance of *XSpanner* and *Chopper* is better than that of *TreeMinerV*. Especially, *TreeMinerV* is halted in 3 hours for memory overflow when $minsup = 0.0006$, while the two algorithms go well. From the figure, we notice that the performance of *XSpanner* is more stable than that of *Chopper*. With threshold decreasing, *XSpanner* surpasses *Chopper* gradually. It should also be noted that, *XSpanner* does not perform excellently until $minsup$ is dropped to 0.0003.

Finally, figure 8(b) shows number of frequent patterns generated by the algorithm, while figure 8(c) shows average number of nodes of frequent patterns generated by the algorithm, where the $minsup$ is set from 0.1 to 0.0003.

6 Conclusions

In this paper, we present two pattern-growth algorithms, *Chopper* and *XSpanner*, for mining frequent tree patterns. In the future, we would like to explore pattern-growth mining of other complex patterns, such as frequent graph patterns.

References

1. R. C. Agarwal, et al. A tree projection algorithm for generation of frequent item sets. *J. of Parallel and Distributed Computing*, 61(3):350–371, 2001.
2. T. Asai, et al. Efficient substructure discovery from large semi-structured data. In *Proc. 2002 SIAM Int. Conf. Data Mining*, Arlington, VA.
3. D. Cook and L. Holder. Substructure discovery using minimal description length and background knowledge. *J. of Artificial Intelligence Research*, 1:231–255, 1994.
4. L. Dehaspe, et al. Finding frequent substructures in chemical compounds. In *KDD'98*, New York, NY.
5. J. Han, et al. Mining frequent patterns without candidate generation. In *SIGMOD'00*, Dallas, TX.
6. M. Kuramochi and G. Karypis. Frequent subgraph discovery. In *ICDM'01*, San Jose, CA.
7. T. Miyahara, et al. Discovery of frequent tree structured patterns in semistructured web documents. In *PAKDD'01*, Hong Kong, China.
8. J. Pei, et al. H-Mine: Hyper-structure mining of frequent patterns in large databases. In *ICDM'01*, San Jose, CA.
9. J. Pei, et al. PrefixSpan: Mining sequential patterns efficiently by prefix-projected pattern growth. In *ICDE'01*, Heidelberg, Germany.
10. R. Srikant and R. Agrawal. Mining generalized association rules. In *VLDB'95*, Zurich, Switzerland.
11. J.T.L. Wang, et al. Automated discovery of active motifs in multiple RNA secondary structures. In *KDD'96*, Portland, Oregon.
12. K. Wang and H. Liu. Schema discovery for semistructured data. In *KDD'97*, Newport Beach, CA.
13. M.J. Zaki. Efficiently mining frequent trees in a forest. In *KDD'02*, Edmonton, Alberta, Canada.

Mining Association Rules from Structural Deltas of Historical XML Documents

Ling Chen, Sourav S. Bhowmick, and Liang-Tien Chia

School of Computer Engineering,
Nanyang Technological University, Singapore, 639798, SINGAPORE

Abstract. Previous work on XML association rule mining focuses on mining from the data existing in XML documents at a certain time point. However, due to the dynamic nature of online information, an XML document typically evolves over time. Knowledge obtained from mining the evolvement of an XML document would be useful in a wide range of applications, such as XML indexing, XML clustering. In this paper, we propose to mine a novel type of association rules from a sequence of changes to XML structure, which we call XML Structural Delta Association Rule (*XSD-AR*). We formulate the problem of *XSD-AR* mining by considering both the frequency and the degree of changes to XML structure. An algorithm, which is derived from the FP-growth, and its optimizing strategy are developed for the problem. Preliminary experiment results show that our algorithm is efficient and scalable at discovering a complete set of *XSD-AR*s.

1 Introduction

XML is rapidly emerging as the *de facto* standard for data representation and exchange on the Web. With the ever-increasing amount of available XML data, the data mining community has been motivated to discover valuable knowledge from collections of XML documents. As one of the most important techniques of data mining, association rule mining has also been introduced into XML repository [2] [4]. Currently, two types of data in XML has been exploited to mine rules: XML content and XML structure. The former aims to discover associations between frequent data values [2]; whereas the latter focuses on discovering associations between frequent substructures [4]. Existing work on XML association rule mining study the data (content or structure) in XML documents at a certain point in time. However, due to the dynamic nature of online information, XML documents typically evolve over time. Changes to XML documents can be divided similarly into changes to XML content (also called *content deltas*, i.e., modifications of element values) and changes to XML structure (also called *structural deltas*, i.e., additions or deletions of elements). From the sequence of historical versions of the XML file, users might be interested in the associations between content or structural deltas. In this paper, we focus on mining association rules from XML structural deltas, which we refer to as *XSD-AR* (XML Structural Delta Association Rule). Given a sequence of historical versions of

H. Dai, R. Srikant, and C. Zhang (Eds.): PAKDD 2004, LNAI 3056, pp. 452–457, 2004.

an XML document, the goal of *XSD-AR* mining is to discover when some substructures of the XML document change, some other substructures change as well with certain probability. The contributions of this paper are summarized as follows: a)A new problem of association rule mining from structural deltas of historical XML documents is formally defined; b)An algorithm, derived from the FP-growth algorithm, and its optimizing strategy are developed to handle this problem; c) Experiments are conducted to evaluate the performance of the algorithm and the effect of the optimizing strategy.

2 Problem Statement

In this section, we first describe basic change operations which result in XML structural deltas. Then, we define several metrics to measure the degree of change and the frequency of change. Finally, the problem of *XSD-AR* is defined formally.

An XML document can be represented as a tree according to Document Object Model (DOM) specification. In this paper, we will focus on unordered XML tree. In the context of *XSD-AR* mining, we consider the following two basic change operations: *Insert(X(name, value), Y)*, which creates a new node X, with node name "name" and node value "value", as a child node of node Y in an XML tree structure and *Delete(X)*, which removes node X from an XML tree structure.

Now we introduce the metrics which measure the degree of change for each single subtree and the frequency of change for a set of subtrees. Interesting substructure patterns and association rules can then be identified based on these metrics.

Degree of Change. Let $<t_i,\ t_{i+1}>$ be two historical versions of a subtree t in an XML tree structure T. Let $|C_i|$ be the number of basic change operations which change the structure of t from the ith version to the $(i+1)$th version. Then the *degree of change* for subtree t from version i to version $(i+1)$ is:

$$DoC(t) = \frac{|C_i|}{|t_i \cup t_{i+1}|}$$

where $t_i \cup\ t_{i+1}$ is the set of unique nodes of tree t in ith version and $(i+1)$th version. □

Frequency of Change. Let $<T_1, T_2, \ldots T_n>$ be a sequence of historical versions of an XML tree structure. Let $<\Delta_1, \Delta_2, \ldots \Delta_{n-1}>$ be the sequence of *deltas* generated by comparing each pair of successive versions, where Δ_i $(1 \leq i \leq n\text{-}1)$ consists of subtrees changed in two versions. Let S be a set of subtrees, $S=\{t_1, t_2, \ldots t_m\}$, where $\forall$ j $(1 \leq j \leq m)$, $\exists$ i $(1 \leq i \leq n\text{-}1)$ s.t. $t_j \in \Delta_i$. Let DoC_{j_i} be the *degree of change* for subtree t_j from ith version to $(i+1)$th version. The *FoC* of the set S is:

$$FoC(S) = \frac{\sum_{i=1}^{n-1} V_i}{n-1} \ where\ V_i = \prod_{j=1}^{m} V_{j_i} \ and\ V_{j_i} = \begin{cases} 1, \text{if } DoC_{j_i} \neq 0 \\ 0, \text{if } DoC_{j_i} = 0 \end{cases} \quad 1 \leq j \leq m \square$$

Weight. Furthermore, in order to measure how significantly subtrees in a set usually changes, we define *Weight* of a set, which compares the *DoC* of subtrees against some user-defined interesting value α:

$$Weight(S) = \frac{\sum_{i=1}^{n-1} D_i}{(n-1) * FoC(S)},$$

$$where\ D_i = \prod_{j=1}^{m} D_{j_i}\ and\ D_{j_i} = \begin{cases} 1, \text{if DoC}_{j_i} \geq \alpha \\ 0, otherwise \end{cases} \quad 1 \leq j \leq m \square$$

Frequent Subtree Pattern. Then, we define a set of subtrees $S=\{t_1, t_2, ..., t_m\}$ as a *frequent subtree pattern* if subtrees in the set frequently change together and they usually change significantly when they change together. That is,

- *FoC* of the set is no less than some user defined minimum *FoC* β, $FoC(S) \geq \beta$.
- *Weight* of the set is no less than some user defined minimum *Weight* γ, $weight(S) \geq \gamma$. □

Confidence. *XSD-ARs* then can be derived from *frequent subtree patterns*. The metric *Confidence* reflects the conditional probability that significant changes of a set of subtrees lead to significant changes of another set of subtrees:

$$Confidence(X \Rightarrow Y) = \frac{FoC(X \cup Y) * Weight(X \cup Y)}{FoC(X) * Weight(X)} \square$$

Problem Definition. The problem of *XSD-AR* mining can be formally stated as follows: Let $<T_1, T_2, ...\ T_n>$ be a sequence of historical versions of an XML tree structure. Let $<\Delta_1, \Delta_2, ..., \Delta_{n-1}>$ be the sequence of *deltas*. A **structural delta database** *SDDB* can be formed, where each tuple <*DID*, *SID*, *DoC*> comprises of a delta identifer, a subtree identifier and a *degree of change* for the subtree. Let $S=\{t_1, t_2, ... t_m\}$ be the set of changed subtrees s.t. each $\Delta \subseteq S$. An *XSD-AR*: $A \Rightarrow B$ is an implication between two subtree sets A and B where $A \cap B = \emptyset$. Given a *FoC* threshold β, a *Weight* threshold γ and a *Confidence* threshold ξ, the **problem of *XSD-AR* mining** is to find the complete set of *XSD-AR*s: $\{A \Rightarrow B | FoC(A \cup B) \geq \beta,\ Weight(A \cup B) \geq \gamma,\ Confidence(A \Rightarrow B) \geq \xi\}$.

3 Algorithm: Weighted-FPgrowth

In this section, we present the procedure of mining *XSD-AR*s. Given a sequence of historical versions of an XML document, three phases are involved in mining the set of *XSD-AR*s: a) *SDDB* construction; b) *Frequent Subtree Pattern* Discovery and c) *XSD-AR*s Derivation. Since phase *a*) and *c*) can be handled in a straightforward way based on known conditions, subsequent discussion will be focused on phase *b*) to discover the set of *frequent subtree patterns*.

DID	SID	DID	SID	DID	SID
1	t_3'	2	t_9	4	t_6'
1	t_6	3	t_3	4	t_2
1	t_9'	3	t_6'	5	t_3'
2	t_3	3	t_2'	5	t_2
2	t_6	4	t_3	5	t_9

(a)

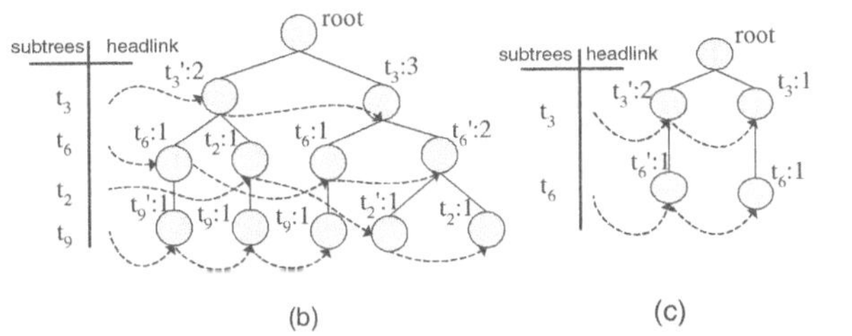

Fig. 1. Weighted FP-tree and Conditional Weighted-FPtree

Data Structure. Consider that each subtree in a *delta* has two states: its *DoC* is either less than the user-defined minimum *DoC* α or no less than α. We use a pair of identifiers to represent the two states of a subtree. Given a subtree t_i, if its *DoC* is less than α, we use t_i' to represent it in this *delta*; otherwise, we use the original identifier t_i. For instance, suppose the threshold of *FoC* and *Weight* are 40% and 50% respectively. *Weighted-FPtree* constructed from the transformed *SDDB* in Figure 1 (a) is presented in Figure 1 (b).

Mining Procedure. We now explain how to mine *frequent subtree patterns* from *Weighted-FPtree*. The critical differences between *Weighted-FPgrowth* and original *FP-growth* are the way we calculate the *FoC* and the *weight* for patterns and the way we construct *conditional Weighted-FPtree*. In the following, we illustrate the algorithm and the differences with an example. Consider the last subtree in header table, t_9. $FoC(t_9)$ is 60% since the total number of occurrence of t_9 and t_9' in *Weighted-FPtree* is three. $Weight(t_9)$ is 66% since t_9 occurs twice. Hence, t_9 is frequent. There are three paths related to t_9: $<t_3'{:}1, t_6{:}1, t_9'{:}1>$, $<t_3'{:}1, t_2{:}1, t_9{:}1>$, and $<t_3{:}1, t_6{:}1, t_9{:}1>$ (the number after colon indicates the number of *deltas* in which the subtrees change together with t_9). Since the minimum *FoC* is 40% and the minimum *weight* is 50%, subtrees t_3 and t_6 may probably be frequent patterns with subtree t_9. Hence, the three prefix paths: $\{t_3'{:}1, t_6{:}1\}$, $\{t_3'{:}1\}$, and $\{t_3{:}1, t_6{:}1\}$, forms t_9*'s* conditional pattern base. We need to construct a *conditional Weighted-FPtree* for subtree t_9 from the three prefix paths. Note that in the first prefix path, subtree t_9 occurs as t_9', which means subtree t_9 does not change significantly with subtree t_3 and t_6 in this path. To record this fact, we need to replace t_6 with t_6'. In fact, when constructing *conditional Weighted-FPtree* for t_i, any node with identifier t_j in paths where subtree t_i occurs as t_i' should be replaced with t_j'. Then the *conditional Weighted-FPtree* of t_9 is shown in Figure 1 (c). Then frequent patterns related to t_9 can be mined in the same way. We consider to optimize the space cost of *Weighted-FPtree* by registering information in edges. Specifically, when a subtree changes significantly in this *delta*, the edge in *Weighted-FPtree* connecting it to its child is labelled as positive. Otherwise, the edge is labelled as negative. Then the number of nodes are reduced because any node in optimized *Weighted-FPtree* will never have more than one child node representing the same subtree, with which it changes significantly (insignificantly).

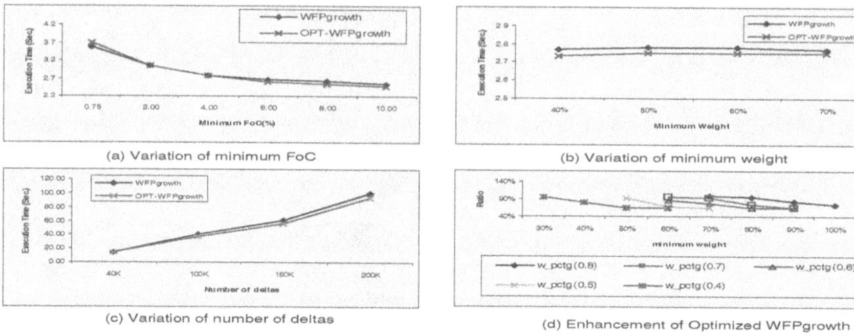

(a) Variation of minimum FoC

(b) Variation of minimum weight

(c) Variation of number of deltas

(d) Enhancement of Optimized WFPgrowth

Fig. 2. Experiment Results

4 Experiment Results

In this section, we study the performance of the proposed algorithms. The algorithms are implemented in Java. Experiments are performed on a Pentium IV 2.8GHz PC with 512 MB memory. The operating system is Windows 2000 professional. We implemented a synthetic *SDDB* generator by extending the one used in [1]. The default value for the set of deltas is 10K, the average size of a delta is 20 and the number of changed subtrees is 1000.

Methodology and Results. We carried out two experiments to evaluate the efficiency of *Wegithed-FPgrowth* by varying thresholds on *FoC* and *Weight* respectively. As shown in Figure 2 (a), *Weighted-FPgrowth* and its optimized version are more efficiency when the threshold turns to be greater. As shown in Figure 2 (b), the efficiency of both versions are similar and insensitive to the variation of minimum *Weight.* The scale-up features of the algorithms are tested against the number of deltas, which is varied from 40K to 200K. As shown in Figure 2 (c), when the number of deltas is larger, the more efficiency the optimized version can obtain by saving the number of nodes in the tree structure. Additionally, we calculate the compression ratio of the size of optimized *Weighted-FPtree* to the size of *Weighted-FPtree* against the mean value (*w_pctg*) of the number of subtrees which has changed greatly in a delta (0.4-0.8). As shown in Figure 2 (d), the compression ratio can be as high as 68%.

5 Conclusions and Future Work

This paper proposed a novel problem of association rule mining based on changes to XML structures: *XSD-AR*. An algorithm, *Weighted-FPgrowth*, and its optimizing strategy are designed to handle it. Experiment results demonstrated the efficiency and scalability of the algorithm. As ongoing work, we would like to collect real life data set to verify the semantic meaning of discovered *XSD-AR*.

References

1. R. Agrawal and R. Srikant. Fast algorithms for mining association rules in large databases. In *Proc. VLDB*, pages 487–499, 1994.
2. D. Braga, A. Campi, S. Ceri, M. Klemettinen, and P. L. Lanzi. A tool for extracting xml association rules from xml documents. In *Proc. IEEE ICTAI*, pages 57–65.
3. J.W. Han, J. Pei, and Y.W. Yin. Mining frequent patterns without candidate generation. In *Proc. ACM SIGMOD*, pages 1–12, 2000.
4. A. Termier, M.C. Rousset, and M. Sebag. Treefinder: A first step towards xml data mining. In *Proc. IEEE ICDE*, pages 450–457, 2002.

Data Mining Proxy: Serving Large Number of Users for Efficient Frequent Itemset Mining

Zhiheng Li[1], Jeffrey Xu Yu[1], Hongjun Lu[2], Yabo Xu[1], and Guimei Liu[2]

[1] Dept. of Systems Engg. & Engg. Mgt.,
The Chinese University Of Hongkong, Hongkong
{zhli,yu,ybxu}@se.cuhk.edu.hk

[2] Dept. of Computer Science,
The Hong Kong University of Science and Technology
{luhj,cslgm}@cs.cuhk.edu.hk

Abstract. Data mining has attracted a lot of research efforts during the past decade. However, little work has been reported on supporting a large number of users with similar data mining tasks. In this paper, we present a data mining proxy approach that provides basic services so that the overall performance of the system can be maximized for frequent itemset mining. Our data mining proxy is designed to fast respond a user's request by constructing the required tree using both trees in the proxy that are previously built for other users' requests and trees stored on disk that are pre-computed from transactional databases. We define a set of basic operations on pattern trees including tree projection and tree merge. Our performance study indicated that the data mining proxy significantly reduces the I/O cost and CPU cost to construct trees. The frequent pattern mining costs with the trees constructed can be minimized.

1 Introduction

Data mining is a powerful technology being widely adopted to help decision makers to focus on the most important nontrivial/predictive information/patterns that can be extracted from large amounts of data they continuously accumulate in their daily business operations. Finding frequent itemsets in a transactional database is a common task for many applications. Recent studies show that pattern-growth method is one of the most efficient methods for frequent pattern mining [1,2,3,4,5,6,7,8,9,10].

In this paper, we study a data mining proxy in a multi-user environment where a large number of users issue similar but different data mining queries from time to time. This work is motivated by the fact that in a large business enterprise a large number of users need to mine patterns according to their needs. Their needs may change from time and time, and a lot of similar queries are registered in a short time. It is undesirable to process these data mining queries one-by-one. In this paper, we focus on how to fast respond a user's data mining query by constructing a smallest but sufficient tree using both trees in a data

H. Dai, R. Srikant, and C. Zhang (Eds.): PAKDD 2004, LNAI 3056, pp. 458–463, 2004.

mining proxy that are previously built for other users' requests and trees stored on disk that are pre-computed for transactional databases. It has significant impacts on data mining. First, the I/O cost can be minimized, because trees do not need always to be constructed by loading data from disk. Second, the data mining cost can be reduced, because the mining cost largely depends on the tree size.

The rest of the paper is organized as follows. Section 2 introduces frequent mining queries and tree building requests. Section 3 gives an overview of the data mining proxy, followed by the definition of tree operations. We will report our experimental results in Section 4, and conclude the paper in Section 5.

2 Preliminaries

A *data mining query* is to mine frequent patterns from a transaction database TDB. Let τ be a given min-support threshold and V be an itemset. We consider three types of data mining queries: *Frequent Itemset Mining Query*, *Frequent super-itemset Mining Query* and *Frequent sub-itemset Mining Query*[1]. All data mining queries can be one of these three types or a combination of them. Each query is processed in two steps: i) constructing an initial *PP*-tree, and ii) mining on top of the *PP*-tree[2] being constructed. We call the former a *frequent tree building request*, which can be done by constructing a tree from either TDB or a previously built tree. The three corresponding types of frequent tree building requests are given as follows.

- **Frequent Itemset Tree Building Request**: constructing a tree in memory which is smallest and sufficient to mine frequent patterns whose support is greater than or equal to τ.
- **Frequent Super-itemset Tree Building Request**: constructing a tree in memory which is smallest and sufficient to mine frequent patterns that include items in V and have a support that greater than or equal to τ.
- **Frequent Sub-itemset Tree Building Request**: constructing a tree in memory which is smallest and sufficient to mine frequent patterns that are included in V and have a support that is greater than or equal to τ.

3 Data Mining Proxy

The data mining proxy is designed and developed to support a large number of users with similar data mining queries by fast responding a user's tree building request. The efficiency of tree building requests is achieved by utilizing both trees in the data mining proxy that are previously built for other users' requests and trees stored on disk that is pre-computed for transactional databases. In addition to efficiency issues, the effectiveness of data mining proxy is achieved by responding a smallest and sufficient tree for a users data mining query. It is

[1] The definitions of the three types of data mining queries are given in [10].

[2] *PP*-tree is introduced in [10].

because the cost of mining all possible patterns in a tree heavily depends on the tree size. The smaller the tree is, the less mining cost it occurs.

Let TDB be a transaction database that includes items in $I = \{x_1, x_2, \cdots, x_n\}$, where x_i is a 1-itemset.

Definition 1. *Given a minimum support τ and a set of items $V \subseteq I$. Let T, T_1 and T_2 be PP-trees. Three tree operations are defined.*

- *A sub-projection operation, denoted $\pi_{\tau,V}(T)$, is to project a subtree from the given tree T. The resulting tree includes all itemsets in V whose minimum support is greater than or equal to τ.*
- *A super-projection operation, denoted $\hat{\pi}_{\tau,V}(T)$, is to project a subtree from the given tree T. The resulting tree includes all itemsets which are a super set of V and have a minimum support that is greater than or equal to τ.*
- *A merge operation, denoted $T_1 \oplus T_2$, is to merge two trees, T_1 and T_2, and results a new tree.*

Given two *PP*-trees, T_i and T_j, we say $T_i \subseteq T_j$ if T_i is a subtree of T_j. More precisely, by $T_i \subseteq T_j$ we mean that every itemset, X, represented in T_i is also in T_j. As examples, if $T_i = \pi_{\tau,V}(T_j)$ then $T_i \subseteq T_j$, and if $T_k = T_i \oplus T_j$, then $T_i \subseteq T_k$ and $T_j \subseteq T_k$. Let T_I be the largest *PP*-tree that include every single $x_i \in I$ appearing in TDB. Obviously, any tree is a subtree of T_I. Consider a data mining query, q, and a mining algorithm, M. We denote the resulting frequent patterns for q as $M_q(\cdot)$. For two *PP*-trees, T_i and T_j, we say $T_i \equiv_q T_j$ if $M_q(T_i)$ is the same as $M_q(T_j)$. In other words, the two trees, T_i and T_j, will give the same frequent patterns for the same query q. The smallest tree for a data mining query is defined below.

Definition 2. *For a data mining query q with a minimum support τ and a set of items V, the smallest tree, T ($\subseteq T_I$), is a tree that satisfies the condition of $T \equiv_q T_I$, but does not satisfy $T \equiv_q T_I$ if any item is removed from T.*

A simple scenario using the tree operations to show the proxy functions is shown in Figure 1. Suppose we have constructed T_1 on disk, which is a *PP*-tree of a transaction database. T_2, T_3 and T_4 are *PP*-trees constructed one by one in memory for three data mining queries. T_2 is the result of sub-projection of c, e on T_1. The merge operation on T_s and $\pi_{4,\{d,g\}}(T_a)$ results in T_3. T_4 is the result of super-projection on T_2.

Remark 1. It is important to know that, for a given data mining query q, a sequence of tree operations can be easily identified that results in a smallest *PP*-tree for q. All the resulting trees shown in Figure 1 are the smallest trees to respond the corresponding data mining query.

4 Performance Studies

Some experiments results are shown in Figure 2 and 3. We conducted our experiments on three datasets: `T25.I20.D100K` with 100K items, 10K items and

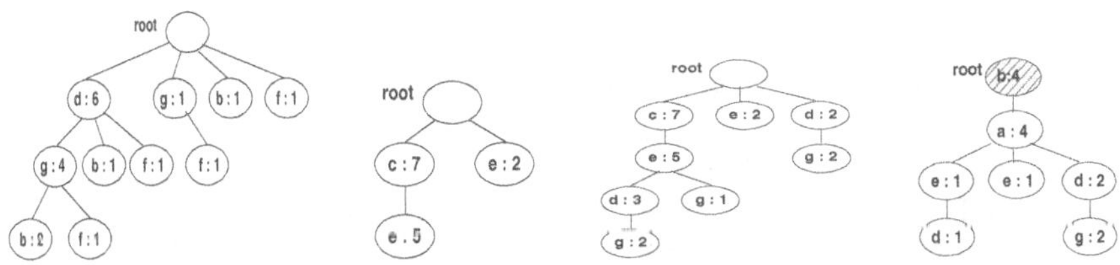

(a) T_1 (b) $T_2 = \pi_{4,\{c,e\}}(T_1)$ (c) $T_3 = \oplus(T_2, \pi_{4,\{d,g\}}(T_1))$ (d) $T_4 = \hat{\pi}_{4,\{b\}}(T_1))$

Fig. 1. Four Tree Projections (Suppose T_1 is the *PP*-tree materialized in disk)

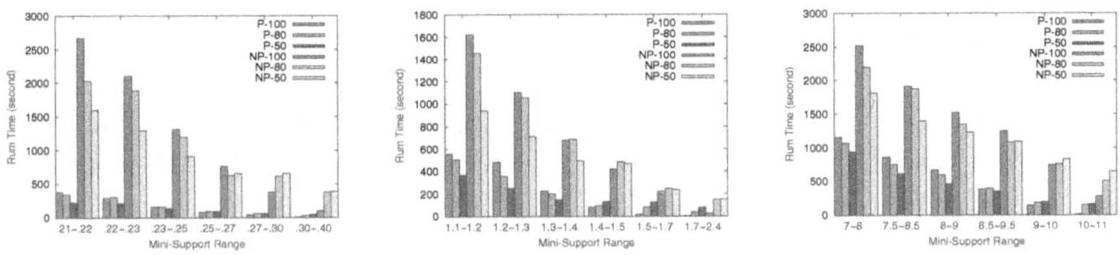

(a) 100K items (Sparse) ($\tau_m = 0.21$, 5MB cache) (b) 10K items (Medium) ($\tau_m = 1.1$, 6.5MB cache) (c) 1K items (Dense) ($\tau_m = 6.8$, 18MB cache)

Fig. 2. Various Queries Patterns (1,000 queries, 60% are mini-support queries, 20% sub-itemset queries (5-10 items), and 20% super-itemset queries (1-5 items)

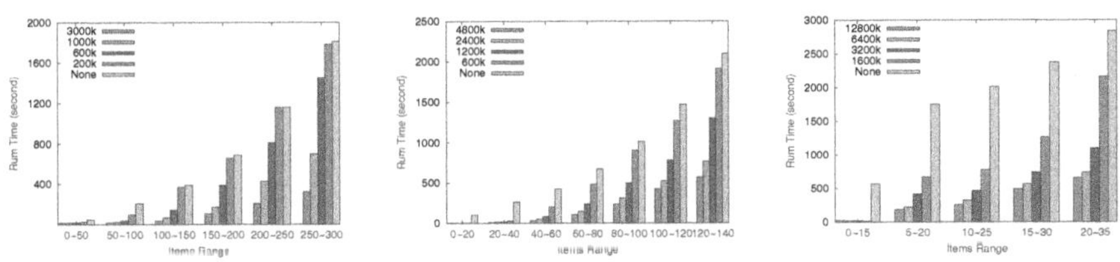

(a) 100K items (Sparse) ($\tau_m = 0.21$, 5-15 subitems) (b) 10K items (Medium) ($\tau_m = 1.1$, 5-15 subitems) (c) 1K items (Dense) ($\tau_m = 6.8$, 5-10 subitems)

Fig. 3. Sub-itemset Queries (non-overlapping sliding window of $[R_{min}, R_{max}]$, 1,000 queries, all sub-itemset queries)

1K items respectively. In the figures, the Y coordinates show the time cost of a test consisting of 1000 randomly generated tree building requests.

In Figure 2, P and NP mean with and without the data mining proxy. For each test we selected a range of minimum supports, $R = [R_{min}, R_{max}]$, which controlled the percentage of the minimum supports in that range. Three cases are considered, 100%, 80% and 50%, the percentage of min-support falling in R, denoted P/NP-100, P/NP-80 and P/NP-50. In all cases, the performance using the proxy outperforms the one without using the proxy with the same setting.

In Figure 3 we focused on sub-itemset building requests. Sliding windows of items range were used to test the proxy. We varied the proxy's size, which is indicated by the label of bar in the figure. As shown in the results, if the memory size was too small, the proxy did not work well, while if the cache size was appreciate, the performance of the proxy is very good.

5 Conclusion

In this paper, we proposed a data mining proxy to support a large number of users' mining queries, and focused on how to build the smallest but sufficient trees in memory efficiently for mining. Three tree operations were proposed: sub-projection, super-projection and merge. The proxy maximize the usage of trees in memory, and minimize the I/O costs. Our experiments showed that the data mining proxy is effective because in-memory tree operations can be processed much faster than loading subtrees from disk.

Acknowledgment. The work described in this paper was supported by grants from the Research Grants Council of the Hong Kong SAR (CUHK4229/01E, HKUST6175/03E).

References

1. Ramesh C. Agarwal, Charu C. Aggarwal, and V. V. V. Prasad. A tree projection algorithm for generation of frequent item sets. *Journal of Parallel and Distributed Computing*, 61:350–371, 2001.
2. Roberto J. Bayardo. Efficiently mining long patterns from databases. In *Proceedigns of 1998 ACM SIGMOD Intl. Conference on Management of Data*, pages 85–93, 1998.
3. Douglas Burdick, Manuel Calimlim, and Johannes Gehrke. MAFIA: A maximal frequent itemset algorithm for transactional databases. In *Proceedings of 2001 Intl. Conference on Data Engineering (ICDE'01)*, pages 443–452, 04 2001.
4. Jiawei Han and Jian Pei. Mining frequent patterns by pattern-growth: Methodology and implications. In *ACM SIGKDD Explorations*. ACM Press, 12 2001.
5. Jiawei Han, Jian Pei, and Yiwen Yin. Mining frequent patterns without candidate generation. In *Proceedings of the 2000 ACM SIGMOD Conference*, pages 1–12, 2000.

6. Guimei Liu, Hongjun Lu, Wenwu Lou, Yabo Xu, and Jeffrey Xu Yu. Efficient mining of frequent itemsets using ascending frequency ordered prefix-tree. In *Proceedings of DASFAA'03*, pages 65–72, 2003.
7. Junqiang Liu, Yunhe Pan, Ke Wang, and Jiawei Han. Mining frequent item sets by opportunistic projection. In *Proceedings of the 8th KDD Conference*, 2002.
8. Jian Pei, Jiawei Han, Hongjun Lu, Shojiro Nishio, and Dongqing Yang S hiwei Tang. H-mine: Hyper-structure mining of frequent patterns in large databases. In *Proceedings of 2001 IEEE Conference on Data Mining*, 2001.
9. Ke Wang, Liu Tang, Jiawei Han, and Junqiang Liu. Top down FP-growth for association rule mining. In *Proceedings of 6th Pacific-Asia conference on Knowledge Discovery and Data Mining*, 2002.
10. Yabo Xu, Jeffrey Xu Yu, Guimei Liu, and Hongjun Lu. From path tree to frequent patterns: A framework for mining frequent patterns. In *Proceedings of IEEE ICDM'02*, pages 514–521, 2002.

Formal Approach and Automated Tool for Translating ER Schemata into OWL Ontologies

Zhuoming Xu[1,2], Xiao Cao[2], Yisheng Dong[1], and Wenping Su[2]

[1] Dept. of Computer Science & Engineering, Southeast University, Nanjing 210096, China
zmxu@acm.org, ysdong@seu.edu.cn
[2] College of Computers & Information Engineering, Hohai University, Nanjing 210098, China
xcao@ieee.org, wpsu@msn.com

Abstract. Ontologies play a key role in creating machine-processable Web content to enable the Semantic Web. Extracting domain knowledge from database schemata can profitably support ontology development and then semantic markup of the instance data with the ontologies. The Entity-Relationship (ER) model is an industrial standard for conceptually modeling databases. This paper presents a formal approach and an automated tool for translating ER schemata into Web ontologies in the OWL Web Ontology Language. The tool can firstly read in an XML-coded ER schema produced with ER CASE tools such as PowerDesigner. Following the predefined knowledge-preserving mapping rules from ER schema to OWL DL (a sublanguage of OWL) ontology, it then automatically translates the schema into the ontology in both the abstract syntax and the RDF/XML syntax for OWL DL. Case studies show that the approach is feasible and the tool is efficient, even to large-scale ER schemata.

1 Introduction and Motivation

Unlike the current Web, where content is primarily intended for human consumption, the *Semantic Web*[3] will represent content that is also machine processable in order to provide better machine assistance for human users in tasks and enable a more open market for information processing and computer services. Ontologies play a key role in creating such machine-processable content by defining shared and common domain theories and providing a controlled vocabulary of terms, each with an explicitly defined and machine-processable semantics[3]. The importance of ontologies to the Semantic Web has prompted the establishment of the normative Web ontology and semantic markup language OWL[5]. Besides, many kinds of ontology tools have been built to help people create ontologies and machine-processable Web content. These include, among others, *ontology development tools*[6] such as Protégé-2000 and OntoEdit that can be used for building or reusing ontologies, and *ontology-based annotation tools*[6] such as OntoMat-Annotizer and SMORE that allow users inserting and maintaining (semi)automatically ontology-based markups in Web pages.

Although many tools exist, acquiring domain knowledge to build ontologies require making great efforts. For this reason, it is necessary to develop methods and

H. Dai, R. Srikant, and C. Zhang (Eds.): PAKDD 2004, LNAI 3056, pp. 464–476, 2004.

techniques that allow reducing the effort necessary for the knowledge acquisition process, being this the goal of *ontology learning*. Maedche and Staab[12] distinguish different ontology learning approaches that focus on the type of input used for learning, i.e., ontology learning from text, from dictionary, from knowledge base, from semi-structured schemata and from database schemata. Database schemata, in particular the conceptual schemata modeled in semantic data models such as the Entity-Relationship (ER) model contain (implicitly) abundant domain knowledge. Extracting the knowledge from them can thus profitably support the development of Web ontologies. This is one aspect which motivates our research.

Another motivation of our work concerns the *semantic markup* of Web content. All existing ontology-based annotation tools can only annotate static HTML pages so far[1]. Nowadays, however, Web-accessible databases are main content sources on the current Web and the majority of Web pages are dynamic ones generated from databases[9]. How to "upgrade" these dynamic Web content to Semantic Web content remains to be an open problem. This is therefore one of the aims of the ongoing IST Project Esperonto Services[2]. The project has imagined a general "semantic wrapper" method which leaves the content in the database and annotates the query that retrieves the concerned content. Other alternatives exist. For instance, Stojanovic *et al*[14] have presented an "instance migration" method which extracts the instance data from a database and stores the content as an RDF file on the Semantic Web. The method has an obvious shortcoming because it assumes that the database is static and no data update occurs. Handschuh *et al*[9] have thus introduced a "deep annotation" method. The method keeps the instance data remaining in the database. Client queries are executed based on the mapping rules between the client ontology and the database with the assumption that the Web site (database owner) is willing to provide the information structure of the underlying database and produce in advance server-side Web page markups according to the information structure. No matter what methods are adopted to "upgrade" a database to Semantic Web content, we argue that a common issue and a precondition here are that we have an ontology at hand that can conceptually capture the knowledge of the domain of discourse of the database. Usually, database construction begins with ER modeling supported by CASE tools. Therefore, it is necessary to develop tools for translating ER schemata into OWL ontologies.

In the paper, we will present an automated tool ER2WO which is based on a formal approach and performs the automatic translation from ER schemata to OWL ontologies. The remainder of the paper is organized as follows. First, we introduce the formal approach in section 2, with the focus on how to define the knowledge-preserving mapping from an ER schema to an OWL ontology. Next, in section 3, we deal with tool implementation and case studies with the intention of proof-of-concept of our approach. Section 4 is the related work, and the last section the conclusions.

---

[1] See Semantic Web - Annotation & Authoring homepage http://annotation.semanticweb.org.

[2] http://www.esperpnto.net/.

2 Formal Approach

To perform the translation form ER schemata to OWL ontologies, we must first establish the correspondence and then define the mapping between the two knowledge representation languages, i.e., the ER model and the OWL language.

Early researches, such as [4], on representing and reasoning on database conceptual schemata have introduced several formalization means to represent an ER schema. The formalized ER schema can then be translated into a knowledge base in *Description Logics* (DLs)[1] (such as *ALUNI*[4]) and it has been proved that the translation preserves the semantics.

The OWL language is developed as a vocabulary extension of RDF and provides three increasingly expressive sublanguages Lite, DL and FULL, designed for use by specific communities of implementers and users[5]. OWL DL can be approximately viewed as the expressive DL *SHOIN(D)* which is more expressive than *ALUNI*, with an OWL DL ontology being equivalent to a *SHOIN(D)* knowledge base, and the set of axioms of the ontology being equivalent to the set of assertions of the knowledge base[11]. Therefore, we believe that there exists a formal and semantics-preserving approach for translating an ER schema into an OWL DL Ontology.

2.1 Formalization of ER Schemata

We adopt the first-order formalization of the ER model introduced by Calvanese *et al* in [4]. This formalization includes the most important features present in the different variants of the ER model supported by existing CASE tools and has a well-defined semantics, which makes it possible to establish a precise correspondence with the Web ontology language OWL DL. In the following, Definition 1 and 2, we give the formal syntax of an ER schema that was presented in [4], with our minor changes of the notation and an augment of key attributes[3] in the model.

Definition 1. For two finite sets X and Y, we call a function from a subset of X to Y an *X-labeled tuple over Y*. The labeled tuple T which maps $x_i \in X$ to $y_i \in Y$, for $i=1, ..., k$, is denoted $[x_1 : y_1, ..., x_k : y_k]$.

Definition 2. An *ER schema* is a tuple $\mathcal{S} = (\mathcal{L}_S, isa_S, att_S, rel_S, card_S)$, where

- $\mathcal{L}_S$ is a finite *alphabet* partitioned into a set $\mathcal{E}_S$ of *entity symbols*, a set $\mathcal{A}_S$ of *attribute symbols*, a set $\mathcal{U}_S$ of *ER-role symbols*, a set $\mathcal{R}_S$ of *relationship symbols*, and a set $\mathcal{D}_S$ of *domain symbols*; each domain symbol $D \in \mathcal{D}_S$ has an associated predefined *basic domain* B^D, and the various basic domains are assumed to be pairwise disjoint.
- $isa_S \subseteq \mathcal{E}_S \times \mathcal{E}_S$ is a binary relation which models the *IS-A relationship* between entities.
- att_S is a function that maps each entity symbol in $\mathcal{E}_S$ to an $\mathcal{A}_S$-labeled tuple over $\mathcal{D}_S$. For an entity $E \in \mathcal{E}_S$ such that $att_S(E) = [..., A : D, ...]$, if there exists an attribute A such that it can have only one (unique) value in D's basic domain B^D for each instance of E, then the attribute A is called a *key attribute* of entity E. The function is used to model *attributes* of entities. For simplicity, we assume here that all attrib-

[3] Here we only consider single-attribute keys and do not consider the individual attributes composed to form a composite-key of an entity.

utes are single-valued and mandatory, but we can also easily handle the situation beyond this assumption.

- rel_S is a function that maps each relationship symbol in $\mathcal{R}_S$ to an $\mathcal{U}_S$–labeled tuple over $\mathcal{E}_S$. We assume without loss of generality that:
  1. Each ER-role is specific to exactly one relationship.
  2. For each ER-role $U \in \mathcal{U}_S$, there is a relationship $R \in \mathcal{R}_S$ and an entity $E \in \mathcal{E}_S$ such that $rel_S(R) = [..., U : E, ...]$.

  The function actually associates a set of *ER-roles* to each *relationship*, determining implicitly also the *arity* of the relationship.
- $card_S$ is a function from $\mathcal{E}_S \times \mathcal{R}_S \times \mathcal{U}_S$ to $N_0 \times (N_1 \cup \{\infty\})$ (where N_0 denotes nonnegative integers, N_1 positive integers) that satisfies the following condition: for a relationship $R \in \mathcal{R}_S$ such that $rel_S(R) = [U_1 : E_1, ..., U_k : E_k]$, $card_S(E, R, U)$ is defined only if $U = U_i$ for some $i = 1, ..., k$, and if $E\ isa_S^*\ E_i$ (where isa_S^* denotes the reflexive transitive closure of isa_S). The first component of $card_S(E, R, U)$ is denoted with $min\_card_S(E, R, U)$ and the second component with $max\_card_S(E, R, U)$. If not stated otherwise, $min\_card_S(E, R, U)$ is assumed to be 0 and $max\_card_S(E, R, U)$ is assumed to be ∞. The function $card_S$ is used to specifies *cardinality constraints*, i.e., constraints on the minimum and maximum number of times an instance of an entity may participate in a relationship via some ER-role.

The semantics of an ER schema can be given by specifying *database states* consistent with the information structure expressed by the schema (see [4]).

2.2 OWL DL Language and Ontology

OWL DL has two types of syntactic form. One is the *exchange syntax*[5], i.e., the RDF/XML syntax, which represents an ontology as a set of RDF triples for the purpose of publishing and sharing the ontology over the Web. Another form is the frame-like style *abstract syntax*[13], where a collection of information about a class or property is given in one large syntactic construct, instead of being divided into a number of atomic chunks (as in most DLs) or even being divided into more triples as when using the exchange syntax. The abstract syntax is abstracted from the exchange syntax and thus facilitates access to and evaluation of the ontologies, being this the reason for presenting our formal approach using this syntax in the paper.

OWL uses a DL style *model theory* to formalize the meaning of the language[13]. The underlying formal foundation of OWL DL is the DL *SHOIN(D)* and the semantics for OWL DL is based on *interpretations*[11], where an interpretation consists of a *domain of discourse* and an *interpretation function*. The domain is divided into two disjoint sets, the *individual domain* Δ^I and the *data-value domain* Δ_D^I. The interpretation function I maps classes into subsets of Δ^I, individuals into elements of Δ^I, datatypes into subsets of Δ_D^I and data values into elements of Δ_D^I. In addition, two disjoint sets of properties are distinguished: *object properties* and *data type properties*. The interpretation function maps the former into subsets of $\Delta^I \times \Delta^I$ and the latter into subsets of $\Delta^I \times \Delta_D^I$. The interpretation function is extended to *concept expressions* in the language.

Information in OWL is gathered into *ontologies*, which can then be stored as documents in the RDF/XML syntax on the Web. The document consists of an optional ontology header plus any number of class axioms, property axioms, and individual axioms (i.e., facts about individuals). An OWL DL ontology in the abstract syntax, started with an optional ontology ID, contains simply a sequence of annotations (optional) and axioms. Regardless of using which syntax form, the formal meaning of the ontology is solely determined by the underlying RDF graph of the ontology, which is interpreted by the model-theoretic semantics for the language[13].

2.3 Mapping from ER Schemata to OWL DL Ontologies

The formal approach for translating an ER schema into an OWL DL ontology follows a set of mapping rules, as specified in Definition 3. In fact, these mapping rules are induced by the semantic-preserving mapping rules from an ER schema to a DL knowledge base introduced in [4] and the semantical correspondence between an OWL DL ontology and a DL knowledge base[11]. Therefore, the translation preserves the semantics of the ER schema.

Definition 3. Let $\mathcal{S} = (\mathcal{L}_\mathcal{S}, isa_\mathcal{S}, att_\mathcal{S}, rel_\mathcal{S}, card_\mathcal{S})$ be an ER schema. The OWL DL ontology O in the abstract syntax is defined by a translation function $O(\mathcal{S}) = (ID_O, axiom_O)$, where

- ID_O is a finite OWL DL *identifier set* partitioned into a subset CID_O of *class identifiers*, a subset $DPID_O$ of *data-valued property identifiers*, a subset $IPID_O$ of *individual-valued property identifiers*, and a subset $DTID_O$ of *datatype identifiers*; each datatype identifier is a predefined XML Schema datatype[4] identifier (that is used in OWL as a local name in the XML Schema canonical URI reference for the datatype), and
- $axiom_O$ is a finite OWL DL *axiom set* partitioned into a subset $caxiom_O$ of *class axioms* and a subset $paxiom_O$ of *property axioms*, and
- ID_O and $axiom_O$ are induced by the elements of $\mathcal{S}$, following the mapping rules from an ER schema to an OWL DL ontology as depicted in Figure 1.

In the table of Figure 1, the left column is the source of the mapping, right column the termination. Starting with the construction of the atomic identifiers (i.e., local names in URI references in the RDF/XML syntax), the approach induces a set of axioms from the ER schema, which forms the body of the target ontology. (To a particular class, multiple class-axioms can be merged for conciseness in practice). The ontology can then be evaluated by the domain expert.

2.4 Ontology Transform from Abstract Syntax to RDF/XML Syntax

For the real use of the resulting ontology over the Web, it should be transformed into the exchange syntax. The W3C has specified a set of semantics-preserving mapping

---

[4] XML Schema datatypes (http://www.w3.org/TR/2001/REC-xmlschema-2-20010502/) are used in OWL as built-in datatypes by means of the XML Schema canonical URI reference for the type, e.g., `xsd:string`, `xsd:integer`, and `xsd:nonNegativeInteger`.

ER Schema Elements	**OWL DL Ontology Elements (in the Abstract Syntax)**	
Alphabet $\mathcal{L}_s$	Identifier Set $\mathcal{ID}_o$ (Symbols E, R, A, and U can be renamed optionally)	
Each entity symbol $E \in \mathcal{E}_s$	A description (exactly, class identifier) $\mathcal{O}(E) \in \mathcal{CID}_o$	
Each relationship symbol $R \in \mathcal{R}_s$	A description (exactly, class identifier) $\mathcal{O}(R) \in \mathcal{CID}_o$	
Each attribute symbol $A \in \mathcal{A}_s$	A data-valued property identifier $\mathcal{O}(A) \in \mathcal{DPID}_o$	
Each domain symbol $D \in \mathcal{D}_s$	A (XML Schema) datatype identifier $\mathcal{O}(D) \in \mathcal{DTID}_o$	
Each ER-role symbol $U \in \mathcal{U}_s$	An individual-valued property identifier $\mathcal{O}(U) \in \mathcal{IPID}_o$	
Other Constructors	Axiom Set $axiom_o$	
Each attribute $A \in \mathcal{A}_s$ such that $att_s(E) = [..., A : D, ...]$	Create a property axiom: `DatatypeProperty(`$\mathcal{O}(A)$` domain(`$\mathcal{O}(E)$`) range(`$\mathcal{O}(D)$`) [Functional]),` where `Functional` occurs only if A is a *key attribute* as defined above	(1)
Each ER-role $U \in \mathcal{U}_s$ such that $rel_s(R) = [..., U : E, ...]$	Create a property axiom: `ObjectProperty(`$\mathcal{O}(U)$` domain(`$\mathcal{O}(R)$`) range(`$\mathcal{O}(E)$`))`	(2)
Each pair of entities $E_1, E_2 \in \mathcal{E}_s$ such that E_1 isa_s E_2	Create a class axiom `subClassOf(`$\mathcal{O}(E_1)$ $\mathcal{O}(E_2)$`)` or `Class(`$\mathcal{O}(E_1)$` partial `$\mathcal{O}(E_2)$`)`	(3)
Each entity $E \in \mathcal{E}_s$ such that $att_s(E) = [A_1 : D_1, ..., A_h : D_h]$	Create a class axiom: `Class(`$\mathcal{O}(E)$` partial restriction(`$\mathcal{O}(A_1)$` allValuesFrom(`$\mathcal{O}(D_1)$`) cardinality(1)) ... restriction(`$\mathcal{O}(A_h)$` allValuesFrom(`$\mathcal{O}(D_h)$`) cardinality(1)))`	(4)

Each relationship $R \in \mathcal{R}_s$ such that $rel_s(R) = [U_1 : E_1, ..., U_k : E_k]$	Create a class axiom:
	`Class(`$\mathcal{O}(R)$` partial restriction(`$\mathcal{O}(U_1)$` allValuesFrom(`$\mathcal{O}(E_1)$`) cardinality(1)) ... restriction(`$\mathcal{O}(U_k)$` allValuesFrom(`$\mathcal{O}(E_k)$`) cardinality(1)))`, and
	For $i = 1 .. k$ do
	Create a property identifier $P_i \in \mathcal{IPID}_o$,
	Create a property axiom:
	`ObjectProperty(`P_i` domain(`$\mathcal{O}(E_i)$`) range(`$\mathcal{O}(R)$`) inverseOf `$\mathcal{O}(U_i)$`)`,
	Create a class axiom:
	`Class(`$\mathcal{O}(E_i)$` partial restriction(`P_i` allValuesFrom(`$\mathcal{O}(R)$`)))`
		(5)
		(6)
		(7)

Each relationship $R \in \mathcal{R}_s$ such that $rel_s(R) = [U_1 : E_1, ..., U_k : E_k]$, for $i = 1, ..., k$, and for each entity $E \in \mathcal{E}_s$, such that E isa_s^* E_i	For $i = 1 .. k$ do (assume P_i has been created with the axiom (6)) If $m = min\_card_s(E, R, U_i) \neq 0$ then create a class axoim: `Class(`$\mathcal{O}(E)$` partial restriction(`P_i` minCardinality(m)))`, If $n = max\_card_s(E, R, U_i) \neq \infty$ then create a class axoim: `Class(`$\mathcal{O}(E)$` partial restriction(`P_i` maxCardinality(n)))`	(8) (9)
For each pair of symbols $X, Y \in \mathcal{E}_s \cup \mathcal{R}_s$ such that $X \neq Y$ and $X \in \mathcal{R}_s$	Create a class axoim: `DisjointClasses(`$\mathcal{O}(X)$ $\mathcal{O}(Y)$`)`	(10)

Fig. 1. Mapping rules from an ER schema to an OWL DL ontology in the abstract syntax

rules from OWL DL abstract syntax to RDF/XML syntax in the OWL Semantics and Abstract Syntax document[13]. In the document, a model-theoretic semantics is given to provide a formal meaning for OWL ontologies written in the abstract syntax. A model-theoretic semantics in the form of an extension to the RDF semantics is also specified to provide a formal meaning for OWL ontologies as RDF graphs. A mapping from the abstract syntax to RDF graphs is then defined in the document and it has been proved that the two model theories are shown to have the same consequences on OWL ontologies that can be written in the abstract syntax. We have simply adopted the mapping rules to perform the syntax transform of the ontology.

3 Prototype Tool and Case Study

Based on the formal approach introduced above, we developed an automated tool ER2WO which can read in an XML-coded ER schema - currently, a conceptual data model (CDM) file produced from CASE tool PowerDesigner 9.5[5] - and translate automatically the schema into an OWL DL ontology using the mapping rules and produce the resulting ontology in both the abstract syntax and the RDF/XML syntax.

3.1 Design and Implementation of the Tool

ER2WO is designed as consisting of 4 modules: *parsing* module which parses an XML-coded ER schema, *translation* module that translates the parsed schema into the ontology in OWL DL abstract syntax, *transformation* module that performs the ontology transformation from the abstract syntax to the RDF/XML syntax, and *output* module which produces the resulting ontology as a text file.

The tool implementation is based on Java 2 v1.4.2 platform. The parsing module uses the SAX API for Java to parse the ER schema file and store the schema data as Java ArrayList classes. The translation module uses Java class methods to implement the mapping from the schema to the OWL DL ontology in the abstract syntax. To the best of our knowledge, there hasn't so far existed any off-the-shelf tool for performing the ontology transformation from the abstract syntax to the RDF/XML syntax. So we developed the transformation module using the XML presentation syntax for OWL specified in the document[10] as an intermediate format between the two syntax formats, and using the XSLT stylesheet (`owlxml2rdf.xsl`[6]) provided in the document for the transformation from the XML presentation syntax to the RDF/XML syntax. The screen snapshot of ER2WO v1.0 is depicted in Figure 2. In the figure, the left list-boxes display the parsed ER schema `univ_dept` in the case study below, the right text-area and the pop window display the resulting ontology in the abstract syntax and in the RDF/XML syntax respectively.

[5] http://www.sybase.com/products/enterprisemodeling/powerdesigner.

[6] http://www.w3.org/TR/owl-xmlsyntax/owlxml2rdf.xsl.

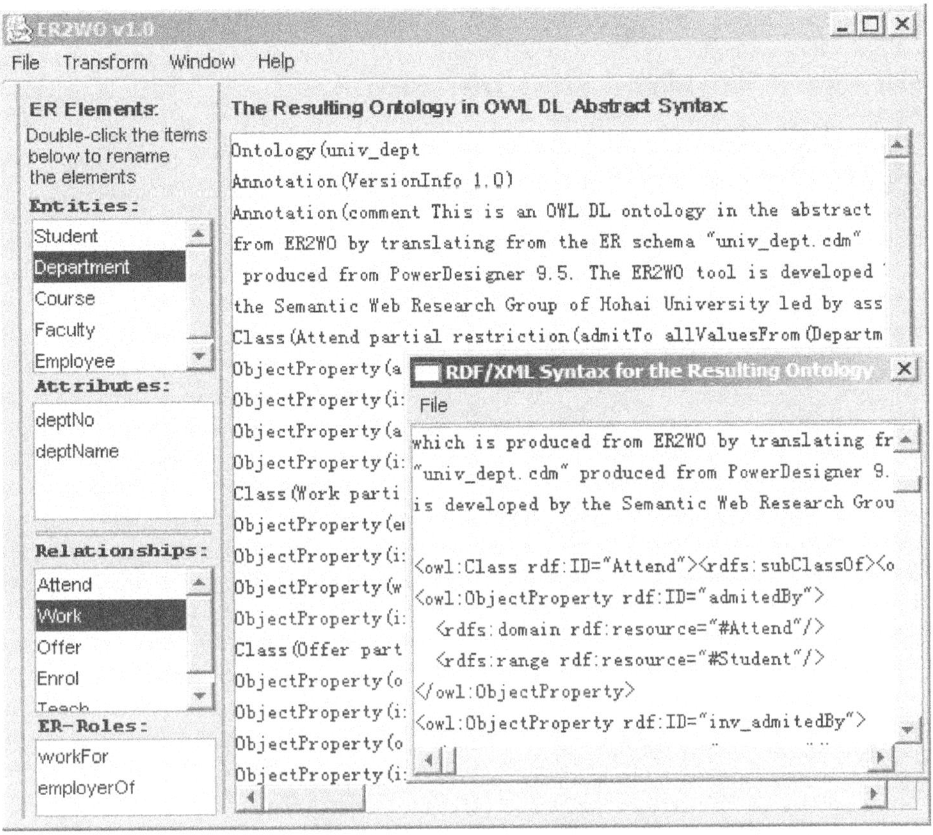

Fig. 2. Screen snapshot of ER2WO

3.2 Case Study

We have carried out more than 10 case studies using ER2WO, with the scale of the ER schema ranging from 57 to 356[7]. Case studies show that the formal approach is feasible and the implemented tool is efficient, even to large-scale ER schemata. All resulting OWL DL ontologies have passed the syntactic validation by the OWL Ontology Validator (http://phoebus.cs.man.ac.uk:9999/OWL/Validator). Four selected cases including *University Department*, *Project Management*, *Book Store* and *Library Management* have been published at the homepage of ER2WO tool[8].

Saving space, Figure 3 shows a small-scale example ER schema `univ_dept` which models a university department with PowerDesigner 9.5. We notice here that PowerDesigner does not currently support *N*-ary relationships ($N>2$), or attributes on

[7] Here we measure the scale of an ER schema by summing up all the numbers of elements in the schema, including entities, (IS-A) relationships, attributes, ER-roles and constraints.

[8] http://cse.seu.edu.cn/people/ysdong/graduates/~zmxu/ER2WO/.

relationships. Hence, the case does not include these ER features. In the figure, 'TXT' denotes datatype string, 'I' denotes integer, and '1,n' denotes that the minimum cardinality is 1 and the maximum cardinality is ∞.

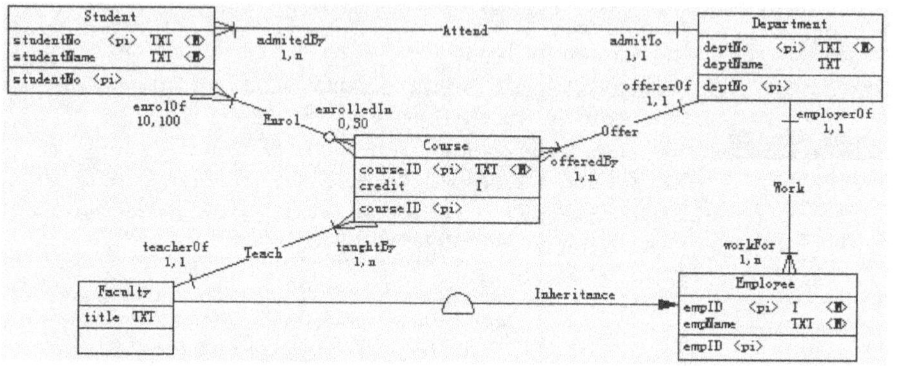

Fig. 3. An example ER schema univ_dept modeled with PowerDesigner 9.5

Following the mapping rules, the ER schema is translated into OWL DL ontology $\mathcal{O}$(univ_dept) = ($\mathcal{ID}_O$, $axiom_O$), where, $\mathcal{ID}_O$ is an OWL DL identifier set partitioned into: $\mathcal{CID}_O$ ={Course, Employee, Student, Department, Faculty, Teach, Attend, Enrol, Offer, Work}; $\mathcal{DTID}_O$={xsd:string, xsd:integer}; $\mathcal{DPID}_O$={studentNo, studentName, deptNo, deptName, courseID, credit, empID, empName, title}; $\mathcal{IPID}_O$={admitTo, admittedBy, enrolOf, enrolledIn, offererOf, offeredBy, workFor, employerOf, teacherOf, taughtBy, inv_admitTo, inv_admittedBy, inv_enrolOf, inv_enrolledIn, inv_offererOf, inv_offeredBy, inv_workFor, inv_employerOf, inv_teacherOf, inv_taughtBy}, where identifiers with prefix inv_ denote the inverse of the corresponding property identifiers (e.g., inv_admitTo is the inverse of property identifier admitTo).

Based on the identifier set, the OWL DL axiom set $axiom_O$ which forms the body of ontology $\mathcal{O}$(univ_dept) is presented in Appendix A. (It can also be found at the homepage of ER2WO tool).

4 Related Work

Two categories of approaches or tools are related to our work. One is the DL-based conceptual modeling approaches such as [4], and tools such as I.COM[7]. These early approaches and tools established the correspondence between DLs and the ER model, which form the formal foundation for our approach and tool. They focus on reasoning about the schema by translating it into a DL knowledge base. Whereas our tool aims at acquiring knowledge from existing ER schemata and then building and publishing OWL ontologies on the Web.

Another category is the tools for ontology learning and instance-data migration from databases. According to recent surveys [2] and [8], there are two ongoing projects aim at developing tools for creating lightweight Web ontologies and ontological instances from databases, i.e., the D2R MAP[9] and KAON REVERSE[14][10] prototype tools. D2R MAP is a declarative language to describe mappings between relational database schemata and RDF(S) ontologies. The mappings can be used by a D2R processor to export data from a database to an RDF file. However the mapping definition process is manual and requires domain-knowledge input from the modeler. Besides, D2R MAP focuses on exposing an RDF description of the relational database, not the conceptual entities which the relational description is attempting to capture[2]. KAON REVERSE is an early prototype for mapping relational database content to RDF(S) ontologies and instances. It is intended to be merged with the Harmonise Mapping Framework tool[11] and the user interface into the KAON OIModeler. The early prototype[14] takes the schema and instance of a domain specific relational database as the source, uses F-Logic predicates and axioms as the intermediate format, and adopts data reverse engineering and mapping approaches to produce the RDF(S) ontology and instance data. However it also needs some degree of human participation and manual check, and because RDF(S) does not have enough expressive power to capture the knowledge and constraints of the domain the tool has inherent drawbacks[2]. Our work seems to be a useful attempt toward developing more formal methods and automated tools used to perform the ontology learning process and support more expressive languages such as OWL.

5 Conclusions

The real power of the Semantic Web will be realized only when people create much machine-readable content, and ontologies play a key role in this effort. Given the fact that many industrial CASE tools can support both database *forward* and *reverse* engineering and produce ER schemata in exchangeable (e.g., in XML) format, extracting knowledge from database schemata and producing OWL ontologies by integrating our ER2WO tool with existing CASE tools can thus profitably support the development and reuse of Web ontologies, and then facilitate the semantic markup of dynamic Web content generated from databases.

In a broad sense, the Semantic Web needs to be able to share and reuse existing knowledge bases. Therefore, interoperability among different knowledge representation systems is essential. In this sense, our ER2WO tool can act as a gap-bridge between existing database applications or legacy systems and the Semantic Web.

Acknowledgements. Research for the paper was funded by the Natural Science Foundation of Jiangsu Province of China as the key project named "Research on Languages and Supporting Software for the Semantic Web", under Grant No. BK2003001. The findings and views expressed in this paper are those of the authors, and not necessarily of the funding organization of the project.

---

[9] http://www.wiwiss.fu-berlin.de/suhl/bizer/d2rmap/D2Rmap.htm.

[10] http://kaon.semanticweb.org/alphaworld/reverse/view.

[11] http://sourceforge.net/projects/hmafra/.

References

1. Franz Baader, Deborah L. McGuinness, Daniele Nardi, Peter F. Patel-Schneider (eds.): *The Description Logic Handbook: Theory, Implementation, and Applications*. Cambridge University Press (2003)
2. Dave Beckett, Jan Grant: Mapping Semantic Web Data with RDBMSes. *IST Project SWAD-Europe Deliverable 10.2*, Feb 18, 2003. Online at: www.w3.org/2001/sw/Europe/reports/scalable_rdbms_mapping_report
3. Tim Berners-Lee, James Hendler, Ora Lassila: The Semantic Web. *Scientific American*, Vol. 284, No. 5, (2001) 34-43
4. Diego Calvanese, Maurizio Lenzerini, Daniele Nardi: Unifying Class-based Representation Formalisms. *Journal-of-Artificial-Intelligence-Research*, Vol. 11, (1999) 199-240
5. Mike Dean, Guus Schreiber (eds.): OWL Web Ontology Language Reference. *W3C Recommendation*, 10 Feb 2004. Online at: www.w3.org/TR/2004/REC-owl-ref-20040210/
6. Dieter Fensel, Asunción Gómes Pérez (eds.): A Survey on Ontology Tools. *IST Project OntoWeb Deliverable 1.3*, May 2002. Online at: ontoweb.aifb.uni-karlsruhe.de/About/Deliverables/D13_v1-0.zip
7. Enrico Franconi, Gary Ng: The i.com Tool for Intelligent Conceptual Modeling. In *Proc. of the 7th Int'l Workshop on Knowledge Representation Meets Databases*. CEUR Electronic Workshop Proceedings (2000) 45-53. Online at: CEUR-WS.org/Vol-29/
8. Asunción Gómez-Pérez, David Manzano-Macho (eds.): A Survey of Ontology Learning Methods and Techniques. *IST Project OntoWeb Deliverable 1.5*, June 2003. Online at: ontoweb.aifb.uni-karlsruhe.de/Members/ruben/Deliverable%201.5
9. Siegfried Handschuh, Steffen Staab, Raphael Volz: On Deep Annotation. In *Proc. of the 12th Int'l World Wide Web Conf.* ACM Press (2003) 431-438
10. Masahiro Hori, Jérôme Euzenat, Peter F. Patel-Schneider: OWL Web Ontology Language XML Presentation Syntax. *W3C Note*, 11 June 2003. Online at: www.w3.org/TR/2003/NOTE-owl-xmlsyntax-20030611/
11. Ian Horrocks, Peter F. Patel-Schneider, Frank van Harmelen, Jason Y. Zienberer: From SHIQ and RDF to OWL: The Making of a Web Ontology Language. *Journal of Web Semantics*, Vol. 1, Issue 1, Dec 2003. Online at: www.websemanticsjournal.org/volume1/issue1/Horrocksetal2003/index.html
12. Alexander Maedche and Steffen Staab: Ontology Learning for the Semantic Web. *IEEE Intelligent Systems*, Vol. 16, No. 2, (2001) 72–79
13. Peter F. Patel-Schneider, Patrick Hayes, Ian Horrocks (eds.): OWL Web Ontology Language Semantics and Abstract Syntax. *W3C Recommendation*, 10 Feb 2004. Online at: www.w3.org/TR/2004/REC-owl-semantics-20040210/
14. Ljiljana Stojanovic, Nenad Stojanovic, Raphael Volz: Migrating Data-intensive Web Sites into the Semantic Web. In *Proc. of the 17th ACM Symposium on Applied Computing*. ACM Press (2002) 1100-1107

Appendix A OWL DL ontology derived from the ER schema `univ_dept`

(For brevity, the disjoint class axioms and annotations are omitted here.)

```
Ontology(univ_dept
Class(Attend partial restriction(admitTo allValuesFrom(Department) cardinality(1))
  restriction(admitedBy allValuesFrom(Student) cardinality(1)))
ObjectProperty(admitedBy domain(Attend) range(Student))
ObjectProperty(inv_admitedBy domain(Student) range(Attend) inverseOf(admitedBy))
ObjectProperty(admitTo domain(Attend) range(Department))
ObjectProperty(inv_admitTo domain(Department) range(Attend) inverseOf(admitTo))
Class(Work partial restriction(workFor allValuesFrom(Employee) cardinality(1))
  restriction(employerOf allValuesFrom(Department) cardinality(1)))
ObjectProperty(employerOf domain(Work) range(Department))
ObjectProperty(inv_employerOf domain(Department) range(Work)
  inverseOf(employerOf))
ObjectProperty(workFor domain(Work) range(Employee))
ObjectProperty(inv_workFor domain(Employee) range(Work) inverseOf(workFor))
Class(Offer partial restriction(offererOf allValuesFrom(Department)
  cardinality(1)) restriction(offeredBy allValuesFrom(Course) cardinality(1)))
ObjectProperty(offeredBy domain(Offer) range(Course))
ObjectProperty(inv_offeredBy domain(Course) range(Offer) inverseOf(offeredBy))
ObjectProperty(offererOf domain(Offer) range(Department))
ObjectProperty(inv_offererOf domain(Department) range(Offer) inverseOf(offererOf))
Class(Enrol partial restriction(enrolOf allValuesFrom(Student) cardinality(1))
  restriction(enrolledIn allValuesFrom(Course) cardinality(1)))
ObjectProperty(enrolledIn domain(Enrol) range(Course))
ObjectProperty(inv_enrolledIn domain(Course) range(Enrol) inverseOf(enrolledIn))
ObjectProperty(enrolOf domain(Enrol) range(Student))
ObjectProperty(inv_enrolOf domain(Student) range(Enrol) inverseOf(enrolOf))
Class(Teach partial restriction(teacherOf allValuesFrom(Faculty) cardinality(1))
  restriction(taughtBy allValuesFrom(Course) cardinality(1)))
ObjectProperty(taughtBy domain(Teach) range(Course))
ObjectProperty(inv_taughtBy domain(Course) range(Teach) inverseOf(taughtBy))
ObjectProperty(teacherOf domain(Teach) range(Faculty))
ObjectProperty(inv_teacherOf domain(Faculty) range(Teach) inverseOf(teacherOf))
Class(Student partial restriction(studentNo allValuesFrom(xsd:string)
  cardinality(1)) restriction(studentName allValuesFrom(xsd:string)
  cardinality(1)) restriction(inv_admitedBy allValuesFrom(Attend) cardinality(1))
  restriction(inv_enrolOf allValuesFrom(Enrol) maxCardinality(30)))
Class(Department partial restriction(deptNo allValuesFrom(xsd:string)
  cardinality(1)) restriction(deptName allValuesFrom(xsd:string) cardinality(1))
  restriction(inv_admitTo allValuesFrom(Attend) minCardinality(1))
  restriction(inv_employerOf allValuesFrom(Work) minCardinality(1))
  restriction(inv_offererOf allValuesFrom(Offer) minCardinality(1)))
Class(Course partial restriction(courseID allValuesFrom(xsd:string)
  cardinality(1)) restriction(credit allValuesFrom(xsd:integer) cardinality(1))
  restriction(inv_offeredBy allValuesFrom(Offer) cardinality(1))
  restriction(inv_enrolledIn allValuesFrom(Enrol) minCardinality(10)
  maxCardinality(100)) restriction(inv_taughtBy allValuesFrom(Teach)
  minCardinality(1)))
Class(Faculty partial Employee restriction(title allValuesFrom(xsd:string)
  cardinality(1)) restriction(inv_teacherOf allValuesFrom(Teach)
  minCardinality(1)))
Class(Employee partial restriction(empID allValuesFrom(xsd:integer)
  cardinality(1)) restriction(empName allValuesFrom(xsd:string) cardinality(1))
  restriction(inv_workFor allValuesFrom(Work) cardinality(1)))
DatatypeProperty(studentNo domain(Student) range(xsd:string) Functional)
DatatypeProperty(studentName domain(Student) range(xsd:string))
DatatypeProperty(deptNo domain(Department) range(xsd:string) Functional)
DatatypeProperty(deptName domain(Department) range(xsd:string))
DatatypeProperty(courseID domain(Course) range(xsd:string) Functional)
DatatypeProperty(credit domain(Course) range(xsd:integer))
DatatypeProperty(title domain(Faculty) range(xsd:string))
DatatypeProperty(empID domain(Employee) range(xsd:integer) Functional)
DatatypeProperty(empName domain(Employee) range(xsd:string)))
```

Separating Structure from Interestingness

Taneli Mielikäinen

HIIT Basic Research Unit
Department of Computer Science
University of Helsinki, Finland
Taneli.Mielikainen@cs.Helsinki.FI

Abstract. Condensed representations of pattern collections have been recognized to be important building blocks of inductive databases, a promising theoretical framework for data mining, and recently they have been studied actively. However, there has not been much research on how condensed representations should actually be represented.
In this paper we propose a general approach to build condensed representations of pattern collections. The approach is based on separating the structure of the pattern collection from the interestingness values of the patterns. We study also the concrete case of representing the frequent sets and their (approximate) frequencies following this approach: we discuss the trade-offs in representing the frequent sets by the maximal frequent sets, the minimal infrequent sets and their combinations, and investigate the problem approximating the frequencies from samples by giving new upper bounds on sample complexity based on frequent closed sets and describing how convex optimization can be used to improve and score the obtained samples.

1 Introduction

Data mining aims to find something interesting from large databases. One of the most important approaches to mine data is *pattern discovery* where the goal is to extract *interesting patterns* (possibly with some *interestingness values* associated to each of them) from data [1,2]. The most prominent example of pattern discovery is the *frequent set mining* problem:

Problem 1 (Frequent set mining). Given a multiset $d = \{d_1, \ldots, d_n\}$ (a *data set*) of subsets (*transactions*) of a set R of *attributes* and a threshold value $\sigma \in [0,1]$, find the collection $\mathcal{F}(\sigma, d) = \{X \subseteq R : fr(X,d) \geq \sigma\}$ where $fr(X,d) = |cover(X,d)|/n$ and $cover(X,d) = \{i : X \subseteq d_i, 1 \leq i \leq n\}$.

The set collection $\mathcal{F}(\sigma, d)$ is called the *collection of σ-frequent sets in d*.

There exist techniques to efficiently compute frequent sets, see e.g. [3]. A major advantage of frequent sets is that they can be computed from data without much domain knowledge: The data set determines an empirical joint probability distribution over the attribute combinations and high marginal probabilities of the joint probability distribution can be considered as a reasonable way to summarize the joint probability distribution. (Note that also a sample from the

H. Dai, R. Srikant, and C. Zhang (Eds.): PAKDD 2004, LNAI 3056, pp. 476–485, 2004.

joint probability distribution is a quite good summary.) However, this generality causes a major problem of frequent sets: the frequent set collections that describe data well tend to be large. Although the computations could be done efficiently enough, it is not certain that the huge collection of frequent sets is very concise summary of the data.

This problem of too large frequent set collections have been tried to solve by computing a small irredundant subcollection of the given frequent set collection such that the subcollection determines the frequent set collection completely. Such subcollections are usually called *condensed representations* of the frequent set collection [4]. The condensed representations of the frequent sets have been recognized to have an important role in *inductive databases* which seems to be a promising theoretical framework for data mining [5,6,7,8].

The condensed representations of frequent sets have been studied actively lately and several condensed representations, such as *maximal sets* [9], *closed sets* [10], *free sets* [11], *disjunction-free sets* [12], *disjunction-free generators* [13], *non-derivable itemsets* [14], *condensed pattern bases* [15], *pattern orderings* [16] and *pattern chains* [17], have been proposed. However, not much has been done on how the condensed representations should actually be represented although it is an important question both for the computational efficiency and for the effectiveness of the data analyst.

In this paper we investigate how the patterns and their interestingness values can be represented separately. In particular, we study how to represent frequent sets and their frequencies: we discuss how the collection of frequent sets can be described concisely by combinations of its maximal frequent and minimal infrequent sets, show that already reasonably small samples determine the frequencies of the frequent sets accurately and describe how a weighting of the sample can further improve the frequency estimates.

The paper is organized as follows. In Section 2 we argue why describing patterns and their interestingness values separately makes sense. In Section 3 we study a representation of interestingness values based on random samples of data and give sample complexity bounds that are sometimes considerably better than the bounds given in [18]. In Section 4 we describe how the samples can be weighted to optimally approximate the frequencies of the set collection w.r.t. a wide variety of loss functions and show experimentally that a considerable decrease of loss can be achieved. The work is concluded in Section 5.

2 Separating Patterns and Interestingness

Virtually all condensed representations of pattern collections represent the collection by listing a subcollection of the interesting patterns. This approach to condensed representations has the desirable closure property that also a subcollection of (irredundant) patterns is a collection of patterns. Another advantage of many condensed representations of pattern collections consisting of a list of irredundant patterns is that the patterns in the original collection and their interestingness values can be inferred conceptually very easily from the subcollection of irredundant patterns and their interestingness values.

The most well-known examples of condensed representations of pattern collections are the collections of *closed σ-frequent sets*:

Definition 1 (Closed σ-frequent sets). *A σ-frequent set $X \in \mathcal{F}(\sigma, d)$ is* closed *if and only if $fr(X, d) > fr(X \cup \{A\}, d)$ for all $A \in R \setminus X$.*
The collection of closed σ-frequent sets is denoted by $\mathcal{C}(\sigma, d)$.

The collection $\mathcal{C}(\sigma, d)$ consists of *closures* of the sets in $\mathcal{F}(\sigma, d)$, i.e., the sets $X \in \mathcal{F}(\sigma, d)$ such that $X = cl(X, d)$ where

$$cl(X, d) = \arg\max\{fr(Y) : Y \supseteq X, Y \in \mathcal{C}(\sigma, d)\}.$$

Clearly, the frequency of the set $X \in \mathcal{F}(\sigma, d)$ is the maximum of the frequencies of the closed frequent supersets of X, i.e., the frequency of its closure

$$fr(X, d) = fr(cl(X, d), d) = \max\{fr(Y) : Y \supseteq X, Y \in \mathcal{C}(\sigma, d)\}.$$

Although there are many positive aspects on representing the pattern collection and their interestingness values by an irredundant subset of the pattern collection, there are some benefits achievable by separating the structure of the collection from the interestingness values. For example, the patterns alone can always be represented at least as compactly as (and most of the time much more compactly than) the same patterns with their interestingness values. As a concrete example, the collection of frequent sets and the collection of closed frequent sets can be represented by their subcollection of maximal frequent sets:

Definition 2 (Maximal σ-frequent sets). *A σ-frequent set $X \in \mathcal{F}(\sigma, d)$ is* maximal *if and only if $Y \supseteq X, Y \in \mathcal{F}(\sigma, d) \Rightarrow X = Y$.*
The collection of maximal σ-frequent sets is denoted by $\mathcal{M}(\sigma, d)$.

Clearly, the frequent sets can be determined using the maximal frequent sets: a set $X \subseteq R$ is in the collection $\mathcal{F}(\sigma, d)$ if and only if it is a subset of some set in $\mathcal{M}(\sigma, d)$.

The collection of maximal frequent sets represents the collection of (closed) frequent sets quite compactly since $|\mathcal{M}(\sigma, d)|$ is never larger than $|\mathcal{C}(\sigma, d)|$ but it can be exponentially smaller than the corresponding closed frequent set collection (and thus the frequent set collection, too): Let the data set d consist of the sets $R \setminus \{A\}$ s.t. $A \in R$. Then the collection of (closed) $(1/n)$-frequent sets consists of all subsets of R except R itself but the collection of maximal frequent sets consists only of the sets $R \setminus \{A\}, A \in R$. Then $|\mathcal{F}(\sigma, d)| / |\mathcal{M}(\sigma, d)| = |\mathcal{C}(\sigma, d)| / |\mathcal{M}(\sigma, d)| > 2^{|R|}/|R|$. Also, the only case when the number of maximal frequent sets is equal to the number of frequent sets is when the collection of frequent sets consists solely of singleton sets, i.e., frequent items.

In addition to the maximal frequent sets, the frequent sets can be described also by the minimal infrequent sets:

Definition 3 (Minimal σ-infrequent sets). *A σ-infrequent set $X \in 2^R \setminus \mathcal{F}(\sigma, d)$ is* minimal *if and only if $Y \subseteq X, Y \in 2^R \setminus \mathcal{F}(\sigma, d) \Rightarrow X = Y$.*
The collection of minimal σ-infrequent sets is denoted by $\mathcal{I}(\sigma, d)$.

Again, obtaining the frequent sets from this representation is straightforward: a set $X \subseteq R$ is σ-frequent if and only if it does not contain any set $Y \in \mathcal{I}(\sigma, d)$.

Minimal infrequent sets for a given collection of frequent sets can be computed quite efficiently since the minimal infrequent sets are the minimal transversals in the complements of the maximal frequent sets [9].

It is not immediate which of the representations – the maximal frequent sets or the minimal infrequent sets – is smaller even in terms of the number of sets: the number $|\mathcal{M}(\sigma, d)|$ of maximal σ-frequent sets is bounded by $(|R| - \sigma n + 1) |\mathcal{I}(\sigma, d)|$ (unless the collection $\mathcal{I}(\sigma, d)$ minimal σ-infrequent sets is empty) but $|\mathcal{I}(\sigma, d)|$ cannot be bounded by a polynomial in $|\mathcal{M}(\sigma, d)|$, in $|R|$ and in n (i.e., the number of transactions in d) [19].

In practice, the representation for a given collection of frequent sets can be chosen to be the one that is smaller for that particular collection. Furthermore, instead of choosing either the maximal frequent sets or the minimal infrequent sets, it is possible to choose a subcollection determining the collection of frequent sets uniquely that contains sets from both collections:

Problem 2 (Smallest representation of frequent sets). Given collections $\mathcal{M}(\sigma, d)$ and $\mathcal{I}(\sigma, d)$, find the smallest subset $\mathcal{T}$ of $\mathcal{M}(\sigma, d) \cup \mathcal{I}(\sigma, d)$ that uniquely determines $\mathcal{F}(\sigma, d)$.

By definition, all subsets of minimal infrequent sets are frequent and all supersets of maximal frequent sets. Thus, Problem 2 can be modeled as a minimum weight set cover problem:

Problem 3 (Minimum weight set cover [20]). Given a collection $\mathcal{S}$ of subsets of a finite set S and a weight function $w : \mathcal{S} \to \mathbb{R}$, find a subcollection $\mathcal{S}'$ of $\mathcal{S}$ with the smallest weight $w(\mathcal{S}') = \sum_{X \in \mathcal{S}'} w(X)$.

The minimum weight set cover problem is approximable within a factor $1 + \ln |S|$ [20].

The set cover instance corresponding to the case of representing the frequent sets by a subcollection of $\mathcal{M}(\sigma, d) \cup \mathcal{I}(\sigma, d)$ is the following. The set S is equal to $\mathcal{M}(\sigma, d) \cup \mathcal{I}(\sigma, d)$. The set collection $\mathcal{S}$ consists of the set $S_X = \{Y \in \mathcal{M}(\sigma, d) \cup \mathcal{I}(\sigma, d) : X \subseteq Y \vee X \supseteq Y\}$ for each $X \in \mathcal{M}(\sigma, d) \cup \mathcal{I}(\sigma, d)$. Clearly, the solution $\mathcal{T} \subset \mathcal{M}(\sigma, d) \cup \mathcal{I}(\sigma, d)$ determines the collection $\mathcal{F}(\sigma, d)$ uniquely if and only if the solution $\mathcal{S}'$ for the corresponding set cover instance covers S. Furthermore, $w(\mathcal{S}') = |\mathcal{T}|$ when $w(X) = 1$ for all $X \in \mathcal{S}$. Due to this reduction and the approximability of Problem 3 we get the following result:

Theorem 1. *Problem 2 is approximable within a factor* $1 + \ln |\mathcal{M}(\sigma, d) + \mathcal{I}(\sigma, d)|$.

A very interesting variant of Problem 2 is the case when user can interactively determine which attributes or frequent sets must be represented. This can be modeled as a minimum set cover problem, too. In the case when user only adds attributes or frequent sets that must be represented this problem can be approximated almost as well as Problem 2 [21].

In addition to knowing which sets are frequent (or, in general, which patterns are interesting), it is usually desirable to know also how frequent each of the frequent sets is. One approach to describe the frequencies of the sets (approximately) correctly is to construct a small representative data set d' from the original data set d [22,23]. One computationally efficient and very flexible way to do this to obtain a random sample from the data set.

In the next two sections we study this approach. In Section 3 we give upper bounds on how many transactions chosen randomly from d suffice to give good approximations for the frequencies of all frequent sets simultaneously with high probability. The bounds can be computed from closed frequent sets which might be beneficial when each randomly chosen transaction is very expensive or the cost of the sample should be bounded above in advance. In Section 4 we show how the frequency estimates computed from the sample can be significantly improved by weighting the transactions using convex programming.

3 Sample Complexity Bounds

If the collection $\mathcal{F}(\sigma, d)$ is known then it can be shown by a simple application of Chernoff bounds that for a sample d' of at least $\ln(2|\mathcal{F}|/\Delta)/2\epsilon^2$ transactions, the absolute error of the frequency estimates for all sets in the collection is at most ϵ with probability at least $1 - \Delta$ [18].

However, the bounds given in [18] can be improved significantly since the frequent set collections have structure that can be useful when estimating the sufficient size for the sample. If the set collection $\mathcal{F}$ is known, the data set d itself is usually a compact representation of the covers of the sets in $\mathcal{F}$. The goal of sampling is to choose a small set of transactions that accurately determine the sizes of the covers for each set in $\mathcal{F}$. This kind of sample (also for arbitrary set collections in addition to the collections of covers) is called an *ϵ-approximation* [24]. The definition of ϵ-approximation can be expressed for covers of a set collection as follows:

Definition 4 (ϵ-approximation). *A finite subset $T \subseteq [n] = \{1, \ldots, n\}$ is an* ϵ-approximation *for the set collection $\mathcal{F}$ w.r.t. d if we have, for all $X \in \mathcal{F}$*

$$\left| \frac{|T \cap cover(X, d)|}{|T|} - \frac{|cover(X, d)|}{n} \right| \leq \epsilon.$$

The sample complexity bound given in [18] is essentially optimal in the general case. However, we can obtain considerably better bounds if we look at the structure of the set collection. One of such structural properties is the *VC*-dimension of the collection:

Definition 5 (*VC*-dimension). *The VC-dimension $VC(cover(\mathcal{F}, d))$ of the set collection $cover(\mathcal{F}, d) = \{cover(X, d) : X \in \mathcal{F}\}$ is*

$$VC(cover(\mathcal{F}, d)) = \max\left\{|T| : |\{T \cap cover(X, d) : X \in \mathcal{F}\}| = 2^{|T|}\right\}.$$

Given the *VC*-dimension of the collection of covers, the number of transactions that form an ϵ-approximation for the covers can be bounded above by the following lemma adapted from the corresponding result for arbitrary set collections [24]:

Lemma 1 (*VC*-**dimension bound**). *Let* $cover(\mathcal{F}, d)$ *be a set system of VC-dimension at most* k, *and let* $\epsilon \leq 1/2$. *Then there exists an* ϵ*-approximation for* $\mathcal{S}$ *of size at most* $\mathcal{O}(k \log(1/\epsilon))/\epsilon^2$

Actually, in addition to mere existence of small ϵ-approximations, an ϵ-approximation of size given by the *VC*-dimension bound can be found efficiently by random sampling [25].

One upper bound for the *VC*-dimension of a frequent set collection can be obtained from the number of closed frequent sets:

Theorem 2. *The VC-dimension of the collection* $cover(\mathcal{F}(\sigma, d), d)$ *is at most* $\log|\mathcal{C}(\sigma, d)|$ *where* $\mathcal{C}(\sigma, d)$ *is the collection of closed frequent sets in* $\mathcal{F}(\sigma, d)$. *This bound is tight in the worst case.*

Proof. The *VC*-dimension of $cover(\mathcal{F}(\sigma, d), d)$ is at most $\log|cover(\mathcal{F}(\sigma, d), d)|$. Clearly, $|cover(\mathcal{C}(\sigma, d), d)| \leq |\mathcal{C}(\sigma, d)|$. Thus it suffices to show that $cover(X, d) = cover(cl(X), d)$. However, this is immediate since, by definition, $cl(X, d)$ is contained in each transaction of d which contains X.

The *VC*-dimension $\log|cover(\mathcal{C}(\sigma, d), d)|$ is achieved by the collection of 0-frequent sets for a data set determined by n attributes $1, 2, \ldots, n$ with $cover(i, d) = [n] \setminus \{i\}$ for each $i \in [n]$. The the $\mathcal{F}(\sigma, d)$ (and $\mathcal{C}(\sigma, d)$, too) consists of all subsets of $[n]$. □

The cardinality of $\mathcal{C}(\sigma, d)$ can be estimated accurately by checking for random subset of frequent sets, how many of them are closed in d. Theorem 2 together with Lemma 1 imply the following upper bound:

Corollary 1. *For the set collection* $\mathcal{F}(\sigma, d)$, *there is an* ϵ*-approximation of* $\mathcal{O}(\log|\mathcal{C}(\sigma, d)| \log(1/\epsilon))/\epsilon^2$ *transactions.*

These bounds are quite general upper bounds neglecting many fine details of the frequent set collections and thus usually smaller samples give approximations with required quality. It is probable that the bounds could be improved by taking into account the actual frequencies of the frequent sets. A straightforward way to improve the upper bounds is to bound the *VC*-dimension more tightly. In practice, the sampling can be stopped right after the largest absolute error between the frequency estimates and the correct frequencies is at most ϵ.

4 Optimizing Sample Weights

Given a random subset d' of transaction in d, the frequencies in d for the sets in a given set collection $\mathcal{F}$ can be estimated from d. If also the correct frequencies are known, the error of the estimates can be measured.

Furthermore, the transactions in the sample can be weighted (in principle) in such a way that the frequency estimates are as good as possible. Let us denote the sample from $d = \{d_1, \ldots, d_n\}$ by $d' = \{d'_1, \ldots, d'_{n'}\}$ and let $w_1, \ldots, w_{n'}$ denote the weights of $d'_1, \ldots, d'_{n'}$. The frequency of X in the weighted sample d' is the sum of the weights w_i of d'_is containing X. Note that due to the weights we can assume that d' is a set although d is a multiset. Furthermore, we assume that all weights are nonnegative, i.e., no transaction in d' can be an anti-transaction.

The search for optimal weights w.r.t. a given loss function ℓ can be formulated as an optimization task as follows:

$$\begin{array}{rl} \text{minimize} & \ell(\mathcal{F}, d, d', w) \\ \text{subject to} & w_i \geq 0, i = 1, \ldots, n'. \end{array}$$

The weight w_i can be interpreted as a measure how representative the transaction d'_i is in the sample d' and thus the weights can be used also to score the transactions: On one hand the transactions with significantly larger weights than the average weight can be considered as very good representatives of the data. On the other hand the distribution of the weights tells about the skewness of the sample (and possibly also the skewness of the set collection $\mathcal{F}$) w.r.t. d. If the loss function ℓ is convex then the optimization task can be solved optimally in (weakly) polynomial time [26]. The most well-known examples of convex loss functions are L_p distances $\left(\sum_{X \in \mathcal{F}} \left| fr(X, d) - \sum_{X \subseteq d'_i \in d'} w_i \right|^p \right)^{1/p}$.

As a concrete example, let us consider the problem of minimizing the maximum absolute error of the frequency estimates. This variation of the problem can be formulated as a linear optimization task which can be solved even faster than the general (convex) optimization task:

$$\begin{array}{rl} \text{minimize} & \varepsilon \\ \text{subject to} & \epsilon \geq fr(X, d) - \displaystyle\sum_{X \subseteq d'_i, i \in [n']} w_i \\ & \epsilon \geq \displaystyle\sum_{X \subseteq d'_i, i \in [n']} w_i - fr(X, d) \\ & w_i \geq 0, i = 1, \ldots, n' \end{array}$$

The instance consists of $n' + 1$ variables and $2 |\mathcal{F}(\sigma, d)| + n'$ inequalities. The number of inequalities in the instance can be further reduced to $2 |\mathcal{C}(\sigma, d)| + |S|$ by recognizing that if $cl(X, d) = cl(Y, d)$ then the corresponding inequalities are equal, i.e., it is enough to have the inequalities for the collection $\mathcal{C}(\sigma, d)$ of closed frequent sets. This can lead to exponential speed-ups.

To evaluate the usability of weighting random samples of transactions, we experimented with the random sampling of transactions (without replacements) and the linear programming refinement using the IPUMS Census data set from UCI KDD Repository:[1] IPUMS Census data set consists of 88443 transactions and 39954 attributes. The randomized experiment were repeated 80 times.

The results are shown in Figure 1. The labels of the curves correspond to the minimum frequency thresholds and the curves with `(lp)` are the corresponding linear programming refinements. The results show that already a very small number of transactions, when properly weighted, suffice to give good approximations for the frequencies of the frequent sets. Similar results were obtained in our preliminary experiments with other data sets.

[1] `http://kdd.ics.uci.edu`

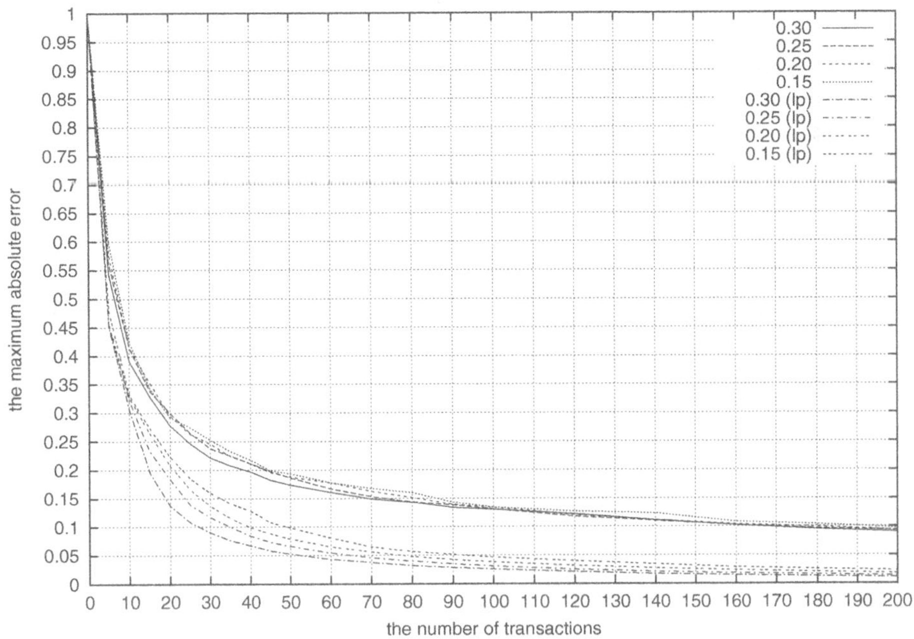

Fig. 1. IPUMS Census data

Note that although it would be desirable to choose the transactions that determine the frequencies best instead of random sample, this might not be easy since finding the best transactions with optimal weights resembles the cardinality constrained knapsack problem that is known to be difficult [27].

5 Conclusions

In this paper we studied how to separate the descriptions of patterns and their interestingness values. In particular, we described how frequent sets can be described in a small space by the maximal frequent sets, the minimal infrequent sets, or their combination. Also, we studied how samples can be used to describe frequencies of the frequent sets: we gave upper bounds for the sample complexity using closed frequent sets and described a practical convex optimization approach for weighting the transactions in a given sample.

Representing pattern collections and their interestingness values separately seems to offer some benefits in terms of understandability, size and efficiency. There are several interesting open problems related to this separation of the structure and the interestingness:

- What kind of patterns and their interestingness values can be efficiently and effectively described by a sample from the data?

- How a small data set that represents the frequencies of frequent sets can be found efficiently?
- What kind of weightings are of interest to determine the costs for the maximal frequent sets and the minimal infrequent sets?
- How the representation of the patterns guides the knowledge discovery process?

References

1. Hand, D.J.: Pattern detection and discovery. In Hand, D., Adams, N., Bolton, R., eds.: Pattern Detection and Discovery. Volume 2447 of Lecture Notes in Artificial Intelligence., Springer-Verlag (2002) 1–12
2. Mannila, H.: Local and global methods in data mining: Basic techniques and open problems. In Widmayer, P., Triguero, F., Morales, R., Hennessy, M., Eidenbenz, S., Conejo, R., eds.: Automata, Languages and Programming. Volume 2380 of Lecture Notes in Computer Science., Springer-Verlag (2002) 57–68
3. Goethals, B., Zaki, M.J., eds.: Proceedings of the Workshop on Frequent Itemset Mining Implementations (FIMI-03), Melbourne Florida, USA, November 19, 2003. Volume 90 of CEUR Workshop Proceedings. (2003) `http://CEUR-WS.org/Vol-90/`.
4. Mannila, H., Toivonen, H.: Multiple uses of frequent sets and condensed representations. In Simoudis, E., Han, J., Fayyad, U.M., eds.: Proceedings of the Second International Conference on Knowledge Discovery and Data Mining (KDD-96), AAAI Press (1996) 189–194
5. De Raedt, L.: A perspective on inductive databases. SIGKDD Explorations **4** (2003) 69–77
6. Imielinski, T., Mannila, H.: A database perspective on knowledge discovery. Communications of The ACM **39** (1996) 58–64
7. Mannila, H.: Inductive databases and condensed representations for data mining. In Maluszynski, J., ed.: Logic Programming, MIT Press (1997) 21–30
8. Mannila, H.: Theoretical frameworks for data mining. SIGKDD Explorations **1** (2000) 30–32
9. Gunopulos, D., Khardon, R., Mannila, H., Saluja, S., Toivonen, H., Sharma, R.S.: Discovering all most specific sentences. ACM Transactions on Database Systems **28** (2003) 140–174
10. Pasquier, N., Bastide, Y., Taouil, R., Lakhal, L.: Discovering frequent closed itemsets for association rules. In Beeri, C., Buneman, P., eds.: Database Theory - ICDT'99. Volume 1540 of Lecture Notes in Computer Science., Springer-Verlag (1999) 398–416
11. Boulicaut, J.F., Bykowski, A., Rigotti, C.: Free-sets: a condensed representation of Boolean data for the approximation of frequency queries. Data Mining and Knowledge Discovery **7** (2003) 5–22
12. Bykowski, A., Rigotti, C.: A condensed representation to find frequent patterns. In: Proceedings of the Twenteenth ACM SIGACT-SIGMOD-SIGART Symposium on Principles of Database Systems, ACM (2001)
13. Kryszkiewicz, M.: Concise representation of frequent patterns based on disjunction-free generators. In Cercone, N., Lin, T.Y., Wu, X., eds.: Proceedings of the 2001 IEEE International Conference on Data Mining, IEEE Computer Society (2001) 305–312

14. Calders, T., Goethals, B.: Mining all non-derivable frequent itemsets. In Elomaa, T., Mannila, H., Toivonen, H., eds.: Principles of Data Mining and Knowledge Discovery. Volume 2431 of Lecture Notes in Artificial Intelligence., Springer-Verlag (2002) 74–865
15. Pei, J., Dong, G., Zou, W., Han, J.: On computing condensed pattern bases. In: Proceedings of the 2002 IEEE International Conference on Data Mining (ICDM 2002), IEEE Computer Society (2002) 378–385
16. Mielikäinen, T., Mannila, H.: The pattern ordering problem. In Lavrac, N., Gamberger, D., Todorovski, L., Blockeel, H., eds.: Knowledge Discovery in Databases: PKDD 2003. Volume 2838 of Lecture Notes in Artificial Intelligence., Springer-Verlag (2003) 327–338
17. Mielikäinen, T.: Chaining patterns. In Grieser, G., Tanaka, Y., Yamamoto, A., eds.: Discovery Science. Volume 2843 of Lecture Notes in Artificial Intelligence., Springer-Verlag (2003) 232–243
18. Toivonen, H.: Sampling large databases for association rules. In Vijayaraman, T., Buchmann, A.P., Mohan, C., Sarda, N.L., eds.: VLDB'96, Proceedings of 22nd International Conference on Very Large Data Bases, Morgan Kaufmann (1996) 134–145
19. Boros, E., Gurvich, V., Khachiyan, L., Makino, K.: On the complexity of generating maximal frequent and minimal infrequent sets. In Alt, H., Ferreira, A., eds.: STACS 2002. Volume 2285 of Lecture Notes in Computer Science., Springer-Verlag (2002) 133–141
20. Ausiello, G., Crescenzi, P., Kann, V., Marchetti-Spaccamela, A., Protasi, M.: Complexity and Approximation: Combinatorial Optimization Problems and Their Approximability Properties. Springer-Verlag (1999)
21. Alon, N., Awerbuch, B., Azar, Y., Buchbinder, N., Naor, J.S.: The online set cover problem. In: Proceedings of the 35th Annual ACM Symposium on Theory of Computing, ACM (2003) 100–105
22. Ioannidis, Y.: Approximations in database systems. In Calvanese, D., Lenzerini, M., Motwani, R., eds.: Database Theory - ICDT 2003. Volume 2572 of Lecture Notes in Computer Science. (2003)
23. Mielikäinen, T.: Finding all occurring sets of interest. In Boulicaut, J.F., Džeroski, S., eds.: 2nd International Workshop on Knowledge Discovery in Inductive Databases. (2003) 97–106
24. Matoušek, J.: Geometric Discrepancy: An Illustrated Guide. Volume 18 of Algorithms and Combinatorics. Springer-Verlag (1999)
25. Chazelle, B.: The Discrepancy Method: Randomness and Complexity. Paperback edn. Cambridge University Press (2001)
26. Ben-Tal, A., Nemirovksi, A.: Lectures on Modern Convex Optimization: Analysis, Algorithms, and Engineering Applications. Volume 2 of MPS-SIAM Series on Optimization. SIAM (2001)
27. de Farias Jr., I.R., Nemhauser, G.L.: A polyhedral study of the cardinality constrained knapsack problem. Mathematical Programming **96** (2003) 439–467

Exploiting Recurring Usage Patterns to Enhance Filesystem and Memory Subsystem Performance

Benjamin Rutt and Srinivasan Parthasarathy

The Ohio State University
Computer Science and Engineering
395 Dreese Laboratories
2015 Neil Ave
Columbus, Ohio 43210-1277
rutt.4@osu.edu, srini@cis.ohio-state.edu

Abstract. In many cases, normal uses of a system form patterns that will repeat. The most common patterns can be collected into a prediction model which will essentially predict that usage patterns common in the past will occur again in the future. Systems can then use the prediction models to provide advance notice to their implementations about how they are likely to be used in the near future. This technique creates opportunities to enhance system implementation performance since implementations can be better prepared to handle upcoming usage.
The key component of our system is the ability to intelligently learn about system trends by tracking file system and memory system activity patterns. The usage data that is tracked can be subsequently queried and visualized. More importantly, this data can also be mined for intelligent qualitative and quantitative system enhancements including predictive file prefetching, selective file compression and and application-driven adaptive memory allocation. We conduct an in-depth performance evaluation to demonstrate the potential benefits of the proposed system.

1 Introduction

System interfaces cleanly separate user and implementor. The user is concerned with invoking the system implementation with legal parameters and in a correct state (e.g. only writing to a valid file handle). The implementor is concerned with providing an efficient and correct solution to a systems-level problem (e.g. completing a file write operation as quickly as possible, returning an error if the operation has failed). Yet these two opposite ends of system interfaces can be designed, built, and deployed without considering the usefulness of information that may not be available until runtime. Systems software components can be optimized by adapting their implementations to how they are being used.

For example, if a file system knows in advance that a particular user will soon open a specific set of files in a specific order, the file system can anticipate those operations and pull data that is about to be opened into filesystem cache, to speed up future operations. Or a dynamic memory manager that is expecting

H. Dai, R. Srikant, and C. Zhang (Eds.): PAKDD 2004, LNAI 3056, pp. 486–496, 2004.

a certain type of allocation requests from a particular program can optimize the memory manager to better serve the needs of the program. Although there is a wealth of data about system usage available at runtime, only a portion of the available information is useful. It remains a challenge to collect, analyze and mine the most useful bits of knowledge from the sea of available data. In this paper, we discuss tools and methods for the collection, analysis and mining of system usage data. The key contributions of our work are:

1. mining user-specific file system traces and enabling personalized prefetching of the user's file system space
2. demonstrating the viability of automatically generating application-specific suballocators and providing tools to automate the process

The rest of this paper is organized as follows. Section 2 discusses related work. Sections 3 and 4 detail the architecture and experimental results for a specialized file system. Sections 5 and 6 detail the architecture and experimental results for a specialized memory allocator. Section 7 discusses conclusions and Sect. 8 discusses future work.

2 Related Work

Kroeger and Long [1] evaluated several approaches to file access prediction. The model they promoted used frequency information about file accesses to develop a probabilistic model which predicted future accesses. However, the downside of storing frequency information about successors is that the memory requirements are high, with potentially little gain if rarely-used successors are stored; in our *N-gram* approaches, we aimed to reduce the state information over these frequency approaches. The *N-gram* prediction model is commonly used in speech-processing research [2], and associates a sequence of events of length N with the event that follows. Su et al. [3] extended the *N-gram* with an *N-gram+* model, a collection of *N-grams* that work together to make predictions. Mowry et al. built a system [4,5] which automatically inserts prefetch and release instructions into programs that work with large (out-of-core) data sets, to reduce virtual memory I/O latency. In [6], the authors develop Markov models based on past WWW accesses to predict future accesses; however, they base predictions on the history of many users, whereas in our work we personalize our predictions per user.

In [7], Grunwald and Zorn develop *CustoMalloc*, a tool that develops a customized allocator to replace `malloc` in a C program, given the program's source code. CustoMalloc inspired the work we have done with dynamic memory management. The primary difference between CustoMalloc and our system is that CustoMalloc requires access to program source-code, while ours does not. In [8], Parthasarathy et al. developed a customized memory allocator for parallel data mining algorithms that reduced program execution time. By observing how data structures were being accessed, they took care to place data structure nodes in a way that reduced false sharing yet increased spatial locality.

3 Prefetching Filesystem Architecture

This section describes an architecture that increases filesystem performance through prefetching. Figure 1 displays an architectural overview; each major component of this architecture will now be discussed.

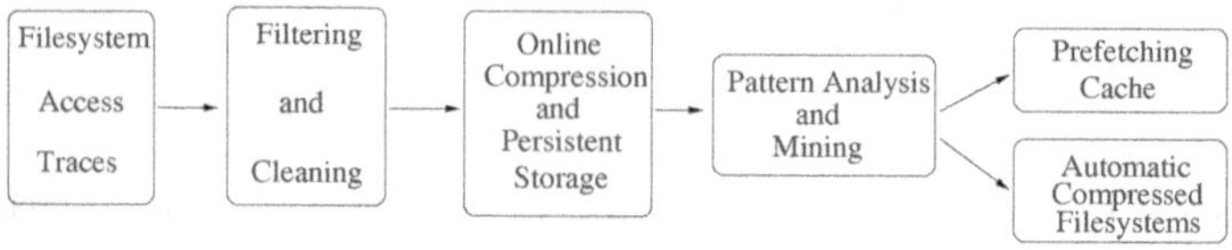

Fig. 1. A specialized prefetching filesystem

3.1 Filesystem Access Traces

In order to apply filesystem access trace data in useful ways, we obviously needed a source of filesystem traces. We decided to generate our own traces. Our test machine was a Solaris 2.8 system with four 296-MHz CPUs and three Gigabytes of RAM. We designed a program that would start up when a login shell was invoked by a user and shutdown when that user logged out. This way, each set of filesystem access traces was generated on a per-user basis, resulting in no interference from other users or system processes. Our users consisted of students who simply performed their normal work on the system for several weeks while the tracing program was enabled. For example, they checked email, edited files, browsed web sites, compiled programs, etc. The attributes we stored per each file access included the name of the file accessed, the date and time of the access, the access mode (read, write, or read+write), the application that accessed the file and the process ID that accessed the file.

3.2 Pattern Analysis and Mining

There are several models presented for usage pattern analysis, including the *1-gram*, *2-gram+*, *p-s-gram+* and *association rule mining* models. Following the discussion of the prediction models, an application of the prediction models will be discussed, a *prefetching cache.*

p-s-gram+ models. As mentioned in Sect. 1, an *N-gram* associates a sequence of events of length N with the event that follows. The *1-gram* approach is an instance of the *N-gram* model where $N = 1$, and all events are file accesses. It has also been described as a *last-successor* model [1], because it records the last successor of each file access. When used in prediction, the *1-gram* simply predicts that the same file to succeed a given file the last time around will also succeed the file the next time around. The *2-gram+* approach is an *N-gram+* where the maximum N is 2. It subsumes both a 2-gram and a 1-gram and maintains tables for both, giving priority to the 2-gram table for predictions.

The *p-s-gram+* model is a generalization of the *N-gram+* model. Rather than predicting only one successor file, it can be configured to predict any number of successors (s). The number of predecessors (p) is configurable as usual. The system supports fall back for p so that if the predictor with the largest predecessor count cannot cannot make a prediction, the predictor with second-largest predecessor count is consulted, and so on. For example, if the file accesses `[A B C A Z Y]` were made previously, a *2-3-gram+* would build the prediction tables at the left in Fig. 2.

When used in prediction, the *p-s-gram+* performs exactly like the *N-gram+*, except that multi-file predictions can be made. For example, if the above tables had been built and the input `BC` was given to the *p-s-gram+*, the output would be `AZY`, meaning that files `A`, `Z` and `Y` are all predicted to be accessed soon.

Although the *p-s-gram+* and similar models are somewhat limited in the sense that they only keep a single collection of successor values compared to an approach that keeps a probability distribution of successor values, one advantage is that the implementation for prediction table maintenance is simplified. Also, outdated successors are removed from the tables automatically when newer successors come along, thus avoiding an expiration policy for rare successors.

Cache simulations revealed that the 1-predecessor and 5-successor combination performed as good as any of the *p-s-gram+* models tested. Therefore, the *1-5-gram+* was chosen to represent the *p-s-gram+* model in this paper.

Example p-s-gram+ table

2-3-gram

key	value
AB	C A Z
BC	A Z Y

1-3-gram

key	value
A	B C A
B	C A Z
C	A Z Y

Example ARM table

Antecedent	Consequent
A	B,C,Z
BC	A,Z,Y
ABC	Z
BCAZ	Y

Fig. 2. Prediction tables

Association Rule Mining. The *Association Rule Mining* approach (*ARM*) builds association rules [9] out of the file accesses that have been seen. The *Apriori* [10] algorithm is applied to the training data to yield a set of association rules that can be used during testing. This approach does not consider the order of file accesses. Rather, it produces rules pertaining to all files accessed within the same *window size* (configured to 5 accesses). For example, if the file accesses `[A B C A Z Y]` were made during training, *ARM* might build the rightmost table of Fig. 2 (for brevity, only a few rules are shown).

By definition, the *ARM* approach subsumes the rules generated by the *p-s-gram+*-based models. In addition, *ARM* will generate rules about files associated a few accesses apart, in any order. Each rule also has a value for *support* (how frequently all files mentioned in the rule appear together, compared to all files that appear together) and *confidence* (what percentage of the time the consequent of the rule appears, if the antecedent of the rule appears).

When used in prediction, *ARM* uses some portion of recent file accesses to generate a list of possible predictions. This list is sorted by support, then by confidence, and finally by length of the antecedent of the rule. Sorting by this order yielded the best results on average during testing than sorting by any other possible sort order. Finally, a configurable number of predictions with highest support are made.

Prefetching Cache. Using these prediction models, a *prefetching cache* can be built. After each file access is made, a cache system could use a prediction model to determine what files are expected to be accessed soon. If the prediction model returned any predictions and there was spare I/O bandwidth, the cache could begin to fill with data from the predicted files, in the hopes that future regular accesses for the predicted files would result in cache hits. Finally, the system could make updates to reflect recent accesses. Instead of just prefetching a file, the system could also uncompress a compressed file. A simulation of prefetching caches is presented in Sect. 4.

4 Results from a Prefetching Cache Simulation

Model	Hit Rate
non-predicting	63%
1-gram	84%
2-gram+	87%
1-5-gram+	90%
ARM	90%
hybrid	91%

Fig. 3. Hit rate comparison

The effectiveness of the prediction models was tested via a filesystem cache simulation. The data set used in this paper is referred to as *vip-fstrace* and contains a single graduate student's file access traces over a 2-week period, totaling around 20,000 file accesses; we have performed the same experiments using data sources from other users as well, with comparable results. All of the file accesses in *vip-fstrace* were captured during normal day-to-day use of a filesystem. In the experiments that follow, *cachesize* defines the number of files the cache can hold. In this paper, we only consider whole-file caching. *Training amount* defines what percentage of the file accesses were used as training data to build the prediction model. During training, no statistics on cache hits were kept.

There were a number of caching strategies that were simulated, including the *non-predicting* model, the *1-gram* model, the *2-gram+* model, the *1-5-gram+* model, and the *association rule mining* model. In addition, we built a *hybrid* model out of the top-performing *ARM* and *1-5-gram+* models. The *hybrid* model simply consults each of its sub-models for predictions, and makes any predictions that either sub-model offers.

The *non-predicting* approach is a baseline cache that has no knowledge of the file access histories. This cache simply adds recently accessed files to the MRU end and removes items from the LRU to make room for new entries in cache. When a file in cache is accessed, it becomes the MRU object. The *non-predicting* approach can still yield a high hit rate, given sufficient re-use of recently accessed files. The other caches were built by augmenting the baseline cache with one of the forms of prediction discussed in this paper. Figure 3 shows a direct comparison of the caching strategies.

The *hybrid* approach offers an advantage in that predictions missed by one model are caught by the other, which explains the slight increase in hit rate for the *hybrid* model over its components.

A closer look at the *1-5-gram+*, *ARM* and *hybrid* hit rates reveals how close to optimal the predictors can be. These caches missed around 10% of the time. However, further investigation showed that 75% of these misses were due to accesses during testing for files that did not exist during training. For these files, it would be impossible to generate any predictions anyway, since no patterns involving these files have been seen. So, the optimal hit rate with this data set would be about 92.5% anyway.

Figure 3 used the *vip-fstrace* data, with *cachesize* set to 70 files and *training amount* set to 55%. For the *ARM* system, *max predictions* were set to 9 (meaning that up to 9 files will be prefetched into filesystem cache), *support threshold* was set to 0.005 and *confidence threshold* was set to 0.5. Unless otherwise specified, the remaining experiments all use this configuration.

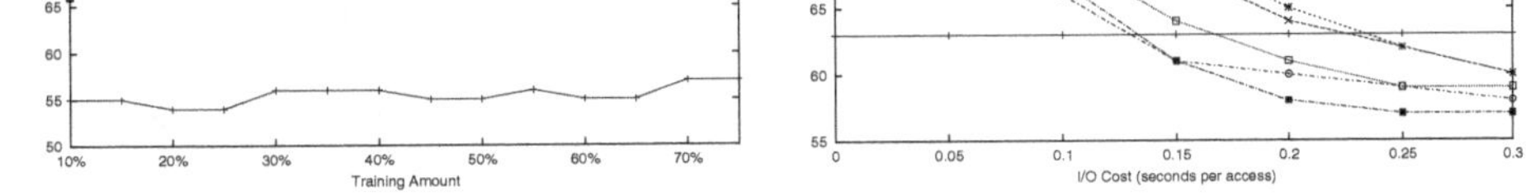

Fig. 4. Hit rates as training data varies, and the effect of varying I/O cost

4.1 Varying Training Data Amount

The left figure of Fig. 4 shows the hit rate performance of the various approaches to caching as training size varies. Not surprisingly, all of the approaches that actually trained performed better with a larger training data amount. The results were obtained by averaging together 5 unique random samples, each using 50%

of the available records for both training and testing purposes. For all training amounts larger than 50%, all predictive approaches yielded hit rates about 1.5 times greater than the *non-predicting* approach.

4.2 Prefetch Arrival Time

When a prefetch is made, time is spent pulling the file into cache to ensure it is available in cache for a later fetch. It is crucial that most of the time, prefetched files arrive in cache before the later fetch is made. If the prefetch is still in progress when the predicted fetch occurs, then the prefetch was not only a waste of time, but also may have caused unrelated I/O to block. On the test system, the cost of each file access was determined to be between 0.001 and 0.002 seconds. However, the cost of performing I/O is system-dependent, and varies with each instance, depending on available I/O bandwidth at access time.

The right figure of Fig. 4 shows the cache hit rate as the cost of I/O varies. As the cost of I/O increases, performance degrades for all of the prefetching approaches. With sufficiently slow I/O, all of the prefetching systems degrade to worse than the non-prefetching approach. For all I/O costs tested, the *ARM*-based approaches (*ARM*, *hybrid*) performed, on average, 1.4 times better than the *p-s-gram+*-based approaches.

5 Application Driven Adaptive Memory Allocation

A general purpose dynamic memory manager has the burden of providing good performance to a variety of applications. In order to achieve this goal, some sacrifices must be made. For example, a general purpose memory manager has to develop a general strategy for exactly how and when to reserve additional heap memory for a process via an expensive system call such as `sbrk`. A general purpose memory manager must also make a decision as to its arrangement of free lists. In order to avoid having to reserve additional heap memory to satisfy every `malloc` request, most memory managers will reserve a large block of heap memory at once and incorporate unused portions of the heap memory block into free lists of various sizes. Then, when a `malloc` request arises of a given size, it may be possible to satisfy the request by retrieving an item from one of the free lists, thus minimizing the number of times that more heap memory must be acquired. The optimal arrangement of these free lists varies per application.

Sometimes programmers develop application-specific memory managers or *suballocators* to try to leverage their own knowledge of their programs. While `malloc` is designed to handle many request sizes, a suballocator could be written to handle the most frequently allocated size in an optimal manner. However, developing these per-application suballocators is time consuming in terms of programmer time and there is no guarantee that the suballocator will improve performance. Ideally, the problem of when and how to optimize dynamic allocation on a per-application basis could be solved generically.

To generically solve the problem of per-application allocation, we developed a system which automatically profiles an application's allocations during initial runs of the program, helps to build a dedicated suballocator for the program, and utilizes the suballocator to enhance performance for future runs of the program. All of this can be done without recompiling the application program.

In order to profile the allocation behavior of a program, special tracing versions of `malloc`, `reallies`, `calloc` and `free` were developed. Using the `LD_PRELOAD` environment variable shared-library hook technique, these special tracing versions replaced the standard C library versions during a program's execution. They recorded each allocation and free that occurred during a program's execution to a history file, along with the allocation size.

We developed an analysis tool to analyze the allocation history file. The analysis tool identified how many allocations were of each size, and which sizes were allocated most frequently, and how often a request for a block of given size was made after the release of a block of the same size (referred to in this paper as a *reuse opportunity*). The tool also reported the maximum number of blocks for each size that were simultaneously freed yet later could be reused; this guided the length of the free lists for each size.

The analysis tool also produced a graph of the total memory in use during each allocation or deallocation event. By viewing this graph, you can quickly see whether or not an application could take advantage of enhanced reuse functionality. For example, the left graph of Fig. 5 shows an example of an application where memory is acquired, released and then reacquired. Each reacquisition of a block of memory represents a reuse opportunity. On the other hand, the right graph of Fig. 5 shows an example of an application with a monotonically increasing memory footprint, which means there are no reuse opportunities.

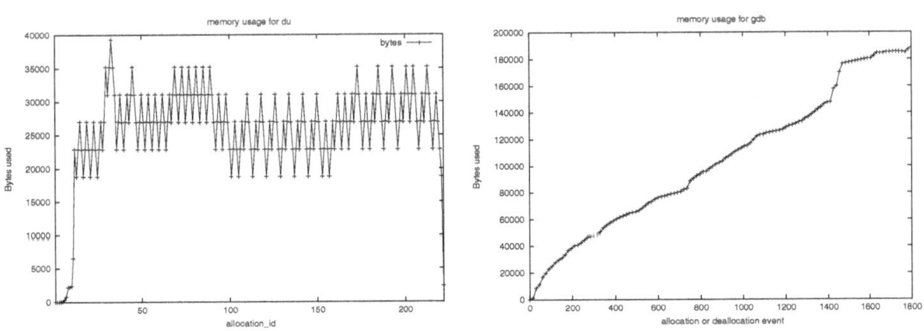

Fig. 5. Different applications have different reuse opportunities

To build a dedicated suballocator for a program, we specified the following properties: the initial size of the heap, the amount the heap should grow when needed, and the size class and length of between zero and four free lists. It has been observed [7] that over 95% of allocation requests could typically be satisfied by using only four free lists, so we only used up to four free lists. For the applications we evaluated, we found this to be true as well. The construction of the suballocator was complete when the above properties were written to a *properties file* corresponding to that application.

Finally, to utilize the suballocator, special versions of `malloc`, `realloc`, `calloc` and `free` were developed. Using the same `LD_PRELOAD` shared-library hook technique that enabled the tracing, these special methods replaced the standard C library versions during the program's execution. Before the program is executed, this special version of `malloc` was initialized. During initialization, the file containing the properties that specified the details of the suballocator was read into memory. The default heap was extended to be as big as the initial size of the heap memory as specified in the file.

If there were any free lists specified in the properties file, then they were initialized at this time. All of the nodes constituting each free list are stored contiguously, starting from the beginning of the heap. If an allocation request cannot be satisfied using one of the free lists, a general purpose first-fit allocation strategy is used.

6 Results from a Specialized Memory Manager

This section reveals some notable performance improvements in program execution time discovered by trying out suballocators on a variety of applications. In each case, at least five runs were executed with and without the suballocator, and the timings for each were averaged. For all of the tests which accessed files, several initial "priming" runs were executed and the results thrown away, to eliminate the effect of file system cache. The test system was a GNU/Linux system with the GNU C library version 2.2.5 installed.

There were actually many applications which showed no benefit during testing; in fact, in some cases, the suballocators slowed down the performance. This reflects the reality that the default dynamic allocator does an excellent job in many cases, and is difficult to beat on average. Since the suballocator can be selectively turned on or off per-application, it can be leveraged where it is needed most, and turned off otherwise.

Figure 6 lists some applications whose performance were enhanced via a suballocator. All of the above performance improvements were realized without requiring source-code access to any of the programs. These performance improvements were discovered after modest amounts of analysis and re-engineering of the suballocators to find the best possible performance.

7 Conclusions

We have several specific conclusions to make related to a prefetching filesystem cache. The *1-5-gram+* model and the *ARM* model are both excellent, near-optimal predictive models for file accesses. A hybrid combination of these appears to do even slightly better, as shown in Fig. 3, as predictions missed by one

application	default	suballocator	improvement
`convert`: convert a 0.5 MB image from JPEG to GIF format	20.70 sec	20.13 sec	2.83%
`tar`: unpacking the linux kernel sources	123.77 sec	114.75 sec	7.87%
`gawk`: update 100,000 key value pairs via gawk's associative arrays	11.83 sec	10.95 sec	8.04%
`zip`: use Info-ZIP to compress 6MB of text files	5.04 sec	4.59 sec	9.68%

Fig. 6. Performance improvements using suballocators

model are caught by the other. The lower space requirements and simpler, more efficient pattern analysis make the *1-5-gram+* model attractive. The *ARM* model performs best when it makes a limited number of predictions; unrestricted, *ARM* would have so many predictions to make that it would end up polluting the cache.

We have explored the idea of using memory usage patterns to enhance dynamic allocation performance. We discovered applications whose overall program execution time improved by nearly 10% by utilizing suballocators, without requiring access to the applications' source code. These per-application improvements indicate that for some applications, there are repetitious memory allocation patterns that are manifested across multiple application runs. When such patterns exist, they can clearly be leveraged to improve performance.

8 Future Work

Future work in the area of file systems includes the full-blown implementation of a prefetching file system cache with automatic decompression. The implementation work will involve replacing the existing file system's cache infrastructure with our own, and measuring changes in performance. The implementation will want to consider the size of prefetched files, to avoid, in some cases, prefetching files that are so large that they will poison the cache.

In the area of dynamic memory management, we would like to integrate into a state of the art dynamic memory manager the ability to generate per-application suballocators. This would avoid the need to use a dynamic library feature of GNU/Linux to enable the suballocators, a feature that is not necessarily available on all platforms. In addition, we would like to fully automate the process of suballocator generation.

References

1. T. M. Kroeger and D. D. E. Long. The case for efficient file access pattern modeling. In *Proceedings of the 1996 USENIX Technical Conference*, January 1996.
2. K. F. Lee and S. Mahajan. *Automatic Speech Recognition: The Development of the SPHINX System.* Kluwer Publishers, 1989.

3. Ye Lu Zhong Su, Qiang Yang and Hong-Jiang Zhang. Whatnext: A prediction system for web requests using n-gram sequence models. In *Proceedings of the First International Conference on Web Information System and Engineering Conference*, 2000.
4. Todd C. Mowry, Angela K. Demke, and Orran Krieger. Automatic compiler-inserted I/O prefetching for out-of-core applications. In *Proceedings of the 1996 Symposium on Operating Systems Design and Implementation*, pages 3–17. USENIX Association, 1996.
5. Angela Demke Brown and Todd C. Mowry. Taming the memory hogs: Using Compiler-Inserted releases to manage physical memory intelligently. In *Proceedings of the 4th Symposium on Operating Systems Design and Implementation (OSDI-00)*, pages 31–44, 2000.
6. I. Zukerman, D. Albrecht, and A. Nicholson. Predicting users' requests on the WWW. In *UM99 – Proceedings of the Seventh International Conference on User Modeling*, 1999.
7. Dirk Grunwald and Benjamin G. Zorn. Customalloc: Efficient synthesized memory allocators. *Software - Practice and Experience*, 23(8):851–869, 1993.
8. Srinivasan Parthasarathy, Mohammed Javeed Zaki, and Wei Li. Memory placement techniques for parallel association mining. In *Knowledge Discovery and Data Mining*, pages 304–308, 1998.
9. R. Agrawal and R. Srikant. Fast algorithms for mining association rules. In *Proceedings of the 20th International Conference on Very Large Databases*, September 1995.
10. Christian Borgelt and Rudolf Kruse. Induction of association rules: A priori implementation. http://citeseer.nj.nec.com/547526.html.

Automatic Text Extraction for Content-Based Image Indexing

Keechul Jung[1] and Eun Yi Kim*[2]

[1] School of Media, College of Information Science, Soongsil University, South Korea
kcjung@ssu.ac.kr

[2] Dept. of Internet and Multimedia Eng., Konkuk University, South Korea

Abstract. This paper proposes an approach for automatic text extraction method using neural networks. Automatic text extraction is a crucial stage for multimedia data mining. We present an artificial neural network(ANNs)-based approach for text extraction in complex images, which uses a combined method of ANNs and non-negative matrix factorization(NMF)-based filtering. An automatically constructed ANN-based classifier can increase recall rates for complex images with small amount of user intervention. NMF-based filtering enhances the precision rate without affecting overall performance. As a result, a combination of two learning mechanism leads to not only robust but also efficient text extraction.

1 Introduction

Content-based image indexing is the process of attaching content-based labels to images and video frames. Based on the fact that texts within an image are very useful for describing the contents of the image and can be easily extracted comparing with other semantic contents, researchers have attempted text-based image indexing using various image processing techniques [1-7, 14, 15]. In the text-based image indexing, automatic text extraction is very important as a prerequisite stage for optical character recognition, and it has been still considered a very difficult problem because of text variations in size, style, and orientation as well as a complex background of images. There are two primary methods for text extraction: connected component (CC) methods (CCMs) and texture-based methods. The CCMs [2,4] are very popular for text extraction thanks to their simplicity in implementation, but are not appropriate for low-resolution and noisy video documents because they depend on the effectiveness of the segmentation method and it is not easy to filter non-text components using only geometrical heuristics. Unlike the CCMs, the texture-based methods [1,3,5,7] regard text regions as textured objects. Although very effective in text extraction, the texture-based methods have some shortcomings: difficulties in manually designing a texture classifier for various text conditions, locality of the texture information, and expensive computation in the texture classification stage.

This paper proposes an approach for attacking several difficulties for text extraction in complex color images, and shows its affirmative results. Based on the

* Corresponding Author (eykim@konkuk.ac.kr)

H. Dai, R. Srikant, and C. Zhang (Eds.): PAKDD 2004, LNAI 3056, pp. 497–507, 2004.

sequential hybrid merging methodology of the texture and CC-based methods, we aim to increase a recall rate with multi-layer perceptrons (MLPs) and a precision rate with CC analysis (Figure 1). MLPs automatically generate a texture classifier that discriminates between text regions and non-text regions on three color bands. Unlike other neural network-based text extraction methods, the MLPs receive raw image pixels as an input feature. Therefore, complex feature designing and feature extraction can be avoided. Bootstrap method is used to force the MLPs to learn a precise boundary between text and non-text classes, which results in reliable detection performance for various conditions of texts in real scenes.

Although we use the bootstrap method, the detection results from the MLPs include many false alarms. In order to tackle this shortcoming of the texture-based method, we use CC-based filtering using a NMF technique: we have applied three stage of filtering on CCs using: features of CCs such as area, fill factor, and horizontal and vertical extents; geometric alignment of text components; and part-based shape information using NMF.

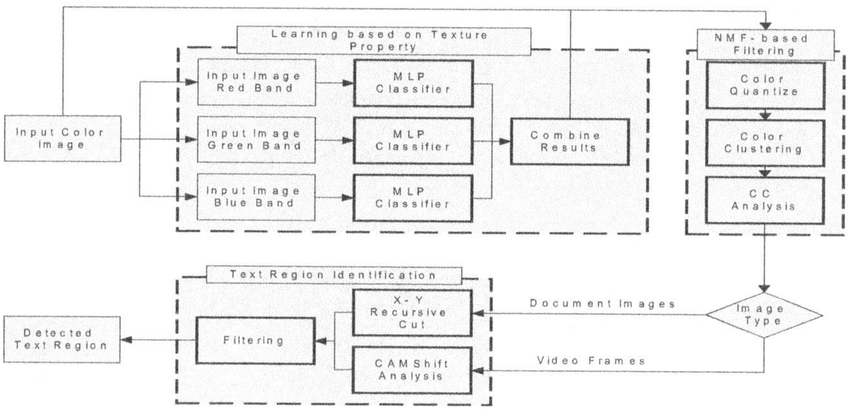

Fig. 1. Overall Structure.

For the last step, we use two different region-marking approaches according to the given image types (video image and document image) both to enhance time performance and to get better results. For video images, we use CAMShift, and for document images, we does X-Y recursive cut algorithm.

2 Neural Networks for Text Detection

We use MLPs to make a texture classifier that discriminates between text pixels and non-text ones. Readers refer the author's previous publication [7]. An input image is scanned by the MLPs, which receives the color values of a given pixel and its neighbors within a small window for each Red, Green, and Blue color band. The three MLPs' outputs are combined to text probability images (TPI), where each pixel's value is in the range [0,1] and represents the probability that the corresponding input pixel is a part of text. The pixel that has a larger value than given a threshold value is considered as a text pixel.

Figure 2 shows the structure of the network for text location. A neural network-based arbitration method is used to describe the influence of each color band. An arbitration neural network produces a final result based on the outputs of the three neural networks. After the pattern passes the network, the value of the output node is compared with a threshold value and the class of each pixel is determined. As a result of this classification, a classified image is obtained.

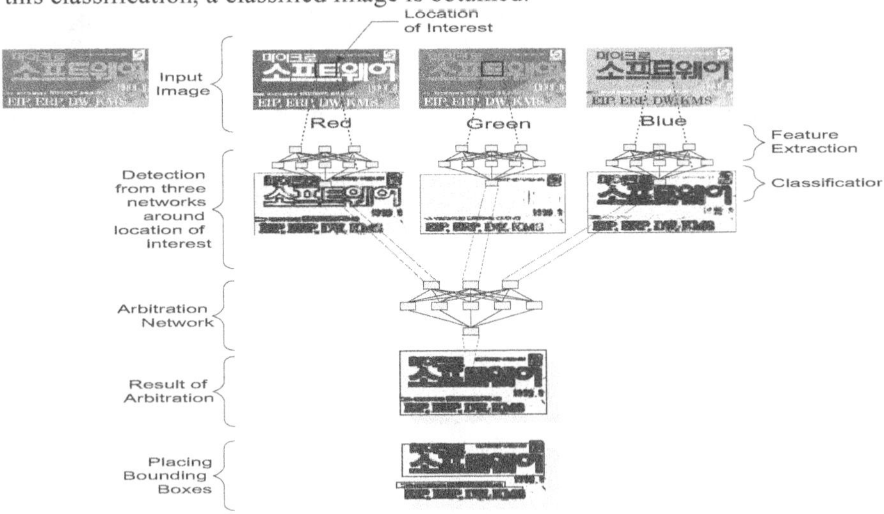

Fig. 2. Architecture of a discrimination network

When using an example-based-learning approach, it would be desirable to make the training set as large as possible in order to attain a comprehensive sampling of the input space. However, when considering real-world limitations, the size of the training set should be moderate. Then the problem is how to build a comprehensive but tractable database. For text samples, we simply collect all text images we can find. However collecting non-text samples is more difficult. Practically infinite images can serve as valid non-text samples. To handle this problem we use the bootstrap method recommended by Sung and Poggio [9], which was initially developed to train neural networks for face detection. Some non-text samples are collected and used for training. Plus, the partially trained system is repeatedly applied to images, which do not contain texts, and then patterns with a positive output are added to the training set as non-text samples. This process iterates until no more patterns are added to the training set. This bootstrap method is used to get non-text training samples more efficiently and to force the MLP to learn a precise boundary between text and non-text classes.

3 Part-Based Component Filtering

Although we use the bootstrap method to make the texture classification MLPs learn precise boundary between a text class and non-text one, the detection result from the MLPs includes many false alarms because we want the MLPs to detect texts as many as possible. In order to tackle this shortcoming of the texture-based method, heuristics such as size and aspect ratio of detected regions are used to filtering non-text regions in previous researches [2,4,6]. However we still have problems in filtering-out high-frequency and high-contrast non-text regions. This is mainly because of the locality of the texture information.

To perform component filtering, the output of the MLPs, or each text region is first passed through a 5×5 median filter, which eliminates spurious text pixels. We then quantize the color of the input image pixels, corresponding to the text regions, into 512 levels (3 bits per color). The different colors present in the image are then grouped together using the single-link clustering algorithm to form clusters of similar colors. To start with, the clustering algorithm defines each color to be in a group of its own. At each step, the single-link clustering algorithm combines the two nearest clusters until the distance between them is above a threshold. The distance, $d(c_1 ; c_2)$, between two colors, $c_1 = (r_1 ; g_1 ; b_1)$ and $c_2 = (r_2 ; g_2 ; b_2)$, is defined to be $d(c_1 ; c_2) = |r_1 - r_2| + |g_1 - g_2| + |b_1 - b_2|$. To efficiently compute the distance between two clusters, we replaced each cluster with the more populous color, while combining two colors. The threshold for terminating the clustering was experimentally determined to be 4. A typical color image in our experiment generates 9 to 15 clusters. Each pixel in text regions of the input image (output of the MLP) is then replaced by the representative color of the group it belongs to. CCs are identified from this color image. The morphological operation of closing (dilation followed by erosion) is applied to the CCs before we compute the attributes of the components such as size and area. We have then applied three-stage filtering on the CCs:

Stage 1: Heuristics using features of CCs such as area, fill factor, and horizontal and vertical extents; The width of the text region must be larger than Min_width, the aspect ratio of the text region must be smaller than Max_aspect, the area of the component should be larger than Min_area and smaller than Max_area, and the fill factor should be larger than Min_fillfactor.

Stage 2: Geometric alignment of text components; We check the number of adjacent text components which has a same color in the same text line. Text components have to be aligned in more than three consecutive components.

Stage 3: Part-based Shape information: It is somewhat difficult to precisely determine the heuristics which are used in Stage 1 and Stage 2. These heuristics have been successfully used in a variety of text detection methods, however, they are limited in their ability to detect texts from images containing multi-segment characters such as Korean. To efficiently represent shape information of the characters, a part-based NMF technique is used.

The NMF algorithm, devised by Lee and Seung [11], decomposes a given matrix into a basis set and encodings, while satisfying the non-negativity constraints. Given an $n \times m$ original matrix V, in which a column is an n-dimensional non-negative vector of the m vectors, NMF decomposes V into an $n \times r$ factorized basis W and $r \times m$ encodings H in order to approximate the original matrix. Then, the NMF algorithm constructs approximate factorizations as in Eq. (1).

$$V_{i\mu} \approx (WH)_{i\mu} = \sum_{a=1}^{r} W_{ia} H_{a\mu} \tag{1}$$

$$F = \sum_{i=1}^{n} \sum_{\mu=1}^{m} V_{i\mu} \log(WH)_{i\mu} - (WH)_{i\mu} \tag{2}$$

$$P(V \mid WH) = \exp(-WH) \frac{(WH)^V}{V!} \tag{3}$$

$$\log P(V \mid WH) = V \log(WH) - WH - \log V! \tag{4}$$

where the rank r of the factorization is generally chosen by $(n+m)r < nm$.

The NMF algorithm starts from a random initialization of W and H, and iteratively updates them until they converge. The NMF algorithm is an iterative algorithm with multiplicative update rules that can be regarded as a special variant of gradient-descent algorithms [12]. The algorithm iteratively updates W and H by multiplicative rules:

$$H_{a\mu} \leftarrow H_{a\mu} \frac{(W^T V)_{a\mu}}{(W^T WH)_{a\mu}} \text{,and } W_{ia} \leftarrow W_{ia} \frac{(VH^T)_{ia}}{(WHH^T)_{ia}},\ W_{ia} \leftarrow W_{ia} \frac{W_{ia}}{\sum_j W_{ja}}. \tag{5}$$

In Figure 5, 4992 character images of six different typefaces are applied to establish the basis and encodings. The character images of each typeface are composed of n = 28 × 28 pixels, in which the intensity values of all the pixels are normalized to the range [0, 1]. As can be seen from Figure 3, the NMF basis and encodings contain a large fraction of vanishing coefficients, so both the basis images and image encodings are sparse.

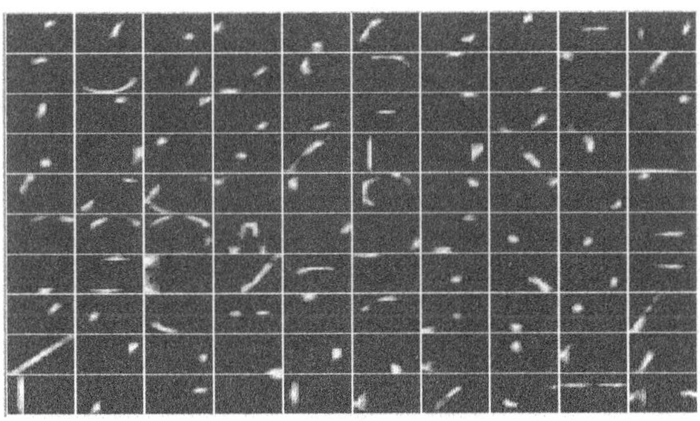

Fig. 3. Basis Images for Text Components.

In the Stage 3, a component image obtained from a document image is projected into the NMF-projected space using the basis obtained in the training phase, resulting in a new feature vector corresponding to the component. The new feature vector is compared with the prototypes, and then classified as the text or non-text.

4 Region Marking

After neural network and NMF analysis, we perform two different region marking approaches according to the given image types. Although lots of works have been done for text detection using texture and CC analysis, post-procesing such as region marking and text extraction is rarely done in previous researches. For gray or color document image we perform X-Y recursive cut based on the assumption that skew correction is done in advance. For video images which have lower text presence rates comparing with document images, we use CAMShift algorithm to increase time performance.

4.1 X-Y Recursive Cut

For document images have larger text portion than non-text regions, we perform X-Y recursive cut on the filtered image. It is a top-down recursive partitioning algorithm. This algorithm takes as input a binary image, where the text pixels are black and the non-text pixels are white (Figure 4). The algorithm projects the text pixels onto the x and y axes and identifies the valleys in the projection. In this paper we assume the document image is skew-corrected in advance. The algorithm recursively divides each region by projecting it, alternately, onto the x and y axes (Figure 5).

4.2 CAMShift Algorithm

For video images, which have lower text presence comparing with non-text region, the problem with the texture-based method is its computational complexity in the texture classification stage, which is the major source of processing time. Especially a texture-based convolution-filtering methods have to exhaustively scan an input image in order to detect the text regions [3, 4], and the convolution operation is computationally expensive and time-consuming.

We use a continuously adaptive mean shift (CAMShift) algorithm for text extraction in video images [10]. To avoid texture analysis of an entire image with the MLP, CAMShift algorithms are invoked at several seed positions on the TPI. This leads to great computational savings when text regions do not dominate the image. At the initial stage of the CAMShift algorithm on each seed position, we decide whether the seed positions contain texts or not using zero-th moment (text detection), and then enlarge detected text regions by merging neighbor pixels within a search window in successive iterations (text extraction).

During the iterations, we estimate the position and size of the text region using 2D moments, change the search window position and size depending on the estimated values, and perform a node-merge operation to eliminate overlapping nodes. If the mean shift is larger than either of the threshold values ε_x (along x-axis) and ε_y (along y-axis), the iteration continues (Readers refer the author's previous publication for detail [7]).

5 Experimental Results

The proposed text extraction method has been tested with several types of images: captured broadcast news, scanned images, web images, and video clips downloaded from the web site of Movie Content Analysis (MoCA) Project [13].

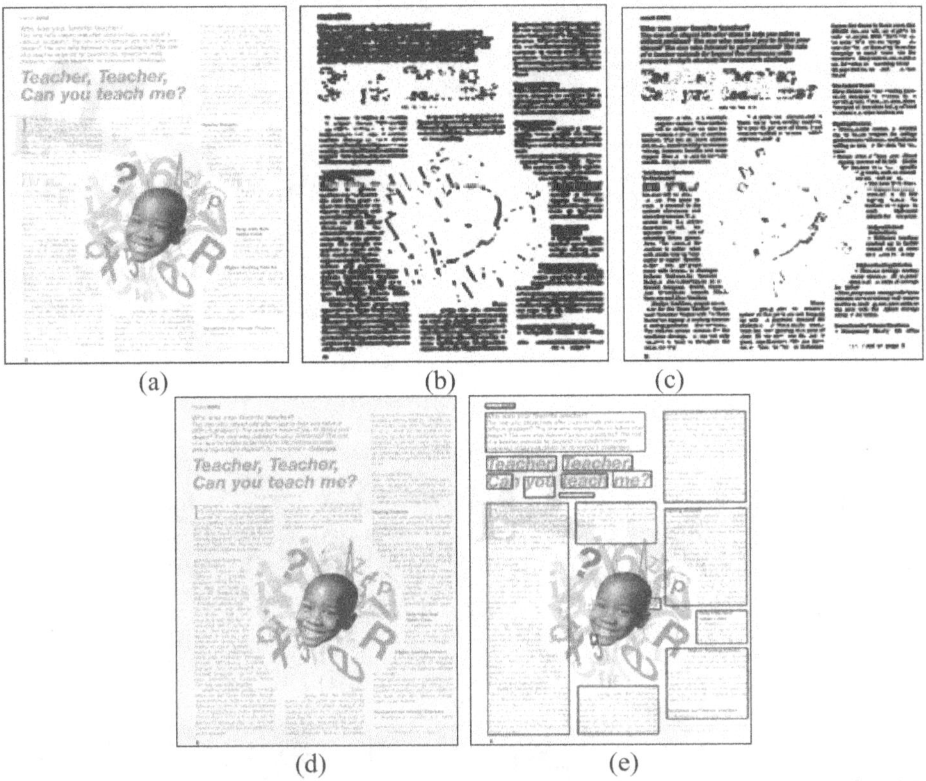

Fig. 4. Intermediate results for X-Y recursive cut: (a) input color image, (b) TPI, (c) after CC-analysis, (d) color quantized image, and (e) text regions.

Fig. 5. Transitions of intermediate text detection results.

Among the training protocols such as on-line, batch, and stochastic trainings, we choose the batch mode. We used a constant learning rate η=0.02, momentum m=0.5 and the sigmoidal activation function f(net)= atanh(b net) with a= 1.7 and b=0.75. We used the number of iteration epochs as a stopping criterion: 1500 epochs were sufficient for convergence in several experiments with varying input window size. For checking the performance of the MLP as a texture classifier, we summarize the precision and recall rates on the pixel level. Without using CAMShift, the MLP is performed at every pixel in the given image.

(a) (b) (c)

Fig. 6. Extraction examples for color document images.

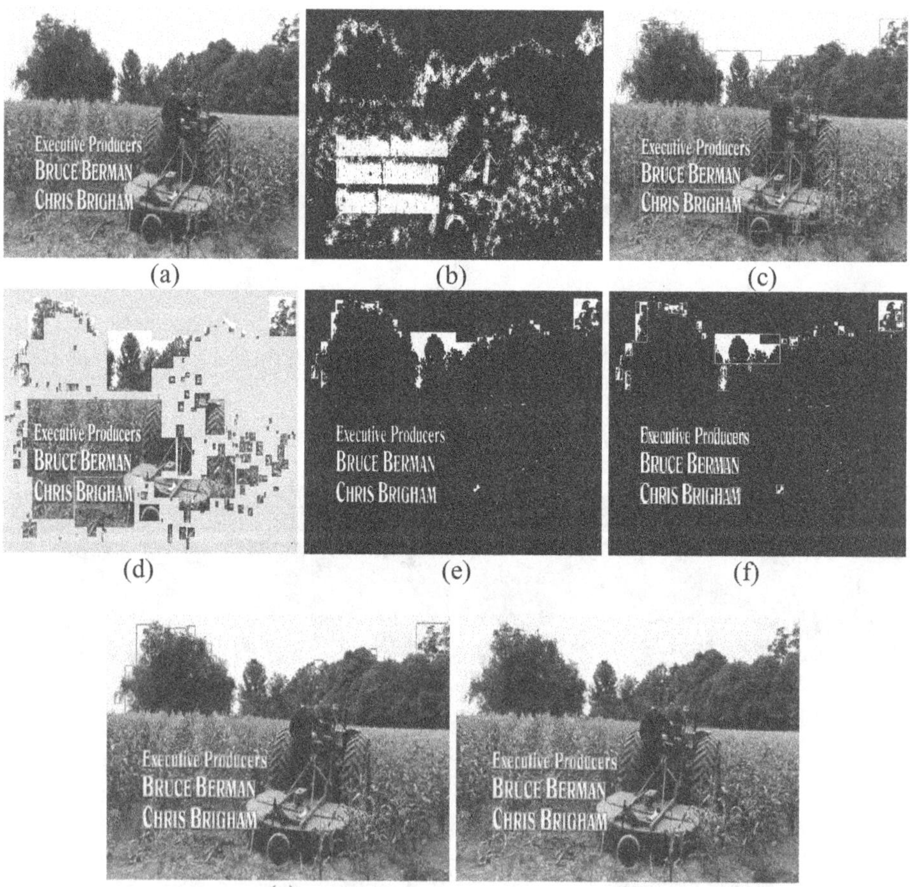

(a) (b) (c) (d) (e) (f) (g) (h)

Fig. 7. CC analysis using NMF: (a) input images, (b) after texture-base text detection, (c) and (d) CCs of (b), (e) binarization based on CCs, (f) CCs of (e), (g) after CC-based filtering (h) after NMF-based filtering.

Figure 6 shows examples of text region detection using X-Y recursive cut. There are some missing errors for relatively large texts in Figure 6(b). They are probably not important for the original purpose of text recognition for image indexing. Figure 7 shows the example images before and after CC analysis based on NMF. Figure 7(c) is the result of CC analysis after the texture-base text detection and Figure 7(e) is a binary image from (c). Figure 7(f) and (g) is the image after CC analysis of (e), and (h) is a result of NMF-based filtering. Non-text CCs that is hard to remove using only CC analysis is removed, and more robust and reliable results are obtained (if it is printed in gray-scale, final results are not clearly visible because of the dark and complex background).

Table 1 shows the pixel-level precision and recall rates after marking bounding boxes. It shows a comparison result with a modified CCM [4] and another texture-based method [7]. The CCM [4] quantized the color spaces into a few prototypes and performed a CC analysis on the resulting gray-scale image. Then, each extracted CC was verified according to its area, ratio, and alignment. The texture-based method [7] adopted an MLP to analyze the textural properties of the texts to identify the text regions and used a top-down X-Y recursive cut technique on projection profiles to generate bounding boxes. It is clear that the performance of the proposed method is superior to the another texture-based method and the CCM. As we generate the text probabilities only for pixels within the search window, the average number of pixels convolved by MLP is reduced to about a tenth of exhaustive search's [7]. So is the processing time. Moreover, the system has given processing time shorter than 0.05 second per a frame with shifting[1] (interval size of 3 pixels for each column and row). The processing time in Table 1 is calculated using video frames in size 320×240.

Table 1. Comparison of precision and recall rates for video images

	CC-only [6]	MLP-only[8]	MLP + CC +CAMShift
Time(seconds)	0.1	4.63	0.47
Precision (%)	78.8	73.1	95.7
Recall (%)	84.5	91.2	89.3

We tried to compare our method's performance with others. However, there is no public comprehensive database on video sequences containing captions, and each researcher used his/her own database. So it is hard to do a comprehensive objective comparison. In this experiment, we can download the sample video clips from MOCA project Web site [13].

6 Conclusions

In this paper, an efficient text extraction technique using a combined method of neural network-based detection and NMF-based filtering is presented. Detection of texts in various conditions can be automatically performed using neural networks without any explicit feature extraction stage. The main drawback of the texture-based method is in

[1] Here, *shifting* means the technique that classifies pixels at regular intervals and interpolates pixels located between the classified pixels.

its locality property, which means that it does not consider the outside of its window. It encourages us to use the hybrid method of texture and CC-based methods. Therefore we aim to increase a recall rate with MLP, and then a precision rate with NMF-filtering-based CC analysis. Moreover, by adopting CAMShift we do not need to analyze the texture properties of an entire image, which results in enhancement of time performance. The proposed method works particularly well in extracting texts from the complex and textured background and shows a better performance than the other proposed methods in open literature. For the future work, incorporating an OCR algorithm with this text extraction method is needed. For this we have to consider enhancing the resolution of the extracted characters and more flexible OCR technology for low-resolution characters.

Acknowledgement. This work was supported by the Soongsil University Research Fund.

References

1. Rainer Lienhart and Frank Stuber, Automatic Text Recognition In Digital Videos, SPIE-The International Society for Optical Engineering, pp. 180-188, 1996.
2. Huiping Li, David Doerman, and Omid Kia, Automatic Text Detection and Tracking in Digital Video, IEEE Transactions on Image Processing, Vol. 9, No. 1, January, pp.147-156, 2000.
3. Yu Zhong, Hongjiang Zhang, and Anil K. Jain, Automatic Caption Extraction in Compressed Video, IEEE Transactions on Pattern Analysis and Machine Intelligence, Vol. 22, No. 4, pp. 385-392, 2000.
4. E.Y. Kim, K. Jung, K.Y. Jeong, and H.J. Kim, Automatic Text Region Extraction Using Cluster-based Templates, International Conference on Advances in Pattern Recognition and Digital Techniques, pp. 418-421, 2000.
5. K. Y. Jeong, K. Jung, E. Y. Kim, and H. J. Kim, Neural Network-based Text Location for News Video Indexing, Proceedings of International Conference of Image Processing, Vol. 3, pp. 319-323, 1999.
6. Victor Wu, Raghavan Manmatha, and Edward M. Riseman, TextFinder: An Automatic System to Detect and Recognize Text in Images, IEEE Transactions on Pattern Analysis and Machine Intelligence, Vol. 21. No. 11, pp. 1224-1229, 1999.
7. K. Jung, Neural Network-based Text Location in Color Images, Pattern Recognition Letters, Vol. 22, No. 14, pp. 1503-1515, 2001.
8. Ullas Gargi, Sameer Antani, and Rangachar Kasturi, Indexing Text Events in Digital Video Database, International Conference on Pattern Recognition, pp. 1481-1483, 1998.
9. K. K. Sung and T. Poggio, Example-based Learning for View-based Human Face Detection, IEEE Transactions on Pattern Analysis and Machine Intelligence, Vol. 20, No. 1, pp. 39-51, 1998.
10. Yizong Cheng, Mean Shift, Mode Seeking, and Clustering, IEEE Transactions on Pattern Analysis and Machine Intelligence, Vol. 17, No. 8, August, pp.790-799, 1995.
11. Lee, D.D, Seung, H.S., 1999. Learning the Parts of Objects by Non-Negative Matrix Factorization. Nature 401, 788-791.
12. Lee, D.D., Seung, H.S., 2001. Algorithms for non-negative matrix factorization. In: In Advances in Neural Information Processing Systems 13, 556–562.

13. http://www.informatik.uni-mannheim.de/informatik/pi4/projects/MoCA/Project-textSegmentationAndRecognition.html
14. Kwang In Kim and Keechul Jung, Texture-based Approach for Text Detection in Images using Support Vector Machines and Continuously Adaptive Mean Shift Algorithm, IEEE Transactions on Pattern Analysis and Machine Intelligence, vol. 25, no. 12, pp. 1631-1639, 2003.
15. Keechul Jung, Kwang In Kim, and Anil K. Jain, Text Information Extraction in Images: A Survey, International Journal of Pattern Recognition, to be published.

Peculiarity Oriented Analysis in Multi-people Tracking Images

Muneaki Ohshima[1], Ning Zhong[1], Y.Y. Yao[2], and Shinichi Murata[3]

[1] Dept. of Infor. Eng., Maebashi Institute of Technology, Japan
[2] Dept. of Computer Science, University of Regina, Canada
[3] Oki Electric Industry Co., Ltd., Japan

Abstract. In the place in which many people gather, we may find a suspicious person who is different from others from a security viewpoint. In other words, the person who takes a *peculiar* action is suspicious. In this paper, we describe an application of our peculiarity oriented mining approach for analysing in image sequences of tracking multiple walking people. A measure of peculiarity, which is called *peculiarity factor*, is investigated theoretically. The usefulness of our approach is verified by experimental results.

1 Introduction

In the place such as a station or an airport in which many people gather, many sacrifices will come out when the terrorism happens. From a security viewpoint, we need to find a suspicious person in such a place. Although the suspicious person can be discovered by using the surveillance camera with a video, not all people can be checked automatically.

In this paper, we describe an application of our peculiarity oriented mining approach for analysing image sequences of tracking multiple walking people. We observed that a suspicious person usually takes action which is different from other people, so-called *peculiar action*, such as coming from and going to the same place and stopping at some place. Hence, our peculiarity oriented mining approach [8, 9] can be used to analyse such data automatically. Such peculiarity oriented analysis is a main difference between our approach and other related work on analysis of human movement [1, 4].

The rest of the paper is organized as follows. Section 2 investigates how to identify peculiar data in our peculiarity oriented mining approach. Section 3 discusses the application of our approach for automatically analysing image sequences of tracking multiple walking people, which were obtained by using the surveillance camera. The experimental results show the usefulness of our approach. Finally, Section 4 gives concluding remarks.

2 Peculiar Data Identification

The main task of peculiarity oriented mining is the identification of peculiar data. An attribute-oriented method, which analyzes data from a new view and

H. Dai, R. Srikant, and C. Zhang (Eds.): PAKDD 2004, LNAI 3056, pp. 508–518, 2004.

is different from traditional statistical methods, is recently proposed by Zhong *et al.* [8, 9].

2.1 A Measure of Peculiarity

Peculiar data are a subset of objects in the database and are characterized by two features:

(1) very different from other objects in a dataset, and
(2) consisting of a relatively low number of objects.

The first property is related to the notion of distance or dissimilarity of objects. Institutively speaking, an object is different from other objects if it is far away from other objects based on certain distance functions. Its attribute values must be different from the values of other objects. One can define distance between objects based on the distance between their values. The second property is related to the notion of support. Peculiar data must have a low support.

At attribute level, the identification of peculiar data can be done by finding attribute values having properties (1) and (2). Table 1 shows a relation with attributes A_1, A_2, ..., A_m. Let x_{ij} be the value of A_j of the i-th tuple, and n the number of tuples. Zhong *et al.* [8, 9] suggested that the peculiarity of x_{ij} can be evaluated by a *Peculiarity Factor*, $PF(x_{ij})$,

$$PF(x_{ij}) = \sum_{k=1}^{n} N(x_{ij}, x_{kj})^{\alpha}, \tag{1}$$

where N denotes the conceptual distance, α is a parameter to denote the importance of the distance between x_{ij} and x_{kj}, which can be adjusted by a user, and $\alpha = 0.5$ as default.

Table 1. A sample table (relation)

A_1	A_2	...	A_j	...	A_m
x_{11}	x_{12}	...	x_{1j}	...	x_{1m}
x_{21}	x_{22}	...	x_{2j}	...	x_{2m}
$\vdots$	$\vdots$		$\vdots$		$\vdots$
x_{i1}	x_{i2}	...	x_{ij}	...	x_{im}
$\vdots$	$\vdots$		$\vdots$		$\vdots$
x_{n1}	x_{n2}	...	x_{nj}	...	x_{nm}

With the introduction of conceptual distance, Eq. (1) provides a more flexible method to calculate peculiarity of an attribute value. It can handle both continuous and symbolic attributes based on a unified semantic interpretation.

Background knowledge represented by binary neighborhoods can be used to evaluate the peculiarity if such background knowledge is provided by a user. If X is a continuous attribute and no background knowledge is available, we use the following distance:

$$N(x_{ij}, x_{kj}) = |x_{ij} - x_{kj}|. \quad (2)$$

If X is a symbolic attribute and the background knowledge for representing the conceptual distances between x_{ij} and x_{kj} is provided by a user, the peculiarity factor is calculated by the conceptual distances [3, 6, 8, 9]. The conceptual distances are assigned to 1 if no background knowledge is available.

Based on peculiarity factor, the selection of peculiar data is simply carried out by using a threshold value. More specifically, an attribute value is peculiar if its peculiarity factor is above minimum peculiarity p, namely, $PF(x_{ij}) \geq p$. The threshold value p may be computed by the distribution of PF as follows:

$$p = mean\ of\ PF(x_{ij}) + \beta \times standard\ deviation\ of\ PF(x_{ij}) \quad (3)$$

where β can be adjusted by a user, and $\beta = 1$ is used as default. The threshold indicates that a data is a peculiar one if its PF value is much larger than the mean of the PF set. In other words, if $PF(x_{ij})$ is over the threshold value, x_{ij} is a peculiar data. By adjusting the parameter β, a user can control and adjust threshold value.

2.2 Analysis of the Peculiarity Factor

A question arises naturally is whether the proposed peculiarity factor reflects our intuitive understanding of peculiarity (i.e. the properties (1) and (2) as mentioned previously). More specifically, whether a high value of Eq. (1) indicates x_{ij} occurs in relatively low number of objects and is very different from other data x_{kj}. Although many experiment results have shown the effectiveness of the peculiarity factor, a detailed analysis may bring us more insights [7].

Table 2. Attribute values and its frequency

attribute value	frequency
x_1	n_1
x_2	n_2
$\vdots$	$\vdots$
x_h	n_h
Total	n

In order to analyze Eq. (1), we adopt a distribution form of attribute value. In Table 2, let $\{x_1, \ldots, x_h\}$ be the set of distinguishing values of an attribute.

With respect to the distribution, the $PF(x_i)$ can be easily computed by:

$$PF(x_i) = \sum_{k=1}^{h} n_k \times N(x_i, x_k)^{\alpha}. \tag{4}$$

Let now consider two special cases, in order to have a better understanding of PF.

Case 1-1. Assume that all attribute values have the same frequency, namely, $n_1 = n_2 = \ldots = n_h = h/n$. In this case, we have:

$$PF(x_i) = \frac{h}{n} \sum_{k=1}^{h} N(x_i, x_k)^{\alpha}. \tag{5}$$

Since h/n is a constant independent of any particular value, the PF value depends only on the total distances of x_i and other values. A value far away from other values would be considered to be peculiar.

Case 1-2. Assume that the distance between a pair of distinguish values are the same, namely, $N(x_i, x_k) = C$ for $i \neq k$ and $N(x_i, x_i) = 0$. In this case, we have:

$$PF(x_i) = (n - n_i)C = nC - n_iC. \tag{6}$$

Since nC is a constant independent of any particular value, the PF value is monotonic decreasing with respect to the value of n_i. A value with low frequency will have a large PF, and in turn, is considered to be peculiar. As expected, the distances between x_i and other values are irrelevant [7].

In general, PF depends on both the distribution n_k and the individual distances $N(x_i, x_k)$. Several qualitative properties can be said about the peculiarity factor based on Eq. (4):

- A value with low frequency tends to have a higher peculiarity value. This follows from the fact $\sum_{k=1}^{k=h} n_k = n$ and there are $n - n_i$ distances to be added for x_i.
- Each term in Eq. (4) is a product of frequency n_k and the distance $N(x_i, x_k)$. This suggests that a value far way from more frequent values is likely to be peculiar. On the other hand, a value for away from less frequent values may not necessarily peculiar, due to a small value of n_k. A value closer to very frequent values may also be considered to be peculiar, due to the large value of n_k. Those latter properties are not desirable properties.
- Eq. (4) can be rewritten as:

$$PF(x_i) = n \sum_{k=1}^{h} \frac{n_k}{n} \times N(x_i, x_k)^{\alpha}. \tag{7}$$

  Thus, the peculiarity factor is in fact a weighted average of distances between x_i and other values. It is the expected distance of $N(x_i, x_k)^{\alpha}$ with respect to probability distribution $(n_1/n, n_2/n, \ldots, n_h/n)$. Under this view, a value is deemed peculiar if it has a large expected distance to other values.

From the above analysis, we can conclude that the peculiarity factor has some desired properties and some undesired properties. The main problem may stem from the fact that average is used in the calculation of peculiarity factor. A best average does not necessarily imply a best choice. Consider the following distribution:

attribute value	frequency
$x_1 = 1$	$n_1 = 10$
$x_2 = 5$	$n_2 = 1$
$x_3 = 10$	$n_3 = 10$
Total	$n = 21$

Assume $\alpha = 1$ and $N(x_i, x_k) = |x_i - x_k|$. We have the following peculiarity values:

$$PF(x_1) = 64, \quad PF(x_2) = 60, \quad PF(x_3) = 62.$$

On the other hand, $x_2 = 5$ seems to be peculiar rather the other two. Furthermore, although a user can adjust the parameter β in the selection of threshold value for peculiarity data selection, its usefulness is limited. The notion of peculiarity, as defined by Eq. (4), mixes together two notions of frequency and distance. Although it is based on a sound theoretic argument, its meaning cannot be simply explained to a non-expert.

Furthermore, α in Eq. (1) can be also considered two special cases with respect to the two cases stated above.

Case 2-1. Assume $\alpha >> n$. This means $N(x_{ij}, x_{kj})^\alpha >> n_i$. Hence,

$$PF(x_i) = \sum_{k=1}^{h} n_k \times N(x_i, x_k)^\alpha \simeq \sum_{k=1}^{h} N(x_i, x_k)^\alpha. \tag{8}$$

This case is the same as Case 1-1 stated above. In other words, the PF value depends only on the total distances of x_i and other values.

Case 2-2. Assume $\alpha \to 0$. Hence,

$$N(x_{ij}, x_{kj}) = \begin{cases} 1 \ (x_{ij} \neq x_{kj}) \\ 0 \ (x_{ij} = x_{kj}) \end{cases} \tag{9}$$

This case is the same as Case 1-2 stated above, when $C = 1$. Based on Eq. (6), we have

$$PF(x_i) = (n - n_i)C = nC - n_iC = n - n_i. \tag{10}$$

In other words, the PF value depends only on the distribution n_k.

Figure 1 shows the relationship between the distance and PF when α changed in Eq. (1). By adjusting the parameter α, a user can control and adjust the degree of PF that depends on both the distribution and the distance. And according to experience, $\alpha = 0.5$ will get a good balance between the distribution and the distance.

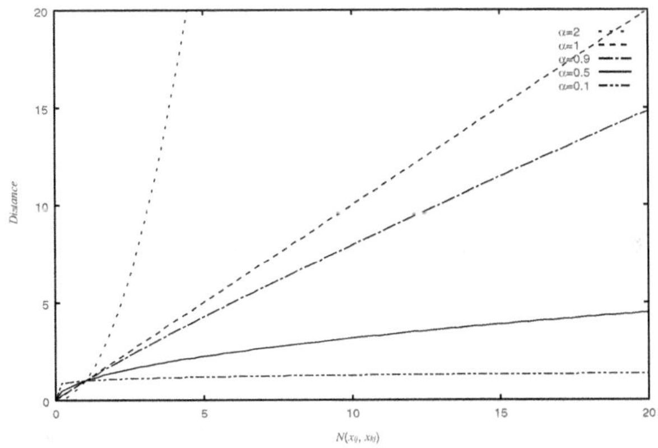

Fig. 1. The relationship between the distance and PF when α changed

2.3 An Algorithm

Based on the above-stated preparation, an algorithm of finding peculiar data can be outlined as follows:

Step 1. Execute attribute oriented clustering for each attribute, respectively.
Step 2. Calculate the peculiarity factor $PF(x_{ij})$ in Eq. (1) for all values in an attribute.
Step 3. Calculate the threshold value in Eq. (3) based on the peculiarity factor obtained in *Step 2*.
Step 4. Select the data that are over the threshold value as the peculiar data.
Step 5. If the current peculiarity level is enough, then goto *Step 7*.
Step 6. Remove the peculiar data from the attribute and thus, we get a new dataset. Then go back to *Step 2*.
Step 7. Change the granularity of the peculiar data by using background knowledge on information granularity if the background knowledge is available.

Furthermore, the algorithm can be done in a parallel-distributed mode for multiple attributes, relations and databases because this is an attribute-oriented finding method.

3 Application in Analysing Image Sequences of Tracking Multiple Walking People

Peculiarity oriented mining has been applied to analyse image sequences of tracking multiple walking people. The path of tracking data of each walking people are used to discover the person action pattern and to detect a person's unique action. The data were obtained by using the surveillance camera with a video, and preprocessed by using person tracking technology developed at OKI Electric

Industry Co., Ltd. In this experiment, we used the video photoed at a station ticket wicket (Fig. 2). The purpose is to discover the person who has taken a suspicious action.

Fig. 2. Multiple walking people at a station ticket wicket

3.1 Data Preparation

The original attributes on tracked image sequences of multiple walking people might not be suitable directly for our peculiarity oriented mining approach. Hence, a key issue is how to generate the attributes to meet our needs from the original data.

At first, the raw data are changed into the coordinates and are given in CSV format for every frame by person tracking technology. Each instance is indicated in the following attributes.

- ID: The unique ID attached to each people under tracking.
- FrameNumber: The frame number in the video.
- Status: The following states are used to describe the state of tracking.
  0: Undefined (unstable state at the time of a tracking start),
  1: Definited (stable tracking), and
  2: Lost (out of the tracking).
- X_1 (coordinate): Raw data under tracking – a left end is 0 and a right end is 256.
- Y_1 (coordinate): Raw data under tracking – a up end is 0 and a down end is 240.
- X_2 (coordinate): The smoothed data – a left end is 0 and a right end is 256.
- Y_2 (coordinate): The smoothed data – a up end is 0 and a down end is 240.

In this experiment, we do not use attributes *Status*, X_1 and Y_1, although *Status* is used in person tracking technology. Further, X_2 and Y_2 will be regarded as the same cases as referred to X_1 and Y_1, respectively. This is because X_1 and Y_1 are unstable, and they are covered by X_2 and Y_2.

The following attributes are used as an attribute for specifying a person's action; *ID*, *In* (the direction included in the photography range), *Out* (the direction left from the photography range), and *seg-n* (the segment number: the number changed the advance direction after going into the photography range before coming out of the range). *In* and *Out* are calculated from the coordinates, respectively, when a person appears for the first time, and the last. *seg-n* is calculated from the number of the divided line segment.

3.2 Linearization of Walking Data

Usually, people goes straight on to the destination from the current position when acting with a goal, if no obstacle prevented him/her from going. In other words, he/she will walk back and forth in a certain range, not to mention a detour, or will stop, if the person is at a unusual state, without a specific goal, and so on. Hence, whether the behavior of a person is usual or not can be analysed by calculating the segment number (i.e. the number of changed direction) in the linearized walking data of each person.

In order to calculate the segment number, linearization of walking data needs to be first performed. An algorithm of linearization can be outlined in Fig. 3. An example of the linearization is shown in Fig. 4.

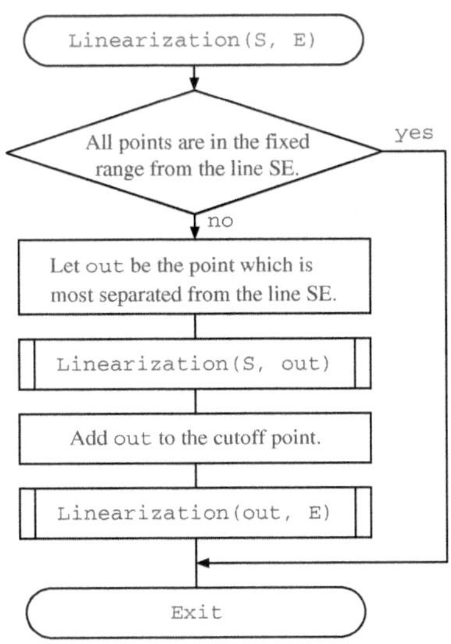

Fig. 3. Linearization flowchart

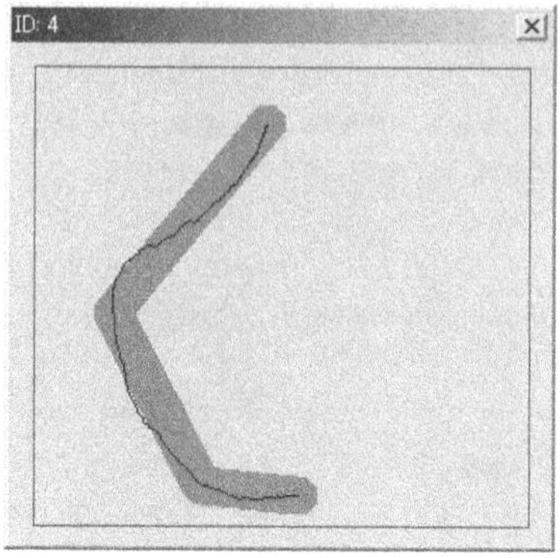

Fig. 4. An example of the linearization

3.3 Simulations

Experiment 1 The following parameters were used to calculate PF, $\alpha = 0.5$, $\beta = 1$. And the linearization of walking data has a 20-point error margin. There are 26 persons who have been judged as taking peculiar actions in this experiment. A part of the result is given Table 3.

Table 3. A part of result 1

ID	In	Out	seg-n
2004	down	up	3
2010	up	left	2
2019	rightup	leftup	1
2039	rightup	leftdown	2
2175	down	leftdown	2
2270	up	left	1
2272	leftdown	up	2
2353	leftdown	center	1

By comparing the result shown in Table 3 with the actual movie, we can see that only 3 persons can be regarded as peculiar ones. The reason why the result is not more exact is that the attributes used in this experiment may be insufficient. Hence, it is necessary to add attributes for describing human's action pattern more specifically.

Experiment 2 Based on the above experiment, we added a new attribute, *frame-n* (i.e. number of frames), which was staying at the photography range.

And the parameters for calculating PF were set to $\alpha = 0.5$, $\beta = 2$. The reason why we set $\beta = 2$ is that points which persons pass in different photography ranges have a bigger unevenness. For example, some people passes along middle and other persons passe along an end of the photography range.

The result of this experiment is shown in Table 4. We can see there are 5 persons judged as taking peculiar action in this experiment. However, the value of attributes *In* or *Out*, "center", means the person who was in the photography range at the time of the start of photography, or the end. Hence, such person cannot be judged to take a peculiar action. As a result, only 3 persons have been considered as taking peculiar actions.

Table 4. Result 2

ID	In	Out	seg-n	frame-n
2004	down	up	3	286
3020	left	up	3	286
4004	center	up	2	286
5024	right	up	2	279
5171	down	center	2	275

Experiment 3 Based on experiments 1 and 2, we added two more attributes *speed* (pixel/frame-n) and *angle* (the average angle of all turnings, i.e., the total angle/# of turnings) for experiment 3. The angle can be calculated by the linearized segment. It was set to 180 degree if not turning. Furthermore, the parameters for calculating PF were set to $\alpha = 0.5$, $\beta = 1$ for attribute *speed*, and to $\alpha = 0.5$, $\beta = 2$ for attribute *angle*, respectively.

As a result as shown in Table 5, we can see there are 16 persons, including the case as shown in Fig. 4, who have been judged as taking peculiar actions.

Table 5. A part of result 3

ID	In	Out	seg-n	frame-n	speed	angle
2004	down	up	3	286	1.0	119
2010	middle	left	2	105	1.7	128
3020	left	leftup	2	286	1.0	136
3031	up	leftdown	3	237	1.1	120
5134	down	rightup	3	223	1.2	134

Although not all the detected persons are peculiar ones, all suspicious persons have been discovered. We observed that the pattern of the whole person's stream will change depending on different time zones. Hence, it is necessary to compare the detected peculiar actions with a more general pattern in each time zone.

4 Conclusions

We presented an application of our peculiarity oriented mining approach for analysing in image sequences of tracking multiple walking people. The strength and usefulness of our approach have been investigated theoretically and demonstrated by experimental results.

In order to increase accuracy, it is necessary to evaluate in multiple stages. Moreover, a general rule (i.e. the pattern of the whole person's stream) needs to be discovered, and will be helpful to recognize suspicious persons quickly.

References

1. Gavrila, D. M. "The Visual Analysis of Human Movement: A Survey", *Computer Vision and Image Understanding*, Vol. 73, No. 1, Academic Press (1999) 82-98.
2. Hilderman, R.J. and Hamilton, H.J. "Evaluation of Interestingness Measures for Ranking Discovered Knowledge", D. Cheung, G.J. Williams, Q. Li (eds) *Advances in Knowledge Discovery and Data Mining*, LNAI 2035, Springer (2001) 247-259.
3. Lin, T.Y. "Granular Computing on Binary Relations 1: Data Mining and Neighborhood Systems", L. Polkowski and A. Skowron (eds) *Rough Sets in Knowledge Discovery*, Vol. 1, Physica-Verlag (1998) 107-121.
4. Song, Y., Goncalves, L., and Perona, P. "Unsupervised Learning of Human Motion", *IEEE Transactions on Pattern Analysis and Machine Intelligence*, Vol. 25, No. 7 (2003) 814-827.
5. Suzuki, E. "Autonomous Discovery of Reliable Exception Rules", *Proc Third Inter. Conf. on Knowledge Discovery and Data Mining (KDD-97)*, AAAI Press (1997) 259-262.
6. Yao, Y.Y. "Granular Computing using Neighborhood Systems", Roy, R., Furuhashi, T., Chawdhry, P.K. (eds) *Advances in Soft Computing: Engineering Design and Manufacturing*, Springer (1999) 539-553.
7. Yao, Y.Y. and Zhong, N. "An Analysis of Peculiarity Factor in Peculiarity Oriented Data Mining", T.Y. Lin and Setsuo Ohsuga (eds) *Proc. ICDM workshop on Foundation of Data Mining and Knowledge Discovery (FDM02)* (2002) 185-188.
8. Zhong, N., Yao, Y.Y., and Ohsuga, S. "Peculiarity Oriented Multi-Database Mining", J. Zytkow and J. Rauch (eds) *Principles of Data Mining and Knowledge Discovery*, LNAI 1704, Springer (1999) 136-146.
9. Zhong, N., Ohshima, M., and Ohsuga, S. "Peculiarity Oriented Mining and Its Application for Knowledge Discovery in Amino-acid Data", D. Cheung, G.J. Williams, Q. Li (eds) *Advances in Knowledge Discovery and Data Mining*, LNAI 2035, Springer (2001) 260-269.
10. Zhong, N., Yao, Y.Y., Ohshima, M., and Ohsuga, S. "Interestingness, Peculiarity, and Multi-Database Mining", Proc. 2001 IEEE International Conference on Data Mining (ICDM'01), IEEE Computer Society Press (2001) 566-573.

AutoSplit: Fast and Scalable Discovery of Hidden Variables in Stream and Multimedia Databases*

Jia-Yu Pan[1], Hiroyuki Kitagawa[2], Christos Faloutsos[1], and Masafumi Hamamoto[2]

[1] Carnegie Mellon University, Pittsburgh PA 15213, USA
[2] University of Tsukuba, Tennohdai, Tsukuba, Ibaraki 305-8573, Japan

Abstract. For discovering hidden (latent) variables in real-world, non-gaussian data streams or an n-dimensional cloud of data points, SVD suffers from its orthogonality constraint. Our proposed method, "AutoSplit", finds features which are mutually independent and is able to discover non-orthogonal features. Thus, (a) finds more meaningful hidden variables and features, (b) it can easily lead to clustering and segmentation, (c) it surprisingly scales linearly with the database size and (d) it can also operate in on-line, single-pass mode. We also propose "Clustering-AutoSplit", which extends the feature discovery to multiple feature/bases sets, and leads to clean clustering. Experiments on multiple, real-world data sets show that our method meets all the properties above, outperforming the state-of-the-art SVD.

1 Introduction and Related Work

In this paper, we focus on discovering patterns in (multiple) data streams like stock-price streams and continuous sensor measurement, and multimedia objects such as images in a video stream. Discovery of the essential patterns in data streams is useful in this area, for it could lead to good compression, segmentation, and prediction. We shall put the related work into two groups: dimensionality reduction and streaming data processing.

Dimensionality reduction/feature extraction. Given a cloud of n points, each with m attributes, we would like to represent the data with fewer attributes/features but still retain most of the information. The standard way of doing this dimensionality reduction is through SVD (Singular Value Decomposition). SVD finds the best set of axes to project the cloud of points, so that the sum of squares of the projection errors is minimized (Figure 1(a)). SVD has

* Supported in part by Japan-U.S. Cooperative Science Program of JSPS; grants from JSPS and MEXT (#15017207, #15300027); the NSF No. IRI-9817496, IIS-9988876, IIS-0113089, IIS-0209107, IIS-0205224, INT-0318547, SENSOR-0329549, EF-0331657; the Pennsylvania Infrastructure Technology Alliance No. 22-901-0001; DARPA No. N66001-00-1-8936; and donations from Intel and Northrop-Grumman.

H. Dai, R. Srikant, and C. Zhang (Eds.): PAKDD 2004, LNAI 3056, pp. 519–528, 2004.

been used in multiple settings: for text retrieval [1], under the name of Latent Semantic Indexing (LSI); for face matching in the eigenface project [2]; for pattern analysis under the name of Karhunen-Loeve transform [3] and PCA [4]; for rule discovery [5]; and recently for streams [6] and online applications [7]. Recently, approximate answers of SVD for timely response to online applications have also been proposed [8,9,10,11].

Streaming data processing. Finding hidden variables is useful in time series indexing and mining [12], modeling [13,14], forecasting [15] and similarity search [16,17]. Fast, approximate indexing methods [18,19] have attracted much attention recently. Automatic discovery of "hidden variables" in, for example, object motions would enable much better extrapolations, and help the human analysts understand the motion patterns.

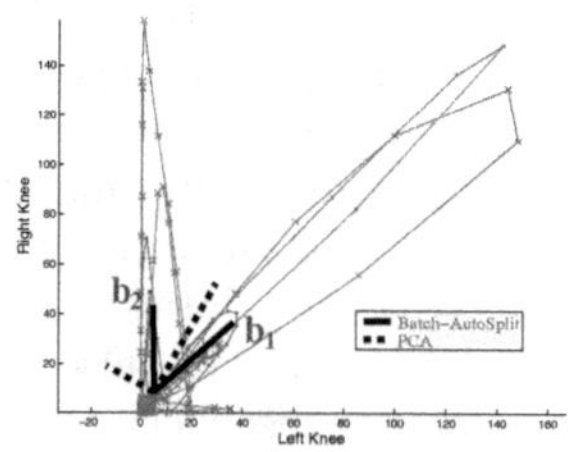

(a) (Energy) Right knee vs left knee

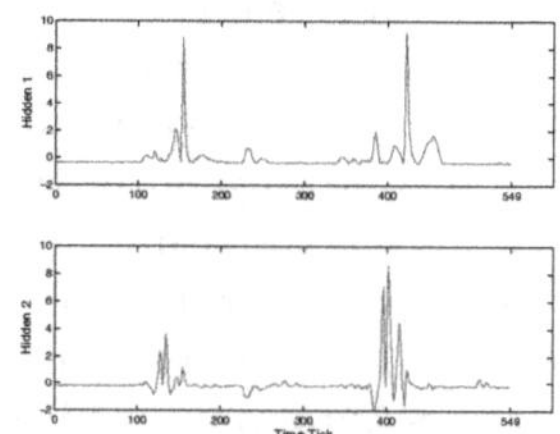

(b) Hidden variables $\mathbf{h_1}$ and $\mathbf{h_2}$

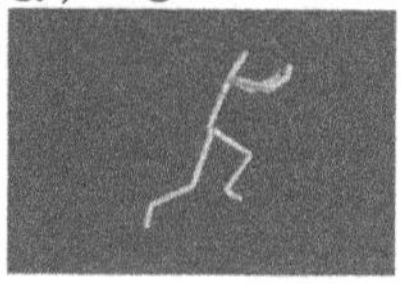

(c) Take off

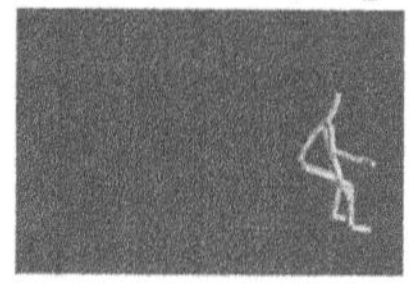

(d) Landing

Fig. 1. (AutoSplit versus SVD/PCA: "Broad jumps") (a): the right knee energy versus left knee energy during the jumps: take off (c) and landing (d). Two jumps are performed at time ticks 100 and 380. In (a), (Batch-)AutoSplit vectors $\mathbf{b_1}$, slope 1:1, corresponds to "landing"; $\mathbf{b_2}$, slope -1:60, for "take off". (b) the hidden variables $\mathbf{h_1}$ (top) and $\mathbf{h_2}$ (bottom) of $\mathbf{b_1}$ and $\mathbf{b_2}$, respectively.

Although popularly used, SVD suffers from its orthogonality requirement for real world data whose distribution is not gaussian. For example, in Figure 1(a), SVD proposes the two *dash* orthogonal vectors as its basis vectors, while completely missing the "natural" ones ($\mathbf{b_1}$, $\mathbf{b_2}$). Is there a way to automatically find the basis vectors $\mathbf{b_1}$ and $\mathbf{b_2}$? Generally, we would like to have a method which (a) finds meaningful feature vectors and hidden variables (better coincide with the true unknown variables which generate the observed data streams), (b) can work in an unsupervised fashion, (c) scales linearly with the database size, (d) is able to operate in on-line, single-pass mode (to cope with continuous, unlimited data streams). Experiments on multiple, real data sets, from diverse settings (motion capture data for computer animation, stock prices, video frames) show that the proposed AutoSplit method and its variants achieve the above properties.

2 Proposed Method

There are two major concepts behind AutoSplit: the *basis vectors*, and the *hidden variables*. The basis vectors are the analog of the eigenvectors of SVD, while the hidden variables are the sources/variables controlling the composition weights of these basis vectors when generating the observed data. We use the "broad jumps" data set in section 1 to illustrate these concepts.

Basis vectors and hidden variables. The "broad jumps" data set (Figure 1) is a motion capture data set from [20]. The actor performed two broad jumps during the recording period. Our data set is a n-by-m data matrix $\mathbf{X}=[x_{i,j}]$ with n=550 rows (time-ticks) and m=2 columns (left and right knee energy). Figure 1(a) shows the scatter plot of the data points: $x_{i,1}$ versus $x_{i,2}$, or, informally: right-knee(i) versus left-knee(i), for time ticks $i = 1, \ldots, n$. For visualization purposes only, data points at successive time-ticks are connected with lines - neither SVD nor AutoSplit use this sequencing information.

For the "broad jumps" data, we would like to (a) *discover the structures of the action* (take-off, landing), and (b) *partitions the action sequences into homogeneous segments.* Notice that the majority of points are close to the origin, corresponding to the time when the knees are not moving much (idle and "flying"). However, there are two pronounced directions, one along the 45° degree line, and one almost vertical. As shown, SVD fails to spot either of the two pronounced directions. On the other hand, AutoSplit clearly locks on to the two "interesting" directions, unsupervisedly.

(Observation 1) *Playing the animation and keeping track of the frame-numbers (= time-ticks), we found that the points along the* 45° *degree line (*$\mathbf{b_1}$*) are from the landing stage of the action (2 knees exert equal energy, Figure 1(d)), while the ones on the near-vertical AutoSplit basis (*$\mathbf{b_2}$*) are from the take-off stage (only right knee is used, Figure 1(c)).*

Exactly because AutoSplit finds "good" basis vectors, most of the data points lie along the captured major axes of activities. Thus, the encoding coefficients (hidden variables) of any data point are all closed to zero, except for the one which controls the axis on which the data point lies (Figure 1(b)). Inspecting the data found that $\mathbf{h_1}$ controls the landing, while $\mathbf{h_2}$ controls the take off. The distinct "fire-up" periods of the hidden variables could lead to clean segmentation.

2.1 AutoSplit: Definitions and Discussion

As shown in Figure 1(a), the failure of SVD partly comes from the orthogonal constraint on its basis vectors. Since our goal is to find clean features, therefore, instead of constraining on the orthogonality, we search for basis vectors which maximize the mutual independence among the hidden variables. The idea is that *independence leads to un-correlation, which implies clean, non-mixed features.*

(Definition 1: Batch-AutoSplit) *Let* $\mathbf{X}_{[\mathbf{nxm}]}$ *be the* ***data matrix*** *representing n data points (rows) with m attributes (columns). We decomposes* $\mathbf{X}_{[\mathbf{nxm}]}$ *as a linear mixture of l bases (rows of the* ***basis matrix*** $\mathbf{B}_{[\mathbf{lxm}]}$*), with weights in the columns of the* ***hidden matrix*** $\mathbf{H}_{[\mathbf{nxl}]}$ *(l: the number of hidden variables,* $l \leq m$*).*

$$\mathbf{X}_{[\mathbf{nxm}]} = \mathbf{H}_{[\mathbf{nxl}]}\mathbf{B}_{[\mathbf{lxm}]},$$

$$\begin{bmatrix} -\boldsymbol{x}_1- \\ -\boldsymbol{x}_2- \\ \vdots \\ -\boldsymbol{x}_n- \end{bmatrix} = \begin{bmatrix} | & | & \cdots & | \\ \boldsymbol{h}_1 & \boldsymbol{h}_2 & \cdots & \boldsymbol{h}_l \\ | & | & \cdots & | \end{bmatrix} \begin{bmatrix} -\boldsymbol{b}_1- \\ -\boldsymbol{b}_2- \\ \vdots \\ -\boldsymbol{b}_l- \end{bmatrix} = \begin{bmatrix} h_{11} & h_{12} & \dots & h_{1l} \\ h_{21} & h_{22} & \dots & h_{2l} \\ \vdots & \vdots & & \vdots \\ h_{n1} & h_{n2} & \dots & h_{nl} \end{bmatrix} \begin{bmatrix} -\boldsymbol{b}_1- \\ -\boldsymbol{b}_2- \\ \vdots \\ -\boldsymbol{b}_l- \end{bmatrix} . \blacksquare$$

We model each data point $\mathbf{x_i}$ as a linear combination of the basis vectors $\mathbf{b_k}$ (features). The value $b_{k,j}$ indicates the weight of the j-th attribute for the k-th hidden variable, where hidden variables represent the unknown data generating factors (e.g., the economic events to a share price series). In the "broad jumps" example, there are n=550 data points $\mathbf{x_i}$, each is m=2 dimensional. The l=2 basis vectors were 2-d vectors, with values $\mathbf{b_1}$=(15.51, 14.12) ($\sim 45^o$ degree line), and $\mathbf{b_2}$=(−0.29, 17.65) (vertical line).

Figure 2 gives the outline of Batch-AutoSplit. The analysis behind Batch-AutoSplit is based on ICA (Independent Component Analysis) [21]. There is a large literature on ICA, with several alternative algorithms. The one most related to Batch-AutoSplit is [22].

Batch-AutoSplit assumes the number of hidden variables is the same as the number of attributes, i.e., **B** is a square matrix. The number of hidden variables, l, is controlled by the *whitening* step of the algorithm. The whitening of matrix **A** is obtained by first making **A** zero mean (and we get $\mathbf{A_0}$), and then apply SVD on $\mathbf{A_0}=\mathbf{U}\Lambda\mathbf{V^T}$. The whitening result is $\hat{\mathbf{A}}=\mathbf{U_l}$, where $\mathbf{U_l}$ is the first l columns of **U**. Also, the Frobenius norm of the matrix $\mathbf{A}_{[\mathbf{nxm}]}$ is defined as $||\mathbf{A}||_F = \sqrt{\sum_{i=1}^{n}\sum_{j=1}^{m} a_{i,j}^2}$.

- Given: The data matrix **X**, and a randomly initialized **B**.
- Step 1: Whiten the data matrix **X**, and get $\hat{\mathbf{X}}$. Initialize $\Delta\mathbf{B}$, such that $||\epsilon\Delta\mathbf{B}||_F > \delta$, where ϵ controls the size of each gradient step, and is usually reduced as more iterations are done, and δ is some pre-defined threshold.
- Step 2: While $||\epsilon\Delta\mathbf{B}||_F > \delta$.
  - Step 2.1: Compute $\mathbf{H}=\hat{\mathbf{X}}\mathbf{B}^{-1}$.
  - Step 2.2: Compute the gradient $\Delta\mathbf{B} = -\mathbf{B^T}\mathbf{Z^T}\mathbf{H} - n\mathbf{B^T}$, where $\mathbf{Z} = -\text{sign}(\mathbf{H})$.
  - Step 2.3: Update $\mathbf{B}=\mathbf{B}+\epsilon\Delta\mathbf{B}$.

Fig. 2. Batch-AutoSplit algorithm: $\mathbf{B_{out}}$=Batch-AutoSplit(**X**,**B**)

2.2 Proposed Method: AutoSplit and Clustering-AutoSplit

Figure 3 extends the basic Batch-AutoSplit algorithm to process online data streams. AutoSplit takes into a infinite stream of data items (grouped into windows $\mathbf{X_0}$,$\mathbf{X_0}$,...) , and continuously output the estimated bases $\mathbf{B_0}$,$\mathbf{B_1}$,.... The memory requirement of the AutoSplit algorithm is tunable, by setting the number of data items (n) to be processed at each loop iteration. The computation time at each update is *constant (O(1))*, given fixed number of data points at each update. In fact, the actual time for each update is tiny, for only a couple of small matrix multiplications and additions are required at each update. Note that AutoSplit is capable of adapting to gradual changes of hidden variables, which is desirable for long-term monitoring applications, where the underlying hidden variables may change over time.

- Given: A infinite stream of data items, every n items are grouped as a data batch $\mathbf{X_l}$, ($l = 1, 2, \dots$). Data patches could be overlapped.
- Step 1: When $\mathbf{X_0}$ is available, initialize $\mathbf{B_0}$ randomly. Let $l = 0$.
- Step 2: While $\mathbf{X_{l+1}}$ is not available, do
  - $\mathbf{B_l}$ = Batch-AutoSplit($\mathbf{X_l}$, $\mathbf{B_l}$).
- Step 3: $\mathbf{B_{l+1}}$=$\mathbf{B_l}$. Goto Step 2.

Fig. 3. AutoSplit algorithm: ($\mathbf{B_0}$,$\mathbf{B_1}$,...) = AutoSplit($\mathbf{X_0}$,$\mathbf{X_1}$,...)

Real world data is not always generated from a single distribution, instead, they are from a mixture of distributions. Many studies model this mixture by a set of Gaussian distributions. Since real world data is often non-Gaussian, we propose "Clustering-AutoSplit", which fits data as a mixture of AutoSplit bases.

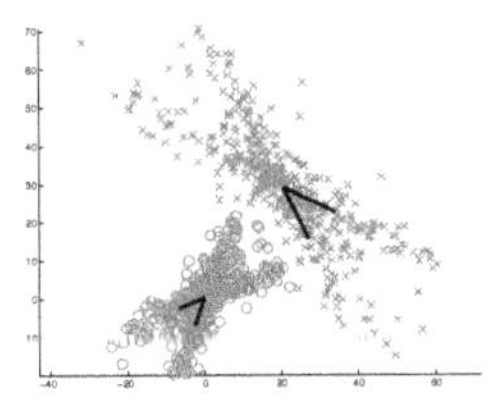

Fig. 4. Clustering-AutoSplit. The number of clusters is set to $k = 2$. Each set of AutoSplit bases is centered at the mean of data points in a cluster.

Figure 4 shows an synthetic example of 2 clusters on the 2-dimensional plane. Each cluster is specified by a set of 2 bases. Note that by fitting the data into the mixture model, the data items are automatically clustered into different classes. Figure 5 gives the outline of the Clustering-AutoSplit algorithm. Note that we can also use AutoSplit algorithm, instead of Batch-AutoSplit, in Step 2.5 of the algorithm, yielding an online clustering algorithm.

- Given: k is the number of clusters to be found, and $\mathbf{X}$ has the data items.
- Step 1: Initialize $\mathbf{B_j}$'s and $\mathbf{c_j}$'s randomly, $j = 1, \ldots, k$. $\mathbf{B_j}$ and $\mathbf{c_j}$ are the bases and the mean of the j data cluster, respectively.
- Step 2: While the changes on $\mathbf{c_j}$'s remain large (above some threshold δ),
  - Step 2.1: For each data point $\mathbf{x}_i$, compute its likelihood of belonging to the j-th cluster,

$$f_{ij} = \frac{\prod_{r=1}^{l} f_h(h_{ir})}{|det(\mathbf{B}_j)|},$$

    where $\mathbf{x}_i = \sum_{r=1}^{l} h_{ir}\mathbf{b}_{jr}$, l is the number of bases for cluster j, and $f_h(h_{ir}) \propto exp(-|h_{ir}|)$.
  - Step 2.2: Compute the relative weight of x_i to cluster j, $p_{ij} = f_{ij}/(\sum_{k=1}^{n} f_{kj})$.
  - Step 2.3: Update $\mathbf{c}_j$'s: $\mathbf{c}_j = \sum_{i=1}^{n} p_{ij}\mathbf{x}_j$.
  - Step 2.4: Cluster each point $\mathbf{x}_i$ to the cluster of maximum likelihood. Let the k clusters be $\mathbf{C_1}, \ldots, \mathbf{C_k}$.
  - Step 2.5: For each cluster j, update $\mathbf{B_j}$=Batch-AutoSplit($\mathbf{C_j}$,$\mathbf{B_j}$) (Figure 2).

Fig. 5. Clustering-AutoSplit: ($\mathbf{B_1}$,$\mathbf{c_1}$, ... ,$\mathbf{B_k}$,$\mathbf{c_k}$)=Clustering-AutoSplit(k,$\mathbf{X}$)

3 Experimental Results

In this section, we show the experimental results of applying (a) AutoSplit to the real world share price sequences and (b) Clustering-AutoSplit to the video frames. We also empirically examine the quality and scalability of AutoSplit.

3.1 Share Price Sequences

The share price data set (DJIA) contains the weekly closing prices of the m=29 companies in the Dow Jones Industrial Average, starting from the week of January 2, 1990 to that of August 5, 2002, and gives data at n=660 time ticks per company. Closing prices collected at the same week/time-tick are grouped into a 29-D vector, i.e., a company is an attribute. The resulting data matrix $\mathbf{X}$ is 660-by-29.

Before doing AutoSplit, we preprocess the data matrix $\mathbf{X}$ and make the value of each company/attribute zero-mean and unit-variance. To extract l=5 hidden variables, the dimensionality is first reduced to 5 from 29 using whitening (section 2.1). Figure 6(b) shows the 2 most influential hidden variables ($\mathbf{h_1}$, $\mathbf{h_2}$). We would like to understand *what do these hidden variables stand for*?

Table 1(a) lists the top 5 companies with largest and smallest contributions $b_{1,j}$ to the hidden variable $\mathbf{h_1}$ (top of Figure 6(b)), where index j is the index

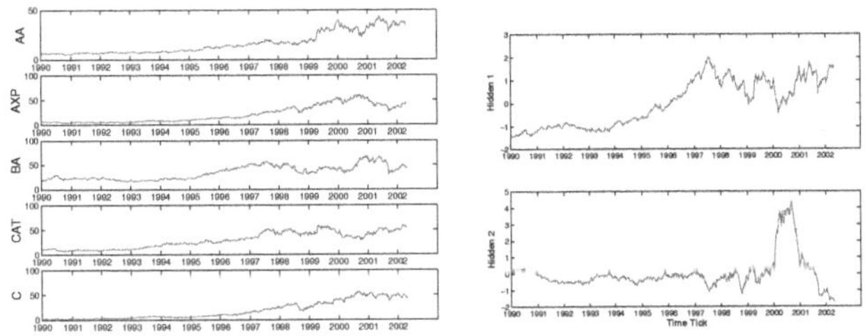

(a) 5 members of DJIA (b) Hidden variables of DJIA

Fig. 6. (Share closing price (DJIA, 1990-2002)) (a) AA: Alcoa, AXP: America Express, BA: Boeing, CAT: Caterpillar, C: CitiGroup. (b) (Top) Probably the general trend of share prices. (Bottom) Probably the Internet bubble.

Table 1. Company contributions according to the hidden variables: $\mathbf{h_1}$, $\mathbf{h_2}$. j is the index to companies. (INTC: Intel, AXP:American Express, DIS:Disnet, MO:Philip Morris, PG:Procter and Gamble, DD: Du Pont)

(a) $b_{1,j}$: Contribution to $\mathbf{h_1}$				(b) $b_{2,j}$: Contribution to $\mathbf{h_2}$			
Highest		Lowest		Highest		Lowest	
CAT	0.938512	T	0.021885	INTC	0.641102	MO	-0.194843
BA	0.911120	WMT	0.624570	HWP	0.621159	IP	-0.089569
MMM	0.906542	INTC	0.638010	GE	0.509164	CAT	0.031678
KO	0.903858	HD	0.647774	AXP	0.504871	PG	0.109576
DD	0.900317	HWP	0.658768	DIS	0.490529	DD	0.133337

to the different companies. As shown, all companies have strong positive contributions (about 0.6 to 0.9) except AT&T (symbol: T). The regularity of contributions among all companies suggests that $\mathbf{h_1}$ represents the general trend of share price series.

On the other hand, the hidden variable $\mathbf{h_2}$ (bottom of Figure 6(b)) is mostly silent (near zero-value) except a sharp rise and drop in year 2000. This seems to correspond to the "Internet bubble". To verify this, we can check the companies' contributions $b_{2,j}$ on this hidden variable $\mathbf{h_2}$. Table 1(b) lists the companies having the 5 highest and 5 lowest contributions $b_{2,j}$ on $\mathbf{h_2}$. Companies having big contributions are mostly technical companies and service (financial, entertainment) providers, while those of near-zero contributions are bio-chemical and traditional industry companies. Since the technical companies are more sensitive to $\mathbf{h_2}$, we suggest that $\mathbf{h_2}$ corresponds to the "Internet bubble" which largely affected technical companies during year 2000-2001.

(Observation 2) *AutoSplit automatically discovered the meaningful underlying factors, namely, the general trend and the Internet bubble (Figure 6(b)).*

(Observation 3) *We found rules like "companies in financial and traditional industry grow steadily during 1990-2002, while technical companies suffered from the event around late-2000", and detected outliers like AT&T, which does not follow the general growth trend during the period 1990-2002 (Table 1(a)).*

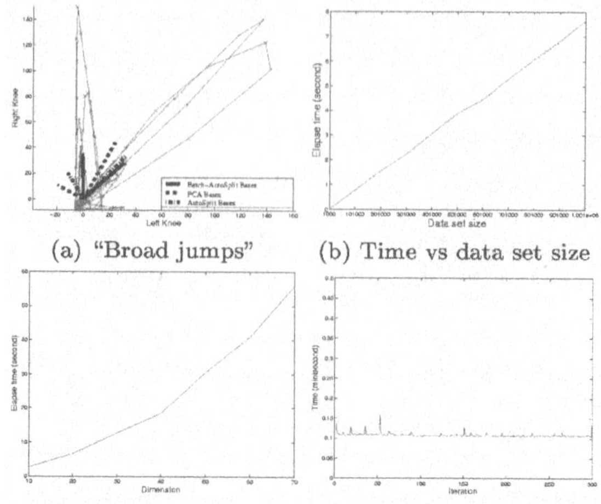

(a) "Broad jumps" (b) Time vs data set size

(c) Time vs dimensionality (d) Time per iteration

Fig. 7. (a): Bases found by Batch-AutoSplit (solid), PCA (dash), and AutoSplit (dash-dot) on the "broad jumps" data set. (b)(c)(d): Scalability AutoSplit.

3.2 AutoSplit : Quality and Scalability

Can we operate AutoSplit on a continuous data stream? If yes, at what accuracy loss? Here we compare our "AutoSplit", the online processing algorithm, with the Batch-AutoSplit. Figure 7(a) shows the bases generated by AutoSplit, along with those by Batch-AutoSplit and PCA. Batch-AutoSplit has access to the complete data, so it gives near perfect bases (solid vectors). AutoSplit gives bases (dash-dot vectors) which are very close to the true bases.

We also studied the scalability of AutoSplit with respect to the data set size and to the data dimensionality. To study the effect of data set size, we generate 2-D synthetic data set similar to our "broad jumps" data set, but with more data points (vary from n=10^3 to 10^6). Figure 7(b) shows the total running time of AutoSplit (until convergence) versus the data set size. The total running time of AutoSplit is linear (O(n)) to the data set size (n), as expected, for the computational cost per iteration is constant (several matrix multiplications). Figure 7(d) shows the small constant computational cost per iteration ($\approx$ 0.12 msec), when AutoSplit is applied to the "broad jumps" data set.

To study the effect of dimensionality, we fixed the data set size to n=35,000, while varies the dimensionality from m=10 to 70. Synthetic data of higher dimensionality are generated to have a non-orthogonal distribution similar to a high-dimensional version of the "broad jumps" data set with more "spikes". The total running time (Figure 7(c)) of AutoSplit is super-linear, and probably quadratic (O(m^2)).

3.3 Experiment with Clustering-AutoSplit

We apply Clustering-AutoSplit to separate two texture classes in an image: overlay text and background. The idea is to find two different sets of bases for image patches of the two classes. Figure 8(a1)(a2) show Clustering-AutoSplit (with $k=2$) gives good separation of the overlay text from the background in a video frame [23]. The data items (each is a 36-dimensional row vector in $\mathbf{X}$) are the 6-by-6 pixel blocks taken from the frame. Figure 8(b1)(b2) show the failure of the PCA-based mixture model (MPPCA, Mixture of Probabilistic PCA [24]) on this task. MPPCA fails to differentiate the background edges with the real texts.

(a1) Background (a2) Overlay text || (b1) Background (b2) Overlay text

Fig. 8. Texture segmentation. Result from (a) Clustering-AutoSplit (b) MPPCA.

4 Conclusions

We propose AutoSplit, a powerful, incremental method for processing streams as well as static, multimedia data. The proposed "Clustering-AutoSplit" extends the feature discovery to multiple feature/bases sets and shows a better performance than the PCA-based method in texture segmentation (Figure 8). The strong points of AutoSplit are:

- It finds *bases* which better capture the natural trends and correlation of the data set (Figure 1(a)).
- It finds rules, which are revealed in the basis matrix $\mathbf{B}$ (Observation 1,3).
- It scales linearly with the number n of data points (Figure 7(b)).
- Its incremental, single-pass algorithm (Figure 3) makes it readily suitable for processing on streams.

References

1. Deerwester, S., Dumais, S.T., Furnas, G.W., Landauer, T.K., Harshman, R.A.: Indexing by latent semantic analysis. Journal of the American Society of Information Science **41** (1990) 391–497
2. Turk, M., Pentland, A.: Eigenfaces for recognition. Journal of Cognitive Neuroscience **3** (1991) 72–86

3. Duda, R.O., Hart, P.E., Stork, D.G.: Pattern Classification, 2nd Edition. New York: Wiley (2000)
4. Jolliffe, I.T.: Principal Component Analysis. Springer-Verlag (1986)
5. Korn, F., Labrinidis, A., Kotidis, Y., Faloutsos, C.: Ratio rules: A new paradigm for fast, quantifiable data mining. In: VLDB. (1998)
6. Garofalakis, M., Gehrke, J., Rastogi, R.: Querying and mining data streams: You only get one look. In: VLDB. (2002)
7. Guha, S., Gunopulos, D., Koudas, N.: Correlating synchronous and asynchronous data streams. In: SIGKDD 2003. (2003)
8. Kanth, K.V.R., Agrawal, D., Singh, A.K.: Dimensionality reduction for similarity searching in dynamic databases. In: SIGMOD. (1998) 166–176
9. Garofalakis, M., Gibbons, P.B.: Wavelet synopses with error guarantees. In: SIGMOD 2002. (2002)
10. Achlioptas, D.: Database-friendly random projections. In: PODS. (2001) 274–281
11. Indyk, P., Koudas, N., Muthukrishnan, S.: Identifying representative trends in massive time series data sets using sketches. In: Proc. VLDB. (2000) 363–372
12. Gunopulos, D., Das, G.: Time series similarity measures and time series indexing. In: SIGMOD. (2001) 624
13. Jensen, C.S., Snodgrass, R.T.: Semantics of time-varying information. Information Systems **19** (1994) 33–54
14. Teng, W.G., Chen, M.S., Yu, P.S.: A regression-based temporal pattern mining scheme for data streams. In: VLDB 2003. (2003) 93–104
15. Yi, B.K., Sidiropoulos, N.D., Johnson, T., Jagadish, H., Faloutsos, C., Biliris, A.: Online data mining for co-evolving time sequences. In: ICDE. (2000)
16. Jagadish, H., Mendelzon, A., Milo, T.: Similarity-based queries. In: PODS '95. (1995)
17. Moon, Y.S., Whang, K.Y., Han, W.S.: General match: a subsequence matching method in time-series databases based on generalized windows. In: SIGMOD 2002. (2002) 382–393
18. Keogh, E., Chakrabarti, K., Mehrotra, S., Pazzani, M.: Locally adaptive dimensionality reduction for indexing large time series databases. In: SIGMOD. (2001) 151–162
19. Korn, F., Jagadish, H.V., Faloutsos, C.: Efficiently supporting ad hoc queries in large datasets of time sequences. In: Proc. SIGMOD. (1997) 289–300
20. Lee, J., Chai, J., Reitsma, P.S.A., Hodgins, J.K., Pollard, N.S.: Interactive control of avatars animated with human motion data. In: SIGGRAPH 2002. (2002)
21. Hyvarinen, A., Karhunen, J., Oja, E.: Independent Component Analysis. John Wiley & Sons (2001)
22. Lewicki, M.S.: Estimating sub- and super-gaussian densities using ica and exponential power distributions with applications to natural images. Unpublished Manuscript (2000)
23. Wactlar, H., Christel, M., Gong, Y., Hauptmann, A.: Lessons learned from the creation and deployment of a terabyte digital video library. IEEE Computer **32** (1999) 66–73
24. Tipping, M., Bishop, C.: Mixture of probabilistic principal component analyzers. Neural Computation (1998)

Semantic Sequence Kin: A Method of Document Copy Detection

Jun-Peng Bao, Jun-Yi Shen, Xiao-Dong Liu, Hai-Yan Liu, and Xiao-Di Zhang

Department of Computer Science and Engineering, Xi'an Jiaotong University, Xi'an 710049, People's Republic of China
baojp@mail.xjtu.edu.cn

Abstract. The string matching and global word frequency model are two basic models of Document Copy Detection, although they are both unsatisfied in some respects. The String Kernel (SK) and Word Sequence Kernel (WSK) may map string pairs into a new feature space directly, in which the data is linearly separable. This idea inspires us with the Semantic Sequence Kin (SSK) and we apply it to document copy detection. SK and WSK only take into account the gap between the first word/term and the last word/term so that it is not good for plagiarism detection. SSK considers each common word's position information so as to detect plagiarism in a fine granularity. SSK is based on semantic density that is indeed the local word frequency information. We believe these measures diminish the noise of rewording greatly. We test SSK in a small corpus with several common copy types. The result shows that SSK is excellent for detecting non-rewording plagiarism and valid even if documents are reworded to some extent.

1 Introduction

In this paper we propose a novel Semantic Sequence Kin (SSK) that is based on the local semantic density, not on the common global word frequency. And we apply it to Document Copy Detection (DCD), not the Text Classification (TC). DCD is to detect whether some part or the whole of the given document is the copy of other documents (it means plagiarism). However, the word frequency based kernel is not suitable for DCD though it is popular in TC. The word frequency model takes mainly global semantic features of a document but loses the detailed local features and structural information. For example, TF-IDF (Term Frequency - Inverse Document Frequency) vector is a basic document representation in TC. But we cannot use TF-IDF vector to distinguish two sentences (or sections) that are just the different arrangements of the same words, which usually have different meanings.

By means of matching string, we can exactly find out the plagiarized sentences. Indeed, many DCD prototypes [4-6] prefer it. This method first gets some strings called fingerprints as text features and then matches the fingerprints to detect plagiarism. The string matching model exploits local features of a document mainly. It can hardly resist noise, and rewording sentences may impair the detection precision

H. Dai, R. Srikant, and C. Zhang (Eds.): PAKDD 2004, LNAI 3056, pp. 529–538, 2004.

heavily. Therefore, it is better to take account of both global and local feature in order to detect plagiarism in certain detail and against rewording noise.

In SSK, We first find out the semantic sequences based on the concept of semantic density, which represents the locally frequent semantic features, and then we collect all of the semantic sequences to imply the global features of the document. When we calculate the similarity between document features, we absorb the ideas of word-sequence kernel and string kernel.

In the next section, we introduce some related work on string kernel and DCD. We present our Semantic Sequence Kin in detail in Section 3 and release experimental results in Section 4. We discuss some aspects of SSK in Section 5. Finally we draw conclusions in Section 6.

2 Related Work

Joachims [1] first applied SVM to TC. He used VSM (Vector Space Model) to construct the text feature vector, which contains only word global frequency information without any structural (sequence) information. Lodhi et al.[3] proposed the string kernel method that classifies documents by the common subsequences between them. The string kernel exploits the structural information (i.e. gaps between terms) instead of word frequency. Before long, Cancedda et al.[2] introduced the word sequence kernel that extends the idea of string kernel. They greatly expand the number of symbols to consider, as symbols are words rather than characters.

Now the kernel methods are popular in TC, but we have not found its application to DCD. Brin et al.[4] proposed the first DCD prototype (i.e. COPS) that detects overlap based on sentence and string matching, but it has some difficulties in detecting sentences and cannot find partial sentence copy. In order to improve COPS, Shivakumar and Garcia-Molina [7] developed SCAM (Stanford Copy Analysis Method), which measures overlap based on word frequency.

Heintze [6] developed a KOALA system for plagiarism detection. Broder et al.[5] proposed a *shingling* method to determine the syntactic similarity of files. These 2 systems are similar to COPS. Monostori et al.[8] proposed the MDR (Match Detect Reveal) prototype to detect plagiarism in large collections of electronic texts. It is also based on string matching, but it uses suffix tree to find and store strings.

Si et al.[9] built a copy detection mechanism CHECK that parsed each document to build an internal indexing structure, which is called structural characteristic (SC), used in document registration and comparison modules. Song et al.[10] presented an algorithm (CDSDG) to detect illegal copy and distribution of digital goods, which indeed combined CHECK and SCAM to discover plagiarism.

3 Semantic Sequence Kin

In the following we first introduce some concepts about the semantic sequence, and then we propose SSK in detail.

3.1 Semantic Density and Semantic Sequence

Definition 1 Let S be a sequence of words, i.e. $S = S_1S_2...S_n$. We denote the word at the position i in S by S_i. The *word distance* of position i $(1 \le i \le n)$, denoted by $\sigma(i)$, is the number of words between S_i and its preceding occurrence S_h:

$$\sigma(i) = i - h \tag{1}$$

where $S_h=S_i$ and $S_k \ne S_i$ $(1 \le h<k<i \le n)$, that is S_h and S_i are the same word and no other words is the same with S_i between S_h and S_i. If no S_h exists, i.e. S_i occurs for the first time, then $\sigma(i)=\infty$.

Definition 2 Let S be a sequence of words, i.e. $S=S_1S_2...S_n$. The *semantic density* of position i $(1 \le i \le n)$, denoted by $\rho(i)$, is the reciprocal of $\sigma(i)$:

$$\rho(i) = 1/\sigma(i) \tag{2}$$

The semantic density is a kind of *word density*, but we believe that a sequence of word should imply certain semantic information. In fact, $\sigma(i)$ is the distance of S_i to its preceding occurrence in the sequence S, and $\rho(i)$ reflects its local frequency. A document is a long sequence of words such that in a given range the small distance means the high density of words in a local section. That is to say the smaller distance leads to the higher density of words in the local section. We believe that the high-density words in some section indicate the local semantic feature of the section.

Definition 3 Let S be a sequence of words, i.e. $S=S_1S_2...S_n$. A *semantic sequence* of S is a subsequence of S, denoted by $S[\boldsymbol{i}]=S_{i_1}S_{i_2}...S_{i_r}$ with $\boldsymbol{i}=[i_1,i_2,...,i_r]$ $(1<i_1<i_2<i_r \le n)$, which satisfies the following conditions:

(1) $|S[\boldsymbol{i}]| > 1$

(2) $0 < i_{k+1} - i_k \le \varepsilon, \quad 1 \le k < r$

(3) $\rho(i_k) \ge \delta, \quad 1 \le k \le r$

(4) $(0 < i_1 - i_h \le \varepsilon) \rightarrow (\rho(i_h) < \delta)$

(5) $(0 < i_j - i_r \le \varepsilon) \rightarrow (\rho(i_j) < \delta)$

where δ and ε are user defined parameters.

In fact, a semantic sequence in S is a *continual* word sequence after the low density words in S are omitted. In definition 3, the condition 1 guarantees $S[\boldsymbol{i}]$ is a non-trivial sub-sequence of S. The condition 2 ensures that we sample with text adequately. The condition 3 ensures that each word in $S[\boldsymbol{i}]$ must be dense, i.e. the word must be locally frequent. The condition 4 and 5 ensure that the sub-sequence cannot be extended in either direction.

A long S may have several semantics. We denote all of the semantic sequences in a document S by $\Omega(S)$, which then includes the global and local semantic features as well as local structural information. However, a single semantic sequence may not represent the global feature of the document. For example, we discuss the question q in the paragraph P of a document, and q is never mentioned in paragraph other than P. If the proportion of P in the whole document is very low, then words about q may seldom appear in the global features. But they can be caught in the semantic sequences of P. Hence we believe the semantic sequence can detect plagiarism in a fine granularity so that we can find n to 1 partial copy well (we explain the n to 1 partial copy in the Section 4.1).

There is a fact for DCD that if one section is shared between two documents, then we think they are a copy pair. It is enough for DCD to compare similarity in several most likely common sequences. We know that the larger number of common words there are between two strings the more similar they are. Therefore, we select candidate semantic sequences in *Ω(S)* and *Ω(T)* according to the number of common words between them.

Let $S[\boldsymbol{i}]\cap T[\boldsymbol{j}]$ be the set of common words between $S[\boldsymbol{i}]$ and $T[\boldsymbol{j}]$. Let $CL(S,T)=[(S[\boldsymbol{i}_1],T[\boldsymbol{j}_1]),\ldots,(S[\boldsymbol{i}_n],T[\boldsymbol{j}_n])]$ be the list of semantic sequence pairs on document S and T, which is sorted by $|S[\boldsymbol{i}_k]\cap T[\boldsymbol{j}_k]|,(1\le k\le n)$ descendingly. We denote the first η semantic sequence pairs by $CL_\eta(S,T)$. We denote the set of all the common words between semantic sequence pairs in $CL_\eta(S,T)$ by $CP(S,T)$:

$$CP(S,T)=\bigcup_{k=1}^{\eta}(S[\boldsymbol{i}_k]\cap T[\boldsymbol{j}_k]),\quad (S[\boldsymbol{i}_k],T[\boldsymbol{j}_k])\in CL_\eta(S,T) \tag{3}$$

The words in $CL_\eta(S,T)$ are not only common words between S and T but also locally frequent words. We believe that $CL_\eta(S,T)$ reflects the local identity between 2 documents in some detail. According to the fact that the more common words there are between documents, the more similar they are, $|CP(S,T)|$ can measure document similarity. However, when two documents share the same words list, they must be very similar but may not be identical. The different arrangements of the same words list often represent different documents on the same topic, namely, they are similar and belonging to the same category, but not identical. $|CP(S,T)|$ regards all possible arrangements of one words list as the same so that it makes a high positive error. In order to discriminate the different arrangement of the same words list, we add structure information of strings in SSK.

3.2 Semantic Sequence Kin

String kernel and word sequence kernel calculate the dot product of two strings based on gaps between terms/words, i.e. $l(\boldsymbol{i})$. The $l(\boldsymbol{i})$ cannot exactly reflect different arrangement of the same words list because it considers only the first and the last term/word in the list, not the others. If we take into account the position of each word, then we can detect plagiarism more precisely. This is the basic idea of SSK.

The Semantic Sequence Kin of two semantic sequences $S[\boldsymbol{i}]$ and $T[\boldsymbol{j}]$ is defined as:

$$K(S[\boldsymbol{i}],T[\boldsymbol{j}])=\sum_{k=1}^{|S[\boldsymbol{i}]\cap T[\boldsymbol{j}]|}\lambda^{x_k},\quad \lambda=\frac{|S[\boldsymbol{i}]\cap T[\boldsymbol{j}]|}{|S[\boldsymbol{i}]|+|T[\boldsymbol{j}]|} \tag{4}$$

where x_k is the difference of a common word's word distances between $S[\boldsymbol{i}]$ and $T[\boldsymbol{j}]$, that is:

$$x_k=|\sigma(i_u)-\sigma(j_v)|,\quad S_{i_u}=T_{j_v}\in S[\boldsymbol{i}]\cap T[\boldsymbol{j}] \tag{5}$$

where $\sigma(i_u)$ is the word distance of i_u in S and $\sigma(j_v)$ is the word distance of j_v in T. If a common word occurs twice or more times in $S[\boldsymbol{i}]$ (or $T[\boldsymbol{j}]$), then in $S[\boldsymbol{i}]\cap T[\boldsymbol{j}]$ we only save the word that occurs first with its position in $S[\boldsymbol{i}]$ (or $T[\boldsymbol{j}]$), and omit the other occurrences. For example, there are six semantic sequences as follows:

S_1: ....A B C.....; S_2: ....A B C.....A B C.....; S_3: ....A B C.....B C A.....;

T_1: ....C B A.....; T_2: ....C B A.....C B A.....; T_3: ....C B A.....A B C.....;

Then $K(S_1,T_1)=K(S_2,T_2)=K(S_3,T_3)$. This could disturb SSK trivially. However, we ignore it and consider it as a noise.

In order to keep the kernel values comparable and independent from the length of the strings, we normalize the kernel as follows:

$$\hat{K}(S[i],T[j])=\frac{K(S[i],T[j])}{\sqrt{K(S[i],S[i])K(T[j],T[j])}} \tag{6}$$

Indeed, $\because \quad \forall S_{i_p} \in S[i]\cap S[i]=S[i], \quad x_k=|\sigma(i_p)-\sigma(i_p)|=0$

$$\therefore \quad K(S[i],S[i])=\sum_{k=1}^{|S[i]\cap S[i]|}\lambda^{x_k}=\sum_{k=1}^{|S[i]\cap S[i]|}\lambda^{0}=|S[i]\cap S[i]|=|S[i]|$$

similarly, $K(T[j],T[j]) = /T[j]/$

$$\therefore \quad \hat{K}(S[i],T[j])=\frac{K(S[i],T[j])}{\sqrt{|S[i]||T[j]|}} \tag{7}$$

We give the Semantic Sequence Kin of a single semantic sequence pair above. For a document pair S and T, we may select several semantic sequence pairs in order to improve accuracy, i.e. $CL_\eta(S,T)$, $\eta \geq 1$. Thus, we define Semantic Sequence Kin of document pair (S,T) as:

$$K(S,T)=\frac{1}{\eta}\sum_{k=1}^{\eta}\hat{K}(S[i_k],T[j_k]), \quad (S[i_k],T[j_k])\in CL_\eta(S,T) \tag{8}$$

It is obvious that $K(S,T)\in[0,1]$. For document pair (S,T), we use the discriminant below to make decision.

$$f(S,T)=sign(aK(S,T)+b), \quad a,b\in R, a\neq 0 \tag{9}$$

4 Experimental Results

In this section, we do some experiments to test our Semantic Sequence Kin and compare it with Relative Frequency Model (RFM) [7], which was successfully used to track real life instances of plagiarism in several conference papers and journals [12]. RFM uses an asymmetric metric to find out subset copy, but it is not so valid for n to 1 partial copy. In order to contrast error trend of SSK with that of RFM, we vary the discriminating threshold manually, that is:

$$f(S,T)=sign(aK(S,T)+b)=sign(a)sign(K(S,T)+b/a), \quad a,b\in R, a\neq 0$$

If we let $a>0$, then $f(S,T) = sign(K(S,T)+b/a)$. Let $\tau = b/a$, called *discriminating threshold*. When we vary the value of τ, we will get different detection error.

4.1 Test Corpus

We collect 75 full text literatures in PDF format from Internet, which can be downloaded from ftp://202.117.15.227/dcd_corpora.zip. These are original documents from which we make our plagiarized documents. (1) Exact self-copy(*D-copy*): these copies are just renamed source files. (2) Self cross copy(*O-copy*): first we divide each source file into 10 blocks and then we randomly reassemble these blocks into a new copy. (3) Subset copy(*Y-copy*): we select 2 files from the source files and divide each of them into 10 blocks, at last we randomly reassemble these 20 blocks into a new copy. It is obvious that the 2 source files are subsets of the copy. (4) N to 1 partial copy(*X-copy*): we select n (we set it to 5 in our experiment) files from the source files and divide each of them into 10 blocks, then we randomly select k (it is 2 in our experiment) blocks from each 10 blocks, at last we randomly reassemble all selected blocks into a new copy. The copy contains small parts of several source files but none of them is subset of the copy. (5) The 4 types above are non-rewording copy types. We reword each word of the non-rewording copy files in certain probability to get the respective rewording files. We use Martin Porter's stemming algorithm and procedure [11] to stem words and then remove the stop words[1]. Finally, we make a document pair with a copy file and a source file.

We define the positive error as the proportion of non-copy document pairs (no plagiarism) above the threshold τ in the whole non-copy document pairs, i.e.

$$E_{\Phi}^{+}(\tau) = \frac{|\{\Phi_N\}_{\geq\tau}|}{|\Phi_N|} \tag{10}$$

where Φ is some type of document pairs, and $\{\Phi_N\}_{\geq\tau}$ is those non-copy pairs of type Φ whose plagiarism score is greater than or equal to τ. The negative error is the proportion of copy document pairs (containing plagiarism) below the threshold τ in the whole copy pairs, i.e.

$$E_{\Phi}^{-}(\tau) = \frac{|\{\Phi_C\}_{<\tau}|}{|\Phi_C|} \tag{11}$$

where $\{\Phi_C\}_{<\tau}$ is those copy pairs of type Φ whose score is less than τ.

4.2 Contrasting Experiments

We have mentioned that we can use $|CP(S,T)|$ to distinguish plagiarism, and we call this approach as Semantic Sequence Model (SSM). The SSM plagiarism score of document S and T is $q_{SSM}(S,T)$.

$$q_{SSM}(S,T) = \min\{\frac{|CP(S,T)|}{a\eta}, 1\} \tag{12}$$

Figure 1 shows the error trends of SSM, RFM and traditional VSM on the whole non-rewording corpus. We see that the negative error of SSM is very low and flat but its positive error is high. Figures 2 show their error trends on X-copy corpus. The negative errors of RFM and VSM increase rapidly while that of SSM is still small. It illustrates that RFM and VSM are futile on the X-copy corpus whereas SSM is valid. SSM gains low negative error at the expense of high positive error.

Figure 3 shows the error trends of Semantic Sequence Kin (SSK), SSM and RFM on the whole non-rewording corpus. We see that the positive errors of SSK are always

[1] Available at http://ls6-www.informatik.uni-dortmund.de/ir/projects/freeWAIS-sf/

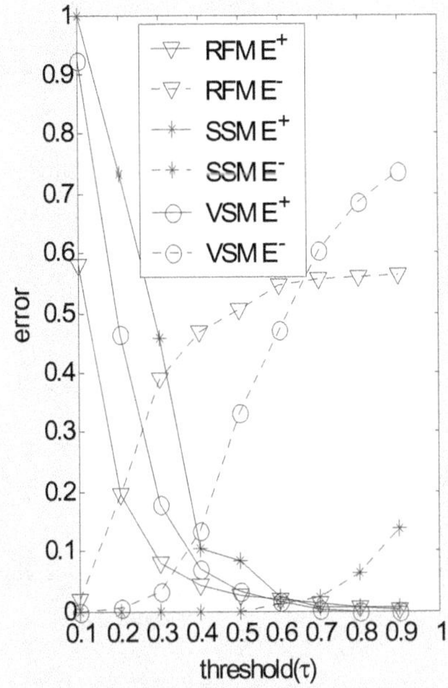

Fig. 1. VSM, RFM and SSM error on the whole non-rewording corpus

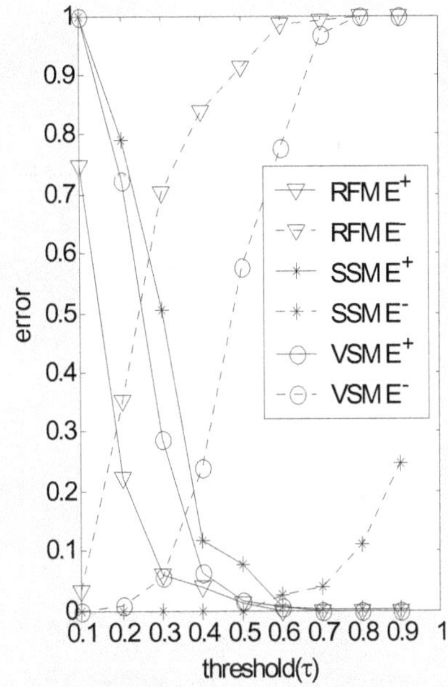

Fig. 2. VSM, RFM and SSM error on the X-copy corpus

near 0 while its negative errors increase slowly. The negative errors of SSK are a little bigger than that of SSM. SSK gains very low positive error at the expense of a little increase of negative error. Whatsoever, SSK is superior to RFM.

Figure 4 shows the error trends of SSM and RFM on the rewording corpus with each word rewording probability[2] θ=0.7. From Figure 4, we see that the negative error of RFM on rewording corpus is far larger than that of SSM. It implies that RFM detects fewer plagiarisms than SSM, that is, RFM will lose more cases of plagiarism than SSM. Figure 5 shows the error trends of SSK with the same rewording probability. Figure 6 shows SSK error trends with different rewording probabilities (θ=0.2,0.4,0.6,0.8). We find that when the rewording probability increases, the positive errors of SSK are stable and the negative errors of SSK increase a little. Whatsoever, we can get the appropriate values of τ to keep both the positive and negative error in a tolerable range. It proves that SSK is valid for DCD even if the document is reworded to some degree.

5 Discussions

From the experiments we find that the positive error and the negative error are a pair of contradiction. That is to say a low positive error will lead to a high negative error

[2] We use JWNL (available on http://sourceforge.net/projects/jwordnet) to reword each word in document. Because the synonym set of a word contains the word itself, the real rewording probability is lower than the labeled value.

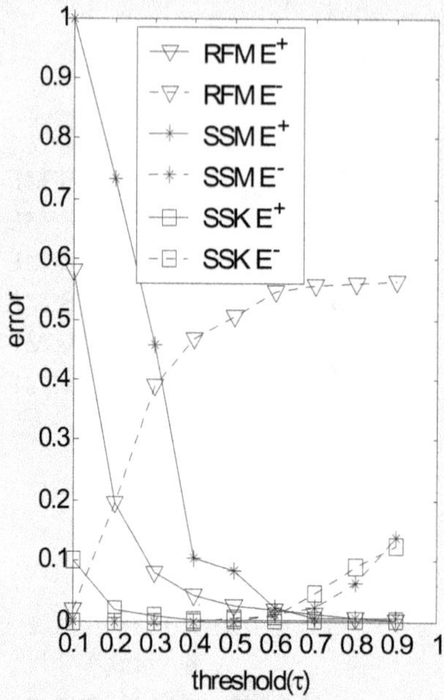

Fig. 3. SSK, RFM and SSM errors on the whole non-rewording corpus

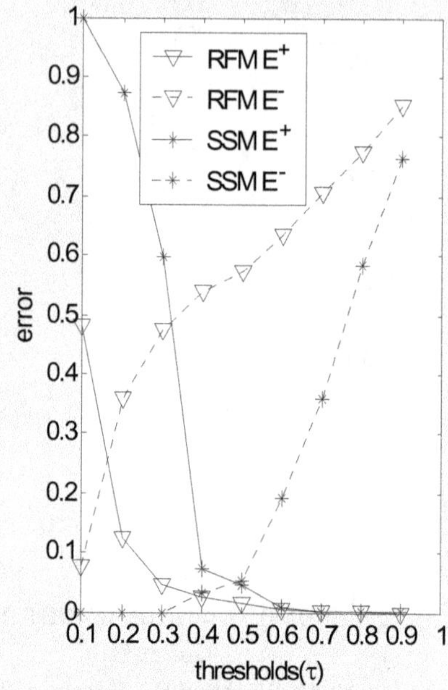

Fig. 4. RFM and SSM errors on the rewording corpus with rewording

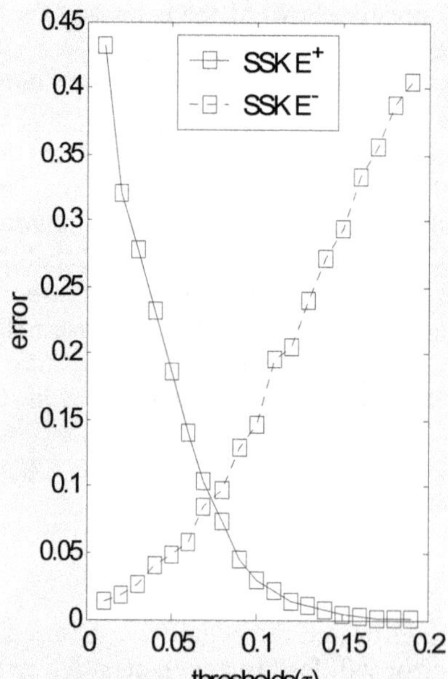

Fig. 5. SSK errors on the rewording corpus with rewording probability θ=0.7

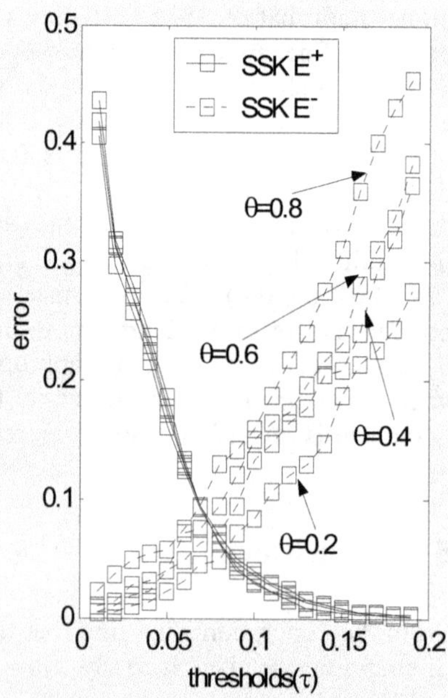

Fig. 6. SSK errors on the rewording corpus with θ=0.2, 0.4, 0.6, 0.8

and vice versa. Therefore we have to make a trade-off between the positive and negative error. The negative error of SSM is the lowest, which leads to its highest positive error. We use SSK to make the positive error very low at the expense of increasing negative error a little. Whatsoever, SSM and SSK are superior to RFM and traditional VSM.

Both SSK and SSM conform to the principle that the more common words there are between documents, the more similar they are. Additionally, SSK satisfies stronger structural condition, i.e. the common words' word distance must be similar, otherwise the word will be penalized. On the one hand, SSM is superior to SSK on negative error because it misses out fewer plagiarisms than SSK, although many documents are mistakenly detected to involve plagiarism. On the other hand, the high SSK score means that one or more word sequence(s) must be almost the same between the documents, which is just what plagiarism means. Thus, SSK seldom mistakes non-plagiarism for plagiarism so that its positive errors are very low.

However, the rewording action may not only reduce the common words but also disorder the words' sequence, which raises the negative error of SSK and SSM, so they may miss some plagiarized documents. Interestingly, the rewording action decreases the probability of mistaking some documents for plagiarism such that the positive error declines while rewording probability rises.

In another respect, while the discriminating threshold (τ) is increasing, the positive error trend is declining and negative error trend is increasing. The higher value of τ means that we need more common words to confirm plagiarism, so that we may decrease mistakes in plagiarism detection. Consequently, when the value of τ is low, we will get low negative error and high positive error, and when the value of τ is high, we will get high negative error and low positive error, which causes the optimal value of τ of SSM larger than that of SSK. Altogether, the low τ makes system radical, which misses out fewer plagiarism cases but easily misjudges the similar documents as plagiarism. In contrast, the high τ makes system conservative, which catches serious plagiarism but may fail to detect many light or partial copy documents.

6 Conclusions

We extend the idea of SK and WSK to make SSK. SSK compares two semantic sequences according to their common words and position information so that we can detect plagiarism in a fine granularity. SSK combines the features of word frequency model and string matching model, which are commonly used in traditional DCD. So SSK is superior to traditional DCD models, such as VSM and RFM et al. Especially in non-rewording corpus, both the positive errors and the negative errors are lower than traditional models. For a given model, if we decrease the positive error by some means, then the negative error must increase and vice versa. In order to get the optimal result, we have to make a trade-off between positive error and negative error. The rewording action may not only change the possible common words between documents, but also influence the local structural information, which impairs the SSK's performance greatly. Hence our next goal is to increase the detection accuracy in rewording corpus by means of adding some rewording probability in SSK.

Acknowledgements

Our study is sponsored by NSFC (National Natural Science Foundation of China, ID: 60173058) and Xi'an Jiaotong University Science Research Fund (ID: 573031).

References

1. Joachims T.: Text categorization with support vector machines: Learning with many relevant features. In Proceedings of the European Conference on Machine Learning. Lecture Notes in Computer Science, Vol. 1398. Springer-Verlag, Berlin Heidelberg New York (1998)137–142
2. Cancedda N., Gaussier E., Goutte C., Renders J. M.: Word-Sequence Kernels. Journal of Machine Learning Research. 3 (2003)1059-1082
3. Lodhi H., Saunders C., Shawe-Taylor J., Cristianini N., Watkins C.: Text Classification using String Kernels. Journal of Machine Learning Research. 2 (2002)419-444
4. Brin S., Davis J., Garcia-Molina H.: Copy detection mechanisms for digital documents. In Proceedings of the ACM SIGMOD Annual Conference, San Francisco, USA (1995)
5. Broder A.Z., Glassman S.C., Manasse M.S.: Syntactic Clustering of the Web. In Proceedings of the Sixth International Web Conference, Santa Clara, California (1997)
6. Heintze N.: Scalable Document Fingerprinting. In Proceedings of the Second USENIX Workshop on Electronic Commerce, Oakland, California (1996)
7. Shivakumar N. and Garcia-Molina H.: SCAM: A copy detection mechanism for digital documents. In Proceedings of 2nd International Conference in Theory and Practice of Digital Libraries, Austin, Texas (1995)
8. Monostori K., Zaslavsky A., Schmidt H.: MatchDetectReveal: Finding Overlapping and Similar Digital Documents. In Proceedings of Information Resources Management Association International Conference, Alaska (2000)
9. Si A., Leong H.V., Lau R.W.H.: CHECK: A Document Plagiarism Detection System. In Proceedings of ACM Symposium for Applied Computing, (1997)70-77
10. Song Qinbao, Shen Junyi. On illegal coping and distributing detection mechanism for digital goods. Journal of Computer Research and Development. 38(2001) 121-125
11. Porter M.F.: An algorithm for suffix stripping, Program. 14(1980) 130-137
12. Denning P.J., Editorial: Plagiarism in the web. Communications of the ACM, 38(1995)

Extracting Citation Metadata from Online Publication Lists Using BLAST

I-Ane Huang[1], Jan-Ming Ho[1], Hung-Yu Kao[1], Wen-Chang Lin[2]

[1]Institute of Information Science, Academia Sinica, Taiwan
{ia@,hoho,bobby}iis.sinica.edu.tw
[2]Institute of Biomedical Sciences, Academia Sinica, Taiwan
wenlin@ibms.sinica.edu.tw

Abstract. Scientific research reports usually contain a list of citations on previous related works. Therefore an automatic citation tool is an essential component of a digital library of scientific literatures. Due to variations in formats, it is difficult to automatically transform semi-structured citation data into structured citations. Some digital library institutes, like ResearchIndex (CiteSeer) or OpCit, have attempted automatic citation parsing. In order to recognize metadata, e.g., authors, title, journal, etc., of a citation string, we present a new methodology based on protein sequence alignment tool. We also develop a template generating system to transform known semi-structured citation strings into protein sequences. These protein sequences are then saved as templates in a database. A new semi-structured citation string is also translated it into a protein sequence. We then use BLAST (Basic Local Alignment Search Tool), a sequence alignment tool, to match for the most similar template to the new protein sequence from the template database previously constructed. We then parse metadata of the citation string according to the template. In our experiment, 2,500 templates are generated by our template generating system. By parsing all of these 2,500 citations using our parsing system, we obtain 89% precision rate. However, using the same template database to train ParaTools, 79% precision rate is obtained. Note that the original ParaTools using its default template database, which contains about 400 templates, only obtains 30% precision rate.

1 Introduction

It is difficult for a computer to automatically parse citations because there are a lot of different citation formats. Citations always include author, title, and publication information. Publication information format varies according to publication type, e.g., books, journals, conference papers, research reports, and technical reports. Publication information can include publication name, volume, number, page number, year published, month published, and publisher's address. Citations can be presented in either structured or semi-structured form. Semi-structured citation form is more flexible, so bibliographies created by different people may have different citation forms. Metadata order may be different as well as their attributes. Bibliographies on the

H. Dai, R. Srikant, and C. Zhang (Eds.): PAKDD 2004, LNAI 3056, pp. 539–548, 2004.

Internet are usually in semi-structured form. If we want to use their data, we must first transform the semi-structured bibliography into structured bibliography. We have to analyze the metadata of each citation, and build up an index for bibliography searches and citing statistics. In this paper, we discuss how to transform semi-structured bibliographies into uniform structured data, the core problem of citation data processing.

CiteSeer [1][2][3][4], which use heuristics to extract certain subfields, can "find the titles and authors in citations roughly 80% of the time and page numbers roughly 40% of the time" [1]. Another system, ParaTools [5][6][7][8] (short for ParaCite Toolkit) is a collection of Perl modules used for reference parsing. It uses a template-based reference parser to extract metadata from references. Our approach is similar to ParaTools in that we also use a template-based reference parser, but we found a better alignment from BLAST [9][10].

There are about 30 billion nucleotides in a human genome, and for about every 1,000 base pairs, there will be a nucleotide difference in genomes. We can use BLAST to compare sequence to identify whether the sequences belong to the same person. We realized that we could use this method to identify citation. In our system, we use a form translation program to translate citations into a form that we can process more easily. The form we use is a protein sequence because we can use BLAST, a well-developed protein sequence matching program, to process it. BLAST needs a scoring table to search for most similar sequences in a protein sequence database. This database stores the templates of known citations that have been translated into protein form. After finding the most similar sequence template, we use a pattern extraction program to parse the citation metadata according to the template. Once the metadata are correctly parsed, we manually validate them and add them into our knowledge database.

The rest of the paper is organized as following. Section 2 is system architecture and design. Section 2.1 introduces the BLAST. Section 2.2 defines the tokens that we use. Section 2.3 tells how the tokens translated to amino acids. Section 2.4 and 2.5 are the template generating system and citation parsing system. In section 3, we do some experiment here. Section 4 is conclusion and future works.

2 System Architecture and Design

As shown figure 1, we can divide our system into two subsystems: a template generating system and a citation parsing system. For the template generating system, we manually crawl BibTeX files on the web to obtain our template generating data. From the BibTeX file, our system can find out the titles, and then use these titles automatically obtain semi-structured citations from the CiteSeer web site. Now we have both semi-structured data from CiteSeer and structured data from BibTeX files. Since the semi-structured and structured data are the same but presented in different forms, this system can use these two forms to automatically make a template database. After the template database is constructed, the system begins the parsing process. In the parsing system, BLAST is used to compare strings. We transform a citation into protein form, and use BLAST to search for the most similar sequence in the template database. We can then parse the citation according to the template that BLAST finds.

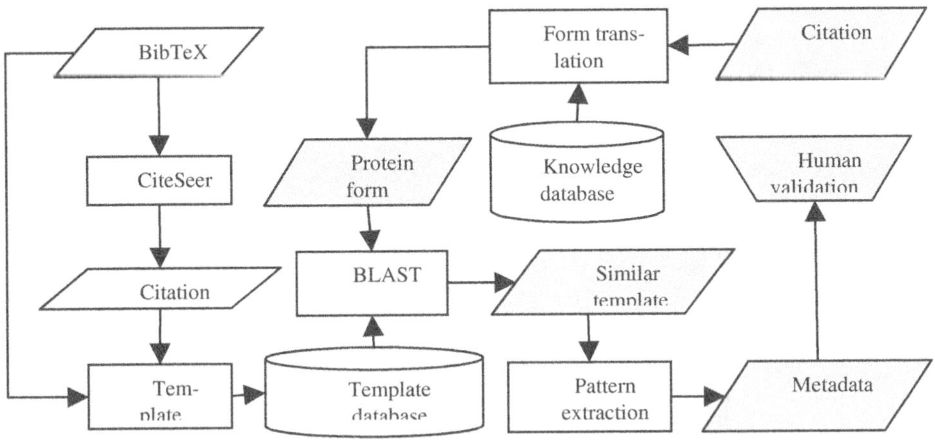

Fig. 1. The architecture of this system.

2.1 BLAST

BLAST is a similarity search tool developed by Altschul et al. (1990), it is based on dynamic programming. It is used to search for optimal local alignments between sequences. BLAST breaks the query and database sequences into fragments (words). It searches the matches for the word of length W that scores at least T. Matches are extended to generate an MST(maximal sequence pair) with a score exceeding the threshold of S. The quality of each pair-wise alignment is represented as a score. Scoring matrices are used to calculate the score of the alignment amino acid by amino acid. The significance of each alignment is computed as a P-value or an E-value [10]. P-value is the likelihood that two random sequences will have an MST with a score greater than or equal to S. E-value is the expected number of MSTs to be found with score S or higher in two random sequences.

2.2 Tokens

Before explaining the architectures of the template generating and citation parsing systems, we have to identify the tokens they use. We use regular expression to precisely define the tokens. Tokens are classified into three types: numerical, general, and punctuation as follows:

- Numerical tokens: [0-9]+

  We reclassify the numerical token again as a year number, or general number. The regular expressions are as follows:

  - Year: [12][0-9][0-9][0-9]
  - General number: Otherwise

- General tokens: [a-zA-Z]+

  We reclassify the general token to key words that often appear in citations, like page, number, volume, month, name or unknown. The regular expressions are as follows:

  - Number:
    [Nn][Oo]
    |[Nn][Nn]
    | [Nn][Uu][Mm][Bb][Ee][Rr]
  - Page:
    [Pp][Pp]
    | [Pp][Aa][Gg][Ee]([Ss])?
  - Volume:
    [Vv][Oo]
    | [Vv][Oo][Ll]([Uu][Mm][Ee])?
  - Month:
    [Jj][Aa][Nn]([Uu][Aa][Rr][Yy])?
    | [Ff][Ee][Bb]([Rr][Uu][Aa][Rr][Yy])?
    | [Mm][Aa][Rr]([Cc][Hh])?
    | [Aa][Pp][Rr]([Ii][Ll])?
    | [Mm][Aa][Yy]
    | [Jj][Uu][Nn]([Ee])?
    | [Jj][Uu][Ll]([Yy])?
    | [Aa][Uu][Gg]([Uu][Ss][Tt])?
    | [Ss][Ee][Pp][Tt]([Ee][Mm][Bb][Ee][Rr])?
    | [Oo][Cc][Tt]([Bb][Ee][Rr])?
    | [Nn][Oo][Vv]([Ee][Mm][Bb][Ee][Rr])?
    | [Dd][Ee][Cc]([Ee][Mm][Bb][Ee][Rr])?
  - Name: We have a database that stores about 2,000 name tokens. If the token appears in the name database, it is identified as a name token.
  - Unknown: If the general token is not classified above, it is classified as unknown.
- Punctuation tokens: [\"\'.\(),:;\-!\?]

2.3 Protein Sequence Translation

In order to identify tokens, we use regular expression to describe the tokens in the last section. We can translate the token into a protein sequence according the following rules:

Y: Match to Year
N: Match to General number
S: Match to Number
P: Match to Page
V: Match to Volume
M: Match to Month
A: Match to Name

NULL: Match to Unknown
G: Match to Punctuation token " or '
D: Match to Punctuation token . or ; or ?
I: Match to Punctuation token (
K: Match to Punctuation token)
R: Match to Punctuation token ,
Q: Match to Punctuation token : or – or !

When the tokens translate to the amino acids, the combinations of the amino acids become a protein sequence. We call this sequence a prototype protein sequences, and use it to represent the citation. We can use this sequence to search for the most similar template in the database. We also use it to construct the template database. The citation in Figure 2(a) can transform into Figure 2(b) by translating the tokens according to the rules described above.

2.4 Template Generating System

In the template generating system, we construct a template database that contains the templates that represent most citation formats. Because BibTeX files are widely used in bibliographies, we retrieve BibTeX files from the Internet. BibTeX format was designed by Oren Patashnik and Leslie Lamport. It is field based, so we can parse the data of each field easily. We use the title field to search the citations in CiteSeer. We get the metadata of the citations found on the web by parsing the corresponding BibTeX files to construct the templates. We begin with the method described in section 2.3 to create the prototype protein sequence. Since we have found the metadata of each field, we can find the data in the CiteSeer citation and then modify the sequence by adding A (author), L (journal), and T (title) into the correct position in the sequence. We then change N to their corresponding amino acids. An amino acid N may become F (value of volume) or W (value of number). This is illustrated in Figure 2(c). It is now a completed protein sequence, and the template is finally constructed. We store this template in the template database.

2.5 Parsing System

After transforming the citation into a prototype protein sequence, we search for the most similar sequence in the template database by using BLAST. The parsing system is like a reverse process of the template generating system. After we find a template, we use it to extend the prototype protein sequence into a complete protein sequence. Now the metadata can be parsed. If we want to parse the citation shown in Figure 2(a), we must transform it into prototype protein sequence as figure 2(b). By entering this sequence into BLAST, we find the template AAAAAAAAARGTTTTTTTTGRLLLLLLLRVDFRSDWRMDYD. In this template, author is in the first position. The field of author and the field of title are separated by RG. (punctuation marks), as are the title and journal fields. Modifying Figure 2(b) according this template, we get Figure 2(c). Then, by checking the original citation, we can parse out all the metadata correctly.

Yieh-Ran Huang and Jan-Ming Ho, "Distributed Call Admission Control for a Heterogeneous PCS Network" , to appear IEEE Trans. On Computers, Vol. 51, no. 11, Nov. 2002.

Fig. 2(a). The original citation we want to parse.

	Q		A		M	Q		A	R	G										G
R					D			R	V	D	N	R	S	D	N	R	M	D	Y	D

Fig. 2(b). The citation in the Fig. 2(a) transforms into its protein sequence as QAMQARGGRDRVDNRSDNRMDYD. Each blank in the table represents a token in the citation.

A	A	A	A	A	A	A	A	A	R	G	T	T	T	T	T	T	T	T	T	G
R	L	L	L	L	L	L	L	R	V	D	F	R	S	D	W	R	M	D	Y	D

Fig. 2(c). We transform the citation sequence into its protein sequence. The template of the citation is AAAAAAAAARGTTTTTTTTTGRLLLLLLLRVDFRSDWRMDYD

3 Experiment Results

Ideally, all of the metadata in the BibTeX should be consistent with the metadata in the citations searched from CiteSeer. Unfortunately, only a few data are consistent. We experimented with 2,500 article citations, but only 100 citations were consistent with BibTeX. To overcome this problem, we create a precision evaluation method to test whether the data is correctly parsed. We define the precision of each subfield as:

$$\text{Subfield precision} = \frac{\#[(Token_{Number}^{subfield} + Token_{word}^{subfield}) \cap Token_{BibTex}^{subfield}]}{\#[(Token_{Number}^{subfield} + Token_{word}^{subfield}) \cap Token_{BibTeX}]} \quad (1)$$

$Token_{Number}^{subfield}$: denotes the number tokens that appear in the parsed subfield.

$Token_{word}^{subfield}$: denotes the general word tokens that appear in the parsed subfield.

$Token_{BibTex}^{subfield}$: denotes the tokens that appear in the specific subfield in the BibTeX file.

$Token_{BibTeX}$: denotes all of the tokens that appear in the BibTeX file.

The denominator of subfield precision represents the number of the tokens which exist both in the citation file and in the BibTeX file. The numerator of the subfield precision represents the number of the tokens which are correctly parsed. We then define the total precision of a citation as the average of subfield precisions. Using the 2,500 templates generated by our template generating system as the template database and parsing all of the 2,500 citations using our parsing system, we obtain 89% precision rate. Using cross-validation to validate the precision of our system, we divide the citations into 10 subsets, and get an average precision rate of 84.8%. The precision distribution chart of the first 100 citations is shown in Figure 3. Parsing results are adequate because 85% of the citations achieve a precision rate of 70% or higher. Although the precision calculated here is not actually precision, the disparity between this precision here and actual precision is small. The actual precision of the parsed result is roughly 80%. BLAST needs a scoring table to evaluate the alignment result. In our system, we use a scoring shown in Figure 4. We had also tried a lot of different scoring table to parse the citations. The diagonals of all the various scoring matrices were always positive, and the variation in the precision of the parsing results because of a particular choice of scoring table was less than 3%. Both ParaTools and our system are template-based reference parsers. The completeness of the template database is an important factor to template-based parsing systems. We illustrate the effect of template completeness on precision in Figure 5. Our system performs better than ParaTools for all tested template completeness. In Table 1, we present the quality of parsing result. Our quality is better than ParaTools'. The parsing results will be used in the future. If the quality is not good, it is insignificant for reusing.

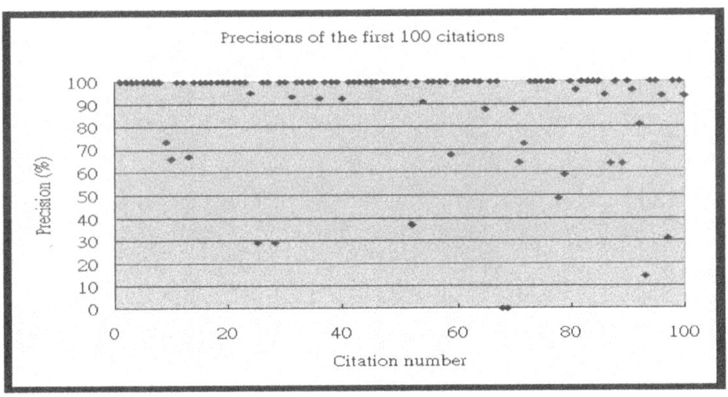

Fig. 3. Precision rates of the first 100 citations. The X axis is the assigned number of one citation, and the Y axis is the precision of that citation.

	A	R	N	D	C	Q	E	G	H	I	L	K	M	F	P	S	T	W
A	9	-9	-9	-9	-9	-9	-9	-9	-9	-9	-9	-9	-9	-9	-9	-9	-9	-9
R	-9	4	-1	-1	-1	-1	-1	-1	-1	-1	-1	-1	-1	-1	-1	-1	-1	-1
N	-9	-1	9	-1	-1	-1	-1	-1	-1	-1	-1	-1	-1	-1	-1	-1	-1	-1
D	-9	-1	-1	4	-1	-1	-1	-1	-1	-1	-1	-1	-1	-1	-1	-1	-1	-1
C	-9	-1	-1	-1	6	-1	-1	-1	-1	-1	-1	-1	-1	-1	-1	-1	-1	-1
Q	-9	-1	-1	-1	-1	4	-1	-1	-1	-1	-1	-1	-1	-1	-1	-1	-1	-1
E	-9	-1	-1	-1	-1	-1	4	-1	-1	-1	-1	-1	-1	-1	-1	-1	-1	-1
G	-9	-1	-1	-1	-1	-1	-1	9	-1	-1	-1	-1	-1	-1	-1	-1	-1	-1
H	-9	-1	-1	-1	-1	-1	-1	-1	4	-1	-1	-1	-1	-1	-1	-1	-1	-1
I	-9	-1	-1	-1	-1	-1	-1	-1	-1	9	-1	-1	-1	-1	-1	-1	-1	-1
L	-9	-1	-1	-1	-1	-1	-1	-1	-1	-1	4	-1	-1	-1	-1	-1	-1	-1
K	-9	-1	-1	-1	-1	-1	-1	-1	-1	-1	-1	9	-1	-1	-1	-1	-1	-1
M	-9	-1	-1	-1	-1	-1	-1	-1	-1	-1	-1	-1	6	-1	-1	-1	-1	-1
F	-9	-1	-1	-1	-1	-1	-1	-1	-1	-1	-1	-1	-1	6	-1	-1	-1	-1
P	-9	-1	-1	-1	-1	-1	-1	-1	-1	-1	-1	-1	-1	-1	6	-1	-1	-1
S	-9	-1	-1	-1	-1	-1	-1	-1	-1	-1	-1	-1	-1	-1	-1	6	-1	-1
T	-9	-1	-1	-1	-1	-1	-1	-1	-1	-1	-1	-1	-1	-1	-1	-1	9	-1
W	-9	-1	-1	-1	-1	-1	-1	-1	-1	-1	-1	-1	-1	-1	-1	-1	-1	4

Fig. 4. The scoring table we use this scoring table to evaluate the alignment score of the protein sequences

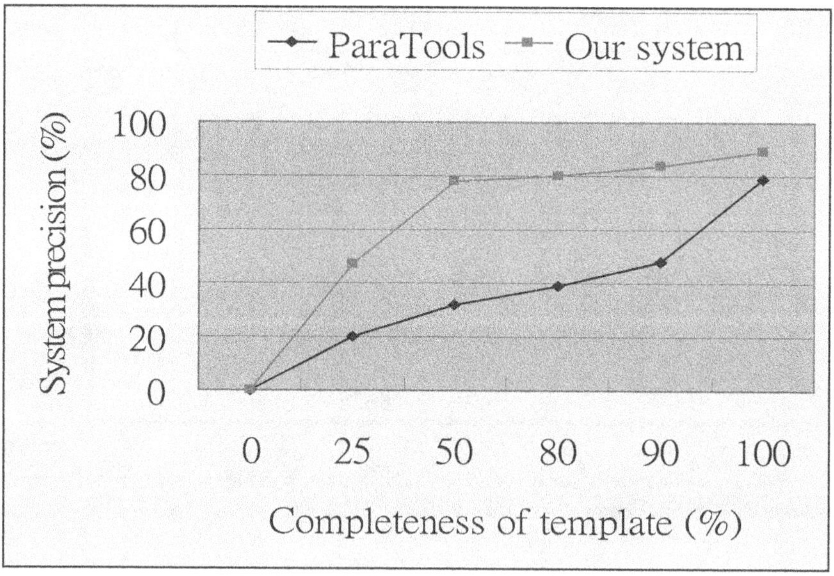

Fig. 5. For varied template database, different precision rates are achieved.

Table 1. Divide the data into k sets, and do cross-validation. The quality of the parsing result is defined good if the precision rate of the parsing result is better than 70%. The good results also contain perfect results which precisions rates are 100%.

Quality of the parsing result

Value of k	Description	Percent OpCit	Percent Our system
10			
	Perfect	21	71
	Good (>70%)	26	85
	Not good(<70%)	74	15
5			
	Perfect	21	59
	Good(>70%)	26	76
	Not good(<70%)	74	24
2			
	Perfect	6	37
	Good(>70%)	6	52
	Not good(<70%)	94	48

4 Conclusion and Future Works

It is flexible for a template-based system to deal with citations. We not only can add new citation templates easily, but also can search the most similar template to rapidly parse the metadata. It is also shown that precision of parsing result is different for various levels of completeness of template database. ParaTools contains about 400 templates in the system, but it does not fit the 2,500 data well. The precision rate for ParaTools to run the data is only 30%. We also demonstrate that our system still performs better in the same set of templates. This is contributed by the power of the string comparison tool, BLAST, and our design of rewriting rules to transform the citation parsing problem to a protein sequence alignment problem. In the future, we will generate more templates to match a broader range of citation formats. Our study also suggests that it is a promising research direction to develop BLAST-based solutions for some other template matching problems.

Acknowledgment. The comments of Shian-Hua Lin and Ming-Jing Hwang[2] during the preparation of this paper were very helpful. We thank for their help.

References

[1] C. Lee Giles, Kurt D. Bollacker, and Steve Lawrence, "CiteSeer: An Automatic Citation Indexing System," Digital Libraries 98 Pittsburgh PA USA.

[2] Kurt D. Bollacker, Steve Lawrence, and C. Lee Giles, "CiteSeer: An Autonomous Web Agent for Automatic Retrieval and Identification of Interesting Publications," 2nd International ACM Conference on Autonomous Agents, pp. 116-123, ACM Press, May, 1998.

[3] Steve Lawrence, C. Lee Giles and Kurt D. Bollacker, "Autonomous Citation Matching," Proceedings of the Third International Conference on Autonomous Agents, Seattle, Washington, May 1-5, ACM Press, New York, NY, 1999.

[4] Steve Lawrence, C. Lee Giles and Kurt D. Bollacker, "Digital Libraries and Autonomous Citation Indexing," IEEE Computer, Vol. 32, No. 6, pp. 67-71, 1999.

[5] Harnad, Stevan and Carr, Leslie, "Integrating, Navigating and Analyzing Eprint Archives Through Open Citation Linking (the OpCit Project)," Current Science (special issue honour of Eugene Garfield), Vol. 79, pp. 629-638, 2000.

[6] Donna Bergmark, "Automatic Extraction of Reference Linking Information from Online Documents," Cornell University Technical Report, TR 2000-1821, November, 2000.

[7] Donna Bergmark and Carl Lagoze, "An Architecture for Automatic Reference Linking," Cornell University Technical Report, TR 2000-1820, October, 2000.

[8] Mike Jewell, "ParaTools Reference Parsing Toolkit - Version 1.0 Released," D-Lib Magazine, Vol. 9, No.2, Feb., 2003.

[9] S. F. Altschul, W. Gish, W. Miller, E. Myers and D. Lipman, "A basic local alignment search tool," J. Mol. Biol., Vol. 215, pp. 403-410, 1990.

[10] http://www.ncbi.nlm.nih.gov/Education/BLASTinfo/similarity.html

Mining of Web-Page Visiting Patterns with Continuous-Time Markov Models

Qiming Huang[1], Qiang Yang[2], Joshua Zhexue Huang[3], and Michael K. Ng[4]

[1]College of Computer Science & Technology, Beijing University of Post & Telecom
qm_huang@yahoo.com
[2]Department of Computer Science, Hong Kong University of Science & Technology
[3]E-Business Technology Institute, The University of Hong Kong
[4]Department of Mathematics, The University of Hong Kong

Abstract. This paper presents a new prediction model for predicting when an online customer leaves a current page and which next Web page the customer will visit. The model can forecast the total number of visits of a given Web page by all incoming users at the same time. The prediction technique can be used as a component for many Web based applications . The prediction model regards a Web browsing session as a continuous-time Markov process where the transition probability matrix can be computed from Web log data using the Kolmogorov's backward equations. The model is tested against real Web-log data where the scalability and accuracy of our method are analyzed.

Keywords: Web mining, Continuous Time Markov Chain, Kolmogorov's backward equations, Sessions, Transition probability

1 Introduction

Web mining is a thriving technology in the practice of Web-based applications. By applying data mining technologies such as clustering, association rules and discrete Markov models, Web mining has been successfully applied to Web personalization and Web-page pre-fetching. Study has shown that the next page an online customer is going to visit can be predicted with statistical models built from Web sessions [4, 5]. However, an open problem is *when* an online customer will click on a predicted next page. A related question is *how many* customers will click the same page at the same time. In this paper, we answer these questions from a perspective of Markov models.

Markov models are one of major technologies for studying the behaviors of Web users. In the past, discrete Markov models are widely used to model sequential processes, and have achieved many practical successes in areas such as Web-log mining. The transition matrix based on the Markov process can be computed from visiting user-session traces, and the Frobenius norm of the differences of two transition probability matrices can show the difference of the two corresponding sequences [4]. Users can be clustered by learning a mixture of the first-order Markov models with an Expectation-Maximization algorithm [7]. The relational Markov model (RMMs) makes effective learning possible in domains of a very large and heterogeneous state space with only sparse data [11].

H. Dai, R. Srikant, and C. Zhang (Eds.): PAKDD 2004, LNAI 3056, pp. 549–558, 2004.

In this paper, we present a continuous-time Markov model for the prediction task. In contrast with the previous work which uses the prediction rules [5], we use the Kolmogorov's backward equations to compute the transition probability from one Web page *A* to another Web page *B* according to the following steps: first, we preprocess the Web-log data to build user sessions; second, we compute the transition rate of a user leaving a Web page *A* and obtain a transition rate matrix; third, we compute the transition probability matrix using the Kolmogorov's backward equations. From the transition probability matrix we can predict Web page a user will visit next, and when the user will visit the page. Furthermore, we can find the total transition count from all other Web pages to the predicted Web page at the same time.

The main contribution of this work is to put forward two hypotheses for computing a transition probability model from one Web page to another. The first hypothesis treats Web browsing sessions as a continuous time Markov process. The second hypothesis regards the probability of leaving a Web page as having an exponential distribution over the time. We can compute the transition rate for leaving the current Web page with the second hypothesis, and then according to the first hypothesiswe compute the transition probability matrix by the Kolmogorov's backward equations.

The paper is organized as follows. Section 2 presents the prediction model. Section 3 presents the experiment results. Section 4 concludes the paper.

2 Continuous Time Markov Chains for Prediction Models

A Web log often contains millions of records, where each record refers to a visit by a user to a certain Web page served by a Web server. A session is a set of ordered Web pages visited in one visit by the same visitor at a given time period.

A sequence of pages in a user session can be modeled by a Markov chain with a finite number of states [5, 9]. In discrete Markov chains, we need to consider the minimal time interval between page transitions, which is not easy to predict. In this paper, we propose to use continuous-time Markov chains for predicting the next visiting web page and when, and the total transition count from all other Web pages to the predicted Web page, instead of using discrete Markov chains. As a result, we can manage different time intervals in which to visit Web pages.

2.1 Continuous Time Markov Chains

A stochastic process is called a *continuou`time Markov chain at* state *i*

- The amount of time it spends in state *i*, before making a transition to a different state, is *exponentially distributed* with rate v_i, and
- It enters the next state *j* from state *i* with probability P_{ij}, where $P_{ii} = 0$ and ${}_j P_{ij} =$ 1 for every possible state *j* in the state space.

Let { $X(t), t \geq 0$ } be a continuous-time Markov chain of a user browsing Web pages. Here the state of a continuous-time Markov chain refers to a Web page and the state space contains all possible Web pages, which is finite but large. The main characteristics of a continuous-time Markov chain is that the conditional distribution of the next Web page to be visited by a user at the time $t+s$ is only dependent on the present Web page at the time s and is independent of the previous Web pages. For simplicity, we let

$$P_{ij} = P[\, X(t+s) = j \mid X(s) = i \,].$$

In a small time interval h,

- $P_{ii}(h)$ is the probability that the process in state i at time 0 will not be in state i at time h. $P_{ii}(h) = v_i\, h + o(h)$ is equal to the probability that a transition occurs in $(0, h)$.
- $P_{ij}(h) = h v_i P_{ij} + o(h)$ is the probability that a transition occurs in $(0, h)$ from state i to state j.

The limits of $(1 - P_{ij}(h)) / h$ and $P_{ij}(h) / h$ are equal to v_i *and* $v_i P_{ij}$ respectively as h approaches zero.

In order to predict the time at which a user will visit the next Web page, we make use of the well-known equation *Kolmogorov's Backward Equations* [1] in the continuous-time Markov chain:

$$P_{ij}'(t) = v_i \times \sum_{k \neq i} P_{ik}(t) \times P_{kj}(t) - v_i \times P_{ij}(t)$$

The transition rate matrix $R\{ r_{ij} \}$ with its entries is defined as

$$r_{ij} = v_i \times P_{ij} \quad \text{if } i \neq j$$
$$r_{ij} = -v_i \quad \text{if } i = j.$$

The Kolmogorov backward equations can be rewritten as

$$P'_{ij}(t) = \Sigma\, r_{ik} P_{kj}(t).$$

We can write it in the matrix form as

$$P'(t) = RP(t).$$

This is a system of ordinary differential equations and the general solution is given by

$$P(t) = e^{Rt}.$$

The above formula gives the probability of the next Web pages to be visited at time t. We need to compute the matrix R in order to use the continuous-time Markov chain for the prediction.

2.2 Computation of the Transition Rate Matrix *R*

According to the definition of P_{ij} *and* v_i in the previous subsection, the term v_i is the rate at which the Markov process leaves Web page i and P_{ij} is the probability that the Markov process enters Web page j.

Let us use the popular NASA Web-log data set to demonstrate how to compute v_i *and* P_{ij}. To illustrate, we extract a URL which index is 2327. Firstly, we compute the time t_{ij} ($j=1,2,\ldots,n$) for all people who visited the Web page i and entered another page j after the visit, where n is the number of visiting next Web pages from Web page i. Then we sort all t_{ij} (t_{i1}<t_{i2}<...t_{il}). Next we compute the accumulative frequency N_{im} of visits of Web page i up to time t_{im}. We note that N_i is the accumulative frequency of visits of Web page i at the time the last visitor leaves. Therefore, we can calculate the probability of leaving aWeb page i as N_{im}/N_i at time t_{im}.

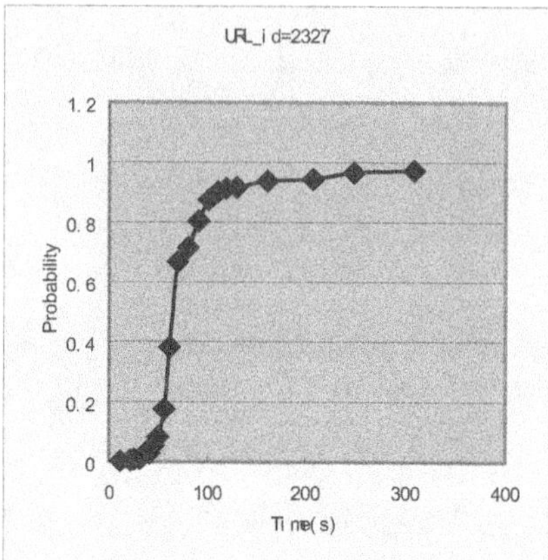

Fig. 1. The probability distribution of Leaving a Web page with time

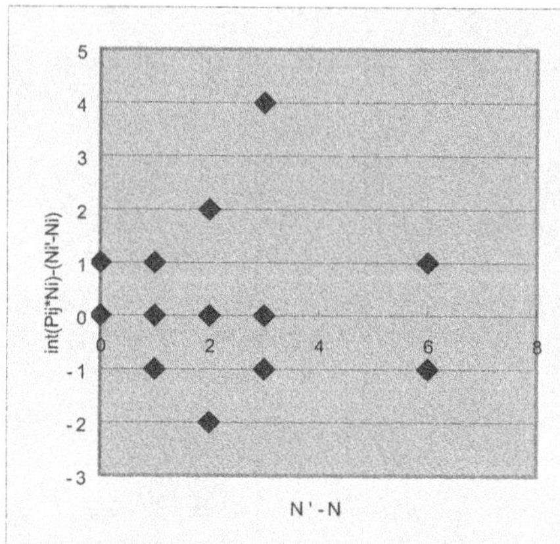

Fig. 2. The Deviation of the Actual Count Visiting Page 2334 from the predicted count

Figure 1 shows that the probability leaving from a Web page is exponentially distributed. By considering the statistical hypothesis in the continuous-time Markov chain, we model the curve in Figure 1 by an exponential distribution function as follows:

$$F(t)=1-e^{-\lambda t},\quad t\ \ 0 \ \ and \ \ F(t)=0,\quad t<0$$

Where λ is equivalent to the transition rate v_i of the continuous-time Markov chain [1]. λ can be determined as follows:

$$\lambda = -(\ln(1-F(t))/t.$$

Based on this formula, we estimate the transition rate v_i as follows:

$$v_{im} = -(\ln(1-N_{im}/N_i))/t_{im} \qquad m=1,2,\ldots,n-1$$
$$v_i = (v_{i1} + v_{i2} + \ldots + v_{i(n-1)})/(n-1).$$

For each Web page, we employ the same procedure to obtain the estimates of the transition rate v_i. Finally, the probability P_{ij} of leaving Web page i to Web page j can be estimated from the data by counting the relative frequency of the next visit to Web page j from Web page i. Finally, we obtain the transition rate matrix R from P_{ij} and v_i by the follow formula.

$$r_{ij} = v_i \times P_{ij} \quad if\ i \neq j$$
$$r_{ij} = -v_i \quad if\ i=j.$$

3 Empirical Analyses

3.1 Experimental Setup

The experiment was conducted on the NASA data set, which contains one month worth of all HTTP requests to the NASA Kennedy Space Center WWW server in Florida. The log was collected from 00:00:00 August 1, 1995 through 23:59:59 August 31, 1995. We filtered out documents that were not requested in this experiment. These were image requests or CSS requests in the log that were retrieved automatically after accessing requests to a document page containing links to these files and some half-baked requests [5].

We consider the Web log data as a sequence of distinct Web pages, where subsequences, such as user sessions can be observed by unusually long gaps between consecutive requests. To save memory space, we use number IDs to identify Web pages and users. To simplify the comparing operation, the time has been transformed to seconds starting from 1970.

In deciding on the boundary of the sessions, we make up two rules as follows. In a user session, if the time interval between two consecutive visiting is larger than 1800 seconds, we consider the next visit starting a new session. If a user has two consecutive records visiting the same Web page in 1800 seconds, we consider the next visit to be a new session. We loaded all sessions into a session data cube and operated the cube to extract the set of sessions for building the continuous-time Markov chain as described in Section 3.

3.2 Computing the Transition Probability from One Web Page to Others

After the continuous-time Markov chain for visiting Web pages was built, we used formula $P(t) = e^{Rt}$ to calculate the probability of entering another Web page from the current page. The Matrix $P(t)$ is estimated by:

$$P(t) = e^{Rt} = \lim_{n \to \infty} (I + Rt/n)^n$$

To obtain the limiting effect, we raised the power of the matrix $I+Rt/n$ to the nth power for sufficiently large n.

3.3 Predicting Which Next Pages People Will Visit, and When

We ran experiments to test the validity of the prediction model. Tracing through the NASA data set, a person visiting the Web page 2359 will decide where to go next. We make a prediction on when and where this visitor will go next, using our model.

Let us set the start time when the person was visiting the Web page 2359 as zero second. We count the transition count from the page 2359 to other pages before the start time of prediction. A total of 19 pages were visited by the person who left the page 2359 before the zeroth second; these pages are listed in Table1, where N_{ij} is the transition count from Web page $i=2359$ to Web page j that actually happened in the data set before the zeroth second.

Table 1. The transition parameter from the Web page 2359 to other Web pages in 600 seconds.

URL_id_j	2324	2442	2383	2364	2334	2362	2969	2333	2358	2496
N_{ij}	1	1	2	1	32	1	5	7	1	2
P_{ij}	0.0093	0.0013	0.0493	0.0111	0.0367	0.0343	0.0049	0.0193	0.0650	0.0714
N'_{ij}	1	1	2	1	33	1	5	7	1	2

URL_id_j	2375	2344	2325	2372	2657	2374	2393	2370	2327	
N_{ij}	2	1	3	1	2	1	1	1		
P_{ij}	0.0237	0.0130	0.0234	0.0321	0.0018	0.0168	0.0011	0.0026	0.0259	
N'_{ij}	2	1	3	1	2	1	1	1		

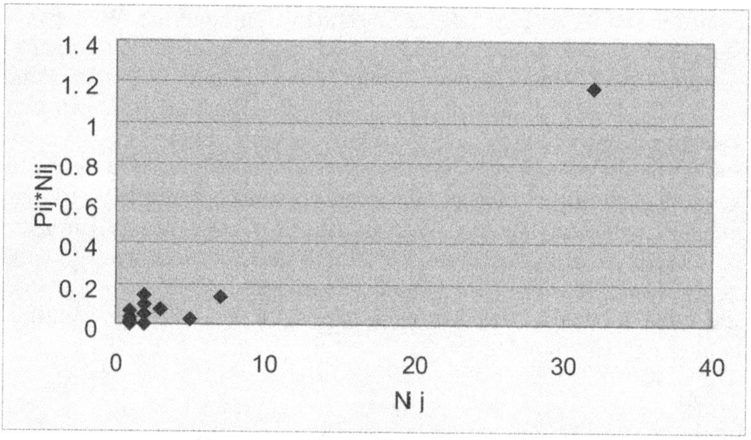

Fig. 3. The $P_{ij} \times N_{ij}$ value at 600 second

With our model, we can compute the transition probability values P_{ij} from Web page 2359 to all other Web pages at time *t*. When the time *t* is 600 seconds, Figure 3 shows the value of $P_{ij} \times N_{ij}$, where the integer part is the predicted transition count from Web page i to Web page j during a given time interval. The value of int($P_{ij} \times N_{ij}$) at Web page 2334 increases up to one, and we can predict that the next page to go from Web page 2359 is Web page 2334 and the transition time is within 600 seconds.

We use N'_{ij} to denote the total transition count from Web page *2359* to another Web page *j* that actually happened in the NASA data set before the end of prediction time. In Table 1, N'_{ij} - N_{ij} is one at Web page 2334. The person left Web page 2359 to Web page 2334 in 600 seconds in reality.

3.4 Predicting the Visiting Count of a Web Page

With the prediction time increasing from zero second and up, we can predict the next Web page visited by an online person, when int($P_{ij} \times N_{ij}$) value first becomes one. We note the corresponding time visiting the next Web page as t_1 seconds. N_i is the total transition count from Web page *i* to other Web pages before the start of prediction. Int($P_{ij} \times N_i$) is the predicted count of visitors leaving Web page *i* and entering Web page *j* during the prediction time. Thus, we can compute the total transition count from all Web pages to the designation Web page in the interval of zero to t_1 seconds.

Table 2. The transition parameters from all Web pages to the Web page 2334 in 600 secs.

URL_id$_i$	2324	2539	2352	2327	2442	2356	2383	2388	2440	2387
P_{ij}	0.036	0.045	0.035	0.038	0.011	0.038	0.037	0.043	0.024	0.047
N_i	214	4	74	187	25	9	108	39	2	19
int(P_{ij}*N_i)	7	0	2	7	0	0	3	1	0	0
N_i'	220	4	75	190	26	9	111	40	2	21
N_i'-N_i	6	0	1	3	1	0	3	1	0	2

URL_id$_i$	2362	2333	2358	2496	2375	2322	2379	2349	2441	2438
P_{ij}	0.037	0.033	0.029	0.037	0.04	0.044	0.041	0.03	0.039	0.043
N_i	60	38	26	15	101	46	38	67	14	13
int(P_{ij}*N_i)	2	1	0	0	4	2	1	2	0	0
N_i'	62	38	27	15	103	47	39	70	15	13
N_i'-N_i	2	0	1	0	2	1	1	3	1	0

URL_id$_i$	2669	2325	2372	2359	2357	2374	2473	2329	2423	3131
P_{ij}	0.028	0.035	0.021	0.037	0.047	0.046	0.031	0.042	0.035	0.044
N_i	6	150	22	65	9	38	16	25	25	7
int(P_{ij}*N_i)	0	5	0	2	0	1	0	1	0	0
N_i'	6	156	22	66	9	38	16	25	26	7
N_i'-N_i	0	6	0	1	0	0	0	0	1	0

URL_id$_i$	2990	2338	2400	2389	2506	2517	2505	2376	2419	2682
P_{ij}	0.035	0.015	0.033	0.052	0.042	0.033	0.035	0.040	0.050	0.046
N_i	7	10	28	3	3	7	14	24	14	12
int(P_{ij}*N_i)	0	0	0	0	0	0	0	1	0	0
N_i'	7	10	29	3	3	8	16	25	14	13
N_i'-N_i	0	0	1	0	0	1	2	1	0	1

URL_id$_i$	2336	2405	2439	2854	2447	2600	2953	2411	2462	2504
P_{ij}	0.049	0.044	0.020	0.023	0.018	0.038	0.022	0.049	0.029	0.026
N_i	26	20	6	5	9	3	5	4	9	4
int(P_{ij}*N_i)	1	0	0	0	0	0	0	0	0	0
N_i'	26	20	6	5	9	3	5	4	9	4
N_i'-N_i	0	0	0	0	0	0	0	0	0	0

URL_id$_i$	2561	2918	2558	2578	2468	2672	2767	2637	2699	3129
P_{ij}	0.033	0.081	0.024	0.032	0.029	0.061	0.029	0.026	0.022	0.040
N_i	2	2	1	4	16	2	5	3	4	2
int(P_{ij}*N_i)	0	0	0	0	0	0	0	0	0	0
N_i'	2	2	3	5	16	2	5	3	4	2
N_i'-N_i	0	0	2	1	0	0	0	0	0	0

Table 2 shows the result of all previous 60 pages from which visitors entered the page 2334 within 600 seconds. The second row lists the transition probability values (P_{ij}) from another page to page 2334 in 600 seconds. N_i (or N_i') is the total visiting count from the previous Web page (URL_id$_i$) to Web page 2334 before the start (or the end) of the prediction time. N'_j - N_i is the actual visiting count from the Web page (URL_id$_i$) to Web page 2334 during the prediction time. The total count of visiting Web page 2334 in 600 seconds was calculated as 45 using $\Sigma \text{abs}(P_{ij} \times N_i)$ and the actual visiting count was 42 according to $\Sigma \text{abs}(N_i' - N_i)$.

Figure 3 shows the prediction deviation value (int($P_{ij} \times N_i$)-($N_i' - N_i$)), which mostly vary in the scope of [-1,1]. The error rate of the total predicted visiting count to a Web page is computed as follows.

$$ER_j = \left(\sum_{i=1}^{m} ER_{ij}\right)/m$$

where

$$ER_{ij} = \begin{cases} 0 & \text{if } \operatorname{int}(P_{ij} \times N_i) = (N_i' - N_i) \\ 1 & \text{if } \operatorname{int}(P_{ij} \times N_i) \neq 0 \text{ and } (N_i' - N_i) = 0 \\ \dfrac{abs(\operatorname{int}(P_{ij} * N_i) - (N_i' - N_i))}{N_i' - N_i} & \text{if } \operatorname{int}(P_{ij} \times N_i) \neq (N_i' - N_i) \text{ and } \operatorname{int}(P_{ij} \times N_i) \neq 0 \\ & \text{and } 0 < abs(\operatorname{int}(P_{ij} \times N_i) - (N_i' - N_i)) < \operatorname{int}(P_{ij} \times N_i) \\ 1 & \text{if } \operatorname{int}(P_{ij} \times N_i) \neq (N_i' - N_i) \text{ and } \operatorname{int}(P_{ij} \times N_i) \neq 0 \\ & \text{and } abs(\operatorname{int}(P_{ij} \times N_i) - (N_i' - N_i)) \geq (N_i' - N_i) \end{cases}$$

The error rate of the predicted count in visiting Web page 2334 in Table 2 was calculated as 34.4%. The main reason for the error rate is that some people do not visit the Web page with the pattern in the training data set. In the prediction model, the computation of the transition probability P_{ij} as shown in the following formula may be unstable when the parameter n is not large enough. This causes an error in the prediction.

$$P(t) = e^{Rt} = \lim_{n \to \infty} (I + Rt/n)^n$$

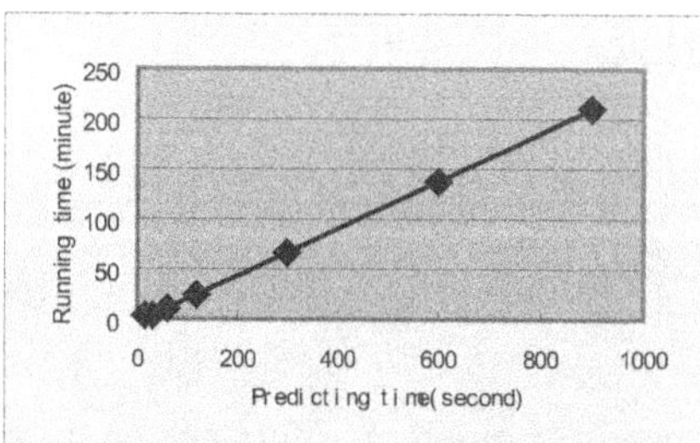

Fig. 4. The relation of the run time and the predicting time.

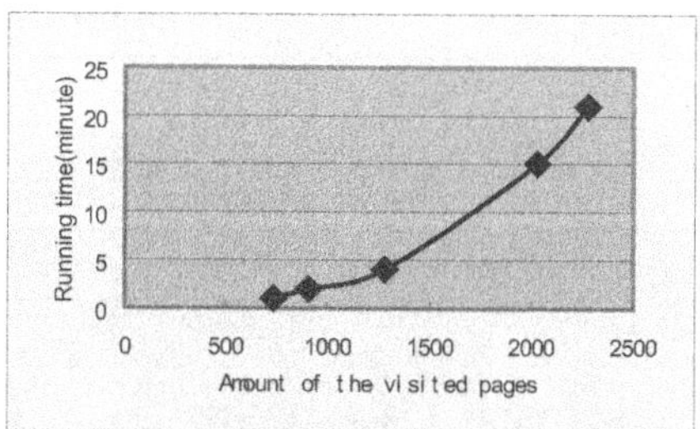

Fig. 5. The run time to the amount of the visited pages

3.5 Scalability Experimental Results

We conducted some experiments to test the scalability of the continuous Markov chain prediction method. The experiments were carried out on a Pentium III, 662 MHz with 196MB RAM. The NASA data in 39608 seconds from 00:00:00 August 1, 1995 was used. We computed the transition probability in different predicting time length (T). Figure 4 shows the linear relation of the predicting time length and the computing time. For a certain precision value, the relation of the predicting time length (T) and the running time is linear too.

Using a different size of training data sets which included different numbers of visited Web pages, we computed the transition probability within the next ten seconds. Figure 5 shows the relation of the number of visited Web pages and the running time.

4 Conclusions

In this paper, we explored using the continuous-time Markov chain to predict not only the next Web page a person will visit and the time when it will happen, but also the visiting count of the Web page in the same time. The transition probability, computed from the continuous-time Markov chain, gives us rich information from which we can compute the transition count from one Web page to another, as well as the total number of visits to a Web page by all people within a certain period of time. The prediction model is validated by an experiment. In the future, we plan to continue to explore the application of continuous time Markov models in Web page prediction. We will also consider how to apply the prediction result to prefetching applications.

Acknowledgement. Michael Ng, Joshua X. Huang, and Qiang Yang are supported by grants from HK ITC and RGC.

References

[1] William J. Anderson (1991) Continuous Time Markov Chains: An Applications-Oriented Approach. Springer-Verlag, New York.

[2] C.M.Leung and J.A.Schormans (2002) Measurement-based end to end latency performance prediction for SLA verification. 41st European Telecommunications Congress, Genoa

[3] J.M.Pitts and J.A.Schormans (2000) Introduction to IP and ATM design and performance. 2nd edition, Wiley.

[4] Qiang Yang, Joshua Zhexue Huang and Michael Ng (2002) A Data Cube Model for Prediction-based Web Prefetching. Journal of Intelligent Information Systems, Vol. 20 (2003), 11-30.

[5] Qiang Yang, Hui Wang and Wei Zhang (2002) Web-log Mining for Quantitative Temporal-Event Prediction. IEEE Computer Society, Computational Intelligence Bulletin, Vol. 1, No. 1, December 2002.

[6] James Pitkow and Peter Pirolli (1999) Mining Longest Repeating Subsequences to Predict World Wide Web Surfing. In Proceedings of USENIX Symposium on Internet Technologies and Systems. 1999.

[7] Igor Cadez, David Heckerman, Christopher Meek, Padhraic Smyth, Steven White (2000) Visualization of Navigation Patterns on a Web Site Using Model-Based Clustering. In Proceedings of the sixth ACM SIGKDD International Conference on Knowledge Discovery and Data Mining, Boston, USA.

[8] Mukund Deshpande and George Karypis (2001) Selective Markov Models for Predicting Web-Page Accesses. In Proceedings SIAM Int. Conference on Data Mining.

[9] Taha, H. (1991) Operations Research, 3rd Edition, Collier Macmillan, N.Y., USA.
[10] Sheldon M. Ross (1996) Stochastic Process. Wiley.
[11] Corin R. Anderson, Pedro Domingos, Daniel S. Weld (2002) Relational Markov Models and Their Application to Adaptive Web Navigation. In Proceedings of SIGKDD 2002.
[12] Alan Jennings and J.J. Mckeown (1992) Matrix Computation. John Wiley & Sons.

Discovering Ordered Tree Patterns from XML Queries

Yi Chen

Institute of Software, Chinese Academy of Sciences

Abstract. Recently, several researches have investigated the discovering of frequent XML query patterns using frequent structure mining techniques. All these works ignore the order properties of XML queries, and therefore are limited in their effectiveness. In this paper, we consider the discovering of ordered query patterns. We propose an algorithm for ordered query pattern mining. Experiments show that our method is efficient.

1 Introduction

Recently, several researchers have investigated the discovering of frequent XML query patterns using frequent structure mining (FSM) techniques [YLH03a, YLH03b, CW03]. Given a set of XML queries, they model them as unordered trees with special XML query constructs like descendant edges or wildcards, and FSM techniques are extended to extract frequent subtrees based on the semantics of XML queries.

In contrast to conventional semi-structured data, elements in XML documents are ordered, and queries about the order of elements are supported in most popular XML query languages. The existing works on the discovering of query patterns ignore the order properties of XML queries, and therefore are limited in their effectiveness. In this paper, we propose an algorithm for ordered query pattern mining. Our algorithm is very efficient since to count the supports of pattern trees, it need only match those single branch pattern trees against QPTs in the transaction database, and the supports of multi-branch ones can be figured out through reusing intermediate results.

2 Preliminaries

A *query pattern tree* is a labeled ordered tree QPT = $<V, E, \leq>$, where V is the vertex set, E is the edge set, and $\leq$ is the order relationship between sibling nodes. Each vertex v has a label, denoted by v.label, whose value is in {"*"} $\cup$ tagSet, where the tagSet is the set of all element names in the context. The root of the query pattern tree is denoted by root(QPT). A distinguished subset of edges representing ancestor-descendant relationships is called descendant edges.

Figure 1 (a)-(d) shows four QPTs QPT_1, QPT_2, QPT_3 and QPT_4. Descendant edges are shown with dotted lines in diagrams, and tree nodes are shown with circles, where the symbols inside the circle indicate their labels. The tuple (n:i) outside of the circle indicates the pre-order number and identifier of the corresponding node respectively.

H. Dai, R. Srikant, and C. Zhang (Eds.): PAKDD 2004, LNAI 3056, pp. 559–563, 2004.

Given any tree node v of a QPT, its node number, denoted by v.number, is assigned according to its position in a pre-order traversal of the QPT, and the meaning of its identifier will be explained in the following sections.

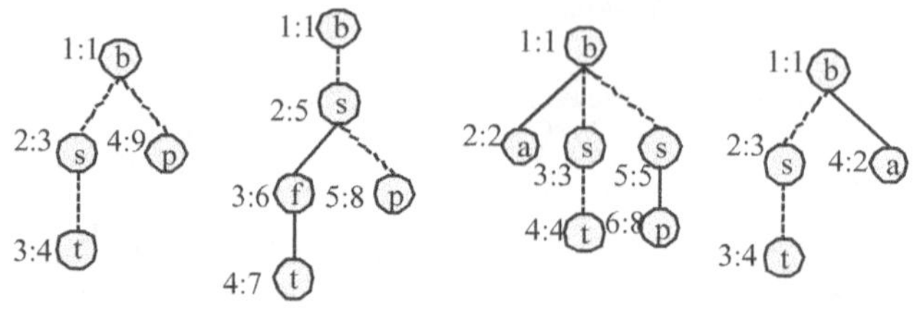

(a)QPT_1 (b)QPT_2 (c) QPT_3 (d) QPT_4

Fig. 1. Query Pattern Trees.

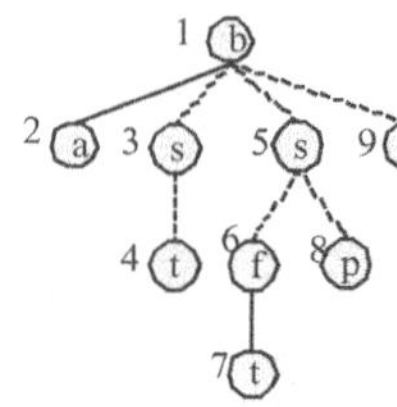

Fig. 2. G-QPT.

In what follows, given a query pattern tree QPT = <V, E, ≤>, sometimes we also refer to V and E with QPT if it's clear from the context. Given an edge $e = (v_1, v_2) \in$ QPT where v_2 is a child of v_1, sometimes v_2 will be denoted as a d-child of v_1 if e is a descendant edge, and as a c-child otherwise. For simplification of representation, in this paper we don't consider duplicate siblings, however, our result is applicable to more general cases.

Given a query pattern tree QPT, a *rooted subtree* RST of QPT is a subtree of QPT such that root(RST)=root(QPT) holds. Let QPT be a pattern tree, the size of QPT is defined by the number of its nodes |QPT|. An RST of size k+1 will be denoted as a k-edge RST sometimes. An RST will also be denoted as a single branch RST if it has only one leaf node, and as a multi-branch RST otherwise.

To discover frequent query patterns, one important issue is how to test the occurrence of a pattern tree in the transaction database. In this paper, we use the concept of tree subsumption [MD02] for the occurrence test.

Given a transaction database $D = \{QPT_i | \ i = 1,\ldots, n\}$, we say RST occurs in D if RST is subsumed in a query pattern tree $QPT_i \in D$. The *frequency* of RST, denoted as Freq(RST), is the total occurrence of RST in D, and supp(RST)=Freq(RST)/|D| is its *support rate*. Given a transaction database D and a positive number $0<\sigma\leq 1$ called the minimum support, mining the frequent query patterns of D means to discover the set of RSTs of D, $F_D = \{RST_1, \ldots, RST_m\}$, such that for each RST $\in F_D$, supp(RST) $\geq \sigma$.

3 Discovering Frequent Query Pattern Trees

Given a transaction database $D=\{QPT_i | \ i = 1,\ldots, n\}$, we construct its *global query pattern tree* G-QPT as follows: At first, a root is created, its label value will always be the name of the root element of the related DTD. Next, for each QPT in the transaction database, we merge it with G-QPT as follows: the root of the QPT is always merged with the root of the G-QPT; and for each other node u of the QPT that is a c-child (or d-child respectively) of its parent, if its parent is merged with a tree node p of the G-QPT, and there exists a c-child (or d-child respectively) q of p such that q.label = u.label holds, then u is merged with q, otherwise, a new node q is inserted as a c-child (or d-child respectively) of p, and q.label is set to u.label. After all the QPTs have been processed, we assign each node p of the G-QPT a number, denoted as its identifier or *p.id*, through a pre-order travesal. Figure 2 shows an example of G-QPT obtained from the QPTs in Figure 1 (a), (b), (c) and (d).

Because each node of QPT$\in D$ is merged with a unique node of the G-QPT, each node of QPT has the same identifier as the corresponding node in the G-QPT (see Figure 1). After labeling each node with an identifier, the representation of QPTs can be simplified to string format. For example, QPT_4 can be simplified as "1, 3, 4, -1, -1, 2, -1". We will call such encoding strings as *string encodings* of QPTs.

As in [YLH03b], we only try to discover pattern trees that are rooted subtrees of some QPTs in the transaction database. In [YLH03b], a schema-guided rightmost expansion method is proposed to enumerate all unordered rooted subtrees of the G-QPT. In our settings, the schema-guided rightmost expansion method can't be directly used because the G-QPT doesn't preserve the order relationship between sibling nodes. For example, "1, 3, -1, 2, -1" is rooted subtree of QPT_4, but it's not a rooted subtree of the G-QPT, and we can't generate it with the method used in [YLH03b].

To handle this issue, we modify the schema-guided rightmost expansion method to generate RSTs as follows: Given a k-edge $RST^k \in [RST^{k-1}]=\{RST_1, RST_2, \ldots, RST_n\}$ sorted in ascending order of their string encodings, let $rmlne(RST^k)=\{RST^{k+1} | RST^{k+1}$ is RMLNE of $RST^k\}$ and $JR(RST^k)= \{RST^{k+1}|RST^{k+1}=RST_i \bowtie RST_j, j=i+1,\ldots,N\} \cup \{RST^{k+1}|RST^{k+1}=RST_i \bowtie RST_j$, where $j=1,\ldots,i-1$, $diff(s_i,s_j) = 1$, s_i and s_j are string encodings of RST_i and RST_j respectively$\}$, then $[RST^k] = rmlne(RST^k) \cup JR(RST^k)$ holds. Definition of equivalence class $[RST^{k-1}]$ and string comparison function diff() can be found in [YLH03b].

The main idea of *QPMiner* is similar to [CW03]. It uses the schema-guided rightmost enumeration method to enumerate candidate RSTs level-wise, counts the frequency of each candidate RST, and prunes infrequent RSTs based on the anti-monotone property of tree subsumption. However, *QPMiner* uses different method to count the frequency of candidate RSTs.

Given an rooted subtree RST and a query pattern tree QPT, where the list $L=<v_1, \ldots, v_m>$ is the set of nodes of RST sorted in pre-order, assume that RST $\subseteq$ QPT holds and *sim* is the simulation relation between their nodes, then the *Proper Combinations* of *sim* is the set PC = $\{<v'_1, \ldots, v'_m>| v'_i \in$ QPT, $i = 1, \ldots, m\}$ such that for each $<v'_1, \ldots, v'_m> \in$ PC: if v_j is a c-child of v_k, v'_j must be a c-child of v'_k; if v_j is a d-child of v_k, v'_j must be a proper descendant of v'_k, where j, k are any integers such that $1 \leq j, k \leq m$ holds. A proper combination is called a strong proper combination iff the following condition holds: given any two sibling nodes v_j and v_k of RST, $v_j \leq v_k$ iff v'_j.number $\leq v'_k$.number.

Given an rooted subtree RST and a query pattern tree QPT where the list $L=<v_1,\ldots, v_m,\ldots, v_k>$ is the set of nodes of RST sorted in pre-order, v_m and v_k is the second rightmost leaf and the rightmost leaf of RST respectively, assume that RST $\subseteq$ QPT holds, *sim* is the simulation relation between their nodes, then the *rightmost occurrence* of RST in QPT is the set rmo(RST, QPT) = $\{<v'_m, v'_k>| <v'_m, v'_k>$ is a sub-list of some proper combination $<v'_1,\ldots, v'_m, \ldots, v'_k>$ of *sim*$\}$. The rightmost occurrence rmo(RST, QPT) is called a strong rightmost occurrence iff $<v'_1,\ldots, v'_m, \ldots, v'_k>$ is a strong proper combination. The rightmost occurrence rmo(RST, QPT) will also be denoted as srmo(RST, QPT) if it's a strong rightmost occurrence.

Given an RST, a transaction database D and its global query pattern tree G-QPT, the *strong rightmost occurrence* of RST in D is the set srmo(RST, D) = $\{<u, v, \{<\text{QPT.tid}, u.\text{number}, v.\text{number}>| \text{ for all QPT} \in D \text{ such that } <u, v> \in \text{srmo(RST, QPT)}\}>| <u, v> \in \text{rmo(RST, G-QPT)}\}$.

Given a rooted subtree RST and a query pattern tree QPT, where $<v_1,\ldots, v_i, \ldots, v_m>$ is the set of nodes of RST sorted in pre-order, and v_i is a node of the rightmost branch of RST, assume that RST $\subseteq$ QPT holds, *sim* is the simulation relation between their nodes, and $<v_i, v'> \in sim$, we define the *Conditional Rightmost Occurrence* satisfying $<v_i,v'> \in sim$ as the set $\{v'_m \mid$ there exists a proper combination of $sim < v'_1,\ldots, v'_i, \ldots, v'_m>$ such that $v'_i = v'\}$. We denote it as $\text{rmo}_{<vi, v'> \in \text{SIM}}$(RST, QPT).

Theorem 1: Given a transaction database D, its global query pattern tree G-QPT, and two k-edge RSTs $RST_1, RST_2 \in [RST^{k-1}]$, let $RST^{k+1} = RST_1 \bowtie RST_2$, p is the node of RST^{k+1} not present in RST_1 (i.e., the rightmost leaf of RST_2), and the junction node q is parent of p, then we have:

1. If RST_1 is a single branch RST, then $\text{srmo}(RST^{k+1}, D) = \{<u, u', \{<\text{tid}, u.\text{number}, u'.\text{number}>| <\text{tid}, v.\text{number}, u.\text{number}> \in List_1, <\text{tid}, v'.\text{number}, u'.\text{number}> \in List_2, u.\text{number} \le u'.\text{number}\}| <v, u, List_1> \subseteq \text{srmo}(RST_1, D), <v', u', List_2> \subseteq \text{srmo}(RST_2, D)$, where $u \in \text{rmo}_{<q, q'> \in \text{SIM}}(RST_1, \text{G-QPT})$, $u' \in \text{rmo}_{<q, q'> \in \text{SIM}}(RST_2, \text{G-QPT})$, and $q' \in \text{G-QPT}\}$.
2. If both RST_1 and RST_2 are multi-branch RSTs, and the parent of the rightmost leaf of RST_1 has only one child, then $\text{srmo}(RST^{k+1}, D) = \{<u, u', \{<\text{tid}, u.\text{number}, u'.\text{number}>| <\text{tid}, v.\text{number}, u.\text{number}> \in List_1, <\text{tid}, v'.\text{number}, u'.\text{number}> \in List_2, u.\text{number} \le u'.\text{number}\}> \mid <v, u, List_1> \subseteq \text{srmo}(RST_1, D), <v', u', List_2> \subseteq \text{srmo}(RST_2, D), u << v'$, where $u \in \text{rmo}_{<q, q'> \in \text{SIM}}(RST_1, \text{G-QPT})$, $u' \in \text{rmo}_{<q, q'> \in \text{SIM}}(RST_2, \text{G-QPT})$, and $q' \in \text{G-QPT}\}$. Here $u << v'$ means v' is the parent (or ancestor respectively) of u if the rightmost leaf of RST_1 is a c-child (or d-child respectively) of its parent.
3. Otherwise, $\text{srmo}(RST^{k+1}, D) = \{<u, u', \{<\text{tid}, u.\text{number}, u'.\text{number}>| <\text{tid}, v.\text{number}, u.\text{number}> \in List_1, <\text{tid}, v'.\text{number}, u'.\text{number}> \in List_2, u.\text{number} \le u'.\text{number}\}| <v, u, List_1> \subseteq \text{srmo}(RST_1, D), <v', u', List_2> \subseteq \text{srmo}(RST_2, D), u = v'$, where $u \in \text{rmo}_{<q, q'> \in \text{SIM}}(RST_1, \text{G-QPT})$, $u' \in \text{rmo}_{<q, q'> \in \text{SIM}}(RST_2, \text{G-QPT})$, and $q' \in \text{G-QPT}\}$.

Theorem 2: For any multi-branch k+1-edge rooted subtree RST^{k+1} generated through rightmost leaf node expansion of a k-edge rooted subtree RST_1, there must be another k-edge rooted subtree RST_2 which is formed by cutting off the second rightmost leaf node of RST^{k+1} such that the join of RST_1 and RST_2 will produce RST^{k+1} itself. If RST_2 exists in F_k, then we have: $\text{rmo}(RST^{k+1}, D) = \{<v, u', \{<\text{tid}, u.\text{number}, u'.\text{number}>| <\text{tid}, v.\text{number}, u.\text{number}> \in List_1, <\text{tid}, v'.\text{number}, u'.\text{number}> \in List_2\}>| <v, u, List_1> \subseteq \text{rmo}(RST_1, D), <v', u', List_2> \subseteq \text{rmo}(RST_2, D), u' << u\}$, otherwise, RST^{k+1} must be infrequent.

Based on Theorem 1 and Theorem 2, only those single-branch RSTs need to be matched with QPTs. The frequencies of other RSTs can be computed through reusing intermediate results.

4 Conclusion

We performed three sets of experiments to evaluate the performance of the *QPMiner* algorithm. We implement an algorithm, *baseMiner*, as the base algorithm to compare *QPMiner* against. By associating each RST with a TIDList attribute that contains transaction IDs of QPTs in which the RST occurs, *baseMiner* need only match 1-edge RSTs against each QPT in the transaction database. To count the frequency of a k-edge RST RST^k (k>1), if RST^k is a single-branch RST, *baseMiner* need only to match RST^k against QPTs whose transaction IDs are in the set RST^{k-1}.tidlist, where RST^{k-1} is the k-1-edge RSTs that is a rooted subtree of RST^k. If RST^k is a multi-branch k-edge RST, then *baseMiner* need only to match RST^k against QPTs whose transaction IDs are in the set $RST_1.tidlist \cap RST_2.tidlist$, where RST_1 and RST_2 are k-1-edge RSTs that are rooted subtrees of RST^k.

The experiments shows that *QPMiner* is 5-7 times faster than *baseMiner*. The reason is that *baseMiner* will match each candidate RST against QPTs in the transaction database while *QPMiner* processes the majority of candidate RSTs without matching operation. In addition, matching time of single-branch RSTs is much less than that of multi-branch RSTs. Consequently, *QPMiner* takes much less time than *baseMiner*.

Future work includes incremental computation of frequent RSTs. To incorporate the result of this paper into caching system, an important issue is to guarantee the freshness of the mining result. However, if the pattern of user activity changes at a relatively high rate, the accuracy of the mining result may deteriorate fast. Because re-computation will incur a high overhead, finding a method to discover frequent RSTs incrementally becomes very important.

Acknowledgments. We'd like to thank Dr. Liang Huai Yang in National University of Singapore for his help in sharing some experimental data sets and clarifying some concepts.

References

[CW03] Yi Chen, Yu Guo Wang, Discovering Frequent Tree Patterns from XML Queries, submitted, 2003.

[MD02] Gerome Miklau, Dan Suciu: Containment and Equivalence for an XPath Fragment. PODS 2002: 65-76.

[YLH03a] Liang Huai Yang, Mong Li Lee, Wynne Hsu, Sumit Acharya. Mining Frequent Query Patterns from XML Queries. DASFAA, 2003, pp:355-362, 2003.

[YLH03b] L.H. Yang, M.L. Lee, W. Hsu. Efficient Mining of Frequent Query Patterns for Caching. In Proc. Of 29th VLDB Conference, 2003.

Predicting Web Requests Efficiently Using a Probability Model

Shanchan Wu and Wenyuan Wang

Department of Automation, Tsinghua University, Beijing 100084, P.R.China
wushanchan@tsinghua.org.cn, wwy-dau@mail.tsinghua.edu.cn

Abstract. As the world-wide-web grows rapidly and a user's browsing experiences are needed to be personalized, the problem of predicting a user's behavior on a web-site has become important. We present a probability model to utilize path profiles of users from web logs to predict the user's future requests. Each of the user's next probable requests is given a conditional probability value, which is calculated according to the function presented by us. Our model can give several predictions ranked by the values of their probability instead of giving one, thus increasing recommending ability. The experiments show that our algorithm and model has a good performance. The result can potentially be applied to a wide range of applications on the web.

1 Introduction

Web mining is the application of data mining technology to huge Web data repositories. The purpose of this paper is to explore ways to exploit the information from web logs for predicting users' actions on the web. There has been an increasing amount of related work. For example, Syskill & Webert [4] is designed to help users distinguish interesting web pages on a particular topic from uninteresting ones. WebTool, an integrated system [5], is developed for mining either association rules or sequential patterns on web usage mining to provide an efficient navigation to the visitor. In [6], the authors proposed to use Markov chains to dynamically model the URL access patterns.

In this paper, we present a probability model to predict the user's next request. From a sequence of previous requests, instead of giving only one prediction, we can give several predictions ranked by the values of their probability. We present a function to calculate the values of their conditional probability and present an efficient algorithm to implement it.

2 Construct the WAP-Tree

There are three main tasks for performing Web Usage Mining or Web usage Analysis: Preprocessing, Pattern Discovery, Pattern Analysis [2].

H. Dai, R. Srikant, and C. Zhang (Eds.): PAKDD 2004, LNAI 3056, pp. 564–568, 2004.

After preprocessing is applied to the original Web log files, pieces of Web logs can be obtained. Let E be a set of events. A Web log piece or (Web) access sequence $S = e_1e_2 \cdots e_n$ $(e_i \in E)$ for $(1 \le i \le n)$ is a sequence of events.

Our algorithm is based on the compact web access pattern tree (WAP-tree) [1] structure. Due to different purpose, our WAP-Tree has some difference. In order not to lose information, we do not do any truncation to WAP-Tree during construction.

Each node in a WAP-Tree registers two pieces of information: label and count, denoted as label: count. The root of the tree is a special virtual node with an empty label and count 0. Auxiliary node linkage structures are constructed to assist node traversal in a WAP-Tree as follows. All of the nodes in the tree with the same label are linked by shared-label linkages into a queue, called event-node queue. The event-node queue with label e_i is also called e_i-queue. There is one header table H for a WAP-Tree, and the head of each event-node queue is registered in H.

The specific process of the algorithm of construction of WAP-Tree can be referred in [1]. We can also build our WAP-tree from frequent sequences which are mined from original sequences with minimum support. We will discuss this in session 4.

3 Predict Future Requests

Using the previous requests to predict the next can be treated as a problem to find the maximum conditional probability. Let E be the set of web pages, and $e_1e_2 \cdots e_{n-1}$ be the previous requests. Our target is to find e_m which satisfy the following equation: $p(e_m \mid e_{n-1}e_{n-2} \cdots e_1) = Max\{p(e_i \mid e_{n-1}e_{n-2} \cdots e_1)\}$, $e_i \in E$, where $p(e_i \mid e_{n-1}e_{n-2} \cdots e_1)$ is the conditional probability of request for page e_i at the next step.

Practically, it is not easy to find the ideal maximum probability $P_{\max}(e_i)$ for a certain user, because there are many facts which affect the probability and some of which are difficult to get and describe in math form. Here we use the user's previous requests in the same session to predict the next request, without taking personal information into account, which is needed for Syskill & Webert [5]. Comparing to well known WebWatcher [3], we also do not require the web site link structure. We only use the logs of web site. This greatly simplifies the process of prediction and without losing much accuracy, and even sometimes may increase it.

The previous requests $e_1, e_{n-2}, \cdots, e_{n-1}$ have different influences to the prediction of the future request e_n. We suppose that the most recent request has the strongest influence, and in most the fact is so. To extend this, we give a coefficient to weigh the influence of every item in the user's previous request. Let $B = b_1b_2 \cdots b_{n-1}b_n$ denote the n-sequence gotten from web logs, f_B denote the support of sequence B, C_k denote the coefficient called weight here of k's event in the user's previous requests, and Ω_i denote the collection of all n-sequences in web logs which satisfy $b_n = e_i$. We define:

$$\omega(b_k) = \begin{cases} 0 & \exists b_j \neq e_j, j \geq k \\ 1 & b_j = e_j, \forall j \geq k \end{cases}$$

We calculate the probability value of e_i to appear at the next step :

$$Q(e_i \mid e_{n-1}e_{n-2}\cdots e_1) = \sum_{B\in\Omega_i}\left\{\frac{f_B}{f_B+M}\cdot\left(\sum_{k=1}^{n-1}\omega(b_k)C_k\right)\right\}, \ M \text{ is a constant} \quad (1)$$

$$P(e_i \mid e_{n-1}e_{n-2}\cdots e_1) = \frac{Q(e_i \mid e_{n-1}e_{n-2}\cdots e_1)}{\sum_{k\in E} Q(e_k \mid e_{n-1}e_{n-2}\cdots e_1)} \quad (2)$$

We use the following rule to C_k.

$$C_{k-1} = \alpha \cdot C_k \qquad k = 2,3,\cdots,n-1 \ , \ \alpha \text{ is a constant.} \quad (3)$$

Function (3) guarantees that no matter what C_{n-1} is, the result will not change if only α is the same. Simply, we set $C_{n-1} = 1$, then $C_k = \alpha^{n-1-k}$.

In order to find the maximum value of $P(e_i \mid e_{n-1}e_{n-2}\cdots e_1)$, we only need to find the maximum value of $Q(e_i \mid e_{n-1}e_{n-2}\cdots e_1)$. We also mark it as Q_{e_i} and calculate $Q(e_i \mid e_{n-1}e_{n-2}\cdots e_1)$ efficiently basing on the WAP-tree [1].

Algorithm 1 (Predicting users' future requests with WAP-tree)

Input: a WAP-tree [1] constructed from web logs, a user's previous request sequence $e_1, e_{n-2}, \cdots, e_{n-1}$, constant M and α, and number n.

Output: the n events of the user's most probable requests at the next step.

Method:

(1). Initialize optional events set $L = \phi$.

(2). Following e_{n-1}'s node-queue of the Header Table of the tree, for each node (marked as Θ) in the e_{n-1}'s node queue.

(a) initialize $\lambda = 0$, $\beta = 1$, node $\Theta' = \Theta$, event $e' = e_{n-1}$ .mark the parent node of node Θ' as *parent* (Θ'), and mark the event exactly before event e' as *parent* (e').

(b) following node Θ's links in the tree from child to parent.

while (the label of node Θ' is the same as event e')

$\lambda \leftarrow \lambda + \beta$, $\beta \leftarrow \beta \cdot \alpha$

$\Theta' \leftarrow parent(\Theta')$, $e' \leftarrow parent(e')$

end **while**

(c) for each child node of node Θ,

mark the count value of the child node as f, then $\Delta Q = \frac{f}{f+M} \cdot \lambda$

If the label of the child node hasn't existed in the optional events set L, create a new item with the label that is the same as that of the child, and set it's $Q \leftarrow \Delta Q$. Otherwise, update Q value of the item with the same label in the set L, $Q \leftarrow Q + \Delta Q$.

(3) For all items in the optional set L, select the n top ones with largest Q values, return their labels, which denote the events.

The algorithm shows that we only need to scan part of the tree once. In addition, the tree isn't needed to be constructed again when making another prediction. Generally, we can put the tree in the memory, and it can be updated easily. We just need to put the user's new sequence to the tree according to the rule of construction.

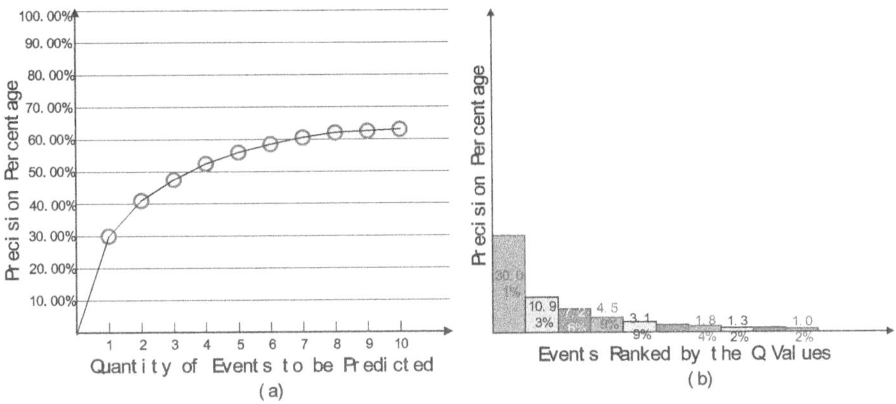

Fig. 1. (a) A curve showing the change of precision according to different quantity of returning events. (b) Precision of top events ranked by Q values. Each histogram represents the precision percentage of the event with a certain rank

4 Experimental Evaluation and Conclusions

In the evaluation of the algorithm, we use the following measures. Let $S = \{S_1 S_2, \cdots, S_n\}$ be the set of sequences in a log file. We build WAP-tree Models on a subset of these sequences, known as the training sequences. The remaining is used as testing sequences. If the returned n top events from the algorithm contain the factual next request, we say that it is a correct prediction. Let P^+ and P^- be set of correct and incorrect predictions separately, and R be the set of all requests. For some requests our algorithm can't give prediction (in the case that in the end the optional set L is empty), so normally $P^+ + P^- \neq |R|$. We use the following measures [6]:

$$precision = \frac{P^+}{P^+ + P^-}, \quad applicability = \frac{P^+ + P^-}{|R|}$$

We use Microsoft anonymous web data as experimental data. The data records the use by 38000 anonymous users. It can be downloaded from the website [7].

As we use the original sequences to build our WAP-tree, almost all the testing requests can be given predictions. So in our experiments, *applicability* = 1. If our WAP-tree is built from frequent sequences, then *applicability* will decrease, but $precision$ will increase.

As *applicability* = 1, here we only need to analyze $precision$.

Let $M = 50$, $\alpha = 2$, and we change parameter n which denotes the returning quantity of predicting events from 1 to 10. The value of $precision$ percentage is shown as figure 1(a):

Further, we get different events predicting ability as figure 1(b).Figure 1(b) explains that for the n returned events, the greater Q value of one event, the greater probable precision the event can be. This means that Q can approximately represent the actual conditional probability.

As we can see, the event with the greatest Q value still hasn't very great precision percentage. There are some reasons. The important one is that, different users have different request sequences. We can reduce applicability to increase precision. As we say previously, we can first mine the web logs to collect frequent sequences with certain minimum support and build WAP-tree from the frequent sequence instead of from original sequences. This method not only increase precision, but also decrease the time consumed in prediction. One shortage of this method is that for some previous sequences, it can not give predictions, or we say that applicability will decrease to less than 1. To let applicability equal 1 or approximately equal 1 and still need high precision, we can use n top ones instead of the greatest one. In some cases, we need to compromise between *precision* and *applicability*. This is very useful in recommendation systems.

References

1. J.Pei, J.Han, H.Zhu and B.Mortazavi-asl. Mining Access Patters Efficiently from Web Logs.PAKDD'00,pages 396-407, Kyoto, Japan, April 2000.
2. J.Srivasta, R.Cooley, M.Deshpande and P.Tan. Web Usage Mining: Discovery and Applications of Usage Patterns from Web Data. In SIGKDD Explorations. (1)2, 2000.
3. T. Joachirms, D. Freitag and T. Mitchell. WebWatcher. A Tour Guide for the World Wide Web. In Proceedings of 15th International Joint Conference on Artificial Intelligence, pages 770-775. Morgan Kaufrnann, Aug. 1997.
4. M. Pazzani, J. Muramatsu and D. Billsus. Syskill&Webert: Identifying interesting web sites. In Proceedings of the 13th National Conference on Artificial Intelligence, Portland, 1996.
5. F. Masseglia, P. Poncelet and M. Teisseire. Using Data Mining Techniques on Web Access Logs to Dynamically Improve Hypertext Structure. In ACM Sib Web Letters, 8(3):13-19, October 1999.
6. R. R. Sarukkai. Link Prediction and Path Analysis Using Markov Chains. In the 9th International WWW Conference, 2000.
7. http://kdd.ics.uci.edu/databases/msweb/msweb.html.

CCMine: Efficient Mining of Confidence-Closed Correlated Patterns

Won-Young Kim*, Young-Koo Lee**, and Jiawei Han

Department of Computer Science
University of Illinois at Urbana-Champaign, Illinois, U.S.A.
{wykim, yklee, hanj}@uiuc.edu

Abstract. Correlated pattern mining has become increasingly important recently as an alternative or an augmentation of association rule mining. Though correlated pattern mining discloses the correlation relationships among data objects and reduces significantly the number of patterns produced by the association mining, it still generates quite a large number of patterns. In this paper, we propose **closed correlated pattern mining** to reduce the number of the correlated patterns produced without information loss. We first propose a new notion of the *confidence-closed* correlated patterns, and then present an efficient algorithm, called CCMine, for mining those patterns. Our performance study shows that confidence-closed pattern mining reduces the number of patterns by at least an order of magnitude. It also shows that CCMine outperforms a simple method making use of the the traditional closed pattern miner. We conclude that confidence-closed pattern mining is a valuable approach to condensing correlated patterns.

1 Introduction

Though association rule mining has been extensively studied in data mining research, its popular adoption and successful industry application has been hindered by a major obstacle: *association mining often generates a huge number of rules, but a majority of them either are redundant or do not reflect the true correlation relationship among data objects.* To overcome this difficulty, interesting pattern mining has become increasingly important recently and many alternative interestingness measures have been proposed [1,2,3,4].

While there is still no universally accepted best measure for judging interesting patterns, *all_confidence* [5] is emerging as a measure that can disclose true correlation relationships among data objects [5,6,7,8]. One of important properties of all_confidence is that it is not influenced by the co-absence of object pairs in the transactions—such an important property is called *null-invariance* [8]. The

* Current address: On Demand Service Team, Digital Home Research Division, ETRI, Korea

** Current address: School of Electronics and Information, Kyung Hee University, Korea

H. Dai, R. Srikant, and C. Zhang (Eds.): PAKDD 2004, LNAI 3056, pp. 569–579, 2004.

co-absence of a set of objects, which is normal in large databases, may have unexpected impact on the computation of many correlation measures. All_confidence can disclose genuine correlation relationships without being influenced by object co-absence in a database while many other measures cannot. In addition, all_confidence mining can be performed efficiently using its downward closure property [5].

Although the all_confidence measure reduces significantly the number of patterns mined, it still generates quite a large number of patterns, some of which are redundant. This is because mining a long pattern may generate an exponential number of sub-patterns due to the downward closure property of the measure. For frequent itemset mining, there have been several studies proposed to reduce the number of items mined, including mining closed [9], maximal [10], and compressed (approximate) [11] itemsets. Among them, the closed itemset mining, which mines only those frequent itemsets having no proper superset with the same support, limits the number of patterns produced without information loss. It has been shown in [12] that the closed itemset mining generates orders of magnitude smaller result set than frequent itemset mining.

In this paper, we introduce the concept of *confidence closed correlated pattern*, which plays the role of reducing the number of the correlated patterns produced without information loss. All_confidence is used as our correlation measure. However, the result can be easily extended to several other correlation measures, such as coherence [6]. First, we propose the notion of the *confidence-closed* correlated pattern. Previous studies use the concept of *support-closed* pattern, i.e., the closed pattern based on the notion of support. However, support-closed pattern mining fails to distinguish the patterns with different confidence values. In order to overcome this difficulty, we introduce *confidence-closed* correlated pattern which encompasses both confidence and support. Then we propose an efficient algorithm, called CCMine, for mining confidence-closed patterns. Our experimental and performance study shows that confidence-closed pattern mining reduces the number of patterns by at least an order of magnitude. It also shows that superiority of the proposed algorithm over a simple method that mines the confidence-closed patterns using the patterns generated by the support-closed pattern miner.

The rest of the paper is organized as follows. Section 2 introduce basic concepts of frequent itemset mining and all_confidence. Section 3 defines the notion of the confidence-closed patterns and discusses its properties. Section 4 presents an efficient algorithm for mining confidence-closed correlated patterns. Section 5 reports our experimental and performance results. Finally, Section 6 concludes the paper.

2 Background

We first introduce the basic concepts of frequent itemset mining, and then present a brief review of the all_confidence measure.

Let $I = \{i_1, i_2, \dots, i_m\}$ be a set of items, and DB be a database that consists of a set of transactions. Each transaction T consists of a set of items such that $T \subseteq I$. Each transaction is associated with an identifier, called *TID*. Let A be a set of items, referred to as an *itemset*. An itemset that contains k items is a *k-itemset*. A transaction T is said to *contain* A if and only if $A \subseteq T$. The *support* of an itemset X in DB, denoted as $sup(X)$, is the number of transactions in DB containing X. An itemset X is *frequent* if it occurs no less frequent than a user-defined minimum support threshold. In most data mining studies, only the frequent itemsets are considered as significant and will be mined.

The all_confidence of an itemset X is the minimal confidence among the set of association rules $i_j \rightarrow X - i_j$, where $i_j \in X$. Its formal definition is given as follows. Here, the max_item_sup of an itemset X means the maximum (single) item support in DB of all the items in X.

Definition 1. *(all-confidence of an itemset) Given an itemset* $X = \{i_1, i_2, \dots, i_k\}$, *the all_confidence of* X *is defined as,*

$$max\_item\_sup(X) = max\{sup(i_j) | \forall i_j \in X\} \tag{1}$$

$$all\_conf(X) = \frac{sup(X)}{max\_item\_sup(X)} \tag{2}$$

Given a transaction database DB, a minimum support threshold min_sup and a minimum all_confidence threshold min_α, a frequent itemset X is ***all_confident*** or ***correlated*** if $all\_conf(X) \geq min\_\alpha$ and $sup(X) \geq min\_sup$.

3 Confidence-Closed Correlated Patterns

It is well known that closed pattern mining has served as an effective method to reduce the number of patterns produced without information loss in frequent itemset mining. Motivated by such practice, we extend the notion of closed pattern so that it can be used in the domain of correlated pattern mining. We present the formal definitions of the original and extended ones in Definitions 2 and 3, respectively. In this paper, we call the former *support-closed* and the latter *confidence-closed*.

Definition 2. *(Support-Closed Itemset) An itemset* Y *is a* ***support-closed (correlated) itemset*** *if it is frequent and correlated and there exists no proper superset* $Y' \supset Y$ *such that* $sup(Y') = sup(Y)$.

Since the support-closed itemset is based on support, it cannot retain the confidence information—notice that in this paper *confidence* means *the value of all_confidence*. In other words, support-closed causes information loss.

Example 1. Let itemset $ABCDE$ be a correlated pattern with support 30 and confidence 30% and itemset CDE be one with support 30 and confidence 80%. Suppose that we want to get a set of non-redundant correlated patterns when *min_sup* = 20 and *min_α* = 20%. Support-closed pattern mining generates *ABCDE* only eliminating *CDE* since *ABCDE* is superset of *CDE* with the same support. We thus lose the pattern *CDE*. However, *CDE* might be more interesting than *ABCDE* since the former has higher confidence that the latter.

We thus extend the support-closed itemset to encompass the confidence so that it can retain the confidence information as well as support information.

Definition 3. *(Confidence-Closed Itemset) An itemset Y is a* ***confidence-closed itemset*** *if it is correlated and there exists no proper superset* $Y' \supset Y$ *such that* $sup(Y') = sup(Y)$ *and* $all\_conf(Y') = all\_conf(Y)$.

By applying mining of confidence-closed itemsets to Example 1, we can obtain not only itemset *ABCDE* but also *CDE* as confidence-closed itemsets since they have different confidence values and therefore no information loss occurs. In the rest of our paper, we call the support-closed pattern as SCP and the confidence-closed pattern as CCP, respectively.

4 Mining Confidence-Closed Correlated Patterns

In this section, we propose two algorithms for mining CCPs: CCFilter and CCMine. CCFilter is a simple algorithm that makes use of the existing support-closed pattern generator. CCFilter consists of the following two steps: First, get the complete set of SCPs using the previous proposed algorithms [13]. Second, check each itemset and its all possible subsets in the resulting set whether it is confidence-closed. If its confidence satisfies *min_α* and it has no proper superset with the same confidence, it is generated as a confidence-closed itemset. CCFilter is used as a baseline algorithm for comparison in Section 5.

CCFilter has a shortcoming: It generates SCPs with less confidence than *min_α* during the mining process. At the end, these patterns are removed. In order to solve this problem, CCMine integrates the two steps of CCFilter into one. Since all_confidence has the downward closure property, we can push down the confidence condition into the process of the confidence-closed pattern mining.

CCMine adopts a pattern-growth methodology proposed in [14]. In the previous studies (e.g., CLOSET+ [13] and CHARM [15]) for mining SCPs, two search space pruning techniques, *item merging* and *sub-itemset merging*, have been mainly used. However, if we apply these techniques directly into confidence-closed pattern mining, we cannot obtain a complete set of CCPs. This is because if there exists a pattern, these techniques remove all of its sub-patterns with the same support without considering confidence. We modify these optimization techniques so that they can be used in confidence-closed pattern mining.

Lemma 1. *(confidence-closed item merging) Let X be a correlated itemset. If every transaction containing itemset X also contains itemset Y but not any proper superset of Y, and* $all\_conf(XY) = all\_conf(X)$, *then XY forms a confidence-closed itemset and there is no need to search any itemset containing X but no Y.* ∎

Lemma 2. *(confidence-closed sub-itemset pruning) Let X be a correlated itemset currently under construction. If X is a proper subset of an already found confidence-closed itemset Y and $all\_conf(X) = all\_conf(Y)$ then X and all of X's descendants in the set enumeration tree cannot be confidence-closed itemsets and thus can be pruned.*

Lemma 1 means that we have to mine the X-conditional database and the XY-conditional database separately if $all\_conf(X) \neq all\_conf(XY)$. However, though $all\_conf(X)$ and $all\_conf(XY)$ are different, the X- and XY- conditional databases are exactly the same if $sup(X) = sup(XY)$. Using this property, we can avoid the overhead of building conditional databases for the prefix itemsets with the same support but different confidence. We maintain a list *candidateList* of the items that have the same support with the size of the X-conditional database but are not included in the item merging because of their confidence. The list is constructed as follows. For X-conditional database, let Y be the set of items in f_list such that they appear in every transaction. Do the following: for each item Y_i in Y, if $sup(Y_i) \leq max\_item\_sup(X)$, $X = X \cup Y_i$; otherwise insert Y_i to *candidateList*. When we check whether an itemset Z containing $X(Z \supset X)$ is confidence-closed, we also check whether the itemset $Z \cup Y'(Y' = Y_1 \ldots Y_k, Y_i \in candidateList)$ could be confidence-closed. Using this method, we compute CCPs without generating the two conditional databases of X and of XY when $all\_conf(X) > all\_conf(XY)$ and $sup(X) = sup(XY)$.

Algorithm 4 shows the CCMine algorithm, which is based on the extension of CLOSET+ [13] and integrates the above discussions into the CLOSET+. Among a lot of studies for support-closed pattern mining, CLOSET+ is the fastest algorithm for a wide range of applications.

CCMine uses another optimization technique to reduce the search space by taking advantage of the property of the all_confidence measure. Lemma 3 describes the pruning rule.

Lemma 3. *(counting space pruning rule) Let $\alpha = i_1 i_2 \ldots i_k$. In the α-conditional database, for item x to be included in an all_confident pattern, the support of x should be less than $sup(\alpha)/min\_\alpha$.* ■

Proof. In order for αx to be an all_confident pattern, $max\_item\_sup(\alpha x) \leq sup(\alpha x)/min\_\alpha$. Moreover, $|sup(\alpha)| \geq |sup(\alpha x)|$. Thus, $max\_item\_sup(\alpha x) \leq sup(\alpha)/min\_\alpha$. Hence the lemma. ■

With this pruning rule, we can reduce the set of items I_β to be counted and, thus, reduce the number of nodes visited when we traverse the FP-tree to count each item in I_β.

Example 2. Let us illustrate the confidence-closed mining process using an example. Figure 1 shows our running example of the transaction database DB. Let $min\_sup = 2$ and $min\_\alpha = 40\%$. Scan DB once. We find and sort the list of frequent items in support descending order. This leads to f_list = $\langle a{:}9, b{:}7, c{:}6, e{:}6, g{:}5, f{:}4, d{:}3, i{:}3, k{:}3, j{:}2, h{:}1\rangle$. Figure 2 shows the global FP-tree. For lack of space, we only show two representative cases: mining for prefix $j{:}2$ and $eg{:}5$.

Algorithm 41 CCMine: Mining confidence-closed correlated patterns

Input: a transaction database DB; a support threshold min_sup
a minimum all_confidence threshold min_α
Output: The complete set of confidence-closed correlated patterns.

Method:

1. Let CCP be the set of confidence-closed patterns. Initialize $CCP \leftarrow \emptyset$
2. Scan DB once to find frequent items and compute frequent list $f\_list(=\langle f_0, f_1, \ldots \rangle)$.
3. Call CCMine($\emptyset$, DB, f_list, CCP, $\emptyset$).

ProcedureCCMine(α, CDB, f_list, CCP, $candidateList$)

1: For each item Y in f_list such that it appears in every transaction of CDB, delete Y from f_list and set $\alpha \leftarrow Y \cup \alpha$ if $all\_conf(Y\alpha) \geq all\_conf(\alpha)$, otherwise insert Y into $candidateList$ in the support increasing order; {confidence-closed item merging}
2: call GenerateCCP(α, $candidateList$, CCP);
3: build FP-tree for CDB using f_list, which excludes all the items Ys in the previous step;
4: **for** each a_i in f_list (in reverse descending support order) **do**
5:   set $\beta = \alpha \cup a_i$;
6:   call GenerateCCP(β, $candidateList$, CCP);
7:   get a set I_β of items to be included in β-projected database; {counting space pruning rule}
8:   for each item in I_β, compute its count in β-projected database;
9:   **for** each b_j in I_β **do**
10:     if $sup(\beta b_j) < min\_sup$, delete b_j from I_β; {pruning based on min_sup}
11:     if $all\_conf(\beta b_j) < min\_\alpha o$, delete b_j from I_β;{pruning based on min_α}
12:   **end for**
13:   call FP-mine(β, CDB, f_list, CCP, $candidateList$);
14:   delete the items that was inserted in step 1 from $candidateList$;
15: **end for**

Procedure GenerateCCP(α, $candidateList$, CCP)
for k-itemset $Y = Y_1 \ldots Y_k (Y_i \in candidateList)$ **do**
  add $\alpha \cup Y$ into CCP if $all\_conf(\alpha \cup Y) \geq min\_\alpha$ if $\alpha \cup Y$ is not a subset of X (in CCP) with the same support and confidence; {confidence-closed sub-itemset pruning}
end for

1. After building the FP-tree we mine the confidence-closed patterns with prefix j:2. **Computing counts:** We compute the counts for items a, c, e, f, and i to be included in the j-projected database by traversing the FP-tree shown in Fig. 2. First, we use Lemma 3 to reduce items to be counted. The support of item $z (z \in \{a, c, e, f, i\})$ should be less than or equal to $sup(j)/min\_\alpha = 2/0.4 = 5$. With this pruning, items a, c and e are eliminated. Now, we compute counts of items f and i and construct j-projected database. They are 2 and 1, respectively. **Pruning:** We conduct pruning based on min_sup and min_α. Item i is pruned since its support is less than min_sup. Item

TID	items bought
10	a, b, c, d, e, g
20	a, b, c, e, g, k
30	a, b, c, e, g
40	a, b, d, f, h
50	a, e, f, i, j
60	a, b, e, g, k
70	a, i, k
80	a, c
90	b, c, f
100	a, d, e, g
110	c, f, j
120	b, i

Fig. 1. An transaction database DB.

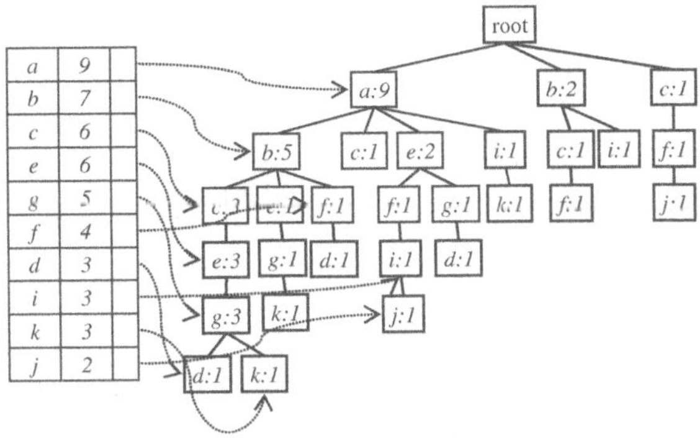

Fig. 2. FP-tree for the transaction database DB.

f is not pruned since and its confidence(2/4) is not less than min_α. Since f is the only item in j-conditional database, we do not need to build the corresponding FP-tree. And fj:2 is a CCP.

2. After building conditional FP-tree for prefix g:5 and we mine g:5-conditional FPtree with $f_list = \langle$a:5, e:5, b:4, c:3$\rangle$. **Confidence Item Merging:** We try confidence-closed item merging of a and e. We delete a and e from f_list. Since $all\_conf(ag) < all\_conf(g)$, we insert a into *candidateList*. Then, we extend the prefix from g to eg by the confidence-closed item merging. **GenerateCCP:** we generate eg:5 as a CCP. In addition, we also generate aeg:5, in which item a come from *candidateList*. Now, in f_list, only two item b:4 and c:3 left. We mine the CCPs with prefix ceg:3. First, we generate ceg as a CCP. However we can not generate $aceg$ as CCP since $all\_conf(aceg) < min\_\alpha$. Since items b the only item in f_list, $bceg$ is a CCP. Again, $abceg$ can not be CCP, since it also does not satisfy min_α. In this way, we mine the beg:4-conditional database and generate beg and $abeg$ as a CCP. After returning mining beg:4-conditonal FP-tree, item a is removed from *candidateList*.

5 Experiments

In this section, we report out experimental results on the performance of CCMine in comparison with CCFilter algorithm. The result shows that CCMine always outperforms CCFilter especially at low min_sup. Experiments were performed on a 2.2GHz Pentium IV PC with 512MB of memory, running Windows 2000. Algorithms were coded with Visual C++.

Our experiments were performed on two real datasets, as shown in Table 1. Pumsb dataset contains census data for population and housing and is obtained from http://www.almaden.ibm.com/software/quest. Gazelle, a transactional data set comes from click-stream data from Gazelle.com. In the table, ATL/MTL represent average/maximum transaction length. The gazelle dataset is rather sparse in comparison with pumsb datasets, which is very dense so that it produce many long frequent itemsets even for very high values of support.

Table 1. Characteristics of Real Datasets.

Dataset	#Tuples	#Items	ATL/MTL
gazelle	59602	497	2.5/267
pumsb	49046	2113	74/74

We first show that the complete set of CCPs is much smaller in comparison with both that of correlated patterns and that of SCPs. Figure 3 shows the number of CCPs, correlated patterns, and SCPs generated from the gazelle data set. In this figure, the number of patterns is plotted on a log scale. Figure 3(a) shows the number of patterns generated when *min_sup* varies and *min_α* is fixed while Figure 3(b) shows those when *min_α* varies and *min_sup* is fixed. We first describe how many we can reduce the number of correlated patterns with the notion of CCPs. Figures 3(a) and 3(b) show that CCP mining generates a much smaller set than that of correlated patterns as the support threshold or the confidence threshold decreases, respectively. It is a desirable phenomenon since the number of correlated patterns increases dramatically as either of the thresholds decreases. These figures also show that the number of SCPs is quite bigger than that of CCPs over the entire range of the support and confidence threshold. These results indicate that CCP mining generates quite a smaller set of patterns even at the low minimum support threshold and low minimum confidence threshold.

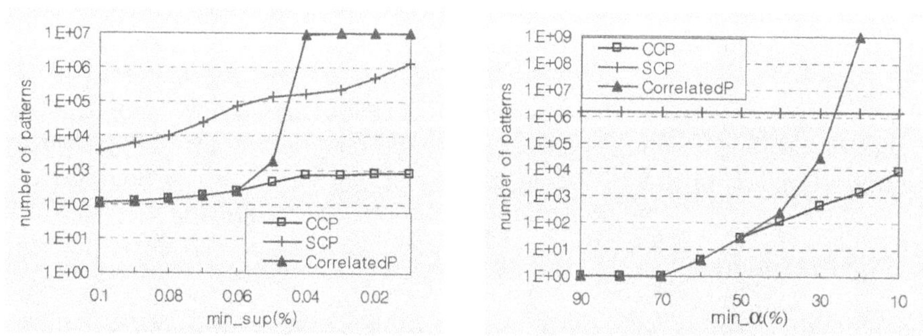

(a) *min_α* = 25% (b) *min_sup* = 0.01%

Fig. 3. Number of patterns generated from the gazelle data set.

Let us then compare the relative efficiency and effectiveness of the CCMine and CCFilter methods. Figure 4 (a) shows the execution time of the two methods on the gazelle dataset using different minimum support threshold while *min_α* is fixed at 25%. Figure 4(a) shows that CCMine always outperforms CCFilter over the entire supports of experiments. When the support threshold is low, CCMine is faster more than 100 times compared with CCFilter, e.g., with *min_sup* 0.05%, CCFilter uses 20 seconds to finish while CCMine only uses 0.2. The reason why CCMine is superior to CCFilter is that CCFilter has to find all of the support closed patterns although many of them do not satisfy the minimum confidence threshold and the number of these patterns increases a lot as the minimum support threshold decreases. Figure 4(b) shows the performance on the gazelle dataset when *min_sup* is fixed at 0.01% and *min_α* varies. As shown in the figure, CCMine always outperforms CCFilter and the execution times of CCMine increases very slowly while *min_α* decreases. CCFilter almost does not change while *min_α* varies, which means it does not take any advantage from *min_α*. This is because it spends most of processing time on mining SCP.

Now, we conduct the experiments on the pumsb dataset, which is a dense dataset. Figure 5(a) shows the execution time on the pumsb dataset when *min_α* varies while *min_sup* is fixed at 60%. Figure 5(a) shows that CCMine method outperforms CCFilter method when *min_sup* is less than 60%. When *min_sup* becomes less then 50%, CCFilter run out of memory and cannot finish. Figure 5(b) shows that CCMine method always outperforms CCFilter method over entire range of *min_α*.

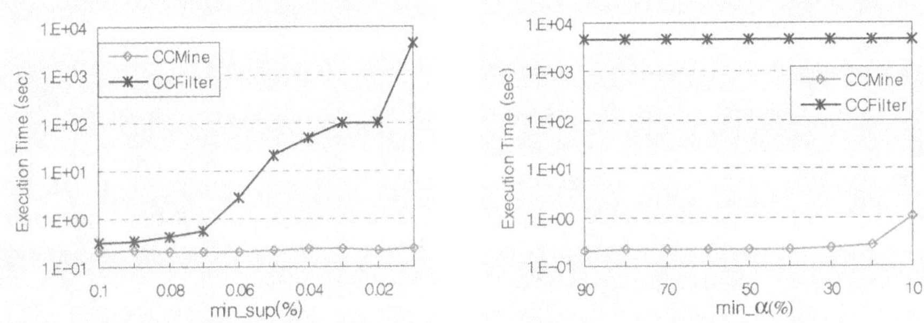

(a) $min\_\alpha = 25\%$. (b) $min\_sup = 0.01\%$.

Fig. 4. Execution time on gazelle data set

In summary, experimental results show that the number of confidence closed correlated patterns are quite small in comparison with that of the support-closed patterns. The CCMine method outperforms CCFilter especially when the support threshold is low or the confidence threshold is high.

6 Conclusions

In this paper, we have presented an approach that can effectively reduce the number of correlated patterns to be mined without information loss. We proposed a new notion of confidence-closed correlated patterns. Confidence-closed correlated patterns are those that have no proper superset with the same support

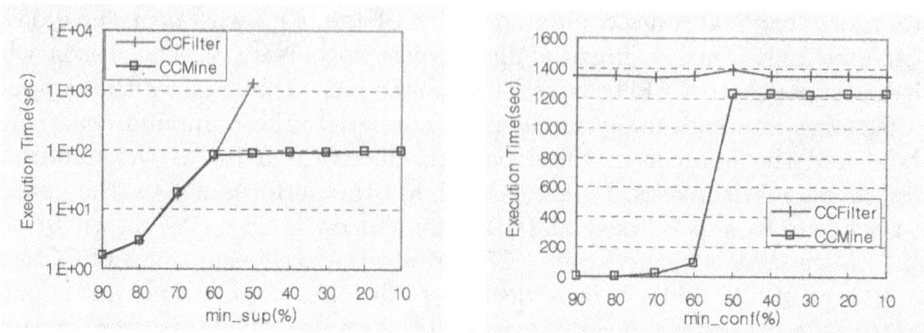

(a) when $min\_\alpha = 60\%$. (b) when $min\_sup = 50\%$.

Fig. 5. Execution time on the pumsb dataset.

and the same confidence. For efficient mining of those patterns, we presented the CCMine algorithm. Several pruning methods have been developed that reduce the search space. Our performance study shows that confidence-closed, correlated pattern mining reduces the number of patterns by at least an order of magnitude in comparison with correlated (non-closed) pattern mining. It also shows that CCMine outperforms CCFilter in terms of runtime and scalability. Overall, it indicates that confidence-closed pattern mining is a valuable approach to condensing correlated patterns.

As indicated in the previous studies of mining correlated patterns, such as [6,5,8], all_confidence is one of several favorable correlation measures, with null-invariance property. Based on our examination, CCMine can be easily extended to mining some correlation measures, such as coherence or bond [6,5,8]. It is an interesting research issue to systematically develop other mining methodologies, such as constraint-based mining, approximate pattern mining, etc. under the framework of mining confidence-closed correlated patterns.

References

1. C. C. Aggarwal and P. S. Yu. A new framework for itemset generation. In *Proc. 1998 ACM Symp. Principles of Database Systems (PODS'98)*, pages 18–24, Seattle, WA, June 1999
2. S. Brin, R. Motwani, and C. Silverstein. Beyond market basket: Generalizing association rules to correlations. In *Proc. 1997 ACM-SIGMOD Int. Conf. Management of Data (SIGMOD'97)*, pages 265–276, Tucson, Arizona, May 1997.
3. S. Morishita and J. Sese. Traversing itemset lattice with statistical metric pruning. In *Proc. 2000 ACM SIGMOD-SIGACT-SIGART Symp. Principles of Database Systems (PODS'00)*, pages 226–236, Dallas, TX, May 2001.
4. P.-N. Tan, V. Kumar, and J. Srivastava. Selecting the right interestingness measure for association patterns. In *Proc. 2002 ACM SIGKDD Int. Conf. Knowledge Discovery in Databases (KDD'02)*, pages 32–41, Edmonton, Canada, July 2002.
5. E. Omiecinski. Alternative interest measures for mining associations. *IEEE Trans. Knowledge and Data Engineering*, 15:57–69, 2003.

6. Y.-K. Lee, W.-Y. Kim, D. Cai, and J. Han. CoMine: Efficient Mining of Correlated Patterns. In *Proc. 2003 Int. Conf. Data Mining (ICDM'03)*, pages 581–584, Melbourne, FL, Nov. 2003.
7. S. Ma and J. L. Hellerstein. Mining mutually dependent patterns. In *Proc. 2001 Int. Conf. Data Mining (ICDM'01)*, pages 409–416, San Jose, CA, Nov. 2001.
8. H. Xiong, P.-N. Tan, and V. Kumar. Mining Strong Affinity Association Patterns in Data Sets with Skewed Support Distribution. In *Proc. 2003 Int. Conf. Data Mining (ICDM'03)*, pages 387–394, Melbourne, FL, Nov. 2003.
9. N. Pasquier, Y. Bastide, R. Taouil, and L. Lakhal. Discovering frequent closed itemsets for association rules. In *Proc. 7th Int. Conf. Database Theory (ICDT'99)*, pages 398–416, Jerusalem, Israel, Jan. 1999.
10. R. J. Bayardo. Efficiently mining long patterns from databases. In *Proc. 1998 ACM-SIGMOD Int. Conf. Management of Data (SIGMOD'98)*, pages 85–93, Seattle, WA, June 1998.
11. J. Pei, G. Dong, W. Zou, and J. Han. On computing condensed frequent pattern bases. In *Proc. 2002 Int. Conf. on Data Mining (ICDM'02)*, pages 378–385, Maebashi, Japan, Dec. 2002.
12. M. Zaki. Generating Non-redundant Association Rules. In *Proc. 2000 ACM SIGKDD Int. Conf. Knowledge Discovery in Databases (KDD'00)*, Aug. 2000.
13. J. Wang, J. Han, and J. Pei. Closet+: Searching for the best strategies for mining frequent closed itemsets. In *Proc. 2003 ACM SIGKDD Int. Conf. Knowledge Discovery and Data Mining (KDD'03)*, Washington, D.C., Aug. 2003.
14. J. Han, J. Pei, and Y. Yin. Mining frequent patterns without candidate generation. In *Proc. 2000 ACM-SIGMOD Int. Conf. Management of Data (SIGMOD'00)*, pages 1–12, Dallas, TX, May 2000.
15. M. Zaki and C. Hsiao. CHARM: An Efficient Algorithm for Closed Itemset Mining, In *Proc. SDM'02*, Apr. 2002.

A Conditional Probability Distribution-Based Dissimilarity Measure for Categorial Data

Le Si Quang and Ho Tu Bao

School of Knowledge science,
Japan Advanced Institute of Science and Technology
Tatsunokuchi, Ishikawa 923-1292 Japan
{quag, bao}@jaist.ac.jp

Abstract. Measuring the similarity between objects described by categorical attributes is a difficult task because no relations between categorical values can be mathematically specified or easily established. In the literature, most similarity (dissimilarity) measures for categorical data consider the similarity of value pairs by considering whether or not these two values are identical. In these methods, the similarity (dissimilarity) of a non-identical value pair is simply considered 0 (1). In this paper, we introduce a dissimilarity measure for categorical data by imposing association relations between non-identical value pairs of an attribute based on their relations with other attributes. The key idea is to measure the similarity between two values of a categorical attribute by the similarities of the conditional probability distributions of other attributes conditioned on these two values. Experiments with a nearest neighbor algorithm demonstrate the merits of our proposal in real-life data sets.

Keywords: Dissimilarity measures, categorical data, conditional probability, hypothesis testing.

1 Introduction

Dissimilarity measures between data objects are crucial notions in data mining, and measuring dissimilarities between data objects is important in detecting patterns or regularities in databases. It is compulsory for distance-based data mining methods such as distance-based clustering [1,2,3], nearest neighbor techniques [4,5,6,7], etc.

Measuring (did)similarities between categorical data objects is a difficult task since relations between categorical values can not be mathematical specified or easily established. This is often the case in many domains in which data are described by a set of descriptive attributes, which are neither numerical nor ordered in any way.

In real-life data sets, categorical attributes are considered as discrete random variables and often depend on each other. For example, the independence confidence level of the attributes *changes in Lym* and *block of affere* in the data set Lymgraphy is less than 1% when applying the χ^2 test on their contingency table (Table 1).

H. Dai, R. Srikant, and C. Zhang (Eds.): PAKDD 2004, LNAI 3056, pp. 580–589, 2004.

Table 1. Contingency table of the attributes *change in node* and the *lymphatic*

	block of affere					*block of affere*		
changes in lym	*no*	*yes*	sum		*changes in lym*	*no*	*yes*	sum
bean	5	1	6	→	*bean*	.83	.17	1
oval	42	35	77		*oval*	.55	.45	1
round	19	46	65		*round*	.29	.71	1

Dependencies among attributes lead to relations between a value v_i of a categorical attribute A^i and the conditional probability distribution (*cpd*) of other attributes when the attribute A^i takes the value v_i. Hence, relations between values of a categorical attribute can be obtained from dissimilarities between the *cpds* of other attributes corresponding to the values. For example, the *cpds* of the attribute *block of affere* are (83% :*yes*, 17%: *no*), (55% :*yes*, 45%: *no*), and (29%: *yes*, 71% :*no*) when the attribute *changes in node* holds the values *bean*, *oval*, and *round*, respectively. The *cpd* of *block of affere* when the attribute *changes in node* holds the value *bean* is more similar to its *cpd* when the attribute *changes in node* holds the value *oval* than when it holds the value *round*. Hence, the dissimilarity between *bean* and *oval* should be smaller than the dissimilarity between *bean* and *round*.

In this paper, we introduce a dissimilarity measure for categorical data by imposing relations between a value v_i of a categorical attribute A^i and *cpds* of another attribute when the attribute A^i takes the value v_i. The main idea here is to define the dissimilarity between values v_i and v_i' of the attribute A^i as the total sum of dissimilarities between *cpds* of other attributes when $A^i = v_i$ and $A^i = v_i'$. Then, the dissimilarity between two data objects is considered as the total sum of dissimilarities of their attribute value pairs.

This paper is organized as follow: related works are briefly summarized in Section 2. Conditional probability distribution-based dissimilarity (*pd*) and its characteristics are presented in Section 3, while our experiments are described in Section 4. Conclusions and recommendations for further work are discussed in the last section.

2 Related Works

A common way to measure dissimilarities between categorical data objects is to transform them into binary vectors. Then, dissimilarities between data objects are considered as dissimilarities between corresponding binary vectors.

(Did)similarity measures between binary vectors have been studied for many years, [8,9,10,11,12]. Most of these measures are based on the number of identical and non-identical value pairs. The similarity (dissimilarity) of a non-identical value pair is simply 0 (1) and the similarity (dissimilarity) of an identical value pair is simply 1 (0).

Table 2. Some well-known similarity measures for binary vectors

	Definition	Range	
Kendall, Sokal-Michener (1958)	$\frac{a+d}{p}$	[0,1]	S
Rogers and Tanimoto (1960)	$\frac{a+d}{p+b+c}$	[1,0]	S
Sokal and Sneath (1963), un_3^{-1}, S	$\frac{b+c}{a+d}$	$[0, \infty]$	S
Hamman (1961)	$\frac{a+d-b-c}{l}$	[-1,1]	T
Jaccard (1900)	$\frac{a}{a+b+c}$	[1,0]	T
Dice (1945), Czekanowski (1913)	$\frac{a}{a+\frac{1}{2}(b+c)}$	[1,0]	T
Sokal and Sneath	$\frac{a}{a+2(b+c)}$	[1,0]	T

Let $\Omega = \{v_1, \ldots, v_l\}$ be a value set of all attributes. A data object $\mathbf{x}$ is transformed into a binary vector $\mathbf{X} = (x_1, \ldots, x_l)$, $x_i \in \{0,1\}$, where $x_i = 1$ if the object $\mathbf{x}$ holds the value v_i, and $x_i = 0$ otherwise.

Denote X and Y as two binary vectors, $XY = \sum_i x_i y_i$, $\overline{X}$ as the complementary vector of X: $\overline{X} = 1 - X = [1 - x_i]$, and the following counters: $a = XY, b = X\overline{Y}, c = \overline{X}Y$ and $d = \overline{XY}$. Obviously, $a + b + c + d = l$ and $a+b+c = m$ where m is the number of attributes. Using these notions, several measures on binary vectors have been defined. A brief summary is given in Table 2.

Most of the popular practical similarity measures in the literature (Jaccard, Dice, etc) belong to one of these two families

$$S_\theta = \frac{a+d}{a+d+\theta(b+c)} \quad \text{and} \quad T_\theta = \frac{a}{a+\theta(b+c)}$$

as introduced by J.C Gower and P.Legendre (1986) [12].

3 Conditional Probability Distribution-Based Dissimilarity

3.1 Notations

Let $A^1, \ldots, A^m$ be m categorical attributes and a data set $D \subseteq A^1 \times \ldots \times A^m$. Let us denote

- $dom(A^i)$: The domain of the attribute A^i.
- $\mathbf{x} = (x_1, \ldots, x_m), x_i \in dom(A^i)$: $\mathbf{x} \in D$
- $p(A^i = v_i)$: the probability of the attribute A^i takes value v_i.

$$p(A^i = v_i) = \frac{|\{\mathbf{x} : x_i = v_i\}|}{|D|}$$

- $p(A^j = v_j | A^i = v_i)$: the conditional probability of the attribute A^j taking the value v_j given that the attribute A^i holds the value v_i.

$$p(A^j = v_j | A^i = v_i) = \frac{p(A^j = v_j, A^i = v_i)}{p(A^i = v_i)}$$

- $cpd(A^j|A^i = v_i)$: the conditional probability distribution of the attribute A^j given that the attribute A^i holds the value v_i:

$$cpd(A^j|A^i = v_i) \equiv \{p(A^j = v_j|A^i = v_i) : v_j \in dom(A^j)\}$$

- $\phi(.,.)$: a dissimilarity function of two probability distributions.

3.2 Conditional Probability Distribution-Based Dissimilarity

Given a value pair (v_i, v_i') of the attribute A^i, the conditional probability distribution-based dissimilarity (*cpd*) between v_i and v_i' is defined as the total sum of dissimilarities between the *cpd* pairs of other attributes, given that the attribute A^i holds the values v_i and v_i':

$$pd(v_i, v_i') = \sum_{j, j \neq i} \phi(cpd(A^j|A^i = v_i), cpd(A^j|A^i = v_i')) \tag{1}$$

The conditional probability distribution-based dissimilarity (*pd*) from the data object $\mathbf{x}$ to the data object $\mathbf{y}$ is defined as the total sum of *pd*s between their attribute value pairs.

$$pd(\mathbf{x}, \mathbf{y}) = \sum_i pd(x_i, y_i) \tag{2}$$

For example, consider the instance in Table 1 and the Kullback-Leiber divergence [13,14] as the dissimilarity function $\phi(.,.)$.

$$\phi_{KL}(P||P') = \sum_x p(x) \log \frac{p(x)}{p'(x)}$$

*pd*s from *bean* to *oval* and from *bean* to *round* are computed as:

$$pd(bean, oval) = 0.83 \log(\frac{0.83}{0.55}) + 0.17 \log(\frac{0.17}{0.45}) = 0.08.$$
$$pd(bean, round) = 0.83 \log(\frac{0.83}{0.29}) + 0.17 \log(\frac{0.17}{0.71}) = 0.27.$$

3.3 Characteristics of *pd*

In this subsection, we discuss a property and two important notes of *pd* in practice.

Proposition 1. *If all attribute pairs are independent, then conditional probability distribution-based dissimilarities between data object pairs are all equal to 0.*

Proof. Since all attribute pairs are independent, we know that:

$$\begin{aligned}
&\forall A^i, A^j, v_i, v_i' \in dom(A^i), v_j \in dom(A^j): \\
&\quad p(A^j = v_j | A^i = v_i) = p(A^j = v_j) = p(A^j = v_j | A^i = v_i') \\
&\Rightarrow cpd(A^j | A^i = v_i) \equiv cpd(A^j | A^i = v_i') \\
&\Rightarrow \phi(cpd(A^j | A^i = v_i), cpd(A^j | A^i = v_i')) = 0 \\
&\Rightarrow pd(v_i, v_i') = 0 \\
&\Rightarrow pd(\mathbf{x}, \mathbf{y}) = 0 \;\; \forall \mathbf{x}, \mathbf{y} \in D.
\end{aligned}$$

The following two notes regarding *pd* are important in practice.

1. *pd* follows the properties of the $\phi(.,.)$ function.
2. Since the number of attribute values of a data set is usually small, it is possible to store temporal data such as $cpd(A^j|A^i = v_i)$, $P(A^i = v_i)$, etc., in memory after one scan for the whole data set. Hence, the computing time for a value pair of an attribute depends mainly on the number of attribute values and is linear with data set sizes.

4 Experiments

In this section, we conduct two experiments on *pd*. First is to analyze *pd* on two data sets, Vote and Monks, from UCI [15]. The second experiment is to employ *pd* in a nearest neighbor classifier (NN)[4], where dissimilarities between data objects play a crucial role, in order to show its usefulness in practice. To evaluate performance of the proposed similarity measure, we compare the accuracies of NN using *pd* with those of NN with Jaccard, employing 10-trial 10-fold cross-validation strategy. In our experiments, we employ the Kullback-Leiber divergence as the distribution similarity function $\phi(.,.)$ in Equation 1.

4.1 Analysis of *pd* on the UCI Data Sets

pd on the Data Set Vote

Table 3 presents dissimilarities between value pairs of the data set Vote. As can be seen from this table,

- Dissimilarities between value pairs are nonnegative, zero on diagonals, and asymmetrical. These properties are attributed to the nature of KL divergence.
- Dissimilarities between value pairs of an attribute are not simply 0 or 1. For example, on the attribute *immigration*, the dissimilarity between *yes* and *nc* is 16.26, the dissimilarity between *yes* and *no* is 0.20, the dissimilarity between *nc* and *no* is 15.11, etc.

Table 3. Dissimilarities between attribute values of Vote

	handicapped			*water-project*			*budget-resolution*			*physician-fee*		
	nc	*no*	*yes*	*nc*	*no*	*yes*	*nc*	*no*	*yes*	*nc*	*no*	*yes*
nc	0	9.77	7.89	0	2.34	1.86	0	13.22	10.20	0	10.99	17.21
no	7.68	0	3.68	1.27	0	0.71	7.77	0	10.17	9.50	0	13.38
yes	4.63	3.47	0	1.32	0.75	0	7.42	10.83	0	16.08	11.54	0

	mx-missile			*immigration*			*synfuels cutback*			*education-spending*		
	nc	*no*	*yes*	*nc*	*no*	*yes*	*nc*	*no*	*yes*	*nc*	*no*	*yes*
nc	0	8.25	2.71	0	15.11	17.67	0	6.69	5.78	0	4.37	7.32
no	8.03	0	11.42	16.15	0	0.21	3.68	0	0.88	3.06	0	10.60
yes	1.97	10.60	0	16.26	0.20	0	5.29	9.79	0	4.45	8.98	0

....

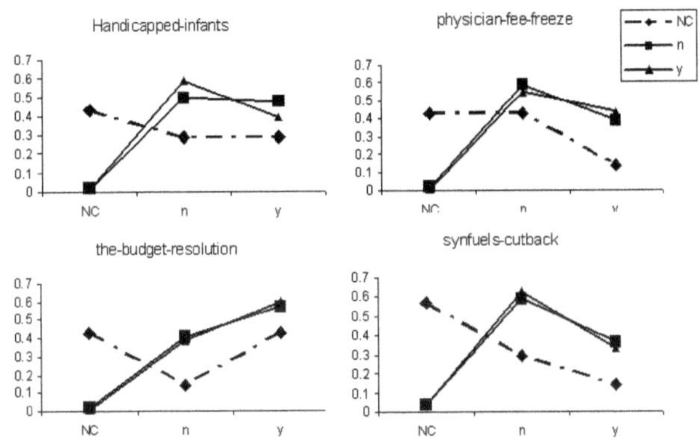

Fig. 1. Some conditional probability distributions when given the attribute immigration values *nc*, *yes*, and *no*

- Dissimilarities between value pairs of an attribute are different from pair to pair. The differences are due to various differences between the *cpds* of other attributes corresponding to values of this attribute. For example, (on the attribute *immigration*) *cpds* of other attributes when the attribute *immigration* takes the value of *yes* are similar to *cpds* when it takes the value of *no* and also different from those when it takes the value of *nc* (some examples are given in Fig. 1). Hence, the dissimilarity between the values *yes* and *no*, 0.20, is much smaller than the dissimilarity between the values *yes* and *nc*, 16.26, or the dissimilarity between the values *no* and *nc*, 16.15.

pd on the Data Set Monks

Table 4 shows dissimilarities between value pairs of the data set Monks. As shown in this table, dissimilarities between value pairs of an attribute are all zero. Since all attribute pairs are independent (an independent test of this data set is shown in Subsection 4.2), the above observation can be explained by Proposition 1.

Table 4. Dissimilarities between attribute values of Monk

A1				A2				A3			A4				A5					A6		
	1	2	3		1	2	3		1	2		1	2	3		1	2	3	4		1	2
1	0	0	0	1	0	0	0	1	0	0	1	0	0	0	1	0	0	0	0	1	0	0
2	0	0	0	2	0	0	0	2	0	0	2	0	0	0	2	0	0	0	0	2	0	0
3	0	0	0	3	0	0	0				3	0	0	0	3	0	0	0	0			
															4	0	0	0	0			

4.2 Nearest Neighbor with *pd*

To analyze how well *pd* boots the performance of the nearest neighbor algorithm [4] compared with Jaccard, we use the hypothesis testing:

$$H_0 : \mu_1 = \mu_2$$
$$H_1 : \mu_1 > \mu_2$$

where μ_1 and μ_2 are the average accuracies of NN with *pd* and NN with Jaccard, respectively.

Since each 10-trial 10-fold cross-validation result contains 100 trials, the difference between two means μ_1 and μ_2 follows the normal distribution:

$$Z = \frac{\mu_1 - \mu_2}{\sqrt{\frac{\delta_1^2}{100} + \frac{\delta_2^2}{100}}}$$

where δ_1 and δ_2 are the deviations of the accuracy test results of NN with *pd* and Jaccard.

Since *pd* bases strongly on dependencies between attributes, we analyze dependencies of attribute pairs by the χ^2 test with the 95% significant level.

$$\chi^2(A^i, A^j) = \sum_{v_i \in dom(A^i)} \sum_{v_j \in dom(A^j)} \left(\frac{\left[p(A^i = v_i, A^j = v_j) - |D|\, p(A^i = v_i)\, p(A^j = v_j)\right]^2}{|D|\, p(A^i = v_i)\, p(A^j = v_j)} \right)$$

with the degree of freedom: $|dom(A^i)|\ |dom(A^j)| - 1$.

The dependent factor of D, denoted $\rho(D)$, is defined as the proportion between the number of dependent attribute pairs and the total number of attribute pairs.

$$\rho(D) = \frac{|\{(A^i, A^j) : A^i \text{ and } A^j \text{ are dependent}\}|}{m(m-1)}$$

4.3 Results

We employ the nearest neighbor algorithm (NN) with *pd* and with Jaccard on 15 data sets from UCI [15]. Discretization for numeric attributes of the data sets is done automatically by the data mining system CBA [16].

Table 5. The accuracies of nearest neighbor algorithm with *pd* and with Jaccard similarity

No.	Name	*pd* $\rho(\%)$	$\mu_1(\%)$	δ_1	Jaccard $\mu_2(\%)$	δ_2	Z
1	splice	100	87.28	0.0167	75.43	0.0260	38.30
2	tictactoe	100	96.61	0.0185	81.01	0.0382	36.76
3	waveform	81	77.11	0.0148	71.58	0.0196	22.56
4	crx	99	83.04	0.0401	78.38	0.0441	7.81
5	anneal	63	99.60	0.0081	98.69	0.0127	6.05
6	pima	53	71.62	0.0472	67.33	0.0566	5.82
7	wine	100	99.83	0.0095	98.15	0.0276	5.76
8	cmc	97	45.90	0.0389	43.59	0.0383	4.24
9	hypo	98	98.97	0.0055	98.61	0.0071	4.09
10	vote	100	94.03	0.0360	92.14	0.0425	3.39
11	labor	99	97.53	0.0719	94.47	0.0880	2.70
12	zoo	98	98.02	0.0399	96.43	0.0500	2.49
13	post-operative	100	60.22	0.1695	55.78	0.1787	1.80
14	sonar	11	79.38	0.0856	75.56	0.0884	-3.10
15	monks	0	49.56	0.0772	76.86	0.0576	-28.34

Table 5 shows the results of NN with *pd* and NN with Jaccard for the 15 data sets. The first column presents the indices of the data sets in the decreasing order of Z values. The second and third columns show the names and the dependent factors of the data sets. The fourth and fifth columns present accuracies and deviations of NN with *pd*. The next two columns present accuracies and deviations of NN with Jaccard. The last column shows Z values.

As can be seen from Table 5:

- The accuracies of NN with *pd* are higher than the accuracies of NN with Jaccard on the first 13 data sets. The confidence levels are more than 99% ($NormDist(Z_{\alpha=2.49}) = 0.994$) for the first 12 data sets and more than 96% ($NormDist(Z_\alpha = 1.80) = 0.964$) for the last one.
- The accuracies of NN with *pd* are lower than the accuracies of NN with Jaccard with a more than 99% confidence level ($NormDist(Z_\alpha = -3.1) < 0.01$) for the last two data sets,.

From Table 5, it is noteworthy that:

- The first 13 data sets, for which the accuracies of NN with *pd* are higher than with Jaccard, have also high dependent factors (i.e., Vote 100%, Splice 100%, waveform 81%).
- The last two data sets, in which the average accuracies of NN with *pd* are lower than with Jaccard, have low dependent factors (sonar 11%, monks 0%).

The note indicates that, in practice, *pd* is suitable for dependent attribute data sets which have large dependent factors, and unsuitable or useless for independent attribute data sets those with small dependent factors. This is because relations between the attribute value pairs of an attribute are obtained from the dependencies of other attributes when given the values. In other words, the relations are based upon the dependencies of other attributes on this attribute. Hence, when there are weak dependencies or no dependencies between attributes, the obtained relations must be very poor and lead to the inefficiency or poor performance of *pd* in practice.

5 Conclusion

In this paper we introduce a new dissimilarity measure for categorical data based on dependencies between attributes. The main idea here is to define the dissimilarity between values (v_i and v_i') of the attribute A^i as the total sum of dissimilarities between *cpds* of other attributes when the attribute A^i holds the values v_i and v_i'. The dissimilarity between two data objects is considered as the total sum of dissimilarities of their attribute value pairs. Since *pd* strongly relies on dependencies between attributes, independent attribute pairs can be considered as 'noises' in *pd*. Thus, in future studies, we will try to limit the effects of independent attribute pairs as much as possible. Moreover, *pd* does not directly reveal (dis)similarities between value pairs but rather, relations between partitions (clusters) of data generated by the value pairs. This led to our approach of using this idea for clustering categorical data.

References

1. MacQueen J. Some methods for classification and analysis of multivariate observation. In *Proceedings 5th Berkeley Symposium on Mathematical Statistics and Probability*, pages 281–297, 1967.
2. Huang Z. Extensions to the k-means algorithm for clustering large data sets with categorical values. *Data Mining Knowledge Discovery II*, pages 283–304, 1988.
3. Kaufmann L. and Rousseeuw P.J. Clustering by means of medoids. *Statistical Data Analysis based on the L1 Norm*, pages 405–416, 1987.
4. Cover T.M. and Hart P.E. Nearest neighbor pattern classification. *IEEE Transactions on Information Theory*, 13:21–27, 1967.
5. Nene S. and Nayar S. A simple algorithm for nearest neighbor search in high dimensions. *IEEETPAMI: IEEE Transactions on Pattern Analysis and Machine Intelligence*, 19, 1997.
6. Aha D. W., Kibler D., and Albert M. Instance-based learning algorithms. *In Machine Learning*, 6:37–66, 1991.
7. Farag'o A., Linder T., and Lugosi G. Fast nearest-neighbor search in dissimilarity spaces. *IEEE Trans. on Pattern Analysis and Machine Intelligence*, 15(9):957–962, 1993.
8. Hubálek Z. Coefficients of association and similarity, based on binary (present-absence) data: an evaluation. *Biological review*, (57):669–689, 1982.
9. Baulieu F.B. Classification of presence/absence based dissimilarity coefficients. *Journal of Classification*, (6):233–246, 1989.
10. Batagelj V. and Bren M. Comparing resemblance measures. *Journal of Classification*, 12(1), 1995.

11. Albert M. *Measures of Association*, volume 32 of *Quantitative Applications in the Social Sciences*. Sage publications, 1983.
12. Gower J.C and Legendre P. Metric and euclidean properties of dissimilarity coefficients. *Journal of Classification*, (3):5–48, 1986.
13. Kullback S. *Information theory and statistics*. John Wiley and Sons., New York, 1959.
14. Kullback S. and Leibler R.A. On information and sufficiency. *Annals of Mathematical Statistics*, 22:79–86, 1951.
15. Blake C.L. and Merz C.J. (uci) repository of machine learning databases, 1998.
16. Bing Liu, Wynne Hsu, and Yiming Ma. Integrating classification and association rule mining. In *Knowledge Discovery and Data Mining*, pages 80–86, 1998.

Learning Hidden Markov Model Topology Based on KL Divergence for Information Extraction

Kwok-Chung Au and Kwok-Wai Cheung*

Hong Kong Baptist University, Kowloon Tong, Hong Kong
{henryau, william}@comp.hkbu.edu.hk

Abstract. To locate information embedded in documents, information extraction systems based on rule-based pattern matching have long been used. To further improve the extraction generalization, hidden Markov model (HMM) has recently been adopted for modeling temporal variations of the target patterns with promising results. In this paper, a state-merging method is adopted for learning the topology with the use of a localized Kullback Leibler (KL) divergence. The proposed system has been applied to a set of domain-specific job advertisements and preliminary experiments show promising results.

1 Introduction

Information extraction (IE), originated from the natural language processing community, has a history of more than a decade, with the goal of extracting target information (normally in the form of some short text segments) from a large collection of documents. Many related pattern matching systems have been built [1] and the extraction rules can automatically be induced [2]. The pattern matching approach only works well for extracting information with some distinctive attributes or a limited vocabulary size. To further improve the extraction generalization, hidden Markov model (HMM) has recently been adopted for modeling and extracting the temporal variations of the target patterns with promising results.

Hidden Markov model (HMM) is a prevalent method for modeling stochastic sequences with an underlying finite-state structure. To apply HMM to information extraction, the model topology should first be determined before the corresponding model parameters can be estimated. In this paper, we adopt a state-merging method to learn the HMM topology, with the learning process guided by the Kullback Leibler (KL) divergence. As learning the optimal structure is known to be computationally expensive, the use of a local KL divergence as well as some domain-specific constraints have been explored. The proposed system has been applied to extracting target items from a set of domain-specific job advertisements and preliminary experiments show promising results regarding its generalization performance.

* This research work is jointly supported by HKBU FRG/00-01/II-8 and HKBU FRG/02-03/II-69.

H. Dai, R. Srikant, and C. Zhang (Eds.): PAKDD 2004, LNAI 3056, pp. 590–594, 2004.

```
<city>Austin</city> company is seeking <area>multimedia</area>
and <area>Internet</area> programmers for exciting web site and
<area>multimedia</area> software production! Successful candidates
will have <req_degree>B.S.</req_degree> degree, preferably in Computer
Science, and <req_years_experience>2 </req_years_experience> years of
professional experience.
```

Fig. 1. A tagged job ad.

2 HMM for Information Extraction

HMM has recently been applied to information extraction and found to work well in a number of application domains, including gene names and locations extraction from scientific abstracts [3], extracting relevant information from research papers [4], etc. The HMM we are studying is a discrete first-order one with four types of states, namely background states (BG), prefix states (PR), target states (TG) and suffix states (SU). Only the target states are intended to produce the tokens we want to extract. Prefix and suffix states are added because the context immediately before and after the targets may help identifying the target items. Background states are for producing the remaining tokens in the document. Figure 1 shows a typical training example under the context of job advertisements. Based on the formulation, we can hand-draft a HMM with the four types of states as mentioned above and learn its parameters based on the tagged information.

Automatically constructing an optimal model topology is known to be challenging. In this paper, a bottom-up approach is adopted (see [5,6] for related works). It starts with a complex HMM as a disjunction of all the training examples (see Figure 2) and induces the model structure with better generalization by merging states, where some measure to determine whether to merge two states is typically required.

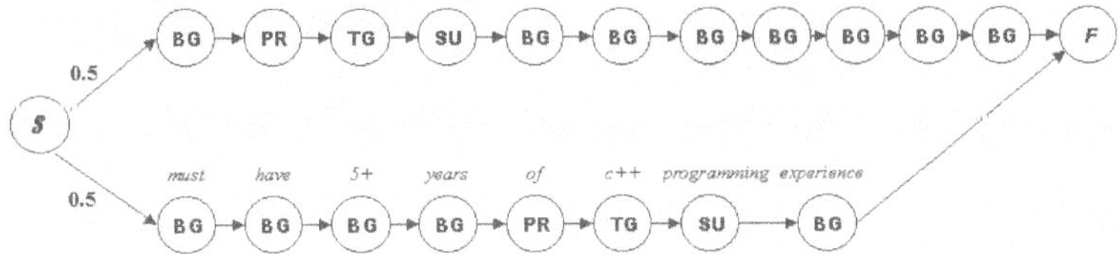

Fig. 2. An initial model constructed by two training examples (S - start state, F - final state, TG - target state, BG - background state, PR - prefix state and SU - suffix state).

3 KL Divergence Based State Merging

3.1 KL Divergence between HMMs

To determine whether two states should be merged, a distance measure between the model before and after modeling is needed. *Kullback Leibler (KL) divergence* $D(A||B)$ (also called *relative entropy* $H_{A,B}(X)$) can be used. The measure has the property that $D(A||B) = 0$ if and only if the model A and B are identical. It is defined as

$$D(A\|B) = H_{A,B}(\mathbf{X}) = -\sum_{\mathbf{x}_i \in \mathbf{X}} p_A(\mathbf{x}_i) \log \frac{p_A(\mathbf{x}_i)}{p_B(\mathbf{x}_i)}. \tag{1}$$

To compute the KL divergence between two HMMs, one can generate a long data sequence using each of the models and compute the divergence empirically according to Eq.(1). This empirical approach is known to be time-consuming, and yet not accurate. An alternative method is to compute directly the KL divergence between two models based on the model parameters. Carrasco et al. [7] have derived an iterative solution for computing the KL divergence between two probablistic deterministic finite state automata (PDFA), given as

$$D(A\|B) = \sum_{q_i \in Q_A} \sum_{q_j \in Q_B} \sum_{\sigma \in \Sigma} c_{ij} p_A(q_i, \sigma) \log \frac{p_A(q_i, \sigma)}{p_B(q_j, \sigma)} \tag{2}$$

where Q_A and Q_B are the sets of states of PDFA A and PDFA B respectively. The coefficients c_{ij} are evaluated iteratively with guaranteed convergence. Related KL divergence evaluation method is used by Tholland *et al.* [8] to infer the structure of probablistic deterministic finite automata (PDFA).

Inspired by the MDI algorithm, we have attempted to apply a similar approach for inducing HMM's topology. However, it turns out that a similar iterative solution does not exists for HMM. The main reason which makes Carrasco's [7] work on PDFA but not applicable to HMM is that every PDFA can generate a stochastic *deterministic* regular language (SDRL). An HMM, however, can generate an observed sequence via multiple possible paths of state transitions and this makes it non-deterministic. So, it is still an open research issue for computing the KL divergence between all kinds of HMMs efficiently. In the following subsection, we propose a learning algorithm that compares the entropies of HMMs in an indirect manner.

3.2 The Proposed Learning Algorithm

To compute the KL divergence between the two HMMs, we first convert them to probabilistic deterministic finite state automata (PDFA) using the well-known subset construction algorithm. Then, the KL divergence between the two converted PDFAs can be computed (using Carrasco's solution [7]) as the approximated value of the KL divergence of the two HMMs. In addition, based on the observation that the change in the structure of the consecutive model candidates during the state merging process are normally localized, we suggest to further restrict the KL divergence comparsion to only the local structure of the models (see Figure 3). With that arrangement, further speed-up is resulted, which however has to trade off with the sacrificed accuracy.

The searching strategy in the topology space greatly affects the computational complexity of the corresponding algorithm. In our proposed algorithm, instead of randomly picking two states to merge as practised by Stolcke in [5], we adopt a search strategy similar to that of MDI [8], which is more efficient via the use of heuristics. Furthermore, an orthogonal direction to address the search complexity is by adding domain-specific constraints to narrow down the search space. For information extraction, the states of the HMM are corresponding to the different target information. Constraints of allowing only the states with same labels to be merged can be incorporated. to pursue [6]. Lastly, as the actual emission vocabulary is much larger than that of the training set, absolute discounting smoothing is adopted to alleviate the problem.

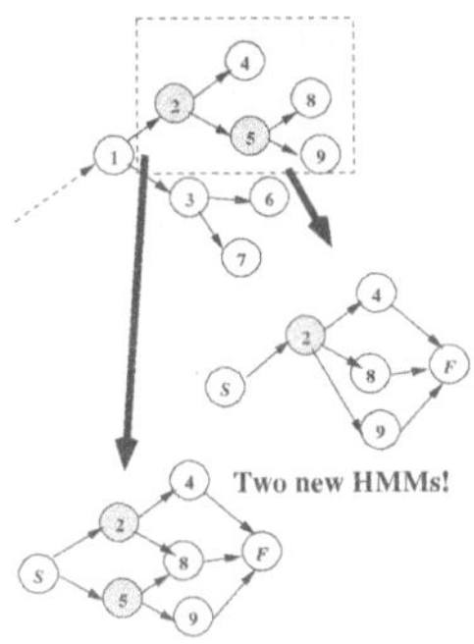

Fig. 3. Two new HMMs constructed by merging two candidate states.

4 Experimental Results

In order to evaluate the proposed topology learning algorithm, we have tested it using a data set containing 100 computer-related job ads. postings Each of them contains more than 200 words. A ten-fold cross-validation is adopted to reduce the sampling bias. According to Table 1, the proposed HMM topology learning system outperforms the well-known RAPIER system [2] as well as a baseline HMM model which merges ALL states of the same type in the initial model. Also, the results show that even a simple baseline HMM model can perform much better than the rule-based system. This demonstrates the advantage of the high representation power of HMM. Also, the results show that the use of HMM topology learning algorithm instead of the baseline HMM model and that the proposed local KL divergence is and effective measure for guiding the learning process. Regarding stemming, according to our experimental results, it makes no significant difference to the overall performance. We guess the effect of stemming may reveal when the training size increases.

Table 1. Performance comparision in extracting the field "language". (# of training = 10, # of validation = 10, # of testing = 80)

System	Initial no. of states	Final no. of states	Stem ?	Precision (%)	Recall (%)	F-Measure (%)
Baseline HMM	530	4	Y	66.3	44.9	53.4
Baseline HMM	530	4	N	66.7	44.7	53.4
Proposed System (α=0.35)	530	21	Y	89.7	55.3	68.4
Proposed System (α=0.35)	530	14	N	87.6	56.4	68.6
RAPIER*	-	-	-	50.8	25.0	33.5

5 Conclusions

An HMM topology learning algorithm using a local KL divergence has been proposed in this paper. Preliminary experimental results are encouraging for applying the methodology to information extraction tasks. In particular, it outperforms a simple HMM model and also the well-known RAPIER system. For future work, more extensive experiments should be conducted for performance comparision. Also, how to further speed up the learning time is also crucial for making the proposed algorithm to be able to scale up. One important issue in the proposed learning algorithm is how to set the threshold for accepting a candidate merge. More research effort along the direction is needed.

References

1. Muslea, I.: Extraction patterns for information extraction tasks: A survey. In: Proceedings of AAAI-99 Workshop on Machine Learning for Information Extraction, Orlando, Florida (1999)
2. Califf, M.: Relational Learning Techniques for Natural Language Information Extraction. PhD thesis, Department of Computer Science, University of Texas, Austin, TX. (1998)
3. Leek, T.R.: Information extraction using hidden Markov models. Master's thesis, UC San Diego (1997)
4. Seymore, K., McCallum, A., Rosenfeld, R.: Learning hidden markov model structure for information extraction. In: AAAI Workshop on Machine Learning for Information Extraction. (1999)
5. Stolcke, A., Omohundro, S.M.: Best-first model merging for hidden Markov model induction. Technical Report TR-94-003, International Computer Science Institue, 1947 Center Street, Berkeley, CA (1994)
6. Freitag, D., McCallum, A.: Information extraction with HMM structures learned by stochastic optimization. In: AAAI/IAAI. (2000) 584–589
7. Carrasco, R.C.: Accurate computation of the relative entropy between stochastic regular grammars. Informatique Theorique et Applications **31** (1997) 437–444
8. Thollard, F., Dupont, P., de la Higuera, C.: Probabilistic DFA inference using Kullback-Leibler divergence and minimality. In: Proc. 17th International Conf. on Machine Learning, Morgan Kaufmann, San Francisco, CA (2000) 975–982

A Non-parametric Wavelet Feature Extractor for Time Series Classification

Hui Zhang[1], Tu Bao Ho[1], and Mao Song Lin[2]

[1] Japan Advanced Institute of Science and Technology, Tatsunokuchi, Ishikawa, 923-1292 Japan,
[2] Southwest University of Science and Technology, Mianyang, Sichuan, 621002 China
{zhang-h,bao}@jaist.ac.jp lms@swust.edu.cn

Abstract. Many representation schemes for time series have been proposed and most of them require predefined parameters. In case of classification, the accuracy is considerably influenced by these predefined parameters. Also, the users usually have difficulty in determining the parameters. The aim of this paper is to develop a representation method for time series that can automatically select the parameters for the classification task. To this end, we exploit the multi-scale property of wavelet decomposition that allows us to automatically extract features and achieve high classification accuracy. Two main contributions of this work are: (1) selecting features of a representation that helps to prevent time series shifts, and (2) choosing appropriate features, namely, features in an appropriate wavelet decomposition scale according to the concentration of wavelet coefficients within this scale.

1 Introduction

Many algorithms have been proposed for mining time series data [14]. Among the different time series mining domains, time series classification is one of the most important domains. Time series classification has been successfully used in various application domains such as medical data analysis, sign language recognition, speech recognition, etc. For efficiency and effectiveness, most of the proposed methods classify time series on high level representations of time series instead of classifying the time series directly. These representations include Fourier Transforms [1], Piecewise Linear Representation (PLR) [15], Piecewise Aggregate Approximation [13,17], Regression Tree Representation [9], Haar Wavelets [4], Symbolic Representation [16], etc.

Most of the proposed representation schemes in the literature require predefined parameters. In case of classification, the accuracy is considerably influenced by these predefined parameters. The selection is not trivial or easy for the users, and as a consequence, the users usually have difficulty in determining the parameters.

In this paper we introduce a time series representation by Haar wavelet decomposition. The problem of time series classification is tackled by means of

H. Dai, R. Srikant, and C. Zhang (Eds.): PAKDD 2004, LNAI 3056, pp. 595–603, 2004.

selecting features of the representation. We propose a novel feature extractor to extract the approximation and change amplitudes of time series. The Euclidean distance between the features of shifted time series and original time series is smaller than those between the raw data. The appropriate features, i.e., features within the appropriate wavelet decomposition scale, are chosen by the concentration of features. The appropriate features also have been used for noise reduction. Corresponding time series classification algorithms will also be addressed.

The rest of this paper is organized as follows. Section 2 briefly discusses background material on time series classification and Haar wavelet decomposition. Section 3 introduces our time series representation and classification algorithms. Section 4 contains a comparative experimental evaluation of the classification accuracy with the proposed approach and other approaches. Section 5 introduces the related work. Section 6 gives some conclusions and suggestions for future work.

2 Background

In order to frame our contribution in the proper context we begin with a review of the concept of time series classification and the process of Haar wavelet decomposition.

2.1 Time Series Classification

A time series is a sequence of chronological ordered data. Time series can be viewed as a series of values of a variable that is a function of time t, that is, $X_t = \{x_0, x_1, \ldots, x_{n-1}\}$. For simplicity, this paper assumes the observed values are obtained with equal time interval 1.

Given a training time series data set $D = \{d_1, d_2, \ldots, d_n\}$, where each sample d_i is a time series associated with a value of class attributes $c \in C$. The task of time series classification is to classify a new testing time series x by a model constructed on D.

2.2 Haar Wavelet Decomposition

Wavelet transform is a domain transform technique for hierarchically decomposing sequences. It allows a sequence to be described in terms of an approximation of the original sequence, plus a set of details that range from coarse to fine. The property of wavelets is that the broad trend of the input sequence is preserved in approximation part, whereas the localized changes are kept in detail parts. No information will be gained or lost during the decomposition process. The original signal can be fully reconstructed from the approximation part and the detail parts.

Haar wavelet is the simplest and most popular wavelet given by Haar. The benefit of Haar wavelet is its decomposition process has low computational complexity. Given a time series with length n, where n is an integral power of 2, the complexity of Haar decomposition is $O(n)$. The concrete mathematical foundation of Haar wavelets can be found in [3].

The length of input time series is restricted to an integer power of 2 in the process of wavelet decomposition. The series will be extended to an integer power of 2 by padding zeros to the end of time series if the length of input time series doesn't satisfy this requirement. Therefore, the number of decomposing scales for the input time series is $log_2(n)$, here n is the length of zero-padded input series.

The structure of decomposed hierarchical coefficients are shown in Table 1.

Table 1. The hierarchical wavelet coefficients

Scale	Coefficients				
0	A_0				
1	A_1				D_1
2	A_2			D_2	D_1
...	...			D_2	D_1
$\log_2(n)$	A_k	D_k	...	D_2	D_1

Scale 0 is the original time series. To calculate the approximation coefficients $A_j = \{a_{0,j}, a_{1,j}, \ldots, a_{n-1,j}\}$ and detail coefficients $D_j = \{d_{0,j}, d_{1,j}, \ldots, d_{n-1,j}\}$ within scale j, the approximation part of scale $j-1$ is divided into an even part including even indexed samples: $even(A_{j-1}) = \{a_{0,j-1}, a_{2,j-1}, \ldots, a_{2n-2,j-1}\}$ and an odd part including odd indexed samples: $odd(A_{j-1}) = \{a_{1,j-1}, a_{3,j-1}, \ldots, a_{2n-1,j-1}\}$. Approximation coefficients A_j are calculated as

$$A_j = \frac{1}{\sqrt{2}}(even(A_{j-1}) + odd(A_{j-1})).$$

The detail coefficients within scale j are calculated as

$$D_j = \frac{1}{\sqrt{2}}(even(A_{j-1}) - odd(A_{j-1})).$$

3 Appropriate Feature Extraction for Wavelet Based Classification

We propose a scheme of wavelet based classification algorithm shown as follows:

1. Decomposing the training data set and a testing data with wavelets;
2. Extracting features from the wavelet coefficients;
3. Constructing a model on the features of training data set;
4. Classifying the testing data by the model;

The performance of classification will depend on the feature extractor and the classification algorithm. We introduce a time series representation and corresponding feature extraction in section 3.1 and section 3.2. We suggest two different classification algorithms in section 3.3 and section 3.4.

3.1 Time Series Representation and Feature Extraction

The concatenation of decomposed wavelet coefficients of a time series $X = \{x_0, x_1, \ldots, x_{n-1}\}$ to a particular scale $k \in \{1, 2, \ldots, log_2(n)\}$ shown in (1) is a representation of X.

$$W_k^X = [A_k^X, D_k^X, \ldots, D_2^X, D_1^X] \tag{1}$$

The Euclidean distance of two time series $X = \{x_0, x_1, \ldots, x_{n-1}\}$ and $Y = \{y_0, y_1, \ldots, y_{n-1}\}$ is

$$Disc(X, Y) = \sqrt{\sum_{i=0}^{n-1}(x_i - y_i)^2},$$

and the Euclidean distance $Disc(W_k^X, W_k^Y)$ between corresponding wavelet coefficients in a particular scale k transformed by X and Y is

$$\sqrt{\sum_i (a_{i,k}^X - a_{i,k}^Y)^2 + \sum_{j=1}^k \sum_i (d_{i,j}^X - d_{i,j}^Y)^2}.$$

The Euclidean distance is preserved through Haar wavelet transform, i.e., $Disc(X, Y) = Disc(W_k^X, W_k^Y)$ [4].

Note that the detail coefficients $D_j^X = \{d_{0,j}, d_{1,j}, \ldots\}$ contain local changes of time series. Thus $|D_j^X| = \{|d_{0,j}|, |d_{1,j}|, \ldots\}$ denote the amplitude of local changes. The concatenation of decomposed wavelet approximation coefficients A_k^X and the absolute value of decomposed wavelet detail coefficients $|D_j^X|, \forall j = 1, 2, \ldots, k$ to a particular scale k, $k \in \{1, 2, \ldots log_2(n)\}$ of a time series X are defined as features shown as following.

$$F_k^X = [A_k^X, |D_k^X|, \ldots, |D_2^X|, |D_1^X|] \tag{2}$$

This definition helps to overcome the well-known problem posed by the fact that wavelet coefficients are sensitive to shifts of series.

The Euclidean distance $Disc(F_k^X, F_k^Y)$ between the features of two time series X and Y can be defined as

$$\sqrt{\sum_i (a_{i,k}^X - a_{i,k}^Y)^2 + \sum_{j=1}^k \sum_i (|d_{i,j}^X| - |d_{i,j}^Y|)^2}.$$

Because $|x| - |y| \leq x - y$, we obtain $Disc(F_k^X, F_k^Y) \leq Disc(W_k^X, W_k^Y)) = Disc(X, Y)$. If X and Y denote the original time series and shifted time series respectively, this inequation is still tenable.

3.2 Appropriate Scale Selection

Notice that with wavelet decomposition, we have multiple choices of features by the multi-scale property of wavelets. It is intuitive to find out which features by a given scale should better than others for classification. If the energy of a scale is concentrated in a few coefficients then only some important coefficients can represent the whole coefficients with low error. This scale may give valuable information for classification.

The Shannon entropy which is a measure of impurity within a set of instances can describe the concentration of coefficients within a scale. The Shannon entropy is defined in (3)

$$H_F = -\sum_i p_i \log_2 p_i \tag{3}$$

The appropriate decomposing scale is defined as the scale with the lowest entropy. The appropriate features of a time series is defined as the wavelet coefficients within appropriate decomposing scale.

$$\text{Appropriate scale} = \operatorname*{argmin}_k \left(-\sum_i p_{i,k} \log_2 p_{i,k}\right) \tag{4}$$

here $p_{i,k} = \frac{|F_{i,k}^X|}{\sum_{i=1}^n |F_{i,k}^X|}$ is the proportion between the absolute value of a coefficient in a feature and the sum of absolute values of whole feature. $p_{i,k}$ has the property $\sum_i p_{i,k} = 1$ and $p_{i,k} \geq 0$.

3.3 Classification Algorithm

For two series with the same length, their corresponding appropriate scales may not be equal. We can't compare the similarity of two set of appropriate features directly because the meaning of each data entry is different. For example, given a time series $X = \{x_0, x_1, x_2, x_3\}$ with appropriate scale 1 and a time series $Y = \{y_0, y_1, y_2, y_3\}$ with appropriate scale 2. The features of series X are $F_1(X) = \{a_{0,1}, a_{1,1}, |d_{0,1}|, |d_{1,1}|\}$ and the features of series Y are $F_2(Y) = \{a_{0,2}, |d_{0,2}|, |d_{0,1}|, |d_{1,1}|\}$. Comparing detail coefficients with approximation coefficients will induce errors and meaningless.

To avoid this problem, we merge the distance of features within different appropriate scales. The classification algorithm will be implemented by means of 1-nearest neighbor algorithm, which we called WCANN (Wavelet Classification Algorithm based on 1-Nearest Neighbor).

Table 2 shows an outline of the classification algorithm. The input is S (the training time series set) and x (a new emerged testing time series). The output is x_c, the label of x.

x will be labeled as a member of a class if and only if, the distance between x and one instance of the class is smaller than between other instances.

The distance of two features is replaced by the average of distance computed on two features with different appropriate scale respectively. The distance between features of time series X and features of time series Y is denoted as

$$Disc(F^X, F^Y) = (Disc(F_m^X, F_m^Y) + Disc(F_n^X, F_n^Y))/2,$$

where m is the appropriate scale of features X and n is the appropriate scale of features Y.

Table 2. The WCANN algorithm (S, x)

For each training example S_i, calculating its appropriate scale m_i and corresponding appropriate features $F^{S_i}_{m_i}$;
Given a testing instance x, calculating its appropriate scale n and appropriate features F^x_n;
best-so-far = inf;
for i = 1 to length(S) **do**
  $Disc(F^{S_i}_{m_i}, F^x_{m_i})$ = distance between F^{S_i} and F^x on scale m_i;
  $Disc(F^{S_i}_n, F^x_n)$ = distance between F^{S_i} and F^x on scale n;
  $Disc(F^{S_i}, F^x) = (Disc(F^{S_i}_{m_i}, F^x_{m_i}) + Disc(F^{S_i}_n, F^x_n))/2$
  if $Disc(F^{S_i}, F^x) <$ best-so-far **then**
    pointer-to-best-series $k = i$;
    best-so-far = $Disc(F^{S_i}, F^x)$;
  end if
end for
return x_c = the label of k;

3.4 Noise Reduction on Features

The idea of wavelet noise shrinkage is based on the assumption that the amplitude of the spectra of the signal to be as different as possible for that of noise. If a signal has its energy concentrated in a small number of wavelet coefficients, these coefficients will be relatively large compared to the noise that has its energy spread over a large number of coefficients. This allows thresholding of the amplitude of the coefficients to separate signals or remove noise. The thresholding is based on a value τ that is used to compare with all the detail coefficients. The definition of appropriate scale defined in section 3.2 helps to reduce noise because the energy of signal gets concentrated in a few coefficients and the noise remains spread out in that scale. Hence it is convenient to separate the signal from the noise by keeping large coefficients (which represent signal) and delete the small ones (which represent noise).

Donoho and Johnstone [8] gave the threshold $\tau = \sigma_n\sqrt{2logN}$, here the σ_n is the standard variation of noise, and N is the length of the time series. Because we don't know the σ_n of the time series in advance, we estimate it by robust median estimation of noise introduced in [8]. The robust median estimation is the median absolute deviation of the detail wavelet coefficients at scale one, divided by 0.6745.

Usual hard thresholding algorithm is used in this paper that is a process of setting the value of detail coefficients whose absolute values are lower than the threshold to zero [7]. The hard thresholding algorithm for features defined in (4) is illustrated in (5).

$$Thre(|D_{i,j}|) = \begin{cases} |D_{i,j}|, & |D_{i,j}| > \tau \\ 0, & |D_{i,j}| \le \tau \end{cases} \tag{5}$$

The classification algorithm for noise reduced coefficients WCANR (Wavelet Classification Algorithm with Noise Reduction) is similar with WCANN algo-

Table 3. The error rates for various feature extraction

Approach	CBF	CC
WCANN	0.0026	0.0067
WCANR	0.0026	0.0067
WC-NN	0.0026	0.0133
1-NN	0.0026	0.0133
Euclidean Distance	0.003	0.013

Table 4. The error rates for various similarity measures

Approach	CBF	CC
Euclidean Distance	0.003	0.013
Aligned Subsequence	0.451	0.623
Piecewise Normalization	0.130	0.321
Autocorrelation Functions	0.380	0.116
Cepstrum	0.570	0.458
String (Suffix Tree)	0.206	0.578
Important Points	0.387	0.478
Edit Distance	0.603	0.622
String Signature	0.444	0.695
Cosine Wavelets	0.130	0.371
Hölder	0.331	0.593
Piecewise Probabilistic	0.202	0.321

rithm. The only difference is the noise within appropriate features will be reduced before classification.

4 Experimental Results

For a fair comparison of methods, we followed the benchmarks introduced in [14]. We also used 1-Nearest Neighbor algorithm, evaluated by "leaving-one-out", and experimented on the same data sets as [14] used:

- Cylinder-Bell-Funnel (CBF): This data set has been used for classification by Kadous [12] , Geurts [9], Lin [16] and Keogh [14]. The aim is to discriminate three types of time series: cylinder(c), bell(b) and funnel(f). We generated 128 examples for each class with length 128 and time step 1 as in [14].
- Control Chart Time Series (CC): This data set has 100 instances for each of the six different classes of control charts. This data set has been used to validate clustering [2], classification [9,16,14]. The data set we used was downloaded from the UCI KDD Archive [10].

We evaluated 4 algorithms using different features extracted from the two data sets described above. 1-NN is the 1-nearest neighbor algorithm that uses the raw data [14]. WC-NN is the algorithm that uses 1-nearest neighbor algorithm with decomposed wavelet coefficients within highest scale [5]. WCANN is our proposed algorithm described in section 3.3. WCANR is our proposed algorithm described in section 3.4.

An unclassified instance is assigned to the same class as its closest match in the training set. The classification error rates about different feature extraction algorithms on the above two data sets are presented in Table 3. We only used Euclidean distance as the distance measure since it superiors other distance measures in classification on the above two data sets [14]. The error rates of various similarity measures with the raw data given by [14] are shown in Table 4. The 1-NN algorithm is actually the same as Euclidean distance algorithm in Table 3. The Euclidean distance algorithm shown in Table 3 is taken from Table 4.

On both data sets WCANN algorithm performs the same with WCANR algorithm and 1-NN algorithm performs the same with WC-NN algorithm. The WCANN and WCANR algorithms achieve the same accuracy compared with 1-NN and WC-NN algorithms on CBF data set and higher accuracy on CC data set.

5 Related Work

There are a large number of techniques which have been proposed for efficiently time series querying based on wavelets. However, no proposed work concern selecting an appropriate scale. Our work is similar in spirit to the wavelet packet algorithm introduced by Coifman et al. [6]. The authors used entropy to select the best basis of wavelet packets. Our solution is different with the solution in [6] on two aspects. We use wavelets not wavelet packets to decompose data and we calculate entropy by defined features not by the energy of wavelet coefficients.

6 Conclusions and Future Work

We defined features on Haar wavelet coefficients that helps to overcome their sensitivity to shifted time series. We selected the appropriate scale of Haar wavelet decomposition for classification. We proposed a nearest neighbor classification algorithm using the derived appropriate scale.

We conducted experiments on two widely used test time series data sets and compared the accuracy of our algorithms and those of two other algorithms on Euclidean distance. Our algorithms outperform other two algorithms on one data set and take the same results on another data set.

We intend to extend this work to two directions. First, applying this approach to a real time series data set. We are in the process of working on hepatitis data set [11]. This hepatitis data set contains irregular time series data measured on nearly one thousand examinations. The target of mining this data set includes time series classification. Second, combining this work with distance-based clustering algorithm. In fact, the main idea of this work is selecting appropriate features of a time series for better similarity comparison. Clustering on these features instead of clustering on the time series data directly may get better results.

References

1. R. Agrawal, C. Faloutsos, and A. Swami. Efficient similarity search in sequence databases. In *Proceedings of the 4th Conference on Foundations of Data Organization and Algorithms*, pages 69–84, October 1993.
2. R. J. Alcock and Y. Manolopoulos. Time-series similarity queries employing a feature-based approach. In *Proceedings of the 7th Hellenic Conference on Informatics*, pages 1–9, August 1999.
3. C. S. Burrus, R. A. Gopinath, and H. Guo. *Introduction to Wavelets and Wavelet Transforms, A Primer*. Prentice Hall, Englewood Cliffs, NJ, 1997.
4. F. K. Chan and A. W. Fu. Harr wavelets for efficient similarity search of time-series: with and without time warping. *IEEE Trans. on Knowledge and Data Engineering*, 15(3):686–705, 2003.
5. K. P. Chan and A. W. Fu. Efficient time series matching by wavelets. In *Proceedings of the 15th Internation Conference on Data Engineering*, pages 126–133, March 1999.
6. R. R. Coifman and M. V. Wickerhauser. Entropy-based algorithms for best basis selection. *IEEE Trans. on Information Theory*, 38(2):713–718, 1992.
7. D. L. Donoho. De-noising by soft-thresholding. *IEEE Trans. on Information Theory*, 41(3):613–627, 1995.
8. D. L. Donoho and I. M. Johnson. Ideal spatial adaptation via wavelet shrinkage. *Biometrika*, 81:425–455, 1994.
9. P. Geurts. Pattern extraction for time series classification. In *Proceedings of the Principles of Data Mining and Knowledge Discovery,5th European Conference*, pages 115–127, September 2001.
10. S. Hettich and S. D. Bay. The uci kdd archive. http://kdd.ics.uci.edu, 1999.
11. T. B. Ho, T. D. Nguyen, S. Kawasaki, S. Q. Le, D. D. Nguyen, H. Yokoi, and K. Takabayashi. Mining hepatitis data with temporal abstraction. In *Proceedings of the 9th ACM International Conference on Knowledge Discovery and Data Mining*, pages 369–377, August 2003.
12. M. W. Kadous. Learning comprehensible descriptions of multivariate time series. In *Proceedings of the 6th International Conference on Machine Learning*, pages 454–463, September 1999.
13. E. Keogh, K. Chakrabarti, M. Pazzani, and S. Mehrotra. Dimensionality reduction of fast similarity search in large time series databases. *Journal of Knowledge and Information System*, 3:263–286, 2000.
14. E. Keogh and S. Kasetty. On the need for time series data mining benchmarks: A survey and empirical demonstration. *Data Mining and Knowledge Discovery*, 7(4):349–371, 2003.
15. E. Keogh and M. Pazzani. An enhanced representation of time series which allows fast and accurate classification, clustering and relevance feedback. In *Proceedings of the 4th International Conference of Knowledge Discovery and Data Mining*, pages 239–241, August 1998.
16. J. Lin, E. Keogh, S. Lonardi, and B. Chiu. A symbolic representation of time series, with implications for streaming algorithms. In *Proceedings of the 8th ACM SIGMOD Workshop on Research Issues in Data Mining and Knowledge Discovery*, pages 2–11, June 2003.
17. B. K. Yi and C. Faloustos. Fast time sequence indexing for arbitrary l_p norms. In *Proceedings of the 26th International Conference on Very Large Databases*, pages 385–394, September 2000.

Rules Discovery from Cross-Sectional Short-Length Time Series*

Kedong Luo[1], Jianmin Wang[2], and Jiaguang Sun[1,2]

[1] Department of Computer Science and Technology
Tsinghua University, Beijing, China
lkd99@mails.tsinghua.edu.cn
[2] School of Software, Tsinghua University, Beijing, China

Abstract. The cross-sectional time series data means a group of multivariate time series each of which has the same set of variables. Usually its length is short. It occurs frequently in business, economics, science, and so on. We want to mine rules from it, such as `"GDP rises if Investment rises in most provinces"` in economic analysis. Rule mining is divided into two steps: events distilling and association rules mining. This paper concentrates on the former and applies Apriori to the latter. The paper defines event types based on relative differences. Considering *cross-sectional property*, we introduce an ANOVA-based event-distilling method which can gain proper events from cross-sectional time series. At last, the experiments on synthetic and real-life data show the advantage of ANOVA-based event-distilling method and the influential factors, relatively to the separately event-distilling method.

1 Introduction

The cross-sectional time series data means a group of multivariate time series each of which have the same set of variables. Usually actual cross-sectional time series have a short length. For example, macroeconomic yearly time series is composed of time series from all provinces (Figure 1) within a few years; each province has yearly data of GDP, Investment and so on. More examples occur frequently in large-scale company, organization, and government, including quarterly marketing data from all branch of country-wide company within several years, monthly taxation data from every county substation of a provincial revenue department within twelve months. The same attributes in all sections have the same meaning and behave similarly, which is called *cross-sectional property*.

Thus, the question is to mine association rules from cross-sectional time series. We decompose the question into two steps: events distillation and association rule mining. In the first step, events are distilled from original time series with respect to cross-sectional property. The step is the emphasis of the paper. In

* This work is supported by both the National High Technology Research and Development Program of China (863 Program) under Grant $No.2002AA444120$ and the National Key Basic Research and Development Program of China (973 Program) under Grant $No.2002CB312006$.

H. Dai, R. Srikant, and C. Zhang (Eds.): PAKDD 2004, LNAI 3056, pp. 604–614, 2004.

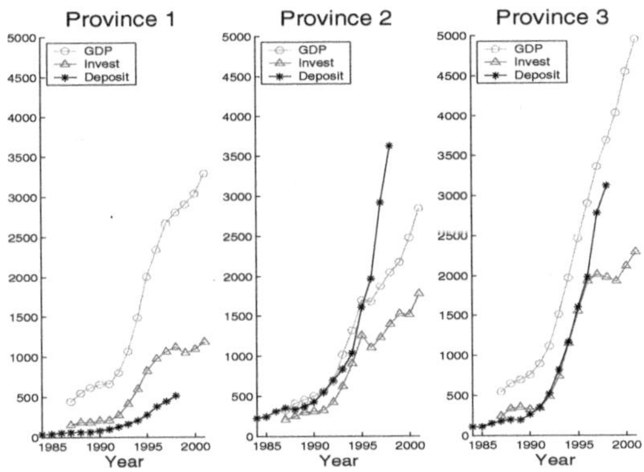

Fig. 1. An Example from Cross-Province Time Series of Chinese Macroeconomic

the second step, interesting rules are mined from the distilled events. We adopt the existing algorithm Apriori in this step.

This paper employs an event definition based on relative differences. This definition is fit for short-length time series and is easy to explanation. In addition, it is convenient to distill events with the relatively simple definition.

Let us take into account *cross-sectional property* via Figure 1 intuitively. Each attribute behaves similarly across provinces. Considering relative differences, or called increasing ratios, some sections may similar ratios on some attribute, e.g. `Province 1` and `Province 2` on `GDP`; but the ratios may not be similar across all sections, e.g. faster increasing of `Province 3` on `GDP`.

So that, we introduce an ANOVA-based algorithm to do event distillation by considering cross-sectional property. More useful information about an attribute will be distilled if we consider the same attributes across sections holistically rather than if we consider them separately. In order to do comparison, the paper gives the separately distilling method at first, and then the ANOVA-based method.

Consider ANOVA-based event-distilling algorithm. Firstly the relative differences of all attributes in all sections are computed. Secondly we use ANOVA to group sections in a greedy manner. At last, for each group of sections, we estimate the distribution parameters of relative differences and distill the events via discretization on probabilistic distribution.

The remainder of the paper is organized as follows. Section 2 defines this problem exactly. Section 3 describes how to pick up proper events from original cross-sectional data. In Section 4, experiments on synthetic and real-life data show the advantage of ANOVA-based algorithm. Section 5 gives the conclusion.

1.1 Related Work

Statistics has introduced some cross-sectional time series analysis methods, such as Dummy Variable Model and Error Components Model [5]. These approaches analyze inter-attribute or intra-attribute relationships via global regression models. Unlike them, the paper does not seek global model and analyze relationship in a relatively non-parametric manner. The difference is as the same as the one between statistical and data mining methods of generic time series analysis.

There are some data mining approaches on rule mining from time series, such as [3,6]. These papers mine rules from time series by two steps: event distillation and rule mining, which our method follows. Event distillation is also called *time series discretization*. But each paper has different methods on the two steps, according to its own problem.

The event definition in the paper is based on relative differences, so does [1]. [3] has different event definition that takes special shapes with fixed length as event types. But, by the definition, we might get too few events from short-length time series. For example, for 15-length time series, we get *6* events with 10-length shape in one time series, but *14* events based on first-order difference.

2 Notions

An **event type** is a specified discrete data type, which usually describes a certain character of time series. For instance, a set $\mathcal{E}$ of event types may be {**up**, **down**, **stable**,**UP**, **DOWN**, **STABLE**} as Table 1 shows.

Given a cross-sectional time series, let $\mathcal{T}$ be the **time domain**, $\mathcal{S}$ be the set of **sections**, and $\mathcal{A}$ be the set of **attributes**. T is a limited set, since time is discrete. For example, in cross-province macroeconomic time series, T is $\{1978, 1979, ..., 2003\}$, S is {`Anhui`, `Beijing`, ..., `Zhejiang`}, and A is {`GDP`, `Investment`, `Payout`, ...}

An event type set $\mathcal{E}$, an attribute set $\mathcal{A}$, a time set $\mathcal{T}$, and a section set $\mathcal{S}$ are given. Any **event** belongs to $A \times \mathcal{E}$, such as "`GDP` **up**" or "`Investment` **stable**". All **events** from one section in certain time form an **event set** $E = \{e_1, e_2, ..., e_n\}$ where $e_i \in \mathcal{A} \times \mathcal{E}$. Thus the number of the **event sets** is $|\mathcal{S}| * |\mathcal{T}|$.

Association Rules Mining. A association rule, can be formulated as an implication of $A \Rightarrow B$, where $A, B \subset \mathcal{A} \times \mathcal{E}$ and $A \cap B = \emptyset$. The first illustration of mining association rules is shopping basket analysis, where [2,7] give efficient algorithm called Apriori. An association rule, such as `"if GDP increases, then Investment increases"`, can be formulated as an implication of $A \Rightarrow B$, where $A, B \subset \mathcal{A} \times \mathcal{E}$ and $A \cap B = \emptyset$. There are two measures: *support* and *confidence*. Given an association rule $A \Rightarrow B$, the *support* is $P(A \cup B)$ and the *confidence* is $P(A \cup B)/P(A)$. In addition, *support of sections* is introduced to measure how many sections the rule covers. That is, *support of sections* = *Number of Rule-Covered Sections / Number of Total Sections.*

Analysis of Variance or ANOVA for short, is a statistical method to decide whether more than two samples can be considered as representative of the same population [4]. The paper adopts the basic One-Way ANOVA among many variations. ANOVA returns *p-value* which is between 0 and 1. The critical *p-value* is usually 0.05 or 0.01. That is, if *p-value* is less than the critical value, data are not likely to be sampled from the same population.

Independently and Identically Distributed or i.i.d. for short. That more than two samples are i.i.d. means that those are sampled from the same population and those are not correlated to one another.

Table 1. Event Types

Symbol	Description	Condition
up	increasing transition is high enough	$d_t^{(1)} > \mu^{(1)} + \sigma^{(1)} T$
down	decreasing transition is high enough	$d_t^{(1)} < \mu^{(1)} - \sigma^{(1)} T$
stable	increasing or decreasing is slight	others
UP	increasing is accelerated or decreasing is decelerated	$d_t^{(2)} < \mu^{(2)} + \sigma^{(2)} T$
DOWN	increasing is decelerated or decreasing is accelerated	$d_t^{(2)} < \mu^{(2)} - \sigma^{(2)} T$
STABLE	increasing or decreasing changes slightly	others

Comment: $d^{(n)}$ means n-order relative difference of x ($n \geq 1$), that is $d_t^{(1)} \equiv (x_t - x_{t-1})/x_{t-1}$, $d_t^{(2)} \equiv d_t^{(1)} - d_{t-1}^{(1)}$. Then $\mu^{(1)}$ and $\sigma^{(1)}$ are mean and standard deviation of the population generating relative differences of some attribute. Similarly $\mu^{(2)}$ and $\sigma^{(2)}$ are for 2-order relative differences. T is the significant threshold.

3 Events Distillation

Table 1 gives two groups of event types. The first three are derived from 1-order relative differences, and the last three are derived from 2-order ones. The event types have intuitive meanings as the descriptions show.

Here the relative differences are assumed in accord with some Gaussian distribution. For generic time series, we can assume relative differences in *one* section are i.i.d., which is the assumption of separately event-distilling method. On the other hand, considering *cross-sectional property*, we can assume relative differences in *some* sections are i.i.d., which is the assumption of ANOVA-based event-distilling method. The following subsections give these two methods, with discussion of assumptions. The ANOVA-based one is the emphases of the paper, and the separately one is mainly for comparison.

3.1 Separately Distilling

[1] has tried to distill events by relative differences. But its events are discriminated by uniform threshold which is set manually. Manual setup of threshold is not proper for multivariate time series. It sounds reasonable that the relative difference threshold of each attribute should be set separately. That is, the threshold of each attribute should be computed automatically according to its Gaussian distribution of relative differences and the significant threshold T.

This paper utilizes *relative difference* rather than *difference* because of cross-sectional diversity. The value of change of an attribute in a section is strongly depended on its base.

Algorithm. Given x, time series data of one attribute in ***one*** section, List 1.1 gives the algorithm to distill events via (1-order) relative differences. Distilling events of 2-order relative difference is alike. E is the array of event sets.

Listing 1.1. Separately Distilling Algorithm

```
Set T;
Compute d[2...Length];
mu:=mean(d);
sigma:=std_deviation(d);
for l:=2 to Length do begin
  if d[l]>mu+sigma*T then
    E[l].addEvent(up);
  else if d[l]<mu-sigma*T then
    E[l].addEvent(down);
  else
    E[l].addEvent(stable);
  end;
end;
```

3.2 ANOVA-Based Distilling

The separately distilling algorithm performs badly when the length of time series is short. It is the reason that Gaussian parameters are difficult to estimate accurately because of small sample size. Section 4 shows this issue using synthetic data.

But lots of cross-sectional time series are short-length. Fortunately the *cross-sectional property* is helpful. That is, we can assume that relative differences of one attribute of *some* sections are i.i.d., and then use holistically these sections to estimate the distribution of relative differences.

But how to determine which sections have i.i.d. relative differences? The statistical method ANOVA can do us a favor. Section 1.1 gives a brief introduction if you are unfamiliar with ANOVA. For example, we can choose `GDP` attribute of any sections, and compute *p-value* by ANOVA on grouped relative differences. If *p-value* is large enough (e.g. > 0.05), these relative differences are significantly i.i.d.

The assumption is also consistent with real-life data. Figure 2 illustrates this assumption intuitively by one attribute from yearly macroeconomic time series of China. The *p-value* of ANOVA test for all provinces is around $5e10^{-6}$, which means significantly *not* i.i.d. If we take Province 3, 5, 7, 23, and 24 as one group, other provinces as the other group, the two groups both have greater-than-0.05 *p-value*. In this way proper events can be distilled.

Given the method of ANOVA, we can use full-search method to group sections. That is, test all possible sections combinations and choose the best one. Obviously it can not avoid the combinational explosion.

Fortunately, ANOVA is concerning about the means, so that a greedy algorithm may be applied considering the sequential order of means of relative

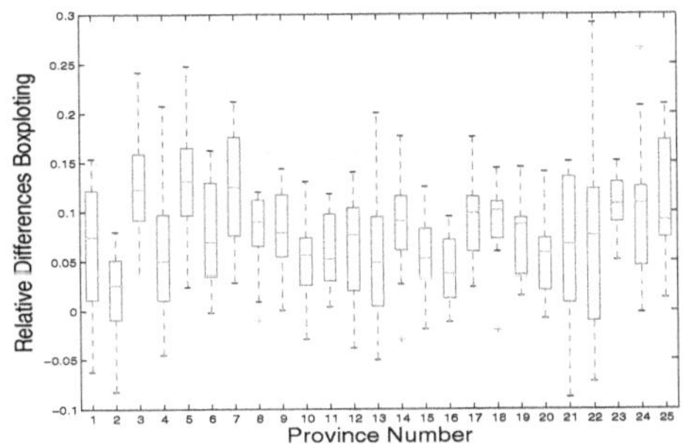

Fig. 2. Differences boxploting of some attribute in Chinese macroeconomic time series.

differences. In the first step, we choose the section with the littlest mean of relative differences. In the second step, we add the section with the littlest mean within the sections left; and then, test ANOVA on relative differences of the chosen sections. If they are significantly i.i.d., we repeat the second step; otherwise we take these sections as an group without the one added newly, and go to the first step. When all sections have been grouped, we can estimate the Gaussian parameters and distill events of relative difference group by group.

Algorithm. Given x, time series data of one attribute in ***all*** section, List 1.2 gives the ANOVA-based algorithm to distill events via (1-order) relative differences. Distilling events of 2-order relative difference is alike. E is the array of event sets. The function GenerateEvents is similar to List 1.1 then omitted.

Listing 1.2. ANOVA-based Distilling Algorithm

```
Set T and Pmin;
Compute d[1...Areas][2...Length];
/*Sort Sections by Mean*/
for a:=1 to NumArea do begin
  mu[a]:=mean(d[a]);
end;
idx=getSortedIndex(mu);
/*A Greedy Algorithm with ANOVA*/
i1:=1; i2:=1;
while(i2<NumArea) begin
  pvalue=ANOVA(d[idx[i1,i2+1]]);
  if pvalue>Pmin then
    i2:=i2+1;
  else
    GenerateEvents(i1,i2);
    i1:=i2+1; i2:=i1;
  end;
end.
if i1<i2 then GenerateEvents(i1,i2);
end.
```

3.3 Remarks on Advantage of ANOVA-Based Method

The previous subsection said that the ANOVA-based distilling algorithm behaves better than the separately one on cross-sectional time series. But how much advantage the ANOVA-based method has? And which factors can affect the advantage?

Regardless of characters in concrete applications, there are three factors in cross-sectional time series: number of sections, number of attributes, and length of time series. ANOVA-based method wants to use cross-sectional property to solve the small sample issue caused by short-length. So that, the cross-sectional property is not helpful yet when there are too few sections. On the other hand, the small sample issue might disappear when the length of time series is long enough. Thus the advantage of ANOVA-based distilling algorithm, relatively to separately one, will be significant if the length is short enough and number of sections is large enough. But it is not a rigorous condition as synthetic experiment in Section 4 shows.

4 Experimental Results

Both synthetic and real-life data are used in experiments. The experiments using synthetic data concerns about the performance of event-distilling algorithms and correlative factors. The experiments using real-life data gives an interesting practice in detail.

4.1 Synthetic Data

It is known that mining association rule in time series has 2 steps: event distillation and rule mining. The latter step has been studied well since we adopt the basic Apriori. So we do the performance analysis on event distillation step, which will make our analysis simpler and clearer.

We are interested in not only the absolute performance of ANOVA-based method, but also its advantage relative to separately method. Discussed in Section 3.3, the number of sections and the length of discrete time will affect the advantage of ANOVA-based method. Thus we will do controlled experiments respectively on these two factors. The absolute performance is also tested in the controlled experiments.

The first experiment concentrates on the effect of the number of sections. The synthetic cross-sectional time series data includes 5 attributes and has 10 time periods, which is long enough that the second experiment can testify. The number of sections is the variable. We generate the start value and relative differences of each attribute to form time series. And relative differences of each attribute in all sections are generated in accordance with the same Gaussian distribution. Different from the original assumption, relative differences of all sections share the same distribution. The reason is that we want to find how much number of sections affects distilling performance. In actual applications, after grouping the sections by ANOVA, it is number of sections in each group

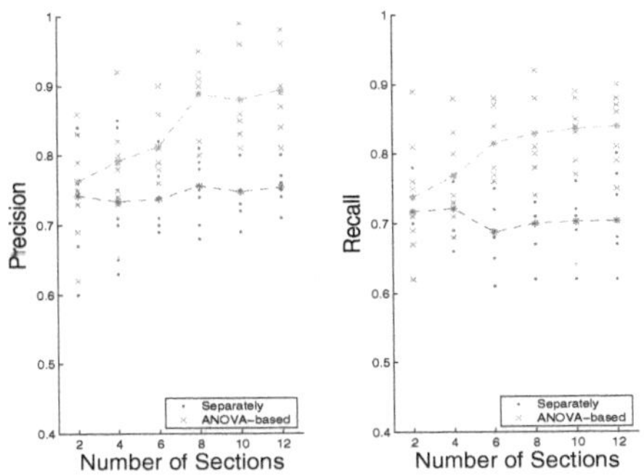

Fig. 3. Comparison by number of sections

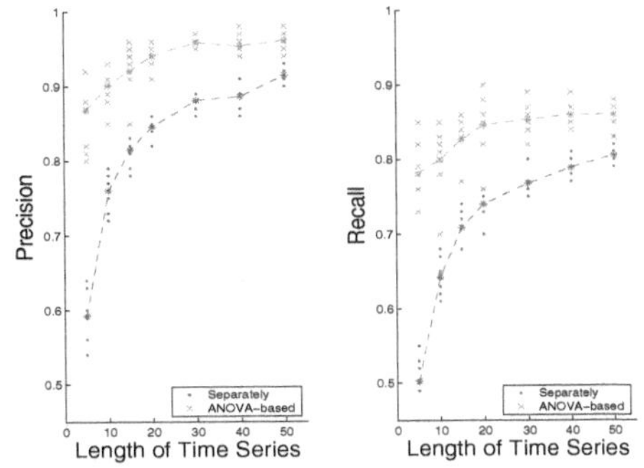

Fig. 4. Comparison by length of time series.

that affects distilling performance. In order to simulate really, we add some rules to these time series: the attribute 2 and 3 increase and decrease dependently on the attribute 1; the attribute 5 increases and decreases dependently on the attribute 4.

The performance of algorithms is the *precision-recall* measure. The *right* events are known because we know the real distributions. We call right events set R, distilled events set D. The precision-call is defined as

$$Precision = \frac{|R \cap D|}{|D|}, \quad Recall = \frac{|R \cap D|}{|R|}$$

where $|.|$ means the norm of a set.

The second experiment concentrates on the effect of length of time series. The synthetic cross-sectional time series data includes 5 attributes and has 20 sections, which is large enough that the first experiment can testify. The length of time series is the variable. The data generation in this experiment is similar to the last experiment. In order to simulate really, relative differences of an attributes in the first 10 sections share one distribution, and ones in the last 10 sections share the other distribution. The rules are as the same as ones in the first experiment.

The experimental results have random changes since the data is generated randomly. Thus we use scatter plots, where the means can illustrate performance.

Experimental Results. Consider Figure 3. The results of separately method are not influenced by the number of sections, so its mean holds on with little random change. The precision-recall of ANOVA-based method becomes better while the number of sections becomes larger. And more, the experimental results illustrate that the *cross-sectional property* is so strong even if there are just several sections. Consider Figure 4. The shorter the length of time series is, the more useful the *cross-sectional property* is. And the ANOVA-based method is still better even though the length is more than 50.

Usual cross-sectional time series, such as economic data, marketing data, or science data, may contain tens or more sections and have a short length. So we can usually take advantage of ANOVA-based event-distilling algorithm on the real-life data.

4.2 Real-Life Data

The real-life data is the cross-province yearly macroeconomic time series of China, which composes 25 provinces and more than 150 attributes. And the time lengths of attributes are between 12 and 18 because of lost values. We use just 25 provinces among 31 provinces of China because 6 provinces have too many lost values. There are still some lost values in the remaining 25 provinces, but they can not harm our method greatly and just mean losing some corresponding events.

Here we do comparison on association rules rather than events, because we do not know the *right* events in real-life data and comparison on rules is more understandable than that on events.

As similar as other technical analysis methods, whether a rule makes sense or not is dependent on whether it is meaningful to economics or not. That is, our method can behave as auxiliary tools to the economics.

Table 2 shows some rules with higher support and confidence. We just use **up**, **down**, **UP**, and **DOWN**, which are more meaningful than **stable** and **STABLE** in economics.

Table 2. Association Rules Mined from Chinese Macroeconomic Data

No.	Rules with **ANOVA-based** Distilling	Supp	Conf	Sect
1	Resident Consumption **up** ⇒ Total Consumption **up**	28%	92%	100%
2	Total Consumption **up** ⇒ Resident Consumption **up**	28%	91%	100%
3	GDP **down** ⇒ Total Retail **down**	26%	79%	100%
4	Total Retail **down** ⇒ GDP **down**	26%	77%	100%
5	Total Industrial Production **UP** ⇒ Heavy I. P. **UP**	26%	93%	100%
6	Heavy Industrial Production **UP** ⇒ Total I. P. **UP**	26%	91%	100%
7	Employers in State-Owned **up** ⇒ College Students **down**	24%	68%	100%
8	College Students **down** ⇒ Employers in State-Owned **up**	24%	67%	100%
9	Agricultural Production **down** ⇒ Sugar Output **down**	24%	87%	100%
10	Sugar Output **down** ⇒ Agricultural Production **down**	24%	91%	100%
No.	Rules with **Separately** Distilling	Supp	Conf	Sect
11	Resident Consumption **up** ⇒ Total Consumption **up**	18%	87%	76%
12	Total Consumption **up** ⇒ Resident Consumption **up**	18%	85%	76%
13	Service Industry **down** ⇒ GDP **down**	17%	76%	60%
14	GDP **down** ⇒ Service Industry **down**	17%	73%	60%
15	Total Retail **DOWN** ⇒ GDP **DOWN**	17%	70%	64%
16	GDP **DOWN** ⇒ Total Retail **DOWN**	17%	69%	64%
17	Total Retail **down** ⇒ Government Payout **down**	17%	77%	52%
18	Government Payout **down** ⇒ Total Retail **down**	17%	79%	52%

Comment: Sect is short for Support of Sections.

Considering rules distilled with ANOVA-based method, some are likely trivial or well-known. For example, the first two rules in Table 2 are well-known, because `Resident Consumption` is a major part of `Total Consumption`. Their confidences are commonly greater than 90%. Rule No.5 and No.6 mean that the increasing ratio change of `Total Industrial Production` is determined by that of `Heavy Industrial Production`, which is trivial. Rule No.3 and No.4 shows that `Total Retail` is important to `GDP`, which is trivial too.

Some rules are interesting, or even strange. For example, Rule No.9 and No.10 mean that `Sugar Output` is important to `Agricultural Production`, which has a little interesting. And Rule No.7 and No.8 look strange. The discussion about they can be left to domain experts.

All rules distilled with ANOVA-based method have 100% section cover rate, as well as relatively high support and confidence. In contrast, the rules distilled separately just cover some sections, as well as have relatively low support and confidence. The reason is that ANOVA-based distilling method distills events much more accurate than separately distilling method does.

5 Conclusion

This paper aims to mine interesting rules from cross-sectional time series. The main contribute of the paper is to introduce an ANOVA-based event-distilling algorithm, which is able to distill events accurately from cross-sectional short-length time series. Interesting and useful association rules can be mined on these events. Event types are defined based on relative differences of time series. We used the experiments on synthetic data to illustrate the advantage of ANOVA-based method, as well as the ones on real-life data to present a vivid illustration.

Acknowledgements. We would like to thank Prof. Zhu Yujie of School of Economics and Management of Tsinghua University for his providing the cross-province macroeconomic time series data of China. We are also grateful to Dr. Jin Xiaoming for his suggestions.

References

1. Rakesh Agrawal, Giuseppe Psaila, Edward L. Wimmers, and Mohamed Zaot. Querying shapes of histories. In Umeshwar Dayal, Peter M. D. Gray, and Shojiro Nishio, editors, *Proc. of VLDB'95*, pages 502–514, Zurich, Switzerland, 1995. Morgan Kaufmann Publishers, Inc. San Francisco, USA.
2. Rakesh Agrawal and Ramakrishnan Srikant. Fast algorithms for mining association rules. In Jorge B. Bocca, Matthias Jarke, and Carlo Zaniolo, editors, *Proc. of VLDB'94*, pages 487–499. Morgan Kaufmann, 12–15 1994.
3. Gautam Das, King-Ip Lin, Heikki Mannila, Gopal Renganathan, and Padhraic Smyth. Rule discovery from time series. In *Knowledge Discovery and Data Mining*, pages 16–22, 1998.
4. John E. Freund. *Mathematical Statistics.* Prentice-Hall, 6nd edition, 1999.

5. George G. Judge, R. Carter Hill, William E. Griffiths, Helmut Lütkepohl, and Tsoung-Chao Lee. *Introduction to the Theory and Practice of Econometrics.* John Wiley and Sons Ltd, 2nd edition, 1988.
6. E. Keogh, S. Lonardi, and B. Chiu. Finding surprising patterns in a time series database in linear time and space. In *Proc. of KDD'02*, Edmonton, Alberta, Canada, July 2002.
7. Heikki Mannila, Hannu Toivonen, and A. Inkeri Verkamo. Efficient algorithms for discovering association rules. In Usama M. Fayyad and Ramasamy Uthurusamy, editors, *AAAI Workshop on Knowledge Discovery in Databases (KDD-94)*, pages 181–192, Seattle, Washington, 1994. AAAI Press.

Constraint-Based Mining of Formal Concepts in Transactional Data

Jérémy Besson[1], Céline Robardet[2], and Jean-François Boulicaut[1]

[1] Institut National des Sciences Appliquées de Lyon
LIRIS CNRS FRE 2672, Bâtiment Blaise Pascal
F-69621 Villeurbanne cedex, France
{jeremy.besson,jean-francois.boulicaut}@insa-lyon.fr
[2] Institut National des Sciences Appliquées de Lyon
PRISMA, Bâtiment Blaise Pascal
F-69621 Villeurbanne cedex, France
celine.robardet@insa-lyon.fr

Abstract. We are designing new data mining techniques on boolean contexts to identify a priori interesting concepts, i.e., closed sets of objects (or transactions) and associated closed sets of attributes (or items). We propose a new algorithm D-MINER for mining concepts under constraints. We provide an experimental comparison with previous algorithms and an application to an original microarray dataset for which D-MINER is the only one that can mine all the concepts.

Keywords: Pattern discovery, constraint-based data mining, closed sets, formal concepts.

1 Introduction

One of the most popular data mining techniques concerns transactional data analysis by means of set patterns. Indeed, following the seminal paper [1], hundreds of research papers have considered the efficient computation of a priori interesting association rules from the so-called frequent itemsets. Transactional data can be represented as boolean matrices (see Figure 1). Lines denotes transactions and columns are boolean attributes that enable to record item occurrences. For instance, in Figure 1, transaction t_4 contains the items g_5, g_6, g_7, g_8, g_9, and g_{10}.

The frequent set mining problem concerns the computation of sets of attributes that are true together in enough transactions, i.e., given a frequency threshold. The typical case of basket analysis (huge - eventually millions - number of transactions, hundreds of attributes, but sparse and lowly-correlated data) can be handled by many algorithms, including the various APRIORI-like algorithms that have been designed during the last decade [2]. When the data are dense and highly-correlated, these algorithms fail but the so-called condensed representations of the frequent itemsets can be computed. For instance, efficient algorithms can compute the frequent closed sets from which every frequent set

H. Dai, R. Srikant, and C. Zhang (Eds.): PAKDD 2004, LNAI 3056, pp. 615–624, 2004.

	Items									
	g_1	g_2	g_3	g_4	g_5	g_6	g_7	g_8	g_9	g_{10}
t_1	1	1	1	1	0	1	1	0	0	0
t_2	1	1	1	1	0	0	0	0	1	1
t_3	1	1	1	1	0	0	0	0	1	1
t_4	0	0	0	0	1	1	1	1	1	1
t_5	1	0	1	0	1	1	1	1	0	0

Fig. 1. Example of a boolean context $\mathbf{r}_1$

and its frequency can be derived without accessing the data [10,6,11,3,14]. Other important applications concern datasets with only a few transactions, e.g., for typical gene expression data where items denote gene expression properties in biological situations. It is however possible to use the properties of Galois connection to compute the closed sets on the smaller dimension and derive the closed sets on the other dimension [12].

In this paper, we consider bi-set mining in difficult cases, i.e., when the data is dense and when none of the dimensions is quite small. Bi-sets are composed of a set of lines T and a set of columns G. T and G can be associated by various relationships, e.g., the fact that all the items of G belong to each transaction of T (1-rectangles). It is interesting to constrain further bi-set components to be closed sets (also called maximal 1-rectangles or concepts [13]). Other constraints, e.g., minimal and maximal frequency, can be used as well.
We propose an original algorithm called D-MINER that computes concepts under constraints. It works differently from other concept discovery algorithms (see, e.g., [8,4,9]) and (frequent) closed set computation algorithms. D-MINER can be used in dense boolean datasets when the previous algorithms generally fail. Thanks to an active use of the constraints, it enlarges the applicability of concept discovery for matrices whose none of the dimensions is small. Section 2 contains the needed definitions and a presentation of D-MINER. Section 3 provides an experimental validation. Finally, Section 4 is a short conclusion.

2 D-Miner

Let $\mathcal{O}$ denote a set of objects or transactions and $\mathcal{P}$ denote a set of items or properties. In Figure 1, $\mathcal{O} = \{t_1, \ldots, t_5\}$ and $\mathcal{P} = \{g_1, g_2, \ldots, g_{10}\}$. The transactional data is represented by the matrix $\mathbf{r}$ of relation $R \subseteq \mathcal{O} \times \mathcal{P}$. We write $(t_i, g_j) \in \mathbf{r}$ to denote that item j belongs to transaction i or that property j holds for object i.

The language of bi-sets is the collection of couples from $\mathcal{L}_\mathcal{O} \times \mathcal{L}_\mathcal{P}$ where $\mathcal{L}_\mathcal{O} = 2^\mathcal{O}$ (sets of objects) and $\mathcal{L}_\mathcal{P} = 2^\mathcal{P}$ (sets of items).

Definition 1. *A bi-set (T, G) is a 1-rectangle in* $\mathbf{r}$ *iff $\forall t \in T$ and $\forall g \in G$ then $(t, g) \in \mathbf{r}$. A bi-set (T, G) is a 0-rectangle in* $\mathbf{r}$ *iff $\forall t \in T$ and $\forall g \in G$ then $(t, g) \notin \mathbf{r}$.*

Definition 2. *(Concept) A bi-set (T, G) is a concept in* $\mathbf{r}$ *iff (T, G) is a 1-rectangle and $\forall T' \subseteq \mathcal{O}\backslash T$, $(T \cup T', G)$ is not a 1-rectangle and $\forall G' \subseteq \mathcal{P}\backslash G$, $(T, G \cup G')$ is not a 1-rectangle.*

Notice that, by construction, both sets of a concept are closed sets and any algorithm that computes closed sets can be used for concept discovery [12]. Given Figure 1, $(\{t_1, t_2, t_3\}, \{g_1, g_2\})$ is a 1-rectangle in $\mathbf{r}_1$ but it is not a concept. Twelve bi-sets are concepts in $\mathbf{r}_1$. Two of them are $(\{t_1, t_2, t_3, t_5\}, \{g_1, g_3\})$ and $(\{t_2, t_3\}, \{g_1, g_2, g_3, g_4, g_9, g_{10}\})$. Interesting data mining processes on transactional data can be formalized as the computation of bi-sets whose set components satisfy combinations of primitive constraints.

Definition 3. *(Monotonic and anti-monotonic constraints) Given $\mathcal{L}$ a collection of sets, a constraint $\mathcal{C}$ is said anti-monotonic w.r.t. $\subseteq$ iff $\forall \alpha, \beta \in \mathcal{L}$ such that $\alpha \subseteq \beta$, $\mathcal{C}(\beta) \Rightarrow \mathcal{C}(\alpha)$. $\mathcal{C}$ is said monotonic w.r.t. $\subseteq$ iff $\forall \alpha, \beta \in \mathcal{L}$ such that $\alpha \subseteq \beta$, $\mathcal{C}(\alpha) \Rightarrow \mathcal{C}(\beta)$.*

In A-PRIORI like algorithms, the minimal frequency constraint (on $\mathcal{L}_\mathcal{P}$) is used to prune the search space. This constraint is anti-monotonic w.r.t. $\subseteq$ on $\mathcal{L}_\mathcal{P}$. This constraint can be considered as monotonic on $\mathcal{L}_\mathcal{O}$ because when a set of items is larger, the associated set of transactions is smaller.

Definition 4. *(Specialization relation) Our specialisation relation on bi-sets from $\mathcal{L} = \mathcal{L}_\mathcal{O} \times \mathcal{L}_\mathcal{P}$ is defined by $(T_1, G_1) \leq (T_2, G_2)$ iff $T_1 \subseteq T_2$ and $G_1 \subseteq G_2$.*

We generalize the frequency constraints on this partial order $\leq$.

Definition 5. *(Frequency constraints on concepts) A concept (T, G) satisfies a constraint $\mathcal{C}_t(\mathbf{r}, \sigma_1, T)$ (resp. $\mathcal{C}_g(\mathbf{r}, \sigma_2, G)$) if $|T| \geq \sigma_1$ (resp. $|G| \geq \sigma_2$). These constraints are both monotonic w.r.t. $\leq$ on $\mathcal{L}_\mathcal{O} \times \mathcal{L}_\mathcal{P}$.*

For example, the set of concepts (T,G) satisfying $\mathcal{C}_g(\mathbf{r}_1, 4, G) \wedge \mathcal{C}_t(\mathbf{r}_1, 3, T)$ (a conjonction of monotonic constraints) is $\{(\{g_1, g_2, g_3, g_4\}, \{t_1, t_2, t_3\})\}$.

2.1 D-Miner Principle

D-MINER is a new algorithm for extracting concepts (T, G) under constraints. It builds the sets T and G and it uses monotonic constraints simultaneously on $\mathcal{L}_O$ and $\mathcal{L}_P$ to reduce the search space. A concept (T, G) is such that all its items and objects are in relation by R. Thus, the absence of relation between an item g and an object t generates two concepts, one with g and without t, and another one with t and without g. D-MINER is based on this observation. Let us denote by H a set of 0-rectangles such that it is a partition of the false values (0) of the boolean matrix, i.e., $\forall g \in \mathcal{P}$ and $\forall t \in \mathcal{O}$ such that $(t, g) \notin \mathbf{r}$, it exists one and only one element (X, Y) of H such that $t \in X$ and $g \in Y$. The elements of H are called *cutters*. H must be as small as possible to reduce the depth of recursion and thus execution time. On another hand, one should not waste too much time to compute H. H contains as many elements as lines in the matrix. Each element is composed of the attribute valued by 0 in this line. Time complexity for computing H is in $\mathcal{O}(n \times m)$ where n and m are the dimensions of the matrix. Computing time is negligible w.r.t. the one of the cutting procedure. Furthermore, using this definition makes easier the pruning of 1-rectangles that are not concepts.

D-MINER starts with the couple $(\mathcal{O},\mathcal{P})$ and then splits it recursively using the elements of H until H is empty and consequently each couple is a 1-rectangle. An element (a, b) of H is used to cut a couple (X, Y) if $a \cap X \neq \emptyset$ and $b \cap Y \neq \emptyset$. By convention, one defines the left son of (X, Y) by $(X \backslash a, Y)$ and the right son by $(X, Y \backslash b)$. Recursive splitting leads to all the concepts, i.e., the maximal 1-rectangles (see Example 1) but also some non-maximal ones (see Example 2). We consider in Example 3 how to prune them to obtain all the concepts and only the concepts.

We now provide examples of D-MINER executions. The use of monotonic constraints on $\mathcal{L}_\mathcal{O}$ and $\mathcal{L}_\mathcal{P}$ is presented later. Notice that for clarity, sets like $\{g_1, g_2\}$ are written $g_1 g_2$.

Example 1. Assume $\mathcal{O} = \{t_1, t_2, t_3\}$ and $\mathcal{P} = \{g_1, g_2, g_3\}$. $\mathbf{r}_2$ is defined in Table 1 (left). Figure 2 (left) illustrates D-MINER execution. We get 4 1-rectangles that are the 4 concepts for this boolean context.

Table 1. Contexts $\mathbf{r}_2$ for Example 1 (left) and $\mathbf{r}_3$ for Examples 2 and 3 (right)

	g_1	g_2	g_3
t_1	0	1	1
t_2	1	1	1
t_3	1	0	1

	g_1	g_2	g_3
t_1	0	0	1
t_2	1	0	1
t_3	0	0	1

Example 2. Assume now $\mathbf{r}_3$ as given in Table 1 (right). Computing H provides $\{(t_1, g_1 g_2), (t_2, g_2), (t_3, g_1 g_2)\}$. Figure 2 (right) illustrates D-MINER execution. Some bi-sets are underlined and this will be explained in Example 3.

From Figure 2 (right), we can see that (t_2, g_2) and $(t_3, g_1 g_2)$ from H are not used to cut $(t_1 t_2 t_3, g_3)$ because $\{g_2\} \cap \{g_3\} = \emptyset$ and $\{g_1 g_2\} \cap \{g_3\} = \emptyset$. The computed collection of bi-sets is:

$$\{(t_1 t_2 t_3, g_3), (t_2, g_1 g_3), (\emptyset, g_1 g_2 g_3), \underline{(t_3, g_3), (t_2 t_3, g_3)}\}$$

We see that $(t_3, g_3) \leq (t_1 t_2 t_3, g_3)$ and $(t_2 t_3, g_3) \leq (t_1 t_2 t_3, g_3)$ and thus these 1-rectangles are not concepts.

To solve this problem, let us introduce a new notation. Let $\mathbf{r}[T, G]$ denote the reduction of $\mathbf{r}$ on objects from T and on items from G. When a couple (X, Y) is split by a cutter $(a, b) \in H$, then $(X \backslash a, Y)$ (the left son) and $(X, Y \backslash b)$ (the

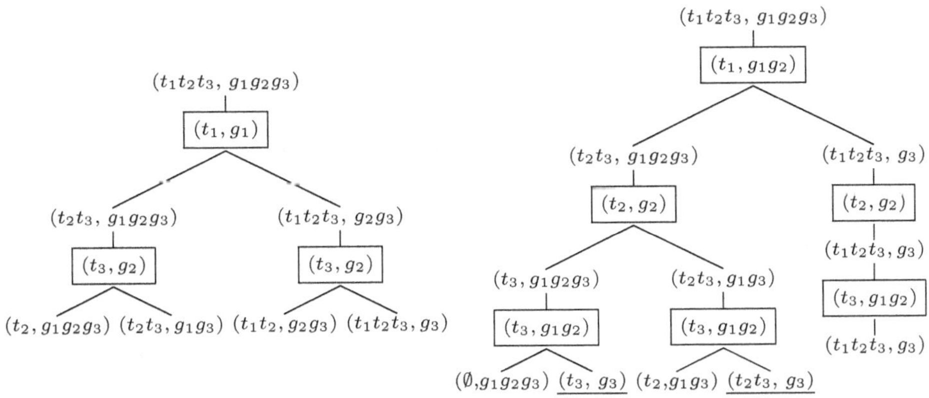

Fig. 2. Concept construction on $\mathbf{r}_2$ (left) and on $\mathbf{r}_3$ (right)

right son) are generated. By construction of H $(a, Y\backslash b)$ is a 1-rectangle which is not necessarily maximal. If a concept (C_X, C_Y) exists in $\mathbf{r}[X\backslash a, Y]$ such that $C_Y \cap b = \emptyset$ then $(C_X \cup a, C_Y)$ is a concept in $\mathbf{r}[X, Y]$. However $(C_X \cup a, C_Y)$ is a concept in $\mathbf{r}[X, Y\backslash b]$ and consequently would be a son of the right son of (X, Y) (see Figure 3). To avoid these non-maximal 1-rectangles, we have to enforce that the property $b \cap Y \neq \emptyset$ is always satisfied for all the previously used left-cutters.

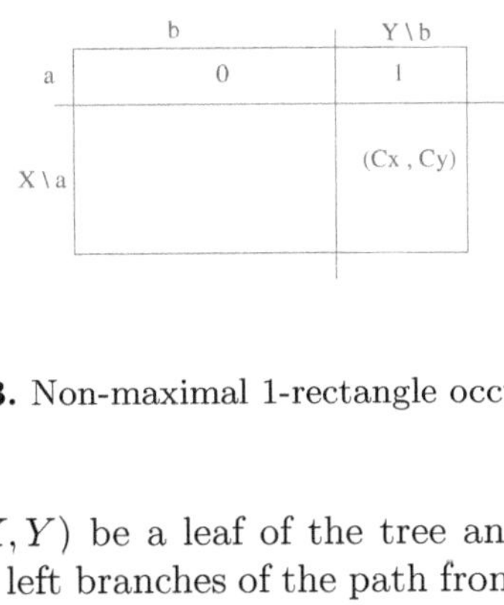

Fig. 3. Non-maximal 1-rectangle occurrence

Property 1: Let (X, Y) be a leaf of the tree and $H_L(X, Y)$ be the set of cutters associated to the left branches of the path from the root to (X, Y). Then (X, Y) is a concept iff it contains at least one item of each element of $H_L(X, Y)$. It means that when trying to build a right son (X, Y) (i.e., to remove some elements from Y), we must check that $\forall(a, b) \in H_L(X, Y)$, $b \cap Y \neq \emptyset$. This is called later the left cutting constraint.
This has been formally studied in [5] that contains correctness and completeness proofs for D-MINER.

Example 3. We take the context used for Example 2 (see Table 1 on the right). 1-rectangles (t_3, g_3) and (t_2t_3, g_3) are pruned using Property 1. (t_3, g_3) comes from the left cutting of $(t_1t_2t_3,g_1g_2g_3)$ and then the left cutting of $(t_3, g_1g_2g_3)$. The items of (t_3, g_3) must contain at least one item of $\{g_1, g_2\}$ and of $\{g_2\}$, i.e., the precedent left cutter set of items. It is not the case and thus (t_3, g_3) is pruned. (t_2t_3, g_3) comes from just one left cutter: (t_1,g_1g_2). It contains neither g_1 nor g_2. Nodes that are underlined in Figure 2 (right) are pruned.

2.2 Algorithm

Before cutting a couple (X, Y) by a cutter (a, b) in two couples $(X\backslash a, Y)$ and $(X, Y\backslash b)$, two types of constraints must be checked, first the monotonic constraints and then the left cutting constraint. Closeness property (maximality) is implied by the cutting procedure.

D-MINER is a depth-first method which generates couples ordered by relation $\leq$. Monotonic constraints w.r.t. either $\mathcal{O}$ or $\mathcal{P}$ are used to prune the search space: if (X, Y) does not satisfy a monotonic constraint $\mathcal{C}$ then none of its sons satisfies $\mathcal{C}$ and it is unnecessary to cut (X, Y). For instance, we can push the constraint $\mathcal{C}((T, G)) \equiv \mathcal{C}_t(T) \wedge \mathcal{C}_g(G)$ where (T, G) is a bi-set, $\mathcal{C}_t(T) \equiv |T| \geq 5$, and $\mathcal{C}_g(G) \equiv |G| \geq 4$.

Algorithms 1 and 2 contain the pseudo-code of D-MINER. First, the set H of cutters is computed. Then the recursive function *cutting()* is called.

Function *cutting* cuts out a couple (X, Y) with the first cutter $H[i]$ that satisfies the following constraints. First, (X, Y) must have a non empty intersection with $H[i]$. If it is not the case, *cutting* is called with the next cutter. Before cutting (X, Y) in $(X\backslash a, Y)$, we have to check the monotonic constraint on $X\backslash a$ (denoted $\mathcal{C}_t(X\backslash a)$) to try to prune the search space. (a, b) is inserted into H_L, the set of cutters in the left cutting. Then *cutting* is called on $(X\backslash a, Y)$ and (a, b) is removed from H_L. For the second cutting of (X, Y), two constraints have to be checked. First the monotonic constraint on $Y\backslash b$ (denoted $\mathcal{C}_g(Y\backslash b)$) is checked. Therefore, D-MINER constructs first an element (X, Y) and then reduces simultaneously X and Y to have the collection of concepts derived from (X, Y). Secondly, monotonic constraints can be applied on X and Y to prune the search space: if $\alpha \leq \beta$ and $\neg\, \mathcal{C}(\beta)$ then $\neg\, \mathcal{C}(\alpha)$.

It is possible to optimize this algorithm. First, the order of the elements of H is important. The aim is to cut as soon as possible the branches which generate non-maximal 1-rectangles. H must be sorted by decreasing order of size of the object components. Moreover, to reduce the size of H, the cutters which have the same items are gathered: $\forall (a_1, b_1),\ (a_2, b_2) \in H,\ if\ b_1 = b_2\ then\ H = H\backslash\{(a_1, b_1), (a_2, b_2)\} \cup (a_1 \cup a_2, b_1)$. If $|\mathcal{P}| > |\mathcal{O}|$, we transpose the data matrix to obtain a set H of minimum size. The symmetry of our extractor on $\mathcal{L}_O$ and $\mathcal{L}_P$ allows to transpose the matrix without loosing the possibility of using the constraints. Indeed, in some contexts where there are few objects and many items, we first perform a simple transposition like in [12].

Algorithm 1: D-MINER

Input : Database **r** with n lines and m columns, $\mathcal{O}$ the set of objects, $\mathcal{P}$ the set of items, $\mathcal{C}_t$ and $\mathcal{C}_g$ are monotonic constraints on $\mathcal{O}$ and $\mathcal{P}$.
Output : $\mathcal{Q}$ the set of concepts that satisfy $\mathcal{C}_t$ and $\mathcal{C}_g$
$H_L \leftarrow$ empty()
H and $H_{size} = |H|$ are computed from **r**;
$\mathcal{Q} \leftarrow$ cutting(($\mathcal{O}$, $\mathcal{P}$), H, 0, H_{size}, H_L);

Algorithm 2: cutting

Input: (X, Y) a couple of $2^{\mathcal{O}} \times 2^{\mathcal{P}}$, H the list of cutters, i the number of iterations, H_{size} the size of H, H_L a set of precedent cutters in left cuttings, $\mathcal{C}_t$ monotonic constraint on $\mathcal{O}$, $\mathcal{C}_g$ monotonic constraint on $\mathcal{P}$.
Output: $\mathcal{Q}$ the set of concepts that satisfy $\mathcal{C}_t$ and $\mathcal{C}_g$
$(a, b) \leftarrow$ H[i]
If $(i \leq H_{size} - 1)$ // i-th cutter is selected
  If $((a \cap X = \emptyset)$ or $(b \cap Y = \emptyset))$
    $\mathcal{Q} \leftarrow \mathcal{Q} \cup cutting((X, Y), H, i + 1, H_{size}, H_L)$
  Else
    If ($\mathcal{C}_t(X \backslash a)$ is satisfied)
      $H_L \leftarrow H_L \cup (a, b)$
      $\mathcal{Q} \leftarrow \mathcal{Q} \cup\ cutting((X \backslash a, Y), H, i + 1, H_{size}, H_L)$
      $H_L \leftarrow H_L \backslash (a, b)$
    If ($\mathcal{C}_g(Y \backslash b)$ is satisfied $\wedge\ \forall (a', b') \in H_L,\ b' \cap Y \backslash b \neq \emptyset$)
      $\mathcal{Q} \leftarrow \mathcal{Q} \cup cutting((X, Y \backslash b), H, i + 1, H_{size}, H_L)$
Else
  $\mathcal{Q} \leftarrow (X, Y)$
Return $\mathcal{Q}$

---

3 Experimental Validation

We compare the execution time of D-MINER with those of CLOSET [11], AC-MINER [6] and CHARM [14] in three datasets. CLOSET, CHARM and AC-MINER compute closed sets under a minimal frequency constraint, i.e., the frequent closed sets. Due to Galois connection, in a given dataset, the number of closed sets in $\mathcal{L}_O$ is the number of closed sets in $\mathcal{L}_P$. For a fair comparison, we transpose the matrices to have a smaller number of columns for CLOSET, AC-MINER and CHARM and to have a smaller number of lines for D-MINER. In the three first experiments, we compare the effectiveness of the four algorithms when computing the collection of closed sets under a minimal frequency constraint. We used Zaki's implementation of CHARM and Bykowski's implementations of AC-MINER and CLOSET [7]. In all the following figures, the minimal frequencies are relative frequencies.

First, we have studied the performance of D-Miner for computing the frequent closed sets from benchmark datasets available on line at IBM Almaden[1] and the UCI repository. All extractions have been performed on a Pentium III (450 MHz, 128 Mb). We have used the benchmark "Mushroom". Its derived boolean context contains 8 124 lines and 120 columns. The needed execution time (in seconds) to obtain the frequent closed sets is shown on Figure 4 (left).

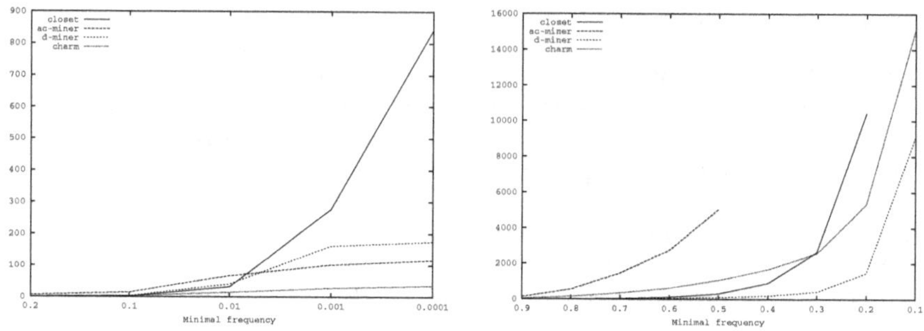

Fig. 4. Mushroom (left) and Connect4 (right)

The four algorithms can compute every closed set and thus all concepts (minimum frequency 0.0001) within a few minutes. Indeed, the lowest frequency threshold corresponds to at least 1 object. Once every closed set on one dimension is computed, the associated concept is obtained easily. The execution time of Closet increases very fast compared to the three others.

Next, we considered the benchmark "Connect4". The derived boolean context contains 67 557 lines and 149 columns. The execution time to obtain frequent closed sets is shown on Figure 4 (right). Only Charm and D-Miner can extract concepts with minimal frequency equals to 0.1 (10%). D-Miner is almost twice faster than Charm on this dataset.

We now provide an experimental validation on an original biological dataset that contains 104 lines and 304 columns. For the purpose of this paper, an important information is that its density is high: 17 % of the cells contain the value true. This is a gene expression dataset that can not be described further due to space limitation (see [5] for details). The execution time (in seconds) for computing frequent closed sets with Closet, Ac-Miner, Charm and D-Miner is shown on Figure 5 (left).

D-Miner is the only algorithm which succeeds in extracting all the concepts. These data are in fact very particular: there are very few concepts before the frequency 0.1 (5 534 concepts) and then the number of concepts increases very fast (at the lowest frequency threshold, there are more than 5 millions of concepts). In this context, extracting putative interesting concepts needs for a

[1] See www.almaden.ibm.comcsquestdemos.html.

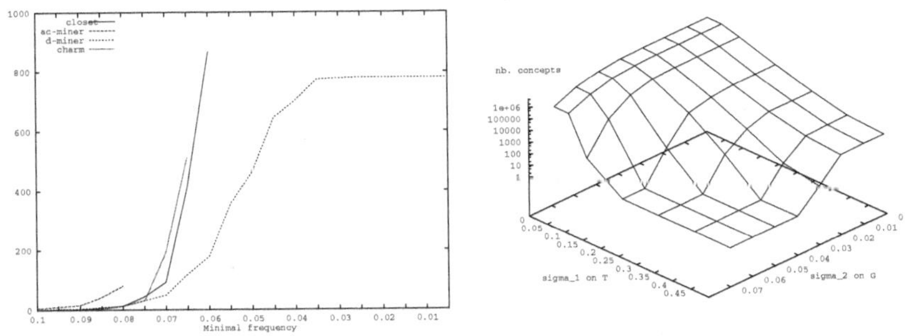

Fig. 5. A microarray dataset analysis

very low frequency threshold, otherwise almost no concept is provided. Consequently D-MINER is much better than the other algorithms because it succeeds to extract concepts when the frequency threshold is lower than 0.06 whereas it is impossible with the others.

End users, e.g., biologists, are generally interested in a rather small subset of the extracted collections. These subsets can be specified by means of user-defined constraints that can be checked afterwards (post-processing) on the whole collection of concepts. It is clear however that this approach leads to tedious or even impossible post-processing phases when, e.g., more than 5 millions of concepts are computed (almost 500M bytes of patterns). It clearly motivates the need for constraint-based mining of concepts.

Let us consider the computation of the bi-sets that satisfy two minimal frequency constraints on $\mathcal{L}_\mathcal{O}$ and $\mathcal{L}_\mathcal{P}$: one on item (gene) sets and the other one on objects (biological situations). In Figure 5 (right), we plot the number of concepts obtained using both $\mathcal{C}_t(\mathbf{r}, \sigma_1, T)$ and $\mathcal{C}_g(\mathbf{r}, \sigma_2, G)$ when σ_1 and σ_2 vary. It appears that using only one of the two constraints does not reduce significantly the number of extracted concepts (see values when $\sigma_1 = 0$ or $\sigma_2 = 0$). However, when we use simultaneously the two constraints, the size of the concept collection decreases strongly (the surface of the values forms a basin). For example, the number of concepts verifying $|G| \geq 10$ and $|T| \geq 21$ is 142 279. The number of concepts verifying $|G| \geq 10$ and $|T| \geq 0$ is 5 422 514. The number of concepts verifying $|G| \geq 0$ and $|T| \geq 21$ is 208 746. The gain when using simultaneously both constraints is significant.

4 Conclusion

Computing formal concepts has been proved useful in many application domains but remains extremely hard from dense boolean datasets like the one we have to process nowadays for gene expression data analysis. We have described an original algorithm that computes concepts under monotonic constraints. First, it can be used for closed set computation and thus concept discovery. Next, for difficult contexts, i.e., dense boolean matrices where none of the dimension is small, the analyst can provide monotonic constraints on both set components of desired concept and the D-MINER algorithm can push them into the extraction process. Considering one of our applications that is described in [5], we are now working on the biological validation of the extracted concepts. We have also to compare D-MINER with new concept lattice construction algorithms like [4].

Acknowledgments. We thank M. Zaki who has kindly provided his CHARM implementation. J. Besson is funded by INRA. This work has been partially funded by the EU contract cInQ IST-2000-26469 (FET arm of the IST programme).

References

1. R. Agrawal, T. Imielinski, and A. Swami. Mining association rules between sets of items in large databases. In *ACM SIGMOD Conference on Management of Data SIGMOD'93*, pages 207–216, 1993.
2. R. Agrawal, H. Mannila, R. Srikant, H. Toivonen, and A. I. Verkamo. Fast discovery of association rules. In *Advances in Knowledge Discovery and Data Mining*, pages 307–328. AAAI Press, 1996.
3. Y. Bastide, R. Taouil, N. Pasquier, G. Stumme, and L. Lakhal. Mining frequent patterns with counting inference. *SIGKDD Explorations*, 2(2):66 – 75, Dec. 2000.
4. A. Berry, J.-P. Bordat, and A. Sigayret. Concepts can not afford to stammer. In *Proceedings JIM'03*, pages 25–35, Metz, France, September 2003.
5. J. Besson, C. Robardet, J.-F. Boulicaut, and S. Rome. Constraint-based bi-set mining for biologically relevant pattern discovery in microarray data. *Intelligent Data Analysis*. Accepted for publication in February 2004.
6. J.-F. Boulicaut, A. Bykowski, and C. Rigotti. Approximation of frequency queries by mean of free-sets. In *PKDD'00*, volume 1910 of *LNAI*, pages 75–85, 2000.
7. A. Bykowski. *Condensed representations of frequent sets: application to descriptive pattern discovery*. PhD thesis, Institut National des Sciences Appliquées de Lyon, LIRIS, F-69621 Villeurbanne cedex, France, Oct. 2002.
8. B. Ganter. Two basic algorithms in concept analysis. Technical report, Technisch Hochschule Darmstadt, Preprint 831, 1984.
9. L. Nourine and O. Raynaud. A fast algortihm for building lattices. *Information Processing Letters*, 71:190–204, 1999.
10. N. Pasquier, Y. Bastide, R. Taouil, and L. Lakhal. Efficient mining of association rules using closed itemset lattices. *Information Systems*, 24(1):25–46, Jan. 1999.
11. J. Pei, J. Han, and R. Mao. CLOSET an efficient algorithm for mining frequent closed itemsets. In *ACM SIGMOD Workshop DMKD'00*, 2000.
12. F. Rioult, J.-F. Boulicaut, B. Crémilleux, and J. Besson. Using transposition for pattern discovery from microarray data. In *Proceedings ACM SIGMOD Workshop DMKD'03*, pages 73–79, San Diego, USA, June 2003.
13. R. Wille. Restructuring lattice theory: an approach based on hierarchies of concepts. In I. Rival, editor, *Ordered sets*, pages 445–470. Reidel, 1982.
14. M. J. Zaki and C.-J. Hsiao. CHARM: An efficient algorithm for closed itemset mining. In *Proceedings SIAM DM'02*, Arlington, USA, April 2002.

Towards Optimizing Conjunctive Inductive Queries*

Johannes Fischer** and Luc De Raedt

Institut für Informatik
Albert-Ludwigs-University Freiburg
Georges Koehler Allee 79
D-79110 Freiburg, Germany
fischer@bio.informatik.uni-muenchen.de
deraedt@informatik.uni-freiburg.de

Abstract. Inductive queries are queries to an inductive database that generate a set of patterns in a data mining context. Inductive querying poses new challenges to database and data mining technology. We study conjunctive inductive queries, which are queries that can be written as a conjunction of a monotonic and an anti-monotonic subquery. We introduce the conjunctive inductive query optimization problem, which is concerned with minimizing the cost of computing the answer set to a conjunctive query. In the optimization problem, it is assumed that there are costs c_a and c_m associated to evaluating a pattern w.r.t. a monotonic and an anti-monotonic subquery respectively. The aim is then to minimize the total cost needed to compute all solutions to the query. Secondly, we present an algorithm that aims at optimizing conjunctive inductive queries in the context of the pattern domain of strings and evaluate it on a challenging data set in computational biology.

1 Introduction

Many data mining problems address the problem of finding a set of patterns that satisfy a constraint. Formally, this can be described as the task of finding the set of patterns $Th(Q, \mathcal{D}, \mathcal{L}) = \{\phi \in \mathcal{L} \mid Q(\phi, \mathcal{D})$, i.e. those patterns ϕ satisfying query Q on database $\mathcal{D}\}$. Here $\mathcal{L}$ is the language in which the patterns or rules are described and Q is a predicate or constraint that determines whether a pattern ϕ is a solution to the data mining task or not [19]. This framework allows us to view the predicate or the constraint Q as an *inductive query* to an *inductive database system*. It is then the task of the inductive database management system to efficiently generate the answers to the query [7]. Within this framework data mining becomes an (inductive) querying process that puts data mining on the same methodological grounds as databases. This view of data mining raises several new challenges for database and data mining technology.

* An early version of this paper was presented at the 2nd ECML/PKDD Workshop on Knowledge Discovery with Inductive Querying, Dubrovnik, 2003.

** Now at the Institut für Informatik, Ludwig-Maximilians-Universität München

H. Dai, R. Srikant, and C. Zhang (Eds.): PAKDD 2004, LNAI 3056, pp. 625–637, 2004.

In this paper, we address one of these challenges, the optimization of *conjunctive* inductive queries. These queries can be written as the conjunction $Q_a \wedge Q_m$ of an anti-monotonic and a monotonic subquery. An example query could ask for molecular fragments that have frequency at least 30 per cent in the active molecules and frequency at most 5 per cent in the inactive ones [5,18]. Conjunctive inductive queries of this type have been studied in various contexts, cf. [5, 6,18,4]. One important result is that their solution space $Th(Q, \mathcal{D}, \mathcal{L})$ is convex, which is related to the well-known concept of version spaces [20] and boundary sets [19], a fact that is exploited by several pattern mining algorithms. The key contribution of this paper is that we introduce an algorithm for computing the set of solutions $Th(Q, \mathcal{D}, \mathcal{L})$ to a conjunctive inductive query that aims at minimizing the cost of evaluating patterns w.r.t. the primitive constraints in the inductive query. More precisely, we assume that there is a cost c_m and c_a associated to testing whether a pattern satisfies Q_m, resp. Q_a, and we aim at minimizing the total cost of computing $Th(Q, \mathcal{D}, \mathcal{L})$. The algorithm that we introduce builds upon the work by [6] that has introduced an effective data structure, called the version space tree, and algorithms for computing $Th(Q, \mathcal{D}, \mathcal{L})$ in a level wise manner for string patterns. In the present paper, we modify this data structure into the partial version space tree and we also present an entirely different approach to computing the answer set to a conjunctive inductive query. Even though the algorithm and data structure are presented in the context of string patterns and data, we believe that the principles also apply to other pattern domains. The approach is also empirically evaluated on the task of finding patterns in a computational biology data set.

The paper is organized as follows. Section 2 introduces the problem of conjunctive query optimization. Section 3 presents a data structure and algorithm for tackling it. Section 4 reports on an experimental evaluation and finally, Sect. 5, concludes and touches upon related work.

2 Conjunctive Inductive Queries

In this section, we define conjunctive inductive queries as well as the pattern domain of strings. Our presentation closely follows that of [6].

A pattern language $\mathcal{L}$ is a formal language for specifying patterns. Each pattern $\phi \in \mathcal{L}$ matches (or covers) a set of examples ϕ_e, which is a subset of the universe $\mathcal{U}$ of possible examples. One pattern ϕ is *more general* than a pattern ψ, written $\phi \succeq \psi$, if and only if $\phi_e \supseteq \psi_e$. E.g., let Σ be a finite alphabet, $\mathcal{U}_\Sigma = \Sigma^*$ the universe of all strings over Σ and denote the empty string by ϵ. The traditional pattern language in this domain is $\mathcal{L}_\Sigma = \mathcal{U}_\Sigma$. A pattern $\phi \in \mathcal{L}_\Sigma$ covers the set $\phi_e = \{\psi \in \Sigma^* \mid \phi \sqsubseteq \psi\}$, where $\phi \sqsubseteq \psi$ denotes that ϕ is a substring of ψ. For this language, $\phi \succeq \psi$, if and only if $\phi \sqsubseteq \psi$.

A pattern *predicate* defines a primitive property of a pattern, usually relative to some data set $\mathcal{D}$ (a set of examples), and sometimes other parameters. For any given pattern, it evaluates to either *true* or *false*.

Inspired by the domain specific inductive database MolFea [18], we now introduce a number of pattern predicates that will be used throughout this paper:

- minfreq(ϕ,n,$\mathcal{D}$) evaluates to true iff ϕ is a pattern that occurs in database $\mathcal{D}$ with frequency at least $n \in \mathbb{N}$. The frequency $f(\phi, \mathcal{D})$ of a pattern ϕ in a database $\mathcal{D}$ is the (absolute) number of examples in $\mathcal{D}$ covered by ϕ. Analogously, the predicate maxfreq($\phi, n, \mathcal{D}$) is defined.
- ismoregeneral(ϕ,ψ) is a predicate that evaluates to true iff pattern ϕ is more general than pattern ψ.

We say that m is a *monotonic* predicate, if for all possible parameter values *params* and data sets $\mathcal{D}$: $\forall\ \phi, \psi \in \mathcal{L}$ such that $\phi \succeq \psi$: m($\psi, \mathcal{D}$, *params*) $\rightarrow$ m($\phi, \mathcal{D}$, *params*). The class of *anti-monotonic* predicates is defined dually. Thus, minfreq, and ismoregeneral are monotonic; their duals are anti-monotonic.

A pattern predicate pred(ϕ,$\mathcal{D}$,*params*) defines the *solution set* *Th*(pred($\phi, \mathcal{D}$, *params*), $\mathcal{L}$) $= \{\psi \in \mathcal{L} \mid$ pred($\psi, \mathcal{D}$, *params*) $=$ *true*$\}$.

We are interested in computing solution sets *Th*($Q, \mathcal{D}, \mathcal{L}$) for conjunctive queries Q, i.e. boolean queries Q that can be written as a conjunction of a monotonic and an anti-monotonic predicate $Q_a \wedge Q_m$. Observe that in a conjunctive query $Q_a \wedge Q_m$, Q_a and Q_m need not be atomic expressions. Indeed, it is well-known that both the disjunction and conjunction of two monotonic (resp. anti-monotonic) predicates are monotonic (resp. anti-monotonic). Furthermore, the negation of a monotonic predicate is anti-monotonic and vice versa. We will also assume that there are cost-functions c_a and c_m associated to the anti-monotonic and monotonic subqueries Q_a and Q_m. The idea is that the cost functions reflect the (expected) costs of evaluating the query on a pattern. E.g., $c_a(\phi)$ denotes the expected cost needed to evaluate the anti-monotonic query Q_a on the pattern ϕ. The present paper does not contribute specific concrete cost functions but rather an overall and general framework for working with such cost functions. Even though it is clear that some predicates are more expensive than other ones, more work seems needed in order to obtain cost estimates that are as reliable as in traditional databases. It is worth mentioning that several of the traditional pattern mining algorithms, such as Agrawal et al.'s Apriori [2] and the level wise algorithm [19], try to minimize the number of passes through the database. Even though this could also be cast within the present framework, the cost functions introduced below better fit the situation where one attempts to minimize the number of covers test, i.e. the number of tests whether a pattern matches or covers a given example. Even though for simple representation languages such as item sets and strings this covers test can often be evaluated efficiently, there exist important applications, such as mining graphs and molecules [18,17], where covers testing corresponds to the subgraph isomorphism problem, which is known to be NP-complete. When mining such data, it is more important to minimize the number of covers test than to minimize the number of passes through the data. Furthermore, for applications in mining graphs or molecules, the data often fit in main memory. Thus the present framework is directly applicable to (and motivated by) molecular feature mining.

By now, we are able to formulate the *conjunctive inductive query optimization problem* that is addressed in this paper:

Given

- a language $\mathcal{L}$ of patterns,
- a conjunctive query $Q = Q_a \wedge Q_m$
- two cost functions c_a and c_m from $\mathcal{L}$ to $\mathbb{R}$

Find the set of patterns $Th(Q, \mathcal{D}, \mathcal{L})$, i.e. the solution set of the query Q in the language $\mathcal{L}$ with respect to the database $\mathcal{D}$, in such a way that the total cost needed to evaluate patterns is as small as possible.

One useful property of conjunctive inductive queries is that their solution space $Th(Q, \mathcal{D}, \mathcal{L})$ is a version space (sometimes also called a convex space).

Definition 1. *Let $\mathcal{L}$ be a pattern language, and $I \subseteq \mathcal{L}$. I is a version space, if $\forall \phi, \phi', \psi \in \mathcal{L} : \phi \preceq \psi \preceq \phi'$ and $\phi, \phi' \in I \implies \psi \in I$.*

3 Solution Methods

In this section, we first introduce partial version space trees (an extension of the data structure in [6]) and then show how it can be used for the optimization problem.

3.1 Partial Version Space Trees

A partial version space tree $\mathcal{T}$ is an extension of a *suffix trie*:

- Each edge is labelled with a symbol from Σ, and all outgoing edges from a node have different labels.
- Each node $n \in \mathcal{T}$ uniquely represents a string $s(n) \in \Sigma^\star$ which is the concatenation of all labels on the path from the root to n. We define $s(root) = \epsilon$. If it is clear from the context, we will simply write n instead of $s(n)$.
- For each node $n \in \mathcal{T}$ there is also a node $n' \in \mathcal{T}$ for all suffixes n' of n. Furthermore, if $n \neq root$, there is a *suffix-link* to the longest proper suffix of n, which we denote by $suffix(n)$. We write $suffix^2(n)$ for $suffix(suffix(n))$ etc. and define $suffix^i(root) = \perp \; \forall\, i \in \mathbb{N}$, where $\perp$ is a unique entity.

To obtain a partial version space tree, we augment each node $n \in \mathcal{T}$ with:

- There are two different labels l_m and l_a, one for the monotonic and one for the anti-monotonic constraint. Each label may obtain one of the values $\oplus$ or $\ominus$, indicating that the string $s(n)$ satisfies the constraint or not. If the truth value of a constraint has not been determined yet, the corresponding label gets a ⓘ.
- There is a link to the father of n, denoted by $parent(n)$. For example, with $n =$ `abc` we have $parent(n) =$ `ab`. As for $suffix(n)$ we write $parent^2(n)$ for $parent(parent(n))$ etc. and define $parent^i(root) = \perp \; \forall\, i \in \mathbb{N}$.
- There is a list of links to all incoming suffix-links to n which we denote by $isl(n)$. For example, if $n =$ `ab` and `aab`, `cab` are the only nodes in $\mathcal{T}$ that have n as their suffix, $isl(n) = \{$`aab`, `cab`$\}$.

The following conditions are imposed on partial version space trees:

(C1) for all leaves n in $\mathcal{T}$, either $l_m(n) = \ominus$ or $l_m(n) = ⓘ$, and all nodes with $l_m = \ominus$ are leaves.
(C2) all expanded nodes n have $l_m(n) = \oplus$

The first condition is motivated by the monotonicity of our query Q_m: if n does not satisfy Q_m, none of its descendants can satisfy the monotonic constraint either, so they need neither be considered nor expanded. A consequence of these requirements is that nodes n with $l_m(n) = \oplus$ must always be expanded. An example of a partial version space tree can be seen in Fig. 1, where the upper part of a node stands for the monotonic and the lower part for the anti-monotonic label. The second condition is imposed because the pattern space is infinite in the downwards direction.

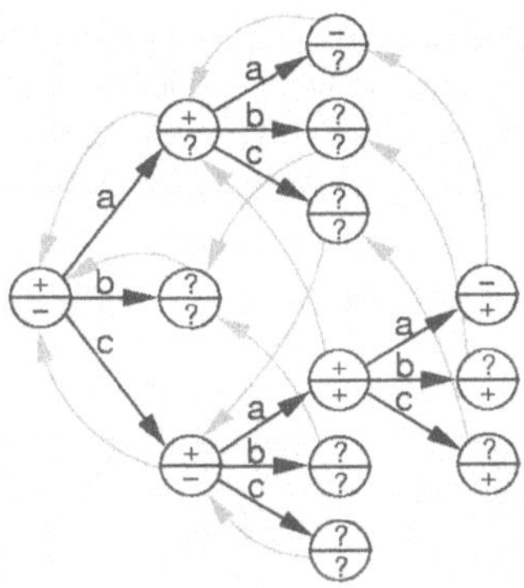

Fig. 1. An example of a partial version space tree. Here and in the rest of this paper, suffix-links are drawn lighter to distinguish them from the black parent-to-child links.

The algorithm given in the next subsection computes the version space tree starting from the tree containing only the root and then expanding the tree until no ⓘ-labels are found. Then, the solution to the original query consists of all nodes n in $\mathcal{T}$ that have a $\oplus$ in both of their labels, i.e. they are of type ⊕. As described in [6], it is also possible to construct the boundary sets S and G from the partial version space tree. The differences with the previous definition of a version space tree given by [6] are that 1) previously each node had only one label that indicated membership of the solution space, and 2) that neither *parent* nor *isl* links were present.

3.2 An Algorithmic Framework

We are now ready to present an algorithmic framework for computing $Th(Q_m \wedge Q_a, \mathcal{D}, \mathcal{L}_\Sigma)$. The key idea is that instead of constructing the version space tree in a top-down, Apriori-like manner, we allow for more freedom in selecting the

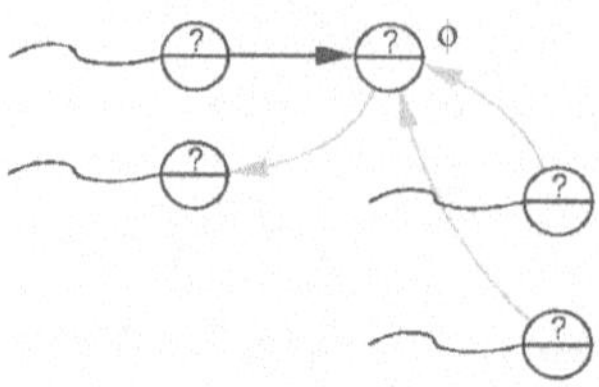

(a) Node ϕ is tested against Q_m. Note that this scheme also applies when condition (C2) is not enforced.

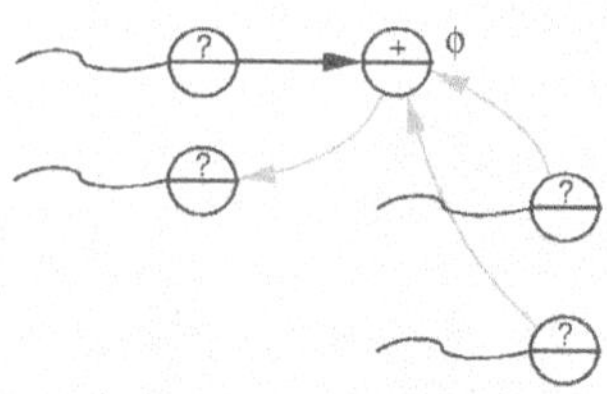

(b) If it is positive, we mark it accordingly. . .

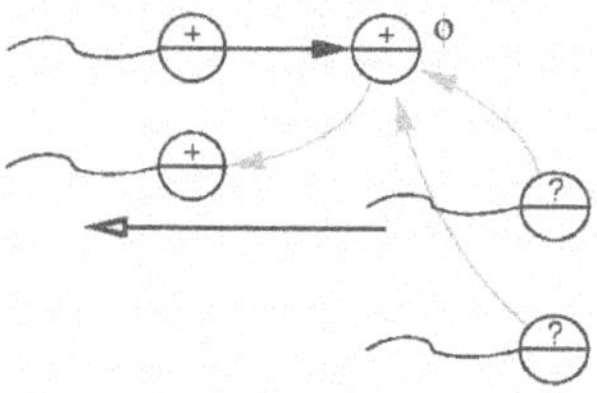

(c) . . . and propagate this label up to the root until we reach a node that has already been marked positive.

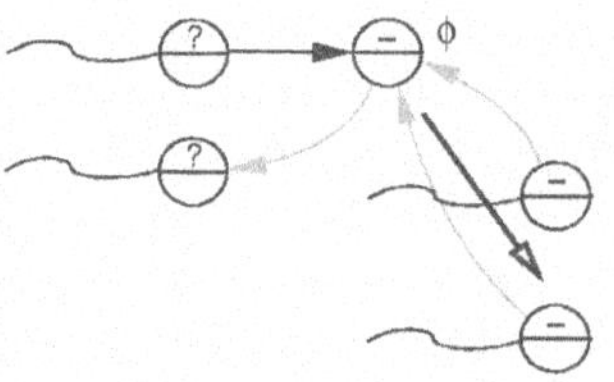

(d) If it is negative, the label is propagated down to the leaves by recursively following the incoming suffix-links.

Fig. 2. How monotone labels are propagated in the tree.

pattern ϕ and the query $Q_a(\phi)$ or $Q_m(\phi)$ to be evaluated. By doing so, we hope to decrease the total cost of obtaining the solution space. As a motivating example, assume our alphabet is $\Sigma = \{\mathtt{a}, \mathtt{b}, \mathtt{c}\}$ and pattern $\phi = \mathtt{abc}$ turns out to satisfy Q_m. Then, by the monotonicity of Q_m, we know that all patterns more general than ϕ satisfy Q_m as well, so ϵ, a, b, c, ab and bc need no longer be tested against Q_m. Thus, by evaluating ϕ, we also obtain the truth values (w.r.t. Q_m) of six other patterns, which would all have been tested using a levelwise strategy. If, on the other hand, ϕ does not satisfy Q_m, all patterns more specific than ϕ cannot satisfy Q_m, so the node representing ϕ need not be expanded.

This suggests the following approach: whenever a pattern ϕ is positive w.r.t. Q_m, we propagate the monotonic ⊕-label up to the root by recursively following ϕ's parent- and suffix-links, until we reach a node that has already been marked positive. Furthermore, ϕ will be expanded and all of its children are labelled appropriately. If ϕ does not satisfy Q_m, we stop the expansion of this node and propagate the monotonic ⊖ down to the leaves by following the incoming suffix-links of ϕ. See Fig. 2 for a schematic overview of these operations.

For the anti-monotonic query Q_a, we propagate the labels in opposite directions. That is, a ⊕ is propagated down to the leaves (by following the children- and incoming suffix-links) and a ⊖ up to the root. The corresponding algorithm is shown in Fig. 3. We use a priority queue P to store those nodes whose truth value has not been fully determined, i.e. all nodes of types ⊕, ⊕, ⊕ and ⊕. The queue not only returns the next pattern ϕ to be evaluated, but also a variable *pred* that tells us which of the predicates Q_m or Q_a should be evaluated for ϕ.

Input: a query $Q = Q_m \wedge Q_a$ and a database $\mathcal{D}$
Output: a version space tree $\mathcal{T}$ representing $Th(Q, \mathcal{D}, \mathcal{L})$
VSTree $\mathcal{T} \leftarrow \{(\epsilon, \overset{?}{\underset{?}{\ominus}})\}$ // insert empty string and mark it as unseen
PriorityQueue $P \leftarrow \{\epsilon\}$
while $(|P| > 0)$
  $(\phi, pred) \leftarrow P$.next // get next pattern and predicate to be evaluated
  if $(pred = antimonotone)$ // evaluate Q_a by accessing $\mathcal{D}$
    if $(Q_a(\phi, \mathcal{D}))$ // a pattern satisfying Q_a
      propagate $\underset{+}{\ominus}$ down to the leaves and remove determined patterns from P
    else // a pattern not satisfying Q_a
      propagate $\underset{-}{\ominus}$ up to the root and remove determined patterns from P
  else if $(pred = monotone)$ // evaluate Q_m by accessing $\mathcal{D}$
    if $(Q_m(\phi, \mathcal{D}))$ // a pattern satisfying Q_m
      propagate $\overset{+}{\ominus}$ up to the root and remove determined patterns from P
      expand ϕ in $\mathcal{T}$
      for all children ψ of ϕ // set children's labels
        if $(l_m(suffix(\psi)) = \ominus)$ $l_m(\psi) \leftarrow \ominus$ else $l_m(\psi) \leftarrow$ ⑦
        if $(l_a(\phi) = \oplus$ or $l_a(suffix(\psi)) = \oplus)$ $l_a(\psi) \leftarrow \oplus$ else $l_a(\psi) \leftarrow$ ⑦
        insert ψ in P if it is not fully determined
    else // a pattern not satisfying Q_m
      propagate $\overset{-}{\ominus}$ down to the leaves and remove determined patterns from P
return $\mathcal{T}$

Fig. 3. An algorithmic framework

Whenever a change of labels results in a label of the other type ($\overset{+}{\underset{+}{\ominus}}$, $\overset{+}{\underset{-}{\ominus}}$, $\overset{-}{\underset{+}{\ominus}}$, $\overset{-}{\underset{-}{\ominus}}$ or $\overset{-}{\underset{?}{\ominus}}$), we remove the node from P. Note that nodes of type $\overset{-}{\underset{?}{\ominus}}$ are also deleted from P although their anti-monotonic part is undetermined; cf. the above discussion.

The choice of priorities for nodes determines the search strategy being used. By assigning the highest priorities to the most shallow nodes (i.e. nodes that are close to the root), a level wise search is simulated as in [6]. On the other hand, by assigning the highest priorities to the deepest nodes, we are close to the idea of *Dualize & Advance* [11,10], since we will go deeper and deeper into the tree until we encounter a node that has only negative children. Somewhere in the middle between these two extremes lies a completely randomized strategy, which assigns random priorities to all nodes.

3.3 Towards an Optimal Strategy

Let us first assign four counters to each node ϕ in the partial version space tree $\mathcal{T}$: $z_m(\phi)$, $z_{\neg m}(\phi)$, $z_a(\phi)$ and $z_{\neg a}(\phi)$. Each of them counts how many labels of nodes in $\mathcal{T}$ would be marked if ϕ's label were changed (including ϕ itself). For example, $z_m(\phi)$ counts how many monotone labels would be marked $\oplus$ if ϕ turned out to satisfy Q_m. In the tree in Fig. 1, we have $z_m(\texttt{cab}) = 3$, because marking cab with a $\overset{+}{\ominus}$ would result in marking ab and b as well, whereas $z_m(\texttt{ac}) = 1$, because no other nodes could be marked in their monotonic part. Likewise, $z_{\neg m}(\phi)$ counts how many monotone labels would change to $\ominus$ if $Q_m(\phi, \mathcal{D})$ turned out to be false. The $z_a(\phi)$- and $z_{\neg a}(\phi)$-counters form the anti-monotonic counterpart. We define $z_m(\phi) = z_{\neg m}(\phi) = 0$ if ϕ's monotonic label is $\neq$?, and likewise for z_a and $z_{\neg a}$. If there is no confusion about which node we talk, we will simply write z_m instead of $z_m(\phi)$ etc. Assume further that we know the following values:

- $P_m(\phi, \mathcal{D})$, the probability that ϕ satisfies the predicate Q_m in database $\mathcal{D}$,
- $P_a(\phi, \mathcal{D})$, the dual of P_m for the anti-monotonic predicate Q_a,
- $c_m(\phi, \mathcal{D})$, the costs for evaluating the monotonic predicate $Q_m(\phi, \mathcal{D})$ and
- $c_a(\phi, \mathcal{D})$, the dual of c_m for the anti-monotonic predicate Q_a.

Now

$$P_m(\phi, \mathcal{D}) \cdot z_m(\phi) + (1 - P_m(\phi, \mathcal{D})) \cdot z_{\neg m}(\phi)$$

is the expected value of the number of monotone labels that get marked by evaluating Q_m for pattern ϕ. Since the operation of evaluating $Q_m(\phi, \mathcal{D})$ has costs $c_m(\phi, \mathcal{D})$, the average number of marked labels per cost unit are

$$\tfrac{1}{c_m(\phi,\mathcal{D})}(P_m(\phi, \mathcal{D}) \cdot z_m(\phi) + (1 - P_m(\phi, \mathcal{D})) \cdot z_{\neg m}(\phi)).$$

A similar formula holds for the average number of marked anti-monotone labels, so the optimal node in the partial version space tree is the one where

$$\max\{\tfrac{1}{c_m}(P_m z_m + (1 - P_m) \cdot z_{\neg m}), \tfrac{1}{c_a} \cdot (P_a z_a + (1 - P_a) \cdot z_{\neg a})\} \qquad (1)$$

is maximal.

The question now is how to determine the probabilities and costs. For certain types of queries and databases it is conceivable that costs grow with increasing complexity of the patterns. For example, testing the coverage of a string becomes more expensive as the pattern becomes longer. On the other hand, short patterns are more likely to satisfy Q_m than long ones (and vice versa for Q_a). Therefore, length could be taken into account when approximating the the above costs and probabilities, but let us for now assume that there is no prior knowledge about those values, so we simply take $P_m = P_a = \frac{1}{2}$ and $c_m = c_a = 1$. With uniform costs and probabilities (1) breaks down to $\frac{1}{2}\max\{z_m + z_{\neg m}, z_a + z_{\neg a})\}$.

3.4 Calculating the Counter-Values

Next, we show how the four counter-values can be computed. Let us start with the $z_m(\phi)$-counter. Since a monotone ⊕-label for ϕ would be propagated up to the root by following ϕ's parent- and suffix-link, we basically have that $z_m(\phi)$ is the sum of $z_m(\mathit{parent}(\phi))$ and $z_m(\mathit{suffix}(\phi))$ plus 1 for ϕ itself. But, due to the fact that $\mathit{parent}(\mathit{suffix}(\phi)) = \mathit{suffix}(\mathit{parent}(\phi))$, we have that this z_m-value has been counted twice; we thus need to subtract it once (see Fig. 4):

$$z_m(\phi) \approx z_m(\mathit{parent}(\phi)) + z_m(\mathit{suffix}(\phi)) - z_m(\mathit{parent}(\mathit{suffix}(\phi))) + 1 \qquad (2)$$

There are some exceptions to this equation: The easiest case is when $\mathit{suffix}(\phi) = \mathit{parent}(\phi)$, which happens if $\phi = \gamma^n$ for $\gamma \in \Sigma, n \in \mathbb{N}$. We then have

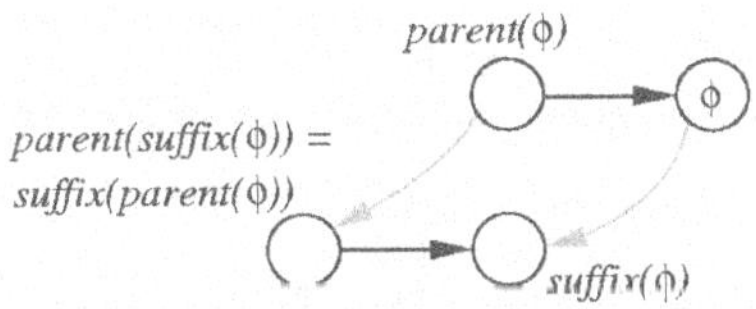

Fig. 4. Calculating z_m: When summing up $z_m(\mathit{parent}(\phi))$ and $z_m(\mathit{suffix}(\phi))$, $z_m(\mathit{parent}(\mathit{suffix}(\phi)))$ has been counted twice.

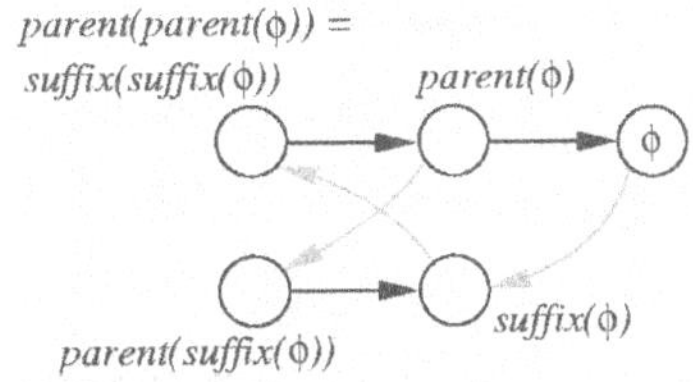

Fig. 5. If $\phi = \gamma_1 \dots \gamma_n$ $(\gamma_i \in \Sigma)$ and $\mathit{suffix}^2(\phi) = \mathit{parent}^2(\phi)$, we have $\gamma_1 \dots \gamma_{n-3} = \gamma_3 \dots \gamma_n$, i.e. $\gamma_1 = \gamma_3 = \gamma_5 = \dots, \gamma_2 = \gamma_4 = \dots$

$z_m(\phi) = z_m(\mathit{parent}(\phi)) + 1$ because $\mathit{parent}(\phi)$ is the only immediate generalization of ϕ. A slightly more complicated exception is when $\mathit{suffix}^2(\phi) = \mathit{parent}^2(\phi)$, which happens when $\phi = \gamma\delta\gamma\delta\gamma\dots$ for $\gamma, \delta \in \Sigma$ (see Fig. 5). Then $z_m(\phi) = z_m(\mathit{parent}(\phi)) + 2$, because all patterns that are more general than ϕ (apart from $\mathit{suffix}(\phi)$) are already counted by $z_m(\mathit{parent}(\phi))$. Similar rules for calculating z_m hold for exceptions of the type $\mathit{suffix}^3(\phi) = \mathit{parent}^3(\phi)$:

Lemma 1. *The counter-value for z_m is given by*

$$z_m(\phi) = \begin{cases} z_m(\mathit{parent}(\phi)) + n & \text{if } \mathit{suffix}^n(\phi) = \mathit{parent}^n(\phi) \\ z_m(\mathit{parent}(\phi)) + z_m(\mathit{suffix}(\phi)) - \\ \quad z_m(\mathit{parent}(\mathit{suffix}(\phi))) + 1 & \text{otherwise} \end{cases}$$

where we take the smallest value of n for which the "exceptional" case applies.

In practical implementations of this method it is advisable to "cut off" the search for the exceptional cases at a fixed depth and take the value of the "otherwise"-case as an approximation to the true value of z_m.

Since anti-monotone $\ominus$-labels are propagated in the same direction as monotone $\oplus$-labels, lemma 1 holds for $z_{\neg a}$ as well. For the remaining two counters, we have to "peek" in the other direction, i.e. we have to consider the children and incoming suffix-links of ϕ. In a similar manner as we did for z_m, we have to consider the values that have been counted twice when summing over the z_a's of ϕ's children and incoming suffix-links. These are the children of all incoming suffix-links, because their suffix-links point to the children of ϕ.

$$z_a(\phi) \approx \sum_{\psi \in \mathit{children}(\phi)} z_a(\psi) + \sum_{\psi \in \mathit{isl}(\phi)} z_a(\psi) - \sum_{\psi \in \mathit{children}(\mathit{isl}(\phi))} z_a(\psi) + 1. \quad (3)$$

Again, we need to consider some special cases where the above formula does not hold. Due to space limitations, we only sketch some of these cases, see [8] for more details. The first is when one of ϕ's children has ϕ as its suffix, which happens iff $\phi = \gamma^n$ for $\gamma \in \Sigma, n \in \mathbb{N}$, because one of ϕ's sons is γ^{n+1}. In this case, we just sum *once* over this node and do *not* subtract z_a of γ^{n+1}'s children, because they were counted only once. The second exception arises when ψ, one of ϕ's grandchildren, has one of ϕ's incoming suffix-links as its suffix.

4 Experimental Results

We implemented the algorithm from Fig. 3 with two different queuing-strategies. The first, called *Random*, uses random priorities for the nodes in queue P. The second strategy, called *CounterMax*, works with the counter-values from Sect. 3.3, where we chose uniform costs and probabilities. We checked for exceptional cases up to suffix^2, as explained in Sect. 3.4. According to the algorithm and to (1), each pattern is tested either against Q_m or Q_a, depending on which of the subqueries yields the maximum. We compared the results to an implementation of algorithm VST of [6] which constructs the version space tree in two passes (called DESCEND and ASCEND). The DESCEND algorithm is a straightforward adaptation of the Apriori and levelwise algorithm for use with strings and version space trees. It computes the set of all solutions w.r.t. Q_m. ASCEND starts from this result working bottom up and starting from the leaves of the tree. For each leaf, ASCEND tests whether it satisfies Q_a, if it does, the parent of the leaf will be tested; if it does not, the pattern is labelled $\ominus$ and the labels are propagated towards the parents and suffixes, more details can be found in [6].

We used a nucleotide database to compare the three algorithms, so our alphabet was $\Sigma = \{\mathtt{a}, \mathtt{c}, \mathtt{g}, \mathtt{t}\}$. Working with a large alphabet significantly increases the number of nodes with a ?⃝. The first dataset $\mathcal{D}_1$ was used for a minfrequency query and consisted of the first hundred nucleotide sequences from the *Hepatitis C* virus of the NIH genetic sequence database *GenBank* [23]. The second dataset $\mathcal{D}_2$ held the first hundred sequences from the *Escherichia coli* bacterium and was used for a maxfrequency query. The average length of the entries in $\mathcal{D}_1$ was about 500, the maximum 10,000. For $\mathcal{D}_2$ we had the values 2,500 and 30,000, respectively. We do not pretend that our results have any biological relevance; we simply used these datasets as a testbed for the different methods. We ran each algorithm several times for the query $\mathsf{minfreq}(\phi; min; \mathcal{D}_1) \wedge \mathsf{maxfreq}(\phi; max; \mathcal{D}_2)$, where each of the variables *min* and *max* could take one of the values $\{2, 3, 4, 5, 6, 7, 8, 9, 10, 15, 20, \ldots, 95\}$. Some of the results are listed in Fig. 6.

Figures 6.(a) and 6.(b) show how the number of database accesses grows with decreasing values for *min* when the maximum frequency is constant. Although *max* has been fixed to 1, similar graphs could be shown for other values of *max*. Note that in the region where the hard instances lie ($min \in \{2, \ldots, 10\}$), *CounterMax* performs significantly better than VST. This is in particular true for the number of evaluations of Q_a (Fig. 6.(b)). For the easy instances, our method takes slightly more evaluations of both predicates. This is obvious; if the length of the longest pattern satisfying Q_m is small, it is very unlikely to beat a levelwise method. The random strategy lies somewhere in the middle between those two other methods.

Figures 6.(c) and 6.(d) show the performance when *min* is fixed and *max* changes. The first thing to note is that in Fig. 6.(c) the number of evaluations of Q_m is constant for VST. This is a simple consequence of how the algorithm works. Again, *CounterMax* takes less evaluations than VST, and *Random* is in between. In Fig. 6.(d) we can see that for VST, the number of evaluations of Q_a levels off when *max* decreases, whereas the other two methods behave conversely. The reasons for this are clear: VST treats the anti monotonic query

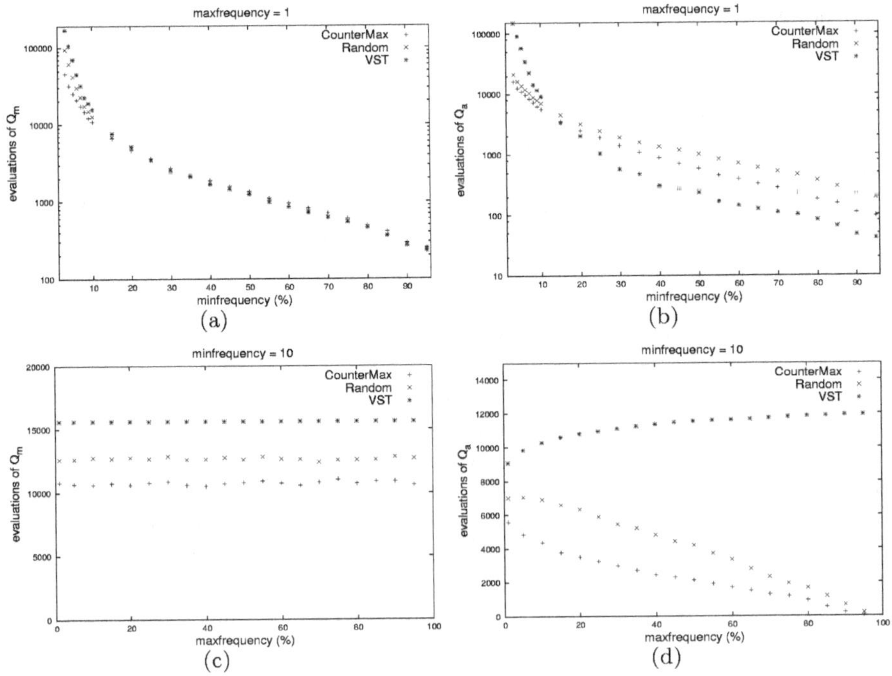

Fig. 6. A comparison of the three algorithms. Note that in figures (a) and (b) a logarithmic scale has been used.

in a bottom-up manner by starting at the leaves. When it encounters a negative pattern w.r.t. Q_a, it propagates this label up to the root. This is more likely to happen at the leaves for small *max*, so in these cases it saves a lot of evaluations of the anti monotonic predicate. For methods *CounterMax* and *Random* it is better when positive patterns (w.r.t. Q_a) are close to the root, which happens for large *max*, because then all newly expanded nodes will "automatically" be marked with a $\ominus$ and need never be tested against Q_a.

5 Related Work and Conclusions

Algorithms that try to minimize the total number of predicate evaluations have been around for several years, most notably Gunopulos et al.'s *Dualize & Advance*-algorithm [11,10] that computes $S(\mathsf{minfreq}(\phi, \cdot, \mathcal{D}), \mathcal{L})$ in the domain of itemsets. This works roughly as follows: first, a set *MS* of maximal specific sentences is computed by a randomized depth-first search. Then the negative border of *MS* is constructed by calculating a *minimum hypergraph transversal* of the complements of all itemsets in *MS*. This process is repeated with the elements from the hypergraph transversal until no more maximal specific sentences can be found. The result is then set of *all* maximal interesting itemsets.

Although Gunopulos et al. work with itemsets and only consider monotonic predicates, there is a clear relation to our approach. Whereas the former method needs to compute the minimum hypergraph transversals to find the candidates for new maximal interesting sentences, these can be directly read off the partial version space tree. In fact, all nodes whose monotonic part is still undetermined are the only possible patterns of $S(\mathsf{minfreq}(\phi, \cdot, \mathcal{D}), \mathcal{L})$ that have not been found so far. These are exactly the nodes that are still waiting in the priority queue. So by performing a depth-first expansion until all children are negative, our algorithm's behaviour is close to that of *Dualize & Advance.* It should be noted that the two strategies are not entirely equal: if a node ϕ has negative children only, it is not necessarily a member of S because there could still be more specific patterns that have ϕ as their *suffix* and satisfy Q_m.

One of the fastest algorithms for mining maximal frequent sets is Bayardo's *Max-Miner* [3]. This one uses a special set-enumeration technique to find large frequent itemsets before it considers any of their subsets. Although this is completely different from what we do at first sight, *CounterSum* also has a tendency to test long strings first because they will have higher z_m-values. By assigning higher values to P_m for long patterns this behaviour can even be enforced.

Finally, the present work is a significant extension of that by [6] in that we have adapted and extended their version space tree and also shown how it can be used for optimizing the evaluation of conjunctive queries. The presented technique has also shown to be more cost effective and flexible than the ASCEND and DESCEND algorithms proposed by [6], which employ a traditional level wise search which minimizes the number of passes through the data rather than a more informative cost-function.

Acknowledgements. This work was partly supported by the European IST FET project cInQ. The authors are grateful to Manfred Jaeger, Sau Dan Lee and Heikki Mannila for contributing the theoretical framework on which the present work is based, to Amanda Clare and Ross King for the yeast database on which we also evaluated our method and to Sau Dan Lee and Kristian Kersting for feedback on this work.

References

1. R. Agrawal and R. Srikant. Fast algorithms for mining association rules. In *Proc. VLDB*, 1994.
2. R. Agrawal, T. Imielinski, A. Swami. Mining association rules between sets of items in large databases. In *Proc. SIGMOD*, pp. 207-216, 1993.
3. R. Bayardo. Efficiently mining long patterns from databases. In *Proc. SIGMOD*, 1998.
4. C. Bucila, J. Gehrke, D. Kifer, W. White. DualMiner: A dual pruning algorithm for itemsets with constraints. In *Proc. of SIGKDD*, 2002.
5. L. De Raedt, S. Kramer. The levelwise version space algorithm and its application to molecular fragment finding. In *Proc. IJCAI*, 2001.
6. L. De Raedt, M. Jäger, S. D. Lee, H. Mannila: A Theory of Inductive Query Answering. In *Proceedings IEEE ICDM*, 2002.

7. L. De Raedt, A perspective on inductive databases. *SIGKDD Explorations*, Vol. 4 (2), 2002.
8. J. Fischer. Version Spaces im Contraint-Based Data Mining. Diplomarbeit. Albert-Ludwigs-University Freiburg, 2003.
9. B. Goethals, J. Van den Bussche. On supporting interactive association rule mining. In *Proc. DAWAK*, LNCS Vol. 1874, Springer Verlag, 2000.
10. D. Gunopulos, H. Mannila, S. Saluja. Discovering All Most Specific Sentences by Randomized Algorithms. In *Proc. ICDT*, LNCS Vol. 1186, Springer Verlag, 1997.
11. D. Gunopulos, R. Khardon, H. Mannila, H. Toivonen: Data mining, Hypergraph Transversals, and Machine Learning. In *Proceedings PODS*, 1997.
12. J. Han, Y. Fu, K. Koperski, W. Wang, and O. Zaiane. DMQL: A Data Mining Query Language for Relational Databases.In *Proc. SIGMOD'96 Workshop on Research Issues on Data Mining and Knowledge Discovery*, 1996.
13. J. Han, L. V. S. Lakshmanan, and R. T. Ng. Constraint-Based, Multidimensional Data Mining, *Computer*, Vol. 32(8), pp. 46-50, 1999.
14. J. Han, J. Pei, and Y. Yin. Mining frequent patterns without candidate generation. In *Proc. SIGMOD*, 2000.
15. H. Hirsh. Generalizing Version Spaces. *Machine Learning*, Vol. 17(1): 5-46 (1994).
16. H. Hirsh. Theoretical underpinnings of versionspaces. In *Proc. IJCAI*, 1991.
17. Akihiro Inokuchi, Takashi Washio, Hiroshi Motoda: Complete Mining of Frequent Patterns from Graphs: Mining Graph Data. *Machine Learning*, 50(3): 321-354 (2003)
18. S. Kramer, L. De Raedt, C. Helma. Molecular Feature Mining in HIV Data. In *Proc. SIGKDD*, 2001.
19. H. Mannila and H. Toivonen. Levelwise search and borders of theories in knowledge discovery, *Data Mining and Knowledge Discovery*, Vol. 1, 1997.
20. T. Mitchell. Generalization as Search, *Artificial Intelligence*, Vol. 18 (2), pp. 203-226, 1980.
21. T. Mitchell. Machine Learning. McGraw Hill. 1997.
22. R. T. Ng, L. V.S. Lakshmanan, J. Han, and A. Pang. Exploratory mining and pruning optimizations of constrained associations rules. In *Proc. SIGMOD*, 1998.
23. *NIC Genetic Sequence Database*. Available at `www.ncbi.nlm.nih.gov/Genbank/GenbankOverview.html`
24. E. Ukkonen. On-line construction of suffix trees. *Algorithmica*, 14(3):249–260, 1995.
25. P. Weiner. Linear pattern matching algorithm. In *Proc. 14th IEEE Symposium on Switching and Automata Theory*, pages 1–11, 1973.

Febrl – A Parallel Open Source Data Linkage System

http://datamining.anu.edu.au/linkage.html

Peter Christen[1]*, Tim Churches[2], and Markus Hegland[3]

[1] Department of Computer Science, Australian National University, Canberra ACT 0200, Australia, peter.christen@anu.edu.au
[2] Centre for Epidemiology and Research, New South Wales Department of Health, Locked Mail Bag 961, North Sydney NSW 2059, Australia, tchur@doh.health.nsw.gov.au
[3] Centre for Mathematics and its Applications, Mathematical Sciences Institute, Australian National University, Canberra ACT 0200, Australia, markus.hegland@anu.edu.au

Abstract. In many data mining projects information from multiple data sources needs to be integrated, combined or linked in order to allow more detailed analysis. The aim of such linkages is to merge all records relating to the same entity, such as a patient or a customer. Most of the time the linkage process is challenged by the lack of a common unique entity identifier, and thus becomes non-trivial. Linking todays large data collections becomes increasingly difficult using traditional linkage techniques. In this paper we present an innovating data linkage system called *Febrl*, which includes a new probabilistic approach for improved data cleaning and standardisation, innovative indexing methods, a parallelisation approach which is implemented transparently to the user, and a data set generator which allows the random creation of records containing names and addresses. Implemented as open source software, *Febrl* is an ideal experimental platform for new linkage algorithms and techniques.

Keywords: Record linkage, data matching, data cleaning and standardisation, parallel processing, data mining preprocessing.

1 Introduction

Data linkage can be used to improve data quality and integrity, to allow re-use of existing data sources for new studies, and to reduce costs and efforts in data acquisition for research studies. In the health sector, for example, linked data might contain information which is needed to improve health policies, information that is traditionally collected with time consuming and expensive survey methods. Linked data can also help in health surveillance systems to enrich data that is used for pattern detection in data mining systems. Businesses routinely

* Corresponding author

H. Dai, R. Srikant, and C. Zhang (Eds.): PAKDD 2004, LNAI 3056, pp. 638–647, 2004.

deduplicate and link their data sets to compile mailing lists. Another application of current interest is the use of data linkage in crime and terror detection.

If a unique entity identifier or key is available in all the data sets to be linked, then the problem of linking at the entity level becomes trivial, a simple *join* operation in *SQL* or its equivalent in other data management systems is all that is required. However, in most cases no unique key is shared by all of the data sets, and more sophisticated linkage techniques need to be applied. These techniques can be broadly classified into *deterministic* or rules-based approaches (in which sets of often very complex rules are used to classify pairs of records as *links*, i.e. relating to the same person or entity, or as *non-links*), and *probabilistic* approaches (in which statistical models are used to classify record pairs). Probabilistic methods can be further divided into those based on *classical* probabilistic record linkage theory as developed by *Fellegi & Sunter* [6], and newer approaches using maximum entropy, clustering and other machine learning techniques [2,4, 5,10,12,14,19].

Computer-assisted data linkage goes back as far as the 1950s. At that time, most linkage projects were based on *ad hoc* heuristic methods. The basic ideas of probabilistic data linkage were introduced by *Newcombe & Kennedy* [15] in 1962 while the theoretical foundation was provided by *Fellegi & Sunter* [6] in 1969. The basic idea is to link records by comparing common attributes, which include person identifiers (like names, dates of birth, etc.) and demographic information. Pairs of records are classified as *links* if their common attributes predominantly agree, or as *non-links* if they predominantly disagree. If two data sets **A** and **B** are to be linked, record pairs are classified in a product space $\mathbf{A} \times \mathbf{B}$ into M, the set of true matches, and U, the set of true non-matches. *Fellegi & Sunter* [6] considered ratios of probabilities of the form

$$R = \frac{P(\gamma \in \Gamma | M)}{P(\gamma \in \Gamma | U)}$$

where γ is an arbitrary agreement pattern in a comparison space Γ. For example, Γ might consist of six patterns representing simple agreement or disagreement on (1) given name, (2) surname, (3) date of birth, (4) street address, (5) suburb and (6) postcode. Alternatively, some of the γ might additionally account for the relative frequency with which specific values occur. For example, a surname value *"Miller"* is normally much more common than a value *"Dijkstra"*, resulting in a smaller agreement value. The ratio R or any monotonically increasing function of it (such as its logarithm) is referred to as a *matching weight*. A decision rule is then given by

if $R > t_{upper}$, then	designate a record pair as *link*
if $t_{lower} \leq R \leq t_{upper}$, then	designate a record pair as *possible link*
if $R < t_{lower}$, then	designate a record pair as *non-link*

The thresholds t_{lower} and t_{upper} are determined by a-priori error bounds on false links and false non-links. If $\gamma \in \Gamma$ mainly consists of agreements then the ratio R would be large and thus the record pair would more likely to be designated as a link. On the other hand for a $\gamma \in \Gamma$ that primarily consists of disagreements the ratio R would be small. The class of *possible links* are those record pairs for which human oversight, also known as *clerical review*, is needed to decide their final linkage status (as often no additional information is available the clerical review process becomes one of applying human intuition, experience or common sense to the decision based on available data).

In this paper we present some key aspects of our parallel open source data linkage system *Febrl* (for "Freely extensible biomedical record linkage"), which is implemented in the object-oriented language *Python*[1] (which is open source itself) and freely available from the project web page. Due to the availability of its source code, *Febrl* is an ideal platform for the rapid development and implementation of new and improved data linkage algorithms and techniques.

2 Related Work

The processes of data cleaning, standardisation and data linkage have various names in different user communities. While statisticians and epidemiologists speak of *record* or *data linkage* [6], the same process is often referred to as *data* or *field matching*, *data scrubbing*, *data cleaning*, *preprocessing*, or as the *object identity problem* [7,11,18] by computer scientists and in the database community, whereas it is sometimes called *merge/purge processing* [9], *data integration* [4], *list washing* or *ETL* (extraction, transformation and loading) in commercial processing of customer databases or business mailing lists. Historically, the statistical and the computer science community have developed their own techniques, and until recently few cross-references could be found.

Improvements [19] upon the classical *Fellegi & Sunter* [6] approach include the application of the expectation-maximisation (EM) algorithm for improved parameter estimation [20], and the use of approximate string comparisons [16] to calculate partial agreements when attribute values have typographical errors. Fuzzy techniques and methods from information retrieval have recently been used to address the data linkage problem [2]. One approach is to represent records as document vectors and to compute the *cosine distance* [4] between such vectors. Another possibility is to use an *SQL* like language [7] that allows approximate joins and cluster building of similar records, as well as decision functions that decide if two records represent the same entity. Other methods [11] include statistical outlier identification, pattern matching, clustering and association rules based approaches.

In recent years, researchers have also started to explore the use of machine learning and data mining techniques to improve the linkage process. The authors of [5] describe a hybrid system that in a first step uses unsupervised clustering on a small sample data set to create data that can be used in the second step to classify record pairs into links or non-links. Learning field specific string-edit distance weights [14] and using a binary classifier based on support vector machines (SVM) is another approach. A system that is capable to link very large data sets with hundreds of millions of records – using special sorting and preprocessing techniques – is presented in [21].

[1] `http://www.python.org`

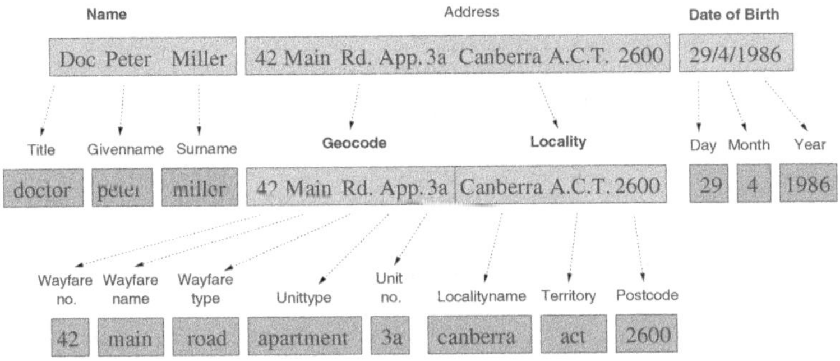

Fig. 1. Example name and address standardisation.

3 Probabilistic Data Cleaning and Standardisation

As most real world data collections contain noisy, incomplete and incorrectly formatted information, data cleaning and standardisation are important preprocessing steps for successful data linkage, and before such data can be loaded into data warehouses or used for further analysis [18]. Data may be recorded or captured in various, possibly obsolete, formats and data items may be missing, out of date, or contain errors. The cleaning and standardisation of names and addresses is especially important for data linkage, to make sure that no misleading or redundant information is introduced (e.g. duplicate records).

The main task of data cleaning and standardisation is the conversion of the raw input data into well defined, consistent forms and the resolution of inconsistencies in the way information is represented or encoded. The example record shown in Figure 1 consisting of three input components is cleaned and standardised into 14 output fields (the dark coloured boxes). Comparing these output fields individually with the corresponding output fields of other records results in a much better linkage quality than just comparing the whole name or the whole address as a string with the name or address of other records.

Rule-based data cleaning and standardisation as currently done by many commercial systems is cumbersome to set up and maintain, and often needs adjustments for new data sets. We have recently developed (and implemented within *Febrl*) new probabilistic techniques [3] based on hidden Markov models (HMMs) [17] which showed to achieve better standardisation accuracy and are easier to set-up and maintain compared to popular commercial linkage software.

Our approach is based on the following three steps. First, the input strings are *cleaned*. This involves converting input into lower-case characters, and replacing certain words and abbreviations with others (these replacements are listed in look-up tables that can be edited by the user). In the second step, the input strings are split into a list of words, numbers and characters, which are then *tagged* using look-up tables (mainly for titles, given- and surnames, street names and types, suburbs, postcodes, states, etc.) and some hard-coded rules (e.g. for numbers, hyphens or commas). Thirdly, these tagged lists are *segmented* into

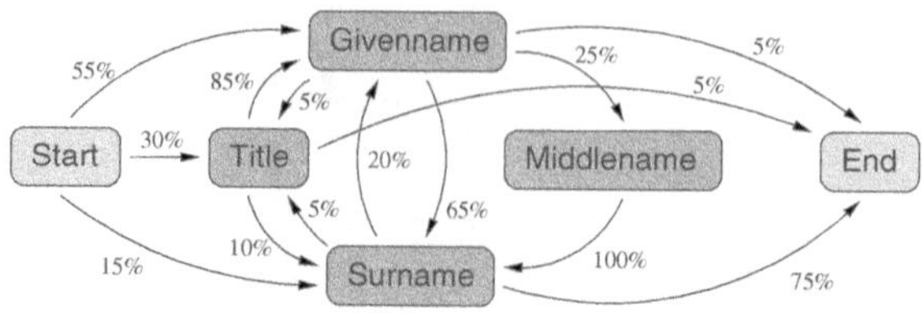

Fig. 2. Simple example hidden Markov model for names.

output fields using a HMM, i.e. by using the *Viterbi* algorithm [17] the most likely path through the HMM gives the corresponding output fields (the states of the HMM).

Details about how to efficiently train the HMMs for name and address standardisation, and experiments with real-world data are given in [3]. Training of the HMMs is quick and does not require any specialised skills. For addresses, our HMM approach produced equal or better standardisation accuracies than a widely-used rule-based system. However, accuracies were slightly worse when used with simpler name data [3].

4 Blocking, Indexing, and Classification

Data linkage considers the distribution of record pairs in the product space $\mathbf{A} \times \mathbf{B}$ and determines which of these pairs are links. The number of possible pairs equals the product of the sizes of the two data sets **A** and **B**. The straight-forward approach would consider all pairs and model their distribution. As the performance bottleneck in a data linkage system is usually the expensive evaluation of a similarity measure between pairs of records [1], this approach is computationally not feasible for large data sets, it is *non-scalable*. Linking two data sets each with $100,000$ records would result in ten billion potential links (and thus comparisons). On the other hand, the maximum number of links that are possible corresponds to the number of records in the smaller data set (assuming a record can be linked to only one other record). Thus, the space of potential links becomes sparser with increasing number of records, while the computational efforts increase exponentially.

To reduce the huge amount of possible record comparisons, traditional data linkage techniques [6,19] work in a blocking fashion, i.e. they use one or more record attributes to split the data sets into blocks. Only records having the same value in such a *blocking variable* are then compared (as they will be in the same block). This technique becomes problematic if a value in a blocking variable is recorded wrongly, as the corresponding record is inserted into a different block. To overcome this problem, several iterations (passes) with different blocking variables are normally performed.

While such blocking (or indexing) techniques should reduce the number of comparisons made as much as possible by eliminating comparisons between records that obviously are not links, it is important that no potential link is overlooked because of the indexing process. Thus there is a trade-off between (a) the reduction in number of record pair comparisons and (b) the number of missed true matches (accuracy).

Febrl currently contains three different indexing methods, with more to be included in the future. In fact, the exploration of improved indexing methods is one of our major research areas [1]. The first indexing method is the standard blocking method [6,19] applied in traditional data linkage systems. The second indexing method is based on the *sorted neighbourhood* [10] approach, where records are sorted alphabetically according to the values of the blocking variable, then a sliding window is moved over the sorted records, and record pairs are formed using all records within the window. The third method uses *n-grams* (substrings of length n) and allows for fuzzy blocking (in the current implementation we use bigrams, i.e. $n = 2$). The values of the blocking variable are converted into lists of bigrams, and permutations of sub-lists are built using a threshold (a value between 0.0 and 1.0) of all possible permutations. The resulting bigram sub-lists are converted back into strings and used as keys in an inverted index, which is then used to retrieve the records in a block.

Initial experiments [1] showed that innovative indexing methods can improve upon the traditional blocking used in data linkage, but further research needs to be conducted. Other research groups have also investigated the use of *n-grams* [2] as well as high-dimensional approximate distance metrics to form overlapping clusters [12].

For each record pair in the index a vector containing *matching weights* is calculated using field comparison functions (for strings, numbers, dates and times). This vector is then used to classify the pair as either a *link*, *non-link*, or *possible link* (in which case the decision should be done by a human review). While the classical *Fellegi & Sunter* [6] simply sums all the weights in the vector into one matching weight, alternative classifiers are possible which improve upon this. For example, separate weights can be calculated for names and addresses, or machine learning techniques can be used for the classification task. Classifiers currently implemented in *Febrl* are the classical *Fellegi & Sunter* [6] classifier described earlier, and a *flexible classifier* that allows the calculation of the matching weights using various functions.

5 Parallelisation

Although computing power has increased tremendously in the last few decades, large-scale data cleaning, standardisation and data linkage are still resource-intensive processes. There have been relatively few advances over the last decade in the way in which probabilistic data linkage is undertaken, particularly with respect to the tedious clerical review process which is still needed to make decisions about pairs of records whose linkage status is doubtful. In order to be able to link large data sets, parallel processing becomes essential. Issues that have to

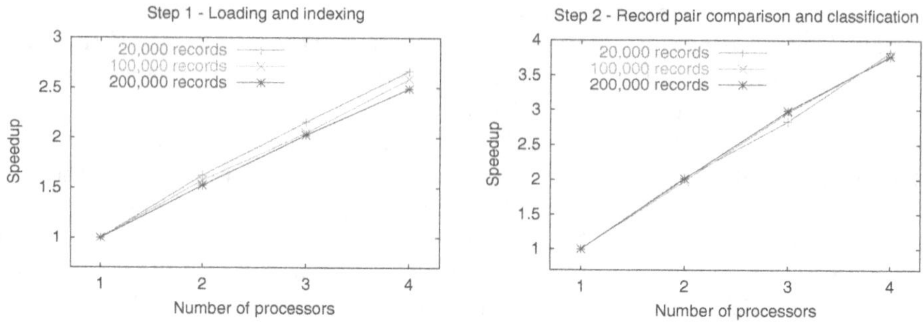

Fig. 3. Speedups for parallel deduplication (internal linkage).

be addressed are efficient data distribution, fault tolerance, dynamic load balancing, portability and scalability (both with the data size and the number of processors used).

Confidentiality and privacy have to be considered as data linkage deals with partially identified data, and access restrictions are required. The use of high-performance computing centers (which traditionally are multi-user environments) becomes problematic. An attractive alternative are networked personal computers or workstations which are available in large numbers in many businesses and organisations. Such office based clusters can be used as virtual parallel computing platforms to run large scale linkage tasks over night or on weekends.

Parallelism within *Febrl* is currently in its initial stages. Based on the well known *Message Passing Interface* (MPI) [13] standard, and the Python module *Pypar*[2] which provides bindings to an important subset of the MPI routines, parallelism is implemented transparently to the user of *Febrl*.

To give an idea on the parallel performance of *Febrl* some initial timing results of experiments made on a parallel computing platform (a *SUN Enterprise 450* shared memory (SMP) server with four 480 MHz *Ultra-SPARC II* processors and 4 Giga Bytes of main memory) are presented in this section. Three internal linkage (deduplication) processes were performed with 20,000, 100,000 and 200,000 records, respectively, from a health data set containing midwife data records provided by the *NSW Department of Health*. Six field comparison functions were used and the classical blocking index technique with three indexes (passes) was applied. The standard *Fellegi & Sunter* [6] classifier was used to classify record pairs.

These deduplication processes were run using 1, 2, 3 and 4 processors, respectively. In Figure 3 *speedup* (which is defined as the time on one processor divided by the time on 2, 3, or 4 processors, respectively) results are shown scaled by the number of records (i.e. the total elapsed run times divided by the number of records were used to calculate the speedups). The results show that

[2] `http://datamining.anu.edu.au/pypar/`

```
       rec_id, streetnum,            address1,            address2,  suburb, postcode
   rec-0-org,         11,         wylly place, inverpine ret vill,   taree,     4860
 rec-0-dup-0,         11,          wyllyplace, inverpine ret vill,   taree,     4860
 rec-0-dup-1,         11,  inverpine ret vill,        wylly place,   taree,     4860
 rec-0-dup-2,          3,         wylly place, inverpine ret vill,   tared,     4860
 rec-0-dup-3,         11,        wylly parade, inverpine ret vill,   taree,     4680
   rec-1-org,         15,       stuart street,           hartford,  menton,     3135
   rec-2-org,         20,    griffiths street,             myross,   kilda,     2756
 rec-2-dup-0,         20,    griffith sstreet,             myross,   kilda,     2576
 rec-2-dup-1,         20,     griffith street,            mycross,   kilda,
   rec-3-org,          5,  ellenborough place,  kalkite homestead,  sydney,     2212
```

Fig. 4. Randomly generated example data set.

the record pair comparison and classification step (step 2, which takes between 94% and 98% of the total run times) is scalable, while loading and building of the blocking indexes (step 1) results in lower speedup values and is not scalable. As most time is spent in step 2, the overall parallel performance is quite scalable. Communication times can be neglected as they were less than 0.35% of the total run times in all experiments.

6 Data Set Generation

As data linkage is dealing with data sets that contain partially identified data, like names and addresses, it can be very difficult to acquire data for testing and evaluating newly developed linkage algorithms and techniques. For the user it can be difficult to learn how to apply, evaluate and customise data linkage algorithms effectively without example data sets where the linkage status of record pairs is known.

To overcome this we have developed a database generator based on ideas by *Hernandez & Stolfo* [9]. This generator can create data sets that contain names (based on frequency look-up tables for given- and surnames), addresses (based on frequency look-up tables for suburbs, postcodes, street numbers, types and names, and state/territory names), dates (like dates of birth), and randomly created identifier numbers (for example for social security numbers).

To generate a data set, a user needs to provide the number of *original* and *duplicate* records to be created, the maximal number of duplicates for one original record, a probability distribution of how duplicates are created (possible are *uniform*, *Poisson* and *zipf*), and the probabilities for introducing various random modifications to create the duplicate records. These modifications include inserting, deleting, transposing and substituting characters; swap a field value (with another value from the same look-up table); inserting or deleting spaces; setting a field value to missing; or swapping the values of two fields (e.g. surname with given name). Each created record is given a unique identifier, which allows the evaluation of accuracy and error rates for data linkage procedures (false linked record pairs and un-linked true matches).

Figure 4 shows a small example data set containing 4 originals and 6 duplicate records, randomly created using Australian address frequency look-up tables.

7 Conclusions and Future Work

Written in an object-oriented open source scripting language, the *Febrl* data linkage system is an ideal experimental platform for researchers to develop, implement and evaluate new data linkage algorithms and techniques. While the current system can be used to perform smaller data cleaning, standardisation and linkage tasks, further work needs to be done to allow the efficient linkage of large data sets. We plan to improve *Febrl* in several areas.

For data cleaning and standardisation, we will be improving upon our recently developed probabilistic techniques [3] based on hidden Markov models (HMMs), by using the *Baum-Welch* forward-backward algorithm [17] to re-estimate the probabilities in the HMMs, and we will explore techniques that can be used for developing HMMs without explicitly specifying the hidden states.

We aim to further explore alternative techniques for indexing based on high-dimensional clustering [12], inverted indexes, or fuzzy *n-gram* indexes [1,2], in terms of their applicability for indexing as well as their scalability both in data size and parallelism.

Current methods for record pair classification based on the traditional *Fellegi & Sunter* [6] approach apply the semi-parametric mixture models and the EM algorithm [20] for the estimation of the underlying densities and the clustering and classification of links and non-links [19]. We will investigate the performance of data mining or non-parametric techniques including tree-based classifiers, sparse grids [8] and Bayesian Nets. An advantage of these methods is that they allow adaptive modelling of correlated data and can deal with complex data and the curse of dimensionality. For all these methods current fitting algorithms will be adapted and new ones developed.

We will continue to improve upon the parallel processing functionalities of *Febrl* with an emphasis on running large linkage processes on clusters of personal computers (PCs) and workstations as available in many businesses and organisations. Such office based PC clusters (with some additional software installed) can be used as virtual parallel computing platforms to run large scale linkage tasks over night or on weekends. Confidentiality and privacy aspects will need to be considered as well, as data linkage in many cases deals with identified data.

Acknowledgments. This project is funded by the *Australian National University (ANU)* and the *New South Wales Department of Health* under an *AICS (ANU-Industry Collaboration Scheme)* grant *AICS #1-2001*. Additional funding is provided by the *Australian Partnership for Advanced Computing (APAC)*.

References

1. Baxter, R., Christen, P. and Churches, T.: A Comparison of Fast Blocking Methods for Record Linkage. ACM SIGKDD '03 Workshop on Data Cleaning, Record Linkage, and Object Consolidation, August 27, 2003, Washington, DC, pp. 25-27.
2. Chaudhuri, S., Ganjam, K., Ganti, V. and Motwani, R.: Robust and efficient fuzzy match for online data cleaning. Proceedings of the 2003 ACM SIGMOD International Conference on on Management of Data, San Diego, USA, 2003, pp. 313-324.

3. Churches, T., Christen, P., Lim, K. and Zhu, J.X.: Preparation of name and address data for record linkage using hidden Markov models. BioMed Central Medical Informatics and Decision Making, Dec. 2002. Available online at: http://www.biomedcentral.com/1472-6947/2/9/
4. Cohen, W.: Integration of heterogeneous databases without common domains using queries based on textual similarity. Proceedings of SIGMOD, Seattle, 1998.
5. Elfeky, M.G., Verykios, V.S. and Elmagarmid, A.K.: TAILOR. A Record Linkage Toolbox. Proceedings of the ICDE' 2002, San Jose, USA, 2002.
6. Fellegi, I. and Sunter, A.: A theory for record linkage. In Journal of the American Statistical Society, 1969.
7. Galhardas, H., Florescu, D., Shasha, D. and Simon, E.: An Extensible Framework for Data Cleaning. Proceedings of the Inter. Conference on Data Engineering, 2000.
8. Hegland, M.: Adaptive sparse grids. ANZIAM J., vol. 44, 2003, pp. C335-C353.
9. Hernandez, M.A. and Stolfo, S.J.: The Merge/Purge Problem for Large Databases. Proceedings of the ACM-SIGMOD Conference, 1995.
10. Hernandez, M.A. and Stolfo, S.J.: Real-world data is dirty: Data cleansing and the merge/purge problem. In Data Mining and Knowledge Discovery 2, Kluwer Academic Publishers, 1998.
11. Maletic, J.I. and Marcus, A.: Data Cleansing: Beyond Integrity Analysis. Proceedings of the Conference on Information Quality (IQ2000), Boston, October 2000.
12. McCallum, A., Nigam, K. and Ungar, L.H.: Efficient clustering of high-dimensional data sets with application to reference matching. Knowledge Discovery and Data Mining, pp. 169-178, 2000.
13. Gropp, W., Lusk, E. and Skjellum, A.: Using MPI – 2nd Edition, Portable Parallel Programming with the Message Passing Interface, MIT Press, 1999.
14. Nahm, U.Y, Bilenko M. and Mooney, R.J.: Two Approaches to Handling Noisy Variation in Text Mining. Proceedings of the ICML-2002 Workshop on Text Learning (TextML'2002), pp. 18-27, Sydney, Australia, July 2002.
15. Newcombe, H.B. and Kennedy, J.M.: Record Linkage: Making Maximum Use of the Discriminating Power of Identifying Information. Communications of the ACM, vol. 5, no. 11, 1962.
16. Porter, E. and Winkler, W.E.: Approximate String Comparison and its Effect on an Advanced Record Linkage System. RR 1997-02, US Bureau of the Census, 1997.
17. Rabiner, L.R.: A Tutorial on Hidden Markov Models and Selected Applications in Speech Recognition. Proceedings of the IEEE, vol. 77, no. 2, Feb. 1989.
18. Rahm, E. and Do, H.H.: Data Cleaning: Problems and Current Approaches. IEEE Data Engineering Bulletin, 2000.
19. Winkler, W.E.: The State of Record Linkage and Current Research Problems. RR 1999-04, US Bureau of the Census, 1999.
20. Winkler, W.E.: Using the EM algorithm for weight computation in the Fellegi-Sunter model of record linkage. RR 2000-05, US Bureau of the Census, 2000.
21. Yancey, W.E.: BigMatch: A Program for Extracting Probable Matches from a Large File for Record Linkage. RR 2002-01, US Bureau of the Census, March 2002.

A General Coding Method for Error-Correcting Output Codes

Yan-huang Jiang, Qiang-li Zhao, and Xue-jun Yang

School of Computer Science, National University of Defense Technology, Changsha, HuNan Province, 410073, P.R. China
jiang-yh@163.com, quentinbox@21.com, and xjyang@nudt.edu.cn

Abstract. ECOC approach can be used to reduce a multiclass categorization problem to multiple binary problems and to improve the generalization of classifiers. Yet there is no single coding method that can generate ECOCs suitable for any number of classes. This paper provides a search-coding method that associates nonnegative integers with binary strings. Given any number of classes and an expected minimum hamming distance, the method can find out a satisfied output code through searching an integer range. Experimental results show that, as a general coding method, the search-coding method can improve the generalization for both stable and unstable classifiers efficiently

1 Introduction

One-per-class[1], *all-pairs*[2], *meaning-distributed code*[3] *and error-correcting* output code (ECOC)[4] are four approaches to reduce a multiclass problem to multiple binary classification problems[5]. The advantage of *error-correcting* output code strategy over other approaches is that: ECOC approach can recover the error results of several binary functions, which improves the predictive accuracy of the supervised classifiers. Some researches show that ECOCs can reduce both variance and bias errors for multiclass problems[6]. But there are no general methods to construct effective error-correcting output codes. Dietterich and Bakiri[5] introduced four coding methods, while all of them have disadvantages: Exhaustive codes are unsuitable for the classification tasks with many classes since it will increase training time greatly; Both column selection and randomized hill climbing are uncertain methods; BCH codes can only adapt to the problems where the number of classes is a power of two. Finding a single method suitable for any number of classes is an open research problem.

This paper explores a search-coding method that associates nonnegative integers with binary strings. Given any number of classes and an expected minimum hamming distance, the method can find out a satisfied output code through searching an integer range. The search-coding method can be used as a general coding method to construct error-correcting output codes.

H. Dai, R. Srikant, and C. Zhang (Eds.): PAKDD 2004, LNAI 3056, pp. 648–652, 2004.

2 Search-Coding Method for Supervised Learning

2.1 Search-Coding Method

Our search-coding method gets error-correcting output codes through searching an integer range. Before searching process, a table named *CodeTable* must be created. Each item of the table $item(d,n)$ saves the maximum number of codewords that satisfy the code length $n(n \geq 1)$ and the minimum hamming distance $d(d \geq 1)$. *CodeTable* can be saved as permanent information after being created.

Table 1. *CodeTable* with $3 \leq d \leq 9$ and $3 \leq n \leq 16$

d	*n*													
	3	**4**	**5**	**6**	**7**	**8**	**9**	**10**	**11**	**12**	**13**	**14**	**15**	**16**
3	2	2	4	8	16	16	32	64	128	256	512	1024	2048	2048
4	0	2	2	4	8	16	16	32	64	128	256	512	1024	2048
5	0	0	2	2	2	4	4	8	16	16	32	64	128	256
6	0	0	0	2	2	2	4	4	8	16	16	32	64	128
7	0	0	0	0	2	2	2	2	4	4	8	16	32	32
8	0	0	0	0	0	2	2	2	2	4	4	8	16	32
9	0	0	0	0	0	0	2	2	2	2	2	4	4	4

Fig. 1(a) depicts the pesudo code of creating one item for *CodeTable*. In function $CreateTableItem(d,n)$, integer "0" is the only one element in the initialized set *A*. Beginning from "1", we search the whole integer range $[1,2^n - 1]$ in sequence and add more integers to *A*. If the hamming distance between the corresponding binary string of an integer *x* and that of any integer in set *A* is equal to or larger than *d*, then add *x* to *A*, and test the next integer in turn. Function $DiffBit(G,H)$ compares two binary strings and returns the number of different bits. Function $Bin(x,n)$ converts integer *x* to an *n*-bit binary string whose $j^{th}(1 \leq j \leq n)$ bit is $\lfloor x/2^{j-1} \rfloor \bmod 2$. After searching through the range $[1,2^n - 1]$, the number of the elements in set *A* is the value of $item(d,n)$. Table 1 lists each item of *CodeTable* where $3 \leq d \leq 9$ and $3 \leq n \leq 16$.

During searching process, the code length must be decided at first according to the saved *CodeTable*, the number of classes *m*, and the expected minimum hamming distance *d*. The length will be *n* if two neighbored table items $item(d,n-1)$ and $item(d,n)$ satisfy $item(d,n-1) < m \leq item(d,n)$. Then we construct an output code with *m* codewords whose code length is *n* and minimum hamming distance is *d*. Fig. 1(b) gives the pesudo code of searching an output code. Given any values of *d* and *m*, function $SearchCode(d,m)$ will return a satisfied output code. In $SearchCode(d,m)$, function $FindCodeLen(CodeTable,d,m)$ is used to decide the length of the output code. Then in a similar way as $CreateTableItem(d,n)$, we find *m* integers recorded in set A. At last, each element of A is translated into an n-bit binary string by $Bin(x,n)$. All m binary strings are saved in code matrix B, and B is the resulted output code. Of cause, if the value of d satisfies $d \geq 3$, the resulted B is an error-correcting output codes.

```
CreateTableItem(d,n)
1. If n<d Then Return 0;
2. Initialization: A={0};
3. For Each Integer x in [1,2^n-1]
    3.1 Tag=True;
    3.2 For Each Integer y in A
            If DiffBit(Bin(x,n),Bin(y,n))<d
                Then Tag=False;
    3.3 If Tag=True Then A={x}∪A ;
4. Return |A|.
```

(a)

```
SearchCode(d,m)
1. Initialization: i=0,A={0},x=1;
2. n=FindCodeLen(CodeTable,d,m);
3. while |A|<m && x<2^n do
    3.1 Tag=True;
    3.2 For Each Integer y in A
            If DiffBit(Bin(x,n),Bin(y,n))<d
                Then Tag=False;
    3.3 If Tag=True Then A={x}∪A ;
    3.4 x=x+1;
4. For Each Element y in A
    B[i]=Bin(y,n),i=i+1;
5. Return B.
```

(b)

Fig. 1. Pesudo code of the search-coding method: (a) Pesudo code of creating one item for *CodeTable*; (b) pesudo code of searching an output code

2.2 Properties of the Output Codes

We can prove that an output code generated by the search-coding method has following properties:

1. The output code has no all-zeros columns;
2. The output code has no all-ones columns;
3. The output code has no complementary columns;
4. If there are identical columns in the output code, they must be consecutive;
5. If there are k identical columns in the output code, then $k \leq d$.

The advantages of our search-coding method are: (1). It can generate output codes for a classification problem with any number of classes. (2). Given an expected minimum hamming distance, the length of the resulted output code is short. This will save training time since the number of binary functions for learning is small relatively. (3). The process of the method is certain. Therefore our search-coding method can be used as a general method to construct error-correcting output codes.

3 Experimental Results

Naïve Bayesian and Back Propagation Neural Network (BPNN) are two representative algorithms in supervised learning field. Naïve Bayesian classifier is relatively stable with respect to small changes of training data, while BPNN classifier is not. We apply our search-coding method to these two algorithms to evaluate its effect on the predictive results of stable and unstable classifiers. In the experiment, both hamming distance (HD) and absolute distance (AD) are selected as the distance measure. All experimental data sets come from the UCI machine learning repository[7]. The expected minimum hamming distance for the search-coding method is 5. The reported results are the means and deviations of 10-fold cross validation.

Table 2. Error results of different naïve Bayesian classifiers

Datasets	NB-normal	NB-D	SCNB (HD)	SCNB (AD)
Austra	19.13 ± 4.52	14.20 ± 4.03	13.91 ± 4.49	14.20 ± 4.36
Bupa	43.53 ± 10.26	40.59 ± 6.01	35.88 ± 5.27	35.58 ± 6.05
Cancer	4.06 ± 1.65	2.90 ± 1.93	2.75 ± 1.93	2.60 ± 1.74
Cleveland	43.67 ± 7.45	42.00 ± 5.02	41.00 ± 3.87	40.33 ± 3.67
Glass	50.48 ± 7.84	29.04 ± 7.26	32.38 ± 7.71	30.47 ± 6.83
Heart	15.56 ± 4.88	16.67 ± 6.59	15.18 ± 6.06	14.07 ± 5.53
Iris	6.47 ± 3.22	6.00 ± 5.83	6.00 ± 5.83	4.00 ± 3.44
Pima	24.34 ± 5.16	25.13 ± 4.49	23.29 ± 3.62	23.03 ± 3.89
Wine	2.35 ± 3.04	2.94 ± 5.00	2.35 ± 4.11	2.94 ± 5.00
Average	23.09	19.94	19.19	18.58

Table 3. Experimental results of error rate and training time for BPNN and SCBP

Data set	BPNN		SCBP		
	error (%)	time (s)	HD error (%)	AD error(%)	time (s)
Austra	15.79 ± 4.29	23.82	14.35 ± 3.64	13.91 ± 4.17	97.67
Bupa	27.06 ± 4.96	19.21	25.59 ± 5.55	26.76 ± 5.08	86.14
Cancer	3.19 ± 1.50	11.19	3.19 ± 1.33	3.19 ± 1.33	23.25
Cleveland	46.67 ± 8.16	9.30	42.67 ± 7.50	43.67 ± 7.10	153.17
Glass	29.52 ± 7.37	11.75	28.57 ± 9.52	29.05 ± 9.90	38.73
Heart	21.11 ± 7.66	8.47	15.19 ± 6.16	15.92 ± 7.21	55.52
Iris	4.67 ± 4.50	0.17	4.00 ± 4.66	3.33 ± 4.71	1.98
Pima	23.15 ± 4.77	22.19	22.50 ± 4.97	23.28 ± 4.80	102.77
Wine	1.76 ± 2.84	0.16	1.76 ± 2.84	1.18 ± 2.48	0.60
Average	19.21	11.81	17.53	17.81	62.20

Table 2 is the predictive results of different naïve Bayesian classifiers. NB-normal is the classifier that uses normal distribution as the conditional probability distribution of each class given any continuous attribute. NB-D uses the discretization for each continuous attribute. SCNB denote the naïve Bayesian classifiers with the search-coding method. Since deciding the probability distribution of either class for the binary functions in SCNB is almost impossible, we discretize the value range of each continuous attribute into several intervals and treat the attribute as a discrete one. From Table 2, we can see that NB-normal gets the worst results for 5 data sets, and also for the average results. Both SCNB (HD) and SCNB (AD) get better results than the other two classifiers for at least 7 data sets, and their average results are also the better ones. For SCNB classifiers, absolute distance measure seems better than hamming distance measure.

Table 3 gives the error rate and training time results of the BPNN and SCBP, where SCBP denotes the BPNN classifiers based on the search-coding method. During training process, the learning rate is 0.5 and the momentum rate is 0.9. As a convergence criterion, we required a mean square error smaller than 0.02. Table 3 shows that, for the Cancer data set, BPNN and SCBP get the same error rate, for the other 8 data sets, SCBP gets less error rate than BPNN. SCBP (HD) has a little better result than SCBP (AD). Yet for all of the data sets, SCBP need much longer training time than BPNN, this indicates that the *error-correcting* output codes constructed by the search-coding method generate more complex binary functions than *one-per-class* output codes.

All above results show that the search-coding method can be used to improve the generalization for both stable and unstable supervised learning classifiers. Yet for two distance measures, it is hard to decide which one is better in our experiments.

4 Conclusion

To address the disadvantages of the existing coding methods for ECOCs, this paper proposes a general coding strategy: search-coding method. Given any number of classes and an expected minimum hamming distance, the method can find out a satisfied output code through searching an integer range. By applying the method to supervised learning algorithms, experimental results show that this coding method can improve the generalization for both stable and unstable classifiers.

Acknowledgements. This research was supported by the National Natural Science Foundation of China, under Grant No. 69825104.

References

1. Rumelhart, D.E., Hinton, G.E., Williams, R.J. Learning internal representations by error propagation. In Parallel Distributed Processing-Explorations in the Microstructure of Cognition, MIT Press, chap. 8 (1986) 318-362
2. Hastie, T., Tibshirani, R. Classification by pairwise coupling. The Annals of Statistics, Vol.26.2 (1998) 451-471.
3. Sejnowski, T.J., Rosenberg, C.R. Parallel networks that learn to pronounce english text. Journal of Complex Systems, Vol.1.1(1987) 145-168
4. Dietterich, T.G., Bakiri, G. Solving multiclass learning problems via error-correcting output codes. Journal of Artificial Intelligence Research, Vol.3. 2(1995) 263-286
5. Allwein, E., Schapire, R. E., Singer, Y. Reducing multiclass to binary: A unifying approach for margin classifier. Journal of Machine Learning Research, Vol.1. (2000) 113-141
6. Kong, E.B., Dietterich, T.G. Error-Correcting Output Coding Corrects Bias and Variance. In proceedings of the twelfth International Conference on Machine Learning, California, USA, (1995) 313-321
7. Bay S.D. UCI KDD Archive [http://kdd.ics.uci.edu] (1999)

Discovering Partial Periodic Patterns in Discrete Data Sequences

Huiping Cao, David W. Cheung, and Nikos Mamoulis

Department of Computer Science and Information Systems
University of Hong Kong
{hpcao, dcheung, nikos}@csis.hku.hk

Abstract. The problem of partial periodic pattern mining in a discrete data sequence is to find subsequences that appear periodically and frequently in the data sequence. Two essential subproblems are the efficient mining of frequent patterns and the automatic discovery of periods that correspond to these patterns. Previous methods for this problem in event sequence databases assume that the periods are given in advance or require additional database scans to compute periods that define candidate patterns. In this work, we propose a new structure, the abbreviated list table (ALT), and several efficient algorithms to compute the periods and the patterns, that require only a small number of passes. A performance study is presented to demonstrate the effectiveness and efficiency of our method.

1 Introduction

A *discrete data sequence* refers to a sequence of discrete values, e.g., events, symbols and so on. The problem of partial periodic pattern mining on a discrete data sequence is to find subsequences that appear periodically and frequently in the data sequence. E.g., in the symbol sequence "*abababab*", the subsequence "*ab*" is a periodic pattern. Since periodic patterns show trends in time series or event sequences, the problem of mining partial periodic patterns has been studied in the context of time series and event sequence databases ([1]-[3]).

Two essential sub-problems are the automatic discovery of periods and the efficient mining of frequent patterns. Given a period value, an Apriori-like algorithm is introduced in [3] to mine the frequent patterns. Han *et al.* in [2] propose a novel structure, max-subpattern tree, to facilitate counting of candidate patterns. This method outperforms the Apriori-like algorithm, but it assumes that the periods are given in advance, which limits its applicability. Berberidis *et al.* [1] has proposed a method that finds periods for which a data sequence may contain patterns. However, this method may miss some frequent periods and it requires a separate pass to scan the data sequence to compute the periods. What's more, for each symbol, it needs to compute(at a high cost) the circular autocorrelation value for different periods in order to determine whether the period is frequent or not.

We observe that a frequent pattern can be approximately expressed by an arithmetic series together with a support indicator about its frequency. In this

H. Dai, R. Srikant, and C. Zhang (Eds.): PAKDD 2004, LNAI 3056, pp. 653–658, 2004.

paper, we propose a novel structure, the *abbreviated list table* (ALT), that maintains the occurrence counts of all distinct elements (symbols) in the sequence and facilitates the mining of periods and frequent patterns. Simultaneously, we present a fast $O(n)$ algorithm to identify periods from ALT.

The paper is organized as follows. Section 2 defines the mining problem formally. Section 3 presents the newly proposed approach. Section 4 includes performance evaluation of our methods. Finally, section 5 concludes the paper and proposes the future work.

2 Problem Definition

Let **Domain** D be the set of elements that can be symbols, events, discretized locations, or any categorical object type. A discrete data sequence S is composed of elements from D and can be expressed as $S = e_0, e_1, ..., e_{n-1}$, where i denotes the relative order of an element and n is the length of S. Given a period T, a **periodic fragment** $s_i = e_{iT}, e_{iT+1}, ..., e_{(i+1)T-1}$, $(0 \leq i \leq \lfloor \frac{n}{T} \rfloor)$, is a subsequence of S, and $\lfloor \frac{n}{T} \rfloor$ is the number of fragments in S with respect to period T. Element e_{iT+j}, $(0 \leq j < T)$, in fragment s_i is at the j-th **period position.** There are T period positions, $0, 1, ..., T-1$, for a given period T.

Given a period T, a T-**period pattern** P is a sequence of elements p_0,p_1,...,p_{T-1}, $(0 \leq j < T)$, where p_j can be the wild card '$*$' or an element from D. If $p_j = *$, then any element from D can be matched at the j-th position of P. A periodic fragment $s_i = e_{iT}, e_{iT+1}, ..., e_{(i+1)T-1}$, $(0 \leq i \leq \lfloor \frac{n}{T} \rfloor)$, of a sequence S **matches** pattern $P = p_0, p_1, ..., p_{T-1}$ if $\forall j$, $0 \leq j < T$, (1) $p_j = *$, or (2) $p_j = e_{iT+j}$.

A pattern is a **partial pattern** if it contains the element '$*$'. The **length** L of a pattern is the number of non-'$*$' elements in the pattern. We will call a length-L T-period pattern P an L-pattern if the period T is clear in the context. P' is a **subpattern** of a pattern P if it is generated from P by replacing some non-'$*$' elements in P by '$*$'. E.g.,'$a * c$' is a subpattern of the 3-pattern 'abc'. Similar to the period position of an element in a fragment, the **period position** of an element in pattern P is also the relative position of this element in the pattern. In the 3-period pattern '$a * *$', the period position of 'a' is 0.

The **support** of a T-period pattern P, denoted as *sup*(P), in a sequence S is the number of periodic fragments that match P. A pattern P is **frequent** with respect to a support parameter *min_sup*, $(0 < min\_sup \leq 1)$, iff $sup(P) \geq \lfloor \frac{n}{T} \rfloor \times min\_sup$, (support threshold). If there exists a frequent T-period pattern, we say that T is a **frequent period**. Element e is a **frequent element** if it appears in a frequent pattern.

The **problem** of mining partial periodic pattern can now be defined as follows. Given a discrete data sequence S, a minimum support *min_sup* and a period window W, find:

(1) the set of frequent periods T such that $1 \leq T \leq W$; and

(2) all frequent T-period patterns w.r.t. *min_sup* for each T found in (1).

3 Mining Using the Abbreviated List Table

This section describes the method for automatic discovery of frequent periods using the *Abbreviated List Table* (ALT), and a method that performs efficient mining of frequent patterns. In phase one, we scan the input sequence to construct an ALT that records the frequency of every element. In phase two, we use the max_subpattern tree [2] to mine the frequent patterns.

3.1 Abbreviated List Table

If an element appears periodically for a period T, its positions in the sequence will form an arithmetic series, which can be captured by three parameters: period position, period and count(frequency). For example, in sequence S = '*bdcadabacdca*', the occurrences of 'a' for period = 2 are $\{3, 5, 7, 11\}$. We can represent them by $(1, 2, 4)$, where 1 is the period position of the occurrence of 'a' with period = 2, and 4 is the frequency of the corresponding pattern '$*a$'. For a given period T, for every element, we need to keep T representations: (0, T, $count_0$), ... , ($T-1$, T, $count_{T-1}$). We call these representations Abbreviated Lists (AL). The ALs of all the elements with respect to all periods bounded by a period window can be constructed at a single scan of the sequence.

The algorithm for maintaining the ALT is shown in Fig. 1a. ALT is a 3-dimensional table, the first dimension is the element's index in domain D, the second and third dimensions are period and period position, respectively.

```
Algorithm ALT1(ALT, W, S)
while(S still has elements){
    Read element e from S,
    //SeqPos is e's position in S;
    Get the index idx of e in D;
    for(p := 1; p ≤ W; p++){
        pos:=SeqPos mod p;
        ALT[idx][p − 1][pos]++;}}
//truncate sequence for all periods
```

Fig. 1a. ALT Maintenance

```
Algorithm ALT2 (ALT, W, min_sup, n)
for (idx:=0; idx < |D|; idx++){
    get element e whose index equals to idx;
    for (p:=1; p ≤ W; p++){
        threshold:= ⌊n/p⌋ × min_sup;
        for (pos:=0;pos < p; pos++){
            if (ALT[idx][p − 1][pos] ≥ threshold)
                output p, pos and element e;
}}}
```

Fig. 1b. Finding Periods and F_1

We now show an example for this algorithm. Let the data sequence S be "*abaaaccaae*" and period window W be 5. For the first 'a' at position 0, the counters at period position 0 of all the periods are incremented by 1. Opon seeing the second 'a' at position 2, for periods 1 and 2, the counters at period position 0 are incremented. For periods 3, 4, and 5, the counters at period position 2 are incremented. The process continues, until we process all the elements in S and have the values shown in Table 1.

While maintaining the Abbreviated List Table, we can compute the frequent periods and F_1 at any time moment against the length of the data sequence scanned. (F_1 is the set of size-1 frequent patterns). Fig. 1b shows the algorithm to compute the periods and F_1. We still take the ALT in Table 1 as example assuming $min\_sup = 0.8$. For $period = 2$, the threshold is $10/2 \times 0.8 = 4$, and 'a' at period position 0 is frequent because its count is 4. However, 'a' is not frequent at period position 1 because its count is only 2. So '$a*$' is a frequent pattern but '$*a$' is not, and 2 is a frequent period. Similarly, we can find other frequent periods 2, 4, 5 and their related F_1s.

Our method is more efficient than the circular autocorrelation method in [1] because of the following reasons: (1) We compute F_1 during the period discovery process in one pass. (2) Our method works well even when the length of data sequence n is unknown in advance. (3) We can find frequent periods directly from ALT.

3.2 Finding Frequent Patterns

For each frequent period found in step 1, the algorithm constructs the max-subpattern tree in step 2 using F_1 and the inverted lists. The lists of frequent elements are merged to reconstruct the fragments and they in turn are inserted into the max-subpattern tree. Finally, we get frequent patterns by traversing all the trees. We note two reasons for the superiority of the ALT-based algorithm over [2]. (1) This phase does not need to compute F_1. (2) Only the inverted lists for frequent elements are needed to build the tree. This requires less I/O compared with the original max-subpattern tree algorithm.

3.3 Analysis

Space complexity: Assume the period window is W, each element may have frequent periods from 1 to W. For each period, we need to record the occurrences at each period position. The number of counters in the ALT of an element is of $O(W^2)$. Given $|D|$ elements, the space required is $O(|D|W^2)$, which is independent of the sequence length. We expect that W and $|D|$ are in the order of hundreds in practice, thus the ALT can be accommodated in the main memory.

Time complexity: We need one scan on the sequence to build the ALT. Since the locations of ALT entries can be accessed in O(1) time, the time complexity to create the ALT is in the order of $O(n)$. The construction of the max-subpattern tree and the computation of the frequent patterns is also of $O(n)$.

4 Experiments

We compare our method with the algorithm proposed in [1]. We will use "ALT+tree" to denote our method, and use "Circular Autocorrelation+tree" to represent the method combining [1] and [2]. All the experiments were performed on a Pentium III 700MHz workstation with 4GB of memory, running Unix. A synthetic data generator is used to generate periodic object movements, with four parameters: period T, sequence length n, max pattern length l and probability p with which the object complies with the pattern.

Given data sequence with parameters n = 1M, $T = 50$, $p = 0.8$ and $l = 25$, it's obvious to see that our method is much faster than [1] for fixed minimum support 0.6 from Fig. 2a. Moreover, the cost difference rises with the increase of the window size. Table 2 records the cost breakdown of the two methods for $W = 100$. Note that the cost difference is mainly attributed to the excessive cost of circular autocorrelation in discovering the periods. The finding of F_1 also contributes some cost difference as [1] needs one more scan on sequence to get F_1. The cost for building the max-subpattern tree in our method is less than that by scanning the whole sequence since we only need to access the inverted lists for frequent elements. Fig. 2b shows that both methods have linear cost to the size of the database, due to the limited number of database scans, however, our method is much faster than the previous technique. For this experiment, we fix parameters $T = 100$, $p = 0.8$ and $l = 50$.

Table 3 lists the frequent periods found and the number of frequent patterns mined from the experiment of Fig. 2b. Since the parameter T used to generate data sequence is set to 100 for all sequences and the mining parameter for W is 100, only one frequent period 100 is found.

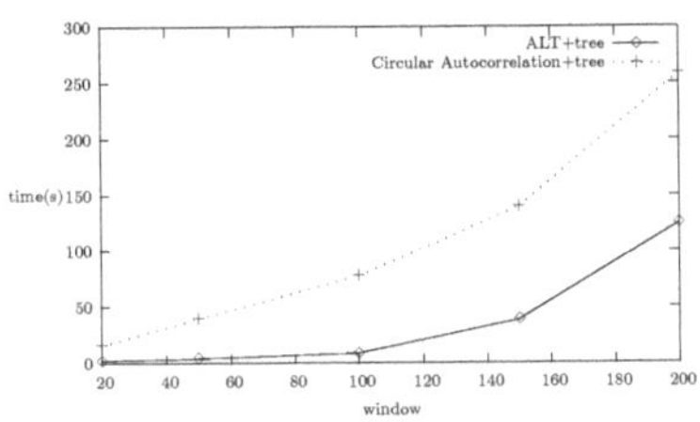

Fig. 2a. Efficiency vs. W

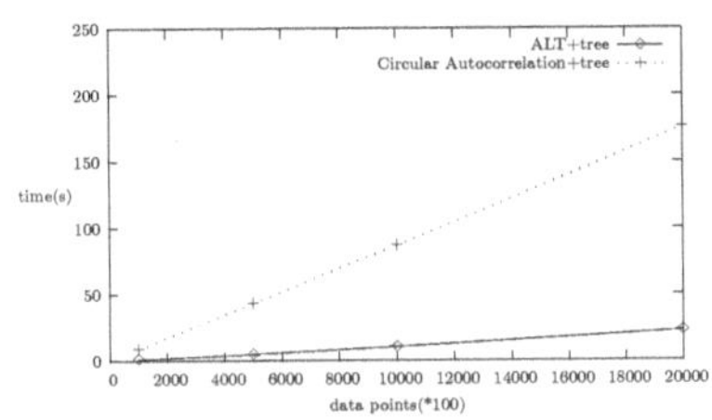

Fig. 2b. Scalability

period pos	0	1	2	3	4
period=1	6				
period=2	4	2			
period=3	2	2	2		
period=4	2	0	1	2	
period=5	1	0	2	2	1

Table 1. ALT for a

time(s)	[1] +tree	ALT+tree
find period	71.34	5.59
find F_1	2.12	0
build trees	2.07	1.11
mine pat.	2.02	2.01

Table 2. Cost comparison

n	period	num of freq. pat.
100K	100	12
500K	100	42
1000K	100	76
2000K	100	133

Table 3.

5 Conclusion

In this paper, we presented a new method to perform partial periodic pattern mining on discrete data sequences. Using the proposed ALT structure, we can find frequent periods and the set of 1-patterns during the first scan on data sequence. This step is much more efficient compared to a previous approach, which is based on circular autocorrelation. Further, frequent patterns are discovered by inserting fragments to the max-subpattern tree, using the inverted lists in order to avoid accessing irrelevant information. Our experiments show that the proposed technique significantly outperforms the previous approaches.

References

1. C. Berberidis, I. P. Vlahavas, W. G. Aref, M. J. Atallah, and A. K. Elmagarmid. On the discovery of weak periodicities in large time series. In *Proc. 6th European Conf. on Principles and Practice of Knowledge Discovery in Databases*, 2002.
2. J. Han, G. Dong, and Y. Yin. Efficient mining of partial periodic patterns in time series database. In *Proc. of 15th Intl. Conf. on Data Engineering*, 1999.
3. J. Han, W. Gong, and Y. Yin. Mining segment-wise periodic patterns in time-related databases. In *Proc. of Intl. Conf. on Knowledge Discovery and Data Mining*, 1998.

Conceptual Mining of Large Administrative Health Data

Tatiana Semenova[1], Markus Hegland[2], Warwick Graco[3], and Graham Williams[4]

[1] CSL, RSISE, Australian National University
tatiana@csl.anu.edu.au
[2] School of Mathematical Sciences, Australian National University
markus.hegland@anu.edu.au
[3] Australian Health Insurance Commission
w.graco@hic.gov.au
[4] Mathematical and Information Sciences, CSIRO
graham.williams@cmis.csiro.au

Abstract. Health databases are characterised by large number of records, large number of attributes and mild density. This encourages data miners to use methodologies that are more sensitive to health undustry specifics. For conceptual mining, the classic pattern-growth methods are found limited due to their great resource consumption. As an alternative, we propose a pattern splitting technique which delivers as complete and compact knowledge about the data as the pattern-growth techniques, but is found to be more efficient.

1 Introduction

In many countries, health care undustry is challenged by growth of costs associated with the use of new treatments or diagnostic techniques and inefficient health care practices where funds are unnecessarily spent with no additional benefits to patients. It has become very important for health service administrators to better understand current health care trends and patterns and associated costs to estimate health costs into the future. The key characteristics of a health system are hospital care, visits to medical practitioners, the consumption of pharmaceuticals calculated with regards to the particular cohorts of patients. One of the measure units for such calculations is episode of care [8], which has a variety of definitions. Episodes take into account various indices of patient care, for instance, patients age, ethnical background, gender, location, medical services provided, information about participating physicians, fees and some other. Aggregating these attributes is important for *Medicare* (Australia's universal health scheme) administrators because they can then produce extensive reports on utilisation. From a data mining point of view, applying some definition of episode is a way to preprocess data according to some temporal principle that is also clinically meaningful. Besides, it is an opportunity to filter out those irrelevant attributes that will not be included in data analyses. Episodic mining of

H. Dai, R. Srikant, and C. Zhang (Eds.): PAKDD 2004, LNAI 3056, pp. 659–669, 2004.

health data is also a method to compress transactional dataset into a collection of health care episodes, that are not so diverse due to the nature of services and limited variance in clinical practice.

We define episode of care as a set of one or more medical services received by an individual during a period of relatively continuous contact with one or more providers of service, in relation to a particular medical problem or situation. Episodes of care should be carefully distinguished from episodes of illness. Care episodes focus on health care delivery whereas illness episodes focus on the patient experience. Episodes of care are the means through which the health care delivery system addresses episodes of illness [8]. Construction of an episode of care begins with the first service for a particular condition and ends when there are no additional claims for a disease-specific number of days.

In the database used for our analyses, for *3,617,556* [9] distinct patients only *368,337* unique patient histories were matched. Applying our definition of a health care episode as a group of tests ordered for a patient by the same doctor on the same day, which in terms of database is the content of all records containing the same *Patient Identification Number*, the same *Referring Provider*, and the same *Date of Reference*, we represented one of the datasets originally containing *13,192,395* transactions as a set of *2,145,864* sequences (episodes). Amongst them only *62,319* sequences were disctinct. Our experience in processing administrative health data has shown that unique health care episodes normally occupy less than 10% of the total size of data, which makes episode-based representation an efficient technique of a database compression. Thus effective pruning of the original data is suggested to be a starting point in handling computations on large datasets. Besides that, the obtained knowledge about diversity and consistency in data is a valuable contribution in understanding the actual meaning of data. This also contributes to the knowledge representation in general [9].

The patterns of practice derived from administrative health data is a way to gain some insights into the clinical side of health care services. *Medicare* transactions do not contain information about any effects of clinical treatments. Neither do they contain information about pre-conditions of the treatments or duration of the disease. *Medicare* items combinations include various mixes of consultation, diagnostic and procedural services provided by health providers to patients for various pathological conditions. Thus, *Medicare* items and possibly other relevant attributes associated within one episode could reveal some clinical side of the event.

One approach to identifying patterns in health data uses association rule mining [1][7]. The *Apriori*-like approaches [5] for discovering frequent associations in data achieve reasonable performance on a variety of datasets, but for large health records collections in particular this method is found limited. Another type of approaches emerged from *Formal Concept Analysis* [10], a field that focuses on the lattices structures extracted from binary data tables, or concepts, which provide a theoretical framework for data mining, conceptual clustering and knowledge representation in general.

In fundamental philosophies, concepts combine things that are different. Implicit in this is the knowledge that the *attributes* to which our concepts apply have many *qualities*. Thus, implicit in the nature of concepts is recognition that each referent of a *concept* differs quantitatively from the others. Like in fundamental philosophies, in *Formal Concept Analysis* a *concept* is constituted by two parts: *extension*, consisting of all *objects* belonging to the concept and *intension*, containing all *attributes* shared by the objects. Such a view at a concept allows to detect all structures in a dataset containing complete information about attributes. All together, these structures present a compact version of the dataset - a *lattice*. Building a lattice can be considered as a conceptual clustering technique because it describes a concept hierarchy. In this context, lattices appear to be a more informative representation comparing with trees, for instance, because they support a multiple inheritance process (e.g. various types of service by the same type of health care provider).

The classic and one of the most efficient techniques for frequent pattern mining is *FP-growth algorithm* and alike [2]. It is an unsupervised learning technique for discovering conceptual structures in data. Its benefits are *completness* and *compactness*, that is, the derived associations contain conclusive information about data and their amount is reduced down to the number of *maximal* frequent patterns. However, on a large scale this technique may face memory problems due to a great *FP-tree* expansion. We suggest an alternative algorithm based on splitting the initial records into two or more sub-records, so that none of the sub-records could carry on irrelevant information. Such an expanded record is in fact a mini-structure (or a *concept*) that already is or will become one of the components of a *formal concept* (or *Galois* lattice) later on.

2 Definition of Formal Concept

Galois connection in its nature is a characteristic of the binary relations that possess structural and logical properties, and can therefore be used as a tool to relate structures. *Galois connection* defines how one structure abstracts another when the relation may be presented as a function [10]. Some of the relations between patterns in health domain can contain functional properties. Health databases typically have many types of relations (*relation* in a database is a set of instances, where *instance* is a vector of attribute values).

Let us denote a database as $\mathcal{D} = (\mathcal{O}, \mathcal{I}, \mathcal{R})$, where $\mathcal{O}$ and $\mathcal{I}$ are the finite sets of objects and items respectively. $\mathcal{R}$ is a binary relation $\mathcal{R} \subseteq \mathcal{O} \times \mathcal{I}$. A triple $\mathcal{D} = (\mathcal{O}, \mathcal{I}, \mathcal{R})$ is called a *formal context*, where elements of $\mathcal{O}$ are *objects*, those of $\mathcal{R}$ are attributes, and $\mathcal{I}$ is the *incidence* of the *context* $\mathcal{D} = (\mathcal{O}, \mathcal{I}, \mathcal{R})$.

Definition 1 (Galois closure operator).
The Galois closure operator $h = f \circ g$ is the composition of the applications f and g, where f associates items common for all objects $o \in O$ with $O \subseteq \mathcal{O}$, and g associates objects related to all items $i \in I$ with an itemset $I \subseteq \mathcal{I}$:

$$f : 2^{\mathcal{O}} \rightarrow 2^{\mathcal{I}} \qquad f(O) = \{i \in \mathcal{I} | \forall o \in O, (o,i) \in \mathcal{R}\}$$
$$g : 2^{\mathcal{I}} \rightarrow 2^{\mathcal{O}} \qquad g(I) = \{o \in \mathcal{O} | \forall i \in I, (o,i) \in \mathcal{R}\}$$

The following properties hold for all $I, I_1, I_2 \subseteq \mathcal{I}$ and $O, O_1, O_2 \subseteq \mathcal{O}$:

1). $I_1 \subseteq I_2 \Rightarrow g(I_1) \supseteq g(I_2)$, $O_1 \subseteq O_2 \Rightarrow f(O_1) \supseteq f(O_2)$
2). $O \subseteq g(I) \Longleftrightarrow I \subseteq f(O)$

The *Galois connection (f,g)* has the following properties:

Extension: $I \subseteq h(I)$
Idempotency: $h(h(I)) = h(I)$
Monotonicity: $I_1 \subseteq I_2 \Rightarrow h(I_1) \subseteq h(I_2)$

Definition 2 (Formal concepts).
An itemset $C \subseteq \mathcal{I}$ *from* $\mathcal{D}$ *is a closed itemset iff* $h(C) = C$. *The minimal closed itemset containing an itemset* I *is obtained by applying* h *to* I. $h(I)$ *is the closure of* I. *The collection of all closed itemsets makes a Galois lattice, or formal concept.* [11][12].

Table 1. Dataset WA/2000 (SEp - size of selection of episodes; UniEp - number of unique episodes; DIt - distinct items)

Table	**Size**	**Episodes**	**SEp**	**UniEp**	**DIt**
WA/2000	6.4Gb	3,220,324	223Mb	212,402	942

3 Data Preprocessing

Table 1 describes a dataset used in our experiments.

This table represents the group of general practitioners and specialists referring both in- and out-hospital patients.

Our testing platform has been the *Sun Enterprise* with 12 *400MHz UltraSPARC* processors each with 8Mb of external cache, 7Gb of main memory, running *Solaris 2.6*. It has 2 *Sun A5100* disk storage arrays (each 250 Gb capacity) which are fibre channel connected and run in a mirrored configuration. A single process user job would normally have a full use of a single processor.

Let us consider some example of episodic presentation. Let us take the following set of transactions where **tid** is transaction ID, **doc** is a doctor ID, **dor** is a date of referral, **dos** is a date of service and **svc** is a code of a medical service provided [4]. A health care episode will be defined by the doctor delivering the initial health care services to an identified patient on the same day.

tid	**pin**	**doc**	**dor**	**dos**	**svc**
1	007	23	9555	9557	65007
2	007	23	9555	9558	73009
3	008	104	10111	10112	10900
4	001	53	9118	9123	65007
5	001	53	9118	9127	73009
6	005	99	9173	9174	65007
7	005	99	9173	9174	73009

The first step in preprocessing our small transactional set will be detecting episodes of care. According to the definition of an episode, we can identify four health episodes and map them into the following structure.

episode id	**content**
(007, 23, 9555)	65007 73009
(001, 53, 9118)	65007 73009
(005, 99, 9173)	65007 73009
(008, 104, 10111)	10900

In the table above, three out of four episodes have the same content. It makes sense to store such a structure in a *hash*-table, where one table will be storing all the unique episodic contents and the other one will be storing their counts.

episode	**count**
(65007, 73009)	3
(10900)	1

Since our purpose is to discover patterns of practice (or concepts) in health data, the episodic-based approach to preprocessing allows to select and store only relevant attributes. Storing only unique episodes in a *hash*-table gives us a compressed presentation of our original set of transactions [3]. Having presented original records in episodic form, we will still treat episodes like records, or a database.

4 PSAlm: Pattern Splitting Algorithm

Let us expand our small collection of episodes from the previous example. A number of items associated with the same object will be an episode. Keeping our notations, the set of objects is $O = \{o_1, o_2, o_3, o_4, o_5\}$ and the set of items is $I = \{i_1, i_2, i_3, i_4, i_5, i_6, i_7, i_8, i_9\}$, where $i_1 = 104, i_2 = 302, i_3 = 304, i_4 = 10900, i_5 = 55853, i_6 = 65007, i_7 = 66716, i_8 = 73009, i_9 = 73011$ [4].

objects	**items**	**count**
o_1	65007, 73009	3
o_2	10900	1
o_3	65007, 66716, 73009, 73011	1
o_4	104, 302, 55853	1
o_5	302, 304, 55853, 65007, 66716	1

Let us set up a support threshold $minsup = 2$. Those patterns whose occurrence in data equal to or exceeds 2 will be considered as frequent. In the technique we propose, the first step is to detect all frequent *2*-itemsets first. These are (65007, 73009), (302, 55853) and (65007, 66716). All other *2*-itemsets are infrequent and are therefore irrelevant. This means that o_3, for instance, can be split into sub-patterns (65007, 66716, 73009) and (65007, 66716, 73011) since (73009, 73011) is irrelevant, and then again, each of those sub-patterns will have to be reduced down to (65007, 66716) because (65007, 73011), (66716, 73011) (and the same operation for the second sub-pattern), are infrequent, therefore their presense in o_3 is unnecessary. Following this principle, we obtain a reduced version of our collection of episodes:

pattern id	items	count
1	65007, 73009	5
2	302, 55853	2
3	65007, 66716	2

With large collections of episodes, this process will involve further splitting operations, with 3-itemsets, 4-itemsets and so forth, depending on the lengths of episodes and their number. In this example, we have detected all frequent patterns in just one step. The derived patterns (65007, 73009), (302, 55853) and (65007, 66716) define a *formal concept* for our dataset. In particular, the binary relation (o_1o_3, i_6i_8) is a concept, (o_4o_5, i_2i_5) is a concept as well, and for instance, (o_5, i_2i_5) is *not* a concept since it does not contain complete information about i_2i_5.

5 Computational Complexity

Let us go through some computations using a small database for example. Presented in a form of a *hash-table*, the dataset is as follows:

episode (key)	count (value)
$I1, I2, I3, I4, I5, I6, I7, I8$	1
$I1, I2, I3, I5, I6$	1
$I3, I4, I6, I7, I8$	1
$I5, I6, I7, I8$	1
$I7$	1
$I5, I6, I8$	1
$I1, I2, I3, I4$	1
$I2, I3, I4$	1

Let us set the *minimal support* threshold equal to *3*. From this table, it is possible to identify frequent *1*-items with linear complexity:

item	I3	I6	I2	I4	I5	I8	I1	I7
count	5	5	4	4	4	4	3	3

The table below lists frequent *2*-itemsets. Other *2*-itemsets composed of the list of frequent *1*-items are infrequent and will be used further on for splitting our original records.

2-itemset	count
$I1, I2$	3
$I1, I3$	3
$I2, I3$	4
$I2, I4$	3
$I3, I4$	4
$I3, I6$	3
$I5, I6$	4
$I5, I8$	3
$I6, I7$	3
$I6, I8$	4
$I7, I8$	3

In order to split the original records into two or more sub-records the *PSAlm* algorithm requires multiple use of the following procedures.

- A record is tested if it contains one or more infrequent pairs.
- If the record does contain such a pair it undergoes a recursive splitting procedure.

Complexity of the first procedure depends mainly on a length of a record. The consecutive matching the record with a key in a dictionary containing infrequent pairs of items takes $O(1)$ time. Let us denote an average records length as l^r. The number of possible pairs of items contained in a record is $l^r * (l^r - 1)/2$. Therefore, to check a record for presence of any of the infrequent pairs takes $S_B = l^r * (l^r - 1)/2$ steps. For an infrequent pattern of length l^w, $S_B = \frac{l^r!}{l^w! * (l^r - l^w)!}$, which is of order $O(m)$. The number of calls for the second procedure depends on l^r and the number of infrequent patterns contained in the record. In the worst case, when a record could be of length m and all possible pairs of items in it were infrequent, then the algorithm would require at most $\frac{m!}{2*(m-2)!} = \frac{m*(m-1)}{2}$ calls for splitting procedure. This makes the first procedure of $O(m^2)$ complexity, for the worst case.

Complexity of the second procedure depends mainly on the length of the record, and as a result of recursive splitting, on the length of the list of sub-records (L^s). Thus, complexity of the whole procedure can be estimated as linear, or $O(L^s)$. L^s depends on l^r and the number of infrequent pairs in the record, which defines the number of calls for the splitting function. But the records can not be split if the resulting sub-records have the length less than l^w. Thus, as l^w increases, the potential maximal number of sub-records decreases. Also, not all the sub-records contain infrequent item combinations, therefore, only some of them undergo further splitting. Since we are not interested in sub-records with length less than l^w, to roughly evaluate the potential maximal number of sub-records L^s we could use the approach suggested in [108], and in our case, such a number could be close to 2^{m-l_w}. But this number would be good only in case if we tried to split the sub-records applying all known infrequent pattern at once. Since we apply only one infrequent pattern at a time, we estimate that L^s may vary from 0 to $\frac{m!}{l_w! * (m-l_w)!}$.

The computations described above are the internal loops of the main computational procedure that includes scanning the original (and then modified) dataset and splitting its records, if necessary. It has two input parameters - a set of records and a set of infrequent patterns of fixed length. For our example, after splitting the original records using the set of infrequent *2*-itemsets, we have to obtain a set of infrequent *3*-itemsets to continue splitting. This procedure requires computations of linear complexity and as a result we obtain a set of frequent *3*-itemsets (below) and an empty set of infrequent *3*-itemsets, which halts the computational process for our example. Otherwise, we could continue with a set of splitted records and the newly computed set of infrequent patterns.

3-itemset	count
$I1, I2, I3$	3
$I2, I3, I4$	3
$I5, I6, I8$	3
$I6, I7, I8$	3

These patterns are supersets of the frequent *2*-itemsets therefore the resulting table of *maximal* frequent patterns is as follows:

concepts	count
$I1, I2, I3$	3
$I2, I3, I4$	3
$I5, I6, I8$	3
$I6, I7, I8$	3
$I3, I6$	3

To evaluate the overall complexity of the process of splitting records and updating the original data, we can consider the following scheme (Figure 1):
Using such a scheme, it is easier to see that the major computational procedure in the *PSAlm* is of the following complexity:

$$S_{BR} \approx k * n * \frac{l^r!}{l^w! * (l^r - l^w)!} * L^s = k * L^s * n * \frac{l^r!}{(k+1)! * (l^r - k - 1)!}$$

It is important to note that n is the number of records in the original dataset. But at every step, records of length less than the current value of l^w get deleted from data. Therefore, n decreases from step to step considerably.

We estimate that the *PSAlm* algorithm is of $O(kL^s m^2 n)$ complexity, in the worst case, where k is the number of data scans, usually just several, L^s is a constant. Thus we could generalise the algorithm's complexity as $O(m^2 n)$. In the worst case, when the number of scans get close to m, or, in other words, when the data is extremely dense , so that l^w gets close to m, then the algorithm's performance will have $O(m^3 n)$ complexity.

The difference in efficiency between *FP-growth* and *PSAlm* becomes especially notable on large data collections (Figure 2), with high number of

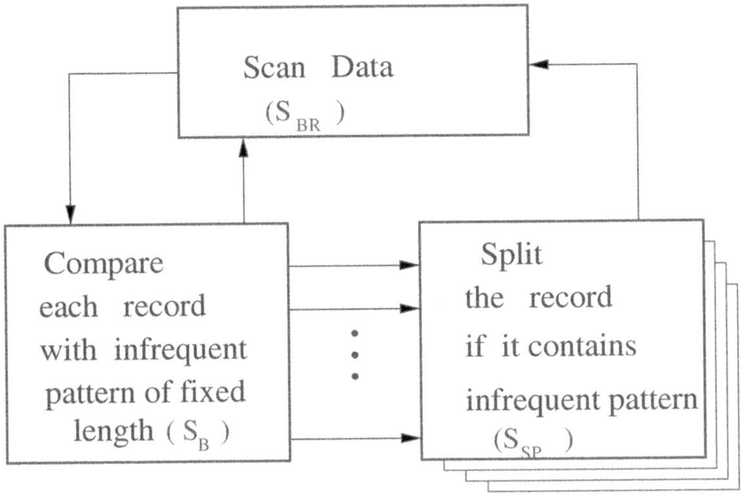

Fig. 1. Main computational process for the PSAlm algorithm

attributes and low support threshold (less than 1%). The most time consuming procedure is the first step of the algorithm, when original records go through recursive splitting into 2 or more sub-records. Since the sub-records are subject to reduction, the dataset gets reduced in size rather significantly, and all further splitting procedures are typically effortless.

Fig. 2. Performance of the Python codes for FP-growth and PSAlm on WA/2000 health data.

We estimate, that the *FP-growth* algorithm requires $O(2^m n)$ steps, in the worst case, where m is the number of attributes and n is the number of records. It is a main memory based mining technique, and its complexity in space consumption may reach $O(2^m)$. Although, each path in the *FP-tree* will be at least partially traversed the number of items existing in that tree path (n in the worst case) times the number of items in the header of the tree (m)[2]. Therefore, the upper bound on complexity of searching through all paths will be $O(m^2n)$. Taking into account that *all* paths containing an itemset with an item that is the *header-item* at the moment and that it is not possible to see, before the traversing along paths, which itemsets turn out to be frequent, the computational complexity rises up to $O(2^m n)$ in the final phase of the algorithm. Trivial examples in the literature on *pattern-growth* techniques do not show these assertions. It only is possible to see in the large scale experiments. In addition, some recursions in counting present an additional computational barrier. Thus, overall we regard the *FP-growth*'s complexity as *exponential* or close to one.

6 Conclusions

Pruning the original data, adequate representation of ordinary transactions as sequences containing specifically selected (or relevant) items, storing and analysing only unique sequences are suggested to be effective techniques for dataset compression and knowledge representation.

The suggested pattern-splitting technique is found an efficient alternative to the *FP-growth* and similar approaches, especially when dealing with large databases.

Formal Concept Analysis (*FCA*) is an adequate learning approach for discovering conceptual structures in data. These structures present conceptual hierarchies that ease the analysis of complex structures and the discovery of regularities within the dataset. *FCA* is also a conceptual clustering technique useful for multipurpose data analysis and knowledge discovery. *FCA* is found to be a suitable theory for the determination of the concept of a *concept*.

References

[1] J. Han and M. Kamber (2001) *Data Mining: Concepts and Techniques.* Morgan Kaufmann.

[2] J. Han, J. Pei, Y. Yin (2000) *Mining Frequent Patterns without Candidate Generation.* In Proc. of the ACM SIGMOD Int. Conf. on Management of Data, pages 1-12, Dallas, Texas, USA.

[3] David M. Beasley (2000) *Python Essential Reference.* New Riders Publishing. Indianapolis, IN.

[4] Commonwealth Department of Human Services and Health (2000) *Medicare Benefits Schedule Book.* Australian Government Publishing Service, Canberra.

[5] Rakesh Agrawal, Tomasz Imielinski, Arun Swami (1994) *Mining Association Rules between Sets of Items in Large Databases.* Proceedings of the 1993 ACM SIGMOD Conference, Washigton DC, USA.

[6] Rakesh Agrawal, Ramakrishnan Srikant (1994) *Fast Algorithms for Mining Association Rules.* IBM Almaden Research Center, San Jose, CA.

[7] Markus Hegland. *Data Mining Techniques* (2000) ANU, Canberra.

[8] B. Edward. *Episodes of Care* (1999) ProfSoft Inc., Cambridge, Massachusets, USA.

[9] T. Semenova, M. Hegland, W. Graco, G. Williams (2001) *Effectiveness of Mining Association Rules in Large Health Databases.* CalPoly: Technical Report No.CPSLO-CSC-01-03. ICDM-01: Workshop on Integrating Data Mining and Knowledge Management. San Jose, California.
[10] B. Ganter and R. Wille (1999) *Formal Concept Analysis: Mathematical foundations.* Springer.
[11] V. Duquenne and J.-L. Guiges (1986) *Famille minimale d'implications informatives résultant d'un tableau de données binaires.* Math. et Sci.Hum., 24(95):5-18.
[12] N. Pasquir, Y. Bastide, R.Taouil and L. Lakhal (1999) *Discovering frequent closed itemsets for association rules.* Proc.ICDT Conf., pp 398-416.

A Semi-automatic System for Tagging Specialized Corpora

Ahmed Amrani[1], Yves Kodratoff[2], and Oriane Matte-Tailliez[2]

[1]ESIEA Recherche, 9 rue Vésale,
75005 Paris, France
amrani@esiea.fr

[2]LRI, UMR CNRS 8623, Bât. 490, Université de Paris-Sud 11,
91405 Orsay, France
{yves.kodratoff, oriane.matte}@lri.fr

Abstract. In this paper, we treat the problem of the grammatical tagging of non-annotated corpora of specialty. The existing taggers are trained on general language corpora, and give inconsistent results on the specialized texts, as technical and scientific ones. In order to learn rules adapted to a specialized field, the usual approach labels manually a large corpus of this field. This is extremely time-consuming. We propose here a semi-automatic approach for tagging corpora of specialty. ETIQ, the new tagger we are building, make it possible to correct the base of rules obtained by Brill's tagger and to adapt it to a corpus of specialty. The user visualizes an initial and basic tagging and corrects it either by extending Brill's lexicon or by the insertion of specialized lexical and contextual rules. The inserted rules are richer and more flexible than Brill's ones. To help the expert in this task, we designed an inductive algorithm biased by the "correct" knowledge he acquired beforehand. By using techniques of machine learning and enabling the expert to incorporate knowledge of the field in an interactive and friendly way, we improve the tagging of specialized corpora. Our approach has been applied to a corpus of molecular biology.

1 Introduction

Knowledge extraction starting from raw and specialized texts is not yet really practical. In order to step forward on this problem, several related steps must be carried out: normalization, grammatical tagging, terminology extraction, and conceptual clustering. Grammatical tagging is a key step in knowledge extraction because its precision strongly influences the results of the following steps in the chain of linguistic treatments. Before doing the grammatical tagging, however, we could observe that it is necessary to normalize the corpus. The normalization of texts consist of several types of treatments [1] highly dependent upon the specialty and the type of information to extract. The goal of the normalization is to limit the errors of the following stage

H. Dai, R. Srikant, and C. Zhang (Eds.): PAKDD 2004, LNAI 3056, pp. 670–681, 2004.

(grammatical tagging) of the text processing and to prepare the ultimate step of knowledge extraction by reducing the complexity of the vocabulary used.

Once the corpus is standardized, the tagging can be carried out. This step consists in associating each word with its grammatical tag, according to its morphology and context. We used the whole labels list of Penn TreeBank. This list is composed of 36 part-of-speech tags and 9 others for the punctuation. A complete list can be downloaded at http://www.cis.upenn.edu/~treebank/home.html.

Some tags carry less semantics than others when they are widely used by all languages, such are the determiners (DT, 'a', 'an', 'the'), tags attached to both kinds of words: very general or very specific ones (such as the adjectives, JJ). Inversely, some tags are associated to highly significant words, such as the nouns.

Table 1. Examples of PoS (Part-of-Speech) tags containing specialized words.

PoS tag	*Signification*	*Examples*
JJ	adjective	*heterologous, mitotic, molecular*
NN	noun, singular or mass	*genome, replication, RNA-splicing*
NNS	noun plural	*genomes, RNAs, strains*
VBG	verb, gerund/present participle	*DNA-damaging, overlapping, oxidizing*
RB	adverb	*evolutionarily, post-translationally, structurally*

2 Grammatical Taggers

Among the data-driven algorithms used for tagging, we can cite Inductive Logic Programming [2, 3, 4], Instance-Based Learning [5, 6], Transformation-Based Learning [7] and statistical approaches [8, 9]. Other sophisticated techniques were used, based on the combination of several kinds of taggers. These techniques are based on the fact that differently designed taggers produce different errors. Thus, the combined tagger shows a higher precision than any of them on its own [10, 11].

Rule-Based Supervised Learning (or *Transformation Based Learning*)
Brill's tagger [7] uses a rule-based supervised learning algorithm which detects and corrects automatically its errors, providing a progressive and automatic improvement of the performance. The tagger functions in two steps. The learning step only takes place once, it is slow and complex. The second step which is the application of the learned rules, is used very often. It is fast and simple.
It will be useful to recall Brill's learning system here, even though many authors already used this approach. It works as follows.

The text is annotated by an expert who assigns grammatical tags to each word of the text on which the training is carried on. At each iteration, the algorithm selects only one rule among all the possible ones, the selected rule is the one which provides the best tagging improvement of the temporary corpus. The new temporary corpus is obtained by applying the rule learned to the current temporary corpus. This process is repeated until no rule can be found with a score higher than a given threshold value. At the end of the process, a list of ordered rules is obtained. Brill's tagger uses a rule-based supervised learning in two successive modules: the lexical module (first module) learns lexical (morphological) rules to label the unknown words. In the contextual module (second module), tags and word morphology of the context are used to improve the accuracy of tagging.

3 Tagging Specialized Corpora

Whatever the technology on which they are based, the current taggers obtain a very satisfactory level; the published results are usually about 98% of correct labels, and even higher figures can be found. These good performances can be explained because test corpora are of type similar to training corpora. Obviously, some specific work has to be done in order to adapt them to a specialized corpus.

The present work is relative to a molecular biology corpus. This corpus was obtained by a Medline request with the key words *'DNA-binding, proteins, yeast'*, resulting in 6119 abstracts (10 megabytes). After having carried out Brill's tagging, we note several problems:

- until many iterations are performed, many cleaning mistakes become obvious
- the technical words are unknown (not listed in the general lexicon)
- Brill's rules are not adapted to a specialized corpus. A possible solution would consist in obtaining an annotated corpus for each specialty. That would require intensive expert work. This is why we could not start from an annotated corpus when we chose a specialized domain.

Another limit of the rule-based taggers is that the rule formats are not sufficiently expressive to take into account the nuances of the specialty language. Here we introduced into our tagging rules the possibility of using an enriched grammar. The specialist can define new labels, using logic descriptions or/and regular expressions to express his knowledge. One of our objectives is to produce methods applicable whatever the specialized domain, with the least possible manual annotation.

4 Description of ETIQ Software

The solution we propose is to adapt a tagger, starting from a "general" corpus, to a corpus of specialty. We thus basically preserve Brill's tagger as the starting point of our system. For the English version, the program was built on the annotated corpus of *Wall Street Journal*, which is of very different nature than our molecular biology corpus. We applied the ordered list of Brill's rules to our corpus. Then ETIQ enabled the expert to display the result of Brill's tagging, to add lexical and contextual rules and to supplement Brill's lexicon by a specialized lexicon. To add the N*th* rule, the expert only had to observe the current state of the corpus after the execution of the (N-1) preceding rules [12] or possibly to look at the results of another tagger. Specialized rules can be written and carried out in a simple way, which significantly reduces the time and the effort of the user.

In a preliminary stage, and to adapt a general tagging to a domain of specialty, the expert inserts a specialized lexicon, i.e., a list of specialized words, where each one is followed by its possible tags. This means that when a word has several possible tags, such *novel* (NN, JJ), one of its tags will be met more often in the corpus. It is also possible that only a single one will be met.

Our strategy is as follows: we start with Brill's lexical stage, then the user visualizes the tagging results and corrects it by adding specialized lexical rules. The corpus is then treated with Brill's contextual stage. Finally the user visualizes and corrects Brill's contextual tagging by adding specialized contextual rules.

4.1 Lexical Tagging

In this module, the goal is to determine the most probable tags of the words unknown to Brill's lexicon by applying specialized lexical rules. These words are then added in a specialized lexicon. The ETIQ software helps in correcting the grammatical tagging while allowing the user to assign new labels. For instance, the word *ABSTRACT* is tagged as a proper noun by Brill's tagger because its rules state that words starting with a capital are proper nouns. An other important feature of ETIQ is that the PoS tag list can be increased by others tag types, specific for the studied field, for example the label *formula* (typed *FRM*) or *citation* (*CIT*). To help the expert identify the tagging errors, the system shows groups of words (and their tags) on the base of any similar morphological feature (For example, words having the same suffix or words corresponding to the same regular expression). According to the detected errors, the expert inserts the adequate lexical rules to correct these errors. An example of lexical rule we had to introduce is: *Assign the tag* ***Adjective*** *to the words having the suffix* ***al***. The used rules are more flexible than the ones in Brill's tagger. Brill's tagger assigns to the word its most probable tag according to simple conditions like the nature of its prefix, the nature of its suffix and its contents. The grammar of our rules enables the combination of the simple conditions used by Brill and the regular expressions. The expert writes precise rules according to given conditions.

The grammar used in ETIQ for the lexical rules is as follows:
<Lexical_rule> ::= if < Sequence_Condition > then <Action>
<Sequence_Condition>::=<Condition><Logical_Operator><Sequence_Condition>
| *NOT <Sequence_Condition>*
| *(<Sequence_Condition>)*
| *<Condition>*
<Condition> ::= The word has the morphological property m.
<Action> ::= Change the tag A to B | *Change the current tag to B*
< Logical_Operator> ::= AND| *OR*

Let us compare two lexical rule examples respectively from Brill's system and from ETIQ:

- Brill's rule: *JJ ery fhassuf 3 NN.* It uses only one morphological condition. This means: if the end of word is *ery* (ery fhassuf) then current tag adjective (JJ) is changed to noun (NN).
- ETIQ's rule : *Nul (char -gene AND hassuf ions) NNS.* This rule combine two morphological conditions. This means: if the word contains *-gene* (char -gene) and its end is *ions* then the current tag is changed to NNS.

The expert visualizes the effect of the rule he just wrote and can modify it before saving it. The writing, the modification, and the checking of each rule are done in a simple way by using a user-friendly interface. The interface is intuitive, and can be used by people who are not computer scientists.

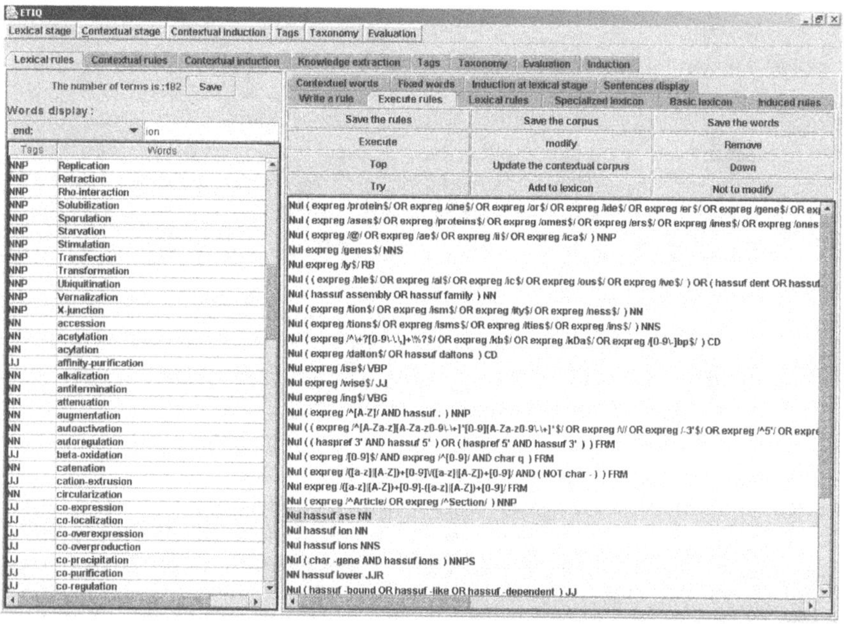

Fig. 1. ETIQ : Lexical step. On the left, a list of words with the suffix *ion*. On the right, lexical rules introduced by the expert. The highlighted line signifies set the tag of all words with suffix *ion* to *NN (noun)*.

Some of the words are sufficiently specialized to be tagged by a unique tag. For example, the word *chromosome* is always tagged as *noun*. The expert can freeze these tags so that they cannot be changed during the contextual phase.

If the expert cannot choose a label for a word during this phase, then this word requires a contextual tagging. These contextual rules are applied to the words having several labels in the corpus (thus, different semantic meanings), or to a group of words (for example with the suffix *ing*) that have different labels according to the context (*NN, JJ, VBG*).

4.2 Contextual Tagging

Once the expert has finished the lexical stage, contextual rules can be used to improve it. The context of the word is defined by the morphology and tags of the neighboring words. In a similar way to the lexical module, the expert can, in an interactive way, carry out requests to look for words, according to words, tags, and morphological criteria. This enables him to visualize the contexts (the target word and its neighbors) and to detect the errors. The user can thus correct these errors by inserting specialized contextual rules. The conditions of the rule can be generated automatically from the request. The expert can refine it if it is necessary.
Let us give two examples needing contextual rules as these words have two possible tags:

- Possible tags of functions are: NNS (noun, common, plural) and VBZ (verb, present tense, 3rd person singular)
- Possible tags of complex are: JJ (adjective) and NN (noun, common, singular)

Compared to Brill's system, which is limited to a context either made of the three preceding words or the three following words, ETIQ explores a context made of both the three preceding and the three following words.

Let us give the contextual rules grammar used in ETIQ:

< Contextual_Rule:> ::= if < Sequence_Condition > then <Action>
<Sequence_Condition>::=<Condition><Logical_Operator><Sequence_Condition>
| NOT <Sequence_Condition>
| (<Sequence_Condition>)
| <Condition>
<Action> ::= Change the tag A to B | Change the current tag to B
<Condition> ::= <Context>
<Context> ::= The nth preceding/following word is tagged with t.
|The nth preceding/following word has the regular expression r.
| The nth preceding/following word is w.
| The current word is tagged with t.
| The current word has the regular expression r
| The current word is w.
<Logical_Operator> ::= AND | OR

4.3 Semi-automated Tagging

Since the tagging task is very cumbersome, most approaches use a supervised learning technique: texts are annotated by humans (supposedly without mistake) and tagging rules are then automatically learned from the annotated corpus. This technique is possible when dealing with the general language, for which annotated corpora already exist.

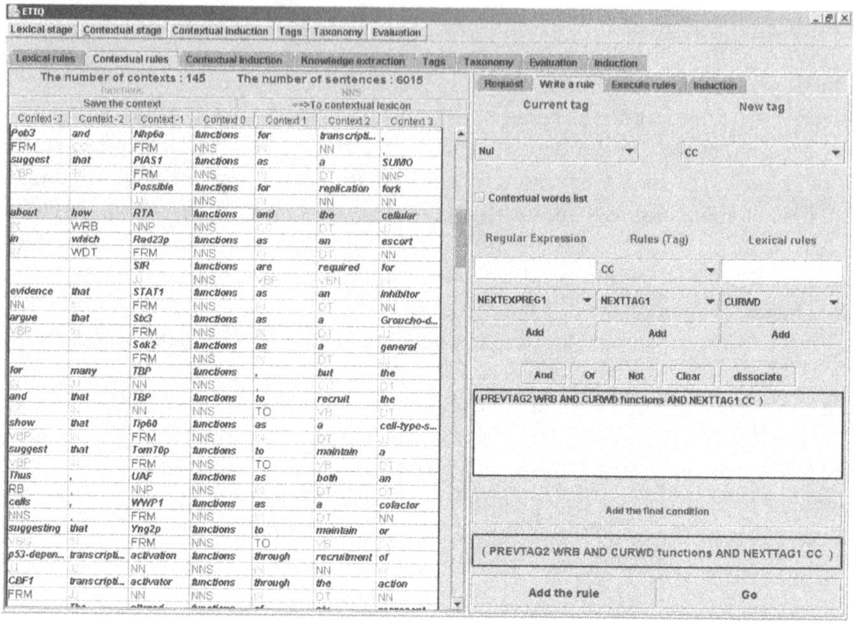

Fig. 2. ETIQ : Contextual step. The left side shows all the contexts of the word *functions* with the tag *NNS (noun common, plural)*. The right side shows a contextual rule for the highlighted *functions*: if the word is *functions* with tag *NNS* and the previous (-2) word tag is *WRB (Wh-adverb)* and the next word tag is *CC (conjunction, coordinating)* then the new tag is *VBZ (verb, present tense, 3rd person singular)*.

Specialty languages, however, ask for expert tagging, a difficult task due to the social fact that all experts are overworked. We thus had to develop a tool speeding the tagging task in order to allow experts to efficiently and rapidly tag the texts. We provide the expert with a user-friendly interface to correct the tagging. The inductive system takes into account these improvements in order to offer new rules to him. In the so-called "lexical" phase (during which the context is limited to relations between the letters inside the word) we use several obvious morphological features such as prefix and suffix. For these features we use values (like *ion* for suffix) suggested by the user or chosen automatically by the system on the basis of their frequency.

Our basic induction algorithm is the classical C4.5 [13] but its optimization measure, the gain ratio, has been modified in order to take into account expert-based biases. To this effect, we introduced an ‘expert gain’, $Gain_{Exp}$. This gain is similar to the one computed by C4.5: the attribute X is applied to the data, and we compute the $Gain_{Exp}$ due to application of X. $Gain_{Exp}$ is not computed on the whole set of tags but only on the set of tags previously changed by the expert. Let T_0 be the current learning set under study, T_{Exp0} the set of tags that have been changed by the expert in T_0, N the number of instances in the set T_0 and c the number of values of the target feature (tag). Applying the feature X, which has *n* possible values, to T_0 returns the classical C4.5 gain. In order to compute $Gain_{Exp}$, we apply X to T_{Exp0} thus generating n subsets $T_{Exp1}, T_{Exp2} \ldots, T_{Expn}$. Let $N_{Exp}Pos(T_{Expi})$ (i = 0, 1, …, n) the number of instances tagged as the majority class of T_{Expi} and $N_{Exp}Neg(T_{Expi})$ the number of instances tagged differently from the majority class of T_{Expi}.

Then, when define

$$M_{Exp}(T_{Exp0}) = N_{Exp}Pos(T_{Exp0}) - N_{Exp}Neg(T_{Exp0}) \quad (1)$$

$$M_{ExpX}(T_{Exp0}) = \sum_{i=1}^{n}(N_{Exp}Pos(T_{Expi}) - N_{Exp}Neg(T_{Expi})) \quad (2)$$

$$Gain_{Exp}(X) = \frac{Log_2(c) * (M_{ExpX}(T_{Exp0}) - M_{Exp}(T_{Exp0}))}{N} \quad (3)$$

$$Gain\ Ratio\ C4.5_{Exp}(X) = \frac{Gain(X) + Gain_{Exp}(X)}{Split\ info(X)} \quad (4)$$

This gain ration (Gain Ratio $C4.5_{Exp}$) selects the feature that favors the classical gain ratio and that contradicts the expert as little as possible. We chose the sum between the C4.5 gain (Gain(X)) and $Gain_{Exp}$ in order to come back to the classical gain ratio when the expert provided no advice. This algorithm proposes new rules to the expert who presently has to choose the best ones. A partial automation of this selection step is presently under study.

4.4 Towards a Perfectly Tagged Corpus

One of the expectations of the system is the semi-automatic acquisition of a perfectly tagged corpus of specialty. On one hand, ETIQ makes it possible to correct easily and manually the few errors introduced by the side effects of the rules. On the other, the expert can manually impose correct labels missed by the rules. All that enabled us to obtain a "perfect" biology sub-corpus of one megabyte with a minimal involvement (half-a-day) of the expert.

5 Experimental Validation

In order to validate our approach, we used a sub-corpus of 600 abstracts.
We generated three tagged corpora starting with our raw corpus, C_0.

→ C_0 annotated (supposedly perfectly) leading to $C_{PERFECT}$.
→ C_0 tagged by standard Brill's tagger leading to C_{BRILL}.
→ C_0 tagged using ETIQ as follows:

C_0 was tagged by Brill's lexical module, yielding C_1. C_1 lexical tagging was then improved by ETIQ, yielding C_2. C_2 was tagged by Brill's contextual module, yielding C_3. Finally, C_3 contextual tagging is improved using ETIQ, yielding C_{ETIQ}.

Table. 2. Tagging results.

PoS Tags	# Tags $C_{PERFECT}$	#errors C_{BRILL}	Precision C_{BRILL}	Recall C_{BRILL}	# errors C_{ETIQ}	Precision C_{ETIQ}	Recall C_{ETIQ}
CC	4347	16	100,00	99,63	5	99,93	99,88
CD	1787	305	65,84	82,93	46	95,45	97,43
DT	11869	56	99,30	99,53	47	99,29	99,60
EX	35	2	100,00	94,29	2	100,00	94,29
FW	57	12	81,82	78,95	15	95,45	73,68
FRM	6417	6417	0,00	0,00	190	97,86	97,04
IN	16559	3	99,54	99,98	3	99,67	99,98
JJ	11044	1016	81,11	90,80	458	98,33	95,85
JJR	116	8	93,10	93,10	7	92,37	93,97
JJS	69	0	90,79	100,00	0	98,57	100,00
MD	490	2	100,00	99,59	3	100,00	99,39
NN	29081	4995	97,42	82,82	2220	97,25	92,37
NNS	7618	153	94,71	97,99	26	98,50	99,66
NNP	4116	239	29,74	94,19	230	66,23	94,41
NNPS	3	0	3,61	100,00	3	0,00	0,00
PDT	12	3	100,00	75,00	0	85,71	100,00
POS	43	39	12,90	9,30	0	61,43	100,00
PRP	1229	0	99,76	100,00	0	99,76	100,00
PRP$	486	0	100,00	100,00	0	100,00	100,00
RB	3555	333	92,69	90,63	33	99,55	99,07
RBR	68	11	100,00	83,82	12	100,00	82,35
RBS	49	6	100,00	87,76	0	100,00	100,00
RP	17	12	100,00	29,41	5	80,00	70,59
SYM	117	66	100,00	43,59	22	100,00	81,20
TO	2242	0	100,00	100,00	0	100,00	100,00
VB	1980	45	93,21	97,73	63	96,62	96,82
VBD	1851	14	94,89	99,24	14	99,19	99,24
VBG	2395	243	90,00	89,85	72	91,93	96,99
VBN	4136	94	99,04	97,73	16	99,66	99,61
VBP	2272	89	99,05	96,08	85	98,20	96,26
VBZ	3518	120	98,87	96,59	70	99,54	98,01
WDT	914	3	99,89	99,67	0	99,78	100,00
WP	5	0	100,00	100,00	0	100,00	100,00
WP$	22	0	100,00	100,00	0	100,00	100,00
WRB	298	126	100,00	57,72	1	99,33	99,66
.	6011	0	99,90	100,00	0	99,90	100,00
"	67	67	0,00	0,00	67	0,00	0,00
(	1634	0	100,00	100,00	0	100,00	100,00
)	1637	0	100,00	100,00	0	100,00	100,00
,	5101	0	100,00	100,00	0	100,00	100,00
--	9	9	0,00	0,00	0	37,50	100,00
'	27	27	0,00	0,00	27	0,00	0,00
:	272	6	91,72	97,79	94	100,00	65,44

Note that we improved mainly the tagging of the adjectives (JJ), of the nouns (NN) and of some of the verbs (VBG and VBN) that are of primary importance for defining a relevant and domain-specific terminology.

Our results can be summarized as follows :
Normalized mean (i.e., a mean weighted by the tag number of a given value) precision of C_{BRILL} : 89.26; Normalized mean recall of C_{BRILL} : 89.12
Normalized mean precision of C_{ETIQ} : 97.48; Normalized mean recall of C_{ETIQ} : 97.12

6 Similar Approaches

Let us now discuss two similar approaches. ANNOTATE is an interactive tool for semi-automatic tagging [14] similar to ours. This system interacts with a statistical tagger [9] and a "parser" to accelerate the tagging. Grammatical tagging works as follows: the tagger determines the tag of each word on the basis of the tag frequency probability, the system then displays the sentences and calls for confirmation or correction of the non reliable labels. In this system, the annotator does not write rules. The internal model is updated gradually with the confirmations and the corrections carried out. These changes are thus exploited immediately.

KCAT [15] is a semi-automatic annotation tool to build a precise and consistent tagged corpus. It combines the use of a statistical tagger, and lexical rules of clarification acquired by an expert. The rules used are limited to word similarity. The method enables the correcting of the words whose labels are not reliable. The expert does not have to repeat the same action on two words in the same context because lexical rules are generated automatically. The expert tags words by lexical rules, the remaining words are tagged by the statistical tagger, then the expert has to correct directly the unreliable tags. These rules are very specific. In order to obtain a well-tagged corpus, the expert must insert a large quantity of tags. This increases human costs and slows down the speed of execution.

In ETIQ, we start with a non annotated corpus of specialty, then we tag the corpus with a general tagger, and, subsequently, the expert inserts the most general possible rules with the help of an inductive module. Finally, when the expert judges that the corpus is correctly tagged, he has the possibility to obtain a perfectly tagged corpus by directly (without rules) correcting the wrong labels.

7 Conclusion and Future Work

Our ultimate goal is domain-specific information extraction, obtained by a complete text-processing chain. Tagging conditions the obtention of a relevant terminology in a given speciality domain which is a key preliminary stage [17] to information extraction. This explains why we give so much importance to the tagging task.

In this paper, we introduced a semi-automatic user-friendly tagger (ETIQ) enabling an expert to deal with non-annotated specialty corpora. Using ETIQ, the expert can easily display the result of a basic tagging (carried out by Brill's tagger). On this basis, the user easily corrects the errors by inserting lexical rules and, subsequently, contextual rules. In the rule writing task, the expert is assisted by an inductive mod-

ule; this module proposes rules learned on the basis built by the expert. To confirm the assumptions upon which our system is built, we carried out its experimental validation: we compared the labels obtained by our system with those obtained by Brill's tagger and those of a 'perfect' corpus.
Progressive induction starting from the behavior of the expert is an approach which is not usually used, but our experience is that the expert finds the proposals of the machine acceptable. At present, we only tested this approach for the lexical rules.

Induction at the contextual stage, and thus a relational one, leads very quickly to a combinatory explosion. We will use progressive induction to solve this problem. The contextual rules come directly from the detection of the erroneous contexts. We will adapt and incorporate (semi-)automatic methods for error detection [16]. These methods will enable the expert to target the contexts where errors are more likely, thus helping him/her to organize and accelerate his/her work.

Our goal is to use "extensional induction" on the various tags, i.e., learning in extension what a tag is. This extension has an extremely large size since it comprises a list of all the strings of letters qualifying for this tag, and of all the contexts that disambiguate this string of letters when it can belong to other tags.

Acknowledgment. The authors thank Mary Felkin for helping in correction of English.

References

1. Kodratoff, Y., Azé, J., Roche, M., Matte-Tailliez, O.: Des textes aux associations entre les concepts qu'ils contiennent. In XXXVièmes Journées de Statistique. RNTI 1 (2003) 171-182
2. Cussens, J.: Part-of-speech tagging using Progol. In S. Dzeroski and N. Lavrac, editors. Proceedings of the 7th International Workshop on Inductive Logic Programming Vol. 1297 Springer-Verlag, (1997) 93-108
3. Lindberg, N., Eineborg, M.: Learning Constraint Grammar-style Disambiguation Rules using Inductive Logic Programming. Proceedings of the Thirty-Sixth Annual Meeting of the Association for Computational Linguistics and Seventeenth International Conference on Computational Linguistics (1998) 775-779
4. Eineborg, M., Lindberg, N.: ILP in Part-of-Speech Tagging - An Overview. In James Cussens and Saso Dzeroski, editors. Learning Language in Logic Vol. 1925 of LNAI. Springer (2000)
5. Daelemans, W., Zavrel, J., Berck, P., Gillis, S.: MBT: A Memory-Based Part of Speech Tagger-Generator. In E. Ejerhed and I. Dagan eds. Proceedings of the Fourth Workshop on Very Large Corpora. Copenhagen (1996) 14-27
6. Zavrel, J., Daelemans, W.: Recent Advances in Memory-Based Part-of-Speech Tagging. In Actas del VI Simposio Internacional de Comunicacion Social. Santiago de Cuba (1999) 590-597
7. Brill, E.: Some Advances in Transformation-Based Part of Speech Tagging. AAAI, 1 (1994) 722-727

8. Cutting, D., Kupiec, J., Pedersen, J., Sibun, P.: A practicial part-of-speech tagger. Proceedings of the Third Conference on Applied Natural Language Processing (1992)
9. Brants, T.: TnT - A Statistical Part- of-Speech Tagger. In Proceedings of the Sixth Conference on Applied Natural Language Processing. Seattle (2000)
10. Halteren, V., Zavrel, J., Daelemans, W.: Improving Accuracy in Word Class Tagging through the Combination of Machine Learning Systems. Computational linguistics 27 (2001) 199-229
11. Brill, E., Wu, J.: Classifier Combination for Improved Lexical Disambiguation. Proceedings of the Thirty-Sixth Annual Meeting of the Association for Computational Linguistics and Seventeenth International Conference on Computational Linguistics (1998)
12. Halteren, V.: Syntactic Wordclass Tagging. Chapter 15. Corpus-Based Rules. E.Brill. Kluwer Academic Publishers (1999)
13. Quinlan, J.R.: C4.5: Programs for Machine Learning. Morgan Kaufmann San Mateo (1993)
14. Plaehn, O., Brants, T.: Annotate - An Efficient Interactive Annotation Tool. In Proceedings of the Sixth Conference on Applied Natural Language Processing. Seattle (2000)
15. Won-Ho, R., Heui-Seok, L., Jin-Dong, K., Hae-Chang R.: KCAT : A Korean Corpus Annotating Tool Minimizing Human Intervention. In Proceedings of the Eighteenth International Conference on Computational Linguistics (2000) 1096-1100
16. Pavel, K., Karel, O.: (semi-)Automatic Detection of Errors in PoS-Tagged Corpora. In Proceedings of the Nineteenth International Conference on Computational Linguistics (2002) 509-515
17. Nenadic, G., Rice, S., Spasic, I., Ananiadou, S., Stapley, B.: Selecting Text Features for Gene Name Classification: from Documents to Terms. In Proceedings of the ACL 2003 Workshop on NLP in Biomedicine. Sapporo (2003) 121-128

A Tree-Based Approach to the Discovery of Diagnostic Biomarkers for Ovarian Cancer

Jinyan Li[1] and Kotagiri Ramamohanarao[2]

[1] Institute for Infocomm Research, 21 Heng Mui Keng Terrace, Singapore 119613
jinyan@i2r.a-star.edu.sg
[2] Dept. of CSSE, The University of Melbourne, VIC 3010, Australia
rao@cs.mu.oz.au

Abstract. Computational diagnosis of cancer is a classification problem, and it has two special requirements on a learning algorithm: perfect accuracy and small number of features used in the classifier. This paper presents our results on an ovarian cancer data set. This data set is described by 15154 features, and consists of 253 samples. Each sample is referred to a woman who suffers from ovarian cancer or who does not have. In fact, the raw data is generated by the so-called mass spectrosmetry technology measuring the intensities of 15154 protein or peptide-features in a blood sample for every woman. The purpose is to identify a small subset of the features that can be used as biomarkers to separate the two classes of samples with high accuracy. Therefore, the identified features can be potentially used in routine clinical diagnosis for replacing labour-intensive and expensive conventional diagnosis methods. Our new tree-based method can achieve the perfect 100% accuracy in 10-fold cross validation on this data set. Meanwhile, this method also directly outputs a small set of biomarkers. Then we explain why support vector machines, naive bayes, and k-nearest neighbour cannot fulfill the purpose. This study is also aimed to elucidate the communication between contemporary cancer research and data mining techniques.

Keywords: Decision trees, committee method, ovarian cancer, biomarkers, classification.

1 Introduction

Contemporary cancer research has distincted itself from the traditional one with the unprecedented large amount of data and tremendous diagnostic and therapeutic innovations. Data are currently generated in high-throughput fashions. Microarray-based gene expression profiling technologies and mass spectrosmetry instruments are two types of such methods in common use [10,18,15,17]. The microarray technologies can measure expression levels (real values) of tens of thousands of genes in a human cell [10,18], while mass spectrosmetry instruments can measure intensities (also real values) of tens or even hundreds of thousands of proteins in a blood sample of a patient [15,17]. As hundreds of cells or blood samples are usually involved in a biomedical study, the resulting data are very complex. In other words, if every gene or protein is considered as a variable, those hundreds of samples are points located at a very high-dimensional Euclidian space with an extremely sparse distribution. These data points are usually labelled with

H. Dai, R. Srikant, and C. Zhang (Eds.): PAKDD 2004, LNAI 3056, pp. 682–691, 2004.

classes, e.g. 'tumor' versus 'normal'. These labelled points can also be converted into a very wide relational table. Such kind of data provide both opportunities and challenges for the data mining field due to the data's high dimensionality and its relatively low volume.

In this paper, we focus on a data set derived from ovarian cancer patients [15]. This disease is the leading cause of death for women who suffer from cancer. Most treatments for advanced ovarian cancer have limited efficacy, and the resulting 5-year survival is just 35%-40%. That is, only a small proportion of advanced ovarian cancer patient can survive 5 years after the treatment. By contrast, if ovarian cancer is detected when it is still at early stage (stage 1), conventional therapy produces a 95% 5-year survival rate [17]. So, early detection—correctly classifying whether a patient has this disease or not—is critical for doctors to save the life of many more patients. The problem is how to use a simple way to make an effective early diagnosis. Mathematically, the problem is how to find such a decision algorithm that uses only a small subset of features but can separate the two classes of samples completely.

Petricoin et al. [15] first attempted to use mass spectrosmetry as basis to identify biomarkers from blood samples for early detection of this disease. Mass spectrosmetry is also called *proteomic profiling*, meaning to measure the mass intensities of all possible proteins or peptides in blood samples. Body fluids such as blood are protein-rich information reservoir that contains traces what the blood has encountered during its circulation through the body. So, a group of diseased ovarian patients should exhibit some distinctive proteomic patterns in their blood samples, compared to those of healthy patients. Petricoin and his colleagues [15] collected 253 blood samples from 91 different healthy patients and 162 different diseased patients. For each blood sample, they measured and obtained the intensities of the same 15154 variables by a so-called PBS-II mass spectrosmetry instrument. They hoped some features could change their intensities in blood sample significantly between the healthy and tumor cases. Petricoin et al. [15] divided the whole data set into a training subset and a test set, and used a genetic computational algorithm to find a subset of features from the training data. Then the identified features were applied to the test set to get an accuracy and other kinds of evaluation indexes such as sensitivity, specificity, and precision. Their results are reported to be perfect: no mis-classifications were made on their test data by their method [15].

However, from data mining point of view, the above computational data analysis is weak. First, 10-fold (or other numbers of folds) cross-validation is commonly used to get an overall reliable accuracy for a learning algorithm on a data set. But, Petricoin and his colleagues evaluated their algorithm only on one small part of pre-reserved test samples. Second, there are many long-studied well-verified learning algorithms such as decision trees [16], support vector machines (SVMs) [3,4], Naive Bayes (NB) [7, 11], and k-nearest neighbour (k-NN) [5] in the field. These algorithms should be also applied to the data set to see whether their performance is perfect and agreeable to each other, and to see whether there indeed exist a small subset of biomarkers in the blood samples of varian cancer patients. So to strengthen the results, in this paper, we conduct the following two aspects of data analysis: (i) Using decision trees (Bagging [1] and Boosting [9]), SVM, NB, and k-NN to get 10-fold (also 7-fold and 5-fold) cross validation performance on this data set; (ii) using a new decision-tree based committee method [12,13] to find biomarkers from the data set.

Over non-linear learning algorithms (SVMs, NB, and k-NN), decision tree based algorithms have advantages for finding a small subset of biomarkers. The induction of a decision tree from a training data set is a recursive learning process [16]. In each iteration, the algorithm selects a most discriminatory feature, and splits the data set into exclusive partitions. If all partitions contain pure or almost pure class of samples, the process stops. Otherwise, the algorithm applies to those un-pure partitions. So such a recursive process can be terminated quickly. Therefore, even a large number of features are input to a tree induction algorithm, the resulting classifier—the tree—may contain just a couple of features. That is, any classification for any new test samples will need only the features in the tree, rather than the whole original features. However, for the non-linear learning algorithms, the classifiers have to use the values of all original features in the test phase. It can be seen that decision tree based algorithms are a more systematic approach to finding a small subset of biomarkers from high-dimensional biomedical data.

The proposal of our new committee method is motivated by the following three reasons: (i) single decision trees do not have high performance; (ii) there exist many diversified decision trees in a high-dimensional data set that have similar performance; (iii) voting by a committee has much potential to eliminate errors made by individual trees. This new committee method has comparable performance with the best of other classifiers. This method also provides small sets of biomarkers that can be directly derived from the trees. This new committee method is different from Bagging [1], AdaBoosting [9], randomized trees [6], and random forest [2].

The remaining of the paper is organized as follows: We give a detailed description of the data set in Section 2. In Section 3, we promote the use of decision tree-based algorithms for finding biomarkers from high-dimensional data, we also introduce a new committee method with technical details. In Section 4, we briefly describe 3 non-linear learning algorithm SVM, NB, and k-nearest neighbour, and explain why these non-linear algorithms cannot fulfill some important biomedical research purpose. In Section 5 and Section 6, we present 10-fold cross-validation results on the ovarian cancer data set. SVM and our new committee method can both achieve the perfect 100% accuracy. We also report the biomarkers identified by our new method from the 253 blood samples. In Section 7, we discuss the difference between our new committee method and state-of-the-art committee algorithms. Finally, we conclude this paper with a summary.

2 Data Set Description

The whole data set consists of two classes of samples: 91 'Normal' samples and 162 'Cancer' samples [15]. A 'Normal' sample refers to a blood (more precisely, serum) sample taking from a healthy patient; while a 'Cancer' sample refers to a blood sample taking from an ovarian cancer patient. Each sample is described by 15154 features. Each feature is a protein or peptide (more precisely, a molecular mass/charge identity) in blood sample, taking continuous real values equal to the intensity of this variable in every particular patient's blood. If a sample is represented by a vector $\langle p_1, p_2, \ldots, p_n \rangle$, where $p_j, 1 \leq j \leq 15154$, is the intensity of the jth feature, then there are total 253 (91 plus 162) such long vectors in this data set.

The raw data are available at a public website `http://clinicalproteomics.steem.com/download-ovar.php` (as of November 12, 2003). As the raw data does not match the common format used in data mining and machine learning, we have performed some pre-processing and transformed the raw data in correspondence with the `.data` and `.names` format. The transformed data are stored at our *Kent Ridge Biomedical Dataset Repository* (`http://sdmc.i2r.a-star.edu.sg/rp/`).

3 A New Decision Tree Based Committee Method for Classification

In this section, we first show that decision trees are a good and systematic approach to finding a small subset of biomarkers. Second, we use an example to show single decision trees are not always accurate in test phase though possessing perfect accuracy on training data. To strengthen the performance, we introduce a new committee method to combine single trees for reliable and accurate classification on test samples.

3.1 Decision Tree: An Ideal Algorithm for Finding Biomarkers

We use a different high-dimensional biomedical data set, a *gene expression* profiling data set, for demonstration. A reason for this is that the ovarian data set does not have an independent test data, but this gene expression data set has. This gene expression data set consists of 215 two-class training samples for differentiation between 14 MLL (mixed-lineage leukemia) subtype patients of childhood leukemia disease and 201 patients suffering any other subtype of this disease [18]. The data are described by 12558 features. Here each feature is a gene, having continuous expression values. This decision tree is constructed by C4.5 [16]. The structure of this tree is depicted in Figure 1. Observe that there are only 4 features in this tree residing at the 4 non-leaf nodes, namely 34306_at, 40033_at, 33244_at, and 35604_at. Compared to the total 12558 input features, the 4 built-in features in the tree is a tiny fraction. For interpretation, each of the five leaf nodes corresponds to a rule, the rule's predictive term is the class label contained in the leaf node.

So, this tree can be decomposed into 5 rules and these rules can be formed a function as follows.

$$f(x_1, x_2, x_3, x_4) = \begin{cases} -1 \text{ if } x_1 \leq a, x_2 \leq b \\ -1 \text{ if } x_1 \leq a, x_2 > b, x_3 \leq c \\ \;\;1 \text{ if } x_1 \leq a, x_2 > b, x_3 > c \\ \;\;1 \text{ if } x_1 > a, x_4 \leq d \\ -1 \text{ if } x_1 > a, x_4 > d \end{cases}$$

where $x_1, x_2, x_3,$ and, x_4 represent gene variables 34306_at, 40033_at, 33244_at, and 35604_at, respectively; $a = 13683.6, b = 3691.4, c = 986.9, d = 846.6$; the two values of this function -1 and 1 represent OTHERS and MLL respectively. So, given a test sample, at most 3 of the 4 genes' expression values are needed to determine $f(x_1, x_2, x_3, x_4)$. If the function value is -1, then the test sample is predicted as OTHERS, otherwise it is predicted as MLL. These 4 genes are biomarkers—simple and effective! We do not necessarily need the entire 12588 original features to make decisions. As seen later, there also exist similar small number of protein biomarkers in the ovarian cancer data set that can be discovered by decision tree based methods.

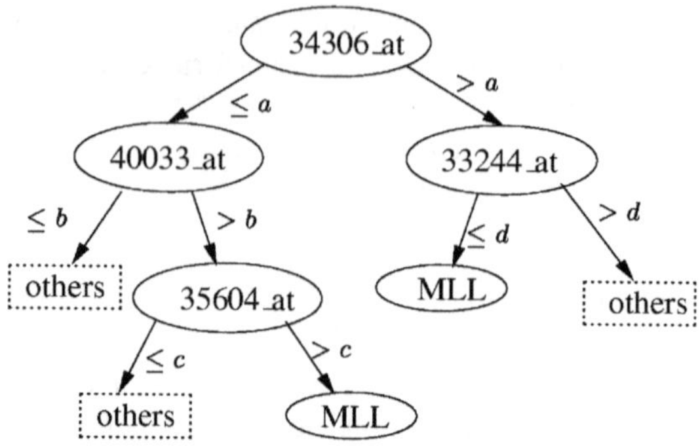

Fig. 1. A decision tree induced by C4.5 from gene expression data set for differentiating the subtype MLL against other subtypes of childhood leukemia. Here $a = 13683.6, b = 3691.4, c = 986.9, d = 846.6$.

3.2 An Introduction to a New Committee Method

Single decision trees usually do not provide good accuracies on test data, especially when handling high-dimensional biomedical data such as gene expression profiling data or proteomic profiling data [14]. For example, on the data set discussed in the previous section, the tree made 4 mistakes on 112 test samples. A possible reason is that the greedy search heuristic confines the capability of the tree induction algorithm, only allowing the algorithm learn well on one aspect of the high-dimensional data.

Next, we introduce our new committee method. It is called CS4[1] [13,12], and its basic idea to discover a committee of trees is to change root nodes using different top-ranked features. For example, to discover 10 trees, we use 10 top-ranked features each as the root node of a tree. Such trees are called *cascading* trees [12]. This method was motivated by the idea of 'second-could-be-the-best', and it is confirmed by the observation that there exist many outstanding features in high-dimensional data that possess similar classification merits with little difference.

Table 1 summarizes the training and test performance, and the numbers of features used in 10 cascading trees induced from the MLL-OTHERS data set. Here, the ith $(1 \leq i \leq 10)$ tree is established using the ith top-ranked feature as root node. Observe that: (a) the 10 trees made similar numbers of errors on the training and test data; (b) the 5th, 8th, or the 9th tree made a smaller number of errors than the first tree made; (c) the rules in these trees were very simple, containing about 2, 3 or 4 features; (d) none of them produced a perfect test accuracy.

Next, we describe how our CS4 method combines these individual trees and how CS4 eliminates those errors. The general method is as follows [13]. Suppose k trees being induced from a data set consisting of only positive and negative training samples. Given a test sample T, each of the k trees in the committee will have a specific rule to tell us a predicted class label for this test sample. Denote the k rules from the tree committee as: $rule_1^{pos}, rule_2^{pos}, \cdots, rule_{k_1}^{pos}$, and $rule_1^{neg}, rule_2^{neg}, \cdots, rule_{k_2}^{neg}$. Here $k_1 + k_2 = k$.

[1] CS4 is an acronym of *Cascading-and-Sharing for* ensembles of decision trees.

Table 1. The performance of 10 cascading trees on the MLL-others data set that consist of 215 training and 112 test samples.

Tree No.	1	2	3	4	5	6	7	8	9	10
Training errors	0	0	0	0	0	0	1	0	0	0
Test errors	4	3	2	3	2	6	7	2	1	6
# of features	4	4	4	4	4	4	4	4	4	6

Each of $rule_i^{pos}$ $(1 \le i \le k_1)$ predicts T to be in the positive class, while each of $rule_i^{neg}$ $(1 \le i \le k_2)$ predicts T to be in the negative class. Sometimes, the k predictions can be unanimous—i.e., either $k_1 = 0$ or $k_2 = 0$. In these situations, the predictions from all the k rules agree with one another, and the final decision is obvious and seemed reliable. Often, the k decisions are mixed with either a majority of positive classes or a majority of negative classes. In these situations, we use the following formulas to calculate two classification scores based on the *coverages* of these rules, i.e., the percentage of a class samples that are satisfied by these rules: $Score^{pos}(T) = \sum_{i=1}^{k_1} coverage(T, rule_i^{pos})$, and $Score^{neg}(T) = \sum_{i=1}^{k_2} coverage(T, rule_i^{neg})$. If $Score^{pos}(T)$ is larger than $Score^{neg}(T)$, we assign the positive class to the test sample T. Otherwise, T is predicted as negative. This weighting method allows the tree committee to automatically distinguish the contributions from the trivial rules and those from the significant rules in the prediction process.

Let's examine the performance of combined trees on the MLL-others data set. When combining the first 2 trees, the committee made 3 mistakes, one mistake less than that made by the sole first tree. When combining the first 4 trees, the committee did not make any errors on the 112 test samples, eliminating all the mistakes made by the first tree. Adding more trees into the committee till the 10th tree, the expanding committee still maintained the perfect test accuracy.

From this example, we can see that the use of multiple trees can much improve the performance of single decision trees. In addition to this perfect accuracy, the committee of trees also provide a small subset of biomarkers. In fact, each decision tree contains 3 or 4 features, so, the committee contains total 30 to 40 features. Compared to the original 12558 features, the features contained in the committee is very small. As seen later, the CS4 learning algorithm can also achieve the two goals for the ovarian cancer data set.

4 Three Non-linear Learning Algorithms

We have seen in the last section that tree based methods are a systematic approach to finding a small subset of features that can be used as biomarkers. In this section, we explain why k-nearest neighbour, support vector machines, and naive bayes cannot fulfill this purpose, and we also explain why genetic algorithm is not most relevant for this purpose.

The k-nearest neighbour [5] classifier is an instance-based algorithm, it does not need a learning phase. The simple intuition behind classification is that the class label of a new sample X should agree with the majority of k nearest points of X if a training data set are viewed as a set of *points* in a high-dimensional Euclidean space. So, this classifier can only make a prediction, but cannot derive any biomarkers because it always depends on the whole input features for the calculation of distance.

Support vector machine is a statistical learning algorithm, it is also called kernel-based learning algorithm [3,4]. Consider a two-class training data set having n number of original features $x_1, x_2, \cdots, x_n$, a classifier induced from this training data by support vector machines is a function defined by:

$$f(x_1, \ldots, x_n) = \begin{cases} 1 \text{ if } \sum_{i \in I} \alpha_i * K(Y_i, X) + b \geq 0 \\ -1 \text{ otherwise} \end{cases}$$

where $X, Y_i \in \mathbf{R}^n$, $\alpha_i, b \in \mathbf{R}$; α_i, Y_i, I and b are parameters and X is the sample to be classified whether it is 1 (normal) or -1 (abnormal). The SVM training process determines the entire parameter set $\{\alpha_i, Y_i, I, b\}$; the resulting $Y_i, i \in I$ are a subset of the training set and are usually called *support vectors*. The kernel function K can have different forms. For example, $K = (X \cdot Y_i)^p$ implements a polynomial SVM classifier; $K = \tanh(X \cdot Y_i + \delta)$ implements a two-layer neural network. The simplest form of the kernel functions is $K = X \cdot Y_i = \sum_{j=1}^{n} x_j * y_{ij}$, namely a n-dimensional linear function.

Regardless of different types of kernels, SVM classifiers always have the same number of variables as that in the input feature space. So, SVM learning algorithms themselves do not derive any small subset of features in the learning phase.

Naive Bayes [7,11] is a probabilistic learning algorithm based on the Bayes theorem and the assumption of class conditional independence. Naive Bayes assumes that the effect of a feature value on a given class is independent of the values of other features. We can also see that it is impossible for Naive Bayes itself to conduct feature selection and derive a small subset of biomarkers because the calculation of the probability crosses the whole feature space.

Due to some randomization mechanism, the output of genetic algorithms is not stable depending machine, time, or even program coding languages. So, biomarkers identified by genetic algorithms may change from time to time simply because of the different randomization seeds. However, tree-based algorithms are stable.

5 Accuracy on the Ovarian Cancer Data Set

This section reports the 10-fold cross-validation results on the ovarian cancer data set using SVM, NB, k-NN, decision trees, and CS4. In each iteration, error numbers on the test data by a learning algorithm is recorded; the overall performance is then the summation of error numbers in the 10 iterations. This evaluation method is much more reliable than the method used by Petricoin et al [15] that only relies on the performance on one small pre-reserved test samples.

The results are shown in Table 2. The Weka software package (`http://www.cs.waikato.ac.nz/ml/weka/`) and our in-house developed CS4 software are used to get the results.

Table 2. Errors in the 10-fold cross validations of 7 learning algorithms on the ovarian cancer data set. The symbol $(x : y)$ stands for x number of errors made in the 'Cancer' class and y number of errors made in the 'Normal' class.

Algorithm	CS4 (20 trees)	C4.5	Bagging (20 trees)	Boosting	SVM	3-NN	NB
Errors	0	10 (4:6)	7 (3:4)	10 (4:6)	0	15 (5:10)	19 (17:2)
Accuracy	100%	96.0%	97.2%	96.0%	100%	94.1%	92.5%

From this table, we can see that both SVM and our CS4 algorithm can achieve the same perfect 100% 10-fold cross validation accuracy on this ovarian cancer data set. But they use different numbers of features. SVM always uses the whole 15154 features in the classification, while CS4 uses less than 100 features. So, for the early detection of ovarian cancer by mass spectrosmetry, our CS4 algorithm is better than SVM for the data analysis including the discovery of biomarkers from the blood samples. More results about biomarkers will be shown in the next section. For other fold (7-fold or 5-fold in this paper) cross validation, CS4 made one mistake, SVM made one mistake in the 7-fold validation, but did not make any mistake in the 5-fold. So, their performance were agreed to each other very much.

Like CS4, Bagging is also a committee method. In addition to using 20 trees in a committee, we also tried other numbers such as 30, 40, or 50 trees for Bagging. However, all these cases did not improve the Bagging's performance (still making the same 7 mistakes).

One possible way for non-linear learning algorithms to find a small subset of biomarkers from high-dimensional data sets is the pre-selection of top-ranked features based on some criteria such as entropy [8]. However, we found that the performance of the non-linear algorithms decreased when only top 10, 20, 25, 30, 35, or 40 ranked features are used. This may be due to the strong bias on the use of only top-ranked features, and this also suggests that opening the whole feature space for consideration, as done by decision tree, is a fair approach to finding biomarkers.

6 Biomarkers Identified from the Blood Samples

Now that CS4 has the perfect 10-fold cross validation performance, we have strong reasons to believe that its identified biomarkers should be very useful for any test samples beyond the 253 samples. In the following, we present detailed results about biomarkers identified from the whole ovarian cancer data set.

We generated 20 cascading trees by CS4. Each of the trees contains 2 to 5 features. In total, there are 72 features in this tree committee. These 72 features are just protein biomarkers that we can use for the early detection of ovarian cancer based on women's blood samples. Mathematically, these biomarkers can form 92 simple rules. We list some of them in the following function:

$$p(x_1, x_2, \ldots, x_7) = \begin{cases} 1 \text{ if } x_1 \le a,\, x_2 \le b \\ -1 \text{ if } x_1 \le a,\, x_2 > b,\, x_4 \le d \\ 1 \text{ if } x_1 \le a,\, x_2 > b,\, x_4 > d \\ -1 \text{ if } x_1 > a,\, x_3 \le c,\, x_5 \le e \\ 1 \text{ if } x_1 > a,\, x_3 \le c,\, x_5 > e \\ -1 \text{ if } x_1 > a,\, x_3 > c \\ 1 \text{ if } x_6 \le f,\, x_7 \le g \\ -1 \text{ if } x_6 \le f,\, x_7 > g \\ -1 \text{ if } x_6 > f \end{cases}$$

where $x_1, \ldots, x_7$ are bio-markers. $a = 0.435461, b = 0.611786, c = 0.277777, d = 0.696293, e = 0.294503, f = 0.251969, g = 0.493934$; the two values of this function -1 and 1 represent NORMAL and CANCER respectively. So, given a new blood sample, at most 3 variables' intensities are needed to determine $p(x_1, x_2, \ldots, x_7)$. If the function value is -1, then the patient is diagnosed as NORMAL, otherwise it is predicted as CANCER.

7 Discussion and Conclusion

CS4 differs fundamentally from the state-of-the-art committee methods such as Bagging [1] and AdaBoosting [9]. Unlike them, our method always uses the original training data instead of *bootstrapped*, or pseudo, training data to construct a sequence of different decision trees. Though a bootstrapped training set is the same size as the original data, some original samples may no longer appear in the new set while others may appear more than once. So, rules produced by the Bagging or Boosting methods may not be correct when applied to the original data, but rules produced by CS4 reflect precisely the nature of the original training data. The bagging or boosting rules should therefore be employed very cautiously, especially in the applications of bio-medicine where such concerns could be critical. In addition to being different from Bagging and Boosting, CS4 also differs from other voting methods such as random forest [2] and randomized decision trees [6]—they randomly select a best feature as root node from a set of candidates.

Finally, we summarize the paper. We have applied a new tree-based committee method to a large ovarian cancer data set for the early detection of this disease and for finding a small subset of biomarkers from blood samples. The new method CS4 has achieved the perfect 100% 10-fold cross validation accuracy on this data set. This method also identified 70 or so biomarkers from the 15154 candidates. Though SVM also achieved the same perfect accuracy, it could not directly derive a small set of biomarkers for the early detection of this disease. Taking this ovarian cancer data set, we have also demonstrated how to bridge the gap between contemporary cancer research and data mining algorithms. We also emphasize that patterns, rules, and mining algorithms should be easily acceptable to the biomedical fields. As a future work, the CS4 algorithm will be extended to data analysis for other specific diseases such as breast cancer, liver cancer, and prostate cancer to identify disease-specific biomarkers.

References

1. Leo Breiman. Bagging predictors. *Machine Learning*, 24:123–140, 1996.
2. Leo Breiman. Random forest. *Machine Learning*, 45:5–32, 2001.
3. C.J. Burges. A tutorial on support vector machines for pattern recognition. *Data Mining and Knowledge Discovery*, 2:121–167, 1998.
4. C. Cortez and V. Vapnik. Support vector networks. *Machine Learning*, 20:273–279, 1995.
5. T. M. Cover and P. E. Hart. Nearest neighbor pattern classification. *IEEE Transactions on Information Theory*, 13:21–27, 1967.
6. Thomas G. Dietterich. An experimental comparison of three methods for constructing ensembles of decision trees: Bagging, boosting, and randomization. *Machine Learning*, 40:139–158, 2000.
7. R. Duda and P. Hart. *Pattern Classification and Scene Analysis*. New York: John Wiley & Sons, 1973.
8. U.M. Fayyad and K.B. Irani. Multi-interval discretization of continuous-valued attributes for classification learning. In Ruzena Bajcsy, editor, *Proceedings of the Thirteenth International Joint Conference on Artificial Intelligence*, pages 1022–1029. Morgan Kaufmann, 1993.
9. Yoav Freund and Robert E. Schapire. Experiments with a new boosting algorithm. In Lorenza Saitta, editor, *ICML*, pages 148–156, Bari, Italy, July 1996. Morgan Kaufmann.
10. T. R. Golub, D. K. Slonim, P. Tamayo, C. Huard, M. Gaasenbeek, J. P. Mesirov, H. Coller, M. L. Loh, J.R. Downing, M. A. Caligiuri, C. D. Bloomfield, and E. S. Lander. Molecular classification of cancer: Class discovery and class prediction by gene expression monitoring. *Science*, 286:531–537, October 1999.
11. P. Langley and S. Sage. Induction of selective bayesian classifiers. In *Proceedings of the Tenth Conference on Uncertainty of Artificial Intelligence*, pages 399–406. Morgan Kaufmann, 1994.
12. Jinyan Li and Huiqing Liu. Ensembles of cascading trees. In *Proceedings of ICDM*, pages 585–588, 2003. IEEE Computer Society.
13. Jinyan Li, Huiqing Liu, See-Kiong Ng, and Limsoon Wong. Discovery of significant rules for classifying cancer diagnosis data. *Bioinformatics*, 19:ii93–102, 2003.
14. Huiqing Liu, Jinyan Li, and Limsoon Wong. A comparative study on feature selection and classification methods using gene and protein expression profiles. In *Genome Informatics 2002*, pages 51 – 60, Tokyo, Japan, 2002. Universal Academy Press.
15. Emanuel F Petricoin, Ali M Ardekani, Ben A Hitt, Peter J Levine, Vincent A Fusaro, Seth M Steinberg, Gordon B Mills, Charles Simone, David A Fishman, Elise C Kohn, and Lance A Liotta. Use of proteomic patterns in serum to identify ovarian cancer. *Lancet*, 359:572–577, 2002.
16. J. R. Quinlan. *C4.5: Programs for Machine Learning*. Morgan Kaufmann, San Mateo, CA, 1993.
17. Julia D. Wulfkuhle, Lance A. Liotta, and Emanuel F. Petricoin. Proteomic applications for the early detection of cancer. *Nature Review: Cancer*, 3:267–275, 2001.
18. Eng-Juh Yeoh, Mary E. Ross, Sheila A. Shurtleff, W. Kent Williams, Divyen Patel, Rami Mahfouz, Fred G. Behm, Susana C. Raimondi, Mary V. Relling, Anami Patel, Cheng Cheng, Dario Campana, Dawn Wilkins, Xiaodong Zhou, Jinyan Li, Huiqing Liu, Ching-Hon Pui, William E. Evans, Clayton Naeve, Limsoon Wong, and James R. Downing. Classification, subtype discovery, and prediction of outcome in pediatric acute lymphoblastic leukemia by gene expression profiling. *Cancer Cell*, 1:133–143, 2002.

A Novel Parameter-Less Clustering Method for Mining Gene Expression Data

Vincent Shin-Mu Tseng and Ching-Pin Kao

Department of Computer Science and Information Engineering,
National Cheng Kung University, Tainan, Taiwan, R.O.C.
tsengsm@mail.ncku.edu.tw

Abstract. Clustering analysis has been applied in a wide variety of fields. In recent years, it has even become a valuable and useful technique for in-silico analysis of microarray or gene expression data. Although a number of clustering methods have been proposed, they are confronted with difficulties in the requirements of automation, high quality, and high efficiency at the same time. In this paper, we explore the issue of integration between clustering methods and validation techniques. We propose a novel, parameter-less, and efficient clustering algorithm, namely CST, which is suitable for analysis of gene expression data. Through experimental evaluation, CST is shown to outperform other clustering methods substantially in terms of clustering quality, efficiency, and automation under various types of datasets.

1 Introduction

In recent years, clustering analysis has become a valuable and useful technique for in-silico analysis of microarray or gene expression data. The main goal of clustering analysis is to partition a given set of objects into homogeneous groups based on given features [1]. Although a number of clustering methods have been proposed [1], [3], [4], [8], [11], they are confronted with some difficulties. First, most clustering algorithms request users to specify some parameters. In real applications, however, it is hard for biologists to determine the suitable parameters manually. Thus an automated clustering method is required. Second, most clustering algorithms aim to produce the clustering results based on the input parameters and their own criterions. Hence, they are incapable of producing optimal clustering result. Third, the existing clustering algorithms may not perform well when the optimal or near-optimal clustering result is enforced from the universal criterions.

On the other hand, a variety of clustering validation measures are applied to evaluate the validity of the clustering results, the suitability of parameters, and the reliability of clustering algorithms. A number of clustering methods have been proposed, such as DB-index [2], Simple matching coefficient, [6], Jaccard coefficient, [6], Hubert's Γ statistic, [6], FOM [10], ANOVA [7], VCV [5], et al. Nevertheless, the roles of them are placed only on the phase of "post-validation", with the exception of

H. Dai, R. Srikant, and C. Zhang (Eds.): PAKDD 2004, LNAI 3056, pp. 692–698, 2004.

Smart-CAST [9]. The study of how validation techniques can help clustering methods performance well has been strangely neglected.

In this paper, we focus on the integration between clustering methods and validation techniques. We propose a novel, parameter-less, and efficient clustering algorithm that is fit for analysis of gene expression data. The proposed algorithm determines the best grouping of genes on-the-fly by using Hubert's Γ statistic as validation measurement during clustering without any user-input parameter. The experiments on synthetic microarray datasets showed that the proposed method can automatically produce "nearly optimal" clustering results in very high speed.

The rest of the paper is organized as follows: In Section 2, we describe the proposed method, namely Correlation Search Technique (CST) algorithm. In Section 3, the empirical evaluation results are presented. Finally, the conclusions are drawn in Section 4.

2 Correlation Search Technique

For the proposed Correlation Search Technique (CST) algorithm, the original dataset must be transformed into a similarity matrix S. The matrix S stores the degree of similarity between every pair of genes in the dataset, with the range of similarity degree in [0, 1]. The similarity can be obtained by various measurements like Euclidean distance, Pearson's correlation coefficient, etc. CST will automatically cluster the genes according to the matrix S without any user input parameters.

2.1 Basic Principles

The CST method integrates clustering method with validation technique so it can cluster the genes quickly and automatically. By embedding the validation technique, CST can produce a "near-optimal" clustering result. The validation measure we used here is Hubert's Γ statistic [6]. Let $\boldsymbol{X}=[X(i,j)]$ and $\boldsymbol{Y}=[Y(i,j)]$ be two $n \times n$ matrices on the same n genes, where $X(i, j)$ indicates the similarity of genes i and j, and $Y(i, j)$ is defined as follows:

$$Y(i,j) = \begin{cases} 1 & \text{if genes } i \text{ and } j \text{ are clustered in the same cluster,} \\ 0 & \text{otherwise.} \end{cases} \tag{1}$$

The *Hubert's Γ statistic* represents the point serial correlation between the matrices $\boldsymbol{X}$ and $\boldsymbol{Y}$, and is defined as follows if the two matrices are symmetric:

$$\Gamma = \frac{1}{M} \sum_{i=1}^{n-1} \sum_{j=i+1}^{n} \left(\frac{X(i,j) - \overline{X}}{\sigma_X} \right) \left(\frac{Y(i,j) - \overline{Y}}{\sigma_Y} \right) \tag{2}$$

where $M = n(n-1)/2$ is the number of entries in the double sum, and σ_X and σ_Y denote the sample standard deviations. $\overline{X}$ and $\overline{Y}$ denote the sample means of the entries of matrices $\boldsymbol{X}$ and $\boldsymbol{Y}$. The value of Γ is between [-1, 1] and a higher value of Γ represents the better clustering quality.

To reduce the time cost in computing Γ statistic, we simplify it as follows:

$$\Gamma = \frac{\left(M\sum_{i=1}^{n-1}\sum_{j=i+1}^{n} X(i,j)Y(i,j) - \sum_{i=1}^{n-1}\sum_{j=i+1}^{n} X(i,j)\sum_{i=1}^{n-1}\sum_{j=i+1}^{n} Y(i,j)\right)}{\sqrt{M\sum_{i=1}^{n-1}\sum_{j=i+1}^{n} X(i,j)^2 - \left(\sum_{i=1}^{n-1}\sum_{j=i+1}^{n} X(i,j)\right)^2}\sqrt{M\sum_{i=1}^{n-1}\sum_{j=i+1}^{n} Y(i,j)^2 - \left(\sum_{i=1}^{n-1}\sum_{j=i+1}^{n} Y(i,j)\right)^2}} \tag{3}$$

Here, the square of $Y(i, j)$ can be ignored since the value of $Y(i, j)$ is either zero or one. Furthermore, similarity $X(i, j)$ of same dataset is invariable regardless that the clustering result is variable. Accordingly, the measurement Γ' indicating the quality of clustering result is

$$\Gamma' = \frac{\left(M\sum_{i=1}^{n-1}\sum_{j=i+1}^{n} X(i,j)Y(i,j) - \sum_{i=1}^{n-1}\sum_{j=i+1}^{n} X(i,j)\sum_{i=1}^{n-1}\sum_{j=i+1}^{n} Y(i,j)\right)}{\sqrt{M\sum_{i=1}^{n-1}\sum_{j=i+1}^{n} Y(i,j) - \left(\sum_{i=1}^{n-1}\sum_{j=i+1}^{n} Y(i,j)\right)^2}} \tag{4}$$

2.2 CST Method

The input of CST is a symmetric similarity matrix $\boldsymbol{X}$, where $X(i, j) \in [0, 1]$. CST is a greedy algorithm that constructs clusters one at a time, and the currently constructed cluster is denoted by C_{open}. Each cluster is started by a seed and is constructed incrementally by adding (or removing) elements to (or from) C_{open} one at a time. The temporary clustering result in each addition (or removal) of x is computed by simplistic Hubert's Γ statistic, i.e. equation (4), and is denoted by $\Gamma_{add}(x)$ (or $\Gamma_{remove}(x)$). In addition, the current maximum of simplistic Hubert's Γ statistic is denoted by Γ_{max}. We say that an element x has high positive correlation if $\Gamma_{add}(x) \geq \Gamma_{max}$, and x has high negative correlation if $\Gamma_{remove}(x) \geq \Gamma_{max}$. CST takes turns between adding high positive correlation elements to C_{open}, and removing high negative correlation elements from it. When C_{open} is stabilized by addition and removal procedure, this cluster is complete and next one is started.

To reduce computing time, we simplify equation (4) further. Considering each addition (or removal) stage, the summation of $Y(i, j)$ is equivalent, no matter which element is chosen. Consequently, (4) is abridged in deciding the element to be added to (or removed from) C_{open} and the measurement Γ'' of effect of each added (or removed) element is

$$\Gamma'' = \sum_{i=1}^{n-1}\sum_{j=i+1}^{n} X(i,j)Y(i,j) \tag{5}$$

```
Input: An n-by-n similarity matrix X

0. Initialization:
    M = n (n - 1) / 2
    S_X = Σ_{i=1}^{n-1} Σ_{j=i+1}^{n} X (i, j)
    S_Y = 0
    S_XY = 0
    C = Ø                                  /* The collection of closed clusters */
    U = {1, 2, ..., n}                     /* Elements not yet assigned to any cluster */
    Γ_max = 0
1. while (U  Ø) do
    C_open = Ø
    a (•) = 0
    1.1. SEED: Pick an element u ∈ U with most neighbors
        U = U - {u}                             /* Remove u from U */
        For all i ∈ U set a(i) = X(u, i)        /* Update the affinity */
        C_open = {u}                            /* Insert u into C_open */
    1.2. ADD: while MaxValidaty( )  Γ_max do
        Pick an element u ∈ U with maximum a(•)
        U = U - {u}                             /* Remove u from U */
        S_Y = S_Y + |C_open|
        S_XY = S_XY + a(u)
        For all i ∈ U ∪ C_open set a(i) = a(i) + X(u, i) /* Update the affinity */
        C_open = C_open ∪ {u}                   /* Insert u into C_open */
        Γ_max = MaxValidaty( )
    1.3. REMOVE: while MaxValidaty( ) > Γ_max do
        Pick an element v ∈ C_open with minimum a(•)
        C_open = C_open - {v}                   /* Remove v from C_open */
        S_Y = S_Y - |C_open|
        S_XY = S_XY - a(u)
        For all i ∈ U ∪ C_open set a(i) = a(i) - X(u, i) /* Update the affinity*/
        U = U ∪ {v}                             /* Insert v into U */
        Γ_max = MaxValidaty()
    1.4. Repeat steps ADD and REMOVE as long as there are no elements been
         removed.
    1.5. C = C ∪ {C_open}
  end
2. Done, return the collection of cluster, C.
```

Fig. 1. The pseudo-code of CST

The pseudo-code of CST is as shown in Fig. 1. The subroutine MaxValidaty(•) computes the maximal value of measurement Γ' in adding (or removing) a certain element. In the addition stage, it is equal to

$$\frac{(M * (S_{XY} + \max\{a(u) \mid u \in U\}) - S_X * (S_{Y+} \mid C_{open} \mid))}{\sqrt{M * (S_{Y+} \mid C_{open} \mid) - (S_{Y+} \mid C_{open} \mid)^2}}$$

For the removal stage, it becomes

$$\frac{(M * (S_{XY} - \min\{a(v) \mid v \in C_{open}\}) - S_X * (S_{Y-} \mid C_{open} \mid +1))}{\sqrt{M * (S_{Y-} \mid C_{open} \mid +1) - (S_{Y-} \mid C_{open} \mid +1)^2}}$$

3 Experimental Evaluation

In order to evaluate the performance and accuracy of our CST method, we design a correlation-inclined cluster dataset generator. Initially we generate four seed sets with size three, six, five, and ten, respectively, and the seeds in each set have correlation coefficient less than 0.1 with 15 dimensions. Then, we load these four seed sets into the generator to generate two synthetic datasets for testing, respectively, called Dataset I and Dataset II. The Dataset I contains three main clusters with size 900, 700, and 500, and 400 additional outliers, while the Dataset II contains six main clusters with size 500, 450, 400, 300, 250, and 200, and 400 additional outliers.

We compare the proposed method CST with the well-known clustering method, namely k-means [5], CAST-FI and Smart-CAST [9]. For k-means, the value of k was varied from 2 to 21 and from 2 to 41 in increment of 1, respectively. For CAST-FI, the value of affinity threshold t was varied from 0.05 to 0.95 in fixed increment of 0.05. The quality of clustering results was measured by Hubert's Γ statistic, Simple matching coefficient [6], and Jaccard coefficient [6]. We also use intensity image [5] to exhibit more information produced by the clustering methods.

Table 1 shows the total execution time and the best clustering quality of the tested methods on Dataset I. The notation "*M*" indicates the number of main clusters produced. Here we consider clusters with size smaller than 50 as outliers. We observe that CST, Smart-CAST and CAST-FI outperform k-means substantially in both of execution time and clustering quality. In particular, our approach performs 396 times to 1638 times faster than k-means on Dataset I. In addition, the results also show that the clustering quality generated by CST is very close to that of Smart-CAST and CAST-FI for measurements of Hubert's Γ statistic, Simple matching coefficient, and Jaccard coefficient. It means that the clustering quality of CST is as good as Smart-CAST and CAST-FI even that the computation time of CST is reduced substantially.

Table 1. Experimental results obtained by applying the tested methods to Dataset I

Methods	Time (s)	# Clusters	Γ Statistic	Matching	Jaccard
CST	< 1	65 (*M*=3)	0.800	0.981	0.926
Smart CAST	12	85 (*M*=3)	0.800	0.986	0.944
CAST-FI	54	91 (*M*=3)	0.799	0.986	0.945
k-means (k=2-21)	396	6 (*M*=6)	0.456	0.825	0.427
k-means (k=2-41)	1638	6 (*M*=6)	0.456	0.825	0.427

(a) (b) (c)

Fig. 2. Intensity images on Dataset I. (a) The prior cluster structure. (b) Clustering results of CST. (c) Clustering results of CAST

Table 1 also shows that k-means produces 6 main clusters with size as 580, 547, 457, 450, 352, and 114 for the best clustering result. This does not match the real cluster structure. In contrast, CST produces 65 clusters as the best clustering result with 3 main clusters of size 912, 717, and 503. This matches the prior cluster structure very well. The same observation applies to Dataset II. Moreover, CST also generates a number of clusters with small size, which are mostly outliers. This means that CST is superior to k-means in filtering out the outliers from the main clusters.

The intensity images of the prior cluster structure and the clustering results on Dataset I are shown in Fig. 2. From Fig. 2(a), we observe easily that there are three main clusters and quite a number of outliers in Dataset I. It is also observed that Fig. 2(b) and Fig. 2(c) are very similar to Fig. 2(a), meaning that the clustering results of CST and Smart-CAST are very similar to the real cluster structure of Dataset I. The above observations also apply to Dataset II.

4 Conclusions

In this paper, we focus on the integration between clustering methods and validation techniques. We propose a novel, parameter-less, and efficient clustering algorithm, called CST, for analysis of gene expression data. CST clusters the genes via Hubert's Γ statistic on the fly and produces a "nearly optimal" clustering result. Performance evaluations on synthetic gene expression datasets showed that CST method can achieve higher efficiency and clustering quality than other methods without requesting the users to setup parameters. Therefore, CST can provide high degree of automation, efficiency and clustering quality, which are lacked in other clustering methods for gene expression mining. In the future, we will use real microarray datasets to evaluate the validity and efficiency of the CST method.

References

1. M.-S. Chen, J. Han, and P. S. Yu, "Data mining: An Overview from a Database Perspective," *IEEE Transactions on Knowledge and Data Engineering*, vol. 8, no. 6, pp. 866-883, 1996
2. D. L. Davies and D. W. Bouldin, "A cluster separation measure," *IEEE Transactions on Pattern Analysis and Machine Intelligence*, vol. 1, no. 2, pp. 224-227, 1979.
3. S. Guha, R. Rastogi, K. and Shim, "CURE: An efficient clustering algorithm for large databases," *Proc. of ACM Int'l Conf. on Management of Data*, pp. 73-84, 1998.
4. S. Guha, R. Rastogi, K. and Shim, "ROCK: a robust clustering algorithm for categorical attributes," *Proc. of the 15th Int'l Conf. on Data Eng*, pp. 512-521, 1999.
5. R. J. Hathaway, J. C. Bezdek, "Visual cluster validity for prototype generator clustering models," *Pattern Recognition Letters*, vol. 24, no. 9-10, pp. 1563-1569, 2003.
6. A. K. Jain and R. C. Dubes, *Algorithms for Clustering Data*, Prentice Hall, Englewood Cliffs, N.J, 1988.
7. M. K. Kerr and G. A. Churchill, "Bootstrapping cluster analysis: assessing the reliability of conclusions from microarray experiments," *Proc. Natl Acad. Sci. USA*, vol. 98, no. 16, pp. 8961–8965, 2001.
8. T. Kohonen, "The self-organizing map," *Proceedings of the IEEE*, vol. 78, no. 9, pp. 1464-1479, 1990.
9. S. M. Tseng and C. P. Kao, "Mining and Validating Gene Expression Patterns: an Integrated Approach and Applications," *Informatica*, vol. 27, pp. 21-27, 2003.
10. K. Y. Yeung, D. R. Haynor and W. L. Ruzzo, "Validating Clustering for Gene Expression Data," *Bioinformatics*, vol. 17, no. 4, pp. 309-318, 2001.
11. T. Zhang, R. Ramakrishnan and M. Livny, "BIRCH: An Efficient Data Clustering Method for Very Large Databases," *Proc. of the 1996 ACM SIGMOD Int'l Conf. on Management of Data*, pp. 103-114, 1996.

Extracting and Explaining Biological Knowledge in Microarray Data

Paul J. Kennedy[1], Simeon J. Simoff[1], David Skillicorn[2], and Daniel Catchpoole[3]

[1] Faculty of Information Technology, University of Technology, Sydney, PO Box 123, Broadway, NSW 2007, Australia
{paulk,simeon}@it.uts.edu.au

[2] School of Computing, Queen's University, Kingston, Ontario, Canada
skill@cs.queensu.ca

[3] The Oncology Research Unit, The Children's Hospital at Westmead, Locked Bag 4001, Westmead NSW 2145, Australia
DanielC@chw.edu.au

Abstract. This paper describes a method of clustering lists of genes mined from a microarray dataset using functional information from the Gene Ontology. The method uses relationships between terms in the ontology both to build clusters and to extract meaningful cluster descriptions. The approach is general and may be applied to assist explanation of other datasets associated with ontologies.

Keywords: Cluster analysis, bioinformatics, cDNA microarray.

1 Introduction

Rapid developments in measurement and collection of diverse biological and clinical data offer researchers new opportunities for discovering relations between patterns of genes. The "classical" statistical techniques used in bioinformatics have been challenged by the large number of genes that are analysed simultaneously and the curse of dimensionality of gene expression measurements (in other words we are looking typically at tens of thousands of genes and only tens of patients). Data mining is expected to be able to assist the bio–data analysis (see [1] for brief overview).

The broad goals of this work are to improve the understanding of genes related to a specific form of childhood cancer. Three forms of data are combined at different stages. Patient data include cDNA microarray and clinical data for 9 patients. Usually between 2 and 10 repeat experiments of the same data (ie. patient) are made. For each patient, there are around 9000 genes with between 2 and 10 log ratios (ie. experiment repeats) for each gene. Clinical data describe a patient in detail, as well as the effect of different treatment protocols. Of the nine patients, 4 are labelled as high risk.

The task is to assist in understanding gene patterns in such biodata. Proposed methodology is shown in Fig. 1. It includes 3 stages. Stage 1 ("DM1:

H. Dai, R. Srikant, and C. Zhang (Eds.): PAKDD 2004, LNAI 3056, pp. 699–703, 2004.

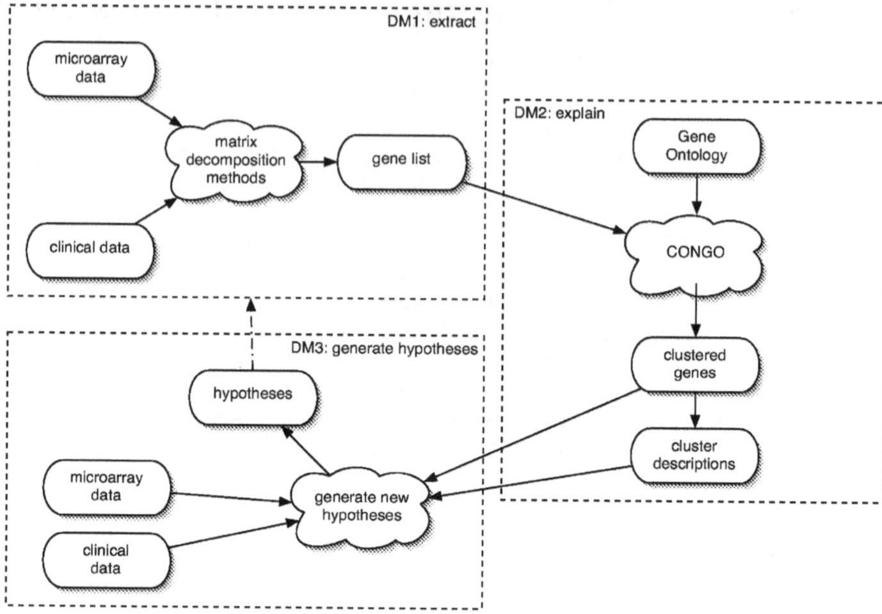

Fig. 1. Diagram showing methodology used to analyse microarray data

extract") is a data mining cycle, which reduces the vast number of genes coming from the microarray experiments to dozens of genes. Techniques used are described in detail in [2]. The output of this stage is interesting from a statistical point of view, however it is difficult for biological interpretation. Stage 2 ("DM2: explain") aims at assisting the interpretation of these outputs. The list of genes is reclustered over a gene ontology [3] into groups of genes with similar biological functionality. Descriptions of the clusters are automatically determined for biological interpretation. Stage 3 ("DM3: generate hypotheses") aims to summarise what is known about the genes and to group them in the context of the microarray measurements. Biologists then can formulate potentially promising hypotheses and may return to Stage 1.

2 DM2: Assisting Biological Explanation

The focus of this paper is on Stage 2. The cluster analysis and visualisation described in this paper takes as input (i) a list of genes highlighted from "DM1: extract" and (ii) data from the Gene Ontology. Clustering data according to an ontology is a new procedure described in [4]. It entails using a special distance measure that considers the relative positions of terms in the ontological hierarchy. The particular clustering algorithm is not as important as the distance measure. Details of the algorithm are presented in [4]. Recent work in [5] takes

Table 1. The first few rows of the dataset for the second step in the methodology.

Gene	GO terms directly associated with gene
AA040427	GO:0004715 GO:0005524 GO:0004674 GO:0006468 GO:0008283 GO:0000074 GO:0005634 GO:0016740
AA046690	GO:0003777 GO:0005524 GO:0007018 GO:0005871
AA055946	GO:0004894 GO:0005057 GO:0004888 GO:0007166 GO:0006968 GO:0005887

Table 2. Discovered clusters. AA*nnnn* are GenBank accession codes.

Cluster Number	Gene Count	Genes
0	6	AA040427 AA406485 AA434408 AA487466 AA609609 AA609759
1	2	AA046690 AA644679
2	6	AA055946 AA398011 AA458965 AA487426 AA490846 AA504272
3	9	AA112660 AA397823 AA443547 AA447618 AA455300 AA478436 AA608514 AA669758 AA683085
4	20	AA126911 AA133577 AA400973 AA464034 AA464743 AA486531 AA488346 AA488626 AA497029 AA629641 AA629719 AA629808 AA664241 AA664284 AA668301 AA669359 AA683050 AA700005 AA700688 AA775874

a similar approach. We use the Gene Ontology [3], a large collaborative public set of controlled vocabularies, in our clustering experiments. Gene products are described in terms of their effect and known place in the cell. Terms in the ontology are interrelated: eg. a "glucose metabolism" *is a* "hexose metabolism". Gene Ontology terms are associated with each gene in the list by searching in the SOURCE database [6]. The list of genes is clustered into groups with similar functionality using a distance measure that explicitly considers the relationship between terms in the ontology. Finally, descriptions of each cluster are found by examining Gene Ontology terms that are representative of the cluster.

Taking the list of genes associated with high risk patients identified in Stage 1 (an example of such genes are shown in the first column in Table 1), we reclustered them using terms in the Gene Ontology (the GO:*nnnnnnn* labels in the right column in Table 1) into groups of similarly described genes.

3 Results of DM2

Five clusters are found as shown in Table 2. Half of the genes have been allocated to one cluster. The rest of the genes have been split into four smaller clusters with one cluster containing only two genes.

Associated GO terms automatically determine functional descriptions of clusters. Starting with all the GO terms directly associated with genes in a particular

Table 3. Principal cluster descriptions for the genes. Last column is the number of genes in the cluster associated with the term.

GO ID	GO Term	Number of Genes
	Cluster 0 — 6 genes	
	20 GO terms but each associated with only one gene	1
	Cluster 1 — 2 genes	
GO:0008092	cytoskeletal protein binding activity	2
GO:0007028	cytoplasm organization and biogenesis	2
GO:0003774	motor activity	2
GO:0005875	microtubule associated complex	2
	5 GO terms but each associated with only one gene	1
	Cluster 2 — 6 genes	
GO:0004871	signal transducer activity	4
GO:0007154	cell communication	4
GO:0005887	integral to plasma membrane	3
GO:0005886	plasma membrane	3
GO:0005194	cell adhesion molecule activity	2
	11 GO terms but each associated with only one gene	1
	Cluster 3 — 9 genes	
GO:0030528	transcription regulator activity	4
GO:0008134	transcription factor binding activity	3
GO:0006366	transcription from Pol II promoter	3
GO:0003700	transcription factor activity	3
GO:0006357	regulation of transcription from Pol II promoter	3
	5 GO terms but each associated with only two genes each	2
	13 GO terms but each associated with only one gene	1
	Cluster 4 — 20 genes	
GO:0003723	RNA binding activity	10
GO:0030529	ribonucleoprotein complex	9
GO:0009059	macromolecule biosynthesis	9
GO:0006412	protein biosynthesis	9
GO:0005829	cytosol	9
GO:0003735	structural constituent of ribosome	8
	2 GO terms but each associated with only four genes each	4
	5 GO terms but each associated with only three genes each	3
	1 GO term associated with only two genes	2
	33 GO terms but each associated with only one gene	1

cluster, we climb the ontology replacing GO terms with their parents. Terms are replaced only if the parent node is *not* associated with genes in another cluster. Cluster descriptions derived in this way are shown in Table 3. Only the *is–a* relationships were followed to build this table. There are far fewer *part–of* relationships in the hierarchies so we do not believe that omitting them affects the results. The terms listed in the table are associated only with genes in each cluster and not in any other cluster. Cluster 0 in Table 3 has no terms that are associated with more than one gene. This suggests that the genes in the cluster are either unrelated or related only in ways that are sufficiently high level that the terms exist in other clusters. This suggests that the quality of the cluster is not good. Cluster 1 contains at least two genes that are related to the cell cytoskeleton and to microtubules (ie. components of the cytoskeleton). Cluster 2 contains three or four genes associated with signal transduction and cell signalling. Cluster 3 contains three or four genes related to transcription of genes and cluster 4 contains genes associated with RNA binding.

4 Conclusions

We present a methodology for extracting and explaining biological knowledge from microarray data. Applying terms from the Gene Ontology brings an understanding of the genes and their interrelationships. Currently biologists search through such lists gene–by–gene analysing each one individually and trying to piece together the many strands of information. Automating the process, at least to some extent, allows biologists to concentrate more on the important relationships rather than the minutiae of searching. Consequently they are enabled to formulate hypotheses to test in future experiments. The approach is general and may be applied to assist explanation other datasets associated with ontologies.

Acknowledgements. We would like to thank the University of Technology, Sydney and The Children's Hospital at Westmead for supporting this project.

References

1. Han, J.: How can data mining help bio–data analysis. In: Proc. 2nd Workshop on Data Mining in Bioinformatics BIOKDD02, ACM Press (2002)
2. Skillicorn, D., et al.: Strategies for winnowing microarray data. In: Proc. SIAM Data Mining Conf. (accepted 2004)
3. Ontology Consortium, T.G.: Gene Ontology: tool for the unification of biology. Nature Genetics **25** (2000) 25–29 PubMed ID:10802651.
4. Kennedy, P.J., Simoff, S.J.: CONGO: Clustering on the Gene Ontology. In: Proc. 2nd Australasian Data Mining Workshop ADM03, University of Technology, Sydney (2003)
5. Lee, S.G., et al.: A graph–theoretic modeling on GO space for biological interpretation of gene clusters. Bioinformatics **20** (2004) 381–388
6. Diehn, M., et al.: SOURCE: a unified genomic resource of functional annotations, ontologies, and gene expression data. Nucleic Acids Research **31** (2003) 219–223

Further Applications of a Particle Visualization Framework[1]

Ke Yin and Ian Davidson

SUNY-Albany, Department of Computer Science
1400 Washington Ave.Albany, NY, USA, 12222.
{ke,davidson}@cs.albany.edu

Abstract. Previous work introduced a 3D particle visualization framework that viewed each data point as a particle affected by gravitational forces. We showed the use of this tool for visualizing cluster results and anomaly detection. This paper generalizes the particle visualization framework and demonstrates further applications such as determining the number of clusters and identifies clustering algorithm biases. We don't claim visualization itself is sufficient in answering these questions. The methods here are best used when combined with other visual and analytic techniques. We have made our visualization software that produces standard VRML available to allow its use for these and other applications.

1 Introduction

Clustering is one of the most popular functions performed in data mining. Applications range from segmenting instances/observations for target marketing, outlier detection, data cleaning and as a general purpose exploratory tool to understand the data. Most clustering algorithms essentially are instance density estimation and thus the results are best understood and interpreted with the aid of visualization. In this paper, we extend our particle based approach to visualizing clustering results [2][3] to facilitate the diagnosis and presentation processes in routine clustering tasks.

Our earlier work describes a general particle framework to display a clustering solution [1] and illustrates its use for anomaly detection and segmentation [2]. The three-dimensional information visualization represents the previously clustered observations as particles affected by gravitational forces. We map the cluster centers into a three-dimensional cube so that similar clusters are adjacent and dissimilar clusters are far apart. We then place the particles (observations) amongst the centers according to the gravitational force exerted on the particles by the cluster centers. A particle's degree of membership to a cluster provides the magnitude of the gravitational force exerted.

---

[1] Our software is available at www.cs.albany.edu/~davidson/ParticleViz. We strongly encourage readers to refer the 3D visualizations at the above address while reading the paper. The visualizations can be viewed by any internet browser with a VRML plug-in (http://www.parallelgraphics.com/products/downloads/).

H. Dai, R. Srikant, and C. Zhang (Eds.): PAKDD 2004, LNAI 3056, pp. 704–710, 2004.

We focus on further applications of the visualization technique in clustering, addressing questions that arises frequently in routine clustering tasks: 1) Deciding the appropriate number of clusters.2)Understanding and visualizing presentational bias and stability of different clustering algorithms.

The rest of the paper is organized as following. We first introduce our clustering visualization methodology and then describe our improvements to achieve better visualization quality for comparing algorithms. We then demonstrate how to utilize our visualization technique to solve the example applications in clustering mentioned above. Finally we define our future research work direction and draw conclusions.

2 Particle Based Clustering Visualization Algorithm

The algorithm takes a $k \times k$ cluster distance matrix C and a $k \times N$ membership matrix P as the inputs. In matrix C, each member c_{ij} denotes the distance between the center of cluster i and the center of cluster j. In the matrix P, each member p_{ij} denotes the probability of instance i belongs to cluster j. The cluster distance matrix may contain the Kullback Leibler (EM algorithm) or Euclidean distances (K-Means) between the cluster descriptions. We believe the degree of membership matrix can be generated by most clustering algorithms, however it must be scaled so that sum of $\sum_j p_{ij} = 1$.

The algorithm first calculates the positions of the cluster centers in the three dimensional space given the cluster distance matrix C. A Multi-Dimensional Scaling (MDS) based simulated annealing method maps the clusters centers from higher dimensions into a three dimensional space while preserving the cluster distances in the higher dimensional instance space. After the cluster centers are placed, the algorithm puts each instance around its closest cluster center according to its degree of membership at a distance of $r_{ij} = f(p_{ij})$. Function f is called the probability-distance transformation function, whose form we will derive in the next section. The exact position of the instance on the sphere shell is finally determined by the remaining clusters gravitational pull on the instance based on the instances degree of membership to them. Precise algorithmic details are provided in earlier work [2].

3 The Probability-Distance Transformation Function

We need a mapping function between p (degree of membership) and r (distance to cluster center) to convey the density of instances correctly in the visualization. Let $N(r)$ be the number of instances assigned to a cluster that are within distance r of its center in the visualization. If p is the degree of membership of an instance to a cluster, then the instance density function Z against the degree of membership is defined by:

$$Z = -\frac{dN(r)}{dp} = -\frac{dN(r)}{dr} \cdot \frac{dr}{dp} . \tag{1}$$

The Z function measures the number of instances that will occupy an interval of size dp. While the D function measures the number of instances that will occupy an interval of size dr in the visualization.

$$D = \frac{1}{4\pi} \frac{dN(r)}{r^2 dr} \tag{2}$$

We wish the two density functions measured by degree of membership and measured by the distance in the visualization to be consistent. To achieve this we equate the two and bound them so they only differ by a positive scaling constant c^3. By solving the differential equation, we attain the probability-distance function f.

$$r = f(p) = c\sqrt{\frac{3}{4\pi}}\sqrt[3]{1-p}\,. \tag{3}$$

The constant c is termed the constricting factor which can be used to zoom in and out. Equation (**3**) can be used to determine how far to place an instance from a cluster center as a function of its degree of membership to the cluster.

4 An Ideal Cluster's Visual Signature

The probability-distance transformation function allows us to convey the instance density in a more accurate and efficient way. By visualizing the clusters we can tell directly whether the density distribution of the cluster is consistent with our prior knowledge or belief. The desired cluster density distribution from our prior knowledge is called an ideal density signature. For example, a mixture model that assumes independent Gaussian attributes will consist of a very dense cluster center with the density decreasing as a function of distance to the cluster center such as in Fig. 1. The ideal visual signature will vary depending on the algorithm's assumptions. These signatures can be obtained by visualizing artificial data that completely abides by the algorithm's assumptions.

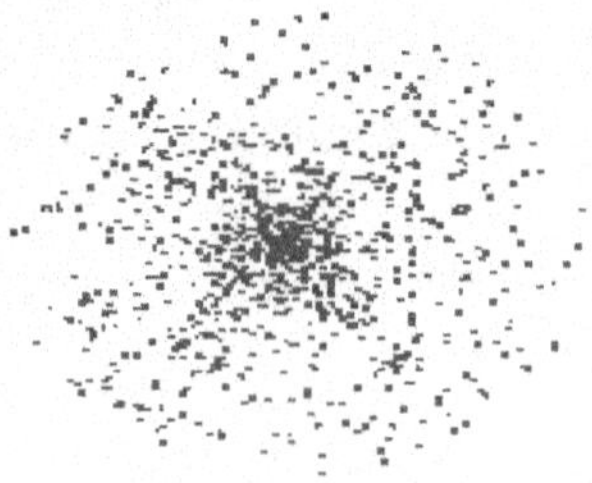

Fig. 1. The Ideal Cluster Signature For an Independent Gaussian Attribute

5 Example Applications

In this section, we demonstrate three problems in clustering we intend to address using our visualization technique. We begin by using three artificial datasets: (A) is generated from three normal distributions, N (-8,1), N (-4,1), N (20,4) with each generating mechanism equally likely. (B) is generated by three normal distributions, N (-9,1), N (-3,1), N (20,4) with the last mechanism twice as likely as the other two. (C) is generated by two equally likely normal distributions, N (-3, 9), N (3, 9).

5.1 Determining the K Value

Many clustering algorithms require the user to *apriori* specify the number of clusters, k, based on information such as experience or empirical results. Though the selection of k can be made part of the problem by making it a parameter to estimate [1], this is not common in data mining applications of clustering. Techniques such as Akaike Information Criterion (*AIC*) and Schwarz's Bayesian Information Criterion (*BIC*) are only applicable in probabilistically formulated problems and often give contradictory results. The expected instance density given by the ideal cluster signature plays a key role in verifying the appropriateness of the clustering solution.

Our visualization technique helps to determine if the current k value is appropriate. As we assumed our data was Gaussian distributed then good clustering results should have a signature density associated with this probability distribution shown Fig. 1.

Dataset (A) is clustered with the K-means algorithm with various values of k. The ideal value of k for this data set is 3. We start with k=2, shown in Fig. 2. The two clusters have quite different densities: The cluster on the left has an almost uniform density distribution that indicates the cluster is not well formed. In contrast, the density of right cluster is indicative of a Gaussian distribution. This suggests that the left-hand-side cluster may be further separable and hence we increase k to 3 as shown in Fig. 3. We can see the density for all clusters approximate the Gaussian distribution (the ideal signature), and that two clusters overlap (bottom left). At this point we conclude that k=3 is a candidate solution as we assumed the instances were drawn from a Gaussian distribution and our visualization gives consistent cluster signature. For completeness, we increased k to 4 the results are shown in Fig. 4. Most of the instances on the right are part of two almost inseparable clusters whose density is not consistent with the Gaussian distribution signature.

Fig. 2. Visualization of dataset (A) clustering results with K-means algorithm (k=2).

Fig. 3. Visualization of dataset (A) clustering results with K-means algorithm (k=3)

Fig. 4. Visualization of dataset (A) clustering results with K-means algorithm (k=4)

5.2 Comparison Clustering Algorithms

In this sub-section we describe how our visualization technique can be used to compare different clustering algorithms on their representational biases and algorithmic stabilities. Most clustering algorithms are sensitive to initial configurations and different initializations lead to different cluster solutions, which is known as the algorithmic stability. Also, different clustering algorithms have different representations of clustering, and the representational biases also contribute to the clustering solutions found. Though analytical studies that compare very similar clustering algorithms such as K-means and EM exist such studies are difficult for fundamentally different classes of clustering algorithms.

Algorithmic Stability. By randomly restarting the clustering algorithm and clustering the clustering solutions and using our visualization technique we can visualize the stability of a clustering algorithm. We represent each solution by the cluster parameter estimates (centroid values for K-Means for example). The number of instances in a cluster indicates the probability of these particular local minima (and its slight variants) being found, while the size of the cluster suggests the basin of attraction associated with the local minima.

We use dataset (B) to illustrate this particular use of the visualization technique. Dataset (B) is clustered using k=3 with three different clustering algorithms: weighted K-means, weighted EM, and unweighted EM. Each algorithm makes 1000 random restarts thereby generating 1000 instances. These 1000 instances represent the different clustering solutions found, are separated into 2 clusters using a weighted K-means algorithm. We do not claim *k=2* is optimal but will serve our purpose of determining the algorithmic stabilities. The results for all three algorithms are shown in Fig. 5, Fig. 6. and Fig. 7. We find the stability of the K-means algorithm is less than that of EM algorithms, and that the stability of unweighted EM is less than the weighted version of the algorithm, which is consistent with the known literature [4].

Fig. 5. Visualization of dataset (B) clustering solutions with unweighted EM

Fig. 6. Visualization of dataset (B) clustering solutions with weighted EM

Fig. 7. Visualization of dataset (B) clustering solutions with weighted K-means

Visualizing Representational Bias. We are going to use dataset (C) to illustrate how to visualize the representational bias effect on the found clustering solutions. We do clustering with both the EM and K-means algorithm and show the visualized results in Fig. 8. and Fig. 9.

The different representational biases of K-means and EM can be inferred from the visualizations. We found that for EM, almost all instances exterior to cluster's main body are attracted to the neighboring cluster. In contrast, only about half of the exterior cluster instances for K-Means are attracted to the neighboring cluster. The other half are too far from the neighboring cluster to show any noticeable attraction. This confirms the well known belief that K-Means finds cluster centers that are further apart, have smaller standard deviations and less well defined than EM for overlapping clusters.

Fig. 8. Visualization of dataset (C) clustering results with EM algorithm (k=2)

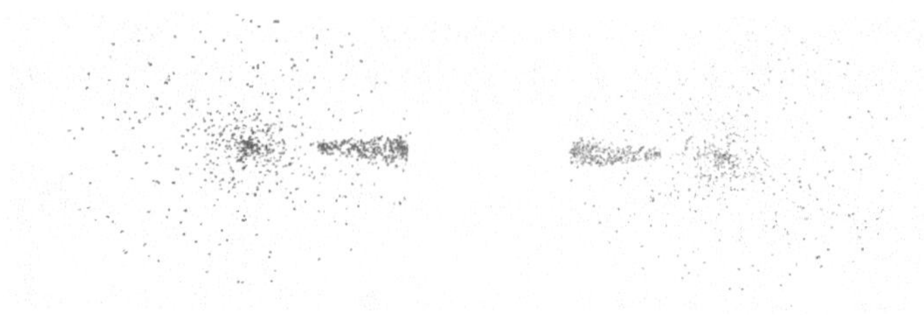

Fig. 9. Visualization of dataset (C) clustering results with K-means algorithm (k=2)

6 Future Work and Conclusions

Visualization techniques provide considerable aids in clustering problems as the problem focuses on instance density estimation. We focused on three visualization applications in clustering.

- The cluster signature indicates the quality of the clustering results thus can be used for clustering diagnosis such as finding the appropriate number of clusters
- Different clustering algorithms have different representational biases and different algorithmic stability. Although sometimes they can be analyzed mathematically, visualization techniques facilitate the process and provide aids in detection and presentation.

We do not claim that visualizations can solely address these problems but believe that in combination with analytic solutions can provide more insight. Similarly we are not suggesting that these are the only problems the technique can address. The software is freely available for others to pursue these and other opportunities.

We can use our approach to generate multiple visualizations of the output of clustering algorithms and visually compare them side by side. We intend to investigate adapting our framework to visually comparing multiple algorithms' output on the one canvas.

References

1. Davidson, I., Ward, M.: A Particle Visualization Framework for Clustering and Anomaly Detection. ACM KDD Workshop on Visual Data Mining (2001)
2. Davidson, I.: Visualizing Clustering Results", SIAM International Conference on Data Mining (2002)
3. Davidson, I.: Minimum Message Length Clustering and Gibbs Sampling. Uncertainty in Artificial Intelligence (2000)
4. Kearns, M., Mansour, Y., Ng, A.: An Information Theoretic Analysis of Hard and Soft Clustering. The 12th International Conference on Uncertainty in A.I. (1996)

Author Index

Lecture Notes in Artificial Intelligence (LNAI)

Vol. 3060: A.Y. Tawfik, S.D. Goodwin (Eds.), Advances in Artificial Intelligence. XIII, 582 pages. 2004.

Vol. 3056: H. Dai, R. Srikant, C. Zhang (Eds.), Advances in Knowledge Discovery and Data Mining. XIX, 713 pages. 2004.

Vol. 3035: M.A. Wimmer, Knowledge Management in Electronic Government. XII, 342 pages. 2004.

Vol. 3034: J. Favela, E. Menasalvas, E. Chávez (Eds.), Advances in Web Intelligence. XIII, 227 pages. 2004.

Vol. 3025: G.A. Vouros, T. Panayiotopoulos (Eds.), Methods and Applications of Artificial Intelligence. XV, 546 pages. 2004.

Vol. 3012: K. Kurumatani, S.-H. Chen, A. Ohuchi (Eds.), Multi-Agnets for Mass User Support. X, 217 pages. 2004.

Vol. 3010: K.R. Apt, F. Fages, F. Rossi, P. Szeredi, J. Váncza (Eds.), Recent Advances in Constraints. VIII, 285 pages. 2004.

Vol. 2980: A. Blackwell, K. Marriott, A. Shimojima (Eds.), Diagrammatic Representation and Inference. XV, 448 pages. 2004.

Vol. 2977: G. Di Marzo Serugendo, A. Karageorgos, O.F. Rana, F. Zambonelli (Eds.), Engineering Self-Organising Systems. X, 299 pages. 2004.

Vol. 2972: R. Monroy, G. Arroyo-Figueroa, L.E. Sucar, H. Sossa (Eds.), MICAI 2004: Advances in Artificial Intelligence. XVII, 923 pages. 2004.

Vol. 2961: P. Eklund (Ed.), Concept Lattices. IX, 411 pages. 2004.

Vol. 2953: K. Konrad, Model Generation for Natural Language Interpretation and Analysis. XIII, 166 pages. 2004.

Vol. 2934: G. Lindemann, D. Moldt, M. Paolucci (Eds.), Regulated Agent-Based Social Systems. X, 301 pages. 2004.

Vol. 2930: F. Winkler (Ed.), Automated Deduction in Geometry. VII, 231 pages. 2004.

Vol. 2926: L. van Elst, V. Dignum, A. Abecker (Eds.), Agent-Mediated Knowledge Management. XI, 428 pages. 2004.

Vol. 2923: V. Lifschitz, I. Niemelä (Eds.), Logic Programming and Nonmonotonic Reasoning. IX, 365 pages. 2004.

Vol. 2915: A. Camurri, G. Volpe (Eds.), Gesture-Based Communication in Human-Computer Interaction. XIII, 558 pages. 2004.

Vol. 2913: T.M. Pinkston, V.K. Prasanna (Eds.), High Performance Computing - HiPC 2003. XX, 512 pages. 2003.

Vol. 2903: T.D. Gedeon, L.C.C. Fung (Eds.), AI 2003: Advances in Artificial Intelligence. XVI, 1075 pages. 2003.

Vol. 2902: F.M. Pires, S.P. Abreu (Eds.), Progress in Artificial Intelligence. XV, 504 pages. 2003.

Vol. 2892: F. Dau, The Logic System of Concept Graphs with Negation. XI, 213 pages. 2003.

Vol. 2891: J. Lee, M. Barley (Eds.), Intelligent Agents and Multi-Agent Systems. X, 215 pages. 2003.

Vol. 2882: D. Veit, Matchmaking in Electronic Markets. XV, 180 pages. 2003.

Vol. 2871: N. Zhong, Z.W. Raś, S. Tsumoto, E. Suzuki (Eds.), Foundations of Intelligent Systems. XV, 697 pages. 2003.

Vol. 2854: J. Hoffmann, Utilizing Problem Structure in Planing. XIII, 251 pages. 2003.

Vol. 2843: G. Grieser, Y. Tanaka, A. Yamamoto (Eds.), Discovery Science. XII, 504 pages. 2003.

Vol. 2842: R. Gavaldá, K.P. Jantke, E. Takimoto (Eds.), Algorithmic Learning Theory. XI, 313 pages. 2003.

Vol. 2838: N. Lavrač, D. Gamberger, L. Todorovski, H. Blockeel (Eds.), Knowledge Discovery in Databases: PKDD 2003. XVI, 508 pages. 2003.

Vol. 2837: N. Lavrač, D. Gamberger, L. Todorovski, H. Blockeel (Eds.), Machine Learning: ECML 2003. XVI, 504 pages. 2003.

Vol. 2835: T. Horváth, A. Yamamoto (Eds.), Inductive Logic Programming. X, 401 pages. 2003.

Vol. 2821: A. Günter, R. Kruse, B. Neumann (Eds.), KI 2003: Advances in Artificial Intelligence. XII, 662 pages. 2003.

Vol. 2807: V. Matoušek, P. Mautner (Eds.), Text, Speech and Dialogue. XIII, 426 pages. 2003.

Vol. 2801: W. Banzhaf, J. Ziegler, T. Christaller, P. Dittrich, J.T. Kim (Eds.), Advances in Artificial Life. XVI, 905 pages. 2003.

Vol. 2797: O.R. Zaïane, S.J. Simoff, C. Djeraba (Eds.), Mining Multimedia and Complex Data. XII, 281 pages. 2003.

Vol. 2792: T. Rist, R.S. Aylett, D. Ballin, J. Rickel (Eds.), Intelligent Virtual Agents. XV, 364 pages. 2003.

Vol. 2782: M. Klusch, A. Omicini, S. Ossowski, H. Laamanen (Eds.), Cooperative Information Agents VII. XI, 345 pages. 2003.

Vol. 2780: M. Dojat, E. Keravnou, P. Barahona (Eds.), Artificial Intelligence in Medicine. XIII, 388 pages. 2003.

Vol. 2777: B. Schölkopf, M.K. Warmuth (Eds.), Learning Theory and Kernel Machines. XIV, 746 pages. 2003.

Vol. 2752: G.A. Kaminka, P.U. Lima, R. Rojas (Eds.), RoboCup 2002: Robot Soccer World Cup VI. XVI, 498 pages. 2003.

Vol. 2741: F. Baader (Ed.), Automated Deduction – CADE-19. XII, 503 pages. 2003.

Vol. 2705: S. Renals, G. Grefenstette (Eds.), Text- and Speech-Triggered Information Access. VII, 197 pages. 2003.

Vol. 2703: O.R. Zaïane, J. Srivastava, M. Spiliopoulou, B. Masand (Eds.), WEBKDD 2002 - MiningWeb Data for Discovering Usage Patterns and Profiles. IX, 181 pages. 2003.

Vol. 2700: M.T. Pazienza (Ed.), Extraction in the Web Era. XIII, 163 pages. 2003.

Vol. 2699: M.G. Hinchey, J.L. Rash, W.F. Truszkowski, C.A. Rouff, D.F. Gordon-Spears (Eds.), Formal Approaches to Agent-Based Systems. IX, 297 pages. 2002.

Vol. 2691: V. Mařík, J.P. Müller, M. Pechoucek (Eds.), Multi-Agent Systems and Applications III. XIV, 660 pages. 2003.

Vol. 2684: M.V. Butz, O. Sigaud, P. Gérard (Eds.), Anticipatory Behavior in Adaptive Learning Systems. X, 303 pages. 2003.

Vol. 2671: Y. Xiang, B. Chaib-draa (Eds.), Advances in Artificial Intelligence. XIV, 642 pages. 2003.

Vol. 2663: E. Menasalvas, J. Segovia, P.S. Szczepaniak (Eds.), Advances in Web Intelligence. XII, 350 pages. 2003.

Vol. 2661: P.L. Lanzi, W. Stolzmann, S.W. Wilson (Eds.), Learning Classifier Systems. VII, 231 pages. 2003.

Vol. 2654: U. Schmid, Inductive Synthesis of Functional Programs. XXII, 398 pages. 2003.

Vol. 2650: M.-P. Huget (Ed.), Communications in Multiagent Systems. VIII, 323 pages. 2003.

Vol. 2645: M.A. Wimmer (Ed.), Knowledge Management in Electronic Government. XI, 320 pages. 2003.

Vol. 2639: G. Wang, Q. Liu, Y. Yao, A. Skowron (Eds.), Rough Sets, Fuzzy Sets, Data Mining, and Granular Computing. XVII, 741 pages. 2003.

Vol. 2637: K.-Y. Whang, J. Jeon, K. Shim, J. Srivastava, Advances in Knowledge Discovery and Data Mining. XVIII, 610 pages. 2003.

Vol. 2636: E. Alonso, D. Kudenko, D. Kazakov (Eds.), Adaptive Agents and Multi-Agent Systems. XIV, 323 pages. 2003.

Vol. 2627: B. O'Sullivan (Ed.), Recent Advances in Constraints. X, 201 pages. 2003.

Vol. 2600: S. Mendelson, A.J. Smola (Eds.), Advanced Lectures on Machine Learning. IX, 259 pages. 2003.

Vol. 2592: R. Kowalczyk, J.P. Müller, H. Tianfield, R. Unland (Eds.), Agent Technologies, Infrastructures, Tools, and Applications for E-Services. XVII, 371 pages. 2003.

Vol. 2586: M. Klusch, S. Bergamaschi, P. Edwards, P. Petta (Eds.), Intelligent Information Agents. VI, 275 pages. 2003.

Vol. 2583: S. Matwin, C. Sammut (Eds.), Inductive Logic Programming. X, 351 pages. 2003.

Vol. 2581: J.S. Sichman, F. Bousquet, P. Davidsson (Eds.), Multi-Agent-Based Simulation. X, 195 pages. 2003.

Vol. 2577: P. Petta, R. Tolksdorf, F. Zambonelli (Eds.), Engineering Societies in the Agents World III. X, 285 pages. 2003.

Vol. 2569: D. Karagiannis, U. Reimer (Eds.), Practical Aspects of Knowledge Management. XIII, 648 pages. 2002.

Vol. 2560: S. Goronzy, Robust Adaptation to Non-Native Accents in Automatic Speech Recognition. XI, 144 pages. 2002.

Vol. 2557: B. McKay, J. Slaney (Eds.), AI 2002: Advances in Artificial Intelligence. XV, 730 pages. 2002.

Vol. 2554: M. Beetz, Plan-Based Control of Robotic Agents. XI, 191 pages. 2002.

Vol. 2543: O. Bartenstein, U. Geske, M. Hannebauer, O. Yoshie (Eds.), Web Knowledge Management and Decision Support. X, 307 pages. 2003.

Vol. 2541: T. Barkowsky, Mental Representation and Processing of Geographic Knowledge. X, 174 pages. 2002.

Vol. 2533: N. Cesa-Bianchi, M. Numao, R. Reischuk (Eds.), Algorithmic Learning Theory. XI, 415 pages. 2002.

Vol. 2531: J. Padget, O. Shehory, D. Parkes, N.M. Sadeh, W.E. Walsh (Eds.), Agent-Mediated Electronic Commerce IV. Designing Mechanisms and Systems. XVII, 341 pages. 2002.

Vol. 2527: F.J. Garijo, J.-C. Riquelme, M. Toro (Eds.), Advances in Artificial Intelligence - IBERAMIA 2002. XVIII, 955 pages. 2002.

Vol. 2522: T. Andreasen, A. Motro, H. Christiansen, H.L. Larsen (Eds.), Flexible Query Answering Systems. X, 383 pages. 2002.

Vol. 2514: M. Baaz, A. Voronkov (Eds.), Logic for Programming, Artificial Intelligence, and Reasoning. XIII, 465 pages. 2002.

Vol. 2507: G. Bittencourt, G.L. Ramalho (Eds.), Advances in Artificial Intelligence. XIII, 417 pages. 2002.

Vol. 2504: M.T. Escrig, F. Toledo, E. Golobardes (Eds.), Topics in Artificial Intelligence. XI, 427 pages. 2002.

Vol. 2499: S.D. Richardson (Ed.), Machine Translation: From Research to Real Users. XXI, 254 pages. 2002.

Vol. 2484: P. Adriaans, H. Fernau, M. van Zaanen (Eds.), Grammatical Inference: Algorithms and Applications. IX, 315 pages. 2002.

Vol. 2479: M. Jarke, J. Koehler, G. Lakemeyer (Eds.), KI 2002: Advances in Artificial Intelligence. XIII, 327 pages. 2002.

Vol. 2475: J.J. Alpigini, J.F. Peters, A. Skowron, N. Zhong (Eds.), Rough Sets and Current Trends in Computing. XV, 640 pages. 2002.

Vol. 2473: A. Gómez-Pérez, V.R. Benjamins (Eds.), Knowledge Engineering and Knowledge Management. Ontologies and the Semantic Web. XI, 402 pages. 2002.

Vol. 2466: M. Beetz, J. Hertzberg, M. Ghallab, M.E. Pollack (Eds.), Advances in Plan-Based Control of Robotic Agents. VIII, 291 pages. 2002.

Vol. 2464: M. O'Neill, R.F.E. Sutcliffe, C. Ryan, M. Eaton, N.J.L. Griffith (Eds.), Artificial Intelligence and Cognitive Science. XI, 247 pages. 2002.

Vol. 2448: P. Sojka, I. Kopecek, K. Pala (Eds.), Text, Speech and Dialogue. XII, 481 pages. 2002.

Vol. 2447: D.J. Hand, N.M. Adams, R.J. Bolton (Eds.), Pattern Detection and Discovery. XII, 227 pages. 2002.

Vol. 2446: M. Klusch, S. Ossowski, O. Shehory (Eds.), Cooperative Information Agents VI. XI, 321 pages. 2002.

GPSR Compliance
The European Union's (EU) General Product Safety Regulation (GPSR) is a set of rules that requires consumer products to be safe and our obligations to ensure this.

If you have any concerns about our products, you can contact us on

ProductSafety@springernature.com

In case Publisher is established outside the EU, the EU authorized representative is:

Springer Nature Customer Service Center GmbH
Europaplatz 3
69115 Heidelberg, Germany

www.ingramcontent.com/pod-product-compliance
Lightning Source LLC
Chambersburg PA
CBHW071405230926
51846CB00001B/11
* 9 7 8 3 5 4 0 2 2 0 6 4 0 *